General Chemistry

William J. Vining
State University of New York, Oneonta

Susan M. Young
Hartwick College

Roberta Day
University of Massachusetts, Amherst

Beatrice Botch
University of Massachusetts, Amherst

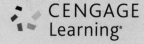

CENGAGE
Learning®

Australia • Brazil • Mexico • Singapore • United Kingdom • United States

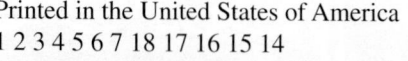

CENGAGE
Learning®

General Chemistry
William J. Vining, Susan M. Young, Roberta Day, and Beatrice Botch

Vice President, MindTap Product: Melissa S. Acuña

MindTap Product Owner: Lisa Lockwood

Director MindTap Operations: Liz Graham

Content Developer: Rebecca Heider

Sr. Media Developer: Lisa Weber

Assoc. Media Developer: Elizabeth Woods

Assoc. Content Developer: Karolina Kiwak

Software Development Manager: Aaron Chesney

Content Implementation Manager: Laura Berger

Digital Content Production Project Manager:
    Madeline Harris

Marketing Director MindTap: Kristin McNary

Marketing Director Sciences: Nicole Hamm

Product Development Specialist: Nicole Hurst

Content Project Manager: Teresa L. Trego

Art Director: Maria Epes

Manufacturing Planner: Judy Inouye

IP Analyst: Christine Myaskovsky

IP Project Manager: John Sarantakis

Production Service/Project Manager: MPS Limited

Photo Researcher: Darren Wright

Text Designer: Ellen Pettengell Designs

Cover Designer: Bartay Studios

Cover Image Credit: ChemDoodle application icon courtesy
    of iChemLabs: all photographs © Charles D. Winters

Compositor: MPS Limited

For product information and technology assistance, contact us at
Cengage Learning Customer & Sales Support, 1-800-354-9706.
For permission to use material from this text or product,
submit all requests online at **www.cengage.com/permissions.**
Further permissions questions can be e-mailed to
permissionrequest@cengage.com.

Library of Congress Control Number: 2014940780

ISBN-13: 978-1-305-27515-7

ISBN-10: 1-305-27515-2

Cengage Learning
20 Channel Center Street
Boston, MA 02210
USA

Cengage Learning is a leading provider of customized learning solutions with office locations around the globe, including Singapore, the United Kingdom, Australia, Mexico, Brazil, and Japan. Locate your local office at **www.cengage.com/global.**

Cengage Learning products are represented in Canada by Nelson Education, Ltd.

To learn more about Cengage Learning Solutions, visit **www.cengage.com.**

Purchase any of our products at your local college store or at our preferred online store **www.cengagebrain.com.**

Printed in the United States of America
1 2 3 4 5 6 7 18 17 16 15 14

Contents

9 Theories of Chemical Bonding 265

10 Gases 293

19 Thermodynamics: Entropy and Free Energy 617

20 Electrochemistry 651

Acknowledgments

A product as complex as ***MindTap for General Chemistry*** could not have been created by the content authors alone; it also needed a team of talented, hardworking people to design the system, do the programming, create the art, guide the narrative, and help form and adhere to the vision. Although the authors' names are on the cover, what is inside is the result of the entire team's work and we want to acknowledge their important contributions.

Special thanks go to the core team at Cengage Learning that guided us through the entire process: Lisa Lockwood, Product Owner; Lisa Weber, Media Producer; and Rebecca Heider, Developmental Editor. Thanks also to Lynne Blaszak, Senior Technology Product Manager; Elizabeth Woods, Associate Media Developer; Gayle Huntress, OWL Administrator and System Specialist; Laura Berger, Content Implementation Manger; Aaron Chesney, Software Development Manger; and Teresa Trego, Senior Content Project Manager.

This primarily digital learning environment would not have been possible without the talents of Bill Rohan, Jesse Charette, and Aaron Russell of Cow Town Productions, who programmed the embedded media activities, and the entire MindTap Engineering Teams. Nor would it have been possible without the continued effort of David Hart, Stephen Battisti, Cindy Stein, Mayumi Fraser, Gale Parsloe, and Gordon Anderson from the Center for Educational Software Development (CESD) team at the University of Massachusetts, Amherst, the creators of OWL and the first OWLBook, who were there when we needed them most. Many thanks also go to Charles D. Winters for filming the chemistry videos and taking beautiful photographs.

We are grateful to Professor Don Neu of St. Cloud State University for his contributions to the nuclear chemistry chapter, and to the many instructors who gave us feedback in the form of advisory boards, focus groups, and written reviews. We also want to thank those instructors and students who tested early versions of the *OWLBook* in their courses, most especially Professors Maurice Odago and John Schaumloffel of SUNY Oneonta and Barbara Stewart of the University of Maine who bravely tested the earliest versions of this product.

MindTap General Chemistry has surely been improved by the hard work of our accuracy checkers, David Shinn, Bette Kreuz, and David Brown.

Bill and Susan would like to thank Jack Kotz, who has been a mentor to both of us for many years. This work would also not have been possible without the support and patience of our families, particularly Kathy, John, John, and Peter.

We are grateful to the many instructors who gave us feedback in the form of advisory boards, focus groups, and written reviews, and most of all to those instructors and students who tested early versions of MindTap General Chemistry in their courses.

Advisory Board

Chris Bahn, *Montana State University*
Christopher Collison, *Rochester Institute of Technology*
Cory DiCarlo, *Grand Valley State University*
Stephen Foster, *Mississippi State University*
Thomas Greenbowe, *Iowa State University*
Resa Kelly, *San Jose State University*
James Rudd, *California State University, Los Angeles*
Jessica Vanden Plas, *Grand Valley State University*

Class Test Participants

Zsuzsanna Balogh-Brunstad, *Hartwick College*
Jacqueline Bennett, *SUNY Oneonta*
Terry Brack, *Hofstra University*
Preston Brown, *Coastal Carolina Community College*
Donnie Byers, *Johnson County Community College*
John Dudek, *Hartwick College*
Deanna (Dede) Dunlavy, *New Mexico State University*
Dan Dupuis, *Coastal Carolina Community College*
Heike Geisler, *SUNY Oneonta*
Victoria Harris, *SUNY Oneonta*
Gary Hiel, *Hartwick College*
Dennis Johnson, *New Mexico State University*
Thomas Jose, *Blinn College*
Kirk Kawagoe, *Fresno City College*
Kristen Kilpatrick, *Coastal Carolina Community College*
Orna Kutai, *Montgomery College—Rockville Campus*
Antonio Lara, *New Mexico State University*
Scott Lefurgy, *Hofstra University*
Barbara Lyons, *New Mexico State University*
Larry Margerum, *University of San Francisco*
Diana Mason, *University of North Texas*
Don Neu, *St. Cloud State University*
Krista Noren-Santmyer, *Hillsborough Community College*
Erik Ruggles, *University of Vermont*
Flora Setayesh, *Nashville State Community College*

Sherril Soman, *Grand Valley State University*
Marjorie Squires, *Felician College*
Paul Tate, *Hillsborough Community College—Dale Mabry Campus*
Trudy Thomas-Smith, *SUNY Oneonta*
John B. Vincent, *University of Alabama*
Mary Whitfield, *Edmonds Community College*
Matthew J. Young, *University of New Hampshire*

Focus Group Participants

Linda Allen, *Louisiana State University*
Mufeed M. Basti, *North Carolina A&T*
Fereshteh Billiot, *Texas A&M University—Corpus Christi*
Kristen A. Casey, *Anne Arundel Community College*
Brandon Cruickshank, *Northern Arizona University*
William Deese, *Louisiana Technical University*
Cory DiCarlo, *Grand Valley State University*
Deanna (Dede) Dunlavy, *New Mexico State University*
Krishna Foster, *California State University, Los Angeles*
Stephen Foster, *Mississippi State University*
Gregory Gellene, *Texas Technical University*
Anita Gnezda, *Ball State University*
Nathaniel Grove, *University of North Carolina at Wilmington*
Bernadette Harkness, *Delta College*
Hongqiu Zhao, *Indiana University—Purdue University at Indianapolis*
Edith Kippenhan, *University of Toledo*
Joseph d. Kittle, Jr., *Ohio University*
Amy Lindsay, *University of New Hampshire*
Krista Noren-Santmyer, *Hillsborough Community College*
Olujide T. Akinbo, *Butler University*
James Reeves, *University of North Carolina at Wilmington*
James Rudd, *California State University, Los Angeles*
Raymond Sadeghi, *University of Texas at San Antonio*
Mark Schraf, *West Virginia University*
Sherril Soman, *Grand Valley State University*
Matthew W. Stoltzfus, *Ohio State University*
Dan Thomas, *University of Guelph*
Xin Wen, *California State University, Los Angeles*
Kurt Winkelmann, *Florida Institute of Technology*
James Zubricky, *University of Toledo*

Reviewers

Chris Bahn, *Montana State University*
Yiyan Bai, *Houston Community College*

Mufeed M. Basti, *North Carolina A&T*
James Beil, *Lorain County Community College*
Fereshteh Billiot, *Texas A&M University—Corpus Christi*
Jeffrey Bodwin, *Minnesota State University Moorhead*
Steven Brown, *University of Arizona*
Phil Brucat, *University of Florida*
Donnie Byers, *Johnson County Community College*
David Carter, *Angelo State University*
Allen Clabo, *Francis Marion University*
Beverly Clement, *Blinn College*
Willard Collier, *Mississippi State*
Christopher Collison, *Rochester Institute of Technology*
Cory DiCarlo, *Grand Valley State University*
Jeffrey Evans, *University of Southern Mississippi*
Nick Flynn, *Angelo State University*
Karin Gruet, *Fresno City College*
Bernadette Harkness, *Delta College*
Carl Hoeger, *University of California, San Diego*
Hongqiu Zhao, *Indiana University—Purdue University Indianapolis*
Richard Jarman, *College of DuPage*
Eric R. Johnson, *Ball State University*
Thomas Jose, *Blinn College*
Kirk Kawagoe, *Fresno City College*
Resa Kelly, *San Jose State University*
Jeffrey A. Mack, *Sacramento State University*
Larry Margerum, *University of San Francisco*
Diana Mason, *University of North Texas*
Donald R. Neu, *St. Cloud University*
Al Nichols, *Jacksonville State University*
Olujide T. Akinbo, *Butler University*
John Pollard, *University of Arizona*
James Reeves, *University of North Carolina at Wilmington*
Mark Schraf, *West Virginia University*
Shawn Sendlinger, *North Carolina Central University*
Duane Swank, *Pacific Lutheran University*
Michael Topp, *University of Pennsylvania*
Ray Trautman, *San Francisco State*
John B. Vincent, *University of Alabama*
Keith Walters, *Northern Kentucky University*
David Wright, *Vanderbilt University*
James Zubricky, *University of Toledo*

institution and arrived at SUNY Oneonta, where he now works with undergraduates, Cow Town Productions, and the UMass OWL team.

Susan M. Young

Hartwick College

Susan Young received her B.S. in Chemistry in 1988 from the University of Dayton and her Ph.D. in Inorganic Chemistry in 1994 from the University of Colorado at Boulder under the direction of Dr. Arlan Norman, where she worked on the reactivity of cavity-containing phosphazanes. She did postdoctoral work with Dr. John Kotz at the State University of New York at Oneonta, teaching and working on projects in support of the development of the first General Chemistry CD-ROM. She taught at Roanoke College in Virginia and then joined the faculty at Hartwick College in 1996, where she is now Professor of Chemistry. Susan maintains an active undergraduate research program at Hartwick and has worked on a number of chemistry textbook projects, including coauthoring an Introduction to General, Organic, and Biochemistry Interactive CD-ROM with Bill Vining.

Roberta Day

Professor Emeritus, University of Massachusetts

Roberta Day received a B.S. in Chemistry from the University of Rochester, Rochester, New York; spent 5 years in the research laboratories of the Eastman Kodak Company, Rochester, New York; and then received a Ph.D. in Physical Chemistry from the Massachusetts Institute of Technology, Cambridge, Massachusetts. After postdoctoral work sponsored by both the Damon Runyon Memorial Fund and the National Institutes of Health, she joined the faculty of the University of Massachusetts, Amherst, rising through the ranks to Full Professor in the Chemistry Department. She initiated the use of online electronic homework in general chemistry at UMass, is one of the inventors of the OWL system, has been either PI or Co-I for several major national grants for the development of OWL, and has authored a large percentage of the questions in the OWL database for General Chemistry. Recognition for her work includes the American Chemical Society Connecticut Valley Section Award for outstanding

Peter W. Samal

William Vining

State University of New York at Oneonta

Bill Vining graduated from SUNY Oneonta in 1981 and earned his Ph.D. in inorganic chemistry at the University of North Carolina-Chapel Hill in 1985, working on the modification of electrode surfaces with polymer-bound redox catalysts. After three years working in industry for S.C. Johnson and Son (Johnson Wax) in Racine, Wisconsin, he became an assistant professor of inorganic chemistry at Hartwick College and eventually department chair. It was here that Bill started working on educational software, first creating the set of simulations called Chemland. This led to work with Jack Kotz on the first General Chemistry CD-ROM and a distance-learning course produced with Archipelago Productions. This work led to a move to the University of Massachusetts, where he served as Director of General Chemistry, which serves 1400 students every semester. He was awarded the University of Massachusetts Distinguished Teaching Award in 1999 and the UMass College of Natural Sciences Outstanding Teacher Award in 2003. At UMass, he also ran a research group dedicated to developing interactive educational software, which included 15 professionals, graduate students, undergraduates, postdoctoral students, programmers, and artists. After nine years at UMass, Bill decided to move back to a primarily undergraduate

contributions to chemistry and the UMass College of Natural Science and Mathematics Outstanding Teacher Award. Her research in chemistry as an x-ray crystallographer has resulted in the publication of more than 180 articles in professional journals. She is now a Professor Emeritus at the University of Massachusetts and continues her work on the development of electronic learning environments for chemistry.

Beatrice Botch

University of Massachusetts

Beatrice Botch is the Director of General Chemistry at the University of Massachusetts. She received her B.A. in Chemistry from Barat College in Lake Forest, Illinois, and her Ph.D. in Physical Chemistry from Michigan State University. She completed her graduate work at Argonne National Laboratory under the direction of Dr. Thom Dunning Jr. and was a post-doctoral fellow at the California Institute of Technology, working in the group of Professor William A. Goddard III. She taught at Southwest State University in Minnesota and Wittenberg University in Ohio before joining the faculty at the University of Massachusetts in 1988. She received the UMass College of Natural Science and Mathematics Outstanding Teacher Award in 1999. She is one of the inventors of OWL, and she authored questions in OWL for General Chemistry. She has been principal investigator and co-investigator on a number of grants and contracts related to OWL development and dissemination and continues to develop learning materials in OWL to help students succeed in chemistry.

To the Student

Welcome to a new integrated approach to chemistry. Chemistry is a continually evolving science that examines and manipulates the world on the atomic and molecular level. In chemistry, it's mostly about the molecules. What are they like? What do they do? How can we make them? How do we even know if we have made them? One of the primary goals of chemistry is to understand matter on the molecular scale well enough to allow us to predict which chemical structures will yield particular properties, and the insight to be able to synthesize those structures.

In this first-year course you will learn about atoms and how they form molecules and other larger structures. You will use molecular structure and the ways atoms bond together to explain the chemical and physical properties of matter on the molecular and bulk scales, and in many cases you will learn to predict these behaviors. One of the most challenging and rewarding aspects of chemistry is that we describe and predict bulk, human scale properties through an understanding of particles that are so very tiny they cannot be seen even with the most powerful optical microscope. So, when we see things happen in the world, we translate and imagine what must be occurring to the molecules that we can't ever see.

Our integrated approach is designed to be one vehicle in your learning; it represents a new kind of learning environment built by making the best uses of traditional written explanations, with interactive activities to help you learn the central concepts of chemistry and how to use those concepts to solve a wide variety of useful and chemically important problems. These readings and activities will represent your homework and as such you will find that your book is your homework, and your homework is your book. In this regard, the interactive reading assignments contain integrated active versions of important figures and tables, reading comprehension questions, and suites of problem solving examples that give you step-by-step tutorial help, recorded "video solutions" to important problems, and practice problems with rich feedback that allow you to practice a problem type multiple times using different chemical examples. In addition to the interactive reading assignments, there are additional OWL problems designed to solidify your understanding of each section as well as end-of-chapter assignments.

The authors of the OWLBook have decades of experience teaching chemistry, talking with students, and developing online chemistry learning systems. For us, this work represents our latest effort to help students beyond our own classrooms and colleges. All in all, we hope that your time with us is rewarding and we wish you the best of luck.

observations on the macroscopic scale and inferring what those observations must mean about atomic scale objects.

For example, careful measurement of the mass of a chemical sample before and after it is heated provides information about the chemical composition of a substance. Observing how a chemical sample behaves in the presence of a strong magnetic field such as that found in a magnetic resonance imaging (MRI) scanner provides information about how molecules and atoms are arranged in human tissues.

An important part of chemistry and science in general is the concept that all ideas are open to challenge. When we perform measurements on chemical substances and interpret the results in terms of atomic scale properties, the results are always examined to see if there are alternative ways to interpret the data. This method of investigation leads to chemical information about the properties and behavior of matter that is supported by the results of many different experiments.

Example Problem 1.1.1 Differentiate between the macroscopic and atomic scales.

Classify each of the following as matter that can be measured or observed on either the macroscopic or atomic scale.
a. An RNA molecule
b. A mercury atom
c. A sample of liquid mercury

Solution:
You are asked to identify whether a substance can be measured or observed on the macroscopic or atomic scales.

You are given the identity of the substance.
a. Atomic scale. An RNA molecule is too small to be seen with the naked eye or with an optical microscope.
b. Atomic scale. Individual atoms cannot be seen with the naked eye or with an optical microscope.
c. Macroscopic scale. Liquid mercury can be seen with the naked eye.

Video Solution

Tutored Practice
Problem 1.1.1

Section 1.1 Mastery

1.2 Classification of Matter

1.2a Classifying Matter on the Atomic Scale

Matter can be described by a collection of characteristics called **properties**. One of the fundamental properties of matter is its composition, or the specific types of atoms or molecules that make it up. An **element**, which is the simplest type of matter, is a pure substance that cannot be broken down or separated into simpler substances. (►Flashforward to Section 2.2 Elements and the Periodic Table) You are already familiar with some of the most common elements such as gold, silver, and copper, which are used in making coins and jewelry, and oxygen, nitrogen, and argon, which are the three most abundant gases in our atmosphere. A total of 118 elements have been identified, 90 of which exist in nature (the rest have been synthesized in the laboratory). Elements are represented by a one- or two-letter element symbol, and they are organized in the periodic table that is shown in Elements and Compounds (Unit 2) and in the Reference Tools. A few common elements and their symbols are shown in Table 1.1.1. Notice that when the symbol for an element consists of two letters, only the first letter is capitalized.

Atoms

An **atom** is the smallest indivisible unit of an element. For example, the element aluminum (Interactive Figure 1.2.1) is made up entirely of aluminum atoms. Although individual

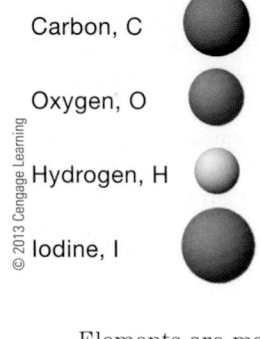

Carbon, C

Oxygen, O

Hydrogen, H

© 2013 Cengage Learning

Iodine, I

atoms are too small to be seen directly with the naked eye or with the use of a standard microscope, methods such as scanning tunneling microscopy (STM) allow scientists to view atoms. Both experimental observations and theoretical studies show that isolated atoms are spherical and that atoms of different elements have different sizes. Thus, the model used to represent isolated atoms consists of spheres of different sizes. In addition, chemists often use color to distinguish atoms of different elements. For example, oxygen atoms are usually represented as red spheres; carbon atoms, as gray or black spheres; and hydrogen atoms, as white spheres.

Elements are made up of only one type of atom. For example, the element oxygen is found in two forms: as O_2, in which two oxygen atoms are grouped together, and as O_3, in which three oxygen atoms are grouped together. The most common form of oxygen is

© 2013 Cengage Learning

Dioxygen, O_2 Ozone, O_3

O_2, dioxygen, a gas that makes up about 21% of the air we breathe. Ozone, O_3, is a gas with a distinct odor that can be toxic to humans. Both dioxygen and ozone are elemental forms of oxygen because they consist of only one type of atom.

Table 1.1.1 Some Common Elements and Their Symbols

Name	Symbol
Hydrogen	H
Carbon	C
Oxygen	O
Sodium	Na
Iron	Fe
Aluminum	Al

Interactive Figure 1.2.1

Explore the composition of elements.

Charles D. Winters

A piece of aluminum

Compounds and Molecules

A **chemical compound** is a substance formed when two or more elements are combined in a defined ratio. Compounds differ from elements in that they can be broken down chemically into simpler substances. You have encountered chemical compounds in many common substances, such as table salt, a compound consisting of the elements sodium and chlorine, and phosphoric acid, a compound found in soft drinks that contains hydrogen, oxygen, and phosphorus.

Water, H_2O

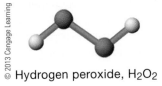

© 2013 Cengage Learning

Hydrogen peroxide, H_2O_2

Molecules are collections of atoms that are held together by chemical bonds. In models used to represent molecules, chemical bonds are often represented using cylinders or lines that connect atoms, represented as spheres. The composition and arrangement of elements in molecules affects the properties of a substance. For example, molecules of both water (H_2O) and hydrogen peroxide (H_2O_2) contain only the elements hydrogen and oxygen. Water is a relatively inert substance that is safe to drink in its pure form. Hydrogen peroxide, however, is a reactive liquid that is used to disinfect wounds and can cause severe burns if swallowed.

Example Problem 1.2.1 Classify pure substances as elements or compounds.

Classify each of the following substances as either an element or a compound.
a. Si   b. CO_2 c. P_4

Solution:

You are asked to classify a substance as an element or a compound.

You are given the chemical formula of the substance.
a. Element. Silicon is an example of an element because it consists of only one type of atom.
b. Compound. This compound contains both carbon and oxygen.
c. Element. Although this is an example of a molecular substance, it consists of only a single type of atom.

Video Solution

Tutored Practice
Problem 1.2.1

1.2b Classifying Pure Substances on the Macroscopic Scale

A **pure substance** contains only one type of element or compound and has fixed chemical composition. A pure substance also has characteristic properties, measurable qualities that are independent of the sample size. The **physical properties** of a chemical substance are those that do not change the chemical composition of the material when they are measured.

Some examples of physical properties include physical state, color, viscosity (resistance to flow), opacity, density, conductivity, and melting and boiling points.

States of Matter

One of the most important physical properties is the physical state of a material. The three physical **states of matter** are solid, liquid, and gas (Interactive Figure 1.2.2).

Interactive Figure 1.2.2

Distinguish the properties of the three states of matter.

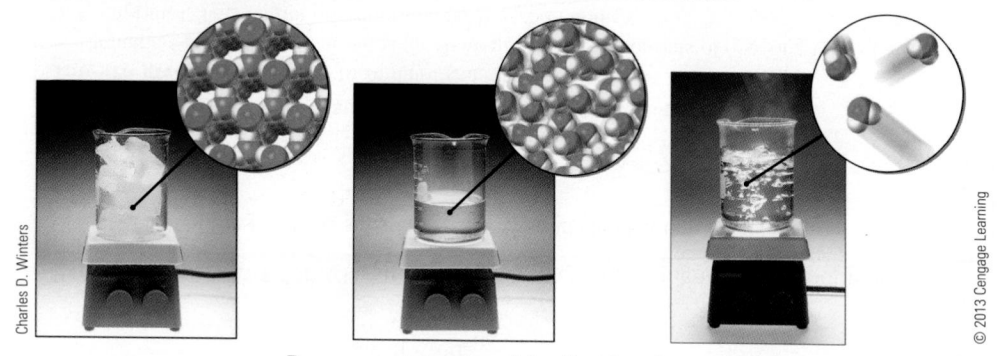

Representations of a solid, a liquid, and a gas

The macroscopic properties of these states are directly related to the arrangement and properties of particles at the atomic level. At the macroscopic level, a **solid** is a dense material with a defined shape. At the atomic level, the atoms or molecules of a solid are packed together closely. The atoms or molecules are vibrating, but they do not move past one another. At the macroscopic level, a **liquid** is also dense, but unlike a solid it flows and takes on the shape of its container. At the atomic level, the atoms or molecules of a liquid are close together, but they move more than the particles in a solid and can flow past one another. Finally, at the macroscopic level, a **gas** has no fixed shape or volume. At the atomic level, the atoms or molecules of a gas are spaced widely apart and are moving rapidly past one another. The particles of a gas do not strongly interact with one another, and they move freely until they collide with one another or with the walls of the container.

The physical state of a substance can change when energy, often in the form of heat, is added or removed. When energy is added to a solid, the temperature at which the solid is converted to a liquid is the **melting point** of the substance. The conversion of liquid to solid occurs at the same temperature as energy is removed (the temperature falls) and is called the **freezing point**. A liquid is converted to a gas at the **boiling point** of a substance. As you will see in the following section, melting and boiling points are measured in Celsius (°C) or Kelvin (K) temperature units.

Not all materials can exist in all three physical states. Polyethylene, for example, does not exist as a gas. Heating a solid polyethylene milk bottle at high temperatures causes it to decompose into other substances. Helium, a gas at room temperature, can be liquefied at very low temperatures, but it is not possible to solidify helium.

A change in the physical property of a substance is called a **physical change**. Physical changes may change the appearance or the physical state of a substance, but they do not change its chemical composition. For example, a change in the physical state of water—changing from a liquid to a gas—involves a change in how the particles are packed together at the atomic level, but it does not change the chemical makeup of the material.

Chemical Properties

The **chemical properties** of a substance are those that involve a chemical change in the material and often involve a substance interacting with other chemicals. For example, a chemical property of methanol, CH_3OH, is that it is highly flammable because the compound burns in air (it reacts with oxygen in the air) to form water and carbon dioxide (Interactive Figure 1.2.3). A **chemical change** involves a change in the chemical composition of the material. The flammability of methanol is a chemical property, and demonstrating this chemical property involves a chemical change.

Charles D. Winters

Interactive Figure 1.2.3

Investigate the chemical properties of methanol.

Methanol is a flammable liquid.

Video Solution

Tutored Practice
Problem 1.2.2

Example Problem 1.2.2 Identify physical and chemical properties and physical and chemical changes.

a. When aluminum foil is placed into liquid bromine a white solid forms. Is this a chemical or physical property of aluminum?
b. Iodine is a purple solid. Is this a chemical or physical property of iodine?
c. Classify each of the following changes as chemical or physical.
  i. Boiling water
  ii. Baking bread

Solution:

You are asked to identify a change or property as chemical or physical.

You are given a description of a material or a change.

a. Chemical property. Chemical properties are those that involve a chemical change in the material and often involve a substance interacting with other chemicals. In this example, one substance (the aluminum) is converted into a new substance (a white solid).

Example Problem 1.2.2 *(continued)*

b. Physical property. A physical property such as color is observed without changing the chemical identity of the substance.

c. i. Physical change. A physical change alters the physical form of a substance without changing its chemical identity. Boiling does not change the chemical composition of water.

   ii. Chemical change. When a chemical change takes place, the original substances (the bread ingredients) are broken down and a new substance (bread) is formed.

1.2c Classifying Mixtures on the Macroscopic Scale

As you can see when you look around you, the world is made of complex materials. Much of what surrounds us is made up of mixtures of different substances. A **mixture** is a substance made up of two or more elements or compounds that have not reacted chemically.

Unlike compounds, where the ratio of elements is fixed, the relative amounts of different components in a mixture can vary. Mixtures that have a constant composition throughout the material are called **homogeneous mixtures**. For example, dissolving table salt in water creates a mixture of the two chemical compounds water (H_2O) and table salt (NaCl). Because the mixture is uniform, meaning that the same ratio of water to table salt is found no matter where it is sampled, it is a homogeneous mixture.

A mixture in which the composition is not uniform is called a **heterogeneous mixture**. For example, a cold glass of freshly squeezed lemonade with ice is a heterogeneous mixture because you can see the individual components (ice cubes, lemonade, and pulp) and the relative amounts of each component will depend on where the lemonade is sampled (from the top of the glass or from the bottom). The two different types of mixtures are explored in Interactive Figure 1.2.4.

Homogeneous and heterogeneous mixtures can usually be physically separated into individual components. For example, a homogeneous mixture of salt and water is separated by heating the mixture to evaporate the water, leaving behind the salt. A heterogeneous mixture of sand and water is separated by pouring the mixture through filter paper. The sand is trapped in the filter while the water passes through. Heating the wet sand to evaporate the remaining water completes the physical separation.

Like pure substances, mixtures have physical and chemical properties. These properties, however, depend on the composition of the mixture. For example, a mixture of 10 grams of table sugar and 100 grams of water has a boiling point of 100.15 °C while a mixture of 20 grams of table sugar and 100 grams of water has a boiling point of 100.30 °C.

Interactive Figure 1.2.5 summarizes how we classify different forms of matter in chemistry.

Interactive Figure 1.2.4

Identify homogeneous and heterogeneous mixtures.

Charles D. Winters

Homogeneous and heterogeneous mixtures

Interactive Figure 1.2.5

Classify matter.

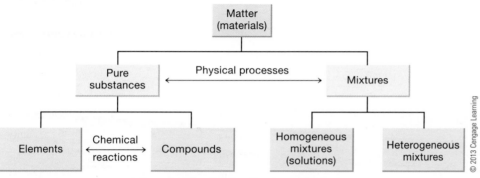

A flow chart for the classification of matter

© 2013 Cengage Learning

Example Problem 1.2.3 Identify pure substances and mixtures.

Classify each of the following as a pure substance, a homogeneous mixture, or a heterogeneous mixture.

a. Copper wire
b. Oil and vinegar salad dressing
c. Vinegar

Solution:

You are asked to classify items as a pure substance, a heterogeneous mixture, or a homogeneous mixture.

You are given the identity of the item.

a. Pure substance. Copper is an element.
b. Heterogeneous mixture. The salad dressing is a mixture that does not have a uniform composition. The different components are visible to the naked eye, and the composition of the mixture varies with the sampling location.
c. Homogeneous mixture. Vinegar is a uniform mixture of water, acetic acid, and other compounds. The different components in this mixture are not visible to the naked eye.

Video Solution

Tutored Practice
Problem 1.2.3

Section 1.2 Mastery

1.3 Units and Measurement

1.3a Scientific Units and Scientific Notation

Chemistry involves observing matter, and our observations are substantiated by careful measurements of physical quantities. Chemists in particular need to make careful measurements because we use those measurements to infer the properties of matter on the atomic scale. Some of the most common measurements in chemistry are mass, volume, time, temperature, and density. Measuring these quantities allows us to describe the chemical and physical properties of matter and study the chemical and physical changes that matter undergoes. When reporting a measurement, we use scientific units to indicate what was measured. **SI units**, abbreviated from the French *Système International d'Unités*, are used in scientific measurements in almost all countries. This unit system consists of seven base units. Other units are called *derived units* and are combinations of the base units (Table 1.3.1).

Metric prefixes are combined with SI units when reporting physical quantities in order to reflect the relative size of the measured quantity. Table 1.3.2 shows the metric prefixes most commonly used in scientific measurements. When making and reporting measurements, it is important to use both the value and the appropriate units. For example, in the United States, speed limits are reported in units of miles per hour (mph).

Table 1.3.1 SI Units

Physical Quantity	SI Unit	Symbol
Base Units		
Length	Meter	m
Mass	Kilogram	kg
Time	Second	s
Electric current	Ampere	A
Temperature	Kelvin	K
Amount of substance	Mole	mol
Luminous intensity	Candela	cd
Some Derived Units		
Volume	—	m^3 (cubic meter)
Density	—	kg/m^3
Energy	—	J (joule)

Table 1.3.2 Common Prefixes Used in the SI and Metric Systems

Prefix	Abbreviation	Factor	
mega	M	1,000,000	(10^6)
kilo	k	1000	(10^3)
deci	d	0.1	(10^{-1})
centi	c	0.01	(10^{-2})
milli	m	0.001	(10^{-3})
micro	m	0.000001	(10^{-6})
nano	n	0.000000001	(10^{-9})
pico	p	0.000000000001	(10^{-12})

A U.S. citizen traveling to Canada might see a speed limit sign reading 100 and assume the units are mph. This could be an expensive mistake, however, because speed limits in Canada are reported in units of kilometers per hour (km/h); a 100 km/h speed limit is the equivalent of 62 mph.

Numbers that are very large or very small can be represented using **scientific notation**. A number written in scientific notation has the general form $N \times 10^x$, where N is a number between 1 and 10 and x is a positive or negative integer. For example, the number 13433 is written as 1.3433×10^4 and the number 0.0058 is written as 5.8×10^{-3} in scientific notation. Notice that x is positive for numbers greater than 1 and negative for numbers less than 1.

To convert a number from standard notation to scientific notation, count the number of times the decimal point must be moved to the right (for numbers less than 1) or to the left (for numbers greater than 1) in order to result in a number between 1 and 10. For the number 13433,

$$13433.$$

the decimal point is moved four places to the left and the number is written 1.3433×10^4. When a number is less than 1, the decimal point is moved to the right and the exponent (x) is negative. For the number 0.0058,

$$0.0058$$

the decimal point is moved three places to the right and the number is written 5.8×10^{-3}. Notice that in both cases, moving the decimal point one place is the equivalent of multiplying or dividing by 10.

To convert a number from exponential notation to standard notation, write the value of N and then move the decimal point x places to the right if x is positive or move the decimal point x places to the left if x is negative.

Example Problem 1.3.1 Write numbers using scientific notation.

a. Write the following numbers in scientific notation:
 i. 0.0000422
 ii. 9700000000
b. Write the following numbers in standard notation:
 i. 7.22×10^6
 ii. 2.5×10^{-3}

Example Problem 1.3.1 *(continued)*

Solution:

You are asked to convert between standard and scientific notation.

You are given a number in standard or scientific notation.

a. i. Moving the decimal point five places to the right results in a number between 1 and 10. The exponent is negative because this number is less than 1.

$$4.22 \times 10^{-5}$$

ii. Moving the decimal point nine places to the left results in a number between 1 and 10. The exponent is positive because this number is greater than 1.

$$9.7 \times 10^{9}$$

b. i. Move the decimal point six places to the right.

$$7220000$$

ii. Move the decimal point three places to the left.

$$0.0025$$

Video Solution

Tutored Practice
Problem 1.3.1

1.3b   SI Base Units and Derived Units

Length

The SI unit of **length**, the longest dimension of an object, is the **meter (m)**. A pencil has a length of about 0.16 m, which is equivalent to 16 centimeters (cm). Atomic radii can be expressed using nanometer (nm) or picometer (pm) units. The definition of the meter is based on the speed of light in a vacuum, exactly 299,792,458 meters per second. One meter is therefore the length of the path traveled by light in a vacuum during 1/299,792,458 of a second.

Mass

The SI unit of **mass**, the measure of the quantity of matter in an object, is the **kilogram (kg)**. This is the only SI base unit that contains a metric prefix. One kilogram is equal to approximately 2.2 pounds (lb). In the chemistry lab, the mass of a sample is typically measured using units of grams (g) or milligrams (mg). The kilogram standard is the mass of a piece of platinum-iridium alloy that is kept at the International Bureau of Weights and Measures.

Temperature

Temperature is a relative measure of how hot or cold a substance is and is commonly reported using one of three temperature scales. In the United States, temperatures are commonly reported using the Fahrenheit temperature scale that has units of degrees

Fahrenheit (°F). In scientific measurements, the Celsius and Kelvin temperature scales are used, with units of degrees Celsius (°C) and kelvins (K), respectively. Notice that for the Kelvin temperature scale, the name of the temperature unit (kelvin) is not capitalized but the abbreviation, K, is capitalized.

As shown in Interactive Figure 1.3.1, the three temperature scales have different defined values for the melting and freezing points of water. In the **Fahrenheit temperature scale**, the freezing point of water is set at 32 °F and the boiling point is 180 degrees higher, 212 °F. In the **Celsius temperature scale**, the freezing point of water is assigned a temperature of 0 °C and the boiling point of water is assigned a temperature of 100 °C. The lowest temperature on the **Kelvin temperature scale**, 0 K, is 273.15 degrees lower than 0 °C. This temperature, known as **absolute zero**, is the lowest temperature possible.

The Celsius and Kelvin temperature scales are similar in that a 1-degree increment is the same on both scales. That is, an increase of 1 K is equal to an increase of 1 °C. Equation 1.1 shows the relationship between the Celsius and Kelvin temperature scales.

$$T(°C) = T(K) - 273.15 \qquad\qquad (1.1)$$

The Fahrenheit and Celsius temperature scales differ in the size of a degree.

$$180 \text{ Fahrenheit degrees} = 100 \text{ Celsius degrees}$$

$$\frac{9}{5} \text{ Fahrenheit degrees} = 1 \text{ Celsius degree}$$

Equation 1.2 shows the relationship between the Fahrenheit and Celsius temperature scales.

$$T(°F) = \frac{9}{5}\left[T(°C)\right] + 32 \qquad\qquad (1.2)$$

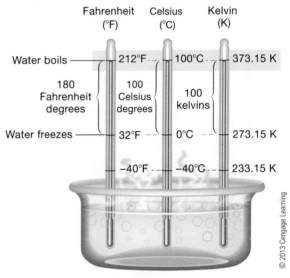

Interactive Figure 1.3.1

Compare different temperature scales.

Fahrenheit, Celsius, and Kelvin temperature scales

Example Problem 1.3.2 Interconvert Fahrenheit, Celsius, and Kelvin temperatures.

The boiling point of a liquid is 355.78 K. What is this temperature on the Celsius and Fahrenheit scales?

Solution:

You are asked to convert a temperature from kelvin to Celsius and Fahrenheit units.

You are given a temperature in kelvin units.

Convert the temperature to Celsius temperature units.

$$T(°C) = T(K) - 273.15$$

$$T(°C) = 355.78 \text{ K} - 273.15 = 82.63 \text{ °C}$$

Example Problem 1.3.2 *(continued)*

Use the temperature in Celsius units to calculate the temperature on the Fahrenheit scale.

$$T(°F) = \frac{9}{5}[T(°C)] + 32$$

$$T(°F) = \frac{9}{5}(82.63\ °C) + 32$$

$$T(°F) = 180.73\ °F$$

Is your answer reasonable? The Celsius temperature should be greater than zero because the Kelvin temperature is greater than 273.15 K, which is equal to 0 °C. The Celsius temperature is close to 100 °C, the boiling point of water, so it is reasonable for the Fahrenheit temperature to be 180.73 °F because this is close to the boiling point of water on the Fahrenheit scale (212 °F).

Video Solution

Tutored Practice
Problem 1.3.2

Volume

Although the SI unit of **volume** (the amount of space a substance occupies) is the cubic meter (m^3), a more common unit of volume is the **liter (L)**. Notice that the abbreviation for liter is a capital L. A useful relationship to remember is that one **milliliter** is equal to one cubic centimeter (1 mL = 1 cm^3).

Energy

The SI unit of **energy**, the capacity to do work and transfer heat, is the **joule (J)**. Another common energy unit is the **calorie (cal)**, and one calorie is equal to 4.184 joules (1 cal = 4.184 J). One dietary calorie (Cal) is equal to 1000 calories (1 Cal = 1000 cal = 1 kcal).

Density

The **density** of a substance is a physical property that relates the mass of a substance to its volume (Equation 1.3).

$$\text{density} = \frac{\text{mass}}{\text{volume}} \tag{1.3}$$

The densities of solids and liquids are reported in units of grams per cubic centimeter (g/cm^3) or grams per milliliter (g/mL), whereas the density of a gas is typically reported in units of grams per liter (g/L). Because the volume of most substances changes with a change in temperature, density also changes with temperature. Most density values are reported at a standard temperature, 25 °C, close to room temperature. The densities of some common substances are listed in Interactive Table 1.3.3. Density can be calculated from mass and volume data as shown in the following example.

Example Problem 1.3.3 Calculate density.

A 5.78-mL sample of a colorless liquid has a mass of 4.54 g. Calculate the density of the liquid and identify it as either ethanol (density = 0.785 g/mL) or benzene (density = 0.874 g/mL).

Solution:

You are asked to calculate the density of a liquid and identify the liquid.

You are given the mass and volume of a liquid and the density of two liquids.

Use Equation 1.3 to calculate the density of the liquid.

$$\text{density} = \frac{\text{mass}}{\text{volume}} = \frac{4.54\ \text{g}}{5.78\ \text{mL}} = 0.785\ \text{g/mL}$$

The liquid is ethanol.

Interactive Table 1.3.3

Densities of Some Common Substances at 25 °C

Substance	Density (g/cm³)
Iron	7.86
Sodium chloride	2.16
Water	1.00
Oxygen	1.33×10^{-3}

Video Solution

Tutored Practice Problem 1.3.3

1.3c Significant Figures, Precision, and Accuracy

The certainty in any measurement is limited by the instrument that is used to make the measurement. For example, an orange weighed on a grocery scale weighs 249 g. A standard laboratory balance, however, like the one shown in Figure 1.3.2, reports the mass of the same orange as 249.201 g. In both cases, some uncertainty is present in the measurement. The grocery scale measurement is certain to the nearest 1 g, and the value is reported as 249 ± 1 g. The laboratory scale measurement has less uncertainty, and the mass of the orange is reported as 249.201 ± 0.001 g. In general, we will drop the ± symbol and assume an uncertainty of one unit in the rightmost digit when reading a measurement. When using a nondigital measuring device such as a ruler or a graduated cylinder, we always estimate the rightmost digit when reporting the measured value. A digital measuring device such as a top-loading laboratory balance or pH meter includes the estimated digit in its readout.

Some measured quantities are infinitely certain, or exact. For example, the number of oranges you have purchased at the grocery store is an exact number. Some units are defined with exact numbers, such as the metric prefixes (1 mm = 0.001 m) and the relationship between inches and centimeters (1 in = 2.54 cm, exactly).

The digits in a measurement, both the certain and uncertain digits, are called **significant figures** or significant digits. For example, the mass of an orange has three significant figures when measured using a grocery scale (249 g) and six significant figures when measured on a laboratory balance (249.201 g). Some simple rules are used to determine the

Figure 1.3.2 A top-loading laboratory balance

number of significant figures in a measurement (Interactive Table 1.3.4). For example, consider the numbers 0.03080 and 728060.

Rules for Determining Significant Figures

1. All non-zero digits and zeros between non-zero digits are significant.
   In 0.03080, the digits 3 and 8 and the zero between 3 and 8 are significant. In 728060, the digits 7, 2, 8, and 6 and the zero between 8 and 6 are significant.

2. In numbers containing a decimal point,
   a. all zeros at the end of the number are significant.
      In 0.03080, the zero to the right of 8 is significant.
   b. all zeros at the beginning of the number are not significant.
      In 0.03080, both zeros to the left of 3 are not significant.

3. In numbers with no decimal point, all zeros at the end of the number are not significant.
   In 728060, the zero to the right of 6 is not significant.

Thus, the number 0.03080 has four significant figures and the number 728060 has five significant figures.

Notice that for numbers written in scientific notation, the number of significant figures is equal to the number of digits in the number written before the exponent. For example, the number 3.25×10^{-4} has three significant figures and 1.200×10^3 has four significant figures.

Example Problem 1.3.4 Identify the significant figures in a number.

Identify the number of significant figures in the following numbers.
a. 19.5400   b. 0.0095   c. 1030

Solution:

You are asked to identify the number of significant figures in a number.

You are given a number.
a. All non-zero digits are significant (there are four), and because this number has a decimal point, the zeros at the end of the number are also significant (there are two). This number has six significant figures.
b. All non-zero digits are significant (there are two), and because this number has a decimal point, the three zeros at the beginning of the number are not significant. This number has two significant figures.
c. All non-zero digits are significant (there are two), and the zero between the non-zero digits is also significant (there is one). Because this number has no decimal, the zero at the end of the number is not significant. This number has three significant figures.

Video Solution

Tutored Practice
Problem 1.3.4

When calculations involving measurements are performed, the final calculated result is no more certain than the least certain number in the calculation. If necessary, the answer is rounded to the correct number of significant figures. For example, consider a density calculation involving a sample with a mass of 3.2 g and a volume of 25.67 cm³. A standard calculator reports a density of 0.124659135 g/cm³, but is this a reasonable number? In this case, the least certain number is the sample mass with two significant figures, and the calculated density therefore has two significant figures, or 0.12 g/cm³.

The final step in a calculation usually involves rounding the answer so that it has the correct certainty. When numbers are rounded, the last digit retained is increased by 1 only if the digit that follows is 5 or greater. When you are performing calculations involving multiple steps, it is best to round only at the final step in the calculation. That is, carry at least one extra significant figure during each step of the calculation to minimize rounding errors in the final calculated result.

When multiplication or division is performed, the certainty in the answer is related to the significant figures in the numbers in the calculation. The answer has the same number of significant figures as the measurement with the fewest significant figures. For example,

$$39.485 \times 6.70 = 264.5495 \quad \text{Round off to 265}$$

Number of significant figures: *5*   *3*   *3*

$$0.029 \div 1.285 = 0.022568093 \quad \text{Round off to 0.023}$$

Number of significant figures: *2*   *4*   *2*

When addition or subtraction is performed, the certainty of the answer is related to the decimal places present in the numbers in the calculation. The answer has the same number of decimal places as the number with the fewest decimal places. For example,

$$12.50 + 6.080 = 18.580 \quad \text{Round off to 18.58}$$

Number of decimal places: *2*   *3*   *2*

$$125.2 - 0.07 = 125.13 \quad \text{Round off to 125.1}$$

Number of decimal places: *1*   *2*   *1*

Exact numbers do not limit the number of significant figures or the number of decimal places in a calculated result. For example, suppose you want to convert the amount of time it takes for a chemical reaction to take place, 7.2 minutes, to units of seconds. There are exactly 60 seconds in 1 minute, so it is the number of significant figures in the measured number that determines the number of significant figures in the answer.

$$7.2 \text{ minutes} \times 60 \text{ seconds/minute} = 430 \text{ seconds}$$

Number of significant figures: *2*   *Exact number*   *2*

Example Problem 1.3.5 Use significant figures in calculations.

Carry out the following calculations, reporting the answer using the correct number of significant figures.

a. $(8.3145)(1.3 \times 10^{-3})$

b. $\dfrac{25 + 273.15}{1.750}$

Solution:

You are asked to carry out a calculation and report the answer using the correct number of significant figures.

You are given a mathematical operation.

a. The numbers in this multiplication operation have five (8.3145) and two (1.3×10^{-3}) significant figures. Therefore, the answer should be reported to two significant figures:

$$(8.3145)(1.3 \times 10^{-3}) = 0.011$$

b. This calculation involves both addition and division, so it must be completed in two steps. First, determine the number of significant figures that result from the addition operation in the numerator of this fraction. The first value, 25, has no decimal places, and the second value, 273.15, has two decimal places. Therefore, the sum should have no decimal places and, as a result, has three significant figures.

$$25 + 273.15 = 298$$

Now the division operation can be performed. The numbers in this operation have three (298) and four (1.750) significant figures. Therefore, the answer should be reported to three significant figures.

$$\frac{25 + 273.15}{1.750} = 170$$

When doing multistep calculations such as this one, it is a good idea to identify the correct number of significant figures that result from each operation but to round to the correct number of significant figures only in the last step of the calculation. This will help minimize rounding errors in your answers.

Along with the number of significant figures in a value, the certainty of a measurement can also be described using the terms *precision* and *accuracy*. **Precision** is how close the values in a set of measurements are to one another. **Accuracy** is how close a measurement or a set of measurements is to a real value. Interactive Figure 1.3.3 demonstrates the relationship between precision and accuracy.

Video Solution

Tutored Practice
Problem 1.3.5

Interactive Figure 1.3.3

Distinguish between precision and accuracy.

Charles D. Winters

These darts were thrown with precision but without accuracy.

Section 1.3 Mastery

1.4 Unit Conversions

1.4a Dimensional Analysis

Quantities of matter, energy, or time can be expressed using different units. To make conversions between different units we use conversion factors derived from equalities. For example, the equality that relates hours and minutes is 1 hour = 60 minutes. Both 1 hour and 60 minutes represent the same amount of time but with different units. When an equality is written as a ratio, it is called a **conversion factor**.

We use conversion factors to change the units of a quantity without changing the amount using the method called **dimensional analysis**. In dimensional analysis, the original value is multiplied by the conversion factor. Because both the numerator and the denominator of a conversion factor are equal, the ratio is equal to 1 and multiplying by this ratio does not change the original value; it changes only the units in which it is expressed.

For example, you can use the conversion factor 1 hour = 60 minutes to express 48 hours in units of minutes, as shown in the following example.

Example Problem 1.4.1 Use dimensional analysis to convert units.

Using dimensional analysis, express 48 hours in units of minutes (1 hour = 60 minutes).

Solution:

You are asked to express a quantity in different units.

You are given a number and an equality.

Step 1. Write the quantity in its current units and identify the desired new units:

48 hours × conversion factor = _____ minutes

Step 2. Identify an equality that relates the current units to the desired units and write it as a ratio.

The equality 1 hour = 60 minutes can be written as a ratio in two ways:

$$\frac{1 \text{ hour}}{60 \text{ min}} \quad \text{or} \quad \frac{60 \text{ min}}{1 \text{ hour}}$$

Because our goal is to convert units of hours to units of minutes we will use the second form of the conversion factor. Remember that we will multiply the original value by the conversion factor. The second form of the conversion factor will allow the hours units to be cancelled in the multiplication step.

Example Problem 1.4.2 Use dimensional analysis with more than one conversion factor.

Using dimensional analysis, express the volume of soda in a soft drink can (12 oz.) in units of milliliters. (1 gallon = 128 oz.; 1 gallon = 3.7854 L; 1 L = 1000 mL)

Solution:

You are asked to express a quantity in different units.

You are given a number and an equality.

Step 1. Write the quantity in its current units and identify the desired new units.

$$12 \text{ oz.} \times \text{conversion factor} = \underline{\hspace{2cm}} \text{ mL}$$

Step 2. Identify equalities that will convert the given units to the desired units and write them as ratios.

It is common to combine conversion factors in sequence to perform a unit conversion if you are not provided with a single conversion factor that converts from the given unit to the desired unit, as shown in the following example.

Example Problem 1.4.1 (*continued*)

Step 3. Multiply the original value by the conversion factor and make sure that the original units are cancelled by the same units in the denominator. The desired units should appear in the numerator.

$$48 \text{ hours} \times \frac{60 \text{ min}}{1 \text{ hour}} = \underline{\hspace{2cm}} \text{ minutes}$$

Notice that the amount of time will not change when we perform this multiplication because the conversion factor is derived from an equality and is therefore equal to 1 (60 minutes is equal to 1 hour).

Step 4. Perform the multiplication and cancel any units that appear in both the numerator and denominator.

$$48 \text{ hours} \times \frac{60 \text{ min}}{1 \text{ hour}} = 2900 \text{ minutes}$$

Notice that the answer is reported to two significant figures. The equality 1 hour = 60 minutes is exact and therefore has an infinite number of significant figures. The initial value, 48 hours, has two significant figures.

Is your answer reasonable? One hour is the equivalent of 60 minutes, so 48 hours is the equivalent of a large number of minutes, much greater than 60 minutes.

Video Solution

Tutored Practice
Problem 1.4.1

Example Problem 1.4.2 *(continued)*

First convert units of ounces (oz.) to units of gallons (gal). The next conversion changes units of gallons (gal) to units of liters (L). Finally, convert units of liters (L) to units of milliliters (mL).

$$\text{Conversion factor 1:} \qquad \frac{1 \text{ gal}}{128 \text{ oz.}}$$

$$\text{Conversion factor 2:} \qquad \frac{3.7854 \text{ L}}{1 \text{ gal}}$$

$$\text{Conversion factor 3:} \qquad \frac{1000 \text{ mL}}{1 \text{ L}}$$

Step 3. Multiply the original value by the conversion factors and make sure that the original units are cancelled by the same units in the denominator. The desired units should appear in the numerator.

$$12 \text{ oz.} \times \frac{1 \text{ gal}}{128 \text{ oz.}} \times \frac{3.7854 \text{ L}}{1 \text{ gal}} \times \frac{1000 \text{ mL}}{1 \text{ L}} = \underline{\qquad} \text{ mL}$$

Step 4. Perform the multiplication and division and cancel any units that appear in both the numerator and denominator.

$$12 \text{ o\!z.} \times \frac{1 \text{ g\!a\!l}}{128 \text{ o\!z.}} \times \frac{3.7854 \text{ \!L}}{1 \text{ g\!a\!l}} \times \frac{1000 \text{ mL}}{1 \text{ \!L}} = 350 \text{ mL}$$

Notice that the initial value limits the answer to two significant figures.

Is your answer reasonable? One liter is approximately equal to one quart, and a can of soda has a volume that is less than half a quart. Thus, it is reasonable for the volume of a can of soda to be less than 500 mL.

1.4b   Unit Conversions Using Density

Density, the relationship between mass and volume, is also an equality that relates the mass and volume of a substance. Whether working in a chemistry lab or a kitchen (Interactive Figure 1.4.1), you can use the density of a material to convert between units of mass and volume.

Example Problem 1.4.3 Convert units using density.

Use the density of diamond ($d = 3.52$ g/cm^3) to determine the volume of a 1.2-carat diamond. (1 carat = 0.200 g)

Video Solution

Tutored Practice
Problem 1.4.2

Interactive Figure 1.4.1

Use density in calculations.

Charles D. Winters

Some household products with different densities.

Example Problem 1.4.3 *(continued)*

Solution:

You are asked to calculate the volume of a solid from its mass.

You are given the mass of the solid and its density.

Step 1. Write the quantity in its current units and identify the desired new units.

$$1.2 \text{ carat} \times \text{conversion factor} = \underline{\hspace{2cm}} \text{ cm}^3$$

Step 2. Identify equalities that will convert the given units to the desired units and write them as ratios.

First convert from carats to units of grams. Next, use density to convert mass units (g) to volume units (cm^3).

Conversion factor 1: $\dfrac{0.200 \text{ g}}{1 \text{ carat}}$

Conversion factor 1: $\dfrac{1 \text{ cm}^3}{3.52 \text{ g}}$

Notice that the density conversion factor is inverted. Density can be expressed as an equality ($3.54 \text{ g} = 1 \text{ cm}^3$), so it can be written with either mass or volume in the numerator. In this example, the desired unit, cubic centimeters (cm^3), appears in the numerator.

Step 3. Multiply the original value by the correct conversion factors and make sure that the original unit is cancelled by the same unit in the denominator. The desired unit should appear in the numerator.

$$1.2 \text{ carat} \times \frac{0.200 \text{ g}}{1 \text{ carat}} \times \frac{1 \text{ cm}^3}{3.52 \text{ g}} = \underline{\hspace{2cm}} \text{ cm}^3$$

Step 4. Perform the multiplication and cancel any units that appear in both the numerator and denominator.

$$1.2 \cancel{\text{ carat}} \times \frac{0.200 \cancel{\text{ g}}}{1 \cancel{\text{ carat}}} \times \frac{1 \text{ cm}^3}{3.52 \cancel{\text{ g}}} = 0.068 \text{ cm}^3$$

Is your answer reasonable? While a 1.2-carat diamond is pretty large in the world of diamonds, it does have a relatively small volume. It is reasonable for the volume of this solid to be less than 1 cm^3.

Video Solution

Tutored Practice Problem 1.4.3

Section 1.4 Mastery

Unit Recap

Key Concepts

1.1 What Is Chemistry?

- Chemistry is the study of matter, its transformations, and how it behaves (1.1a).
- Matter is any physical substance that occupies space and has mass (1.1a).
- The macroscopic scale is used to describe matter that can be seen with the naked eye, and the atomic scale is used to describe individual atoms and molecules (1.1a).

1.2 Classification of Matter

- Matter is described by a collection of characteristics called properties (1.2a).
- An element is a pure substance that cannot be broken down into simpler substances and an atom is the smallest indivisible unit of an element (1.2a).
- A chemical compound is a substance formed when two or more elements are combined in a defined ratio (1.2a).
- Molecules are collections of atoms that are held together by chemical bonds (1.2a).
- A pure substance contains only one type of element or compound and has a fixed chemical composition (1.2b).
- The physical properties of a chemical substance are those that do not change the composition of the material when they are measured, such as physical state (solid, liquid, or gas), melting point, and boiling point (1.2b).
- The chemical properties of a substance are those that involve a chemical change in the material (1.2b).
- A solid is a material with a defined shape in which the atoms or molecules are packed together closely. A liquid is a material that flows and takes on the shape of its container because its particles can flow past one another. A gas is a material that has no fixed shape or volume because its particles are widely spaced and move rapidly past one another (1.2b).
- A change in the physical property of a substance is a physical change, whereas a chemical change involves a change in the chemical composition of the material (1.2b).
- A mixture is a substance made up of two or more elements or compounds that have not reacted chemically and can be homogeneous (the composition is constant throughout the mixture) or heterogeneous (the composition is not uniform throughout the mixture) (1.2c).

1.3 Units and Measurement

- SI units are the standard set of scientific units used to report measurements in chemistry (1.3a).
- Metric prefixes are used with SI units to reflect the relative size of a measured quantity (1.3a).
- Scientific notation is used to represent very large or very small numbers (1.3a).
- Temperature is a relative measure of how hot or cold a substance is and is reported using the Fahrenheit, Celsius, or Kelvin temperature scales (1.3b).
- Absolute zero (0 K) is the lowest temperature possible (1.3b).
- Density is the ratio of the mass of a substance to its volume (1.3b).
- Significant figures are the certain and uncertain digits in a measurement (1.3c).
- When multiplication or division is performed, the number of significant figures in the answer is equal to the number of significant figures in the least certain measurement (1.3c).
- When addition or subtraction is performed, the number of decimal places in the answer is equal to the number of decimal places in the measurement with the fewest decimal places (1.3c).
- Precision, how close the values in a set of measurements are to one another, and accuracy, how close a measurement or set of measurements is to a real value, are ways of describing the certainty of a measurement or set of measurements (1.3c).

1.4 Unit Conversions

- A conversion factor is an equality written in the form of a ratio (1.4a).
- Dimensional analysis is the use of conversion factors to change the units of a quantity without changing its amount (1.4a).

Key Equations

$$T(°C) = T(K) - 273.15 \qquad (1.1)$$

$$T(°F) = \frac{9}{5}[T(°C)] + 32 \qquad (1.2)$$

$$\text{density} = \frac{\text{mass}}{\text{volume}} \qquad (1.3)$$

Key Terms

1.1 What Is Chemistry?
chemistry
matter
macroscopic scale
atomic scale

1.2 Classification of Matter
properties
element
atom
chemical compound
molecule
pure substance
physical properties
states of matter
solid
liquid
gas

melting point
freezing point
boiling point
physical change
chemical properties
chemical change
mixture
homogeneous mixture
heterogeneous mixture

1.3 Units and Measurement
SI units
scientific notation
length
meter (m)
kilogram (kg)
mass
temperature

Fahrenheit temperature scale
Celsius temperature scale
Kelvin temperature scale
absolute zero
volume
liter (L)
milliliter (mL)
energy
joule (J)
calorie (cal)
density
significant figures
precision
accuracy

1.4 Unit Conversions
conversion factor
dimensional analysis

Unit 1 Review and Challenge Problems

2

Elements and Compounds

Unit Outline

In This Unit...

As we learned in Chemistry: Matter on the Atomic Scale (Unit 1), each of the 118 known elements is composed of a unique type of atom. In this unit, we discover that there are actually a number of different variations, or isotopes, of the atoms associated with each element. We explore the structure of the atom in further detail and learn about the composition of isotopes. We also discuss the different ways that molecules and compounds are represented and named.

2.1 The Structure of the Atom

2.1a Components of an Atom

Elements are characterized by the number and type of particles of which they are composed. Atoms, the smallest chemical unit of matter, consist of three **subatomic particles**: protons, neutrons, and electrons. A **proton** carries a relative charge of +1 and has a mass of 1.672622×10^{-24} g. A **neutron** carries no electrical charge and has a mass of 1.674927×10^{-24} g. An **electron** carries a relative charge of −1 and has a mass of 9.109383×10^{-27} g.

Two of the subatomic particles, protons and neutrons, are found in the **atomic nucleus**, a very small region of high density at the center of the atom. Electrons are found in the region around the nucleus. As you will see when we study atomic structure in more detail in Electromagnetic Radiation and the Electronic Structure of the Atom (Unit 6), the precise location of electrons is not determined. (▶Flashforward to Section 6.4 Quantum Theory of Atomic Structure) Instead, we visualize an electron cloud surrounding the nucleus that represents the most probable location of electrons (Interactive Figure 2.1.1).

The atom represented in Interactive Figure 2.1.1 is not drawn to scale. In reality, electrons account for most of the volume of an atom, and the nucleus of an atom is about 1/10,000 the diameter of a typical atom. For example, if an atom had a diameter the same size as a football field, 100 yards, or about 90 meters, the nucleus of the atom would have a diameter of only about 1 cm!

Mass and Charge of an Atom

The mass and charge of an atom affect the physical and chemical properties of the element and the compounds it forms. As shown in Table 2.1.1, the three subatomic particles are easily differentiated by both charge and mass. The mass of an atom is almost entirely

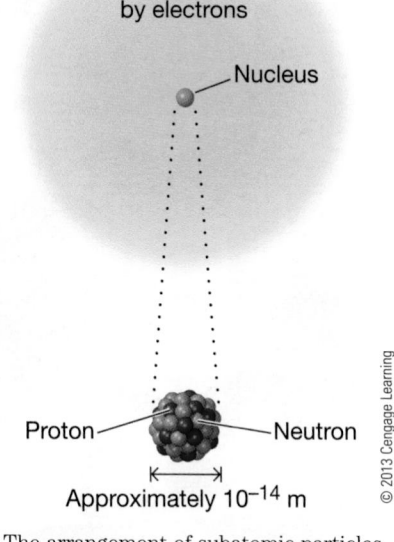

Interactive Figure 2.1.1

Explore the components of an atom.

Approximately 10^{-10} m

Region occupied by electrons

Nucleus

Proton — Neutron

Approximately 10^{-14} m

© 2013 Cengage Learning

The arrangement of subatomic particles in an atom (not drawn to scale)

Table 2.1.1 Properties of Subatomic Particles

	Actual Mass (kg)	Relative Mass	Mass (u)	Actual Charge (C)	Relative Charge
Proton (p)	1.672622×10^{-27}	1836	1.007276	1.602×10^{-19} C	+1
Neutron (n)	1.674927×10^{-27}	1839	1.008665	0	0
Electron (e⁻)	9.109383×10^{-31}	1	5.485799×10^{-4} (≈ 0)	-1.602×10^{-19} C	−1

accounted for by its dense nucleus of protons and neutrons. The actual mass of protons and neutrons is very small, so it is more convenient to define the mass of these particles using a different unit. The **atomic mass unit (u)** is defined as 1/12 the mass of a carbon atom that contains six protons and six neutrons. Because neutrons and protons have very similar masses, both have a mass of approximately 1 u. The mass of an electron is about 2000 times less that of protons and neutrons, and it has a mass of approximately zero on the atomic mass unit scale.

Atoms are neutral because protons and electrons have equal, opposite charges and atoms have equal numbers of positively charged protons and negatively charged electrons. An **ion** is an atom with an unequal number of protons and electrons; because the number of protons and electrons is not equal, the atom has an overall positive or negative charge. When an atom carries more protons than electrons, it carries an overall positive charge and is called a **cation**. An atom with more electrons than protons has an overall negative charge and is called an **anion**. As discussed later in this unit, ions have very different properties than the elements they are derived from.

2.1b Atomic Number, Mass Number, and Atomic Symbols

Atoms of each element can be distinguished by the number of protons in the nucleus. The **atomic number (Z)** of an element is equal to the number of protons in the nucleus. For example, a carbon atom has six protons in its nucleus, and therefore carbon has an atomic number of six ($Z = 6$). Each element has a unique atomic number, and all atoms of that element have the same number of protons in the nucleus. All atoms of hydrogen have 1 proton in the nucleus ($Z = 1$), and all atoms of gold have 79 protons in the nucleus ($Z = 79$). Because protons carry a positive charge ($+1$), in a neutral atom the atomic number also equals the number of electrons (-1 charge) in that atom.

An atom can also be characterized by its mass. Because the mass of electrons is negligible, the mass of an atom in atomic mass units (u) is essentially equal to the number of protons and neutrons in the nucleus of an atom, called the **mass number (A)**. For example, a carbon atom with 6 protons and 6 neutrons in its nucleus has a mass number of 12 ($A = 12$), and a gold atom with 79 protons and 119 neutrons in its nucleus has a mass number of 198 ($A = 198$).

$$\text{mass number} \longrightarrow {}^A_Z X \longleftarrow \text{element symbol}$$
$$\text{atomic number}$$

$$6\,p + 6\,n \longrightarrow {}^{12}_{6}C \qquad 79\,p + 119\,n \longrightarrow {}^{198}_{79}Au$$
$$6\,p \longrightarrow \qquad\qquad\qquad 79\,p \longrightarrow$$

The **atomic symbol** for an element (also called the *nuclear symbol*) consists of the one- or two-letter symbol that represents the element along with the atomic number, written as a subscript number, and the mass number, written as a superscript number. For example, the atomic symbol for a carbon (C) atom with 6 protons and 6 neutrons is ${}^{12}_{6}C$, and the symbol for a gold (Au) atom with 79 protons and 119 neutrons is ${}^{198}_{79}Au$. Note that the number of neutrons in an atom is equal to the difference between the mass number and the atomic number.

Example Problem 2.1.1 **Write atomic symbols.**

Write the atomic symbol for the following atoms.
a. A nitrogen atom containing 7 protons, 8 neutrons, and 7 electrons
b. A uranium atom containing 92 protons, 143 neutrons, and 92 electrons

Solution:

You are asked to write the atomic symbol for an atom.

You are given the number of protons, neutrons, and electrons in the atom.

a. ${}^{15}_{7}N$. The atomic number of nitrogen is equal to the number of protons (7), and the mass number is equal to the number of protons plus the number of neutrons (7 + 8 = 15).

b. ${}^{235}_{92}Au$. The atomic number of uranium is equal to the number of protons (92), and the mass number is equal to the number of protons plus the number of neutrons (92 + 143 = 235).

Video Solution

Tutored Practice
Problem 2.1.1

2.1c Isotopes and Atomic Weight

Although all atoms of a given element have the same number of protons, the number of neutrons found in the nucleus for a particular element can vary. For example, although atoms of carbon always have six protons in the nucleus, a carbon atom might have six, seven, or even eight neutrons in the nucleus (Interactive Figure 2.1.2). Even though the mass number for these atoms differs, each is a carbon atom because each has six protons in the nucleus.

© 2013 Cengage Learning

Interactive Figure 2.1.2

Explore isotopes.

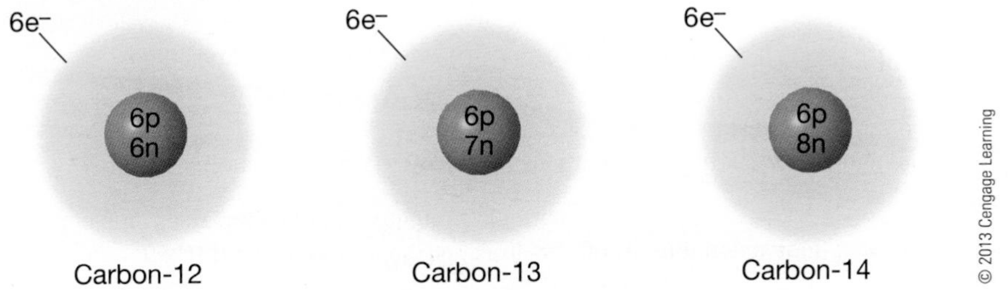

Carbon-12   Carbon-13   Carbon-14
Three isotopes of carbon

Every carbon atom has six protons, and the mass of electrons is negligible; this means we can conclude that the carbon atoms shown in Interactive Figure 2.1.2 have different mass numbers because each has a different number of neutrons. Atoms that have the same atomic number (Z) but different mass numbers (A) are called **isotopes**. Isotopes are named using the element name and the mass number. For example, the isotopes shown in Interactive Figure 2.1.2 are named carbon-12, carbon-13, and carbon-14. The atomic symbols for these elements can be written ^{12}C, ^{13}C, and ^{14}C. Notice that because the atomic number is always the same for a given element, Z is sometimes omitted from the atomic symbol for an isotope.

Atomic Weight

Most samples of an element in nature contain a mixture of various isotopes. Fluorine is an example of an element with only one naturally occurring isotope. Tin, in contrast, has 10 naturally occurring isotopes. When we talk about the mass of an atom of a certain element, therefore, we must take into account that any sample of that element would include different isotopes with different masses. The **atomic weight** for any element is the average mass of all naturally occurring isotopes of that element, taking into account the relative abundance of the isotopes. (Because the atomic weight of an element is actually a mass, not a weight, the term *atomic mass* is often used in its place.) We use **percent abundance**, the percentage of the atoms of a natural sample of the pure element represented by a particular isotope, to describe isotype composition for an element. For example, chlorine ($Z = 17$) has two naturally occurring isotopes, ^{35}Cl and ^{37}Cl. The percent abundance of these two isotopes is 75.78% ^{35}Cl and 24.22% ^{37}Cl. In other words, in any sample of chlorine, about 3/4 of the atoms are ^{35}Cl and about 1/4 are ^{37}Cl. Because there are more ^{35}Cl atoms than ^{37}Cl atoms in the sample, the average mass of chlorine is closer to that of ^{35}Cl than to that of ^{37}Cl. Atomic weight is a weighted average of the atomic masses of all isotopes for a particular element.

Average atomic weight depends on both the mass of each isotope present and the relative abundance of that isotope. To calculate the average atomic weight for an element, the fractional abundance and the exact mass of the isotopes are summed as shown in Equation 2.1 and the example that follows.

$$\text{average atomic weight} = \sum_{\substack{\text{all} \\ \text{isotopes}}} (\text{exact mass})(\text{fractional abundance}) \qquad \textbf{(2.1)}$$

Example Problem 2.1.2 Calculate average atomic weight.

Calculate the average atomic weight for chlorine. Chlorine has two naturally occurring isotopes, chlorine-35 (34.96885 u, 75.78% abundant) and chlorine-37 (36.96590 u, 24.22% abundant).

Solution:

You are asked to calculate the average atomic weight for chlorine.

You are given the relative abundance of the chlorine isotopes and the exact mass of each isotope.

Use Equation 2.1, the exact mass of the isotopes, and the fractional abundance of the isotopes to calculate the average atomic weight of chlorine.

$$\text{average atomic weight (Cl)} = (^{35}\text{Cl exact mass})(^{35}\text{Cl fractional abundance})$$
$$+ (^{37}\text{Cl exact mass})(^{37}\text{Cl fractional abundance})$$

$$\text{average atomic weight (Cl)} = (34.96885 \text{ u})\left(\frac{75.78}{100}\right) + (36.96560 \text{ u})\left(\frac{24.22}{100}\right)$$

$$\text{average atomic weight (Cl)} = 35.45 \text{ u}$$

Is your answer reasonable? The average mass of chlorine should be closer to 35 u than 37 u because the chlorine-35 isotope is more abundant than the chlorine-37 isotope.

Video Solution

Tutored Practice
Problem 2.1.2

Section 2.1 Mastery

2.2 Elements and the Periodic Table

2.2a Introduction to the Periodic Table

The **periodic table of the elements** (Interactive Figure 2.2.1) is the most important tool that chemists use. Not only does it contain information specific to each element, but it also organizes the elements according to their physical and chemical properties.

Each entry in the periodic table represents a single element and contains the element's chemical symbol, atomic number, and average atomic weight. The elements are arranged in vertical columns called **groups** and horizontal rows called **periods**. The elements within each group have similar chemical and physical properties. The periodic, repeating properties of the elements within groups is one of the most important aspects of the periodic table, as you will see in Electron Configurations and the Properties of Atoms (Unit 7). (▶ Flashforward to Section 7.4 Properties of Atoms)

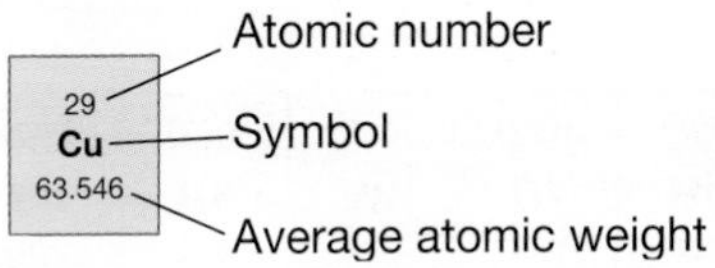

29 — Atomic number
Cu — Symbol
63.546 — Average atomic weight

© 2013 Cengage Learning

Some of the groups in the periodic table are given special names (Table 2.2.1) to reflect their common properties. For example, the Group 1A elements, the **alkali metals**, are all shiny solids that react vigorously with air, water, and **halogens**—the Group 7A elements. Most of the Group 2A elements, the **alkaline earth metals**, react with water to form alkaline solutions, and the **noble gases** (Group 8A) are the least reactive elements in the periodic table.

The 18 groups in the periodic table are numbered according to one of three common numbering schemes. The numbering scheme shown in Interactive Figure 2.2.1 is widely used in North America and consists of a number followed by A or B. The elements in A groups are the **main-group elements**, also called the *representative* elements, and the elements in B groups are **transition metals**. The International Union of Pure and Applied Chemistry (IUPAC) has proposed a simpler numbering scheme, also shown in Interactive Figure 2.2.1, that numbers the groups 1 to 18 from left to right.

Table 2.2.1 Special Names Given to Groups in the Periodic Table

Group	Name
1A	Alkali metals
2A	Alkaline earth metals
6A	Chalcogens
7A	Halogens
8A	Noble gases

Explore the periodic table.

Period	1 / 1A	2 / 2A		3 / 3B	4 / 4B	5 / 5B	6 / 6B	7 / 7B	8	9 / 8B	10	11 / 1B	12 / 2B	13 / 3A	14 / 4A	15 / 5A	16 / 6A	17 / 7A	18 / 8A
1	1 **H** 1.0079																		2 **He** 4.0026
2	3 **Li** 6.941	4 **Be** 9.0122												5 **B** 10.811	6 **C** 12.0107	7 **N** 14.0067	8 **O** 15.9994	9 **F** 18.9984	10 **Ne** 20.1797
3	11 **Na** 22.9898	12 **Mg** 24.3050												13 **Al** 26.9815	14 **Si** 28.0855	15 **P** 30.9738	16 **S** 32.065	17 **Cl** 35.453	18 **Ar** 39.948
4	19 **K** 39.0983	20 **Ca** 40.078		21 **Sc** 44.9559	22 **Ti** 47.867	23 **V** 50.9415	24 **Cr** 51.9961	25 **Mn** 54.9380	26 **Fe** 55.845	27 **Co** 58.932	28 **Ni** 58.6934	29 **Cu** 63.546	30 **Zn** 65.38	31 **Ga** 69.723	32 **Ge** 72.64	33 **As** 74.9216	34 **Se** 78.96	35 **Br** 79.904	36 **Kr** 83.798
5	37 **Rb** 85.4678	38 **Sr** 87.62		39 **Y** 88.9059	40 **Zr** 91.224	41 **Nb** 92.9064	42 **Mo** 95.96	43 **Tc** (98)	44 **Ru** 101.07	45 **Rh** 102.9055	46 **Pd** 106.42	47 **Ag** 107.8682	48 **Cd** 112.411	49 **In** 114.818	50 **Sn** 118.710	51 **Sb** 121.760	52 **Te** 127.60	53 **I** 126.9045	54 **Xe** 131.293
6	55 **Cs** 132.9055	56 **Ba** 137.327		71 **Lu** 174.968	72 **Hf** 178.49	73 **Ta** 180.9479	74 **W** 183.84	75 **Re** 186.207	76 **Os** 190.23	77 **Ir** 192.217	78 **Pt** 195.084	79 **Au** 196.9666	80 **Hg** 200.59	81 **Tl** 204.3833	82 **Pb** 207.2	83 **Bi** 208.9804	84 **Po** (209)	85 **At** (210)	86 **Rn** (222)
7	87 **Fr** (223)	88 **Ra** (226)		103 **Lr** (262)	104 **Rf** (267)	105 **Db** (268)	106 **Sg** (271)	107 **Bh** (272)	108 **Hs** (270)	109 **Mt** (276)	110 **Ds** (281)	111 **Rg** (280)	112 **Cn** (285)	113 **Uut** (284)	114 **Uuq** (289)	115 **Uup** (288)	116 **Uuh** (293)	117 **UUs** (?)	118 **Uuo** (294)

Metal

Metalloid

Nonmetal

Lanthanides

57 **La** 138.9055	58 **Ce** 140.116	59 **Pr** 140.9077	60 **Nd** 144.242	61 **Pm** (145)	62 **Sm** 150.36	63 **Eu** 151.964	64 **Gd** 157.25	65 **Tb** 158.9254	66 **Dy** 162.500	67 **Ho** 164.9303	68 **Er** 167.259	69 **Tm** 168.9342	70 **Yb** 173.054

Actinides

89 **Ac** (227)	90 **Th** 232.0381	91 **Pa** 231.0359	92 **U** 238.0289	93 **Np** (237)	94 **Pu** (244)	95 **Am** (243)	96 **Cm** (247)	97 **Bk** (247)	98 **Cf** (251)	99 **Es** (252)	100 **Fm** (257)	101 **Md** (258)	102 **No** (259)

There are seven horizontal periods in the periodic table. Portions of periods 6 and 7 are placed below the main body of the periodic table to make it fit easily on a single page. These portions of periods 6 and 7 are given special names, the **lanthanides** and **actinides**.

The elements on the left side of the periodic table are **metals**, the elements on the right side are **nonmetals**, and the elements at the interface of these two regions are **metalloids** or **semimetals**. With the exception of mercury (Hg), which is a liquid metal at room temperature, metals are generally shiny solids that are ductile and good conductors of electricity. Nonmetals are generally dull, brittle solids or gases that do not conduct electricity; bromine is the only liquid nonmetal at room temperature. Metalloids have properties of both metals and nonmetals.

Most of the elements in the periodic table are solids. Only 2 elements exist as liquids at room temperature (mercury and bromine) and 11 elements are gases at room temperature (hydrogen, nitrogen, oxygen, fluorine, chlorine, helium, neon, argon, krypton, xenon, and radon). As shown in Interactive Figure 2.2.2, many elements are found as individual atoms at the atomic level (helium [He], sodium [Na], and mercury [Hg], for example). However, many elements exist as molecules consisting of two or more atoms of an element (oxygen [O_2], sulfur [S_8], and white phosphorus [P_4], for example), or as a connected three-dimensional array of atoms (silicon [Si], carbon [C], red phosphorus [P]). Seven elements exist as diatomic molecules in their most stable form: H_2, N_2, O_2, F_2, Cl_2, Br_2, and I_2.

Red phosphorus (P) and white phosphorus (P_4) are examples of **allotropes**, forms of the same element that differ in their physical and chemical properties. Red phosphorus, which consists of long chains of phosphorus atoms, is nontoxic, has a deep red color, and burns in air at high temperatures (above 250 °C). White phosphorus, which is made up of individual molecules of four phosphorus atoms, is a white or yellow waxy solid that ignites in air above 50 °C and is very poisonous. Other examples of elements that exist as different allotropes are oxygen (diatomic oxygen [O_2], and triatomic ozone [O_3]) and carbon (diamond, graphite, and buckminsterfullerene).

The periodic table is used extensively in chemistry, and it is helpful to become familiar with the structure of the table. You should learn the names and symbols for the first 36 elements and some other common elements such as silver (Ag), gold (Au), tin (Sn), iodine (I), lead (Pb), and uranium (U).

Interactive Figure 2.2.2

Explore the composition of elements.

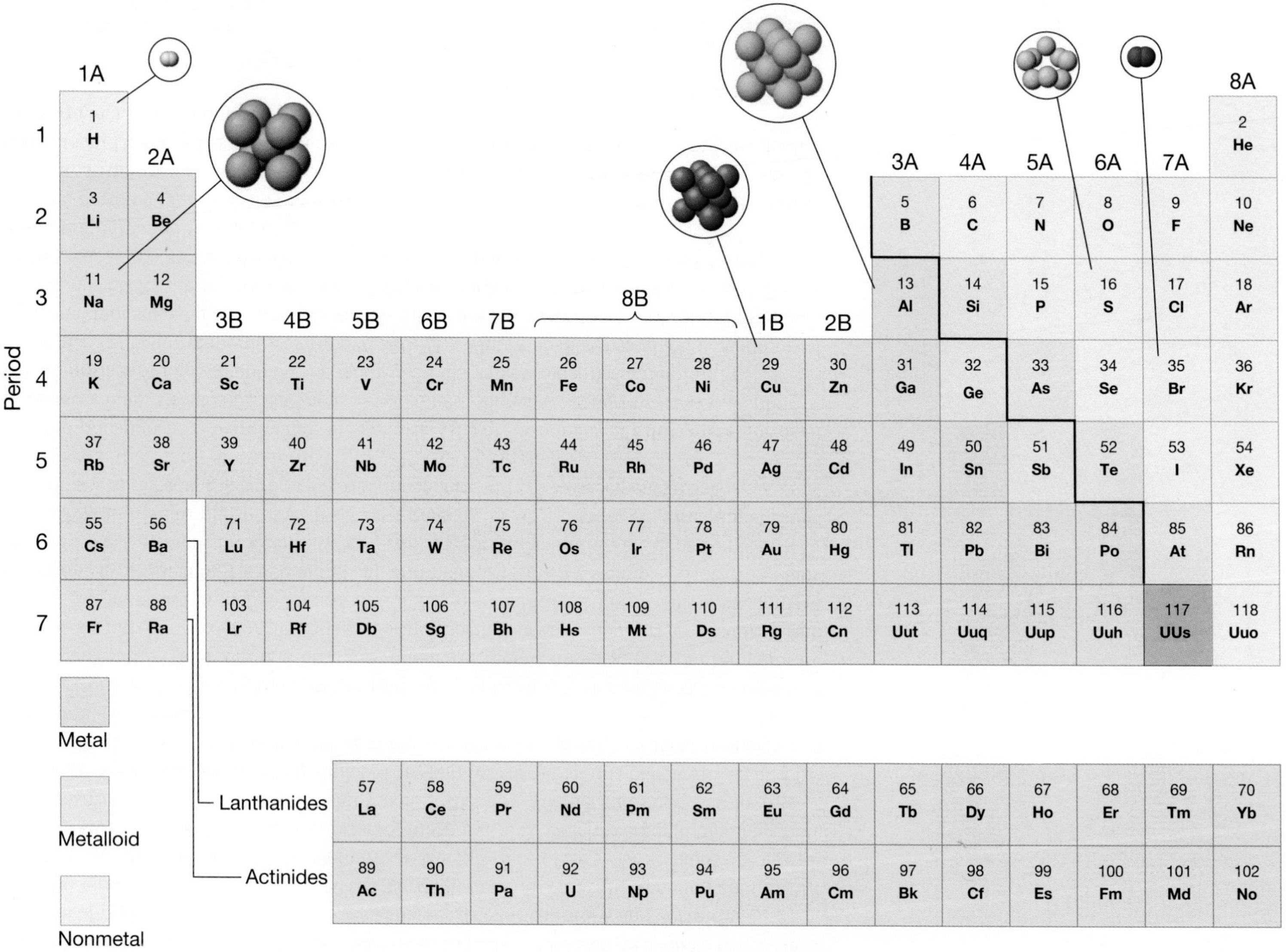

The composition of selected elements

© 2013 Cengage Learning

> **Example Problem 2.2.1** Identify the structure of elements.
>
> Consider the elements phosphorus, bromine, sodium, and hydrogen.
> a. Which of these elements have allotropes?
> b. Which of these elements exist as diatomic molecules?
> c. Which of these elements exist as a metallic lattice?
>
> **Solution:**
>
> **You are asked** to describe the composition of some elements.
>
> **You are given** the identity of the elements.
> a. Phosphorus exists in allotropic forms (red phosphorus and white phosphorus). Both
>   contain only phosphorus atoms, but they differ in the arrangement of atoms in the solid.
> b. Both bromine (Br_2) and hydrogen (H_2) exist as diatomic molecules. Hydrogen is a gas
>   under typical room conditions and bromine is a liquid.
> c. Sodium exists as individual atoms held together in a metallic lattice.

Video Solution

Tutored Practice
Problem 2.2.1

Section 2.2 Mastery

2.3 Covalent Compounds

2.3a Introduction to Covalent Compounds

Covalent compounds consist of atoms of different elements held together by covalent bonds. (▶ Flashforward to Section 8.1 An Introduction to Covalent Bonding) Covalent compounds can be characterized as either molecular covalent compounds or network covalent compounds (Interactive Figure 2.3.1). Water (H_2O) is an example of a **molecular covalent compound**. Water is made up of individual H_2O molecules, with the oxygen and hydrogen atoms in each water molecule held together by covalent bonds (Interactive Figure 2.3.1a).

Silicon dioxide (SiO_2), also known as sand, is an example of a **network covalent compound**. Unlike water, which consists of individual H_2O molecules, silicon dioxide is made up of a three-dimensional network of silicon and oxygen atoms held together by covalent bonds (Interactive Figure 2.3.1b).

2.3b Representing Covalent Compounds with Molecular and Empirical Formulas

Molecules can vary in complexity from only two atoms to many. The simplest way to represent a molecule is through a **molecular formula**. A molecular formula contains the

Distinguish between molecular and network covalent compounds.

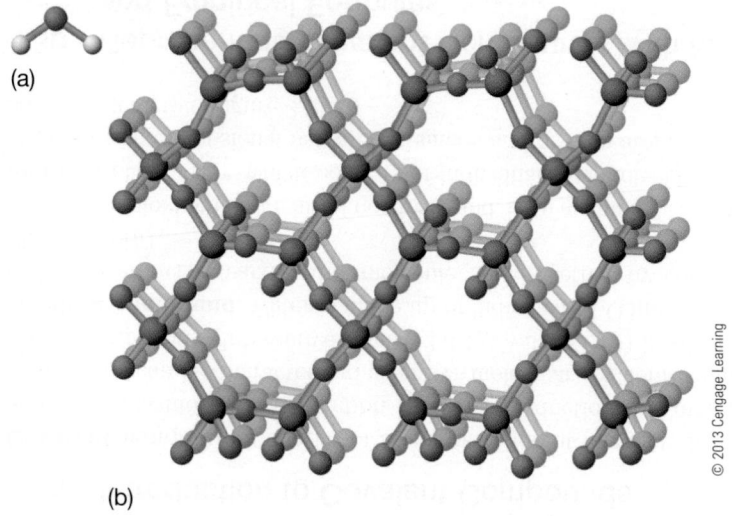

(a)

(b)

(a) Water molecules; (b) Si-O network in quartz (sand)

© 2013 Cengage Learning

symbol for each element present and a subscript number to identify the number of atoms of each element in the molecule. If only one atom of an element is present in a molecule, however, the number 1 is not used. A water molecule, H_2O, is made up of two hydrogen atoms and one oxygen atom. Isopropanol has the molecular formula C_3H_8O, which means that a single molecule of isopropanol contains three carbon atoms, eight hydrogen atoms, and one oxygen atom. Notice that chemical formulas always show a whole-number ratio of elements.

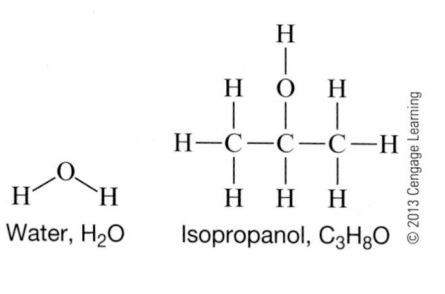

Water, H_2O Isopropanol, C_3H_8O

© 2013 Cengage Learning

Empirical Formulas

Whereas a molecular formula indicates the number of atoms of each element in one molecule of a compound, an **empirical formula** represents the simplest whole-number ratio of elements in a compound. Hydrogen peroxide, for example, is a molecular compound with the molecular formula of H_2O_2. The empirical formula of hydrogen peroxide, HO, shows the simplest whole-number ratio of elements in the compound.

H—O—O—H

Hydrogen peroxide, H_2O_2 © 2013 Cengage Learning

Network covalent compounds are also represented using empirical formulas. Silicon dioxide, for example, does not consist of individual SiO_2 molecules. The simplest ratio of elements in the compound is 1 Si atom:2 O atoms. The empirical formula of silicon dioxide is therefore SiO_2. Carbon (diamond) is an example of a network element. It is made up of carbon atoms held together by covalent bonds in a three-dimensional network, and the element is represented by the empirical formula C.

Example Problem 2.3.1 Write molecular and empirical formulas.

Determine the molecular and empirical formulas for the following substances.

a. Hexane, a laboratory solvent

```
    H   H   H   H   H   H
    |   |   |   |   |   |
H — C — C — C — C — C — C — H
    |   |   |   |   |   |
    H   H   H   H   H   H
```

b. Butyraldehyde, a compound used for synthetic almond flavoring in food

```
    H   H   H   O
    |   |   |   ‖
H — C — C — C — C — H
    |   |   |
    H   H   H
```

Solution:

You are asked to write the empirical and molecular formulas for a compound.

You are given the formula of the compound.

a. A hexane molecule contains 6 carbon atoms and 14 hydrogen atoms. The molecular formula of hexane is C_6H_{14}. The empirical formula, the simplest whole-number ratio of elements in the compound, is C_3H_7.

b. A butyraldehyde molecule contains four carbon atoms, eight hydrogen atoms, and one oxygen atom. The molecular formula of butyraldehyde is C_4H_8O. In this case, the empirical formula, the simplest whole-number ratio of elements in the compound, is the same as the molecular formula, C_4H_8O.

Video Solution

Tutored Practice
Problem 2.3.1

Structural Formulas

A molecular formula identifies the number and types of elements present in a molecule, but it does not provide information on how the atoms are connected. A **structural formula** shows the linkage of all the atoms in the molecule. The covalent bonds are represented by lines between the element symbols. Butanol, a molecular compound made up of 4 carbon atoms, 10 hydrogen atoms, and 1 oxygen atom has the structural formula shown here:

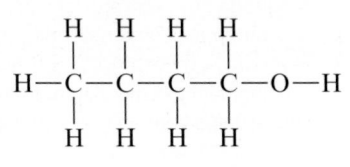

A **condensed structural formula** lists the atoms present in groups to indicate connectivity between the atoms. The condensed structural formula for butanol is $CH_3CH_2CH_2CH_2OH$. Interpretation of this type of formula requires familiarity with commonly encountered groups of atoms, such as the CH_3 or CH_2 groups. Note that although the structural formula does convey information about connectivity, it does not convey information about the three-dimensional shape of the compound.

2.3c Representing Covalent Compounds with Molecular Models

Chemists often need to visualize the three-dimensional shape of a molecule to understand its chemical or physical properties. A variety of models are used to represent the shapes of molecules, each of which has a different purpose. The **wedge-and-dash model** is a two-dimensional representation of a three-dimensional structure that can easily be drawn on paper. In this wedge-and-dash model of butanol, bonds are represented by lines (bonds that lie in the plane of the paper), wedges (bonds that lie in front of the plane of the paper), or dashes (bonds that lie behind the plane of the paper).

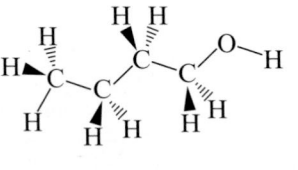

The other two common types of molecular models are created using molecular modeling software, sophisticated computer programs that calculate the spatial arrangement of atoms and bonds. A **ball-and-stick model** shows atoms as colored spheres connected by sticks that represent covalent bonds. This figure shows the same molecule, butanol, represented with balls and sticks. This type of model emphasizes the connections between atoms and the arrangement of atoms in the molecule. A less accurate ball-and-stick model can be created using a commercial molecular modeling kit. This type of model can be held in your hands and the atoms and bonds rotated.

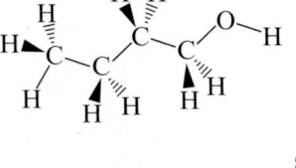

© 2013 Cengage Learning

In a **space-filling model**, interpenetrating spheres represent the relative amount of space occupied by each atom in the molecule. A space-filling model of butanol is shown in Interactive Figure 2.3.2. This type of model is useful when considering the overall shape of molecules and how molecules interact when they come in contact with one another.

2.3d Naming Covalent Compounds

Covalent compounds can be categorized in many ways; two common classes are **binary nonmetals** and **inorganic acids**. Often, a compound will belong to more than one of these categories. Covalent compounds are named according to guidelines created by the Chemical Nomenclature and Structure Representation Division of IUPAC. Some compounds, however, have names that do not follow these guidelines because they have been known by other common names for many years.

Binary Nonmetals

Binary nonmetal compounds consist of only two elements, both nonmetals; some examples include H_2O, CS_2, and SiO_2. Binary nonmetal compounds are named according to the rules in Interactive Table 2.3.1.

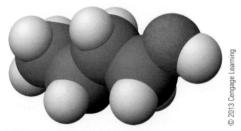
Interactive Table 2.3.1

Rules for Naming Binary Nonmetal Compounds

1. The first word in the compound name is the name of the first element in the compound formula. If the compound contains more than one atom of the first element, use a prefix (Table 2.3.2) to indicate the number of atoms in the formula.
   - CS_2 First word in compound name: carbon
   - N_2O_4 First word in compound name: dinitrogen

2. The second word in the compound name is the name of the second element in the formula that has been changed to end with *-ide*. In all cases, use a prefix (Table 2.3.2) to indicate the number of atoms in the formula.
   - CS_2 Second word in compound name: disulfide
   - N_2O_4 Second word in compound name: tetraoxide

3. The compound is named by combining the first and second words of the compound name.
   - CS_2 carbon disulfide
   - N_2O_4 dinitrogen tetraoxide

Table 2.3.2 Prefixes Used in Naming Binary Nonmetal Compounds

Number	Prefix
1	mono
2	di
3	tri
4	tetra
5	penta
6	hexa
7	hepta
8	octa
9	nona
10	deca
12	dodeca

The names of some common binary nonmetal compounds are shown in Table 2.3.3. Many binary nonmetal compounds have special names that have been used for many years. Examples include water (H_2O), ammonia (NH_3), and nitric oxide (NO).

Table 2.3.3 Names and Formulas of Some Binary Nonmetals

Name	Formula	Name	Formula
Water	H_2O	Sulfur dioxide	SO_2
Hydrogen peroxide	H_2O_2	Sulfur trioxide	SO_3
Ammonia	NH_3	Carbon monoxide	CO
Hydrazine	N_2H_4	Carbon dioxide	CO_2
Nitric oxide	NO	Chlorine monoxide	ClO
Nitrogen dioxide	NO_2	Disulfur decafluoride	S_2F_{10}

Hydrocarbons, binary nonmetal compounds containing only carbon and hydrogen, are also given special names. These compounds are one class of organic compounds, compounds that contain carbon and hydrogen and often other elements such as oxygen and nitrogen. Hydrocarbons are named according to the number of carbon and hydrogen atoms in the compound formula, as shown in Table 2.3.4.

Inorganic Acids

Inorganic acids produce the hydrogen ion (H^+) when dissolved in water and are compounds that contain hydrogen and one or more nonmetals. Inorganic acids can often be identified by their chemical formulas because hydrogen is the first element in the compound formula. Some examples include HCl, H_2S, and HNO_3.

Inorganic acids are named as binary nonmetal compounds but without the use of prefixes (H_2S, hydrogen sulfide), or using common names (HNO_3, nitric acid; H_2SO_4, sulfuric acid). The hydrogen halides (HF, HCl, HBr, and HI) are named as binary nonmetals when in the gas phase, but with common names when dissolved in water (HCl, hydrogen chloride or hydrochloric acid). Groups of acids that differ only in the number of oxygen atoms, **oxoacids**, are named according to the number of oxygen atoms in the formula. Chlorine, bromine, and iodine each form a series of four oxoacids, as shown in Table 2.3.5.

Table 2.3.4 Selected Hydrocarbons with the Formula C_nH_{2n+2}

Hydrocarbon	Name
CH_4	Methane
C_2H_6	Ethane
C_3H_8	Propane
C_4H_{10}	Butane
C_5H_{12}	Pentane
C_6H_{14}	Hexane
C_8H_{18}	Octane
$C_{10}H_{22}$	Decane

Table 2.3.5 Names and Formulas of the Halogen Oxoacids

Formula	Name	Formula	Name	Formula	Name
$HClO_4$	Perchloric acid	$HBrO_4$	Perbromic acid	HIO_4	Periodic acid
$HClO_3$	Chloric acid	$HBrO_3$	Bromic acid	HIO_3	Iodic acid
$HClO_2$	Chlorous acid	$HBrO_2$	Bromous acid	HIO_2	Iodous acid
$HClO$	Hypochlorous acid	$HBrO$	Hypobromous acid	HIO	Hypoiodous acid

When naming oxoacids, the suffix *-ic* is generally used to indicate an acid with more oxygen atoms and the suffix *-ous* is used to indicate an acid with fewer oxygens. For example, HNO_3 is nitric acid and HNO_2 is nitrous acid; H_2SO_4 is sulfuric acid and H_2SO_3 is sulfurous acid. Some common acids are shown in Table 2.3.6.

Table 2.3.6 Names and Formulas of Some Inorganic Acids

Name	Formula	Name	Formula
Hydrochloric acid	HCl	Nitric acid	HNO_3
Hydrobromic acid	HBr	Nitrous acid	HNO_2
Hydrogen sulfide	H_2S	Sulfuric acid	H_2SO_4
Phosphoric acid	H_3PO_4	Sulfurous acid	H_2SO_3

Example Problem 2.3.2 Name covalent compounds.

Name or write the formula for the following covalent compounds:
a. CF_4 b. P_4S_3 c. Hydrogen iodide   d. Hydrazine

Solution:

You are asked to write the name or formula for a covalent compound.

You are given either the formula or the name of the compound.

a. Carbon tetrafluoride. Notice that the name of the first element, carbon, does not include the *mono-* prefix.

b. Tetraphosphorus trisulfide. Both element names include prefixes, and the name of the second element ends in *-ide*.

c. HI. This is the formula of an inorganic acid.

d. N_2H_4. This is a common name that must be memorized.

Video Solution

Tutored Practice
Problem 2.3.2

Section 2.3 Mastery

2.4 Ions and Ionic Compounds

2.4a Monoatomic Ions

Unlike covalent compounds, **ionic compounds** contain ions, species that carry a positive (cation) or negative (anion) charge. Whereas the atoms in covalent compounds are held together by covalent bonds, ionic compounds are held together by strong attractive forces between cations and anions. The different makeup of these two types of compounds results in species with very different physical and chemical properties. For example, many covalent compounds are gases, liquids, or solids with low melting points, whereas most ionic compounds are solids with very high melting points.

Remember that an atom carries no charge because it contains an equal number of positively charged protons and negatively charged electrons. When a single atom gains or loses one or more electrons, the number of electrons and protons is no longer equal and a **monoatomic ion** is formed. The charge on an ion is indicated using a superscript to the right of the element symbol. When the charge is $+1$ or -1, it is written without the number 1. For example, magnesium forms a cation, Mg^{2+}, when it loses two electrons, and bromine forms an anion, Br^-, when it gains one electron (Interactive Figure 2.4.1).

Interactive Figure 2.4.1

Explore ion formation.

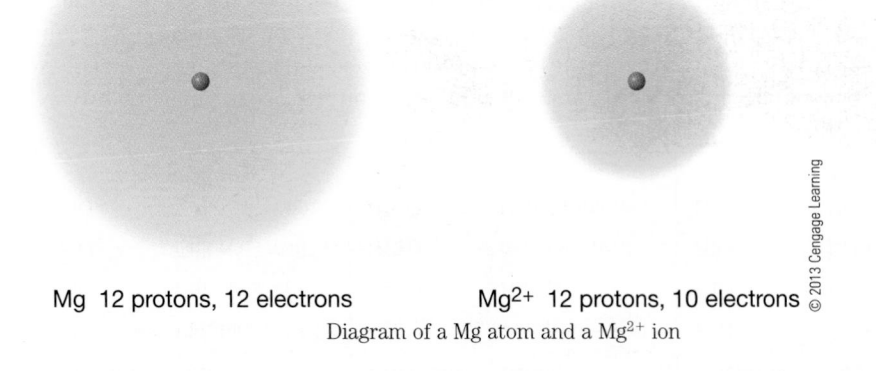

Mg 12 protons, 12 electrons Mg^{2+} 12 protons, 10 electrons

© 2013 Cengage Learning

Diagram of a Mg atom and a Mg^{2+} ion

Notice that when an ion charge is written with an atom symbol, the numeric value is written first, followed by the + or − symbol. When describing the charge on an ion, the + or − symbol is written first, followed by the numeric value. For example, Sr^{2+} has a +2 charge and Br^- has a −1 charge.

Cations and anions have physical and chemical properties that are very different than those of the elements from which they are formed. For example, elemental magnesium is a shiny metal that burns in air with a bright white flame. Magnesium ions are colorless and are found in most drinking water.

Predicting Charge Based on Periodic Group

Most elements in the main groups of the periodic table (Groups 1A–7A) form monoatomic ions that have a charge related to the group number of the element.

- Metals in Groups 1A, 2A, and 3A form cations that have a positive charge equal to the group number of the element.

$$\text{Sodium, Group 1A} \quad Na^+ \qquad \text{Calcium, Group 2A} \quad Ca^{2+}$$

- Nonmetals in Groups 5A, 6A, and 7A form anions that have a negative charge equal to 8 minus the group number of the element.

$$\text{Oxygen, Group 6A} \quad O^{2-} \qquad \text{Bromine, Group 7A} \quad Br^-$$

Other elements form ions with charges that are not easily predicted.

- Hydrogen forms both H^+ and H^- ions.

- Transition metals typically form cations with charges ranging from +1 to +3. Many transition metals form more than one monoatomic ion.

- Group 4A contains both metals and nonmetals, so some elements in this group form cations and others form anions.

- Other than aluminum, the metals in Groups 3A, 4A, and 5A form cations with positive charges that are not easily predicted.

Some of the more common monoatomic ions are shown in Interactive Figure 2.4.2. Practice with predicting charge on monoatomic ions is given in the following example. Notice that in general, metals form cations and nonmetals form anions; in addition, the charges on most monoatomic ions are relatively small, between +2 and −2. The noble gases (Group 8A) are quite unreactive and therefore do not form ions.

Interactive Figure 2.4.2

Explore monoatomic ion formulas.

1A																7A	8A
H^+	2A											3A	4A	5A	6A	H^-	
Li^+														N^{3-}	O^{2-}	F^-	
Na^+	Mg^{2+}	3B	4B	5B	6B	7B	8B		1B	2B	Al^{3+}		P^{3-}	S^{2-}	Cl^-	
K^+	Ca^{2+}		Ti^{4+}		Cr^{2+} Cr^{3+}	Mn^{2+}	Fe^{2+} Fe^{3+}	Co^{2+} Co^{3+}	Ni^{2+}	Cu^+ Cu^{2+}	Zn^{2+}			Se^{2-}	Br^-	
Rb^+	Sr^{2+}									Ag^+	Cd^{2+}		Sn^{2+}	Te^{2-}	I^-	
Cs^+	Ba^{2+}									Hg_2^{2+} Hg^{2+}		Pb^{2+}	Bi^{3+}			

Charges on some monoatomic ions

© 2013 Cengage Learning

Example Problem 2.4.1 Predict charge on monoatomic ions.

a. How many protons and electrons are in a Ca^{2+} ion?
b. Identify the ion formed by phosphorus. How many protons and electrons are in this ion?

Solution:

You are asked to determine the number of protons and electrons in a monoatomic ion.

You are given the identity of the element that forms a monoatomic ion.

a. Calcium is element 20, and the neutral atom has 20 protons and 20 electrons. A Ca^{2+} cation has lost two electrons and has 20 protons and 18 electrons.
b. Phosphorus is in Group 5A and therefore forms an ion with a −3 charge. Phosphorus is element 15 and the P^{3-} ion has 15 protons and 18 electrons.

Is your answer reasonable? A cation (Ca^{2+}) should contain more protons than electrons, while an anion (P^{3-}) has more electrons than protons.

Video Solution

Tutored Practice
Problem 2.4.1

2.4b Polyatomic Ions

Polyatomic ions are groups of covalently bonded atoms that carry an overall positive or negative charge. The formulas, names, and charges of the common polyatomic ions are shown in Interactive Table 2.4.1 and should be memorized. Most polyatomic ions are anions; there is only one common polyatomic cation, the ammonium ion (NH_4^+). You have seen many of the polyatomic ions in the formulas of inorganic acids. For example, nitric acid, HNO_3, contains the monoatomic ion H^+ and the polyatomic nitrate ion, NO_3^-.

Interactive Table 2.4.1

Names and Formulas of Common Polyatomic Ions

Ion	Name	Ion	Name
NH_4^+	Ammonium	NO_2^-	Nitrite
OH^-	Hydroxide	NO_3^-	Nitrate
CN^-	Cyanide	ClO^-	Hypochlorite
$CH_3CO_2^-$	Acetate	ClO_2^-	Chlorite
SO_3^{2-}	Sulfite	ClO_3^-	Chlorate
SO_4^{2-}	Sulfate	ClO_4^-	Perchlorate
HSO_4^-	Hydrogen sulfate (bisulfate)	CO_3^{2-}	Carbonate
$S_2O_3^{2-}$	Thiosulfate	HCO_3^-	Hydrogen carbonate (bicarbonate)
PO_4^{3-}	Phosphate	$C_2O_4^{2-}$	Oxalate
HPO_4^{2-}	Hydrogen phosphate	$Cr_2O_7^{2-}$	Dichromate
$H_2PO_4^-$	Dihydrogen phosphate	CrO_4^{2-}	Chromate
SCN^-	Thiocyanate	MnO_4^-	Permanganate
OCN^-	Cyanate		

2.4c Representing Ionic Compounds with Formulas

Ionic compounds are represented by empirical formulas that show the simplest ratio of cations and anions in the compound. In the formula of an ionic compound, the cation symbol or formula is always written first, followed by the anion symbol or formula. Ionic compounds do not have a positive or negative charge because the total cationic positive charge is balanced by the total anionic negative charge. This means that unlike covalent compounds, it is possible to predict the formula of an ionic compound if the cation and anion charges are known.

For example, consider the ionic compound formed from the reaction between aluminum and bromine (Interactive Figure 2.4.3). We can predict the formula of the ionic compound formed from these elements using the charges on the ions formed from these elements.

- Aluminum is in Group 3A and forms the Al^{3+} ion.

- Bromine is in Group 7A and forms the Br^- ion.

- The $+3$ cation charge must be balanced by a -3 charge for the overall formula to carry no net charge. Three Br^- ions, each with a -1 charge, are needed to balance the Al^{3+} ion charge. The formula of the ionic compound is therefore $AlBr_3$.

When the formula of an ionic compound contains a polyatomic ion, parentheses are used if more than one polyatomic ion is needed to balance the positive and negative charges in the compound. For example, the formula $Ca(NO_3)_2$ indicates that it contains Ca^{2+} ions and NO_3^- ions in a 1:2 ratio.

Example Problem 2.4.2 Write formulas for ionic compounds.

a. What is the formula of the ionic compound expected to form between the elements oxygen and sodium?
b. What is the formula of the ionic compound formed between the ions Zn^{2+} and PO_4^{3-}?
c. What ions make up the ionic compound $Cr(NO_3)_3$?

Solution:

You are asked to write formulas for ionic compounds or identify the ions in an ionic compound.

You are given the identity of the compound or the ions or elements that make up the compound.

a. Na_2O. Sodium is in Group 1A and forms a cation with a $+1$ charge, Na^+. Oxygen is in Group 6A and forms an anion with a -2 charge, O^{2-}. Two Na^+ ions are required to provide a total $+2$ positive charge that balances the -2 charge on O^{2-}.

b. $Zn_3(PO_4)_2$. In this case, more than one of each ion is needed to balance the positive and negative charges. Three Zn^{2+} ions provide a positive charge of $+6$, and two PO_4^{3-} ions provide a negative charge of -6. Parentheses are used to indicate the total number of polyatomic ions in the compound formula.

c. Cr^{3+}, NO_3^-. Cr is a transition metal, and it is impossible to predict its charge when it forms an ion. NO_3^- is a polyatomic ion with a -1 charge. The three NO_3^- ions in the compound formula provide a total negative charge of -3, so the single cation must have a $+3$ charge to balance this negative charge.

Write ionic compound formulas.

Charles D. Winters

Aluminum reacts with bromine to form an ionic compound.

Video Solution

Tutored Practice
Problem 2.4.2

2.4d   Naming Ionic Compounds

Like covalent compounds, ions and ionic compounds are named using guidelines created by IUPAC. Naming ionic compounds involves identifying the charges on the monoatomic and polyatomic ions in a chemical formula, so it is important to memorize the rules for predicting the charges on monoatomic ions and the names, formulas, and charges on polyatomic ions.

Ions and ionic compounds are named according to the rules shown in Table 2.4.2.

Table 2.4.2 Rules for Naming Ions and Ionic Compounds

Monoatomic cations

- The name of a main-group monoatomic cation is the element name followed by the word *ion*.

$$Na^+ \quad \text{sodium ion} \qquad Mg^{2+} \quad \text{magnesium ion}$$

- The name of a transition metal cation is the element name followed by the cation charge in Roman numerals within parentheses and the word *ion*.

$$Fe^{2+} \quad \text{iron(II) ion} \qquad Co^{3+} \quad \text{cobalt(III) ion}$$

Monoatomic anions

The name of a monoatomic anion is the element name changed to include the suffix *-ide*, followed by the word *ion*.

$$Br^- \quad \text{bromide ion} \qquad O^{2-} \quad \text{oxide ion}$$

Polyatomic ions

The names of polyatomic ions are shown in Interactive Table 2.4.1 and must be memorized. Notice that the names also include the word *ion*.

$$NO_3^- \quad \text{nitrate ion} \qquad MnO_4^- \quad \text{permanganate ion}$$

Ionic compounds

The name of an ionic compound consists of the cation name followed by the anion name. The word *ion* is dropped because the compound does not carry a charge. Prefixes are not used to indicate the number of ions present in the formula of an ionic compound.

$$NaNO_3 \quad \text{sodium nitrate} \qquad Co_2O_3 \quad \text{cobalt(III) oxide}$$

Example Problem 2.4.3 Name ionic compounds.

a. What is the name of the compound with the formula CuCN?
b. What is the formula for aluminum nitrite?

Solution:

You are asked to write the name or the formula for an ionic compound.

You are given the formula or the name of the ionic compound.

a. Copper(I) cyanide

The cation is a transition metal, and its name must include the cation charge. The cyanide ion, CN^-, is a polyatomic ion with a -1 charge. The single copper cation therefore has a $+1$ charge. The name of this compound includes the charge on the cation in Roman numerals, within parentheses.

b. $Al(NO_2)_3$

Aluminum is in Group 3A and forms a cation with a $+3$ charge, Al^{3+}. The nitrite ion, NO_2^-, is a polyatomic ion whose name, charge, and formula must be memorized. Three nitrite ions are needed to provide a total -3 charge that balances the $+3$ charge on Al^{3+}. Parentheses are used to indicate the total number of polyatomic ions in the compound formula.

Video Solution

Tutored Practice
Problem 2.4.3

2.4e Identifying Covalent and Ionic Compounds

It is often challenging to determine whether a compound is covalent or ionic. However, the formula or name of a compound contains information that can be used to classify a compound. The following guidelines show how the two classes of compounds are similar and how they differ.

Covalent compounds

- Contain only nonmetals

- Are named using prefixes to indicate the number of each element in a formula

Ionic compounds

- Contain monoatomic and/or polyatomic ions

- Usually contain metals and nonmetals but can also contain only nonmetals

- Are never named using prefixes

- Are sometimes named with the cation charge in Roman numerals within parentheses

> **Example Problem 2.4.4 Identify covalent and ionic compounds.**
>
> Identify each of the following compounds as ionic or covalent.
> a. NH_4NO_3 b. $Na_2S_2O_3$ c. SF_6
>
> **Solution:**
>
> **You are asked** to determine whether a compound is ionic or covalent.
>
> **You are given** the compound formula.
> a. Ionic. Ammonium nitrate contains two polyatomic ions (NH_4^+ and NO_3^-), so it is an ionic compound that consists only of nonmetal elements.
> b. Ionic. Sodium thiosulfate contains the sodium ion (Na^+) and the polyatomic thiosulfate ion ($S_2O_3^{2-}$). It is an ionic compound that consists of a metal ion and a polyatomic ion.
> c. Covalent. Sulfur hexafluoride contains only nonmetals and no polyatomic ions.

Video Solution

Tutored Practice
Problem 2.4.4

Section 2.4 Mastery

Unit Recap

Key Concepts

2.1 The Structure of the Atom

- Atoms consist of a nucleus that contains protons (relative charge = +1) and neutrons (relative charge = 0), and electrons (relative charge = −1), which are found in the region around the nucleus (2.1a).

- Protons and neutrons each have a relative mass of approximately 1 u, whereas the relative mass of an electron is 0 u (2.1a).

- An ion is an atom with an unequal number of protons and electrons. Anions have a negative charge because they contain more electrons than protons; cations carry a positive charge because they contain more protons than electrons (2.1a).

- Atoms are characterized by their atomic number (Z), the number of protons in the nucleus, and their mass number (A), the mass of the protons and neutrons in the nucleus. The mass number for an atom is essentially equal to the number of protons and neutrons in the nucleus (2.1b).

- The atomic symbol for an atom shows the element symbol (X), the atomic number, and the mass number ($^A_Z X$) (2.1b).

- Isotopes are atoms that have the same atomic number but differ in their mass number (2.1c).

- The atomic weight of an element is the weighted average of the isotope masses of that element (2.1c).

2.2 Elements and the Periodic Table

- The periodic table is an organizational chart used to arrange elements in horizontal periods and vertical groups (2.2a).

- Different regions and groups in the periodic table are given special names, such as main-group elements, transition elements, lanthanides, and actinides (2.2a).

- Most elements are metals, a smaller number are nonmetals, and the elements that have properties of both are metalloids (2.2a).

- Allotropes are forms of the same element that differ in their physical and chemical properties.

2.3 Covalent Compounds

- Covalent compounds consist of atoms of different elements held together by covalent bonds (2.3a).

- Molecular covalent compounds consist of individual molecules, whereas network covalent compounds are made up of a three-dimensional network of covalently bonded atoms (2.3a).

- A molecular formula is the simplest way to represent a molecule and consists of element symbols and subscript numbers that indicate the number of atoms of each element in one molecule of the compound (2.3b).

- An empirical formula is the simplest whole-number ratio of elements in a compound (2.3b).

- Structural formulas and condensed structural formulas provide additional information about the atom connectivity in a molecule (2.3b).

- Wedge-and-dash models, ball-and-stick models, and space-filling models give information about the three-dimensional shape of a molecule (2.3c).

- Binary nonmetals, covalent compounds consisting of only two nonmetal elements, are usually named according to a set of simple rules (2.3d).

- Some covalent compounds such as inorganic acids, oxoacids, and hydrocarbons are named according to the composition of the compound or the relative number of atoms of each element in the compound formula (2.3d).

2.4 Ions and Ionic Compounds

- Ionic compounds are made up of cations and anions and have physical and chemical properties that differ significantly from those of covalent compounds (2.4).

- A monoatomic ion is a single atom that carries a positive or negative charge (2.4a).

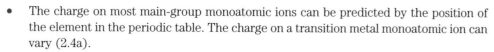

- The charge on most main-group monoatomic ions can be predicted by the position of the element in the periodic table. The charge on a transition metal monoatomic ion can vary (2.4a).

- Polyatomic ions are groups of covalently bonded atoms that carry an overall charge (2.4b).

- Ions and ionic compounds are named according to a set of simple rules (2.4d).

- The formula of a compound can be used to determine whether it is a covalent or an ionic compound (2.4e).

Key Equations

$$\text{average atomic weight} = \sum_{\substack{\text{all} \\ \text{isotopes}}} (\text{exact mass})(\text{fractional abundance})$$ **(2.1)**

Key Terms

2.1 The Structure of the Atom
proton
neutron
electron
atomic nucleus
atomic mass unit (u)
ion
cation
anion
atomic number (Z)
mass number (A)
atomic symbol
isotopes
atomic weight

2.2 Elements and the Periodic Table
periodic table of the elements
groups

periods
alkali metal
halogen
alkaline earth metal
noble gas
main-group elements
transition elements
lanthanides
actinides
metals
nonmetals
metalloids
semimetals
allotropes

2.3 Covalent Compounds
covalent compound
molecular covalent compound

network covalent compound
molecular formula
empirical formula
structural formula
condensed structural formula
wedge-and-dash model
ball-and-stick model
space-filling model
binary nonmetal
inorganic acid
hydrocarbon
oxoacids

2.4 Ions and Ionic Compounds
ionic compound
monoatomic ion
polyatomic ion

Unit 2   Review and Challenge Problems

3 Stoichiometry

Unit Outline

In This Unit...

As you have learned in Units 1 and 2, much of chemistry involves using macroscopic measurements to deduce what happens between atoms and molecules. We now explore the chemical counting unit that links the atomic and macroscopic scales, the mole. The mole allows us to study in greater detail chemical formulas and chemical reactions. Specifically, in this Unit we investigate stoichiometry, the relationship between quantities of materials in chemical reactions. In Unit 4 (Chemical Reactions and Solution Stoichiometry), we will expand our study of stoichiometry to include different types of chemical reactions and focus on reactions that take place in water.

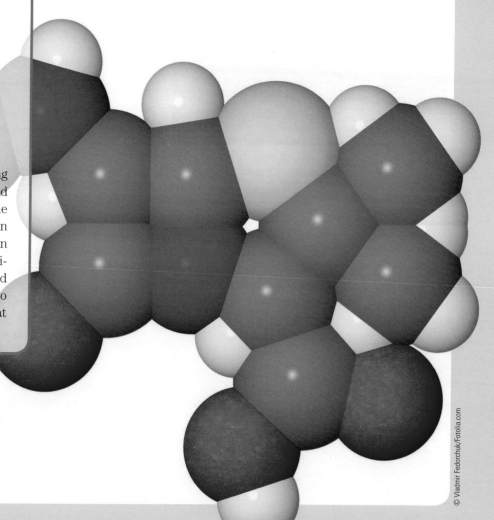

3.1 The Mole and Molar Mass

3.1a Avogadro's Number

As you learned in Unit 1, atoms are so small and have such small masses that any amount of atoms we would work with would be very hard to count. For example, a piece of aluminum about the size of a pencil eraser contains approximately 2×10^{22} aluminum atoms!

The **mole** (abbreviated mol) is the unit chemists use when counting numbers of atoms or molecules in a sample. The number of particles (atoms, molecules, or other objects) in one mole is equal to the number of atoms in exactly 12 g of carbon-12. This number of particles is called **Avogadro's number (N_A)** and has a value of 6.0221415×10^{23}. In most cases we will use 6.022×10^{23} or 6.02×10^{23} for Avogadro's number. One mole of any element contains 6.0221415×10^{23} atoms of that element, and one mole of a molecular compound contains 6.0221415×10^{23} molecules of that compound. Avogadro's number is an extremely large number, as it must be to connect tiny atoms to the macroscopic world. Using the mole counting unit to measure something on the macroscopic scale demonstrates just how big Avogadro's number is. For example, one mole of pencil erasers would cover the Earth's surface to a depth of about 500 meters. Interactive Figure 3.1.1 shows one mole quantities of some elements and compounds. Avogadro's number can be used to convert between moles and the number of particles in a sample, as shown in the following example.

Interactive Figure 3.1.1

Recognize how the mole connects macroscopic and atomic scales.

One mole quantities of (from left to right) the elements copper, aluminum, sulfur and the compounds potassium dichromate, water, and copper(II) chloride dihydrate.

Example Problem 3.1.1 Convert between moles and numbers of atoms.

A sample of titanium contains 8.98×10^{25} Ti atoms. What amount of Ti, in moles, does this represent?

Solution:

You are asked to calculate the amount (moles) of Ti in a given sample of the metal.

You are given the number of atoms of Ti in the sample.

Use the equality 1 mol = 6.022×10^{23} atoms to create a conversion factor that converts from units of atoms to units of moles.

Example Problem 3.1.1 *(continued)*

$$8.98 \times 10^{25} \text{ Ti atoms} \times \frac{1 \text{ mol Ti}}{6.022 \times 10^{23} \text{ Ti atoms}} = 149 \text{ mol Ti}$$

Is your answer reasonable? The sample contains more than Avogadro's number of Ti atoms and therefore contains more than 1 mol of Ti atoms.

Video Solution

Tutored Practice
Problem 3.1.1

3.1b   Molar Mass

Molar mass is the mass, in grams, of one mole of a substance. The molar mass of an element is the mass in grams of one mole of atoms of the element. It is related to the atomic weight of an element, as shown here:

$$1 \ ^{12}\text{C atom} = 12 \text{ u}$$
$$6.022 \times 10^{23} \text{ (1 mole) } ^{12}\text{C atoms} = 12 \text{ grams}$$

Thus, one mole of an element has a mass in grams equal to its atomic weight in atomic mass units (u). For example, according to the periodic table, the element magnesium has an atomic weight of 24.31 u, and one mole of magnesium atoms has a mass of 24.31 g.

Just as Avogadro's number can be used to convert between moles and the number of particles in a sample, molar mass can be used to convert between moles and the mass in grams of a sample, as shown in the following example. Note that the number of significant figures in molar mass and Avogadro's number can vary. In calculations, include one more significant figure than the data given in the problem so that they do not limit the number of significant figures in the answer.

Example Problem 3.1.2 Convert between mass and moles of an element.

What amount of oxygen, in moles, does 124 g O represent?

Solution:

You are asked to calculate the amount (in moles) of oxygen atoms in a given sample.

You are given the mass of the oxygen sample.

Use the molar mass of oxygen (1 mol O = 16.00 g) to create a conversion factor that converts mass (in grams) to amount (in mol) of oxygen.

$$124 \text{ g O} \times \frac{1 \text{ mol O}}{16.00 \text{ g}} = 7.75 \text{ mol O}$$

Example Problem 3.1.2 *(continued)*

Notice that the units of grams cancel, leaving the answer in units of mol. Also, note that the molar mass of oxygen has four significant figures, one more than the number of significant figures in the data given in the problem.

Is your answer reasonable? The mass of oxygen in the sample is greater than the molar mass of oxygen, so there is more than 1 mol O in the sample.

Video Solution

Tutored Practice
Problem 3.1.2

We can now use Avogadro's number and molar mass to relate mass, moles, and atoms of an element, as shown in the following example.

Example Problem 3.1.3 Convert between mass, moles, and atoms of an element.

How many boron atoms are there in a 77.8-g sample of elemental boron?

Solution:

You are asked to calculate the number of boron atoms in a given sample.

You are given the mass of the boron sample.

This calculation involves two conversion factors: the molar mass of boron to convert mass (in grams) to amount (in mol) and Avogadro's number to convert amount (in mol) to amount (atoms).

$$77.8 \text{ g B} \times \frac{1 \text{ mol B}}{10.81 \text{ g}} \times \frac{6.022 \times 10^{23} \text{ atoms B}}{1 \text{ mol B}} = 4.33 \times 10^{24} \text{ atoms B}$$

Notice that the units of grams and mol cancel, leaving the answer in units of atoms. Also, note that the molar mass and Avogadro's number have four significant figures, one more than the number of significant figures in the data given in the problem.

Is your answer reasonable? The mass of the boron sample is greater than the molar mass of boron, so the number of boron atoms in the sample is greater than Avogadro's number.

Video Solution

Tutored Practice
Problem 3.1.3

The molar mass of a compound is the mass in grams of one mole of the compound. It is numerically equal to the compound's formula (or molecular) weight. **Formula weight** is the sum of the atomic weights of the elements that make up a substance multiplied by the number of atoms of each element in the formula for the substance. For substances that exist as individual molecules, the formula weight is called the **molecular weight**. For example, one molecule of water, H_2O, is made up of two hydrogen atoms and one oxygen atom. Therefore, the mass of 1 mol of H_2O molecules (the molar mass of H_2O) is equal to the mass of 2 mol of hydrogen atoms plus the mass of 1 mol of oxygen atoms.

$$\text{molar mass } H_2O = (2 \text{ mol } H)\left(\frac{1.01 \text{ g}}{1 \text{ mol } H}\right) + (1 \text{ mol } O)\left(\frac{16.00 \text{ g}}{1 \text{ mol } O}\right)$$

$$\text{molar mass } H_2O = 18.02 \text{ g/mol}$$

We will generally report the molar mass of an element or compound to two decimal places, unless more significant figures are required in a calculation.

Example Problem 3.1.4 Determine the molar mass of a compound.

Calculate the molar mass for each of the following compounds:

   a. 2-Propanol, $CH_3CH(OH)CH_3$
   b. Iron(II) phosphate, $Fe_3(PO_4)_2$

Solution:

You are asked to calculate the molar mass of a compound.

You are given the compound's formula.

Add the molar masses for the constituent elements, taking into account the number of each element present in the compound.

   a. One mole of 2-propanol, $CH_3CH(OH)CH_3$, contains three moles of C, eight moles of H, and one mole of O.

$$\text{C:} \quad 3 \text{ mol C} \times \frac{12.01 \text{ g}}{1 \text{ mol C}} = 36.03 \text{ g C}$$

$$\text{H:} \quad 8 \text{ mol H} \times \frac{1.01 \text{ g}}{1 \text{ mol H}} = 8.08 \text{ g H}$$

$$\text{O:} \quad 1 \text{ mol O} \times \frac{16.00 \text{ g}}{1 \text{ mol O}} = 16.00 \text{ g O}$$

molar mass 2-propanol = 36.03 g C + 8.08 g H + 16.00 g O = 60.11 g/mol

   b. One mole of iron(II) phosphate, $Fe_3(PO_4)_2$, contains three moles of Fe, two moles of P, and eight moles of O.

$$\text{Fe:} \quad 3 \text{ mol Fe} \times \frac{55.85 \text{ g}}{1 \text{ mol Fe}} = 167.6 \text{ g Fe}$$

$$\text{P:} \quad 2 \text{ mol P} \times \frac{30.97 \text{ g}}{1 \text{ mol P}} = 61.94 \text{ g P}$$

$$\text{O:} \quad 8 \text{ mol O} \times \frac{16.00 \text{ g}}{1 \text{ mol O}} = 128.0 \text{ g O}$$

molar mass iron(II) phosphate = 167.6 g Fe + 61.94 g P + 128.0 g O = 357.5 g/mol

Video Solution

Tutored Practice
Problem 3.1.4

The molar mass of a compound can be used to convert between moles and the mass in grams of a sample, as shown in the following example.

Example Problem 3.1.5 Convert between mass and moles of a compound.

Use the molar mass of 2-propanol, 60.11 g/mol, to calculate the amount of 2-propanol present in a 10.0-g sample of the alcohol.

Solution:

You are asked to calculate the amount (mol) of compound present in a given sample.

You are given the mass of the sample and the molar mass of the compound.

Use the molar mass of 2-propanol to create a conversion factor that converts mass (in grams) to amount (in mol) of 2-propanol.

$$10.0 \text{ g CH}_3\text{CH(OH)CH}_3 \times \frac{1 \text{ mol CH}_3\text{CH(OH)CH}_3}{60.11 \text{ g}} = 0.166 \text{ mol CH}_3\text{CH(OH)CH}_3$$

Note that the molar mass used in the calculation has four significant figures, one more than the data given in the problem.

Is your answer reasonable? The mass of the sample is less than the molar mass of the compound, so the amount of compound present is less than 1 mole.

Video Solution

Tutored Practice
Problem 3.1.5

Section 3.1 Mastery

3.2 Stoichiometry and Compound Formulas

3.2a Element Composition

Stoichiometry is the study of the relationship between relative amounts of substances. The formula of a compound provides information about the relative amount of each element present in either one molecule of the compound or one mole of the compound. For example, one molecule of acetic acid, CH_3CO_2H, contains two atoms of oxygen and one mole of acetic acid and contains 2 mol of oxygen atoms. When working with ionic and other types of non-molecular compounds, the compound formula is still used to describe the stoichiometry of a compound. For example, the ionic compound calcium chloride, $CaCl_2$, is not made up of $CaCl_2$ molecules but rather one mole of $CaCl_2$ contains 1 mol Ca^{2+} ions and 2 mol Cl^- ions.

The compound formula can be used to determine the amount of an element present in a sample of compound, as shown in the following example.

Example Problem 3.2.1 Use compound formulas to determine element composition.

A sample of acetic acid, CH_3CO_2H, contains 2.50 mol of the compound. Determine the amount (in mol) of each element present and the number of atoms of each element present in the sample.

Solution:

You are asked to calculate the amount (mol) and number of atoms of each element present in a given sample of a compound.

You are given the amount of sample (mol) and the chemical formula of the compound.

Use the compound formula to create conversion factors that relate the amount (in mol) of each element to one mole of the compound.

$$2.50 \text{ mol } CH_3CO_2H \times \frac{2 \text{ mol C}}{1 \text{ mol } CH_3CO_2H} = 5.00 \text{ mol C}$$

$$2.50 \text{ mol } CH_3CO_2H \times \frac{4 \text{ mol H}}{1 \text{ mol } CH_3CO_2H} = 10.0 \text{ mol H}$$

$$2.50 \text{ mol } CH_3CO_2H \times \frac{2 \text{ mol O}}{1 \text{ mol } CH_3CO_2H} = 5.00 \text{ mol O}$$

Use Avogadro's number and the amount of each element present to determine the number of atoms of each element present in the 2.50-mol sample of acetic acid.

$$5.00 \text{ mol C} \times \frac{6.022 \times 10^{23} \text{ C atoms}}{1 \text{ mol C}} = 3.01 \times 10^{24} \text{ C atoms}$$

$$10.0 \text{ mol H} \times \frac{6.022 \times 10^{23} \text{ H atoms}}{1 \text{ mol H}} = 6.02 \times 10^{24} \text{ H atoms}$$

$$5.00 \text{ mol O} \times \frac{6.022 \times 10^{23} \text{ O atoms}}{1 \text{ mol O}} = 3.01 \times 10^{24} \text{ O atoms}$$

Is your answer reasonable? The sample contains more than one mole of the compound, so it also contains more than one mole of each element and more than Avogadro's number of atoms of each element in the compound.

Video Solution

Tutored Practice
Problem 3.2.1

3.2b Percent Composition

The mole-to-mole relationships in a chemical formula can also be used to determine the **percent composition** of an element in a compound. The percent composition of an element in a compound is the mass of an element present in exactly 100 g of a compound and is calculated using the following equation:

% element =
$$\frac{(\text{number of atoms of element in formula})(\text{molar mass of element})}{\text{mass of 1 mol of compound}} \times 100\% \qquad \textbf{(3.1)}$$

For example, the percent composition of hydrogen in water, H_2O, is calculated as follows:

$$\% \text{ H in } H_2O = \frac{2 \text{ mol H}\left(\dfrac{1.01 \text{ g}}{1 \text{ mol H}}\right)}{1 \text{ mol } H_2O\left(\dfrac{18.02 \text{ g}}{1 \text{ mol } H_2O}\right)} \times 100\% = 11.2\% \text{ H}$$

Notice that percent composition is calculated by using the mole-to-mole ratio in the chemical formula of water (2 mol H in 1 mol H_2O) and converting it to a mass ratio using molar mass. Because water is made up of only two elements and the sum of all the percent composition values must equal 100%, the percent composition of oxygen in water is

$$\% \text{ O in } H_2O = 100\% - \% \text{ H} = 100\% - 11.2\% = 88.8\% \text{ O}$$

Example Problem 3.2.2 Calculate percent composition from a compound formula.

Determine the percent composition of each element in potassium permanganate, $KMnO_4$, a compound used as an antiseptic in some countries.

Solution:

You are asked to calculate the percent composition of each element in a compound.

You are given the compound formula.

First calculate the molar mass of $KMnO_4$, then use it along with the chemical formula to calculate the percent composition of each element.

$$\text{molar mass } KMnO_4 = (1 \text{ mol K})\left(\frac{39.10 \text{ g}}{1 \text{ mol K}}\right)$$

$$+ (1 \text{ mol Mn})\left(\frac{54.94 \text{ g}}{1 \text{ mol Mn}}\right) + (4 \text{ mol O})\left(\frac{16.00 \text{ g}}{1 \text{ mol O}}\right)$$

$$\text{molar mass } KMnO_4 = 158.0 \text{ g/mol}$$

Example Problem 3.2.2 *(continued)*

$$\% \text{ K in KMnO}_4 = \frac{1 \text{ mol K} \left(\dfrac{39.10 \text{ g}}{1 \text{ mol K}} \right)}{158.0 \text{ g KMnO}_4} \times 100\% = 24.75\% \text{ K}$$

$$\% \text{ Mn in KMnO}_4 = \frac{1 \text{ mol Mn} \left(\dfrac{54.94 \text{ g}}{1 \text{ mol Mn}} \right)}{158.0 \text{ g KMnO}_4} \times 100\% = 34.77\% \text{ Mn}$$

$$\% \text{ O in KMnO}_4 = \frac{4 \text{ mol O} \left(\dfrac{16.00 \text{ g}}{1 \text{ mol O}} \right)}{158.0 \text{ g KMnO}_4} \times 100\% = 40.51\% \text{ O}$$

Is your answer reasonable? Each percentage should be less than 100%, and the sum of the percent composition values for the constituent elements must be equal to 100%. Notice that the oxygen percent composition could have been calculated from the % K and % Mn values.

$$\% \text{ O in KMnO}_4 = 100\% - \% \text{ K} - \% \text{ Mn} = 100\% - 24.75\% - 34.77\% = 40.48\% \text{ O}$$

The slight difference between the values is due to rounding during the calculation steps.

Video Solution

Tutored Practice
Problem 3.2.2

3.2c   Empirical Formulas from Percent Composition

A common practice in the chemical laboratory is to determine the percent composition of each element in a compound and use that information to determine the formula that shows the simplest whole-number ratio of elements present in the compound, the **empirical formula** (◄ Flashback to Section 2.3 Covalent Compounds) of the compound. In a sense, this is the reverse of the process that is used to determine percent composition.

The first step in the process is to understand that the percent composition of any element is equal to the mass of that element in exactly 100 g of compound. For example, a compound containing only phosphorus and chlorine is analyzed and found to contain 22.55% P and 77.45% Cl by mass. The percent composition of phosphorus can be written as

$$22.55\% \text{ P} = \frac{22.55 \text{ g P}}{100 \text{ g compound}}$$

Thus, if we assume that we are working with a sample that has a mass of exactly 100 g, the mass of each element present is equal to the percent composition. For the compound containing only phosphorus and chlorine,

$$22.55\% \text{ P} = 22.55 \text{ g P in 100 g compound}$$

$$77.45\% \text{ Cl} = 77.45 \text{ g Cl in 100 g compound}$$

To determine the empirical formula of the compound, the mass of each element is converted to an amount in moles.

$$22.55 \text{ g P} \times \frac{1 \text{ mol P}}{30.974 \text{ g}} = 0.7280 \text{ mol P}$$

$$77.45 \text{ g Cl} \times \frac{1 \text{ mol Cl}}{35.453 \text{ g}} = 2.185 \text{ mol Cl}$$

Thus, this compound contains phosphorus and chlorine in a mole ratio of $P_{0.7280}Cl_{2.185}$. The empirical formula, however, shows the simplest whole-number ratio of elements. Dividing each amount by the smallest amount present produces integer subscripts without changing the relative amounts of each element present in the compound.

$$\frac{2.185 \text{ mol Cl}}{0.7280} = 3.001 \text{ mol Cl} \approx 3 \text{ mol Cl}$$

$$\frac{0.7280 \text{ mol P}}{0.7280} = 1.000 \text{ mol P} = 1 \text{ mol P}$$

Notice that the amount of Cl can be rounded to the nearest whole number. The empirical formula of the compound, therefore, contains 3 mol of Cl and 1 mol of P, or PCl_3.

Example Problem 3.2.3 Use percent composition to determine an empirical formula.

A solid compound is found to contain K, S, and O with the following percent composition:

K:	41.09%
S:	33.70%
O:	25.22%

What is the empirical formula of this compound?

Solution:

You are asked to determine the empirical formula for a compound.

You are given the percent composition of each element in the compound.

First, assume that you are given a 100-g sample, which would mean that the mass of each element in grams is equal to the percent composition value. Calculate the amount (in mol) of each element present, and then use the relative amounts to determine the empirical formula of the compound.

$$41.09 \text{ g K} \times \frac{1 \text{ mol K}}{39.098 \text{ g}} = 1.051 \text{ mol K}$$

$$33.70 \text{ g S} \times \frac{1 \text{ mol S}}{32.065 \text{ g}} = 1.051 \text{ mol S}$$

$$25.22 \text{ g O} \times \frac{1 \text{ mol O}}{15.999 \text{ g}} = 1.576 \text{ mol O}$$

To determine the simplest whole-number ratio of elements, divide each by the smallest value.

$$\frac{1.051 \text{ mol K}}{1.051} = 1.000 \text{ mol K} = 1 \text{ mol K}$$

$$\frac{1.051 \text{ mol S}}{1.051} = 1.000 \text{ mol S} = 1 \text{ mol S}$$

$$\frac{1.576 \text{ mol O}}{1.051} = 1.500 \text{ mol O} = 1.5 \text{ mol O}$$

Recall that the value 1.5 is the same as the fraction $^3/_2$. Multiply all three amounts by 2, the fraction denominator, to express the ratio of elements as whole numbers.

$$1 \text{ mol K} \times 2 = 2 \text{ mol K}$$

$$1 \text{ mol S} \times 2 = 2 \text{ mol S}$$

$$1.5 \text{ mol O} \times 2 = 3 \text{ mol O}$$

The empirical formula is therefore $K_2S_2O_3$.

Video Solution

Tutored Practice
Problem 3.2.3

3.2d   Determining Molecular Formulas

The empirical formula of a compound gives the simplest whole-number ratio of atoms of each element present in a compound, but it may not be representative of the molecular formula of a compound. For example, the hydrocarbons cyclopropane and cyclobutane have the same empirical formula (CH_2) but different molecular formulas (C_3H_6 and C_4H_8, respectively). Ionic compounds and other compounds that do not exist as individual molecules are always represented using empirical formulas. For molecular compounds, additional information is needed to determine the molecular formula from an empirical formula.

Cyclopropane   Cyclobutane

The **molecular formula** of a compound is a whole-number multiple of its empirical formula. For cyclobutane, the empirical formula (CH_2) has a molar mass of 14.03 g/mol and the molecular formula has a molar mass of 56.11 g/mol.

$$\frac{\text{molar mass of compound}}{\text{molar mass of empirical formula}} = \text{whole number}$$

$$\frac{56.11 \text{ g/mol}}{14.03 \text{ g/mol}} = 3.999 \approx 4$$

The whole-number multiple is equal to the number of empirical formula units in one molecule of the compound. For cyclobutane, there are four empirical formula units per molecule, and the molecular formula is $(CH_2)_4$ or C_4H_8. The molar mass of a compound must be determined, usually through additional experiments, in order to determine the molecular formula of a compound.

Example Problem 3.2.4 Use percent composition and molar mass to determine molecular formula.

Resorcinol is a compound used in the manufacture of fluorescent and leather dyes as well as to treat acne and other greasy skin conditions. Analysis of the compound showed that it is 65.45% C and 5.493% H, with oxygen accounting for the remainder. In a separate experiment, the molar mass of the compound was found to be 110.11 g/mol. Determine the molecular formula of resorcinol.

Solution:

You are asked to determine the molecular formula of a compound.

You are given the percent composition of all but one element in the compound and the compound's molar mass.

First, determine the percent composition of oxygen.

% O = 100.00% − % C − % H = 100.00% − 65.45% C − 5.493% H = 29.06% O

Next, assume a 100-g sample and calculate the amount (in mol) of each element present.

$$65.45 \text{ g C} \times \frac{1 \text{ mol C}}{12.011 \text{ g}} = 5.449 \text{ mol C}$$

$$5.493 \text{ g H} \times \frac{1 \text{ mol H}}{1.0079 \text{ g}} = 5.450 \text{ mol H}$$

$$29.06 \text{ g O} \times \frac{1 \text{ mol O}}{15.999 \text{ g}} = 1.816 \text{ mol O}$$

Example Problem 3.2.4 (continued)

To determine the simplest whole-number ratio of elements, divide each by the smallest value.

$$\frac{5.449 \text{ mol C}}{1.816} = 3.001 \text{ mol C} \approx 3 \text{ mol C}$$

$$\frac{5.450 \text{ mol H}}{1.816} = 3.001 \text{ mol H} \approx 3 \text{ mol H}$$

$$\frac{1.816 \text{ mol O}}{1.816} = 1.000 \text{ mol O} = 1 \text{ mol O}$$

The empirical formula is C_3H_3O. Use the molar mass of the empirical formula and the molar mass of the compound to determine the whole-number multiple that relates the empirical and molecular formulas.

$$\frac{\text{molar mass of compound}}{\text{molar mass of empirical formula}} = \frac{110.11 \text{ g/mol}}{55.06 \text{ g/mol}} = 2$$

There are two empirical formula units in the molecular formula. The molecular formula of resorcinol is $(C_3H_3O)_2$ or $C_6H_6O_2$.

Is your answer reasonable? The molar mass of the compound should be a whole-number multiple of the empirical formula's molar mass. In this case, it is twice the empirical formula's molar mass, which suggests that the empirical formula was determined correctly.

Video Solution

Tutored Practice
Problem 3.2.4

3.2e Hydrated Compounds

A **hydrated ionic compound** is an ionic compound that has a well-defined amount of water trapped within the crystalline solid. The water associated with the compound is called the **water of hydration**. A hydrated compound formula includes the term $\cdot n\text{H}_2\text{O}$, where n is the number of moles of water incorporated into the solid per mole of ionic compound. Prefixes are used in naming hydrated compounds to indicate the number of waters of hydration.

Many solids used in the laboratory are hydrated. For example, reagent-grade copper(II) sulfate is usually provided as the hydrated compound $CuSO_4 \cdot 5H_2O$, copper(II) sulfate pentahydrate. Some common hydrated compounds and their uses are shown in Table 3.2.1. Notice that the molar mass of a hydrated compound includes the mass of the water of hydration.

Table 3.2.1 Some Common Hydrated Ionic Compounds

Molecular Formula	Name	Molar Mass (g/mol)	Common Name	Uses
$Na_2CO_3 \cdot 10H_2O$	Sodium carbonate decahydrate	286.14	Washing soda	Water softener
$Na_2S_2O_3 \cdot 5H_2O$	Sodium thiosulfate pentahydrate	248.18	Hypo	Photography
$MgSO_4 \cdot 7H_2O$	Magnesium sulfate heptahydrate	246.47	Epsom salt	Dyeing and tanning
$CaSO_4 \cdot 2H_2O$	Calcium sulfate dihydrate	172.17	Gypsum	Wallboard
$CaSO_4 \cdot \frac{1}{2}H_2O$	Calcium sulfate hemihydrate	145.15	Plaster of Paris	Casts, molds
$CuSO_4 \cdot 5H_2O$	Copper(II) sulfate pentahydrate	249.68	Blue vitriol	Algicide, root killer

Heating a hydrated compound releases the water in the crystalline solid. For example, heating the compound $CuCl_2 \cdot 2H_2O$ releases two moles of water (in the form of water vapor) per mole of hydrated compound. As shown in the following example problem, we can determine the formula of a hydrated compound by performing this experiment in the laboratory.

Example Problem 3.2.5 Determine the formula of a hydrated compound.

A 32.86 g sample of a hydrate of $CoCl_2$ was heated thoroughly in a porcelain crucible until its weight remained constant. After heating, 17.93 g of the dehydrated compound remained in the crucible. What is the formula of the hydrate?

Solution:

You are asked to determine the formula of an ionic hydrated compound.

You are given the mass of the hydrated compound and the mass of the compound when it has been dehydrated.

First, determine the mass of water lost when the hydrated compound was heated.

$$32.86 \text{ g} - 17.93 \text{ g} = 14.93 \text{ g } H_2O$$

Example Problem 3.2.5 *(continued)*

Next, calculate moles of water and moles of the dehydrated compound.

$$14.93 \text{ g H}_2\text{O} \times \frac{1 \text{ mol H}_2\text{O}}{18.015 \text{ g}} = 0.8288 \text{ mol H}_2\text{O}$$

$$17.93 \text{ g CoCl}_2 \times \frac{1 \text{ mol CoCl}_2}{129.84 \text{ g}} = 0.1381 \text{ mol CoCl}_2$$

Finally, determine the simplest whole-number ratio of water to dehydrated compound.

$$\frac{0.8288 \text{ mol H}_2\text{O}}{0.1381 \text{ mol CoCl}_2} = \frac{6.001 \text{ mol H}_2\text{O}}{1 \text{ mol CoCl}_2} \approx \frac{6 \text{ mol H}_2\text{O}}{1 \text{ mol CoCl}_2}$$

The chemical formula for the hydrated compound is $CoCl_2 \cdot 6H_2O$.

Video Solution

Tutored Practice
Problem 3.2.5

Section 3.2 Mastery

3.3   Stoichiometry and Chemical Reactions

3.3a   Chemical Reactions and Chemical Equations

Stoichiometry is used not only to determine the chemical formula of a compound but also to study the relative amounts of compounds involved in chemical reactions. Chemical reactions are represented by **chemical equations**, in which the reacting species (**reactants**) are shown to the left of a reaction arrow and the species produced during the reaction (**products**) appear to the right of the reaction arrow. In addition, the physical state of each reactant and product is often indicated using the symbols (g), (ℓ), (s), and (aq) for gas, liquid, solid, and aqueous (dissolved in water) solution, respectively. For example, the reaction of gaseous methane (CH_4) with gaseous molecular oxygen to produce gaseous carbon dioxide and water vapor is written as

$$\underbrace{CH_4(g) + 2 O_2(g)}_{\text{reactants}} \rightarrow \underbrace{CO_2(g) + 2 H_2O(g)}_{\text{products}}$$

Reading the equation, we say that one CH_4 molecule and two O_2 molecules react to form one CO_2 molecule and two H_2O molecules. The number that appears to the left of a compound formula in a balanced equation is called a **stoichiometric coefficient**; it indicates the relative number of molecules of that reactant or product in the reaction. (A stoichiometric coefficient of 1 is not written in a chemical equation.) Recall that the subscript to the right of each element symbol indicates the relative number of atoms of that element in the compound. Therefore, multiplying the stoichiometric coefficient of a compound by the subscript for each element in the compound formula gives the total number of atoms of each element that the compound contributes to the reactants or products.

The word *equation*, which means "equal on both sides," reflects the fact that when a chemical reaction takes place, matter is neither created nor destroyed, in accordance with the **Law of Conservation of Matter**. This means that the products of a chemical reaction are made up of the same type and number of atoms found in the reactants—but rearranged into new compounds. In the reaction of methane with oxygen, for example, both the reactants and the products contain one carbon atom, four hydrogen atoms, and four oxygen atoms.

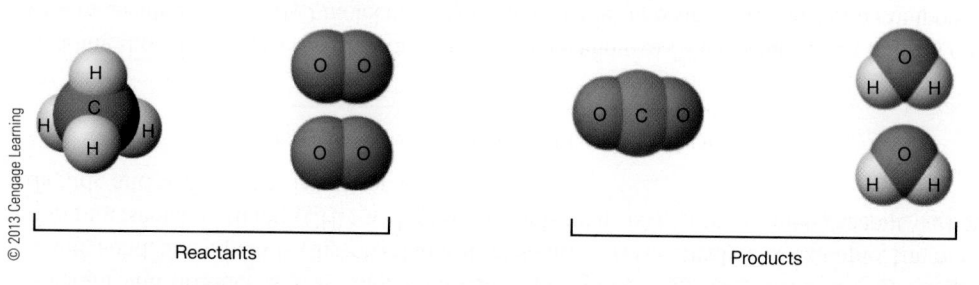

Reactants   Products

The reactants in this reaction are a CH_4 molecule and two O_2 molecules. When the reaction occurs, the chemical bonds between elements in the reactants are broken and new bonds are formed to make products. The same atoms are present both before and after the reaction, but the linkages between them have changed. When an equation meets the conditions of the Law of Conservation of Matter by having the same num-ber of atoms of each element on both sides of the reaction arrow, the equation is **balanced**.

Example Problem 3.3.1 Identify the components of a chemical equation.

Consider the following chemical equation:

$$MgCO_3(s) + 2 HCl(aq) \rightarrow CO_2(g) + MgCl_2(aq) + H_2O(\ell)$$

a. Is CO_2 a reactant or a product in the reaction?
b. How many chlorine atoms are there in the reactants?
c. What is the physical state of H_2O in this reaction?

Solution:

You are asked to answer questions about a chemical equation.

You are given a chemical equation.

a. Product. Chemical products are shown to the right of a reaction arrow.
b. Two. There is one chlorine atom in an HCl molecule. The stoichiometric coefficient to the left of HCl indicates that there are two HCl molecules, and thus two Cl atoms, in the reactants.
c. Liquid. The liquid physical state is indicated by the symbol (ℓ).

Video Solution

Tutored Practice
Problem 3.3.1

3.3b Balancing Chemical Equations

Very often we know what elements and compounds are involved in a reaction but not the relative amounts. To correctly describe the reaction, we must determine the relative amounts of the reactants and products involved in the reaction by balancing the equation (Interactive Figure 3.3.1). For example, consider the reaction that occurs in a gas grill when propane (C_3H_8) reacts with oxygen to form carbon dioxide and water. The unbalanced equation and the number of each element present in the reactants and product are

$$C_3H_8 + O_2 \rightarrow CO_2 + H_2O$$

3 C	1 C
8 H	2 H
2 O	3 O

The equation is not balanced because the number of atoms of each element present in the reactants is not equal to the number present in the products. The equation can be balanced by changing the coefficients to the left of each chemical species from 1 to a whole number that results in equal numbers of atoms of each element on both sides of the reaction arrow. It is important to note that the subscripts in a chemical formula are never changed when balancing an equation. A change in the subscript in a formula changes the chemical identity of the compound, not the amount of compound present in the reaction.

Although balancing chemical equations is usually done by trial and error, it is helpful to follow some general guidelines.

1. *If an element appears in more than one compound in the reactants or products, it is usually best to balance that element last.* In this case, we will begin by balancing carbon and hydrogen, balancing oxygen last because it appears in two compounds in the products.
2. As stated earlier, *only coefficients are changed when balancing equations, never the subscripts within a chemical formula.* For example, we cannot balance carbon by changing the formula of CO_2 to C_3O_2. Although this balances the number of carbon atoms, this change alters the chemical identity of the compound and makes the chemical equation invalid.
3. *Balanced chemical equations are written so that the coefficients are the lowest possible whole numbers.*

Balance Carbon

In the unbalanced equation, there are three carbon atoms in the reactants and only one carbon atom in the products. Changing the coefficient in front of CO_2 from 1 to 3 balances

Relate conservation of mass to balanced equations.

1 CH$_4$	2 O$_2$		1 CO$_2$	2 H$_2$O
Reactants			Products	

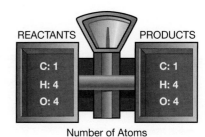

Number of Atoms

The mass of reactants equals the mass of the products

© 2013 Cengage Learning

the carbon atoms in the equation and also increases the number of oxygen atoms in the products from 3 to 7.

$$C_3H_8 + O_2 \rightarrow 3\,CO_2 + H_2O$$

3 C	3 C
8 H	2 H
2 O	7 O

Balance Hydrogen

There are eight hydrogen atoms in the reactants and only two hydrogen atoms in the products. Each increase in the H_2O coefficient increases the number of hydrogen atoms by a factor of 2. Changing the coefficient in front of H_2O from 1 to 4 balances the number of hydrogen atoms in the equation and also increases the number of oxygen atoms in the products from 7 to 10.

$$C_3H_8 + O_2 \rightarrow 3\,CO_2 + 4\,H_2O$$

3 C	3 C
8 H	8 H
2 O	10 O

Balance Oxygen

There are 10 oxygen atoms in the products (6 in the three CO_2 molecules and 4 in the four H_2O molecules) and only 2 in the reactants. Each increase in the O_2 coefficient increases the number of oxygen atoms by a factor of 2, so changing the coefficient in front of O_2 from 1 to 5 balances the oxygen atoms in the equation.

$$C_3H_8 + 5\,O_2 \rightarrow 3\,CO_2 + 4\,H_2O$$

3 C	3 C
8 H	8 H
10 O	10 O

The number of atoms of each element in the reactants and products is the same, so the equation is balanced. Finally, note that this equation can be balanced using a multiple of the coefficients above.

$$2\ C_3H_8 + 10\ O_2 \rightarrow 6\ CO_2 + 8\ H_2O$$

6 C	6 C
16 H	16 H
20 O	20 O

However, this is incorrect because it is possible to balance the equation using simpler coefficients.

Example Problem 3.3.2 Balance equations.

Balance the chemical equation for the neutralization of phosphoric acid by calcium hydroxide to form calcium phosphate and water. The unbalanced equation is
$H_3PO_4 + Ca(OH)_2 \rightarrow Ca_3(PO_4)_2 + H_2O$.

Solution:

You are asked to balance a chemical equation.

You are given an unbalanced equation.

Notice that both hydrogen and oxygen appear in more than one compound in the reactants. Begin by balancing the other elements, phosphorus and calcium, before balancing hydrogen and oxygen.

$$H_3PO_4 + Ca(OH)_2 \rightarrow Ca_3(PO_4)_2 + H_2O$$

5 H	2 H
1 P	2 P
6 O	9 O
1 Ca	3 Ca

Balance phosphorus. Change the coefficient in front of H_3PO_4 from 1 to 2 and note the change in H and O.

$$2\ H_3PO_4 + Ca(OH)_2 \rightarrow Ca_3(PO_4)_2 + H_2O$$

8 H	2 H
2 P	2 P
10 O	9 O
1 Ca	3 Ca

Balance calcium. Change the coefficient in front of $Ca(OH)_2$ from 1 to 3 and note the change in H and O.

$$2\ H_3PO_4 + 3\ Ca(OH)_2 \rightarrow Ca_3(PO_4)_2 + H_2O$$

12 H	2 H
2 P	2 P
14 O	9 O
3 Ca	3 Ca

Example Problem 3.3.2 (continued)

Balance hydrogen and oxygen. Changing the coefficient in front of H_2O from 1 to 6 (each increase in the H_2O coefficient adds two H atoms) balances both hydrogen and oxygen.

$$2 H_3PO_4 + 3 Ca(OH)_2 \rightarrow Ca_3(PO_4)_2 + 6 H_2O$$

12 H	12 H
2 P	2 P
14 O	14 O
3 Ca	3 Ca

The equation is balanced.

Is your answer reasonable? When balancing equations, it is a good idea to check the atom balance for all elements in the equation when you are finished. In this case, there are equal numbers of elements in both reactants and products, so the equation is balanced.

Video Solution

Tutored Practice
Problem 3.3.2

3.3c   Reaction Stoichiometry

A balanced chemical equation shows the relative amounts of reactants and products involved in a chemical reaction on both the molecular and macroscopic scale. For example, the balanced equation for the reaction of methane with oxygen to form carbon dioxide and water describes the reaction at the atomic level and at the macroscopic level.

$$CH_4 + 2 O_2 \rightarrow CO_2 + 2 H_2O$$

$$1 \text{ CH}_4 \text{ molecule} + 2 \text{ O}_2 \text{ molecules} \rightarrow 1 \text{ CO}_2 \text{ molecule} + 2 \text{ H}_2\text{O molecules}$$

$$1 \text{ mol CH}_4 + 2 \text{ mol O}_2 \rightarrow 1 \text{ mol CO}_2 + 2 \text{ mol H}_2\text{O}$$

$$16.04 \text{ g CH}_4 + 64.00 \text{ g O}_2 \rightarrow 44.01 \text{ g CO}_2 + 36.03 \text{ g H}_2\text{O}$$

On the molecular level, one molecule of CH_4 reacts with two molecules of O_2 to produce one molecule of CO_2 and two molecules of H_2O. Scaling up to macroscopic amounts, the equation also represents the reaction of one mole of CH_4 (16.04 g) with two moles of O_2 (64.00 g) to form one mole of CO_2 (44.01 g) and two moles of H_2O (36.03 g). Reaction stoichiometry is the study of the relationships between the amount of reactants and products on the macroscopic scale, and balanced chemical equations are the key to understanding these relationships.

Consider the reaction between butane and oxygen to form carbon dioxide and water. The balanced equation is

$$2 C_4H_{10} + 13 O_2 \rightarrow 8 CO_2 + 10 H_2O$$

According to the balanced equation, 8 mol of CO_2 are produced for every 2 mol of C_4H_{10} that react with oxygen. A typical reaction, however, might involve more or fewer than 2 mol of butane. The balanced equation gives us the information needed to determine the amount of CO_2 produced from any amount of C_4H_{10}.

The mole-to-mole relationships in a balanced chemical equation are used in the form of conversion factors to relate amounts of materials reacting or forming during a chemical reaction. For example, consider a butane lighter that contains 0.24 mol of C_4H_{10}. We can use the coefficients in the balanced equation to determine the amount of carbon dioxide produced in the reaction of this amount of butane with oxygen.

$$2\ C_4H_{10} + 13\ O_2 \rightarrow 8\ CO_2 + 10\ H_2O$$

$$0.24\ \text{mol}\ C_4H_{10} \times \frac{8\ \text{mol}\ CO_2}{2\ \text{mol}\ C_4H_{10}} = 0.96\ \text{mol}\ CO_2$$

The ratio of coefficients from the balanced equation, called the **stoichiometric factor** (or *stoichiometric ratio*), relates the amount of one species to another. In this case, the stoichiometric factor relates the amount of carbon dioxide formed from the reaction of butane with oxygen. It is important to note that *stoichiometric factors are always created from coefficients in a balanced chemical equation.*

The amount of water produced in the reaction can also be determined. Again, a stoichiometric factor is created using the coefficients in the balanced equation that relate moles of H_2O and C_4H_{10}.

$$2\ C_4H_{10} + 13\ O_2 \rightarrow 8\ CO_2 + 10\ H_2O$$

$$0.24\ \text{mol}\ C_4H_{10} \times \frac{10\ \text{mol}\ H_2O}{2\ \text{mol}\ C_4H_{10}} = 1.2\ \text{mol}\ H_2O$$

Stoichiometric relationships exist between all species in a balanced equation, not just between reactants and products. For example, if the reaction of a sample of butane with oxygen produces 2.8 mol of CO_2, it is possible to determine the amount of H_2O produced using a stoichiometric factor.

$$2.8\ \text{mol}\ CO_2 \times \frac{10\ \text{mol}\ H_2O}{8\ \text{mol}\ CO_2} = 3.5\ \text{mol}\ H_2O$$

Example Problem 3.3.3 Use balanced chemical equations to relate amounts of reactants and products.

The unbalanced equation for the reaction between magnesium nitride and sulfuric acid is shown here.

$$Mg_3N_2(s) + H_2SO_4(aq) \rightarrow MgSO_4(aq) + (NH_4)_2SO_4(aq)$$

Balance the equation and determine the amount of H_2SO_4 consumed and the amounts of $MgSO_4$ and $(NH_4)_2SO_4$ produced when 3.5 mol of Mg_3N_2 reacts.

Solution:

You are asked to balance a chemical equation and determine the amount of reactant consumed and products formed by a given amount of reactant.

You are given an unbalanced equation and the amount of one reactant consumed in the reaction.

Step 1. Write a balanced chemical equation.

$$Mg_3N_2(s) + 4\ H_2SO_4(aq) \rightarrow 3\ MgSO_4(aq) + (NH_4)_2SO_4(aq)$$

Step 2. Use the coefficients in the balanced equation to create a stoichiometric factor that will convert moles of Mg_3N_2 to moles of H_2SO_4 consumed.

$$3.5\ \text{mol}\ Mg_3N_2 \times \frac{4\ \text{mol}\ H_2SO_4}{1\ \text{mol}\ Mg_3N_2} = 14\ \text{mol}\ H_2SO_4\ \text{consumed}$$

Step 3. Use the coefficients in the balanced equation to create a stoichiometric factor that will convert moles of Mg_3N_2 to moles of $MgSO_4$ produced.

$$3.5\ \text{mol}\ Mg_3N_2 \times \frac{3\ \text{mol}\ MgSO_4}{1\ \text{mol}\ Mg_3N_2} = 11\ \text{mol}\ MgSO_4$$

Step 4. Use the coefficients in the balanced equation to create a stoichiometric factor that will convert moles of Mg_3N_2 to moles of $(NH_4)_2SO_4$ produced.

$$3.5\ \text{mol}\ Mg_3N_2 \times \frac{1\ \text{mol}\ (NH_4)_2SO_4}{1\ \text{mol}\ Mg_3N_2} = 3.5\ \text{mol}\ (NH_4)_2SO_4$$

Is your answer reasonable? According to the balanced equation, one mole of Mg_3N_2 reacts with four moles of H_2SO_4, producing three moles of $MgSO_4$ and one mole of $(NH_4)_2SO_4$. Here, more than one mole of Mg_3N_2 reacts, which means more than these amounts of reactants and products are consumed and produced, respectively.

Video Solution

Tutored Practice
Problem 3.3.3

Molar mass is combined with stoichiometric factors in order to determine the mass of each species consumed or produced in the reaction. In these types of calculations, it is very important to keep track of the units for each value calculated. For example, the mass of oxygen consumed by 7.8 g of C_4H_{10} is calculated as shown here. First, convert the mass of butane to an amount in units of moles.

$$7.8 \text{ g } C_4H_{10} \times \frac{1 \text{ mol } C_4H_{10}}{58.1 \text{ g}} = 0.13 \text{ mol } C_4H_{10}$$

Next, use the stoichiometric factor from the balanced equation to convert moles of C_4H_{10} to moles of O_2.

$$0.13 \text{ mol } C_4H_{10} \times \frac{13 \text{ mol } O_2}{2 \text{ mol } C_4H_{10}} = 0.87 \text{ mol } O_2$$

Finally, use the molar mass of oxygen to calculate the mass of O_2 consumed by 7.8 g of C_4H_{10}.

$$0.87 \text{ mol } O_2 \times \frac{32.0 \text{ g}}{1 \text{ mol } O_2} = 28 \text{ g } O_2$$

In the calculations above, values were rounded to the correct number of significant figures after each step. This introduces rounding errors that can be eliminated by carrying extra significant figures until the final calculation is performed or by combining the calculations into a single step.

In general, stoichiometric calculations are performed in the following order:

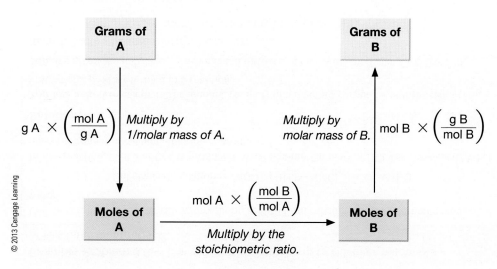

Example Problem 3.3.4 Use reaction stoichiometry to calculate amounts of reactants and products.

Calcium carbonate reacts with hydrochloric acid to form calcium chloride, carbon dioxide, and water.

$$\text{Unbalanced equation:} \quad CaCO_3 + HCl \rightarrow CaCl_2 + CO_2 + H_2O$$

In one reaction, 54.6 g of CO_2 is produced. What amount (in mol) of HCl was consumed? What mass (in grams) of calcium chloride is produced?

Solution:

You are asked to calculate the amount (in mol) of a reactant consumed and the mass (in grams) of a product formed in a reaction.

You are given the unbalanced equation and the mass of one of the products of the reaction.

Step 1. Write a balanced chemical equation.

$$CaCO_3 + 2\,HCl \rightarrow CaCl_2 + CO_2 + H_2O$$

Step 2. Calculate moles of CO_2 produced.

$$54.6 \text{ g } CO_2 \times \frac{1 \text{ mol } CO_2}{44.01 \text{ g}} = 1.24 \text{ mol } CO_2$$

Step 3. Calculate moles of HCl consumed and moles of $CaCl_2$ produced using stoichiometric factors derived from the balanced chemical equation.

$$1.24 \text{ mol } CO_2 \times \frac{2 \text{ mol HCl}}{1 \text{ mol } CO_2} = 2.48 \text{ mol HCl}$$

$$1.24 \text{ mol } CO_2 \times \frac{1 \text{ mol } CaCl_2}{1 \text{ mol } CO_2} = 1.24 \text{ mol } CaCl_2$$

Step 4. Calculate mass (in grams) of $CaCl_2$ produced.

$$1.24 \text{ mol } CaCl_2 \times \frac{111.0 \text{ g}}{1 \text{ mol } CaCl_2} = 138 \text{ g } CaCl_2$$

Notice that it is possible to set up both calculations as a single step.

$$54.6 \text{ g } CO_2 \left(\frac{1 \text{ mol } CO_2}{44.01 \text{ g}} \right) \left(\frac{2 \text{ mol HCl}}{1 \text{ mol } CO_2} \right) = 2.48 \text{ mol HCl}$$

$$54.6 \text{ g } CO_2 \left(\frac{1 \text{ mol } CO_2}{44.01 \text{ g}} \right) \left(\frac{1 \text{ mol } CaCl_2}{1 \text{ mol } CO_2} \right) \left(\frac{111.0 \text{ g}}{1 \text{ mol } CaCl_2} \right) = 138 \text{ g } CaCl_2$$

Is your answer reasonable? More than one mole of CO_2 is produced in the reaction, so the amounts of HCl consumed and $CaCl_2$ produced are greater than one mole.

Video Solution

Tutored Practice
Problem 3.3.4

Section 3.3 Mastery

3.4 Stoichiometry and Limiting Reactants

3.4a Limiting Reactants

A balanced equation represents a situation in which a reaction proceeds to completion and all of the reactants are converted into products. Under nonstoichiometric conditions, the reaction continues only until one of the reactants is entirely consumed, and at this point the reaction stops. Under these conditions, it is important to know which reactant will be consumed first and how much product can be produced.

When a butane lighter is lit, for example, only a small amount of butane is released from the lighter and combined with a much larger amount of oxygen present in air.

$$2 \, C_4H_{10} + 13 \, O_2 \rightarrow 8 \, CO_2 + 10 \, H_2O$$

Under these conditions, the oxygen is considered an **excess reactant** because more is available than is required for reaction with butane. The butane is the **limiting reactant**, which means that it controls the amount of products produced in the reaction. When a nonstoichiometric reaction is complete, the limiting reactant is completely consumed and some amount of the excess reactant remains unreacted (Interactive Figure 3.4.1).

Consider a situation where 12 mol of butane is combined with 120 mol of oxygen. Under stoichiometric conditions, butane and oxygen react in a 2:13 mole ratio. The amount of oxygen needed to react with 12 mol of butane is

$$\text{mol } O_2 \text{ needed} \; = \; 12 \text{ mol } C_4H_{10} \times \frac{13 \text{ mol } O_2}{2 \text{ mol } C_4H_{10}} \; = \; 78 \text{ mol } O_2$$

Because more oxygen is available than is needed to react with all of the available butane (120 mol available versus 78 mol needed), oxygen is the excess reactant. The amount of butane that would be needed to react with 120 mol of O_2 is

$$\text{mol } C_4H_{10} \text{ needed} \; = \; 120 \text{ mol } O_2 \times \frac{2 \text{ mol } C_4H_{10}}{13 \text{ mol } O_2} \; = \; 18 \text{ mol } C_4H_{10}$$

Less butane is available than is needed to react with all of the oxygen available (12 mol available versus 18 mol needed), so butane is the limiting reactant. The amount of products that can be produced in the reaction is determined by the amount of limiting reactant present. For water,

$$\text{mol } H_2O \text{ produced} \; = \; 12 \text{ mol } C_4H_{10} \times \frac{10 \text{ mol } H_2O}{2 \text{ mol } C_4H_{10}} \; = \; 60. \text{ mol } H_2O$$

Determine how much Zn remains after reaction with HCl.

Charles D. Winters

Reaction of Zn with HCl

The limiting reactant concept can be illustrated in the laboratory. For example, consider the reaction of iron(III) chloride with sodium hydroxide.

$$FeCl_3(aq) + 3\ NaOH(aq) \rightarrow Fe(OH)_3(s) + 3\ NaCl(aq)$$

If a student starts with 50.0 g of $FeCl_3$ and adds NaOH in 1-gram increments, the mass of $Fe(OH)_3$ produced in each experiment can be measured. Plotting mass of $Fe(OH)_3$ as a function of mass of NaOH added results in the graph shown in Interactive Figure 3.4.2. As expected, as more NaOH is added to the $FeCl_3$, increasing amounts of $Fe(OH)_3$ are produced. But when more than 37.0 g of NaOH are added, the mass of $Fe(OH)_3$ no longer changes. When 37.0 g (0.925 mol) of NaOH are added to 50.0 g (0.308 mol) of $FeCl_3$, the stoichiometric ratio is equal to 3:1 and all reactants are consumed completely. When more than 37.0 g of NaOH are added to 50.0 g $FeCl_3$, there is excess NaOH available ($FeCl_3$ is limiting) and no additional $Fe(OH)_3$ is produced.

Interactive Figure 3.4.2

Analyze reaction stoichiometry in the laboratory.

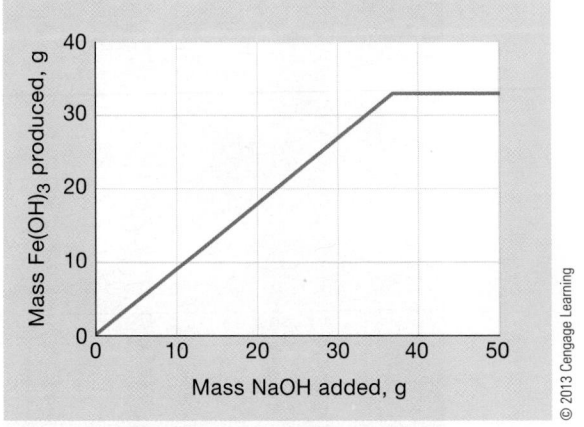

Mass of $Fe(OH)_3$ as a function of mass of NaOH

© 2013 Cengage Learning

Example Problem 3.4.1 Identify limiting reactants (mole ratio method).

Identify the limiting reactant in the reaction of hydrogen and oxygen to form water if 61.0 g of O_2 and 8.40 g of H_2 are combined. Determine the amount (in grams) of excess reactant that remains after the reaction is complete.

Solution:

You are asked to identify the limiting reactant and mass (in grams) of the excess reactant remaining after the reaction is complete.

You are given the mass of the reactants.

Step 1. Write a balanced chemical equation.

$$2\ H_2 + O_2 \rightarrow 2\ H_2O$$

Step 2. Determine the limiting reactant by comparing the relative amounts of reactants available.

Calculate the amount (in mol) of one of the reactants needed and compare that value to the amount available.

$$\text{amount of } O_2 \text{ needed: } 8.40\ \text{g } H_2\left(\frac{1\ \text{mol } H_2}{2.016\ \text{g}}\right)\left(\frac{1\ \text{mol } O_2}{2\ \text{mol } H_2}\right) = 2.08\ \text{mol } O_2$$

$$\text{amount of } O_2 \text{ available: } 61.0\ \text{g } O_2 \times \frac{1\ \text{mol } O_2}{32.00\ \text{g}} = 1.91\ \text{mol } O_2$$

More O_2 is needed (2.08 mol) than is available (1.91 mol), so O_2 is the limiting reactant.

Example Problem 3.4.1 *(continued)*

Alternatively, the amounts of reactants available can be compared to the stoichiometric ratio in the balanced equation in order to determine the limiting reactant.

$$\frac{4.17 \text{ mol } H_2}{1.91 \text{ mol } O_2} = \underset{\substack{\text{ratio of reactants} \\ \text{(available)}}}{\frac{2.18 \text{ mol } H_2}{1 \text{ mol } O_2}} > \underset{\substack{\text{ratio of reactants} \\ \text{(balanced equation)}}}{\frac{2 \text{ mol } H_2}{1 \text{ mol } O_2}}$$

Here, the mole ratio of H_2 to O_2 available is greater than the mole ratio from the balanced chemical equation. Thus, hydrogen is in excess and oxygen is the limiting reactant.

Step 3. Use the amount of limiting reactant (O_2) available to calculate the amount of excess reactant (H_2) needed for complete reaction.

$$1.91 \text{ mol } O_2 \left(\frac{2 \text{ mol } H_2}{1 \text{ mol } O_2}\right)\left(\frac{2.016 \text{ g}}{1 \text{ mol } H_2}\right) = 7.70 \text{ g } H_2$$

Step 4. Calculate the mass of excess reactant that remains when all the limiting reactant is consumed.

$$8.40 \text{ g } H_2 \text{ available} - 7.70 \text{ g } H_2 \text{ consumed} = 0.70 \text{ g } H_2 \text{ remains}$$

Is your answer reasonable? More O_2 is needed than is available, so it is the limiting reactant. There is H_2 remaining after the reaction is complete because it is the excess reactant in the reaction.

Video Solution

Tutored Practice
Problem 3.4.1

An alternative method of performing stoichiometry calculations that involve limiting reactants is to simply calculate the maximum amount of product that could be produced from each reactant. (This method works particularly well for reactions with more than two reactants.) In this method, the reactant that produces the least amount of product is the limiting reactant.

Example Problem 3.4.2 Identify limiting reactants (maximum product method).

Consider the reaction of gold with nitric acid and hydrochloric acid.

$$Au + 3 HNO_3 + 4 HCl \rightarrow HAuCl_4 + 3 NO_2 + 3 H_2O$$

Determine the limiting reactant in a mixture containing 125 g of each reactant and calculate the maximum mass (in grams) of $HAuCl_4$ that can be produced in the reaction.

Solution:

You are asked to calculate the maximum mass of a product that can be formed in a reaction.

You are given the balanced equation and the mass of each reactant available.

In this case there are three reactants, so it is most efficient to calculate the maximum amount of product ($HAuCl_4$) that can be produced by each one. The reactant that produces the least amount of $HAuCl_4$ is the limiting reactant.

Step 1. Write the balanced chemical equation.

$$Au + 3\,HNO_3 + 4\,HCl \rightarrow HAuCl_4 + 3\,NO_2 + 3\,H_2O$$

Step 2. Use the molar mass of each reactant and the stoichiometric factors derived from the balanced equation to calculate the amount of $HAuCl_4$ that can be produced by each reactant.

$$125\text{ g Au}\left(\frac{1\text{ mol Au}}{197.0\text{ g}}\right)\left(\frac{1\text{ mol HAuCl}_4}{1\text{ mol Au}}\right)\left(\frac{339.8\text{ g}}{1\text{ mol HAuCl}_4}\right) = 216\text{ g HAuCl}_4$$

$$125\text{ g HNO}_3\left(\frac{1\text{ mol HNO}_3}{63.01\text{ g}}\right)\left(\frac{1\text{ mol HAuCl}_4}{3\text{ mol HNO}_3}\right)\left(\frac{339.8\text{ g}}{1\text{ mol HAuCl}_4}\right) = 225\text{ g HAuCl}_4$$

$$125\text{ g HCl}\left(\frac{1\text{ mol HCl}}{36.46\text{ g}}\right)\left(\frac{1\text{ mol HAuCl}_4}{4\text{ mol HCl}}\right)\left(\frac{339.8\text{ g}}{1\text{ mol HAuCl}_4}\right) = 291\text{ g HAuCl}_4$$

Gold is the limiting reagent. The maximum amount of $HAuCl_4$ that can be produced is 216 g.

Video Solution

Tutored Practice
Problem 3.4.2

3.4b   Percent Yield

When chemists perform experiments to make a compound, it is rare that the amount of materials produced is equal to that predicted based on stoichiometric calculations. The reasons for this are numerous. For example, side reactions may occur, using some of the reactants to make other, undesired products. Sometimes, the reaction does not go to completion and some reactants remain unreacted. In other cases, some of the product cannot be physically isolated (for example, because it is embedded in filter paper, it does not completely precipitate from solution or is lost via evaporation). Because laboratory results rarely match calculations exactly, we make a distinction between the predicted amount of product, called the **theoretical yield**, and the actual amount of product produced in the experiment, called the **experimental yield** (or *actual yield*).

When reporting the success of a chemical synthesis, the ratio of product mass obtained (the experimental yield) to the maximum mass that could have been produced based on the amounts of reactants used (the theoretical yield) is calculated. This ratio is reported in the form of a percent, called the **percent yield** of a reaction.

$$\text{percent yield} = \frac{\text{experimental yield}}{\text{theoretical yield}} \times 100\% \qquad \textbf{(3.2)}$$

All the calculations we have done to this point allow us to determine the theoretical yield for a reaction. An experimental yield value is never calculated but is instead measured in the chemical laboratory after a reaction is complete.

Example Problem 3.4.3 Calculate percent yield.

Consider the reaction of $Pb(NO_3)_2$ with $NaCl$ to form $PbCl_2$ and $NaNO_3$. If 24.2 g $Pb(NO_3)_2$ is reacted with excess NaCl and 17.3 g of $PbCl_2$ is ultimately isolated, what is the percent yield for the reaction?

Solution:

You are asked to calculate the percent yield for a reaction.

You are given the mass of the limiting reactant and the experimental yield for one of the products in the reaction.

Step 1. Write a balanced chemical equation for the reaction.

$$Pb(NO_3)_2 + 2\, NaCl \rightarrow PbCl_2 + 2\, NaNO_3$$

Step 2. Use the amount of limiting reactant to calculate the theoretical yield of $PbCl_2$.

$$24.2 \text{ g } Pb(NO_3)_2 \left(\frac{1 \text{ mol } Pb(NO_3)_2}{331.2 \text{ g}} \right) \left(\frac{1 \text{ mol } PbCl_2}{1 \text{ mol } Pb(NO_3)_2} \right) \left(\frac{278.1 \text{ g}}{1 \text{ mol } PbCl_2} \right) = 20.3 \text{ g } PbCl_2$$

Step 3. Use the experimental yield (17.3 g $PbCl_2$) and the theoretical yield (20.3 g $PbCl_2$) to calculate the percent yield for the reaction.

$$\frac{17.3 \text{ g } PbCl_2}{20.3 \text{ g } PbCl_2} \times 100\% = 85.2\%$$

Is your answer reasonable? The theoretical yield is greater than the experimental yield, so the percent yield is less than 100%.

Video Solution

Tutored Practice
Problem 3.4.3

Section 3.4 Mastery

3.5 Chemical Analysis

3.5a Determining a Chemical Formula

Chemical analysis is a large field of chemistry that revolves around the need to analyze samples to determine the nature and amount of chemical species present. The types of analysis done in a chemical laboratory typically involve determining the chemical formula of an unknown compound, the composition of a mixture of compounds, or the purity of a chemical sample.

The most common technique used to determine the chemical formula of a compound that contains carbon and hydrogen is **combustion analysis**. This technique involves a **combustion reaction**, the reaction of a chemical species with molecular oxygen that produces energy in the form of heat and light. In combustion analysis, a weighed sample of the compound is burned in the presence of excess oxygen, converting all of the carbon in the sample to carbon dioxide and all of the hydrogen present in the sample to water (Interactive Figure 3.5.1). All the hydrogen in the sample ends up in the form of H_2O, which is absorbed in one chamber. All the carbon in the sample ends up in the form of CO_2, which is absorbed in the other chamber. The two absorption chambers are weighed before and after the reaction, and the chemical formula of the hydrocarbon can be determined.

Interactive Figure 3.5.1

Explore CHO combustion analysis.

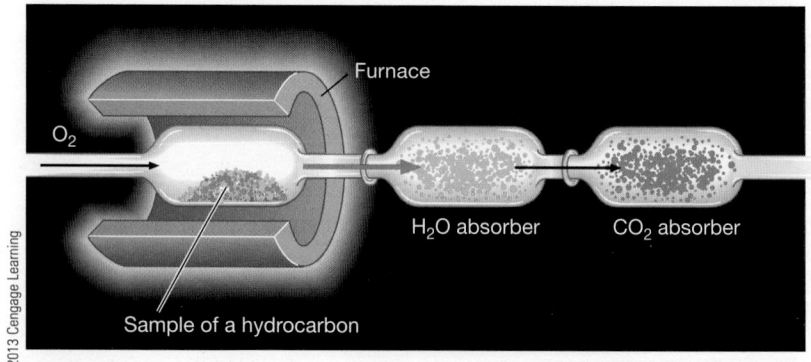

Analysis of an organic compound by combustion

For the general hydrocarbon C_xH_y, the following reaction takes place during combustion analysis.

$$C_xH_y + \text{excess } O_2 \rightarrow x\, CO_2 + \frac{y}{2}\, H_2O$$

Using the stoichiometric relationships developed earlier in this unit, the data produced from combustion analysis can be used to determine the empirical formula of a hydrocarbon, as shown in the following example.

Example Problem 3.5.1 Use combustion analysis to determine empirical and molecular formulas (hydrocarbons).

When 1.827 g of a hydrocarbon, C_xH_y, was burned in a combustion analysis apparatus, 6.373 g of CO_2 and 0.7829 g of H_2O were produced. In a separate experiment, the molar mass of the compound was found to be 252.31 g/mol. Determine the empirical formula and molecular formula of the hydrocarbon.

Solution:

You are asked to determine the empirical and molecular formulas of a compound.

You are given the mass of compound analyzed, data from a combustion analysis experiment (mass of H_2O and CO_2), and the molar mass of the compound.

Step 1. Use the mass of CO_2 and H_2O produced in the combustion analysis to calculate moles of H and moles of C present in the original sample. Note that there are 2 mol of H present in 1 mol of H_2O.

$$6.373 \text{ g } CO_2 \times \frac{1 \text{ mol } CO_2}{44.010 \text{ g}} \times \frac{1 \text{ mol C}}{1 \text{ mol } CO_2} = 0.1448 \text{ mol C}$$

$$0.7829 \text{ g } H_2O \times \frac{1 \text{ mol } H_2O}{18.015 \text{ g}} \times \frac{2 \text{ mol H}}{1 \text{ mol } H_2O} = 0.08692 \text{ mol H}$$

Step 2. Dividing each amount by the smallest value results in a whole-number ratio of elements.

$$\frac{0.1448 \text{ mol C}}{0.08692} = 1.666 \text{ mol C}$$

$$\frac{0.08692 \text{ mol H}}{0.08692} = 1.000 \text{ mol H} = 1 \text{ mol H}$$

In this case, the ratio is not made up of two integers. Writing the relative amounts as a ratio and rewriting the ratio as a fraction results in a whole-number ratio of elements and the correct empirical formula.

$$\frac{1.666 \text{ mol C}}{1 \text{ mol H}} = \frac{1\frac{2}{3} \text{ mol C}}{1 \text{ mol H}} = \frac{\frac{5}{3} \text{ mol C}}{1 \text{ mol H}} = \frac{5 \text{ mol C}}{3 \text{ mol H}}$$

Example Problem 3.5.1 (continued)

The empirical formula of the hydrocarbon is C_5H_3.

Step 3. Compare the molar mass of the compound to the molar mass of the empirical formula to determine the molecular formula.

$$\frac{\text{molar mass of compound}}{\text{molar mass of empirical formula}} = \frac{252.31 \text{ g/mol}}{63.079 \text{ g/mol}} = 4$$

The molecular formula is $(C_5H_3)_4$, or $C_{20}H_{12}$.

Is your answer reasonable? The molar mass of the compound must be a whole-number multiple of the molar mass of the empirical formula, which is true in this case.

Combustion analysis can also be performed on compounds containing carbon, hydrogen, and oxygen, as shown in the following example. In this case, the amount of oxygen must be determined from the mass of the original sample, not directly from the combustion data.

Video Solution

Tutored Practice
Problem 3.5.1

Example Problem 3.5.2 Use combustion analysis to determine empirical and molecular formulas (C, H, and O).

A 1.155-g sample of butyric acid, an organic compound containing carbon, hydrogen, and oxygen, is analyzed by combustion and 2.308 g of CO_2 and 0.9446 g of H_2O are produced. In a separate experiment, the molar mass is found to be 88.11 g/mol. Determine the empirical and molecular formulas of butyric acid.

Solution:

You are asked to determine the empirical and molecular formulas of a compound.

You are given the mass of compound analyzed, data from a combustion analysis experiment (mass of H_2O and CO_2), and the molar mass of the compound.

When combustion analysis involves a compound containing carbon, hydrogen, and oxygen, the amount of oxygen in the compound cannot be determined directly from the amount of CO_2 or H_2O produced. Instead, the amounts of CO_2 and H_2O are used to calculate the mass of carbon and hydrogen in the original sample. The remaining mass of the original sample is oxygen.

Step 1. Use the mass of CO_2 and H_2O produced in the combustion analysis to calculate moles of C and moles of H present in the original sample. Note that there are 2 mol of H present in 1 mol of H_2O.

Example Problem 3.5.2 *(continued)*

$$2.308 \text{ g CO}_2 \times \frac{1 \text{ mol CO}_2}{44.010 \text{ g}} \times \frac{1 \text{ mol C}}{1 \text{ mol CO}_2} = 0.05244 \text{ mol C}$$

$$0.9446 \text{ g H}_2\text{O} \times \frac{1 \text{ mol H}_2\text{O}}{18.015 \text{ g}} \times \frac{2 \text{ mol H}}{1 \text{ mol H}_2\text{O}} = 0.1049 \text{ mol H}$$

Step 2. Use the amount (in mol) of C and H to calculate the mass of C and H present in the original sample.

$$0.05244 \text{ mol C} \times \frac{12.011 \text{ g}}{1 \text{ mol C}} = 0.6299 \text{ g C}$$

$$0.1049 \text{ mol H} \times \frac{1.0079 \text{ g}}{1 \text{ mol H}} = 0.1057 \text{ g H}$$

Step 3. Subtract the amounts of C and H from the mass of the original sample to determine the mass of O present in the original sample. Use this value to calculate the moles of O present in the original sample.

$$1.155 \text{ g sample} - 0.6299 \text{ g C} - 0.1057 \text{ g H} = 0.419 \text{ g O}$$

$$0.419 \text{ g O} \times \frac{1 \text{ mol O}}{16.00 \text{ g}} = 0.0262 \text{ mol O}$$

Step 4. Divide each amount by the smallest value to obtain a whole-number ratio of elements.

$$\frac{0.05244 \text{ mol C}}{0.0262} = 2.00 \text{ mol C} = 2 \text{ mol C}$$

$$\frac{0.1049 \text{ mol H}}{0.0262} = 4.00 \text{ mol H} = 4 \text{ mol H}$$

$$\frac{0.0262 \text{ mol O}}{0.0262} = 1.00 \text{ mol O} = 1 \text{ mol O}$$

The empirical formula of butyric acid is C_2H_4O.

Step 5. Compare the molar mass of the compound to the molar mass of the empirical formula to determine the molecular formula.

$$\frac{\text{molar mass of compound}}{\text{molar mass of empirical formula}} = \frac{88.11 \text{ g/mol}}{44.05 \text{ g/mol}} = 2$$

The molecular formula is $(C_2H_4O)_2$, or $C_4H_8O_2$.

Is your answer reasonable? The molar mass of the compound must be a whole-number multiple of the molar mass of the empirical formula, which is true in this case.

Video Solution

Tutored Practice
Problem 3.5.2

The formula of other types of binary (two-element) compounds can be determined from experiments where the two elements react to form a single compound, as shown in the following example.

Example Problem 3.5.3 Determine the chemical formula of a binary compound.

A 2.64-g sample of Cr is heated in the presence of excess oxygen. A metal oxide (Cr_xO_y) is formed with a mass of 3.86 g. Determine the empirical formula of the metal oxide.

Solution:

You are asked to determine the empirical formula of a binary compound.

You are given the mass of a reactant and the mass of the binary compound formed in the reaction.

Step 1. Use the mass of the metal oxide and the mass of the metal to determine the amount of oxygen present in the metal oxide sample.

$$3.86 \text{ g } Cr_xO_y - 2.64 \text{ g Cr} = 1.22 \text{ g O}$$

Step 2. Use the mass of Cr and mass of O to determine moles of each element present in the compound.

$$2.64 \text{ g Cr} \times \frac{1 \text{ mol Cr}}{52.00 \text{ g}} = 0.0508 \text{ mol Cr}$$

$$1.22 \text{ g O} \times \frac{1 \text{ mol O}}{16.00 \text{ g}} = 0.0763 \text{ mol O}$$

Step 3. To determine the simplest whole-number ratio of elements, divide each by the smallest value.

$$\frac{0.0508 \text{ mol Cr}}{0.0508} = 1.00 \text{ mol Cr} = 1 \text{ mol Cr}$$

$$\frac{0.0763 \text{ mol O}}{0.0508} = 1.50 \text{ mol O}$$

In this case, the ratio is not made up of two integers. Writing the relative amounts as a ratio and rewriting the ratio as a fraction results in a whole-number ratio of elements and the correct empirical formula.

$$\frac{1.50 \text{ mol O}}{1 \text{ mol Cr}} = \frac{\frac{3}{2} \text{ mol O}}{1 \text{ mol Cr}} = \frac{3 \text{ mol O}}{2 \text{ mol Cr}}$$

The empirical formula is Cr_2O_3.

Video Solution

Tutored Practice
Problem 3.5.3

3.5b Analysis of a Mixture

The analysis of a mixture is a common process performed in the chemical laboratory. For example, an environmental sample containing a mixture of compounds can be analyzed for a single compound by first chemically isolating the compound of interest, followed by chemical analysis to determine the amount of compound in the sample. This type of analysis uses the stoichiometric methods developed in this unit to determine the purity of a chemical sample or the chemical makeup of the sample, as shown in the following example.

Example Problem 3.5.4 Use stoichiometric methods to analyze a mixture.

A soil sample is analyzed to determine the iron content. First, the iron is isolated as $Fe(NO_3)_2$. This compound is then reacted with $KMnO_4$.

$$5\ Fe(NO_3)_2 + KMnO_4 + 8\ HNO_3 \rightarrow 5\ Fe(NO_3)_3 + Mn(NO_3)_2 + 4\ H_2O + KNO_3$$

In one experiment, a 12.2-g sample of soil required 1.73 g of $KMnO_4$ to react completely with the $Fe(NO_3)_2$. Determine the percent (by mass) of iron present in the soil sample.

Solution:

You are asked to calculate the mass percent of iron in a soil sample.

You are given a balanced equation, the mass of solid analyzed, and the mass of one of the reactants used to analyze the sample.

Step 1. Use the mass of $KMnO_4$ to determine the amount of $Fe(NO_3)_2$ present in the soil sample.

$$1.73\ g\ KMnO_4 \left(\frac{1\ mol\ KMnO_4}{158.0\ g} \right) \left(\frac{5\ mol\ Fe(NO_3)_2}{1\ mol\ KMnO_4} \right) = 0.0547\ mol\ Fe(NO_3)_2$$

Step 2. Use the stoichiometry of the chemical formula of $Fe(NO_3)_2$ and the molar mass of Fe to calculate the mass of Fe present in the soil sample.

$$0.0547\ mol\ Fe(NO_3)_2 \left(\frac{1\ mol\ Fe}{1\ mol\ Fe(NO_3)_2} \right) \left(\frac{55.85\ g}{1\ mol\ Fe} \right) = 3.06\ g\ Fe$$

Step 3. Use the mass of Fe in the soil sample and the mass of the soil sample to calculate the percent by mass of Fe in the soil.

$$\frac{3.06\ g\ Fe\ in\ soil\ sample}{12.2\ g\ soil\ sample} \times 100\% = 25.1\%\ Fe$$

Is your answer reasonable? Because the soil sample is not pure, the mass of iron in the soil is less than the mass of the soil sample and the weight percent of iron in the sample is less than 100%.

Video Solution

Tutored Practice
Problem 3.5.4

Section 3.5 Mastery

Unit Recap

Key Concepts

3.1 The Mole and Molar Mass

- Avogadro's number (6.022×10^{23}) is the number of particles in one mole of a substance (3.1a).

- Molar mass is the mass in grams of one mole of particles of a substance (3.1b).

- Molar mass is used to convert between moles and grams of a substance (3.1b).

3.2 Stoichiometry and Compound Formulas

- Stoichiometry is the study of the relationship between relative amounts of substances (3.2a).

- Percent composition of an element in a compound is the mass of an element in exactly 100 g of a compound (3.2b).

- The empirical formula of a compound is the simplest whole-number ratio of elements present in the compound (3.2c).

- The molecular formula of a compound is a whole-number multiple of the empirical compound formula and shows the number of atoms of each element in one molecule of a compound (3.2d).

- Hydrated ionic compounds have water trapped in the crystal lattice (3.2e).

- The formula of a hydrated compound can be determined by heating the compound to drive off waters of hydration (3.2e).

3.3 Stoichiometry and Chemical Reactions

- Chemical reactions are represented by chemical equations, in which the reacting species (reactants) are shown to the left of a reaction arrow and the species produced during the reaction (products) appear to the right of the reaction arrow (3.3a).

- Stoichiometric coefficients are numbers that appear to the left of a compound formula in a chemical equation (3.3a).

- A balanced equation, which has equal numbers of atoms of each element present in the reaction on both sides of the reaction arrow, reflects the Law of Conservation of Matter, which states that matter is neither created nor destroyed during a chemical reaction (3.3a).

- The stoichiometric coefficients in a balanced chemical equation can be used to create a stoichiometric factor (or stoichiometric ratio) that relates the amount of one species to another (3.3c).

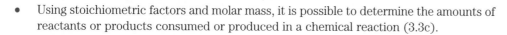

- Using stoichiometric factors and molar mass, it is possible to determine the amounts of reactants or products consumed or produced in a chemical reaction (3.3c).

3.4 Stoichiometry and Limiting Reactants

- The limiting reactant is the reactant in a chemical reaction that limits or controls the amount of product that can be produced (3.4a).
- Excess reactants are present in amounts that exceed the amount required for reaction (3.4a).
- Percent yield is a ratio of the experimental yield (the amount of material produced in the reaction) to the theoretical yield (the predicted amount of product) (3.4b).

3.5 Chemical Analysis

- Combustion analysis is an analytical method used to determine the carbon and hydrogen content, by mass, in a compound (3.5a).

Key Equations

$$\% \text{ element} = \frac{(\text{number of atoms of element in formula})(\text{molar mass of element})}{\text{mass of 1 mol of compound}} \times 100\% \quad \textbf{(3.1)}$$

$$\text{percent yield} = \frac{\text{experimental yield}}{\text{theoretical yield}} \times 100\% \quad \textbf{(3.2)}$$

Key Terms

3.1 The Mole and Molar Mass
mole (mol)
Avogadro's number (N_A)
molar mass
formula weight
molecular weight

3.2 Stoichiometry and Compound Formulas
stoichiometry
percent composition
empirical formula

molecular formula
hydrated ionic compound
water of hydration

3.3 Stoichiometry and Chemical Reactions
chemical equations
reactants
products
stoichiometric coefficient
Law of Conservation of Matter
balanced equation
stoichiometric factor

3.4 Stoichiometry and Limiting Reactants
excess reactant
limiting reactant
theoretical yield
experimental yield
percent yield

3.5 Chemical Analysis
chemical analysis
combustion analysis
combustion reaction

Unit 3   Review and Challenge Problems

4 Chemical Reactions and Solution Stoichiometry

Unit Outline

In This Unit...

The science of chemistry brings three central benefits to society. First, chemistry helps explain how the world works, principally by examining nature on the molecular scale. Second, chemical analysis is used to identify substances, both natural and artificial. This is evident every time a person's blood is tested to determine cholesterol level or when athletes undergo testing for banned drug use. Finally, the science of chemistry is unique in that it involves the ability to create new forms of matter by combining the elements and existing compounds in new but controlled ways. The creation of new chemical compounds is the subject of this unit. Here, we explore different types of chemical reactions, emphasizing the reactions that take place in aqueous solution.

4.1 Types of Chemical Reactions

4.1a Combination and Decomposition Reactions

Chemical **reactions** involve the transformation of matter, the reactants, into different materials, the products. A **combination reaction** is one in which typically two or more reactants, usually elements or compounds, combine to form one product, usually a compound. In one type of combination reaction, two elements combine to form a compound. For example, the elements hydrogen and oxygen combine to form water, giving off large amounts of energy in the process (Figure 4.1.1):

$$2 \ H_2(g) + O_2(g) \rightarrow 2 \ H_2O(g)$$

This combination reaction was used to generate power from the main engines in the space shuttle, which flew its final mission in 2011.

Other types of combination reactions include the combination of an element with a compound and the combination of two different compounds to form a new compound. For example, oxygen reacts with sulfur dioxide to make sulfur trioxide.

$$2 \ SO_2(g) + O_2(g) \rightarrow 2 \ SO_3(g)$$

Sulfur trioxide and water undergo a combination reaction to form sulfuric acid.

$$SO_3(g) + H_2O(\ell) \rightarrow H_2SO_4(aq)$$

Notice that in a combination reaction, the number of reactants is greater than the number of products (Figure 4.1.2).

In a typical **decomposition reaction** (Interactive Figure 4.1.3), the number of products is greater than the number of reactants. The reaction is essentially the reverse of a combination reaction and usually results from the addition of thermal or electrical energy. For example, water is a very stable compound under typical conditions, but it can be made to decompose to its constituent elements by using electrical energy, a process called electrolysis.

$$2 \ H_2O(\ell) \rightarrow 2 \ H_2(g) + O_2(g)$$

Metal carbonates such as calcium carbonate, $CaCO_3$, undergo decomposition when heated, giving off carbon dioxide gas as one of the products.

$$CaCO_3(s) \rightarrow CaO(s) + CO_2(g)$$

Charles D. Winters

Figure 4.1.1 Hydrogen burns in air to form water vapor.

Figure 4.1.2 Zinc reacts with iodine to form zinc iodide.

Interactive Figure 4.1.3

Explore combination and decomposition reactions.

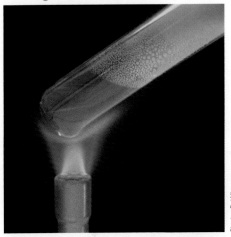

Interactive Figure 4.1.3 The decomposition of mercury(II) oxide

4.1b Displacement Reactions

Many common chemical reactions are classified as **displacement reactions**, reactions where the number of reactants is typically equal to the number of products. In displacement reactions, also called *exchange reactions*, one atom, ion, or molecular fragment displaces another. Displacement reactions can be single or double reactions.

Single Displacement Reactions
In a **single displacement reaction**, one molecular fragment is exchanged for another.

$$AB + X \rightarrow XB + A$$

For example, when solid magnesium is placed in a solution of copper(II) chloride, the magnesium displaces the copper. In this case, the more reactive metal, magnesium, replaces the less reactive metal, copper.

$$CuCl_2(aq) + Mg(s) \rightarrow MgCl_2(aq) + Cu(s)$$

Another common single displacement reaction occurs when a metal reacts with water. In the reaction of sodium metal with water, the metal displaces a hydrogen atom in water, forming sodium hydroxide (Figure 4.1.4).

$$2 \, H_2O(\ell) + 2 \, Na(s) \rightarrow H_2(g) + 2 \, NaOH(aq)$$

Double Displacement Reactions

A **double displacement reaction**, also called a *metathesis reaction*, occurs when two atoms, ions, or molecular fragments exchange.

$$AB + XY \rightarrow AY + XB$$

There are three common types of double displacement reactions, each of which will be discussed in greater detail later in this unit.

Precipitation reactions are double displacement reactions that result in the formation of an insoluble compound. For example, when aqueous solutions of lead nitrate and potassium sulfate are combined, solid lead sulfate is formed (Interactive Figure 4.1.5).

$$Pb(NO_3)_2(aq) + K_2SO_4(aq) \rightarrow 2 \, KNO_3(aq) + PbSO_4(s)$$

Many **acid–base reactions**, also called **neutralization reactions**, are double displacement reactions that result in the formation of water.

$$HCl(aq) + NaOH(aq) \rightarrow H_2O(\ell) + NaCl(aq)$$

The H^+ and Na^+ ions exchange to form the products, water and sodium chloride.

Gas-forming reactions, one example of which is the reaction between an ionic compound and an acid, are double displacement reactions that result in the formation of a gas, often carbon dioxide (Figure 4.1.6). When hydrochloric acid is added to solid nickel(II) carbonate, for example, a double displacement reaction takes place.

$$NiCO_3(s) + 2 \, HCl(aq) \rightarrow NiCl_2(aq) + H_2CO_3(aq)$$

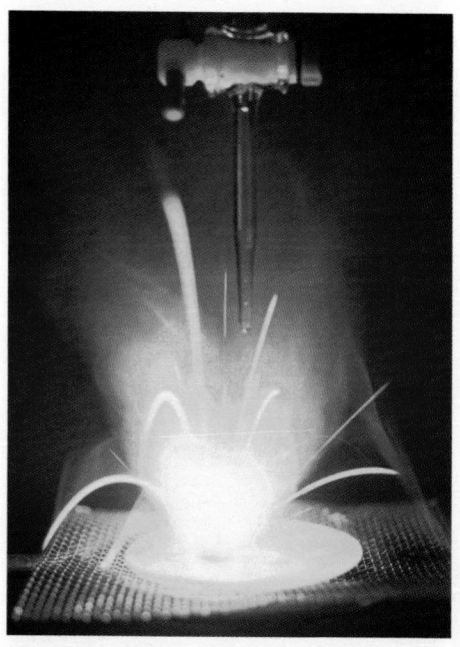

Charles D. Winters

Figure 4.1.4 The displacement reaction between sodium metal and water

Explore displacement reactions.

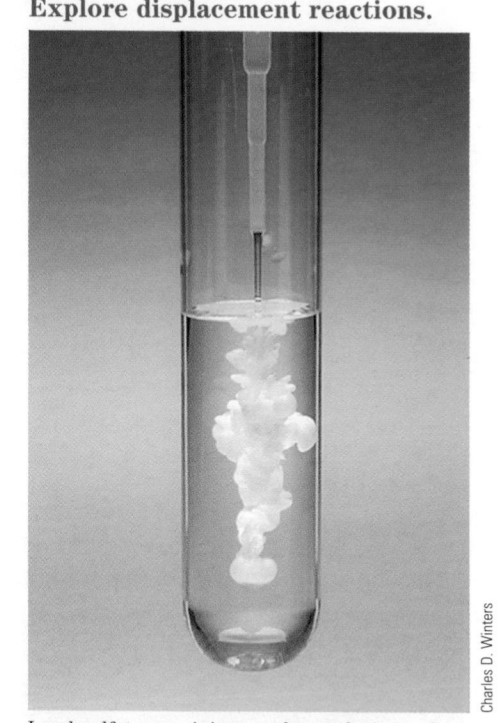

Charles D. Winters

Lead sulfate precipitates when solutions of lead nitrate and potassium sulfate are combined.

The carbonic acid, H_2CO_3, formed in the reaction decomposes into water and carbon dioxide gas.

$$H_2CO_3(aq) \rightarrow H_2O(\ell) + CO_2(g)$$

Thus, the net gas-forming reaction between nickel(II) carbonate and hydrochloric acid is written as

$$NiCO_3(s) + 2\,HCl(aq) \rightarrow NiCl_2(aq) + H_2O(\ell) + CO_2(g)$$

Notice that the decomposition of carbonic acid makes this both a double displacement reaction and a decomposition reaction.

Example Problem 4.1.1 Identify reaction types.

Classify and balance each of the following reactions.

a. $HNO_3(aq) + Ca(OH)_2(s) \rightarrow H_2O(\ell) + Ca(NO_3)_2(aq)$
b. $Al(s) + Br_2(\ell) \rightarrow Al_2Br_6(s)$

Solution:

You are asked to balance a chemical equation and classify the type of chemical reaction.

You are given an unbalanced equation.

a. This is a double displacement reaction, specifically an acid–base reaction. The H^+ and Ca^{2+} ions exchange to form the products of the reaction.

$$2\,HNO_3(aq) + Ca(OH)_2(s) \rightarrow 2\,H_2O(\ell) + Ca(NO_3)_2(aq)$$

b. This is a combination reaction.

$$2\,Al(s) + 3\,Br_2(\ell) \rightarrow Al_2Br_6(s)$$

Charles D. Winters

Figure 4.1.6 The reaction of calcium carbonate with an acid produces carbon dioxide gas.

Video Solution

Tutored Practice
Problem 4.1.1

Section 4.1 Mastery

4.2 Aqueous Solutions

4.2a Compounds in Aqueous Solution

Many chemical reactions take place in a **solvent**, a chemical species in which the chemical reactants are dissolved. A **solution** is a homogeneous mixture formed when a chemical species, a **solute**, is mixed with a solvent, typically a liquid. Much of the chemistry of life as well as much of the chemistry of commerce involves solutions in which water is the solvent. A compound dissolved in water is in an **aqueous** state, indicated with the symbol (aq) in chemical equations.

When an ionic compound such as sodium chloride dissolves in water, its constituent ions separate and become **solvated**, surrounded by solvent molecules. When water is the solvent, the ions are **hydrated**, and after hydration occurs, the individual ions have little contact with one another (Interactive Figure 4.2.1).

Interactive Figure 4.2.1

Explore the dissolution of NaCl.

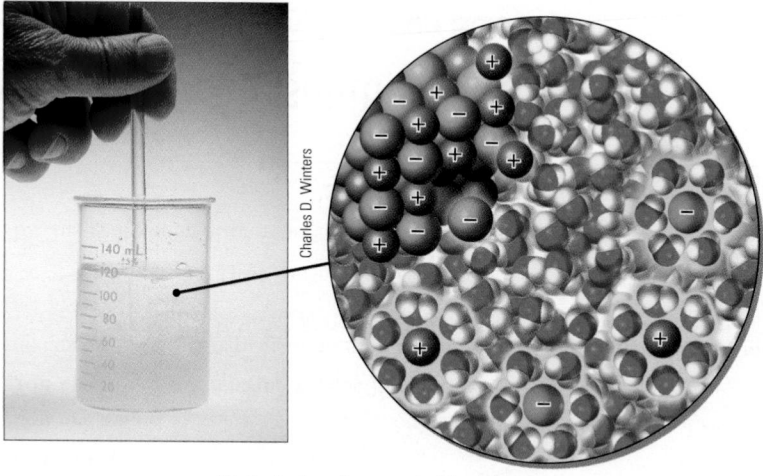

Hydrated sodium and chloride ions

Notice that in Interactive Figure 4.2.1 the water molecules orient themselves so that the oxygen atoms are near the Na^+ cations and the hydrogen atoms are near the Cl^- anions. This is due to the polar nature of water, a result of uneven electron distribution in water molecules. (▶ Flashforward to Section 8.6 Molecular Polarity) Water is a neutral compound, but the electrons in the covalent O—H bonds are distributed unevenly so that they are closer to oxygen than hydrogen. As a result, each H_2O molecule has a buildup of partial negative charge near the oxygen end of the molecule and partial positive charge near the hydrogen atoms. Thus, the negative end of a water molecule is strongly attracted to positively charged cations, and the positive end of each water molecule is attracted to anions.

The dissolution of solid sodium chloride can be written as a chemical equation.

$$NaCl(s) \rightarrow Na^+(aq) + Cl^-(aq)$$

Because the two types of ions are independent once dissolved, it does not matter where they originated. This means there is no difference between a hydrated Na^+ ion that came from dissolving NaCl and one that came from dissolving, for example, Na_2SO_4.

Electrolytes

The presence of hydrated ions in a solution affects the electrical conductivity of the solution. Using the simple conductivity apparatus shown in Interactive Figure 4.2.2, you can see that a solution of NaCl conducts electricity. The hydrated ions present in the aqueous NaCl solution carry the electrical charge from one electrode to the other, completing the circuit.

Compounds that dissolve to form hydrated ions and increase the electrical conductivity of water are called **electrolytes**. Ionic compounds such as sodium chloride that dissociate 100% in water to form hydrated ions are **strong electrolytes**.

$$NaCl(s) \xrightarrow[\text{ionization}]{100\%} Na^+(aq) + Cl^-(aq)$$

There are many examples of compounds that dissolve in water but do not form ions, such as sucrose ($C_{12}H_{22}O_{11}$) and ethanol (CH_3CH_2OH). These compounds are **nonelectrolytes** and form aqueous solutions that do not conduct electricity. **Weak electrolytes**, such as acetic acid, are compounds whose solutions conduct electricity only slightly. These compounds dissolve in water and form ions, but they do not dissociate completely. For example, when acetic acid is added to water, the solution contains non-ionized acetic acid molecules along with hydrated H^+ and $CH_3CO_2^-$ ions.

$$CH_3CO_2H(aq) \xrightarrow[\text{ionization}]{\text{less than } 100\%} CH_3CO_2^-(aq) + H^+(aq)$$

4.2b Solubility of Ionic Compounds

Many ionic compounds are **soluble** in water; that is, they dissolve in water to form solutions containing hydrated ions. Ionic compounds that do not dissolve to an appreciable extent are **insoluble** in water. Recall that ionic compounds consist of individual cations and anions held together by ionic forces in a three-dimensional arrangement. When a soluble ionic compound dissolves in water, strong hydration forces between ions and water molecules replace these ionic forces. If the ionic forces are very strong, however, the compound does not dissociate into hydrated ions and it does not dissolve in water.

A precipitation reaction occurs when an insoluble ionic compound is the product of an exchange reaction. Predicting the solubility of ionic compounds, therefore, allows us to determine whether a solid, called a **precipitate**, might form when solutions containing ionic compounds are mixed. We can use general trends in the solubility of ionic compounds containing commonly encountered ions to predict whether an ionic compound will be soluble in water, as shown in Interactive Table 4.2.1.

Explore the conductivity of aqueous solutions.

Charles D. Winters

An aqueous sodium chloride solution conducts electricity.

Solubility Rules for Ionic Compounds in Water

Soluble Ionic Compounds*	Notable Exceptions
All sodium (Na^+), potassium (K^+), and ammonium (NH_4^+) salts	
All nitrate (NO_3^-), acetate ($CH_3CO_2^-$), chlorate (ClO_3^-), and perchlorate (ClO_4^-) salts	
All chloride (Cl^-), bromide (Br^-), and iodide (I^-) salts	Compounds also containing lead, silver, or mercury(I) (Pb^{2+}, Ag^+, Hg_2^{2+}) are insoluble.
All fluoride (F^-) salts	Compounds also containing calcium, strontium, barium, or lead (Ca^{2+}, Sr^{2+}, Ba^{2+}, Pb^{2+}) are insoluble.
All sulfate (SO_4^{2-}) salts	Compounds also containing calcium, mercury(I), strontium, barium, or lead (Ca^{2+}, Hg_2^{2+}, Sr^{2+}, Ba^{2+}, Pb^{2+}) are insoluble.
Insoluble Ionic Compounds	**Exceptions**
Hydroxide (OH^-) and oxide (O^{2-}) compounds	Compounds also containing sodium, potassium, or barium (Na^+, K^+, Ba^{2+}) are soluble.
Sulfide (S^{2-}) salts	Compounds also containing sodium, potassium, ammonium, or barium (Na^+, K^+, NH_4^+, Ba^{2+}) are soluble.
Carbonate (CO_3^{2-}) and phosphate (PO_4^{3-}) salts	Compounds also containing sodium, potassium, or ammonium (Na^+, K^+, NH_4^+) are soluble.

*Soluble compounds are defined as those that dissolve to the extent of 1 g or more per 100 g water.

Notice that classification of ionic compound solubility is primarily based on the anion in the compound. When classifying an ionic compound as soluble or insoluble, first determine whether the anion present is typically found in soluble or insoluble compounds. Next, see if the cation in the compound results in an exception to the solubility guidelines.

For example, sodium nitrate, $NaNO_3$, is a soluble ionic compound because it contains the NO_3^- ion, an anion found in soluble compounds. Calcium nitrate, $Ca(NO_3)_2$, is also soluble because it too contains the NO_3^- anion. Silver chloride, AgCl, is an insoluble compound even though compounds containing Cl^- are generally soluble. As shown in Interactive Table 4.2.1, when the chloride ion is paired with Ag^+, Hg_2^{2+}, or Pb^{2+}, an insoluble compound results. Calcium carbonate, $CaCO_3$, is an insoluble compound. The presence of the carbonate ion (CO_3^{2-}) places this compound in the group of generally insoluble compounds. The only exceptions to this rule are carbonate compounds containing NH_4^+, K^+, or Na^+ ions.

It is important to note that many common ionic compounds have solubilities that are on the borderline of being considered soluble or insoluble. For example, both Ag_2SO_4 and $PbCl_2$ have solubilities (at 25 °C) that are only slightly below the 1 g compound per 100 g water cutoff for soluble compounds in Interactive Table 4.2.1. It is the concentration of ions in solution that determines whether these borderline soluble compounds precipitate from solution.

Example Problem 4.2.1 Characterize the solubility of ionic compounds.

Classify each of the compounds as soluble or insoluble in water:
a. Lead chloride   b. Magnesium iodide   c. Nickel sulfide

Solution:

You are asked to classify a compound as soluble or insoluble in water.

You are given the name of a compound.

a. Lead chloride, $PbCl_2$, is insoluble in water. As shown in Interactive Table 4.2.1, ionic compounds containing the chloride ion are generally soluble, but lead chloride is an exception to this rule.
b. Magnesium iodide, MgI_2, is soluble in water. As shown in Interactive Table 4.2.1, most ionic compounds that contain the iodide ion are soluble in water, with the exception of PbI_2, AgI, and Hg_2I_2.
c. Nickel sulfide, NiS, is insoluble in water. As shown in Interactive Table 4.2.1, most metal sulfides are insoluble in water.

Video Solution

Tutored Practice
Problem 4.2.1

Section 4.2 Mastery

4.3 Reactions in Aqueous Solution

4.3a Precipitation Reactions and Net Ionic Equations

When solutions containing soluble ionic compounds are mixed, the hydrated ions intermingle. Whether or not one of the three common exchange reactions—precipitation reactions, acid–base reactions, and gas-forming reactions—occurs depends on the identity of the ions in solution.

A precipitation reaction occurs when a solution, originally containing dissolved species, produces a solid, which generally is denser and falls to the bottom of the reaction vessel (Figure 4.3.1).

The most common precipitation reactions occurring in aqueous solution involve the formation of an insoluble ionic compound when two solutions containing soluble compounds are mixed.

Consider what happens when an aqueous solution of NaCl is added to an aqueous solution of $AgNO_3$. The first solution contains hydrated Na^+ and Cl^- ions and the second solution, Ag^+ and NO_3^- ions.

$$NaCl(s) \rightarrow Na^+(aq) + Cl^-(aq)$$
$$AgNO_3(s) \rightarrow Ag^+(aq) + NO_3^-(aq)$$

When mixed, a double displacement reaction takes place, forming the soluble compound $NaNO_3$ and the insoluble compound AgCl. In the reaction vessel the Ag^+ and Cl^- ions combine, and a white solid precipitates from the solution. As the solid precipitates, the Na^+ and NO_3^- ions remain in solution. The overall double displacement reaction is represented by the following balanced equation:

$$NaCl(aq) + AgNO_3(aq) \rightarrow AgCl(s) + NaNO_3(aq)$$

When determining whether a precipitation reaction will occur, first identify the ions present when the two solutions are mixed and then see if an exchange of ions results in an insoluble ionic compound. If an insoluble compound can form, a precipitation reaction occurs. If no combination of ions results in an insoluble compound, no net reaction occurs and the two solutions mix without the formation of a precipitate.

For example, consider the reaction of calcium nitrate with sodium carbonate. Calcium nitrate is a soluble compound (all ionic compounds containing the NO_3^- ion are soluble), and sodium carbonate is a soluble compound (all ionic compounds containing the Na^+ ion are soluble).

$$Ca(NO_3)_2(s) \rightarrow Ca^{2+}(aq) + 2 NO_3^-(aq)$$
$$Na_2CO_3(s) \rightarrow 2 Na^+(aq) + CO_3^{2-}(aq)$$

The exchange of ions results in Ca^{2+} paired with CO_3^{2-} to form $CaCO_3$ and Na^+ paired with NO_3^- to form $NaNO_3$. Calcium carbonate is an insoluble compound (most ionic compounds containing the carbonate ion are insoluble), and $NaNO_3$ is soluble. A precipitation reaction occurs, and the balanced equation is

$$Ca(NO_3)_2(aq) + Na_2CO_3(aq) \rightarrow CaCO_3(s) + 2 NaNO_3(aq)$$

Charles D. Winters

Figure 4.3.1 The precipitation of lead iodide

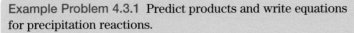

Example Problem 4.3.1 Predict products and write equations for precipitation reactions.

a. When aqueous solutions of $Pb(NO_3)_2$ and KI are mixed, does a precipitate form?
b. Write a balanced equation for the precipitation reaction that occurs when aqueous solutions of copper(II) iodide and potassium hydroxide are combined.

Solution:

You are asked to predict whether a precipitate will form during a chemical reaction and to write a balanced equation for a precipitation reaction.

You are given the identity of two reactants.

a. Yes, a solid precipitate, PbI_2, forms when these solutions are mixed:

$$Pb(NO_3)_2(aq) + KI(aq) \rightarrow PbI_2(s) + 2\ KNO_3(aq)$$

b. The two products of the reaction are insoluble copper (II) hydroxide and soluble potassium iodide.

$$CuI_2(aq) + 2\ KOH(aq) \rightarrow Cu(OH)_2(s) + 2\ KI(aq)$$

Video Solution

Tutored Practice
Problem 4.3.1

In the equation for the formation of solid AgCl shown previously, three of the species, NaCl, $AgNO_3$, and $NaNO_3$, are strong electrolytes that dissociate 100% into ions in aqueous solution. It is therefore possible to write the balanced equation in its completely ionized form:

$$Na^+(aq) + Cl^-(aq) + Ag^+(aq) + NO_3^-(aq) \rightarrow AgCl(s) + Na^+(aq) + NO_3^-(aq)$$

Two of the ionic species, Na^+ and NO_3^-, do not change during the reaction; they are present in exactly the same form in the reactants and the products. These unchanged ions are called **spectator ions** because they do not participate in the chemical reaction. A **net ionic equation** is written by removing the spectator ions from the completely ionized equation and shows only the chemical species that participate in the chemical reaction.

Completely ionized equation:

$$Na^+(aq) + Cl^-(aq) + Ag^+(aq) + NO_3^-(aq) \rightarrow AgCl(s) + Na^+(aq) + NO_3^-(aq)$$
spectator ion   *spectator ion*   *spectator ions*

Net ionic equation:

$$Cl^-(aq) + Ag^+(aq) \rightarrow AgCl(s)$$

For precipitation reactions, net ionic equations are written following the steps in Interactive Table 4.3.1.

Interactive Table 4.3.1

Writing Net Ionic Equations for Precipitation Reactions

Step 1.	Write a balanced chemical equation for the precipitation reaction.
Step 2.	Identify all strong electrolytes (soluble compounds) in the reaction and write them in their completely ionized form.
Step 3.	Identify the spectator ions in the reaction.
Step 4.	Remove spectator ions and write the reaction using only the species that remain.

Example Problem 4.3.2 Write net ionic equations for precipitation reactions.

Write the net ionic equation for the precipitation reaction that occurs when aqueous solutions of sodium sulfide and chromium(III) acetate are combined.

Solution:

You are asked to write a net ionic equation for a chemical reaction.

You are given the identity of the reactants.

Step 1. $3\,Na_2S(aq) + 2\,Cr(CH_3CO_2)_3(aq) \rightarrow 6\,NaCH_3CO_2(aq) + Cr_2S_3(s)$

Step 2. Strong electrolytes (soluble compounds): Na_2S, $Cr(CH_3CO_2)_3$, and $NaCH_3CO_2$

$6\,Na^+(aq) + 3\,S^{2-}(aq) + 2\,Cr^{3+}(aq) + 6\,CH_3CO_2^-(aq)$
$$\rightarrow 6\,Na^+(aq) + 6\,CH_3CO_2^-(aq) + Cr_2S_3(s)$$

Step 3. Spectator ions: Na^+ and $CH_3CO_2^-$

Step 4.

$6\,Na^+(aq) + 3\,S^{2-}(aq) + 2\,Cr^{3+}(aq) + 6\,CH_3CO_2^-(aq) \longrightarrow 6\,Na^+(aq) + 6\,CH_3CO_2^-(aq) + Cr_2S_3(s)$

$$3\,S^{2-}(aq) + 2\,Cr^{3+}(aq) \rightarrow Cr_2S_3(s)$$

Video Solution

Tutored Practice Problem 4.3.2

4.3b Acid–Base Reactions

Acid–base reactions are a second example of double displacement (exchange) reactions that take place in aqueous solution. Acids and bases are common materials, found in foods (citric acid gives lemons their sour taste; acetic acid is one component of vinegar) and household materials (sodium hydroxide and ammonia are bases found in many cleaning products). There are many ways to define acids and bases. (▶ Flashforward to Section 16.1 Introduction to Acids and Bases) In one of the simplest definitions, an **acid** is a species that produces H^+ ions when dissolved in water and a **base** is a species that increases the amount of OH^- ions in a solution. Bases also react with H^+ ions and can therefore also be described as proton (H^+) acceptors. Acids and bases are electrolytes that produce hydrated ions in aqueous solution.

An acid–base reaction is also known as a neutralization reaction because the reaction can result in a solution that is neutral. That is, it no longer has acidic or basic properties. For example, consider the reaction of nitric acid (HNO_3) and sodium hydroxide ($NaOH$).

$$HNO_3(aq) + NaOH(aq) \rightarrow H_2O(\ell) + NaNO_3(aq)$$
$$\quad\ acid \qquad\qquad base \qquad\quad water \qquad\quad salt$$

This reaction between an acid and a base produces water and a **salt**, an ionic compound formed as the result of an acid–base reaction that consists of a cation donated from a base and an anion donated from an acid. Notice that this is also a double displacement reaction. The base in this reaction, $NaOH$, acts as a proton acceptor, combining with H^+ to form water.

Acids

Nitric acid is a **strong acid** because it is a strong electrolyte that ionizes 100% in aqueous solution. Acetic acid is an example of a **weak acid**, a weak electrolyte that does not completely ionize in solution (Table 4.3.2).

$$HNO_3(aq) \xrightarrow[\text{ionized}]{100\%} H^+(aq) + NO_3^-(aq)$$

$$CH_3CO_2H(aq) \xrightarrow[\text{ionized}]{\text{less than } 100\%} H^+(aq) + CH_3CO_2^-(aq)$$

There are six important strong acids in aqueous solution (Table 4.3.2) and numerous weak acids. Many weak acids are **organic acids**, compounds made up mostly of carbon, hydrogen, and oxygen that also contain the —C(O)OH structural group (Figure 4.3.2).

Table 4.3.2 Some Important Acids

Strong Acids		Weak Acids	
HCl	Hydrochloric acid	CH_3CO_2H	Acetic acid
HBr	Hydrobromic acid	$H_3C_6H_5O_5$	Citric acid
HI	Hydroiodic acid	$H_2C_2O_4$	Oxalic acid
HNO_3	Nitric acid	HF	Hydrofluoric acid
H_2SO_4	Sulfuric acid	H_3PO_4	Phosphoric acid
$HClO_4$	Perchloric acid	H_2CO_3	Carbonic acid

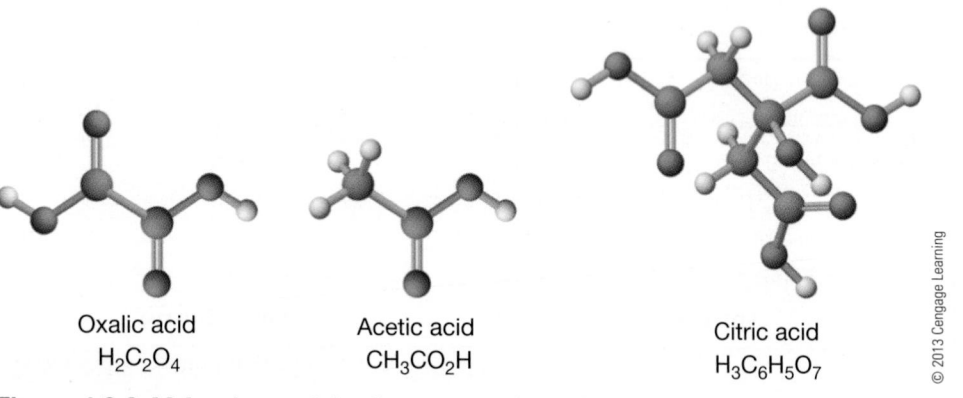

Oxalic acid
$H_2C_2O_4$

Acetic acid
CH_3CO_2H

Citric acid
$H_3C_6H_5O_7$

© 2013 Cengage Learning

Figure 4.3.2 Molecular models of some organic acids

Some acids can produce more than one mole of H^+ ions per mole of acid. An acid that produces only one mole of H^+ ions per mole of acid is a **monoprotic acid**. Sulfuric acid, H_2SO_4, and carbonic acid, H_2CO_3, are examples of **diprotic acids**, acids that can dissociate to form two moles of H^+ ions per mole of acid.

$$H_2SO_4(aq) \rightarrow H^+(aq) + HSO_4^-(aq)$$
$$HSO_4^-(aq) \rightarrow H^+(aq) + SO_4^{2-}(aq)$$

Bases

Bases are also characterized as being either strong or weak (Table 4.3.3). All common strong bases are hydroxide salts such as NaOH and KOH. Weak bases are most often compounds containing a nitrogen atom such as ammonia, NH_3, and trimethylamine, $(CH_3)_3N$.

Table 4.3.3 Some Important Bases			
Strong Bases		**Weak Bases**	
LiOH	Lithium hydroxide	NH_3	Ammonia
NaOH	Sodium hydroxide	$(CH_3)_3N$	Trimethylamine
KOH	Potassium hydroxide	$CH_3CH_2NH_2$	Ethylamine
$Ca(OH)_2$	Calcium hydroxide		
$Ba(OH)_2$	Barium hydroxide		

When added to water, **strong bases** act as strong electrolytes and are ionized completely. For example, when NaOH is added to water, it forms hydrated Na^+ and OH^- ions.

$$NaOH(s) \xrightarrow[\text{ionized}]{100\%} Na^+(aq) + OH^-(aq)$$

When added to water, weak bases also increase the concentration of hydroxide ion (OH^-) but do so by reacting with water. For example, NH_3 is a weak base and reacts with water to form NH_4^+ and OH^- ions.

$$NH_3(g) + H_2O(\ell) \xrightarrow[\text{ionized}]{\text{less than }100\%} NH_4^+(aq) + OH^-(aq)$$

Like a weak acid, a **weak base** is a weak electrolyte that does not completely ionize. In a solution of NH_3, for example, only about 1% of the NH_3 molecules react with water.

Net Ionic Equations for Acid–Base Reactions

The net ionic equation for a strong acid–strong base reaction is written by first writing a completely ionized equation. Consider the reaction between nitric acid and sodium hydroxide.

$$H^+(aq) + NO_3^-(aq) + Na^+(aq) + OH^-(aq) \rightarrow H_2O(\ell) + Na^+(aq) + NO_3^-(aq)$$

Identifying the spectator ions that remain unchanged during the reaction (Na^+ and NO_3^-) and removing them results in the net ionic equation for most strong acid–strong base reactions.

$$H^+(aq) + OH^-(aq) \rightarrow H_2O(\ell)$$

The net ionic equation for the reaction of strong acids or bases with weak acids or bases also involves writing a completely ionized equation and identifying any spectator ions. For example, consider the reaction of hydrochloric acid with the weak base ammonia.

$$HCl(aq) + NH_3(aq) \rightarrow NH_4Cl(aq)$$

In this reaction, the base acts as a proton acceptor, accepting H^+ from the acid to form the ammonium ion, NH_4^+. Writing the equation in its completely ionized form,

$$H^+(aq) + Cl^-(aq) + NH_3(aq) \rightarrow NH_4^+(aq) + Cl^-(aq)$$

Ammonium chloride, NH_4Cl, is a soluble ionic compound that dissociates completely in aqueous solution. Removing the spectator ion (Cl^-) results in the following net ionic equation:

$$H^+(aq) + NH_3(aq) \rightarrow NH_4^+(aq)$$

For acid–base reactions, net ionic equations are written following the steps in Table 4.3.4.

Table 4.3.4 Writing Net Ionic Equations for Acid–Base Reactions	
Step 1.	Write a balanced chemical equation for the acid–base reaction.
Step 2.	Identify all strong electrolytes (strong acids and bases and soluble ionic compounds) in the reaction and write them in their completely ionized form.
Step 3.	Identify the spectator ions in the reaction.
Step 4.	Remove spectator ions and write the reaction using only the species that remain.

Example Problem 4.3.3 Write net ionic equations for acid–base reactions.

Write the net ionic equation for the reaction that occurs when aqueous solutions of potassium hydroxide and phosphoric acid are combined.

Solution:

You are asked to write the net ionic equation for an acid–base reaction.

You are given the identity of the reactants in the acid–base reaction.

Step 1. $3\ KOH(aq) + H_3PO_4(aq) \rightarrow 3\ H_2O(\ell) + K_3PO_4(aq)$

Step 2. KOH is a strong base, and K_3PO_4 is a soluble ionic compound.

$$3\ K^+(aq) + 3\ OH^-(aq) + H_3PO_4(aq) \rightarrow 3\ H_2O(\ell) + 3\ K^+(aq) + PO_4^{3-}(aq)$$

Notice that H_3PO_4 is a weak acid and therefore is not written in an ionized form.

Step 3. K^+ is a spectator ion.

Step 4. $\cancel{3\ K^+(aq)} + 3\ OH^-(aq) + H_3PO_4(aq) \rightarrow 3\ H_2O(\ell) + \cancel{3\ K^+(aq)} + PO_4^{3-}(aq)$

$$3\ OH^-(aq) + H_3PO_4(aq) \rightarrow 3\ H_2O(\ell) + PO_4^{3-}(aq)$$

Video Solution

Tutored Practice
Problem 4.3.3

4.3c   Gas-Forming Reactions

A third double displacement reaction commonly encountered in the laboratory is the reaction of an ionic compound with an acid to form a gas. These gas-forming reactions, introduced previously in this unit, are a special type of acid–base reaction involving less obvious bases such as $CuCO_3$ and Na_2SO_3 (Figure 4.3.3).

When the gas-forming reaction involves a metal carbonate, one product of the reaction is carbon dioxide gas. Consider the reaction between copper(II) carbonate and hydrochloric acid.

$$CuCO_3(s) + 2\ HCl(aq) \rightarrow CuCl_2(aq) + H_2CO_3(aq)$$

As shown previously, one product of the double displacement reaction, H_2CO_3, undergoes a decomposition reaction to produce gaseous carbon dioxide and water.

$$H_2CO_3(aq) \rightarrow H_2O(\ell) + CO_2(g)$$

The net reaction, therefore, is

$$CuCO_3(s) + 2\ HCl(aq) \rightarrow CuCl_2(aq) + CO_2(g) + H_2O(\ell)$$

The procedure for writing net ionic equations for gas-forming reactions is similar to that used for precipitation and acid–base net ionic equations, as shown in Table 4.3.5.

Figure 4.3.3 Alka-seltzer contains an acid (citric acid) and a base (sodium bicarbonate) that react to form carbon dioxide gas.

Charles D. Winters

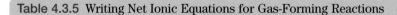

Table 4.3.5 Writing Net Ionic Equations for Gas-Forming Reactions

Step 1.	Write a balanced chemical equation for the gas-forming reaction.
Step 2.	Identify all strong electrolytes (strong acids and bases and soluble ionic compounds) in the reaction and write them in their completely ionized form.
Step 3.	Identify the spectator ions in the reaction.
Step 4.	Remove spectator ions and write the reaction using only the species that remain.

When a gas-forming reaction involves an ionic compound that is not a metal carbonate, it is not always easy to predict the products of the reaction. For example, sodium sulfite, Na_2SO_3, reacts with hydrochloric acid to form gaseous sulfur dioxide.

$$Na_2SO_3(aq) + 2\,HCl(aq) \rightarrow 2\,NaCl(aq) + SO_2(g) + H_2O(\ell)$$

The net ionic equation for this reaction is

$$SO_3{}^{2-}(aq) + 2\,H^+(aq) \rightarrow SO_2(g) + H_2O(\ell)$$

Example Problem 4.3.4 Write net ionic equations for gas-forming reactions.

Write a net ionic equation for the reaction that occurs when aqueous hydroiodic acid is combined with solid zinc carbonate.

Solution:

You are asked to write the net ionic equation for a gas-forming reaction.

You are given the identity of the reactants in the gas-forming reaction.

Step 1. $2\,HI(aq) + ZnCO_3(s) \rightarrow ZnI_2(aq) + H_2O(\ell) + CO_2(g)$

Note that H_2CO_3 is formed in the reaction but undergoes a decomposition reaction to produce gaseous carbon dioxide and water.

Step 2. HI is a strong acid and ZnI_2 is a soluble ionic compound.

$$2\,H^+(aq) + 2\,I^-(aq) + ZnCO_3(s) \rightarrow Zn^{2+}(aq) + 2\,I^-(aq) + H_2O(\ell) + CO_2(g)$$

Step 3. The iodide ion (I^-) is a spectator ion in this reaction.

Step 4.

$$2\,H^+(aq) + \cancel{2\,I^-(aq)} + ZnCO_3(s) \longrightarrow Zn^{2+}(aq) + \cancel{2\,I^-(aq)} + H_2O(\ell) + CO_2(g)$$

$$2\,H^+(aq) + ZnCO_3(s) \longrightarrow Zn^{2+}(aq) + H_2O(\ell) + CO_2(g)$$

Video Solution

Tutored Practice
Problem 4.3.4

Section 4.3 Mastery

4.4 Oxidation–Reduction Reactions

4.4a Oxidation and Reduction

The two simplest particles in chemistry are the proton and the electron. Whereas acid–base chemistry involves a transfer of protons, H^+ ions, between acids and bases, many other reactions involve the transfer of electrons. These electron-transfer reactions are known as **oxidation–reduction reactions**, or *redox reactions*. Most of the combination, decomposition, and single displacement reactions described earlier in this unit are oxidation–reduction reactions.

Oxidation is the loss of one or more electrons from a chemical species, and **reduction** is the gain of one or more electrons. For example, consider the single displacement reaction involving zinc metal and a solution of copper(II) nitrate (Interactive Figure 4.4.1). If a piece of zinc metal is immersed in a $Cu(NO_3)_2$ solution, a thin coating of copper metal quickly forms on the zinc.

Complete equation: $Zn(s) + 2\ Cu(NO_3)_2(aq) \rightarrow Zn(NO_3)_2(aq) + Cu(s)$

Net ionic equation: $Zn(s) + Cu^{2+}(aq) \rightarrow Zn^{2+}(aq) + Cu(s)$

In this oxidation–reduction reaction, zinc gives up two electrons to form Zn^{2+} and the copper (II) ions gain two electrons to form copper metal, Cu. Thus, Zn is oxidized and Cu^{2+} is reduced.

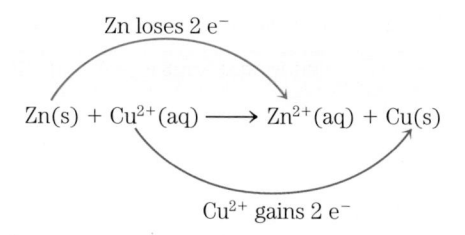

Zn loses 2 e⁻

$Zn(s) + Cu^{2+}(aq) \longrightarrow Zn^{2+}(aq) + Cu(s)$

Cu^{2+} gains 2 e⁻

Notice that in order for a species to be oxidized and give up electrons, some other species must gain the electrons lost. The electron-acceptor species in an oxidation–reduction reaction is called the **oxidizing agent**. Because the oxidizing agent gains electrons, it is reduced. The **reducing agent** donates electrons to the species that is reduced and is therefore oxidized. In summary,

- An oxidizing agent oxidizes another substance and is reduced.

- A reducing agent reduces another substance and is oxidized.

In the reaction of zinc with copper(II) ions,

$Zn(s)$	$+$	$Cu^{2+}(aq) \rightarrow Zn^{2+}(aq) + Cu(s)$
loses 2 e⁻		gains 2 e⁻
oxidized		reduced
reducing agent		oxidizing agent

Investigate oxidation and reduction.

Zinc metal reacts with aqueous Cu^{2+} ions to form copper metal and aqueous Zn^{2+} ions.

Charles D. Winters

The equation for an oxidation–reduction reaction can be broken down in a way that emphasizes the gain and loss of electrons. A **half-reaction** shows only the species involved in oxidation or reduction and also includes the electrons lost or gained by each species. Adding the oxidation and reduction half-reactions results in the net ionic equation. Notice that in the half-reactions that follow, the total number of electrons lost by Zn is equal to the total number of electrons gained by Cu^{2+}.

Oxidation half-reaction: $Zn(s) \rightarrow Zn^{2+}(aq) + 2 e^-$

Reduction half-reaction: $Cu^{2+}(aq) + 2 e^- \rightarrow Cu(s)$

Net ionic equation: $Zn(s) + Cu^{2+}(aq) \rightarrow Zn^{2+}(aq) + Cu(s)$

The term *oxidation* originally referred to the addition of oxygen, the most commonly encountered oxidizing agent, to an element or compound. When molecular oxygen reacts with a metal, for example, each O_2 molecule gains four electrons (two for each O atom) and is reduced to form oxide ions. Halogens, F_2, Cl_2, Br_2, and I_2, are also very good oxidizing agents.

Metals are generally good reducing agents that are easily oxidized to form positively charged ions, and they show a range of reducing ability. For example, alkali and alkaline earth metals are excellent reducing agents, as is aluminum. On the other hand, the coinage metals, Cu, Ag, and Au, are poor reducing agents. This is why they are useful for making coins; they do not oxidize easily and therefore do not corrode.

4.4b Oxidation Numbers and Oxidation States

In the reaction of zinc and copper(II) ions, it is easy to see in the net ionic equation the flow of electrons from Zn to Cu^{2+}. In many other cases involving more complex compounds and ions, the electron transfer is more difficult to discern. To aid in identifying oxidizing and reducing agents, a system of electron counting has been developed to identify species that are either gaining or losing electrons during a reaction. This system involves the assignment of oxidation numbers to elements, ions, and atoms in compounds.

The **oxidation number**, also called *oxidation state* of an atom, represents the number of electrons that have been gained (if the oxidation number is negative) or lost (if the oxidation number is positive) from the neutral atom. Oxidation numbers are primarily used to identify the transfer of electrons in an oxidation–reduction reaction and are not always representative of the actual charge carried by an atom in a compound. Oxidation numbers are assigned according to the rules shown in Table 4.4.1.

Table 4.4.1 Rules for Assigning Oxidation Numbers

Rule	Example
1. Each atom in a pure element has an oxidation number of zero.	Fe in Fe(s) oxidation number = 0 Each O in O_2(g) oxidation number = 0
2. A monoatomic ion has an oxidation number equal to the ion charge.	Cl in Cl^- oxidation number = −1 Mg in Mg^{2+} oxidation number = +2
3. In compounds, halogens (F, Cl, Br, I) have an oxidation number of −1. *Exception:* When halogens are combined with oxygen or fluorine, the oxidation number of Cl, Br, and I is not −1.	Each F in CF_4 oxidation number = −1 Cl in ClF_3 oxidation number = +3
4. In compounds, oxygen has an oxidation number of −2. *Exception:* In compounds containing the peroxide ion (O_2^{2-}), oxygen has an oxidation number of −1.	Each O in CO_2 oxidation number = −2 Each O in H_2O_2 oxidation number = −1
5. When combined with nonmetals, hydrogen is assigned an oxidation number of +1. With metals, hydrogen has an oxidation number of −1.	Each H in CH_4 oxidation number = +1 H in LiH oxidation number = −1
6. The sum of the oxidation numbers for all atoms in an ion is equal to the overall charge on the ion. For a neutral compound, the sum of all of the oxidation numbers is equal to zero.	CO_2 (C oxidation number) + 2 × (O oxidation number) = 0 (C oxidation number) + 2 × (−2) = 0 C oxidation number = +4 ClO_4^- (Cl oxidation number) + 4 × (O oxidation number) = −1 (Cl oxidation number) + 4 × (−2) = −1 Cl oxidation number = +7

Example Problem 4.4.1 Assign oxidation numbers.

Determine the oxidation number of each atom in K_2CrO_4.

Solution:

You are asked to assign oxidation numbers to elements in a compound.

You are given the formula of the compound.

This compound contains two K^+ ions, each of which has an oxidation number of +1. Each of the four O atoms has an oxidation number of −2, for a total of −8. The oxidation number of Cr is found using the K and O oxidation numbers and the overall charge on the compound.

Example Problem 4.4.1 *(continued)*

charge on K_2CrO_4 = sum of K oxidation numbers + chromium oxidation number + sum of oxygen oxidation numbers

$$0 = 2(+1) + \text{oxidation number Cr} + 4(-2)$$

The oxidation number of Cr is therefore +6. This number does not represent the actual charge on Cr in K_2CrO_4.

Video Solution

Tutored Practice
Problem 4.4.1

4.4c Recognizing Oxidation–Reduction Reactions

We can generally predict the products of acid–base and precipitation reactions. This is not the case for oxidation–reduction reactions. In these reactions, predicting the products is beyond the scope of this introductory textbook. That being said, you should be able to recognize an oxidation–reduction reaction and be able to predict when one is likely to occur, if not what products will be formed.

Oxidation–reduction reactions always involve a change in the oxidation number of one or more atoms. Therefore, the most certain means of determining whether a reaction is an oxidation–reduction reaction is to determine the oxidation numbers of all elements in the reaction. If any of the oxidation numbers change when proceeding from reactants to products, the reaction is an oxidation–reduction reaction.

For example, consider the following reaction.

$$Cl_2(g) + H_3AsO_4(aq) \rightarrow HAsO_2(aq) + 2\ HClO(aq)$$

First, assign oxidation numbers for all atoms.

$$Cl_2(g) + H_3AsO_4(aq) \rightarrow HAsO_2(aq) + 2\ HClO(aq)$$
$$\quad(0)\qquad (+1)(+5)(-2)\quad (+1)(+3)(-2)\quad (+1)(+1)(-2)$$

Any element with a change in oxidation number when proceeding from reactants to products has been oxidized or reduced. In the preceding reaction, the oxidation number of chlorine increases from 0 (in Cl_2) to +1 (in HClO). The oxidation number has increased because each chlorine atom has lost an electron. Chlorine, Cl_2, has been oxidized to HClO and is the reducing agent. The oxidation number of arsenic decreases from +5 (in H_3AsO_4) to +3 (in $HAsO_2$). The oxidation number has decreased because each arsenic atom has gained two electrons. H_3AsO_4 has been reduced to $HAsO_2$ and is the oxidizing agent.

Some general guidelines can also help make it easier to recognize oxidation–reduction reactions. For example, oxidation–reduction reactions often involve one or more common oxidizing and reducing agents (Table 4.4.2). As shown in Table 4.4.2, reactions involving a pure element are usually oxidation–reduction reactions. Also, almost all reactions involving

Table 4.4.2 Common Oxidizing and Reducing Agents

Oxidizing Agent	Reaction Product	Reducing Agent	Reaction Product
O_2 (oxygen)	O^{2-} (oxide ion) or an oxygen-containing molecular compound	H_2 (hydrogen) or hydrogen-containing molecular compound	H^+ (hydrogen ion) or H combined in H_2O
H_2O_2 (hydrogen peroxide)	$H_2O(\ell)$	C (carbon) used to reduce metal oxides	CO and CO_2
F_2, Cl_2, Br_2, or I_2 (halogens)	F^-, Cl^-, Br^-, or I^- (halide ions)	M, metals such as Na, K, Fe, or Al	M^{n+}, metal ions such as Na^+, K^+, Fe^{3+}, or Al^{3+}
$Cr_2O_7^{2-}$ (dichromate ion)	Cr^{3+} (chromium(III) ion), in acid solution		
MnO_4^- (permanganate ion)	Mn^{2+} (manganese(II) ion), in acid solution		

compounds with atoms in very high or very low oxidation states are oxidation–reduction reactions. For example, Mn has a very high oxidation number of +7 in the permanganate ion, MnO_4^-. The permanganate ion is a strong oxidizing agent because reactions involving the ion almost always involve it being reduced to a lower oxidation number.

Finally, all combustion reactions and most explosive reactions are oxidation–reduction reactions. In reactions that involve the burning of organic matter in air, the organic material is the reducing agent and O_2 in air is the oxidizing agent.

Example Problem 4.4.2 Identify oxidizing and reducing agents.

Identify the oxidizing agent and the reducing agent in the following reactions:

a. $2 K(s) + 2 H_2O(\ell) \rightarrow 2 KOH(aq) + H_2(g)$
b. $I_2(s) + 2 Br^-(aq) \rightarrow 2 I^-(aq) + Br_2(\ell)$

Solution:

You are asked to identify the oxidizing and reducing agents in a reaction.

You are given a chemical equation.

a. Potassium metal is a common reducing agent, and in this reaction it is oxidized to form the potassium ion, K^+. Water is therefore the oxidizing agent. Hydrogen in water is reduced to elemental H_2.
b. Iodine, I_2, is a halogen and is a common oxidizing agent. Elemental iodine is reduced to form the iodide ion, I^-, and Br^-, the reducing agent, is oxidized to form elemental bromine, Br_2.

Video Solution

Tutored Practice
Problem 4.4.2

Section 4.4 Mastery

4.5 Stoichiometry of Reactions in Aqueous Solution

4.5a Solution Concentration and Molarity

To work with quantities of materials in aqueous solution, first we must measure the amount of solute dissolved in a solvent. We will work with the **concentration** of a solute in a solution, a quantitative measure of the amount of material dissolved in a known quantity of solvent or solution (Figure 4.5.1).

There are many measurements of concentration, including weight percent, parts per million (ppm), and molarity. The most common concentration used in the chemistry lab is **molarity (M)**, the amount of solute (in moles) dissolved in exactly 1 liter of solution (Equation 4.1).

$$\text{molarity (M)} = \frac{\text{moles of solute}}{\text{1 L of solution}} \qquad \textbf{(4.1)}$$

It is helpful to always think of molarity as a ratio because it can be used as a conversion factor in calculations, converting between the amount (in moles) of solute in a solution and the volume (in liters) of a solution.

Square brackets are placed around the formula of a solute to indicate a concentration in molarity units. For example, reporting a concentration as $[NaCl] = 0.15$ M means that the amount of sodium chloride dissolved in a solution is 0.15 mol per liter of solution, or the NaCl concentration is 0.15 molar.

To determine the concentration of a solute in a solution in units of molarity, first calculate the amount of solute present (in moles) and then divide that amount by the volume of the solution, in units of liters (L). For example, the concentration (M) of 250. mL of solution that contains 14.5 g $CaCl_2$ is calculated as follows.

First, determine the amount of solute present.

$$14.5 \text{ g } CaCl_2 \times \frac{1 \text{ mol } CaCl_2}{111.0 \text{ g}} = 0.131 \text{ mol } CaCl_2$$

Next, convert the solution volume to units of liters.

$$250. \text{ mL} \times \frac{1 \text{ L}}{1000 \text{ mL}} = 0.250 \text{ L}$$

Finally, use Equation 4.1 to calculate the concentration of the solution.

$$[CaCl_2] = \frac{0.131 \text{ mol } CaCl_2}{0.250 \text{ L}} = 0.523 \text{ M } CaCl_2$$

Figure 4.5.1 The concentration of hydrogen peroxide in this solution is indicated on the label.

Example Problem 4.5.1 Calculate solution concentration in molarity units.

A student weighs out 5.33 g of $NiBr_2$ and transfers it to a 100-mL volumetric flask, adds enough water to dissolve the solid, and then adds water to the 100-mL mark on the neck of the flask. Calculate the concentration (in molarity units) of nickel bromide in the resulting solution.

Solution:

You are asked to calculate the concentration of a solution in molarity units.

You are given the identity of the solute, the amount of solute in the solution, and the total volume of the solution.

First determine the amount of solute present.

$$5.33 \text{ g NiBr}_2 \times \frac{1 \text{ mol NiBr}_2}{218.5 \text{ g}} = 0.0244 \text{ mol NiBr}_2$$

Use Equation 4.1 to calculate the concentration of the solution.

$$[\text{NiBr}_2] = \frac{0.0244 \text{ mol NiBr}_2}{0.100 \text{ L}} = 0.244 \text{ M NiBr}_2$$

Video Solution

Tutored Practice Problem 4.5.1

The concentration of a solute in a solution can be used along with the compound formula to determine the concentration of a specific ionic species in a solution. For example, the concentration of calcium ions and chloride ions in the 0.523 M $CaCl_2$ solution is calculated here:

$$[\text{Ca}^{2+}] = \frac{0.523 \text{ mol CaCl}_2}{1 \text{ L}} \times \frac{1 \text{ mol Ca}^{2+}}{1 \text{ mol CaCl}_2} = 0.523 \text{ M}$$

$$[\text{Cl}^-] = \frac{0.523 \text{ mol CaCl}_2}{1 \text{ L}} \times \frac{2 \text{ mol Cl}^-}{1 \text{ mol CaCl}_2} = 1.05 \text{ M}$$

As shown in this calculation and in Interactive Figure 4.5.2, the chloride ion concentration is twice that of the calcium ion concentration because the ions are present in a 2:1 ratio in the compound formula, $CaCl_2$.

Interactive Figure 4.5.2

Determine ion concentrations.

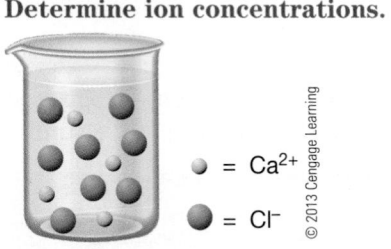

= Ca^{2+}

= Cl^-

© 2013 Cengage Learning

In an aqueous solution of $CaCl_2$, the chloride ion concentration is twice the calcium ion concentration.

Example Problem 4.5.2 Calculate ion concentration.

In the laboratory, a student adds 15.8 g of sodium sulfate to a 500-mL volumetric flask and adds water to the mark on the neck of the flask. Calculate the concentration (in mol/L) of sodium sulfate, the sodium ion, and the sulfate ion in the solution.

Solution:

You are asked to calculate the concentration of a solution and the concentration of the ions in the solution.

You are given the identity of the solute, the amount of solute in the solution, and the total volume of the solution.

First determine the concentration of sodium sulfate.

$$15.8 \text{ g Na}_2\text{SO}_4 \times \frac{1 \text{ mol Na}_2\text{SO}_4}{142.0 \text{ g}} = 0.111 \text{ mol Na}_2\text{SO}_4$$

$$[\text{Na}_2\text{SO}_4] = \frac{0.111 \text{ mol Na}_2\text{SO}_4}{0.500 \text{ L}} = 0.223 \text{ M Na}_2\text{SO}_4$$

Use the compound stoichiometry to determine the ion concentration in the solution.

$$[\text{Na}^+] = \frac{0.223 \text{ M Na}_2\text{SO}_4}{1 \text{ L}} \times \frac{2 \text{ mol Na}^+}{1 \text{ mol Na}_2\text{SO}_4} = 0.445 \text{ M Na}^+$$

$$[\text{SO}_4^{2-}] = \frac{0.223 \text{ M Na}_2\text{SO}_4}{1 \text{ L}} \times \frac{1 \text{ mol SO}_4^{2-}}{1 \text{ mol Na}_2\text{SO}_4} = 0.223 \text{ M SO}_4^{2-}$$

Video Solution

Tutored Practice
Problem 4.5.2

   The concentration of a solution can also be used to determine the volume of a solution that contains a specific amount of solute or the amount of solute in a given volume of solution. That is, a concentration in units of molarity can be used as a conversion factor to convert between the amount of solute in a solution (in moles) and the volume of a solution (in liters). For example, we can calculate the volume of a 0.264 M solution of $AgNO_3$ needed to provide 18.6 g of the ionic compound.

   First, determine the amount of $AgNO_3$ needed in units of moles.

$$18.6 \text{ g AgNO}_3 \times \frac{1 \text{ mol AgNO}_3}{169.9 \text{ g}} = 0.109 \text{ mol AgNO}_3 \text{ needed}$$

Next, use the amount of $AgNO_3$ needed and the solution concentration to calculate the volume of solution needed.

$$0.109 \text{ mol AgNO}_3 \times \frac{1 \text{ L solution}}{0.264 \text{ mol AgNO}_3} = 0.415 \text{ L solution}$$

Notice that in this case, the concentration of the solution is inverted when used as a conversion factor so that the units cancel correctly, resulting in an answer with volume units.

Example Problem 4.5.3 Use molarity as a conversion factor.

Calculate the mass (in grams) of magnesium iodide that must be added to a 250-mL volumetric flask in order to prepare 250 mL of a 0.169 M aqueous solution of the salt.

Solution:

You are asked to calculate the mass of solute required to make a solution of known volume and concentration.

You are given the identity of the solute, the volume of the solution, and the concentration of the solution.

First determine the amount of magnesium iodide needed using the concentration and volume of the desired solution.

$$0.250 \text{ L} \times \frac{0.169 \text{ mol MgI}_2}{1 \text{ L}} = 0.0423 \text{ mol MgI}_2$$

Use the amount of magnesium iodide to determine the mass of solid needed to prepare the solution.

$$0.0423 \text{ mol MgI}_2 \times \frac{278.1 \text{ g}}{1 \text{ mol MgI}_2} = 11.7 \text{ g MgI}_2$$

4.5b Preparing Solutions of Known Concentration

One of the most important skills learned by chemists is how to accurately prepare a solution of known concentration. This process has two parts: First, calculate the quantities that must be combined to make the solution; then correctly prepare the solution in the laboratory. We will describe both parts here, but it is only by practicing in the chemistry lab that you can become proficient in the skills necessary to prepare a solution correctly.

The laboratory equipment necessary for this process must allow you to accurately measure the volume of a solution. **Volumetric glassware** is laboratory glassware that has been carefully calibrated to contain very accurate volumes or allow for the measurement of very accurate volumes (Interactive Figure 4.5.3). Volumetric flasks are typically used to make solutions of known concentration. These flasks are generally available with volumes of 25 mL, 50 mL, 100 mL, 250 mL, 1 L, and 2 L. Although the flasks can hold more than the stated volume, a mark is placed on the neck of the flask that denotes a specific volume; filling the flask to that mark ensures that it holds the stated volume of liquid. Pipets and burets are examples of volumetric glassware used to deliver a specific volume of solution.

Video Solution

Tutored Practice
Problem 4.5.3

Interactive Figure 4.5.3

Select the correct glassware.

A collection of volumetric and nonvolumetric laboratory glassware

Two methods are used to make a solution of known concentration: the **direct addition** method and the **dilution** method.

Direct Addition Method

The direct addition method is generally used to prepare solutions with relatively high concentrations of solute, whereas the dilution method is generally used to prepare more dilute solutions.

For example, the direct addition method can be used to prepare 100. mL of a 0.200 M solution of sucrose, $C_{12}H_{22}O_{11}$. First, calculate the quantity (in grams) of sucrose needed to make the solution. Using the volume of the desired solution and the desired concentration, calculate the amount of sucrose needed (in moles).

$$100. \text{ mL solution} \times \frac{1 \text{ L}}{1000 \text{ mL}} = 0.100 \text{ L solution}$$

$$0.100 \text{ L solution} \times \frac{0.200 \text{ mol sucrose}}{1 \text{ L solution}} = 0.0200 \text{ mol sucrose needed}$$

Next, calculate the quantity of sucrose needed, in units of grams.

$$0.0200 \text{ mol } C_{12}H_{22}O_{11} \times \frac{342.3 \text{ g}}{1 \text{ mol } C_{12}H_{22}O_{11}} = 6.85 \text{ g } C_{12}H_{22}O_{11}$$

To make 100. mL of a 0.200 M solution of sucrose, weigh out 6.85 g of sucrose on a laboratory balance and carefully add it to a 100-mL volumetric flask. Add a small amount of water to the flask and swirl until the sucrose dissolves completely. Add additional water to the flask, mixing as each portion is added. Carefully add water to the volumetric flask until the bottom of the meniscus sits on the mark on the neck of the flask. Invert the flask to mix the contents (Interactive Figure 4.5.4).

Make a solution by direct addition.

Charles D. Winters

The materials required to make a solution by direct addition

Example Problem 4.5.4 Use the direct addition method.

Describe the steps involved in preparing 500. mL of a 0.125 M potassium nitrate solution using solid potassium nitrate, a 500-mL volumetric flask, and deionized water.

Solution:

You are asked to describe the steps in making a solution of known volume and concentration using the direct addition method.

You are given the volume and concentration of the desired solution and the identity of the solute.

Example Problem 4.5.4 *(continued)*

Calculate the amount of KNO_3 required to prepare the solution.

$$0.500 \text{ L} \times \frac{0.125 \text{ mol } KNO_3}{1 \text{ L}} \times \frac{101.1 \text{ g}}{1 \text{ mol } KNO_3} = 6.32 \text{ g } KNO_3$$

To make the solution, weigh out 6.32 g of KNO_3 on a laboratory balance and carefully add it to a 500-mL volumetric flask. Add a small amount of deionized water to the flask and swirl until the solid dissolves completely. Add additional water to the flask, mixing as each portion is added. Carefully add water to the volumetric flask until the bottom of the meniscus sits on the mark on the neck of the flask. Invert the flask to mix the contents.

Video Solution

Tutored Practice
Problem 4.5.4

Dilution Method

The dilution method is used to prepare very dilute solutions because it is difficult to weigh small amounts of solute with great accuracy. In the dilution method, the solution is prepared by diluting a more concentrated existing solution with a known concentration. For example, we can use a 0.600 M $NiCl_2$ stock solution to prepare 50.0 mL of solution that is 0.0125 M in $NiCl_2$.

First, use the concentration and volume of the desired solution to calculate the amount of solute (in moles) needed from the stock solution.

$$50.0 \text{ mL solution} \times \frac{1 \text{ L}}{1000 \text{ mL}} = 0.0500 \text{ L solution}$$

$$0.0500 \text{ L solution} \times \frac{0.0125 \text{ mol } NiCl_2}{1 \text{ L solution}} = 6.25 \times 10^{-4} \text{ mol } NiCl_2 \text{ needed}$$

Next, use the amount of solute needed and the concentration of the stock solution to calculate the volume of stock solution that contains the desired amount of $NiCl_2$.

$$6.25 \times 10^{-4} \text{ mol } NiCl_2 \times \frac{1 \text{ L solution}}{0.600 \text{ mol } NiCl_2} = 1.04 \times 10^{-3} \text{ L}$$

To prepare the solution, transfer 1.04×10^{-3} L (1.04 mL) of the 0.600 M stock solution to a 50.0-mL volumetric flask using a graduated pipette and follow the procedure outlined previously for using a volumetric flask (Interactive Figure 4.5.5).

When using the dilution method to prepare a solution of known concentration, remember that the total amount of solute is the same in both the measured volume of the more concentrated stock solution and in the dilute solution created. This means we can use a simple equation (Equation 4.2) to calculate the volume or concentration of either the stock solution or the more dilute solution. Equation 4.2 is based on the fact that the concentration and volume of any solution can be used to calculate the amount of solute present in a solution.

Interactive Figure 4.5.5

Make a solution by dilution.

Charles D. Winters

The materials required to make a solution by dilution

amount of solute (mol) = solution concentration $\left(\dfrac{\text{mol}}{\text{L}}\right)$ × solution volume (L)

$\qquad\qquad\qquad\quad = C \times V$

Because the amount of solute (in moles) in the measured volume of concentrated solution is equal to the amount of solute (in moles) in the diluted solution,

$$\text{mol solute in concentrated solution} = \text{mol solute in dilute solution}$$

and

$$C_{\text{conc}} \times V_{\text{conc}} = C_{\text{dil}} \times V_{\text{dil}} \qquad\qquad \textbf{(4.2)}$$

where C_{conc} and V_{conc} are the concentration (mol/L) and volume (L) of the concentrated solution and C_{dil} and V_{dil} are the concentration (mol/L) and volume (L) of the diluted solution.

Example Problem 4.5.5 Use the dilution method to make a solution.

In the laboratory, a student dilutes 21.5 mL of a 10.0 M nitric acid solution to a total volume of 125 mL. What is the concentration of the diluted solution?

Solution:

You are asked to calculate the concentration of a diluted solution.

You are given the volume and concentration of the concentrated solution and the volume of the diluted solution.

First, calculate the amount of HNO_3 in the concentrated sample.

$$0.0215 \text{ L} \times \frac{10.0 \text{ mol } HNO_3}{1 \text{ L}} = 0.215 \text{ mol } HNO_3 \text{ in the concentrated sample}$$

Use the amount of HNO_3 and the volume of the diluted sample to calculate the concentration of the dilute solution.

$$\frac{0.215 \text{ mol } HNO_3}{0.125 \text{ L}} = 1.72 \text{ M } HNO_3$$

You can also use the equation $C_{\text{conc}} \times V_{\text{conc}} = C_{\text{dil}} \times V_{\text{dil}}$ to calculate the concentration of the diluted solution.

$$C_{\text{conc}} \times V_{\text{conc}} = C_{\text{dil}} \times V_{\text{dil}}$$

$$(10.0 \text{ M } HNO_3)(21.5 \text{ mL}) = (C_{\text{dil}})(125 \text{ mL})$$

$$C_{\text{dil}} = 1.72 \text{ M } HNO_3$$

Notice that the volume units cancel and do not need to be in units of liters.

Video Solution

Tutored Practice
Problem 4.5.5

4.5c Solution Stoichiometry

The central premise of stoichiometry is that the amount of matter produced in a chemical reaction can be determined using the mole relationship provided by a balanced chemical equation. The amount of reactants and products can be determined from mass (in grams) or, as we have just seen, from the volume and concentration of a solution. As shown in Interactive Figure 4.5.6, the stoichiometric relationships between reactants and products developed in Stoichiometry (Unit 3) can be expanded to include using the volume and concentration of a solution to determine amounts of reactant and product.

Interactive Figure 4.5.6

Explore stoichiometric relationships involving aqueous solutions.

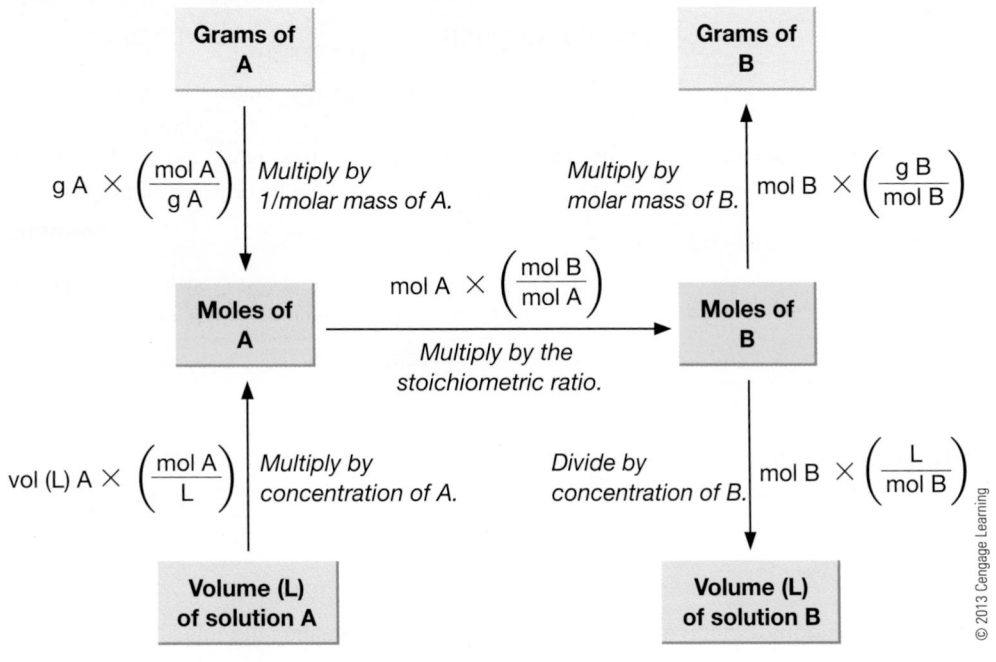

For example, the mass of $PbCrO_4$ produced when 50.0 mL of a 0.400 M $Pb(NO_3)_2$ solution is mixed with an excess of 0.100 M Na_2CrO_4 is calculated as follows:

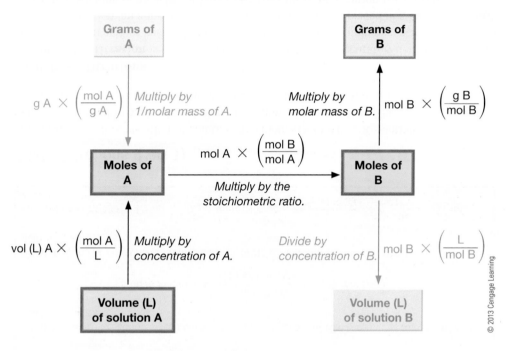

Step 1. Write a balanced chemical equation.

$$Pb(NO_3)_2(aq) + Na_2CrO_4(aq) \rightarrow 2 NaNO_3(aq) + PbCrO_4(s)$$

Step 2. Calculate moles of limiting reactant.

$$\frac{0.400 \text{ mol } Pb(NO_3)_2}{1 \text{ L}} \times \frac{1 \text{ L}}{1000 \text{ mL}} \times 50.0 \text{ mL} = 0.0200 \text{ mol } Pb(NO_3)_2$$

Step 3. Calculate moles of product.

$$0.0200 \text{ mol } Pb(NO_3)_2 \times \frac{1 \text{ mol } PbCrO_4}{1 \text{ mol } Pb(NO_3)_2} = 0.0200 \text{ mol } PbCrO_4$$

Step 4. Calculate mass of product.

$$0.0200 \text{ mol } PbCrO_4 \times \frac{323.2 \text{ g}}{1 \text{ mol } PbCrO_4} = 6.46 \text{ g } PbCrO_4$$

Example Problem 4.5.6 Use solution stoichiometry.

Calculate the volume (in mL) of 0.715 M HNO_3 needed to react completely with 6.35 g of $CaCO_3$ in a gas-forming reaction.

Solution:

You are asked to calculate the volume of a solution required to completely consume a reactant.

You are given the concentration of the solution and the mass of the reactant that will be consumed.

Step 1. Write a balanced chemical equation for the gas-forming reaction.

$$2\ HNO_3(aq) + CaCO_3(s) \rightarrow Ca(NO_3)_2(aq) + H_2O(\ell) + CO_2(g)$$

Step 2. Calculate moles of calcium carbonate available to react with HNO_3.

$$6.35\ \text{g CaCO}_3 \times \frac{1\ \text{mol CaCO}_3}{100.1\ \text{g}} = 0.0634\ \text{mol CaCO}_3$$

Step 3. Use the reaction stoichiometry to calculate moles of HNO_3 required to react with the available $CaCO_3$.

$$0.0634\ \text{mol CaCO}_3 \times \frac{2\ \text{mol HNO}_3}{1\ \text{mol CaCO}_3} = 0.127\ \text{mol HNO}_3$$

Step 4. Use the concentration and amount of HNO_3 solution to calculate the volume of nitric acid needed.

$$\text{volume of HNO}_3\ \text{solution} = 0.127\ \text{mol HNO}_3 \times \frac{1\ \text{L}}{0.715\ \text{mol HNO}_3} \times \frac{10^3\ \text{mL}}{1\ \text{L}} = 177\ \text{mL}$$

Video Solution

Tutored Practice
Problem 4.5.6

4.5d   Titrations (Part 1)

A **titration** is an application of solution stoichiometry in which a solution of known concentration is used to analyze a solution of unknown concentration. The two most common types of titrations are acid–base titrations and oxidation–reduction titrations.

Acid–Base Titrations

An **acid–base titration** is used in analytical laboratories to accurately determine the concentration of an acid or base solution or the molar mass of an unknown acid or base. In an acid–base titration, an aqueous solution of a base (or acid) is placed in a buret and slowly added to a solution of an acid (or base) in a flask until the reaction is complete. An acid–base **indicator**, a dye that shows by a change in color when the acid–base reaction is complete, is added to the solution in the flask before the titration is begun. The most common acid–base indicator used in the general chemistry laboratory is phenolphthalein, an organic dye that is colorless in acidic solutions and pink in basic solutions.

The **equivalence point** in an acid–base titration is the point at which the acid (or base) in the flask has been consumed completely by the base (or acid) that has been added from the buret. That is, $[H^+] = [OH^-]$ at the equivalence point in an acid–base titration.

The steps involved in an acid–base titration calculation are similar to those in any solution stoichiometry problem. Consider, for example, the titration of 30.0 mL of tartaric acid ($H_2C_4H_4O_6$, a diprotic acid) solution with 0.354 M NaOH (Interactive Figure 4.5.7). In this

Interactive Figure 4.5.7

Explore acid–base titrations.

The titration of tartaric acid with sodium hydroxide

titration, 28.79 mL of NaOH is required to reach the equivalence point in the titration. The concentration of the tartaric acid solution is determined as shown in the following steps.

Step 1. Write a balanced equation for the acid–base reaction.

$$H_2C_4H_4O_6(aq) + 2\ NaOH(aq) \rightarrow Na_2C_4H_4O_6(aq) + 2\ H_2O(\ell)$$

Step 2. Use the volume and concentration of the NaOH solution to calculate the amount of base added to the acid solution during the titration.

$$28.79\ \text{mL solution} \times \frac{1\ \text{L}}{1000\ \text{mL}} \times \frac{0.354\ \text{mol NaOH}}{1\ \text{L solution}} = 0.0102\ \text{mol NaOH}$$

Step 3. Use the balanced equation to determine the amount of acid present in the flask.

$$0.0102\ \text{mol NaOH} \times \frac{1\ \text{mol}\ H_2C_4H_4O_6}{2\ \text{mol NaOH}} = 0.00510\ \text{mol}\ H_2C_4H_4O_6$$

Step 4. Use the amount of acid and the volume of the acid solution to determine the concentration of the tartaric acid solution.

$$\frac{0.00510\ \text{mol}\ H_2C_4H_4O_6}{30.0\ \text{mL solution}} \times \frac{1000\ \text{mL}}{1\ \text{L}} = 0.170\ \text{M}$$

Example Problem 4.5.7 Determine an unknown concentration or volume using an acid–base titration.

Calculate the volume of 0.106 M barium hydroxide required to neutralize 18.6 mL of a 0.288 M hydrochloric acid solution.

Solution:

You are asked to calculate the volume of base solution required to neutralize an aqueous solution containing an acid.

You are given the concentration of the base, the identity of the base, the concentration and volume of the acid solution, and the identity of the acid.

Step 1. Write a balanced equation for the acid–base reaction.

$$2\ HCl(aq) + Ba(OH)_2(aq) \rightarrow BaCl_2(aq) + 2\ H_2O(\ell)$$

Step 2. Use the volume and concentration of the HCl solution to calculate the amount of acid neutralized during the titration.

$$18.6\ \text{mL HCl solution} \times \frac{1\ \text{L}}{1000\ \text{mL}} \times \frac{0.288\ \text{mol HCl}}{1\ \text{L solution}} = 0.00536\ \text{mol HCl}$$

Example Problem 4.5.7 *(continued)*

Step 3. Use the balanced equation to determine the amount of base required to neutralize the acid.

$$0.00536 \text{ mol HCl} \times \frac{1 \text{ mol Ba(OH)}_2}{2 \text{ mol HCl}} = 0.00268 \text{ mol Ba(OH)}_2$$

Step 4. Use the amount of base and the concentration of the base solution to determine the volume of barium hydroxide solution required to neutralize the acid solution.

$$\text{volume Ba(OH)}_2 \text{ solution} = 0.00268 \text{ mol Ba(OH)}_2 \times \frac{1 \text{ L}}{0.106 \text{ mol Ba(OH)}_2} \times \frac{1000 \text{ mL}}{1 \text{ L}}$$

$$= 25.3 \text{ mL}$$

This type of calculation is often performed before beginning a titration in order to estimate the amount of solution in the buret required to reach the equivalence point.

Video Solution

Tutored Practice
Problem 4.5.7

It is often important to have an accurate concentration of an acid or base solution. With **standardization**, another type of acid–base titration, the concentration of an acid or base is determined accurately by the use of either a primary standard or another solution whose concentration has already been determined. A **primary standard** is an acid or base that can be obtained in a very pure form, such as potassium hydrogen phthalate, $KHC_8H_4O_4$ (often abbreviated KHP), or sodium carbonate, Na_2CO_3. The steps involved in standardization calculations are very similar to other acid–base titration calculations, as shown in the following example.

Example Problem 4.5.8 Use a primary standard to determine an unknown concentration using an acid–base titration.

Potassium hydrogen phthalate is a solid, monoprotic acid frequently used in the laboratory as a primary standard. It has the unwieldy formula of $KHC_8H_4O_4$. This is often written in shorthand notation as KHP. If 37.1 mL of a barium hydroxide solution is needed to neutralize 1.83 g of KHP, what is the concentration (in mol/L) of the barium hydroxide solution?

Solution:

You are asked to determine the concentration of a solution containing a base.

You are given the mass of primary standard used, the volume of base solution required to neutralize the primary standard, and the identity of the base and the primary standard.

Step 1. Write a balanced equation for the acid–base reaction.

$$2 \text{ KHC}_8\text{H}_4\text{O}_4(\text{aq}) + \text{Ba(OH)}_2(\text{aq}) \rightarrow \text{BaK}_2(\text{C}_8\text{H}_4\text{O}_4)_2(\text{aq}) + 2 \text{ H}_2\text{O}(\ell)$$

Example Problem 4.5.8 *(continued)*

Step 2. Use the mass of the primary standard, KHP, and the balanced equation to determine the amount of $Ba(OH)_2$ in the solution.

$$1.83 \text{ g KHC}_8\text{H}_4\text{O}_4 \times \frac{1 \text{ mol KHC}_8\text{H}_4\text{O}_4}{204.2 \text{ g}} \times \frac{1 \text{ mol Ba(OH)}_2}{2 \text{ mol KHC}_8\text{H}_4\text{O}_4} = 0.00448 \text{ mol Ba(OH)}_2$$

Step 3. Use the amount of base and the volume of the base solution to determine the concentration of the barium hydroxide solution.

$$\frac{0.00448 \text{ mol Ba(OH)}_2}{37.1 \text{ mL solution}} \times \frac{1000 \text{ mL}}{1 \text{ L}} = 0.121 \text{ M Ba(OH)}_2$$

4.5e Titrations (Part 2)

A third type of acid–base titration calculation involves the determination of the molar mass of an acid or base. For example, as shown in Interactive Figure 4.5.8, an acid–base titration can be used to calculate the effective molecular mass of the basic chemicals found in the ashes produced by burning firewood.

Example Problem 4.5.9 Use an acid–base titration to calculate the molar mass of an acid.

A 0.125-g sample of an unknown monoprotic acid is dissolved in water and titrated with standardized sodium hydroxide. The equivalence point in the titration is reached after the addition of 20.59 mL of 0.0193 M NaOH to the sample of the unknown acid. Calculate the molar mass of the acid.

Solution:

You are asked to calculate the molar mass of an unknown monoprotic acid.

You are given the mass of the acid sample, the volume and concentration of base required to neutralize the acid sample, the identity of the base, and the fact that the acid is monoprotic.

Step 1. Write a balanced equation for the acid–base reaction. In this case we do not know the identity of the acid, but we know it is monoprotic; therefore, we can use the general formula HA to represent the unknown acid in an equation.

$$HA(aq) + NaOH(aq) \rightarrow H_2O(\ell) + NaA(aq)$$

Step 2. Use the volume and concentration of the NaOH solution to calculate the amount of base used to neutralize the acid during the titration.

$$20.59 \text{ mL NaOH solution} \times \frac{1 \text{ L}}{1000 \text{ mL}} \times \frac{0.0193 \text{ mol NaOH}}{1 \text{ L solution}} = 3.97 \times 10^{-4} \text{ mol NaOH}$$

Video Solution

Tutored Practice
Problem 4.5.8

Interactive Figure 4.5.8

Determine the molar mass of an unknown using an acid–base titration.

Charles D. Winters

The ashes produced from burning firewood are basic.

Example Problem 4.5.9 (continued)

Step 3. Use the balanced equation to determine the amount of acid in the flask.

$$3.97 \times 10^{-4} \text{ mol NaOH} \times \frac{1 \text{ mol HA}}{1 \text{ mol NaOH}} = 3.97 \times 10^{-4} \text{ mol HA}$$

Step 4. Use the mass of the acid sample and the amount of acid (in mol) in the sample to calculate the molar mass of the unknown acid.

$$\text{molar mass of HA} = \frac{0.125 \text{ g}}{3.97 \times 10^{-4} \text{ mol}} = 315 \text{ g/mol}$$

> **Video Solution**
>
> **Tutored Practice Problem 4.5.9**

Oxidation–Reduction Titrations

An **oxidation–reduction titration** uses an oxidation–reduction reaction to analyze a solution of unknown concentration. The equivalence point in an oxidation–reduction reaction occurs when the reactants are consumed and often is detected by a color change in one of the reactants.

Example Problem 4.5.10 Determine an unknown volume or concentration using an oxidation–reduction titration.

The concentration of a hydrogen peroxide solution is determined by titrating it with a 0.275 M solution of potassium permanganate. The balanced net ionic equation for the reaction is

$$5 \text{ H}_2\text{O}_2(aq) + 6 \text{ H}^+(aq) + 2 \text{ MnO}_4^-(aq) \rightarrow 5 \text{ O}_2(g) + 2 \text{ Mn}^{2+}(aq) + 8 \text{ H}_2\text{O}(\ell)$$

In one experiment, 17.3 mL of the 0.275 M KMnO$_4$ solution is required to react completely with 25.0 mL of the hydrogen peroxide solution. Calculate the concentration of the hydrogen peroxide solution.

Solution:

You are asked to calculate the concentration of a solution using an oxidation–reduction titration.

You are given the balanced oxidation–reduction reaction, the volume and concentration of the titrant, and the volume of the solution of unknown concentration.

Step 1. Calculate the amount of KMnO$_4$ added to the flask during the titration.

$$17.3 \text{ mL KMnO}_4 \text{ solution} \times \frac{1 \text{ L}}{1000 \text{ mL}} \times \frac{0.275 \text{ mol KMnO}_4}{1 \text{ L solution}} = 0.00476 \text{ mol KMnO}_4$$

Example Problem 4.5.10 *(continued)*

Step 2. Use the balanced equation to determine the amount of hydrogen peroxide in the sample.

$$0.00476 \text{ mol KMnO}_4 \times \frac{5 \text{ mol H}_2\text{O}_2}{2 \text{ mol KMnO}_4} = 0.0119 \text{ mol H}_2\text{O}_2$$

Step 3. Use the amount of hydrogen peroxide and the volume of the sample to calculate the solution concentration.

$$\frac{0.0119 \text{ mol H}_2\text{O}_2}{25.0 \text{ mL}} \times \frac{1000 \text{ mL}}{1 \text{ L}} = 0.476 \text{ M H}_2\text{O}_2$$

The equivalence point in this titration is indicated by the color of the solution. As the purple $KMnO_4$ solution is added to the flask containing H_2O_2, the colorless Mn^{2+} ion is formed. At the equivalence point a slight excess of $KMnO_4$ is present in the flask and the solution has a faint purple color.

Video Solution

Tutored Practice
Problem 4.5.10

Section 4.5 Mastery

Unit Recap

Key Concepts

4.1 Types of Chemical Reactions

- In a combination reaction, two or more reactants combine to form one product (4.1a).

- A decomposition reaction is often the reverse of a combination reaction, and the number of products in this type of reaction is typically greater than the number of reactants (4.1a).

- In displacement reactions, also called exchange reactions, one atom, ion, or molecular fragment displaces another (4.1b).

- Double-displacement reactions are also called metathesis reactions. Examples include precipitation reactions and acid–base reactions (4.1b).

4.2 Aqueous Solutions

- A solution is formed when a solute is mixed with a solvent (4.2a).

- A compound dissolved in water is in an aqueous state (4.2a).

- Species dissolved in a solvent are solvated (or hydrated when water is the solvent) (4.2a).

- An electrolyte increases the electrical conductivity of water (4.2a).

- Electrolytes are characterized as strong or weak; a species that does not increase the electrical conductivity of water is a nonelectrolyte (4.2a).

- Ionic compounds can be soluble or insoluble in water; their solubility is predicted using solubility rules for ionic compounds in water (4.2b).

4.3 Reactions in Aqueous Solution

- A precipitation is an exchange reaction that results in the formation of an insoluble ionic compound, a precipitate (4.3a).

- Ionic species that are unchanged during an exchange reaction are called spectator ions (4.3a).

- A net ionic equation is a chemical equation for a reaction written without spectator ions (4.3a).

- Acid–base reactions, which occur between an acid and a base, are typically exchange reactions where a proton is transferred between the acid and base and water and an ionic salt are formed (4.3b).

- Acids and bases can be characterized as strong (strong electrolytes) or weak (weak electrolytes) (4.3b).

- A monoprotic acid produces one mole of H^+ ions per mole of acid, and a diprotic acid produces two moles of H^+ per mole of acid (4.3b).

- A gas-forming reaction is an exchange reaction that has a gas as one of its products (4.3c)

4.4 Oxidation–Reduction Reactions

- An oxidation–reduction reaction involves the exchange of electrons between two or more species (4.4a).

- Oxidation is the loss of one or more electrons from a chemical species and reduction is the gain of one or more electrons (4.4a).

- An oxidizing agent is reduced in an oxidation–reduction reaction, and a reducing agent is oxidized in an oxidation–reduction reaction (4.4a).

- A half-reaction shows only the oxidation or reduction in the oxidation–reduction reaction (4.4a).

- The oxidation number (or oxidation state) of an atom is assigned using a set of rules (4.4b).

- Oxidation numbers are used to identify the species oxidized or reduced in an oxidation–reduction reaction (4.4c).

4.5 Stoichiometry of Reactions in Aqueous Solution

- Concentration is a quantitative measure of the amount of solute dissolved in a solvent (4.5a).
- Molarity is a concentration unit defined as moles of solute dissolved in 1 liter of solution (4.5a).
- Two methods of making solutions of known concentration are the direct addition method and the dilution method (4.5b).
- The concentration of a solution can be used, along with the solution volume, to determine the number of moles of solute in a sample (4.5c).
- In a titration, a solution of known concentration is used to analyze a solution of unknown concentration (4.5d).
- Acid–base titrations use an acid–base reaction to analyze an unknown solution (4.5d).
- The equivalence point in an acid–base titration occurs when all of the acid and base are consumed (4.5d).
- A primary standard can be used, along with an acid–base titration, to determine the concentration of an unknown acid or base solution (4.5e).
- Oxidation–reduction titrations use an oxidation–reduction reaction to analyze an unknown solution (4.5e).

Key Equations

$$\text{molarity (M)} = \frac{\text{moles of solute}}{1\ \text{L of solution}} \qquad (4.1)$$

$$C_{\text{conc}} \times V_{\text{conc}} = C_{\text{dil}} \times V_{\text{dil}} \qquad (4.2)$$

Key Terms

4.1 Types of Chemical Reactions
combination reaction
decomposition reaction
displacement reaction
single displacement reaction
double displacement reaction
precipitation reaction
acid–base reaction
gas-forming reaction

4.2 Aqueous Solutions
solvent
solution
solute
aqueous
solvated
hydrated
electrolyte
strong electrolyte
nonelectrolyte
weak electrolyte

soluble
insoluble
precipitate

4.3 Reactions in Aqueous Solution
spectator ion
net ionic equation
acid
base
salt
strong acid
weak acid
organic acid
monoprotic acid
diprotic acid
strong base
weak base

4.4 Oxidation–Reduction Reactions
oxidation–reduction reaction
oxidation

reduction
oxidizing agent
reducing agent
half-reaction
oxidation number

4.5 Stoichiometry of Reactions in Aqueous Solution
concentration
molarity (M)
volumetric glassware
direct addition
dilution
titration
acid–base titration
indicator
equivalence point
standardization
primary standard
oxidation–reduction titration

Unit 4   Review and Challenge Problems

Thermochemistry

Unit Outline

In This Unit...

This unit begins an exploration of thermochemistry, the study of the role that energy in the form of heat plays in chemical processes. We investigate the energy changes that take place during phase changes and the chemical reactions you have studied previously and learn why some chemical reactions occur while others do not. In Electromagnetic Radiation and the Electronic Structure of the Atom (Unit 6), we will study energy changes at the molecular level and the consequences those energy changes have on the properties of atoms and elements.

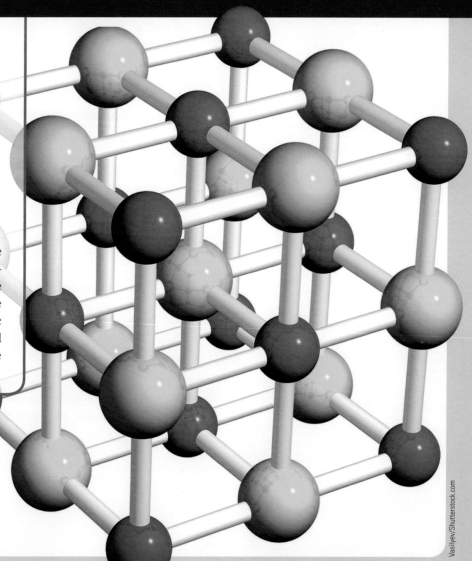

5.1 Energy

5.1a Kinetic and Potential Energy

Chemical reactions involve reactants undergoing chemical change to form new substances, products.

$$\text{reactants} \rightarrow \text{products}$$

What is not apparent in the preceding equation is the role of energy in a reaction. For many reactions, energy, often in the form of heat, is absorbed—that is, it acts somewhat like a reactant. You might write an equation for those reactions that looks like this:

$$\text{energy} + \text{reactants} \rightarrow \text{products}$$

In other reactions, energy is produced—that is, it acts like a product:

$$\text{reactants} \rightarrow \text{products} + \text{energy}$$

Energy is defined most simply as the ability to do work. **Work** is defined in many ways, the simplest definition being the force involved in moving an object some distance. From a chemist's point of view, energy is best viewed as the ability to cause change, and **thermochemistry** is the study of how energy in the form of heat is involved in chemical change.

Energy takes many forms, such as mechanical, electrical, or gravitational. These are categorized into two broad classes: **kinetic energy**, energy associated with motion, and **potential energy**, energy associated with position. **Mechanical energy** is the sum of the kinetic and potential energy of an object as a whole, while **thermal energy** is the sum of the kinetic and potential energies of all the atoms, molecules, or ions within a system. **Chemical energy** is a form of potential energy that can be released when new chemical bonds are formed. Kinetic energy is calculated from the equation

$$KE = \tfrac{1}{2}\, mv^2 \qquad\qquad \textbf{(5.1)}$$

where

m = mass
v = velocity

Potential energy calculations depend on the forces that exist between particles. Because different types of particles experience different types of forces, it is not possible to use a single equation to calculate potential energy. Some common types of kinetic and potential energy are shown in Table 5.1.1.

Table 5.1.1 Types of Energy

Energy	Source	Kinetic or Potential?
Mechanical	Bulk matter	Both
Thermal	Random motion and position of molecules	Both
Electrical	Charged particles	Both
Chemical	Structural arrangement of nuclei and electrons in atoms and molecules	Potential
Nuclear	Structural arrangement of protons and neutrons within the atomic nucleus	Potential
Electromagnetic (radiant)	Disturbance in the electric and magnetic fields of space due to oscillating charged particles	Both

Most of the events we see around us involve *conversion* of energy from one form to another, as shown in Interactive Figure 5.1.1. For example, the photocell in Interactive Figure 5.1.1 absorbs light (radiant energy) and converts it into an electric current. That electric current is then used to drive a fan. The energy conversions occurring are therefore:

radiant (kinetic and potential) → electrical (kinetic and potential) → mechanical (kinetic)

5.1b Measuring Energy: Energy Units

Energy is measured in different units. For example, heating fuel is typically measured in British thermal units, Btu, and food energy content is measured in Calories. Energy associated with most chemical processes is reported in terms of joules (J) and kilojoules (kJ), or calories (cal) and kilocalories (kcal).

One joule is equal to the energy required to accelerate a 1-kg object using a force of one **newton**, the SI unit of force, over a distance of one meter (1 J = 1 kg · m^2/s^2). One calorie is the energy needed to raise the temperature of 1 g of pure water by 1 degree Celsius. The food energy unit, Calorie, is equal to 1 kcal. Table 5.1.2 shows conversion factors for joules, calories, Btu, and kilowatt-hours (kWh), the energy unit used in measuring electrical energy.

Interactive Figure 5.1.1

Recognize different types of energy.

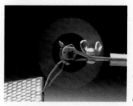

A photocell drives this small fan

Table 5.1.2 Energy Unit Conversion Factors

	J	kJ	cal	kcal	kWh	Btu
1 J =	1	0.001	0.2390	2.390×10^{-4}	2.778×10^{-7}	9.479×10^{-4}
1 kJ =	1000	1	239.0	0.2390	2.778×10^{-4}	0.9479
1 cal =	4.184	4.184×10^{-3}	1	0.001	1.162×10^{-6}	3.968×10^{-3}
1 kcal =	4184	4.184	1000	1	1.162×10^{-3}	3.968
1 kWh =	3.6×10^{6}	3.6×10^{3}	8.604×10^{5}	860.4	1	3413
1 Btu =	1055	1.055	252	0.252	2.93×10^{-4}	1

Example Problem 5.1.1 Use and interconvert energy units.

A barrel contains 42.0 gallons of oil. This is the equivalent of 4.50×10^{10} J of energy. How many kilowatt-hours of electrical energy does this barrel represent?

Solution:

You are asked to calculate the number of kWh of energy in an amount of oil.

You are given an amount of oil and the amount of energy in that oil.

The conversion factor table tells us that $1 \text{ J} = 2.778 \times 10^{-7}$ kWh of energy. The conversion is therefore,

$$4.50 \times 10^{10} \text{ J}\left(\frac{2.778 \times 10^{-7} \text{ kWh}}{1 \text{ J}}\right) = 1.25 \times 10^{4} \text{ kWh}$$

Is your answer reasonable? According to the energy conversion factor table, $1 \text{ kWh} = 3.6 \times 10^{6}$ J. The barrel of oil contains about 10^{4} times more than this amount of energy (in J), so it also contains about 10^{4} kWh of energy.

Video Solution

Tutored Practice
Problem 5.1.1

5.1c Principles of Thermodynamics

Thermochemistry is part of the field of **thermodynamics**, the study of the relationships between heat, energy, and work and the conversion of one into the other. When considering chemical events, it is useful to define the **system**, the item or reaction of interest, and separate that from the **surroundings**, which include everything else. An **isolated system** is one in which neither matter nor energy can be passed to or from the surroundings. A **closed system** is one in which energy but not matter can be passed to or from the surroundings.

In almost all cases in chemistry, the system of interest is closed and the **internal energy**, E_{system}, the energy of the system, changes when energy in the form of heat (q) is added or lost and work (w) is done by or on the system. Although the total internal energy of a system cannot be measured directly, the **change in internal energy**, ΔE_{system}, is calculated from the following equation:

$$\Delta E_{system} = q + w \qquad (5.2)$$

where

q = energy in the form of heat exchanged between system and surroundings
w = work done by or on the system

The **first law of thermodynamics** states that the total energy for an isolated system is constant. That is, the combined amount of energy and matter in an isolated system is constant. Energy is neither created nor destroyed during chemical or physical changes, but it is instead transformed from one form to another. In other words, energy is conserved during a chemical or physical change, or

$$\Delta E_{universe} = 0 \qquad (5.3)$$

As shown in Interactive Figure 5.1.2, positive and negative signs are used in thermodynamics to indicate the direction of heat and work transfer (in or out of a system) and to indicate what is happening to the internal energy of the system.

Interactive Figure 5.1.2

Define system and surroundings.

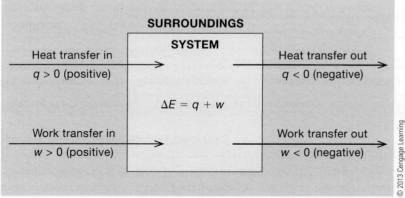

Heat and work sign conventions in thermodynamics

When energy in the form of heat is transferred from the surroundings to the system, q is positive; when heat is transferred from the system to the surroundings, q is negative. Similarly, when work is done by the surroundings on the system, w is positive, and it is negative when work is done by the system on the surroundings. These thermodynamic sign conventions are summarized in Table 5.1.3.

Table 5.1.3 Sign Conventions in Thermodynamics

$+q$ and $+w$	$-q$ and $-w$
• Energy is added to the system.	• Energy is removed from the system.
• Internal energy of the system increases.	• Internal energy of the system decreases.

Example Problem 5.1.2 Calculate internal energy change.

A gas is compressed, and during this process the surroundings do 128 J of work on the gas. At the same time, the gas loses 270 J of energy to the surroundings as heat. What is the change in the internal energy of the gas?

Solution:

You are asked to calculate the change in internal energy (ΔE) of the gas.

You are given the amount of work done, the amount of heat transferred, and the directionality of the work and heat.

Work is done by the surroundings on the system, so w is positive. Heat is transferred from the system to the surroundings, so q is negative.

According to the first law of thermodynamics, $\Delta E_{system} = q + w$.

$$q = -270 \text{ J}$$

$$w = 128 \text{ J}$$

$$\Delta E_{system} = q + w = (-270 \text{ J}) + (128 \text{ J}) = -142 \text{ J}$$

Is your answer reasonable? The amount of heat lost to the surroundings is greater than the amount of work done by the surroundings, so the internal energy change should be negative.

Video Solution

Tutored Practice
Problem 5.1.2

Section 5.1 Mastery

5.2 Enthalpy

5.2a Enthalpy

Although internal energy represents the total energy of a chemical system, it is more common to study the energy of a chemical system under the specific conditions of constant pressure. **Enthalpy**, H, is a thermodynamic quantity that allows us to study heat exchange under constant pressure. It is defined as the sum of the internal energy of a system plus the product of pressure and volume.

$$H = E + PV \qquad \qquad \textbf{(5.4)}$$

In most chemical systems under study, reactions are performed under conditions of constant pressure. Under these conditions, the **change in enthalpy**, ΔH, is equal to the heat exchanged under constant pressure. Note that, like the internal energy of a system, enthalpy cannot be measured directly and it is not possible to know the amount of enthalpy present in a chemical sample. However, enthalpy change, and therefore *relative* enthalpy, can be measured.

Enthalpy is a measure of the total heat content of a system and is related to both chemical potential energy and the degree to which electrons are attracted to nuclei in molecules. When electrons are strongly attracted to nuclei, there are strong bonds between atoms, molecules are relatively stable, and enthalpy is low. In contrast, when electrons are only weakly attracted to nuclei, there are weak bonds between atoms, molecules are relatively unstable, and enthalpy is high. You might wonder if this means that breaking chemical bonds releases energy. It does not. The presence of strong bonds in a molecule does not correlate with a large degree of stored potential energy. A chemical state where bonds could be formed but do not exist is a state of high chemical potential energy.

The sign of ΔH indicates the direction of energy transfer, as shown in Interactive Figure 5.2.1. In an **exothermic process**, heat is transferred from the system to the surroundings. The enthalpy change for an exothermic process has a negative value ($\Delta H < 0$). During exothermic reactions, weakly bonded molecules are converted to strongly bonded

Interactive Figure 5.2.1

Distinguish endothermic and exothermic processes.

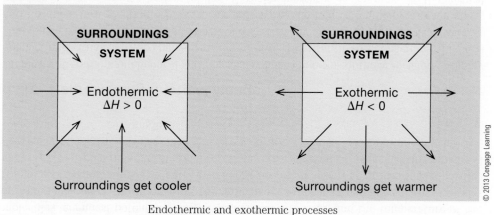

Endothermic and exothermic processes

molecules, chemical potential energy is converted into heat, and the temperature of the surroundings increases. In an **endothermic process**, heat is transferred from the surroundings to the system. The enthalpy change for an endothermic process has a positive value ($\Delta H > 0$). During endothermic reactions, strongly bonded molecules are converted to weakly bonded molecules, heat is converted into chemical potential energy, and the temperature of the surroundings decreases.

5.2b   Representing Energy Change

Chemists often think of chemical and physical changes in terms of the associated enthalpy changes and visualize these changes in an *enthalpy diagram.* In these diagrams, the horizontal axis indicates the different states of a system undergoing change or the reactants and products in a reaction. The vertical axis shows the relative enthalpy of each state, which is indicated using a horizontal line. Enthalpy increases as you move up the vertical axis, so the higher that line occurs on the y-axis, the higher the enthalpy for a given species. Interactive Figure 5.2.2 allows you to explore simple enthalpy diagrams for endothermic and exothermic chemical reactions.

The enthalpy change for the reaction, ΔH, is the difference between the enthalpies of the different states or the reactants and products. In the exothermic enthalpy diagram in Interactive Figure 5.2.2, the products of the reaction are at lower enthalpy than the reactants, so ΔH for the reaction is negative. The reaction is exothermic, the reaction releases heat, and the chemical bonding in the products is stronger than that in the reactants.

Section 5.2 Mastery

Interactive Figure 5.2.2

Interpret energy level diagrams.

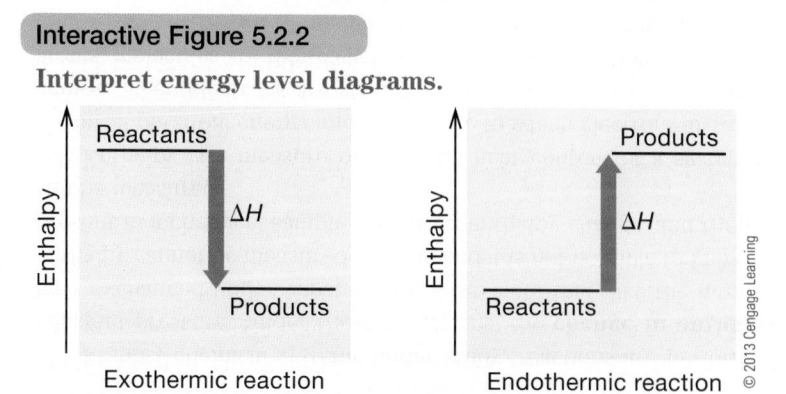

Energy level diagrams for exothermic and endothermic systems

© 2013 Cengage Learning

5.3 Energy, Temperature Changes, and Changes of State

5.3a Heat Transfer and Temperature Changes: Specific Heat Capacity

When an object gains thermal kinetic energy, a variety of things can occur. One of the most common is that the energy raises the object's temperature. Three factors control the magnitude of a temperature change for an object: the amount of heat energy added to the object, the mass of the object, and the material the object is made of. Consider lighting a match and using it to heat a large glass of water. Heat is transferred from the burning match to the water, but the temperature of the water does not increase very much. Now consider using a lit match to heat the tip of a needle. In this case, the needle becomes quite hot. A similar amount of heat energy is added to each object, but the needle gets hotter because it has a smaller mass than the water and is made of metal, which has a lower specific heat capacity than the water. **Specific heat capacity** is the amount of energy needed to raise the temperature of 1 g of a substance by 1 °C.

$$c, \text{specific heat capacity (J/g} \cdot {}^\circ\text{C)} = \frac{q, \text{heat energy absorbed (J)}}{m, \text{mass (g)} \times \Delta T, \text{change in temperature (}^\circ\text{C)}} \quad \textbf{(5.5)}$$

Some specific heat capacity values are given in Table 5.3.1. Specific heat capacity values are reported in units of J/g · °C or J/g · K. The values do not change with the different units because a 1-degree increment is the same on both temperature scales. Notice in Table 5.3.1 that some materials, such as metals, have low specific heat capacities, which means it takes relatively little energy to cause a large temperature increase. Other materials, such as water, have high specific heat capacities, so it takes much more energy to effect the same increase in temperature. For example, the same amount of heat energy will raise the temperature of a 1-g sample of gold over 30 times more than it would a 1-g sample of water.

The value of the specific heat capacity can be determined if the energy, mass, and temperature change are all known for the sample. Consider the experiment shown in Interactive Figure 5.3.1. In this experiment, 5-g samples of copper and glass are heated and 150 J of heat energy is added to each sample. The temperature change of the copper sample is much greater than that of the glass sample, which indicates that copper has a smaller specific heat capacity.

Table 5.3.1 Specific Heat Capacity Values of Common Substances

Substance	Name	Specific Heat (J/g · °C)
Al(s)	Aluminum	0.897
Fe(s)	Iron	0.449
Cu(s)	Copper	0.385
Au(s)	Gold	0.129
$O_2(g)$	Oxygen	0.917
$N_2(g)$	Nitrogen	1.04
$NH_3(g)$	Ammonia	4.70
$C_2H_5OH(\ell)$	Ethanol	2.44
$H_2O(\ell)$	Water (liquid)	4.184
$H_2O(s)$	Water (ice)	2.06
	Wood	1.8
	Concrete	0.9
	Glass	0.8
	Granite	0.8

Interactive Figure 5.3.1

Recognize the effect of heat capacity on temperature change.

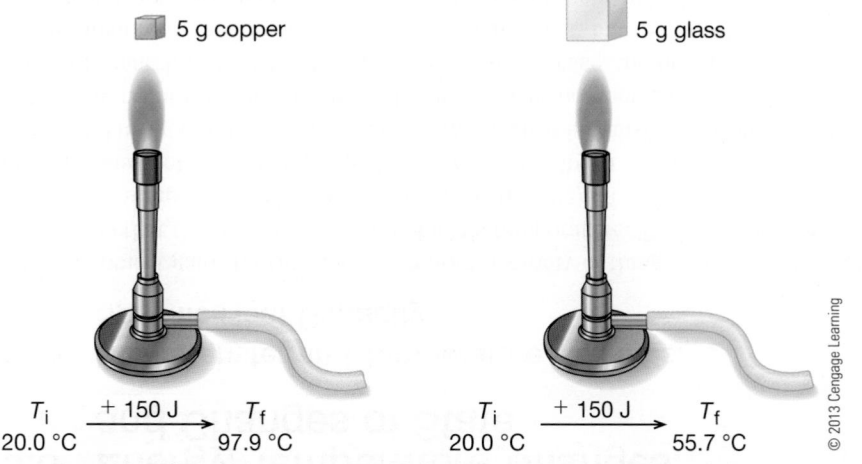

5 g copper

5 g glass

T_i + 150 J T_f
20.0 °C   97.9 °C

T_i + 150 J T_f
20.0 °C   55.7 °C

A total of 150 J of heat energy is added to 5-g samples of copper and glass.

© 2013 Cengage Learning

Example Problem 5.3.1 Determine specific heat capacity.

Using the following data, determine the specific heat capacity of silver.

$$q = 150 \text{ J}$$
$$m = 5.0 \text{ g Ag}$$
$$T_{final} = 145.0 \text{ °C}$$
$$T_{initial} = 20.0 \text{ °C}$$

Solution:

You are asked to determine the heat capacity of silver.

You are given the amount of heat transferred, the mass of silver, and the initial and final temperature of the silver.

Use Equation 5.5 to calculate the specific heat capacity of silver.

$$q = \text{heat energy transferred} = 150 \text{ J}$$
$$m = 5.0 \text{ g}$$
$$\Delta T = T_{final} - T_{initial} = 145.0 \text{ °C} - 20.0 \text{ °C} = 125.0 \text{ °C}$$

$$c_{Ag} = \frac{q}{m \times \Delta T} = \frac{150 \text{ J}}{(5.0 \text{ g})(125 \text{ °C})} = 0.24 \text{ J/g} \cdot \text{°C}$$

Is your answer reasonable? As you can see in Table 5.3.1, many metals have a heat capacity less than 1.0 J/g·°C. This answer is reasonable for the heat capacity of silver.

Calculating the amount of energy associated with temperature changes and predicting the magnitude of temperature changes are essential skills in working with energy and chemical systems. The following example shows how to perform both of these calculations.

Example Problem 5.3.2 Use specific heat capacity.

a. Determine the amount of heat energy that is associated with heating a 154-g iron bar from 20.0 °C to 485 °C.
b. Calculate the final temperature reached when 324 J of heat is added to a 24.5-g iron bar initially at 20.0 °C.

Solution:　.

a. **You are asked** to calculate the amount of heat (q) transferred to an iron bar.

You are given the mass of the iron bar and the initial and final temperature of the iron.

Use Equation 5.5 to calculate the amount of heat transferred to the iron bar.

Video Solution

Tutored Practice
Problem 5.3.1

Example Problem 5.3.2 (continued)

$$c_{Fe} = 0.449 \text{ J/g} \cdot {}^\circ\text{C}$$

$$m = 154 \text{ g}$$

$$\Delta T = T_{final} - T_{initial} = 485 \, {}^\circ\text{C} - 20.0 \, {}^\circ\text{C} = 465 \, {}^\circ\text{C}$$

$$q = m \times c_{Fe} \times \Delta T = (154 \text{ g})(0.449 \text{ J/g} \cdot {}^\circ\text{C})(465 \, {}^\circ\text{C}) = 3.22 \times 10^4 \text{ J}$$

Is your answer reasonable? The temperature change was relatively large, so you would expect a large amount of heat energy was transferred to the iron.

b. **You are asked** to calculate the final temperature when a given amount of heat is added to an iron bar.

You are given the amount of energy, the mass of the bar, and the initial temperature of the bar.

Use Equation 5.5 to calculate the change in temperature of the iron bar.

$$c_{Fe} = 0.449 \text{ J/g} \cdot {}^\circ\text{C}$$

$$m = 24.5 \text{ g}$$

$$q = 324 \text{ J}$$

$$\Delta T = \frac{q}{m \times c_{Fe}} = \frac{324 \text{ J}}{(24.5 \text{ g})(0.449 \text{ J/g} \cdot {}^\circ\text{C})} = 29.5 \, {}^\circ\text{C}$$

The temperature of the iron bar therefore increases from 20.0 °C to 49.5 °C.

Is your answer reasonable? A relatively small amount of heat energy was transferred to the iron bar (less than in part a. of this example problem), so you would expect a relatively small temperature change.

5.3b Heat Transfer Between Substances: Thermal Equilibrium and Temperature Changes

When objects at different temperatures come into contact, the hotter object transfers thermal energy to the cooler object. This causes the hotter object to cool and the cooler object to warm. This process occurs until the two objects reach the same temperature, a state of **thermal equilibrium** (Interactive Figure 5.3.2). At any point in the heat transfer, the quantity of heat energy lost by the hotter object ($-q_{lost}$) is equal to that gained by the cooler object ($+q_{gained}$). That is, $q_{lost} + q_{gained} = 0$.

Video Solution

Tutored Practice
Problem 5.3.2

Interactive Figure 5.3.2

Investigate the principle of thermal equilibrium.

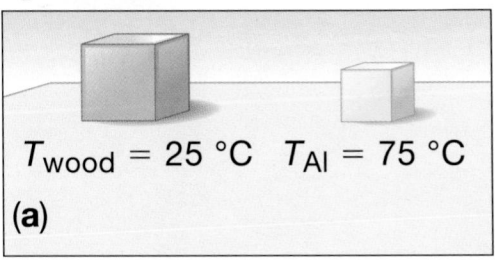

$T_{wood} = 25 \, {}^\circ\text{C} \quad T_{Al} = 75 \, {}^\circ\text{C}$

(a)

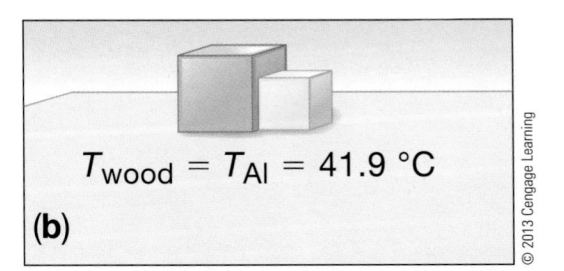

$T_{wood} = T_{Al} = 41.9 \, {}^\circ\text{C}$

(b)

© 2013 Cengage Learning

(a) Separated blocks of wood and aluminum (b) The blocks reach thermal equilibrium when combined.

Interactive Figure 5.3.3

Explore thermal energy transfer.

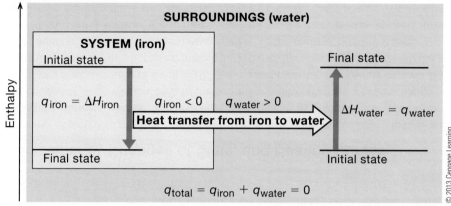

An energy diagram representing energy transfer from iron to water

Consider a heated iron bar plunged into water (Interactive Figure 5.3.3). If we define iron as the system and the water as the surroundings, the process is exothermic as heat energy is lost by the system (iron) to the surroundings (water). As shown in Interactive Figure 5.3.2, when two objects with different initial temperatures are brought into contact, they reach the same temperature at thermal equilibrium. The final temperature is calculated by recognizing that magnitude of energy transferred for each is the same, as shown in the following example.

Example Problem 5.3.3 Predict thermal equilibrium temperatures.

A 12.00-g block of copper at 12.0 °C is immersed in a 5.00-g pool of ethanol with a temperature of 68.0 °C. When thermal equilibrium is reached, what is the temperature of the copper and ethanol?

Solution:

You are asked to calculate the temperature of the copper and ethanol when they reach thermal equilibrium.

You are given the mass of copper and ethanol and the initial temperature of copper and ethanol.

Because the magnitude of energy lost by the ethanol is equal to the energy gained by the copper, $q_{Cu} + q_{ethanol} = 0$.

Example Problem 5.3.3 (continued)

Use Equation 5.5 to calculate the energy lost and gained upon reaching thermal equilibrium and then calculate the final temperature.

$$c_{Cu} = 0.385 \text{ J/g} \cdot \text{°C} \qquad c_{ethanol} = 2.44 \text{ J/g} \cdot \text{°C}$$

$$m_{Cu} = 12.00 \text{ g} \qquad m_{ethanol} = 5.00 \text{ g}$$

$$T_{initial}(Cu) = 12.0 \text{ °C} \qquad T_{initial}(ethanol) = 68.0 \text{ °C}$$

$$q_{Cu} + q_{ethanol} = 0$$

$$[(12.00 \text{ g})(0.385 \text{ J/g} \cdot \text{°C})(T_{final} - 12.0 \text{ °C})] + [(5.00 \text{ g})(2.44 \text{ J/g} \cdot \text{°C})(T_{final} - 68.0 \text{ °C})] = 0$$

$$(4.62 \times T_{final}) - 55.44 + (12.2 \times T_{final}) - 829.6 = 0$$

$$16.82 \times T_{final} = 885.04$$

$$T_{final} = 52.6 \text{ °C}$$

Is your answer reasonable? At thermal equilibrium the two substances in contact with one another have the same temperature, and that temperature lies between the two initial temperatures. Whether the final temperature lies closer to the temperature of the initially hotter or cooler object is related to the mass and specific heat of each object. In this case the final temperature is closer to ethanol's initial temperature because of its significantly higher specific heat (about six times greater than the specific heat of copper).

At the molecular level, heat transfer from a hot object to a cooler object is the result of a transfer of molecular momentum. Explore the energetic changes that take place at the molecular level as two objects with different temperatures come in contact in Interactive Figure 5.3.4.

5.3c Energy, Changes of State, and Heating Curves

When an object is heated, it can get warmer, it can undergo a phase change, and it can undergo a chemical change. We have discussed the energy changes associated with warming and cooling objects. We now turn to the second possible outcome: a phase change.

If you heat a sample of ice at −10 °C at 1 atm pressure, it becomes warmer. If you heat ice at 0 °C (and 1 atm), however, initially it does not become warmer. Instead, the ice melts while maintaining a constant temperature of 0 °C, the melting point of water.

When a solid such as ice is heated and melts, the heat energy is used to overcome the forces holding the water molecules together in the solid phase. The heat energy is therefore changed into chemical potential energy. This potential energy can be reconverted to thermal energy when the forces between water molecules form again, that is, when liquid water freezes to form ice (Interactive Figure 5.3.5).

Video Solution

Tutored Practice
Problem 5.3.3

Interactive Figure 5.3.4

Visualize thermal energy transfer on the molecular scale.

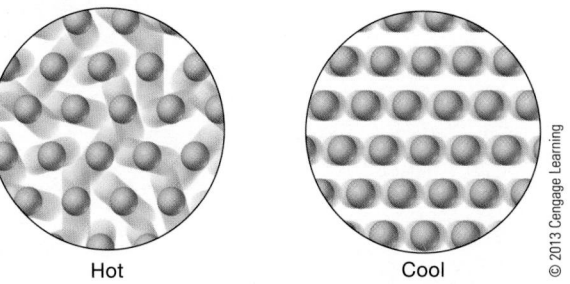

Hot   Cool

Molecular view of a material at different temperatures

© 2013 Cengage Learning

Interactive Figure 5.3.5

Investigate the relationship between energy and phase changes.

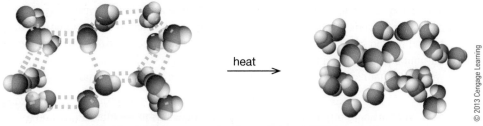

heat →

Melting ice involves overcoming forces between water molecules.

The energy to effect a phase change has been measured for a number of substances. The **enthalpy (heat) of fusion** (ΔH_{fus}) for a substance is the energy needed to melt 1 g or 1 mol of that substance. The **enthalpy (heat) of vaporization** (ΔH_{vap}) is the energy needed to vaporize 1 g or 1 mol of that substance.

Example Problem 5.3.4 Calculate the energy change during a phase change.

Given the following information for mercury, Hg (at 1 atm), calculate the amount of heat needed (at 1 atm) to vaporize a 30.0-g sample of liquid mercury at its normal boiling point of 357 °C.

$$\text{boiling point} = 357 \text{ °C} \qquad \Delta H_{vap}(357 \text{ °C}) = 59.3 \text{ kJ/mol}$$

$$\text{melting point} = -38.9 \text{ °C} \qquad \Delta H_{fus}(-38.9 \text{ °C}) = 2.33 \text{ kJ/mol}$$

$$\text{specific heat (liquid)} = 0.139 \text{ J/g} \cdot \text{°C}$$

Solution:

You are asked to calculate the amount of heat needed to vaporize a sample of liquid mercury at its normal boiling point.

You are given the mass of the mercury, its boiling and melting points, its heat of vaporization and heat of fusion, and the specific heat of the liquid.

This is a constant-temperature process in which the liquid is vaporized at its normal boiling point. Use the heat of vaporization of mercury to calculate the amount of heat needed to vaporize the liquid.

$$30.0 \text{ g} \times \frac{1 \text{ mol Hg}}{200.6 \text{ g}} \times \frac{59.3 \text{ kJ}}{1 \text{ mol Hg}} = 8.87 \text{ kJ}$$

Is your answer reasonable? The quantity of liquid mercury being vaporized is less than one mol, so it should require less energy to vaporize than the heat of vaporization of mercury, the amount of energy required to vaporize one mole of liquid.

Video Solution

Tutored Practice
Problem 5.3.4

Explore heating curves for benzene and water.

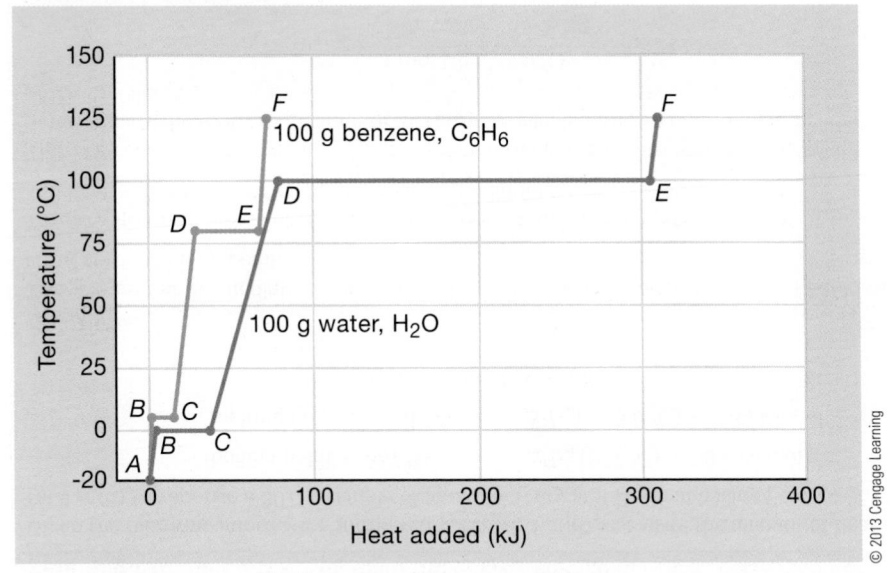

Heating curves for benzene and water

We now have the tools needed to calculate both the amount of thermal energy required to warm an object in a particular physical state and the amount of thermal energy required to change the physical state of a substance. Interactive Figure 5.3.6 shows the energy involved in heating 100-g samples of benzene and water over a large temperature range.

When 100-g samples of the solids at −20 °C are heated, the following processes occur between the indicated points on the heating curves:

A to B: Temperature increases as added heat warms the solid to its melting point.
B to C: Temperature remains constant as solid is converted to liquid at the melting point.
C to D: Temperature increases as added heat warms the liquid to its boiling point.
D to E: Temperature remains constant as liquid is converted to vapor at the boiling point.
E to F: Temperature increases as added heat warms the vapor above its boiling point.

As shown in the following example, the amount of heat required to heat a substance can be calculated when specific heat capacities, enthalpy of fusion, and enthalpy of vaporization values are known.

Example Problem 5.3.5 Calculate the energy change for a pure substance over a temperature range.

The following information is given for ethanol at 1 atm:

boiling point = 78.40 °C ΔH_{vap} (78.40 °C) = 837.0 J/g

melting point = −114.5 °C $\Delta H_{fus}(-114.5\ °C) = 109.0$ J/g

specific heat (gas) = 1.430 J/g · °C   specific heat (liquid) = 2.460 J/g · °C

A 33.50-g sample of liquid ethanol is initially at 13.50 °C. If the sample is heated at constant pressure ($P = 1$ atm), calculate the amount of energy needed to raise the temperature of the sample to 94.50 °C.

Solution:

You are asked to calculate the total energy required to heat a sample of liquid ethanol.

You are given the mass of the ethanol sample, its initial and final temperature, and physical data for ethanol in the liquid and vapor phases.

This is a three step process: (1) raise the temperature of the liquid to its boiling point, (2) vaporize the liquid at its boiling point, and (3) raise the temperature of the resulting gas to the final temperature.

1. Calculate the amount of energy required to heat the liquid from 13.50 °C to 78.40 °C.

$$q(1) = m \times c_{liquid} \times \Delta T = (33.50\ g)(2.460\ J/g · °C)(78.40\ °C - 13.50\ °C) = 5348\ J$$

2. Calculate the amount of energy required to vaporize the liquid at its boiling point.

$$q(2) = m \times \Delta H_{vap} = (33.50\ g)(837.0\ J/g) = 2.804 \times 10^4\ J$$

3. Calculate the amount of energy required to heat the vapor from 78.40 °C to 94.50 °C.

$$q(3) = m \times c_{gas} \times \Delta T = (33.50\ g)(1.430\ J/g · °C)(94.50\ °C - 78.40\ °C) = 771.3\ J$$

The total amount of energy needed to heat the sample of ethanol is the sum of the energy required for the three steps.

$$q_{total} = q(1) + q(2) + q(3) = 5348\ J + (2.804 \times 10^4\ J) + 771.3\ J = 3.416 \times 10^4\ J = 34.16\ kJ$$

Is your answer reasonable? Over the temperature range in this problem, ethanol is heated, converted from liquid to vapor phase and then heated in the vapor phase. All three changes involve a great deal of energy, especially the conversion of a liquid to a vapor. The total amount of energy required to raise the temperature of the ethanol is expected to be large.

Video Solution

Tutored Practice
Problem 5.3.5

Section 5.3 Mastery

5.4 Enthalpy Changes and Chemical Reactions

5.4a Enthalpy Change for a Reaction

We have examined the ways in which energy transferred to or from a sample can change its temperature or phase. Of far greater interest and importance is the energy associated with chemical change.

When elements or compounds combine to form a new compound, bonds between the atoms in the reactants are broken and new bonds form in the products. Breaking or forming chemical bonds leads to a change in enthalpy, which is observed as an absorption or release of thermal energy, respectively.

For example, when zinc and sulfur react, a large amount of thermal energy is released (Interactive Figure 5.4.1).

$$Zn(s) + S(s) \rightarrow ZnS(s) + 205.98 \text{ kJ}$$

When measured at constant pressure, the energy released is the **enthalpy change for a reaction** (ΔH_{rxn}).

$$\Delta H_{rxn} = -205.98 \text{ kJ}$$

The negative sign of ΔH_{rxn} indicates that the reaction is exothermic. The enthalpy of the system (zinc and sulfur) decreases, and energy is released.

In contrast, when nitrogen and oxygen react to form nitrogen dioxide, thermal energy is absorbed.

$$N_2(g) + O_2(g) + 66.36 \text{ kJ} \rightarrow 2 \text{ NO}_2(g)$$

This reaction is endothermic, and the enthalpy change for the reaction has a positive value.

$$\Delta H_{rxn} = +66.36 \text{ kJ}$$

The enthalpy change for a reaction is the energy associated when the reaction occurs as written. That is, 66.36 kJ of energy is absorbed when 1 mol of N_2 reacts with 2 mol of O_2 to form 2 mol of NO_2.

Investigate energy change of zinc and sulfur reaction.

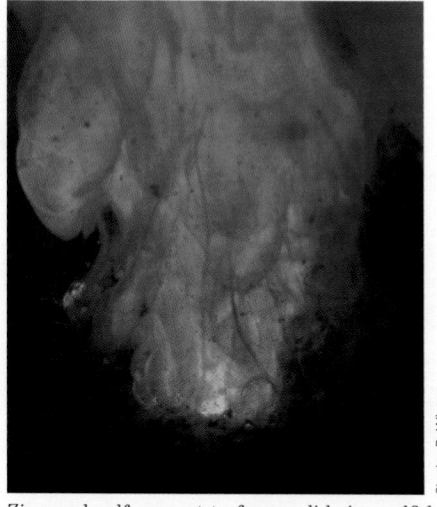

Charles D. Winters

Zinc and sulfur react to form solid zinc sulfide.

Example Problem 5.4.1 Calculate enthalpy change for a chemical reaction.

When CO reacts with NO according to the following reaction, 373 kJ of energy is evolved for each mole of CO that reacts. Use this information to calculate the enthalpy change for the following reaction.

$$2\,CO(g) + 2\,NO(g) \rightarrow 2\,CO_2(g) + N_2(g) \qquad \Delta H_{rxn} = ?$$

Solution:

You are asked to calculate the enthalpy change for a given reaction.

You are given the amount of energy released when a given amount of reactant is consumed.

Energy is evolved in this reaction, so the sign of ΔH_{rxn} is negative. The equation shows the reaction of 2 mol of CO, so the energy change for the reaction is

$$\Delta H_{rxn} = 2\ \text{mol CO} \times \frac{-373\ \text{J}}{1\ \text{mol CO}} = -746\ \text{kJ}$$

Is your answer reasonable? The energy change should be negative because the reaction gives off energy. The amount of energy released is twice that given in the problem because of the stoichiometry of the chemical equation.

Video Solution

Tutored Practice
Problem 5.4.1

5.4b   Enthalpy Change and Chemical Equations

The enthalpy change for a given reaction is a function of the stoichiometry of the chemical reaction. For instance, Example Problem 5.4.1 showed that the reaction of carbon monoxide with nitric oxide releases 373 kJ of energy per mole of CO.

$$2\,CO(g) + 2\,NO(g) \rightarrow 2\,CO_2(g) + N_2(g) \qquad \Delta H_{rxn} = -746\ \text{kJ}$$

Consider, however, the reaction of 0.250 mol carbon monoxide. The amount of energy released can be calculated from the enthalpy of reaction.

$$0.250\ \text{mol CO} \times \frac{746\ \text{kJ}}{2\ \text{mol CO}} = 93.3\ \text{kJ energy released}$$

This calculation is valid because enthalpy is an **extensive variable**, which means that it is dependent on the amount of substance present; therefore, a change in enthalpy is also dependent on the amount of substance that reacts. The reaction of 250. mol of CO with NO, for example, results in the release of 9.33×10^4 kJ of energy.

Knowing that enthalpy is an extensive variable allows us to look at the relationship between the enthalpy change for a reaction and the chemical equation. (◄Flashback to Section 3.3c Reaction Stoichiometry) Consider, for example, the production of ammonia from elemental nitrogen and hydrogen.

$$N_2(g) + 3\,H_2(g) \rightarrow 2\,NH_3(g) \qquad \Delta H_{rxn} = -23.1\ kJ$$

In this reaction, 23.1 kJ of energy is released when 2 mol of NH_3 is formed. The formation of 1 mol of NH_3,

$$\tfrac{1}{2}\,N_2(g) + \tfrac{3}{2}\,H_2(g) \rightarrow NH_3(g) \qquad \Delta H_{rxn} = -11.6\ kJ$$

releases half as much energy, 11.6 kJ. Thus, *when a chemical equation is multiplied by a constant, the enthalpy change is also multiplied by that constant.*

Consider the decomposition of ammonia to form elemental nitrogen and hydrogen.

$$2\,NH_3(g) \rightarrow N_2(g) + 3\,H_2(g) \qquad \Delta H_{rxn} = +23.1\ kJ$$

This is the reverse of the equation for the formation of ammonia from elemental nitrogen and hydrogen. Because the original equation represented an exothermic reaction (heat is evolved to the surroundings), the reverse reaction is endothermic (heat is absorbed from the surroundings). *When the reactants and products in a chemical equation are reversed, the magnitude of ΔH is identical but the sign of ΔH is reversed.*

Example Problem 5.4.2 Use reaction stoichiometry to calculate enthalpy change.

The reaction of HCl with O_2 is exothermic.

$$4\,HCl(g) + O_2(g) \rightarrow 2\,H_2O(\ell) + 2\,Cl_2(g) \qquad \Delta H(1) = -202.4\ kJ$$

Calculate the enthalpy change for the reaction of water with elemental chlorine to produce HCl and O_2.

$$H_2O(\ell) + Cl_2(g) \rightarrow 2\,HCl(g) + \tfrac{1}{2}\,O_2(g) \qquad \Delta H(2) = ?$$

Solution:

You are asked to calculate the enthalpy change for a given reaction.

You are given the enthalpy change for the reaction under a different set of stoichiometric conditions.

The second reaction is related to the first reaction by (1) reversing reactants and products and (2) multiplication by a constant ($\times\ \tfrac{1}{2}$). Thus, the enthalpy change for the second reaction is equal to $-\Delta H(1) \times \tfrac{1}{2}$.

$$\Delta H(2) = -\Delta H(1) \times \tfrac{1}{2} = -(-202.4\ kJ) \times \tfrac{1}{2} = 101.2\ kJ$$

Is your answer reasonable? The second reaction is the reverse of the exothermic first reaction, so it is endothermic ($+\Delta H$). It also involves only half as many reactants and products, so the amount of energy absorbed is half as much as the amount of energy released in the first reaction.

Video Solution

Tutored Practice
Problem 5.4.2

5.4c Constant-Pressure Calorimetry

Tabulated ΔH_{rxn} values are derived from experimental data. Experiments that measure heat exchange are referred to collectively as **calorimetry** experiments, and the measurement device is called a **calorimeter**. There are two important types of calorimetry experiments: constant-pressure calorimetry (also known as coffee cup calorimetry) and constant-volume calorimetry (also known as bomb calorimetry).

In constant-pressure calorimetry, the heat evolved or absorbed during a chemical change is measured under conditions in which the pressure does not change. This typically means that the chemical change is allowed to occur in a reaction vessel that is open to the atmosphere, and thus the pressure remains essentially constant. Under these conditions, the heat evolved or absorbed, q, is equal to the enthalpy change for the reaction, ΔH.

Heat exchange under conditions of constant pressure can be measured using a simple coffee cup calorimeter (Interactive Figure 5.4.2). The reaction vessel consists of two nested Styrofoam coffee cups that are covered with a lid. A thermometer measures the temperature change. Most coffee cup calorimetry experiments involve a solvent such as water and a stirring device ensures thorough mixing in the reaction vessel.

In a constant-pressure calorimetry experiment, the system consists of the chemicals undergoing change and the surroundings consist of the other contents of the calorimeter, the calorimeter itself, and all materials around the calorimeter. However, the calorimeter is designed to minimize energy transfer, so it is possible to assume that the surroundings consist only of the contents of the calorimeter. Therefore, the temperature change measured by the thermometer allows calculation of $q_{surroundings}$ for a chemical change in a coffee cup calorimeter.

$$q_{surroundings} = (\text{mass of calorimeter contents})(\text{specific heat of solution})(\Delta T)$$

Because $0 = q_{system} + q_{surroundings}$, the enthalpy change for a chemical process taking place in a coffee cup calorimeter is

$$\Delta H = q_{system} = -q_{surroundings}$$

In a typical experiment, enthalpy change is reported as kilojoules per mole of reactant (kJ/mol). Note that in constant-pressure calorimetry experiments, an increase in temperature indicates an exothermic reaction ($q_{surroundings}$ is positive, q_{system} and ΔH are negative) and a decrease in temperature indicates an endothermic reaction ($q_{surroundings}$ is negative, q_{system} and ΔH are positive). The sign conventions in constant-pressure calorimetry experiments are summarized in Table 5.4.1.

Interactive Figure 5.4.2

Explore constant-pressure calorimetry.

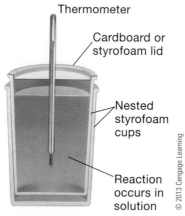

A constant-pressure calorimeter

Table 5.4.1 Sign Conventions in Constant-Pressure Calorimetry

Exothermic Reactions	Endothermic Reactions
Temperature increases $(+\Delta T)$	Temperature decreases $(-\Delta T)$
$+q_{surroundings}$	$-q_{surroundings}$
$-q_{system}$	$+q_{system}$
ΔH is negative	ΔH is positive

As shown in the following example, one application of constant-pressure calorimetry is an experiment to determine **enthalpy of dissolution** ($\Delta H_{dissolution}$), the amount of energy as heat involved in the process of solution formation.

Example Problem 5.4.3 Use constant-pressure calorimetry to determine enthalpy change.

Ammonium chloride is very soluble in water. When 4.50 g NH_4Cl is dissolved in 53.00 g of water, the temperature of the solution decreases from 20.40 °C to 15.20 °C. Calculate the enthalpy of dissolution of NH_4Cl (in kJ/mol).

Assume that the specific heat of the solution is 4.18 J/g · °C and that the heat absorbed by the calorimeter is negligible.

Solution:

You are asked to calculate the enthalpy change for the dissolution of ammonium chloride.

You are given the masses of the solid and water, and the temperature change that occurs when the two are combined.

First calculate the energy change for the surroundings ($q_{solution}$) in the coffee cup calorimeter.

$$c_{solution} = 4.18 \text{ J/g} \cdot {}^{\circ}\text{C}$$

$$m_{solution} = 4.50 \text{ g} + 53.00 \text{ g} = 57.50 \text{ g}$$

$$\Delta T = T_{final} - T_{initial} = 15.20 \text{ }^{\circ}\text{C} - 20.40 \text{ }^{\circ}\text{C} = -5.20 \text{ }^{\circ}\text{C}$$

$$q_{solution} = m \times c_{solution} \times \Delta T = (57.50 \text{ g})(4.18 \text{ J/g} \cdot {}^{\circ}\text{C})(-5.20 \text{ }^{\circ}\text{C}) = -1250 \text{ J}$$

Next calculate q for the dissolution of NH_4Cl, q_{system}.

$$q_{dissolution} + q_{solution} = 0$$

$$q_{dissolution} = -q_{solution} = 1250 \text{ J}$$

Finally, calculate the amount of NH_4Cl dissolved (in mol) and ΔH for the dissolution of NH_4Cl (kJ/mol).

$$4.50 \text{ g } NH_4Cl \times \frac{1 \text{ mol } NH_4Cl}{53.49 \text{ g}} = 0.0841 \text{ mol } NH_4Cl$$

$$\Delta H_{dissolution} = \frac{q_{dissolution}}{\text{mol } NH_4Cl} = \frac{1250 \text{ J}}{0.0841 \text{ mol}} \times \frac{1 \text{ kJ}}{10^3 \text{ J}} = 14.9 \text{ kJ/mol}$$

Is your answer reasonable? The temperature of the solution decreased, indicating an endothermic process and a positive value for $\Delta H_{dissolution}$.

Video Solution

Tutored Practice
Problem 5.4.3

Although most constant-pressure calorimetry experiments are performed in a laboratory, Interactive Figure 5.4.3 shows how you can use this type of experiment to determine practical information about a heat source in your home.

5.4d Constant-Volume Calorimetry

Many constant-pressure calorimetry experiments assume that no heat is transferred to the calorimeter or to the outside surroundings. A more precise experiment is performed using constant-volume calorimetry. Constant-volume experiments can be performed using a bomb calorimeter (Interactive Figure 5.4.4), a reaction vessel designed to measure the energy change involved in combustion reactions.

The chemical reaction studied in a bomb calorimeter takes place inside a sealed steel vessel (the bomb) that is completely surrounded by a water bath, and an insulated jacket surrounds the water bath. The temperature change of the chemical reaction is determined by measuring the temperature change of the water surrounding the bomb.

Because the reaction studied in a bomb calorimetry experiment takes place in a sealed bomb, this is a constant-volume calorimetry experiment, not a constant-pressure experiment. Under constant volume conditions, the heat evolved or absorbed by a chemical change, q, is equal to the change in energy, ΔE, not the change in enthalpy, ΔH. However, the difference between ΔE and ΔH is quite small for most chemical reactions.

The heat evolved in a bomb calorimeter combustion reaction ($q_{reaction}$) is absorbed by the steel bomb (q_{bomb}) and the water in the water bath (q_{water}).

$$0 = q_{reaction} + q_{bomb} + q_{water}$$

The heat absorbed by the steel bomb is calculated from the temperature change of the water bath (ΔT) and the bomb heat capacity (c_{bomb}, J/°C). The heat absorbed by the water bath is calculated from the mass of water, the specific heat of water, and the water bath temperature change.

Example Problem 5.4.4 Use constant-volume calorimetry to determine energy change.

A 0.444-g sample of sucrose ($C_{12}H_{22}O_{11}$) is burned in a bomb calorimeter and the temperature increases from 20.00 °C to 22.06 °C. The calorimeter contains 748 g of water, and the bomb has a heat capacity of 420. J/°C. Calculate ΔE for the combustion reaction per mole of sucrose burned (kJ/mol).

$$C_{12}H_{22}O_{11}(s) + 12\ O_2(g) \rightarrow 12\ CO_2(g) + 11\ H_2O(\ell)$$

Solution:

You are asked to calculate the energy change (in kJ/mol sucrose) for the combustion of sucrose.

Interactive Figure 5.4.3

Measure the temperature of a fire using calorimetry.

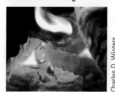

Charles D. Winters

How hot is this fire?

Interactive Figure 5.4.4

Explore constant-volume calorimetry.

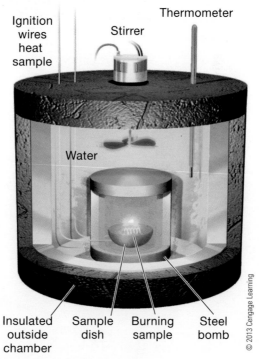

Ignition wires heat sample

Stirrer

Thermometer

Water

Insulated outside chamber

Sample dish

Burning sample

Steel bomb

© 2013 Cengage Learning

A constant-volume calorimeter

Example Problem 5.4.4 *(continued)*

You are given the mass of sucrose and bomb calorimeter data (mass of the water, temperature change of the water, and calorimeter heat capacity).

First, calculate the energy absorbed by the water bath and the bomb.

Water bath:

$$c_{water} = 4.184 \text{ J/g} \cdot \text{°C}$$

$$m_{water} = 748 \text{ g}$$

$$\Delta T = T_{final} - T_{initial} = 22.06 \text{ °C} - 20.00 \text{ °C} = 2.06 \text{ °C}$$

$$q_{water} = m \times c_{water} \times \Delta T = (748 \text{ g})(4.184 \text{ J/g} \cdot \text{°C})(206 \text{ °C}) = 6450 \text{ J}$$

Bomb:

$$c_{bomb} = 420. \text{ J/°C}$$

$$\Delta T = T_{final} - T_{initial} = 22.06 \text{ °C} - 20.00 \text{ °C} = 2.06 \text{ °C}$$

$$q_{bomb} = c_{bomb} \times \Delta T = (420. \text{ J/g} \cdot \text{°C})(2.06 \text{ °C}) = 865 \text{ J}$$

Next, calculate the energy released by the combustion reaction.

$$0 = q_{reaction} + q_{bomb} + q_{water}$$

$$q_{reaction} = -(q_{bomb} + q_{water}) = -(6450 \text{ J} + 865 \text{ J}) = -7320 \text{ J}$$

Finally, calculate the amount (in mol) of sucrose burned in the combustion reaction and the energy change for the reaction.

$$0.444 \text{ g } C_{12}H_{22}O_{11} \times \frac{1 \text{ mol } C_{12}H_{22}O_{11}}{342.3 \text{ g}} = 0.00130 \text{ mol } C_{12}H_{22}O_{11}$$

$$\Delta E = \frac{q_{reaction}}{\text{mol } C_{12}H_{22}O_{11}} = \frac{-7320 \text{ J}}{0.00130 \text{ mol}} \times \frac{1 \text{ kJ}}{10^3 \text{ J}} = -5640 \text{ kJ/mol}$$

Because the difference between ΔE and ΔH is quite small, this energy change can be taken as the enthalpy of combustion (ΔH_{comb}) for sucrose.

Is your answer reasonable? This is an exothermic reaction; temperature increased, and $q_{reaction}$ and ΔE are negative. In addition, the combustion of a hydrocarbon typically produces a large amount of energy per mole of compound.

Video Solution

Tutored Practice
Problem 5.4.4

Section 5.4 Mastery

5.5 Hess's Law

5.5a Hess's Law

Enthalpy is a **state function**, which means its value for a system depends only on the current state of the system and not on its history. Therefore, the change in enthalpy for a process depends only on the initial and final states for a system and not on the path between the two. For example, the altitude of a given location is like a state function: the difference in altitude between the top and bottom of a mountain has a fixed value, but the length of the hike from bottom to top does not, because there is more than one possible route, each with a different length. The state function nature of enthalpy is expressed in **Hess's law**, which states that *if a chemical reaction can be expressed as the sum of two or more chemical reactions, the enthalpy change for the net reaction is equal to the sum of the enthalpy changes for the individual steps.*

Consider the enthalpy changes for three reactions involved in the formation of formic acid (Interactive Figure 5.5.1).

Reaction 1: $C(s) + \frac{1}{2} O_2(g) \rightarrow CO(g)$ $\Delta H_1 = -110.5$ kJ

Reaction 2: $CO(g) + H_2O(\ell) \rightarrow HCO_2H(g)$ $\Delta H_2 = +33.7$ kJ

Reaction 3: $HCO_2H(g) \rightarrow HCO_2H(\ell)$ $\Delta H_3 = -62.9$ kJ

The net reaction is the production of formic acid from carbon, oxygen, and water.

Net reaction: $C(s) + \frac{1}{2} O_2(g) + H_2O(\ell) \rightarrow HCO_2H(\ell)$

To apply Hess's law, sum reactions 1, 2, and 3 and cancel any species common to the products and reactants in the reaction.

1 + 2 + 3:
$C(s) + \frac{1}{2} O_2(g) + CO(g) + H_2O(\ell) + HCO_2H(g) \rightarrow CO(g) + HCO_2H(g) + HCO_2H(\ell)$

Cancel:
$C(s) + \frac{1}{2} O_2(g) + \cancel{CO(g)} + H_2O(\ell) + \cancel{HCO_2H(g)} \rightarrow \cancel{CO(g)} + \cancel{HCO_2H(g)} + HCO_2H(\ell)$

Net reaction: $C(s) + \frac{1}{2} O_2(g) + H_2O(\ell) \rightarrow HCO_2H(\ell)$

Calculate enthalpy change using Hess's law.

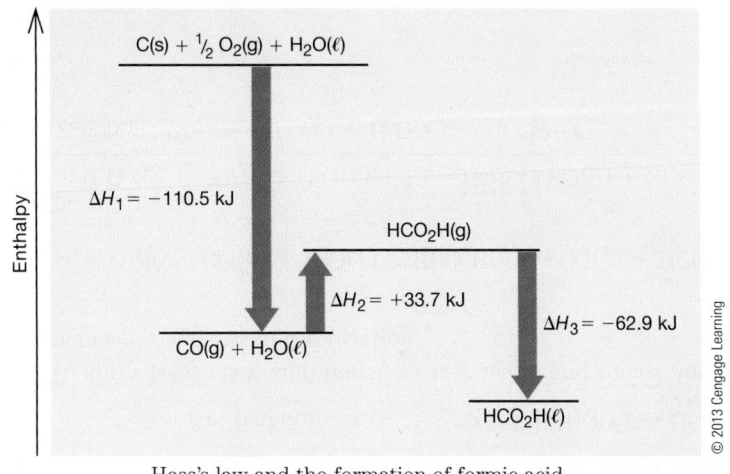

Hess's law and the formation of formic acid

The net reaction is the sum of reactions 1, 2, and 3, and therefore the enthalpy change for the net reaction is equal to the sum of the enthalpy changes for reactions 1, 2, and 3.

$$\Delta H_{rxn} = \Delta H_1 + \Delta H_2 + \Delta H_3 = (-110.5 \text{ kJ}) + 33.7 \text{ kJ} + (-62.9 \text{ kJ}) = -139.7 \text{ kJ}$$

As shown in the following example, reactions might need to be reversed or multiplied by a constant before summing to obtain a net reaction.

Example Problem 5.5.1 Use Hess's law to calculate enthalpy change.

Given the following two reactions,

Reaction 1: $SnCl_2(s) + Cl_2(g) \rightarrow SnCl_4(\ell)$ $\Delta H(1) = -195$ kJ

Reaction 2: $TiCl_2(s) + Cl_2(g) \rightarrow TiCl_4(\ell)$ $\Delta H(2) = -273$ kJ

calculate the enthalpy change for the following chlorine exchange reaction.

Reaction 3: $SnCl_2(s) + TiCl_4(\ell) \rightarrow SnCl_4(\ell) + TiCl_2(s)$ $\Delta H_{net} = ?$

Solution:

You are asked to calculate the enthalpy change for a given reaction.

You are given the enthalpy change for two other reactions.

Example Problem 5.5.1 *(continued)*

In order to use Hess's law, the sum of two or more chemical reactions must result in the net reaction. In this example, reversing the second reaction and adding it to the first reaction results in the net reaction. Notice that reversing the second reaction requires changing the sign of $\Delta H(2)$. Because the net reaction can be expressed as the sum of two chemical reactions, the enthalpy change for the net reaction is equal to the sum of the enthalpy changes for the individual steps.

$$SnCl_2(s) + Cl_2(g) \rightarrow SnCl_4(\ell) \quad \Delta H(1) = -195 \text{ kJ}$$

$$TiCl_4(\ell) \rightarrow TiCl_2(s) + Cl_2(g) \quad \Delta H(2)' = -\Delta H(2) = +273 \text{ kJ}$$

$$SnCl_2(s) + TiCl_4(\ell) \rightarrow SnCl_4(\ell) + TiCl_2(s) \quad \Delta H_{net} = \Delta H(1) + \Delta H(2)' = -195 \text{ kJ} + 273 \text{ kJ}$$
$$= 78 \text{ kJ}$$

Video Solution

Tutored Practice
Problem 5.5.1

Section 5.5 Mastery

5.6 Standard Heats of Reaction

5.6a Standard Heat of Formation

The enthalpy change for a chemical reaction varies with experimental conditions such as temperature, pressure, and solution concentration. Therefore, it is helpful to specify a standard set of conditions under which to tabulate enthalpy change, as well as a standard set of enthalpy change values that can be used to calculate the enthalpy change for a reaction.

Enthalpy change values that are tabulated under standard conditions are indicated with a superscripted ° symbol, as in $\Delta H°_{rxn}$. **Standard state** conditions for enthalpy values are:

- Gases, liquids, and solids in their pure form at a pressure of 1 bar at a specified temperature

- Solutions with concentrations of 1 mol/L at a specified temperature

Notice that standard state conditions do not specify a temperature. Most standard enthalpy values are tabulated at 25 °C (298 K), but it is possible to report standard enthalpy change values at other temperatures.

Creating a list of all possible standard enthalpy values is impossible. Fortunately, tabulating only one type of standard enthalpy change is all that is needed to calculate the standard enthalpy change for almost any chemical reaction. The **standard heat of formation** (or standard enthalpy of formation) for a species is the enthalpy change for the formation of one mole of a species from its constituent elements in their most stable form. Standard heat of formation values are denoted with the symbol ΔH_f°.

For example, the standard heat of formation for solid N_2O_5 is the enthalpy change when one mole of the compound is formed from its constituent elements, nitrogen and oxygen, both in their standard states.

$$N_2(g) + \tfrac{5}{2} O_2(g) \rightarrow N_2O_5(s) \qquad \Delta H^\circ = \Delta H_f^\circ = -43.1 \text{ kJ/mol}$$

Notice that the chemical reaction for the standard heat of formation of N_2O_5 includes a fractional coefficient. It is often necessary to include fractional coefficients in chemical reactions representing ΔH_f° values because the enthalpy change must be for the formation of one mole of a species.

A chemical reaction that does not represent a standard heat of formation is the formation of gaseous N_2O_4 from nitric oxide and oxygen. The enthalpy change for this reaction is not a ΔH_f° value because NO is not a constituent element of N_2O_4.

$$2 \text{ NO}(g) + O_2(g) \rightarrow N_2O_4(g) \qquad \Delta H^\circ = -171.3 \text{ kJ}$$

Some standard heat of formation values are shown in Table 5.6.1 and a more complete table is found in the Reference Tools. There are some important details that should be noted about ΔH_f° values.

- ΔH_f° values have units of kJ/mol because each is the enthalpy change for the formation of one mole of a chemical species.

- The ΔH_f° value for an element in its standard state is equal to 0 kJ/mol. For example, the reaction for ΔH_f° of elemental bromine is written

$$Br_2(\ell) \rightarrow Br_2(\ell)$$

There is no change from reactants to products, so $\Delta H_f^\circ = 0$ kJ/mol.

- Most ΔH_f° values are negative. This indicates that for most species, the formation from elements in their standard states is an exothermic process.

Table 5.6.1 Standard Heat of Formation Values for Selected Compounds (kJ/mol) at 25 °C

Formula	Name	ΔH_f° (kJ/mol)	Formula	Name	ΔH_f° (kJ/mol)
$Al_2O_3(s)$	Aluminum oxide	−1675.7	$HF(g)$	Hydrogen fluoride	−271.1
$BaCO_3(s)$	Barium carbonate	−1219.0	$HCl(g)$	Hydrogen chloride	−92.3
$CaCO_3(s)$	Calcium carbonate	−1206.9	$HBr(g)$	Hydrogen bromide	−36.3
$CaO(s)$	Calcium oxide	−635.1	$KCl(s)$	Potassium chloride	−436.7
$CCl_4(\ell)$	Carbon tetrachloride	−135.4	$KClO_3(s)$	Potassium chlorate	−397.7
$CH_4(g)$	Methane	−74.8	$MgO(s)$	Magnesium oxide	−601.7
$C_2H_5OH(\ell)$	Ethanol	−277.7	$Mg(OH)_2(s)$	Magnesium hydroxide	−924.5
$CO(g)$	Carbon monoxide	−110.5	$NaCl(s)$	Sodium chloride	−411.2
$CO_2(g)$	Carbon dioxide	−393.5	$NaBr(s)$	Sodium bromide	−361.0
$C_2H_2(g)$	Acetylene (ethyne)	226.7	$NaI(s)$	Sodium iodide	−288.0
$C_2H_4(g)$	Ethylene (ethene)	52.3	$NH_3(g)$	Ammonia	−46.1
$C_2H_6(g)$	Ethane	−84.7	$N_2H_4(\ell)$	Hydrazine	50.6
$C_3H_8(g)$	Propane	−103.8	$NO(g)$	Nitrogen monoxide	90.3
$C_6H_6(\ell)$	Hexane	49.0	$NO_2(g)$	Nitrogen dioxide	33.2
$C_6H_{12}O_6(s)$	Glucose	−1275.0	$PCl_3(g)$	Phosphorus trichloride	−287.0
$CuCO_3(s)$	Copper(II) carbonate	−595.0	$SiCl_4(g)$	Silicon tetrachloride	−657.0
$Fe_2O_3(s)$	Iron(III) oxide (hematite)	−824.2	$SiO_2(s)$	Silicon dioxide (quartz)	−910.9
$FeSO_4(s)$	Iron(II) sulfate	−929.0	$SnCl_4(\ell)$	Tin(IV) chloride	−511.3
$H_2O(g)$	Water (vapor)	−241.8	$SO_2(g)$	Sulfur dioxide	−296.8
$H_2O(\ell)$	Water (liquid)	−285.8	$SO_3(g)$	Sulfur trioxide	−395.7
$H_2O_2(\ell)$	Hydrogen peroxide	−187.8	$ZnCl_2(s)$	Zinc chloride	−415.1

Example Problem 5.6.1 Use and interpret standard heats of formation.

a. Write the balanced chemical equation that represents the standard heat of formation of $KClO_3(s)$ at 298 K.
b. The standard enthalpy change for the following reaction is 2261 kJ at 298 K.

$$2\ Na_2CO_3(s) \rightarrow 4\ Na(s) + 2\ C(graphite) + 3\ O_2(g) \qquad \Delta H^\circ_{rxn} = 2261\ kJ$$

What is the standard heat of formation of $Na_2CO_3(s)$?

Solution:

a. **You are asked** to write an equation for the standard heat of formation of a compound.

You are given the compound formula.

The standard heat of formation (or standard enthalpy of formation) of a substance in a specified state at 298 K is the enthalpy change for the reaction in which one mole of the substance is formed from the elements in their stable forms at 1 bar and 298 K. Potassium is a solid and chlorine and oxygen are diatomic gases at 1 bar and 298 K.

$$K(s) + \tfrac{1}{2}\ Cl_2(g) + \tfrac{3}{2}\ O_2(g) \rightarrow KClO_3(s)$$

Is your answer reasonable? The equation shows the formation of only one mole of product and the reactants are all elements in their most stable forms at 1 bar and 298 K.

b. **You are asked** to determine the standard heat of formation of a compound.

You are given the enthalpy change under standard conditions for a reaction involving that compound.

The equation that represents the standard heat of formation of $Na_2CO_3(s)$ is

$$2\ Na(s) + C(graphite) + \tfrac{3}{2}\ O_2(g) \rightarrow Na_2CO_3(s)$$

Reversing the equation in the problem and multiplying by ½ results in the equation for the standard heat of formation of Na_2CO_3, so

$$\Delta H_f^\circ = -\tfrac{1}{2}(\Delta H^\circ_{rxn}) = -\tfrac{1}{2}(2261\ kJ) = -1131\ kJ/mol$$

Is your answer reasonable? The given equation shows sodium carbonate as a reactant, and the standard heat of formation is the energy change for the formation of one mole of a compound. The given reaction is endothermic as written, so the standard heat of formation for this compound is exothermic.

Video Solution

Tutored Practice
Problem 5.6.1

5.6b Using Standard Heats of Formation

Standard heat of formation values, when combined with Hess's law, are used to calculate a **standard enthalpy of reaction** (ΔH°_{rxn}), the enthalpy change for a reaction under standard conditions. Consider the following reaction, at 25 °C.

$$2\,NaHCO_3(s) \rightarrow Na_2CO_3(s) + CO_2(g) + H_2O(\ell)$$

The reaction can be thought to occur in two steps. First, reactants are broken down into constituent elements in their standard states, and second, the elements are recombined to form products.

Step 1: $2 \times [NaHCO_3(s) \rightarrow Na(s) + \frac{1}{2}\,H_2(g) + C(graphite) + \frac{3}{2}\,O_2(g)]$
$\Delta H^{\circ} = 2 \times (-\Delta H_f^{\circ}[NaHCO_3(s)]) = (2\,mol\,NaHCO_3)[-(-950.81\,kJ/mol)]$

Step 2: $2\,Na(s) + C(graphite) + \frac{3}{2}\,O_2(g) \rightarrow Na_2CO_3(s)$
$\Delta H^{\circ} = \Delta H_f^{\circ}[Na_2CO_3(s)] = (1\,mol\,Na_2CO_3)(-1130.68\,kJ/mol)$

$C(graphite) + O_2(g) \rightarrow CO_2(g)$
$\Delta H^{\circ} = \Delta H_f^{\circ}[CO_2(g)] = (1\,mol\,CO_2)(-393.51\,kJ/mol)$

$H_2(g) + \frac{1}{2}\,O_2(g) \rightarrow H_2O(\ell)$
$\Delta H^{\circ} = \Delta H_f^{\circ}[H_2O(\ell)] = (1\,mol\,H_2O)(-285.8\,kJ/mol)$

Both steps involve ΔH_f° values. In step 1, the sign of $\Delta H_f^{\circ}[NaHCO_3(s)]$ is reversed because the standard heat of formation reaction is reversed. The value is also multiplied by 2 because 2 mol $NaHCO_3$ is broken down into constituent elements. Step 2 reactions are standard heat of formation reactions, so the enthalpy change for each is equal to ΔH_f°.

Summing the four reactions in steps 1 and 2 results in the net reaction. Therefore, using Hess's law, the enthalpy change for the net reaction is equal to the sum of the enthalpy changes for the reactions in steps 1 and 2.

$$\Delta H^{\circ}_{rxn} = 2 \times (-\Delta H_f^{\circ}[NaHCO_3(s)]) + \Delta H_f^{\circ}[Na_2CO_3(s)] + \Delta H_f^{\circ}[CO_2(g)] + \Delta H_f^{\circ}[H_2O(\ell)]$$

$$= 1901.62\,kJ + (-1130.68\,kJ) + (-393.51\,kJ) + (-285.8\,kJ)$$

$$= 91.6\,kJ$$

A general formula used to calculate the standard enthalpy change for a reaction from standard heats of formation is

$$\Delta H^{\circ}_{rxn} = \Sigma \Delta H_f^{\circ}(\text{products}) - \Sigma \Delta H_f^{\circ}(\text{reactants}) \qquad \textbf{(5.6)}$$

where each ΔH_f° value is multiplied by the stoichiometric coefficient in the balanced chemical equation. Note that the sign of the ΔH_f° values for the reactants is reversed. As shown in the calculation above, this is due to the need to reverse the standard heat of formation reaction in order to break apart reactants into constituent elements. Notice also that, because each ΔH_f° value is multiplied by the number of moles of the species in the balanced equation, the standard enthalpy change calculated using Equation 5.6 has units of kJ. Using Equation 5.6 to calculate the standard enthalpy change for the reaction above,

$$\Delta H^{\circ}_{rxn} = (1 \text{ mol Na}_2\text{CO}_3)(\Delta H_f^{\circ}[\text{Na}_2\text{CO}_3(s)]) + (1 \text{ mol CO}_2)(\Delta H_f^{\circ}[\text{CO}_2(g)])$$

$$+ (1 \text{ mol H}_2\text{O})(\Delta H_f^{\circ}[\text{H}_2\text{O}(\ell)]) - (2 \text{ mol NaHCO}_3)(\Delta H_f^{\circ}[\text{NaHCO}_3(s)])$$

$$= (1 \text{ mol Na}_2\text{CO}_3)(-1130.68 \text{ kJ/mol}) + (1 \text{ mol CO}_2)(-393.51 \text{ kJ/mol})$$

$$+ (1 \text{ mol H}_2\text{O})(-285.8 \text{ kJ/mol}) - (2 \text{ mol NaHCO}_3)(-950.81 \text{ kJ/mol})$$

$$= 91.6 \text{ kJ}$$

Example Problem 5.6.2 Calculate enthalpy change using standard heats of formation.

Using the standard heats of formation that follow, calculate the standard enthalpy change for the following reaction.

$$3 \text{ Fe}_2\text{O}_3(s) + \text{H}_2(g) \rightarrow 2 \text{ Fe}_3\text{O}_4(s) + \text{H}_2\text{O}(g)$$

Compound	ΔH_f° (kJ/mol)
$\text{Fe}_2\text{O}_3(s)$	-824.2
$\text{Fe}_3\text{O}_4(s)$	-1118.4
$\text{H}_2\text{O}(g)$	-241.8

Example Problem 5.6.2 *(continued)*

Solution:

You are asked to calculate the standard enthalpy change for a reaction.

You are given the chemical equation for the reaction and the standard heats of formation for the compounds in the equation.

The standard enthalpy change for a reaction can be calculated from the standard heats of formation of the products and reactants, each multiplied by the stoichiometric coefficient in the balanced equation. Recall that the standard heat of formation for an element in its stable state at 298 K and 1 bar is zero.

$\Delta H°_{rxn} = \Sigma \Delta H_f°(\text{products}) - \Sigma \Delta H_f°(\text{reactants})$

$\quad = (2 \text{ mol } Fe_3O_4)(\Delta H_f°[Fe_3O_4(s)]) + (1 \text{ mol } H_2O)(\Delta H_f°[H_2O(g)])$
$\quad\quad - (3 \text{ mol } Fe_2O_3)(\Delta H_f°[Fe_2O_3(s)])$

$\quad = (2 \text{ mol } Fe_3O_4)(-1118.4 \text{ kJ/mol}) + (1 \text{ mol } H_2O)(-241.8 \text{ kJ/mol})$
$\quad\quad - (3 \text{ mol } Fe_2O_3)(-824.2 \text{ kJ/mol})$

$\quad = -6.0 \text{ kJ}$

Video Solution

Tutored Practice
Problem 5.6.2

Section 5.6 Mastery

Unit Recap

Key Concepts

5.1 Energy

- Energy is the ability to do work (5.1a).
- Kinetic energy and potential energy are two broad classes of energy (5.1a).
- Thermodynamics is the study of the relationships between heat, energy, and work (5.1c).
- A change in internal energy is the sum of the work and heat added to or removed from a system (5.1c).
- Sign conventions are used to indicate the direction of heat and work flow between a system and the surroundings (5.1c).
- The first law of thermodynamics states that the total energy for an isolated system is constant (5.1c).

5.2 Enthalpy

- Enthalpy is the sum of the internal energy of a system plus the product of pressure and volume (5.2a).

- A change in enthalpy is equal to the heat exchanged between the system and surroundings at constant pressure (5.2a).

- Heat is transferred from the system to the surroundings in an exothermic process (5.2a).

- Heat is transferred from the surroundings to the system in an endothermic process (5.2a).

5.3 Energy, Temperature Changes, and Changes of State

- Specific heat capacity is the amount of energy required to raise the temperature of one gram of a substance by one degree Celsius or one kelvin (5.3a).

- Heat transfer occurs from a hot object to a cooler object until thermal equilibrium is reached (5.3b).

- The heat lost by a hot object is equal to the amount of heat gained by the cooler object (5.3b).

- During a phase change, temperature does not change because heat energy is used to overcome the forces between particles (5.3c).

- Enthalpy of fusion is the energy required to melt a solid and enthalpy of vaporization is the energy required to vaporize a liquid (5.3c).

5.4 Enthalpy Changes and Chemical Reactions

- Enthalpy change for a reaction is the energy released or absorbed during a chemical reaction (5.4a).

- Enthalpy is an extensive variable; it is dependent on the amount of substance present (5.4b).

- Calorimetry is used to determine enthalpy of reaction values (5.4c).

- Constant-pressure (coffee-cup) calorimetry experiments measure enthalpy change for a reaction (5.4c).

- Constant-volume (bomb) calorimetry experiments measure a change in internal energy (5.4d).

5.5 Hess's Law

- Enthalpy is a state function (5.5a).
- Hess's Law states that if a reaction can be carried out in a series of steps, the overall enthalpy change for the reaction is the sum of the enthalpy changes for the series of steps (5.5a).

5.6 Standard Heats of Reaction

- Standard enthalpy change is measured under standard conditions, where substances are in their most stable form at 1 bar and a given temperature, and all solutions have a concentration of 1 mol/L (5.6a).
- Standard heat of formation is the enthalpy change for the formation of one mole of a substance from its constituent elements, all in their standard states (5.6a).
- The standard enthalpy of a reaction can be calculated using standard heats of reaction for the reactants and products (5.6b).

Key Equations

$$KE = \tfrac{1}{2}\,mv^2 \tag{5.1}$$

$$\Delta E_{\text{system}} = q + w \tag{5.2}$$

$$\Delta E_{\text{universe}} = 0 \tag{5.3}$$

$$H = E + PV \tag{5.4}$$

$$c, \text{ specific heat capacity (J/g} \cdot {}^\circ\text{C)} = \frac{q, \text{ heat energy absorbed (J)}}{m, \text{ mass (g)} \times \Delta T, \text{ change in temperature (}^\circ\text{C)}} \tag{5.5}$$

$$\Delta H^\circ_{\text{rxn}} = \Sigma \Delta H_f^\circ(\text{products}) - \Sigma \Delta H_f^\circ(\text{reactants}) \tag{5.6}$$

Key Terms

5.1 Energy
energy
work
thermochemistry
kinetic energy
potential energy
mechanical energy
thermal energy
chemical energy
newton
thermodynamics
system
surroundings
isolated system
closed system

internal energy
change in internal energy
first law of thermodynamics

5.2 Enthalpy
enthalpy
change in enthalpy
exothermic process
endothermic process

5.3 Energy, Temperature Changes, and Changes of State
specific heat capacity
thermal equilibrium
enthalpy of fusion
enthalpy of vaporization

5.4 Enthalpy Changes and Chemical Reactions
enthalpy change for a reaction
extensive variable
calorimetry
calorimeter
enthalpy of dissolution

5.5 Hess's Law
state function
Hess's law

5.6 Standard Heats of Reaction
standard state
standard heat of formation
standard enthalpy of reaction

Unit 5   Review and Challenge Problems

6 Electromagnetic Radiation and the Electronic Structure of the Atom

Unit Outline

In This Unit...

Physical and chemical properties of compounds are influenced by the structure of the molecules that they consist of. Chemical structure depends, in turn, on how electrons are arranged around atoms and how electrons are shared among atoms in molecules. Understanding physical and chemical properties of chemical compounds therefore relies on a detailed understanding of the arrangement of electrons in atoms and molecules. This unit begins that exploration by examining what we know about atomic electronic structure and how we know it. This is the first of a group of units that, in turn, explore the arrangement of electrons in atoms with many electrons (Electron Configurations and the Properties of Atoms, Unit 7), the manner in which chemical bonds form and control molecular structure (Covalent Bonding and Molecular Structure, Unit 8), and two theories of bonding (Theories of Chemical Bonding, Unit 9). In this unit, we examine the ways we learn about the electronic structure of elements. This, for the most part, involves studying how electromagnetic radiation interacts with atoms. We therefore begin with the nature of electromagnetic radiation.

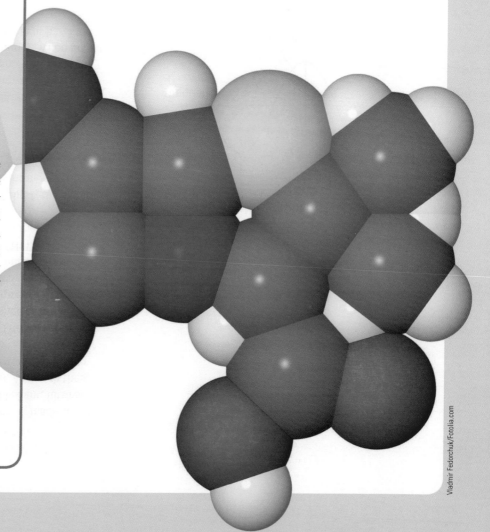

6.1 Electromagnetic Radiation

6.1a Wavelength and Frequency

Electromagnetic radiation, energy that travels through space as waves, is made up of magnetic and electric fields oscillating at right angles to one another. Visible light, ultraviolet radiation, and radio waves are all examples of electromagnetic radiation. Although these forms of electromagnetic radiation have different energies, they all have wavelike properties and travel at the same speed in a vacuum.

Waves are characterized by their wavelength, frequency, and speed. The **wavelength** of a wave (symbolized by the lowercase Greek letter lambda, λ) is the distance between two consecutive peaks or troughs in a wave (Interactive Figure 6.1.1). The **frequency** of a wave (symbolized by the lowercase Greek letter nu, ν) is the number of complete waves that pass a point in space in a given amount of time. Wavelength has units of length (meters) and frequency has units of cycles per second (1/s, s^{-1}) or **hertz** (Hz). Waves also have **amplitude**, the maximum positive displacement from the medium to the top of the crest of a wave.

The wavelength and frequency of a wave are related by the **speed of light**, the speed at which all electromagnetic radiation travels in a vacuum. The speed of light in a vacuum, 2.998×10^8 m/s, is equal to the wavelength (in meters, m) times the frequency (in Hz, 1/s) of the radiation (Equation 6.1)

$$c = \lambda\nu \qquad (6.1)$$

Equation 6.1 and the fixed speed of light in a vacuum allow the calculation of the wavelength or frequency of electromagnetic radiation if the other value is known. Notice that wavelength and frequency are inversely related. When light has a very short wavelength, many waves pass a point in space per second and the light has a high frequency. When light has a long wavelength, fewer waves pass a point in space per second and the light has a low frequency.

> **Example Problem 6.1.1** Calculate wavelength and frequency of waves.
>
> a. A local radio station broadcasts at a frequency of 91.7 MHz (91.7×10^6 Hz). What is the wavelength of these radio waves?
> b. What is the frequency of blue light with a wavelength of 435 nm?

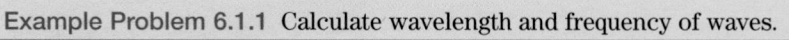

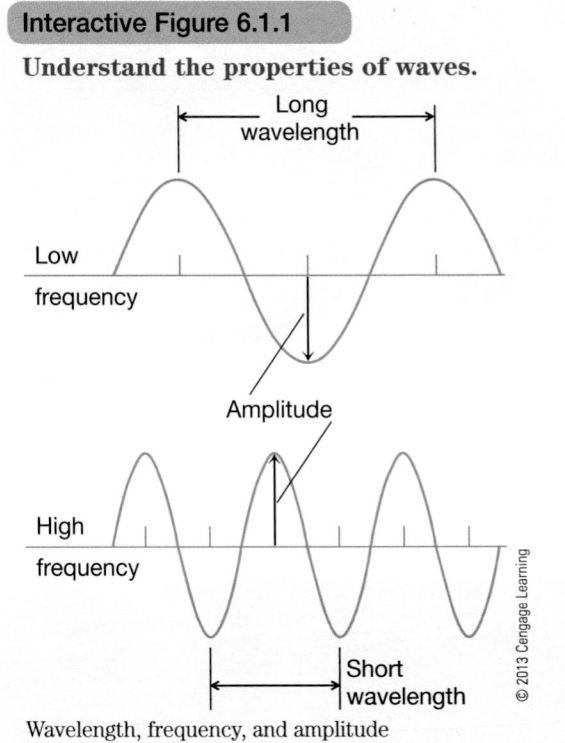

© 2013 Cengage Learning

Interactive Figure 6.1.1

Understand the properties of waves.

Long wavelength

Low frequency

Amplitude

High frequency

Short wavelength

Wavelength, frequency, and amplitude

Example Problem 6.1.1 *(continued)*

Solution:

You are asked to calculate the wavelength or frequency of electromagnetic radiation.

You are given the frequency or wavelength of the radiation.

a. First rearrange Equation 6.1 to solve for wavelength (λ). Then substitute the known values into the equation and solve for wavelength.

$$\lambda = \frac{c}{\nu} = \frac{2.998 \times 10^8 \text{ m/s}}{91.7 \times 10^6 \text{ 1/s}} = 3.27 \text{ m}$$

b. First rearrange Equation 6.1 to solve for frequency (ν). Then substitute the known values into the equation and solve for frequency. Notice that wavelength must be converted to units of meters before using it in Equation 6.1.

$$435 \text{ nm} \times \frac{10^{-9} \text{ m}}{1 \text{ nm}} = 4.35 \times 10^{-7} \text{ m}$$

$$\nu = \frac{c}{\lambda} = \frac{2.998 \times 10^8 \text{ m/s}}{4.35 \times 10^{-7} \text{ m}} = 6.89 \times 10^{14} \text{ Hz}$$

Video Solution

Tutored Practice
Problem 6.1.1

6.1b   The Electromagnetic Spectrum

The **electromagnetic spectrum** shows the different types of electromagnetic radiation arranged by wavelength, from gamma rays with very short wavelengths (in the picometer range) to radio waves with very long wavelengths (from about 1 meter to many kilometers in length). Each type of electromagnetic radiation has a range of wavelengths and frequencies, as shown in Interactive Figure 6.1.2.

Visible light, the electromagnetic radiation that can be observed by the human eye, ranges in wavelength from about 400 to 700 nm. Each color in the visible spectrum has a different wavelength and frequency. Light with a wavelength of 450 nm is blue, for example, and light with a wavelength of 675 nm is red.

The visible spectrum is a very small portion of the entire electromagnetic spectrum. The different types of radiation that make up the entire spectrum are all important to humans. For example, x-rays are used for imaging living tissues, microwave radiation is used in microwave ovens to cause water molecules to rotate and generate heat, and radio waves are used in radio and cell phone communication as well as in television and digital satellite signals.

Identify regions of the electromagnetic spectrum.

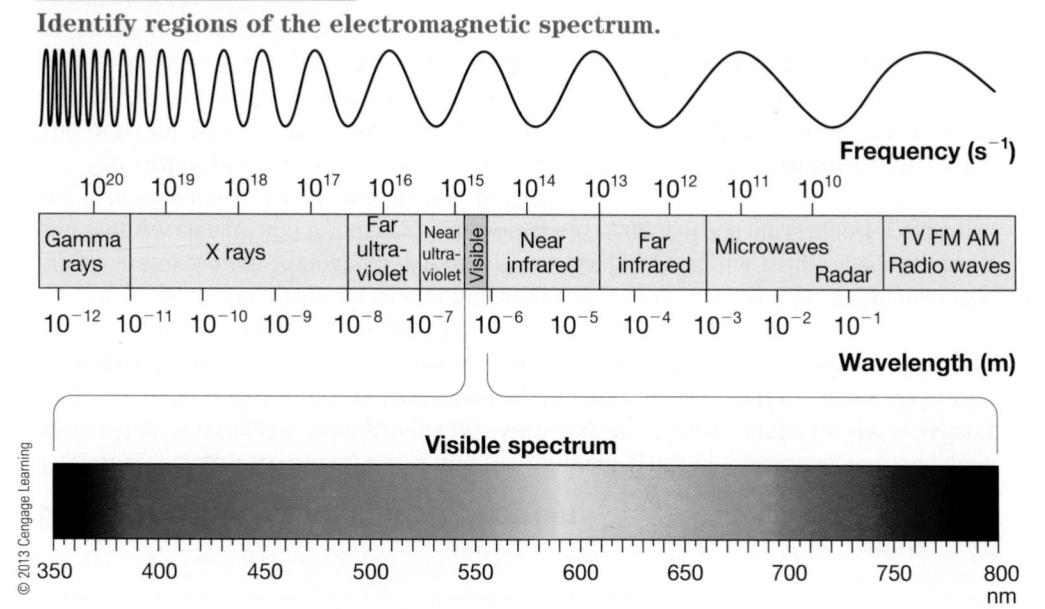

The electromagnetic spectrum

6.2 Photons and Photon Energy

6.2a The Photoelectric Effect

The wavelike properties of electromagnetic radiation are demonstrated by experiments that show wave interference and diffraction. At the beginning of the 20th century, other experiments puzzled the scientific community because they suggested that light acts more like it is composed of particles of energy. That light acts alternately as a wave and as a particle is known as the matter–wave duality of electromagnetic radiation.

One experiment that could not easily be explained by the wavelike properties of electromagnetic radiation is the **photoelectric effect**. The photoelectric effect is exhibited when light is shone on a metal and electrons are ejected from the surface of the metal. In a typical experiment, a piece of metal is placed in a vacuum tube. If light with a long wavelength (low frequency) is directed at the metal surface, nothing happens—even if the light has high intensity. However, if the light has a short wavelength (high frequency), even with low or moderate intensity, electrons are ejected from the metal surface. The ejection of electrons from the metal depends not on the total energy of the light, but only on the wavelength of the light. This experiment suggests that light has particle-like properties.

The explanation for the photoelectric effect is that light travels in packets, called **photons**, and that the energy of a single packet is related to the wavelength of the light. If a photon has low energy, it will be unable to knock an electron out of the metal. Hitting the metal with large numbers of these low-energy photons (very bright light with long wavelength) has no effect because no single photon can do the job (Interactive Figure 6.2.1).

On the other hand, a photon of high energy (short wavelength) can lead to the ejection of electrons. Thus, even low-intensity light with high energy (short wavelength) will lead to a measurable current. This implies that a photon carries an explicit amount of energy, called a quantum of energy, and that energy itself is quantized.

The relationship between frequency and the energy of a photon is given by Planck's equation,

$$E_{photon} = h\nu \qquad \qquad (6.2)$$

where h is **Planck's constant**, 6.626×10^{-34} J·s. Planck's constant is named for Max Planck (1858–1947), the scientist who first proposed the idea that energy is quantized.

Notice that the energy of a photon is directly related to the frequency of the radiation. Also notice that Plank's constant is very small, which means that a single photon of light carries a small amount of energy.

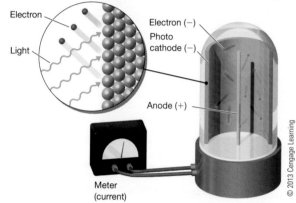

Interactive Figure 6.2.1

Identify the properties of photons.

Electron

Light

Electron (−)

Photo cathode (−)

Anode (+)

Meter (current)

The photoelectric effect

Example Problem 6.2.1 Use Planck's equation to calculate photon energy.

a. Calculate the energy of a single photon of light with a frequency of 8.66×10^{14} Hz.
b. Calculate the energy of a single photon of yellow light with a wavelength of 582 nm.

Solution:

You are asked to calculate the energy of a photon of light.

You are given the frequency or wavelength of the radiation.

a. Use Planck's equation to calculate the energy of this light.

$$E = h\nu = (6.626 \times 10^{-34} \text{ J·s})(8.66 \times 10^{14} \text{ s}^{-1}) = 5.74 \times 10^{-19} \text{ J}$$

b. First, convert wavelength to frequency and then use Planck's equation to calculate the energy of this light.

$$\nu = \frac{c}{\lambda} = \frac{2.998 \times 10^8 \text{ m/s}}{582 \text{ nm}\left(\dfrac{10^{-9} \text{ m}}{1 \text{ nm}}\right)} = 5.15 \times 10^{14} \text{ s}^{-1}$$

$$E = h\nu = (6.626 \times 10^{-34} \text{ J·s})(5.15 \times 10^{14} \text{ s}^{-1}) = 3.41 \times 10^{-19} \text{ J}$$

Video Solution

Tutored Practice
Problem 6.2.1

Section 6.2 Mastery

6.3 Atomic Line Spectra and the Bohr Model of Atomic Structure

6.3a Atomic Line Spectra

When sunlight passes through a prism hanging in a window, the white light is separated into a band that shows the colors of the rainbow, called a **continuous spectrum**. When the light emitted by "excited" gas phase elements is passed through a prism, a different kind of spectrum is seen—one that led scientists to a better understanding of electronic structure.

If a sample of a gas is placed in a sealed glass tube and "excited" with an electric current, the atoms absorb energy and begin to glow, emitting light. A common example of this is the light emitted by sealed tubes containing neon (neon lights), but the same principle is at work in fluorescent lightbulbs. When the radiation emitted by these samples is analyzed by passing it through a prism, a **line spectrum** (Interactive Figure 6.3.1) is observed that shows only very specific wavelengths of light.

Notice that the line spectra for different elements in Interactive Figure 6.3.1 show a finite number of lines (the spectra are not continuous) and that the number of lines and the wavelengths of the lines are different for each element. The fact that only certain wavelengths of light are emitted when atoms of a given element are excited suggests that the energy of an atom is quantized.

Interactive Figure 6.3.1

Explore the hydrogen line spectrum.

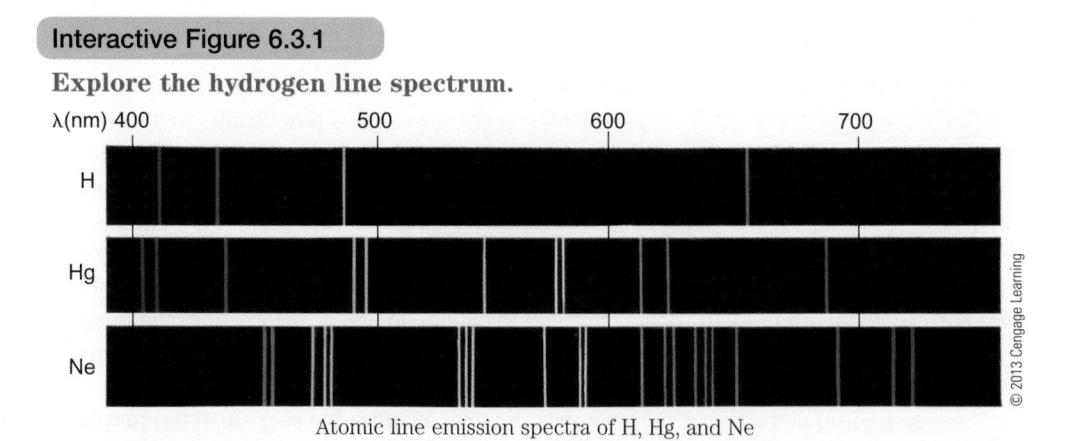

© 2013 Cengage Learning

Atomic line emission spectra of H, Hg, and Ne

6.3b The Bohr Model

Niels Bohr (1885–1962) was the first scientist to explain atomic line spectra. He proposed that electrons in atoms could occupy only certain energy levels. That is, the energy of electrons in atoms is quantized. According to his model of atomic electronic structure, electrons move around the nucleus of an atom in defined energy levels, called orbits. He assigned a number to each energy level, known as the **principle quantum number** (n). Electrons in low-energy levels can absorb energy and be promoted to higher-energy levels. When an electron moves from a high-energy level to a lower-energy level, energy is emitted in the form of light. Hydrogen, for example, is in its **ground state** when its electron is in the lowest energy level, $n = 1$. When a hydrogen atom absorbs energy, the electron moves to an **excited state** where $n > 1$.

Using his model, Bohr was able to explain the energies and wavelengths of the lines observed in the visible portion of the hydrogen spectrum, shown in the table below.

Using Bohr's equation relating the energy (E_n) and energy level (n) for an electron, it is possible to calculate the energy of a single electron in a ground state or excited state, or the energy change when an electron moves between two energy levels.

Color	λ (nm)
Red	656
Blue–green	486
Blue	434
Violet	410

$$E_n = -2.179 \times 10^{-18} \text{ J}\left(\frac{1}{n^2}\right) \tag{6.3}$$

For example, the calculation of the energy of an electron in the $n = 2$ and the $n = 3$ energy levels follows:

$$n = 2 \quad E_2 = -2.179 \times 10^{-18} \text{ J}\left(\frac{1}{2^2}\right) = -5.448 \times 10^{-19} \text{ J}$$

$$n = 3 \quad E_3 = -2.179 \times 10^{-18} \text{ J}\left(\frac{1}{3^2}\right) = -2.421 \times 10^{-19} \text{ J}$$

The energy emitted when an electron moves from the $n = 3$ to the $n = 2$ energy level is therefore

$$\Delta E = -2.179 \times 10^{-18} \text{ J}\left(\frac{1}{2^2} - \frac{1}{3^2}\right) = -3.026 \times 10^{-19} \text{ J}$$

and the frequency and wavelength of light that corresponds to this energy change is

$$\nu = \frac{E}{h} = \frac{3.026 \times 10^{-19} \text{ J}}{6.626 \times 10^{-34} \text{ J} \cdot \text{s}} = 4.567 \times 10^{14} \text{ s}^{-1}$$

$$\lambda = \frac{c}{\nu} = \frac{2.998 \times 10^8 \text{ m/s}}{4.567 \times 10^{14} \text{ s}^{-1}} = 6.564 \times 10^{-7} \text{ m} = 656.4 \text{ nm}$$

The wavelength and energy of the red line in the hydrogen spectrum corresponds to an electron moving from the $n = 3$ energy level to the $n = 2$ energy level.

As shown in Interactive Figure 6.3.2, all of the lines in the visible region of the hydrogen spectrum result from an electron moving from an excited state ($n > 2$) to the $n = 2$ energy level. Notice in the figure that electron energy becomes more negative as it occupies lower energy levels and is closer to the nucleus (as n decreases).

The observed emissions in the ultraviolet region are the result of a transition from an excited state to the $n = 1$ level, and emissions in the infrared region result from transitions from $n > 3$ or 4 to the $n = 3$ or 4 energy level. The energy change for each of these transitions is calculated using the general form of the Bohr equation:

$$\Delta E = -2.179 \times 10^{-18} \text{ J}\left(\frac{1}{n_{\text{final}}^2} - \frac{1}{n_{\text{initial}}^2}\right) \tag{6.4}$$

where n_{final} and n_{initial} are the final energy level and initial energy level, respectively.

It is important to keep in mind that a simple relationship between energy and the integer n is true only for species with a single electron such as hydrogen. The line spectra observed for all other species are more complex (Interactive Figure 6.3.1) and follow no clear, defined mathematical relationship.

Example Problem 6.3.1 Calculate energy change for electron transitions between energy levels.

Calculate the wavelength of the radiation emitted when an electron in a hydrogen atom moves from the $n = 5$ to the $n = 3$ energy level. Is the radiation visible?

Solution:

You are asked to calculate the wavelength of light emitted when an electron moves between energy levels in a hydrogen atom.

You are given the initial and final energy levels for the electron.

Use the general form of the Bohr equation to calculate the energy of the transition and then calculate the wavelength of the radiation.

$$\Delta E = -2.179 \times 10^{-18} \text{ J}\left(\frac{1}{3^2} - \frac{1}{5^2}\right) = -1.550 \times 10^{-19} \text{ J}$$

$$\nu = \frac{E}{h} = \frac{1.550 \times 10^{-19} \text{ J}}{6.626 \times 10^{-34} \text{ J} \cdot \text{s}} = 2.339 \times 10^{14} \text{ s}^{-1}$$

$$\lambda = \frac{c}{\nu} = \frac{2.998 \times 10^8 \text{ m/s}}{2.339 \times 10^{14} \text{ s}^{-1}} = 1.282 \times 10^{-6} \text{ m} = 1282 \text{ nm}$$

This radiation is not visible. It is in the infrared region of the electromagnetic spectrum.

Video Solution

Tutored Practice
Problem 6.3.1

Section 6.3 Mastery

Interactive Figure 6.3.2

Understand the Bohr model of the atom.

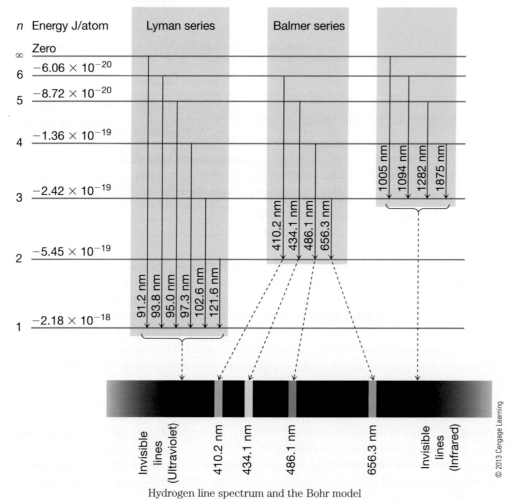

Hydrogen line spectrum and the Bohr model

6.4 Quantum Theory of Atomic Structure

6.4a Wave Properties of Matter

The Bohr model of an electron orbiting around the nucleus like a planet around the sun fails to explain almost all properties of atoms. Indeed, the planetary view of one charged particle orbiting another particle of opposite charge violates some of the best known laws of classical physics. Because of this deficiency, scientists have developed **quantum mechanics**, which presents a different view of how electrons are arranged about the nucleus in the atom. This view depends on two central concepts: the wave behavior of matter and the uncertainty principle. These two ideas combined lead to a mathematical description of electronic structure.

Bohr's model of the electronic structure of the atom failed when it was applied to species with more than a single electron. Although his theory is still useful in thinking about electron transitions between energy levels, it ultimately failed because it addressed only the particle nature of electrons. Similar to the way that light energy can act as a wave or as a particle, matter such as an electron displays **wave-particle duality**, properties of wave motion in addition to its particle behavior.

In the early 20th century, Louis de Broglie proposed that all matter in motion has a characteristic wavelength, according to a relationship now known as the de Broglie equation.

$$\lambda = \frac{h}{mv} \qquad (6.5)$$

where h is Plank's constant, m is the mass of the particle (in kg), and v is the velocity (m/s). For any macroscopic object, the calculated wavelength is vanishingly small and not observable. For an electron, though, the wavelength is significant and measurable.

Example Problem 6.4.1 Calculate the wavelength of moving particles using de Broglie's equation.

Calculate the wavelength of the following:

a. An electron (mass = 9.11×10^{-31} kg) moving at a speed of 2.2×10^6 m/s
b. A golf ball (mass = 45 g) moving at a speed of 72 m/s

Example Problem 6.4.1 *(continued)*

Solution:

You are asked to calculate the wavelength of a moving particle.

You are given the mass and velocity of the particle.

a. $\lambda = \dfrac{h}{mv} = \dfrac{6.626 \times 10^{-34}\,\text{J} \cdot \text{s}}{(9.11 \times 10^{-31}\,\text{kg})(2.2 \times 10^{6}\,\text{m/s})} = 3.3 \times 10^{-10}\,\text{m}$

b. $\lambda = \dfrac{h}{mv} = \dfrac{6.626 \times 10^{-34}\,\text{J} \cdot \text{s}}{(0.045\,\text{kg})(72\,\text{m/s})} = 2.0 \times 10^{-34}\,\text{m}$

Is your answer reasonable? The electron has a wavelength similar to the wavelength of x-rays, whereas the heavier, slower golf ball has a wavelength too small to be measured by any known instrument.

Video Solution

Tutored Practice
Problem 6.4.1

The Heisenberg Uncertainty Principle

A few years after de Broglie proposed that all particles have wavelike properties, Werner Heisenberg (1901–1976) proposed the **uncertainty principle**, which states that it is not possible to know with great certainty both an electron's position and its momentum (which is related to its kinetic energy) at the same time. Equation 6.6 shows the mathematical expression of this principle, where Δx is the uncertainty in the position of the electron and $\Delta \rho$ is the uncertainty in the electron's momentum.

$$\Delta x \Delta \rho \geq \frac{h}{4\pi} \qquad \textbf{(6.6)}$$

As shown earlier with calculations involving line spectra, it is possible to determine with great certainty the energy of an electron. Therefore, according to Heisenberg's uncertainty principle, we cannot know the position of the electron with any certainty. This principle demonstrated another weakness in Bohr's atomic theory. Bohr proposed that electrons moved in defined orbits around the nucleus, implying that both the position and the energy of an electron could be known with great certainty.

Describe wave functions and orbitals.

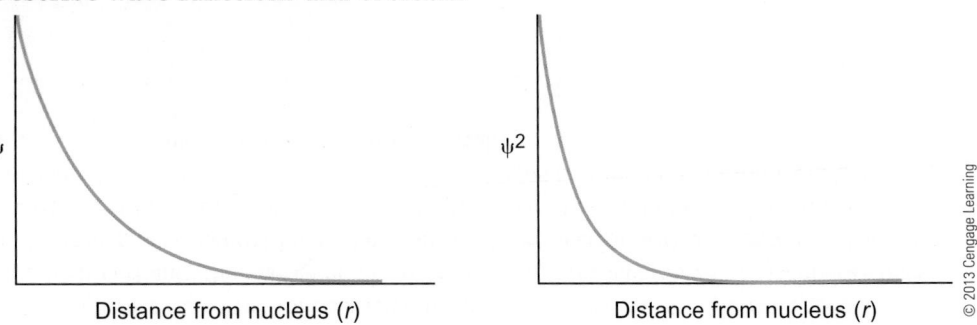

Wave function (ψ) and probability density (ψ²) for the lowest-energy electron in a hydrogen atom

© 2013 Cengage Learning

6.4b The Schrödinger Equation and Wave Functions

In the early 20th century, Erwin Schrödinger (1877–1961) derived a new mathematical equation to describe the wavelike behavior of electrons. Solutions to this equation produced **wave functions** (ψ), mathematical equations that predict the energy of an electron and the regions in space where an electron is most likely to be found. The square of a wave function, ψ^2, is equal to the probability of finding an electron in a given region around the nucleus of an atom. These probability functions, called **orbitals**, are represented using two- or three-dimensional shapes as shown in Interactive Figure 6.4.1 and Figures 6.4.2 and 6.4.3.

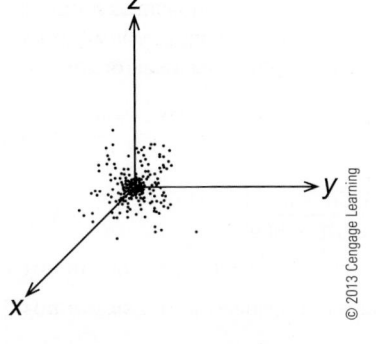

© 2013 Cengage Learning

Figure 6.4.2 Dot picture representation of an electron in a hydrogen atom

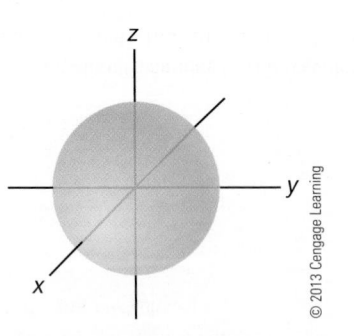

© 2013 Cengage Learning

Figure 6.4.3 Boundary surface representation of an electron in a hydrogen atom

We will not examine the mathematical form of wave functions in detail, but it is worth examining one to make clear how orbital shapes are derived. In Interactive Figure 6.4.1, the wave function (ψ) and the probability density (ψ^2) for the lowest-energy electron in a hydrogen atom are plotted as a function of r, the distance from the nucleus. Notice that as the distance from the nucleus (r) decreases, the value of ψ^2 becomes larger. This suggests that the lowest-energy electron in a hydrogen atom is most likely to be found close to the nucleus.

This probability function is usually visualized in two ways. First, imagine visualizing the position of an electron as a dot, sampling the position once every second over a very long period of time. If this is done for an electron in a hydrogen atom, the "dot picture" would look like that shown in Figure 6.4.2. Notice that the density of dots is greatest close to the nucleus. This is the region of greatest **electron density**, the region of greatest probability of finding an electron.

The second way to visualize the electron in a hydrogen atom is to think about a boundary surface that encloses the region of space where the electron is likely to be found most of the time. In this view, the electron is much more likely to be found within the boundary surface than outside the boundary surface. The boundary surface for the lowest-energy orbital is shown in Figure 6.4.3.

Section 6.4 Mastery

6.5 Quantum Numbers, Orbitals, and Nodes

6.5a Quantum Numbers

The organization of the periodic table is intimately related to the organization of electrons in atoms, which can be expressed in terms of **quantum numbers**. Quantum numbers refers to a series of numbers that results from solving Schrödinger's wave equation for the hydrogen atom.

Three quantum numbers are used to characterize an orbital. The **principal quantum number** (n) describes the size and energy of the shell in which the orbital resides. Recall that Bohr's atomic structure model also included a principal quantum number (n). The allowed values of n are positive integers ($n = 1, 2, 3, 4, ...$). The **angular momentum quantum number** (ℓ) indicates the shape of the orbital. For any given value of n, ℓ can have values that range from 0 to ($n - 1$). For example, in the $n = 4$ shell, ℓ has values of 0, 1, 2, and 3. Each value of ℓ corresponds to an orbital label and an orbital shape. Beyond $\ell = 3$, the

ℓ	Orbital Label
0	s
1	p
2	d
3	f

orbitals are labeled alphabetically ($\ell = 4$ is a g orbital, for example). The **magnetic quantum number** (m_ℓ) is related to an orbital's orientation in space. For a specific value of ℓ, m_ℓ can have values that range from $-\ell$ to $+\ell$. For example, when $\ell = 1$, the possible values of m_ℓ are -1, 0, and $+1$. Thus, there are three p ($\ell = 1$) orbitals (because there are three m_ℓ values when $\ell = 1$) in any shell.

Example Problem 6.5.1 Use quantum numbers.

a. List all possible values for ℓ in the sixth energy level ($n = 6$).
b. List all possible values for m_ℓ for an f orbital.

Solution:

You are asked to list all possible quantum numbers.

You are given a specific orbital or group of orbitals.

a. For any given value of n, ℓ can have values that range from 0 to $(n - 1)$. When $n = 6$, ℓ can have values of 0, 1, 2, 3, 4, and 5. Thus, there are six different types of orbitals in the sixth energy level.
b. For a specific value of ℓ, m_ℓ can have values that range from $-\ell$ to $+\ell$. An f orbital corresponds to an ℓ value of 3. Therefore, the possible values of m_ℓ are -3, -2, -1, 0, 1, 2, and 3. Thus, there are seven unique f orbitals, one for each m_ℓ value.

Video Solution

Tutored Practice
Problem 6.5.1

6.5b Orbital Shapes

The three-dimensional shapes of orbitals play a vital and important role in the behavior of electrons and thus control chemical processes and determine the shapes of molecules. Everything from the physical states of compounds to the selective way biological molecules react with one another is a facet of the way electrons are arranged in orbitals about the atom's nucleus.

Schrödinger's wave equation can be solved to produce wave functions and orbitals for single-electron species such as hydrogen. More complicated mathematical treatment leads to comparable orbitals for multielectron atoms. Orbitals are given designations, such as $1s$, $3p_x$, and $4d_{xz}$, that provide information about the orbital's energy, shape, and orientation in space. The lowest-energy orbitals for a hydrogen atom are shown in Interactive Figure 6.5.1. Each boundary surface contains the volume in space where an electron is most likely to be found. The surfaces also contain regions of high electron density, and some orbitals, such as the $2p_x$ orbital (Figure 6.5.2), contain regions of space called **nodes**, where there is no probability of finding an electron (no electron density).

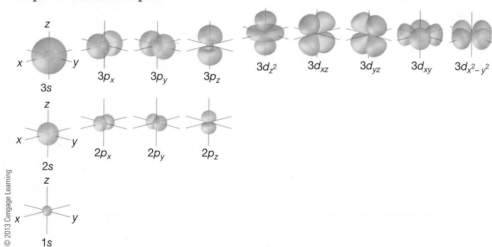

Interactive Figure 6.5.1

Explore orbital shapes.

3s

3p_x   3p_y   3p_z

3d_{z^2}   3d_{xz}   3d_{yz}   3d_{xy}   3d_{x^2-y^2}

2s

2p_x   2p_y   2p_z

1s

© 2013 Cengage Learning

Atomic orbitals (boundary surfaces, $n = 1–3$)

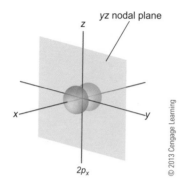

yz nodal plane

$2p_x$

© 2013 Cengage Learning

Figure 6.5.2 The node in a $2p_x$ orbital.

Orbitals are classified by energy, shape, and orientation in space. The main classification is the energy of the orbital, designated by the value of the principal quantum number, n. The principal quantum number indicates the relative size of the orbital. For example, the 2s orbital ($n = 2$) is higher in energy and larger than the 1s orbital ($n = 1$). Orbitals with the same n value are said to be in the same shell. Higher n values indicate orbitals where electrons are located, on average, farther from the nucleus. These electrons are less attracted to the positively charged nucleus (because it is farther away) and have a higher relative energy. Electrons closer to the nucleus have relatively large, negative energy. Electrons farther from the nucleus have energy closer to zero or a smaller negative energy. In summary,

High-energy electrons

- Have a higher n value

- Are farther from the nucleus

- Have a small, negative energy (close to zero)

Low-energy electrons

- Have a lower n value

- Are closer to the nucleus

- Have a large, negative energy

Each shell contains one or more **subshells**, each defined by a different value of ℓ and designated by a different letter (s, p, d, and f are the letters assigned to the first four subshells). The subshell label indicates the shape of the orbitals in that subshell. An s orbital has a spherical shape, p orbitals have two regions of electron density on either side of the nucleus, and most of the d orbitals have four regions of electron density surrounding the nucleus. When an orbital has more than one region of electron density, the different regions are separated by a node.

Each subshell is made up of one or more orbitals. The number of orbitals in a subshell is given by the number of m_ℓ values for that subshell. For subshells with more than one orbital, subscript letters are used to differentiate between orbitals. The subshells and orbitals that exist for the first four shells are shown in Table 6.5.1.

Table 6.5.1 Subshells and Orbitals in $n = 1$–4 Energy Levels

Principal Quantum Number	ℓ	Subshell	Number of m_ℓ Values	Orbitals in the Subshell
1	0	s	1	1 ($1s$ orbital)
2	0	s	1	1 ($2s$ orbital)
	1	p	3	3 ($2p_x$, $2p_y$, $2p_z$ orbitals)
3	0	s	1	1 ($3s$ orbital)
	1	p	3	3 ($3p_x$, $3p_y$, $3p_z$ orbitals)
	2	d	5	5 ($3d_{xy}$, $3d_{xz}$, $3d_{yz}$, $3d_{z^2}$, and $3d_{x^2-y^2}$ orbitals)
4	0	s	1	1 ($4s$ orbital)
	1	p	3	3 ($4p_x$, $4p_y$, and $4p_z$ orbitals)
	2	d	5	5 ($4d_{xy}$, $4d_{xz}$, $4d_{yz}$, $4d_{z^2}$, and $4d_{x^2-y^2}$ orbitals)
	3	f	7	7 ($4f_{y^3}$, $4f_{x^3}$, $4f_{z^3}$, $4f_{xz^2y^2}$, $4f_{yz^2x^2}$, $4f_{zx^2y^2}$, and $4f_{xyz}$ orbitals)

Notice that as n increases, so does the number of subshells in that shell. Also, the number of subshells for any shell is equal to the principal quantum number, n, and the total number of orbitals in a given energy level is equal to n^2.

6.5c Nodes

Nodes are regions of space where there is zero probability of finding an electron. Two types of nodes are found in orbitals: planar nodes (also called angular nodes and sometimes take the shape of a cone) and radial nodes (also called spherical nodes).

Planar nodes are found in all orbitals other than the s orbitals. As shown in Figure 6.5.2, each p orbital has a planar node that is perpendicular to the axis where there are regions of high electron density. The d orbitals generally have two planar nodes that are perpendicular to each other. For example, the $3d_{xy}$ orbital has two planar nodes, one in the xz plane and one in the yz plane. The number of planar nodes in a subshell is given by the value of the angular momentum quantum number, ℓ.

Radial nodes are more difficult to visualize than planar nodes. Consider the $2s$ orbital, which has a single radial node. The boundary surface for a $2s$ orbital looks much like that of a $1s$ orbital. But, if we "slice" the orbital in half, we can see that the $2s$ orbital has a spherical region where there is no electron density. This is a radial node, and it separates two regions of electron density: one near the nucleus and the other farther from the nucleus (on the other side of the node) (Interactive Figure 6.5.3).

The number of radial nodes increases as the value of n increases. For example, a $3s$ orbital has two radial nodes separating three regions of electron density, and a $4s$ orbital has three radial nodes. Likewise, a $2p$ orbital has no radial nodes, and a $3p$ orbital has one. The number of radial nodes for an orbital is related to both n and ℓ and is equal to $(n - \ell - 1)$. The total number of nodes (planar and radial) for an orbital is equal to $n - 1$.

6.5d Orbital Energy Diagrams and Changes in Electronic State

We use energy diagrams as a means for depicting orbital energies (Figure 6.5.4). Notice that the energy of each orbital depends only on its shell, not its subshell. This is true only for hydrogen, and we will see in Electron Configurations and the Properties of Atoms (Unit 7) that the energy of orbitals for all other atoms depends on both the shell and the subshell. The energy of each shell is the same as that seen earlier in the Bohr model of the hydrogen atom,

$$E_n = -2.179 \times 10^{-18} \text{ J} \left(\frac{1}{n^2} \right).$$

ℓ	Subshell	Number of Planar Nodes
0	s	0
1	p	1
2	d	2
3	f	3

Interactive Figure 6.5.3

Relate quantum numbers to nodes.

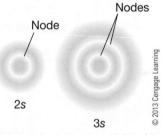

Radial nodes in the $2s$ and $3s$ orbitals

© 2013 Cengage Learning

We can now give a more complete view of what occurs when a hydrogen atom absorbs a photon of light. The ground state for the hydrogen atom finds its electron in the lowest-energy orbital, the $1s$ orbital. When an atom absorbs a photon of light, the electron is promoted to a higher-energy orbital ($n > 1$). The photon of light causes the electron to move farther away from the nucleus, as shown in Interactive Figure 6.5.5.

Note that although Interactive Figure 6.5.5 shows a hydrogen atom absorbing energy and promoting an electron from a $1s$ orbital to a $3p_x$ orbital, absorption of a photon with a wavelength of 102.6 nm can promote the electron to any p orbital in the $n = 3$ shell. Absorption of light with a different wavelength can promote the electron to an orbital with a different energy.

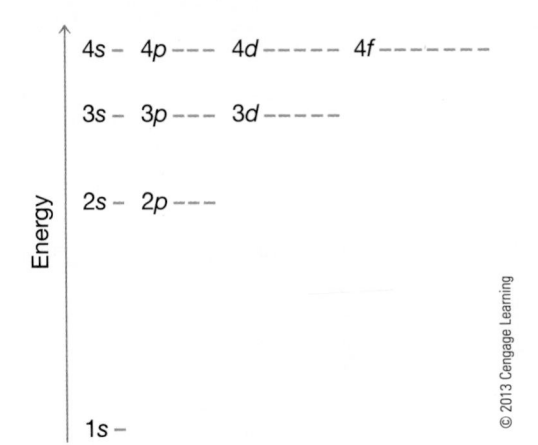

Figure 6.5.4 Energy diagram for hydrogen orbitals ($n = 1-4$)

Interactive Figure 6.5.5

Add energy to a hydrogen atom.

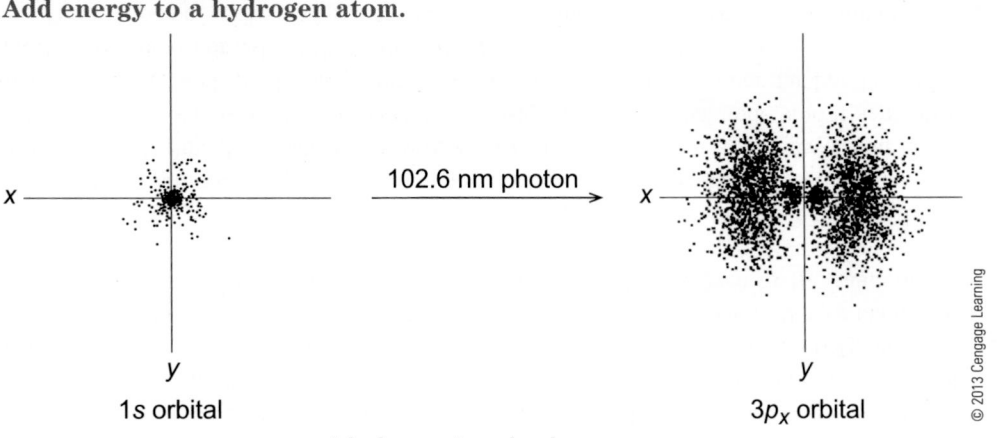

A hydrogen atom absorbs energy.

Section 6.5 Mastery

Unit Recap

Key Concepts

6.1 Electromagnetic Radiation

- Electromagnetic radiation is made up of magnetic and electric fields oscillating at right angles to one another (6.1a).
- Waves are characterized by their wavelength, frequency, and speed (6.1a).
- Wavelength and speed are related by the speed of light (6.1a).
- The electromagnetic spectrum shows the different types of electromagnetic radiation (6.1b).
- Visible light is a small portion of the electromagnetic spectrum (6.1b).

6.2 Photons and Photon Energy

- The photoelectric effect suggests that light has particle-like properties (6.2a).
- Light travels in packets called photons, and the energy of a photon is related to the wavelength and frequency of the light (6.2a).
- Planck's equation relates the energy of a photon to its frequency (6.2a).

6.3 Atomic Line Spectra and the Bohr Model of Atomic Structure

- When light emitted by excited atoms is passed through a prism, the resulting spectrum is called a line spectrum and is characteristic for a given element (6.3a).
- Niels Bohr's atomic theory states that electrons in an atom can occupy only certain energy levels and that they move from lower- to higher-energy levels when they absorb energy (6.3b).
- Bohr's equation relates the energy of an electron to its energy state as indicated by n, the principal quantum number (6.3b).

6.4 Quantum Theory of Atomic Structure

- Louis de Broglie proposed that all matter in motion has a characteristic wavelength (6.4a).
- Werner Heisenberg proposed the uncertainty principle, which states that it is not possible to know with great certainty both an electron's position and its momentum at the same time (6.4a).

- Erwin Schrödinger proposed a mathematical system that produced wave function equations that predict the energy of an electron and the regions in space where it is most likely to be found (6.4b).

- The square of a wave function is equal to the probability of finding an electron in a given region of space around the nucleus of an atom (6.4b).

- An orbital is the probability function visualized as a two- or three-dimensional shape (6.4b).

6.5 Quantum Numbers, Orbitals, and Nodes

- The quantum numbers n, ℓ, and m_ℓ are a series of numbers that result from solving Schrödinger's wave equation for the hydrogen atom and are used to characterize orbitals (6.5a).

- Each energy level (shell) contains one or more subshells, and each subshell is made up of one or more orbitals (6.5b).

- Orbitals can contain planar and radial nodes, regions of space where there is no probability of finding an electron (6.5c).

- The number of nodes in an orbital is related to the quantum numbers n and ℓ (6.5c).

- Energy diagrams are used to depict relative orbital energies (6.5d).

Key Equations

$$c = \lambda \nu \qquad \textbf{(6.1)}$$

$$\Delta E = -2.179 \times 10^{-18} \text{ J}\left(\frac{1}{n_{\text{final}}^2} - \frac{1}{n_{\text{initial}}^2}\right) \qquad \textbf{(6.4)}$$

$$E_{\text{photon}} = h\nu \qquad \textbf{(6.2)}$$

$$\lambda = \frac{h}{mv} \qquad \textbf{(6.5)}$$

$$E_n = -2.179 \times 10^{-18} \text{ J}\left(\frac{1}{n^2}\right) \qquad \textbf{(6.3)}$$

$$\Delta x \Delta \rho \geq \frac{h}{4\pi} \qquad \textbf{(6.6)}$$

Key Terms

6.1 Electromagnetic Radiation
electromagnetic radiation
wavelength
frequency
hertz
amplitude
speed of light
electromagnetic spectrum

6.2 Photons and Photon Energy
photoelectric effect
photons
Planck's constant

6.3 Atomic Line Spectra and the Bohr Model of Atomic Structure
continuous spectrum
line spectrum
principle quantum number
ground state
excited state

6.4 Quantum Theory of Atomic Structure
quantum mechanics
wave-particle duality
uncertainty principle

wave functions
orbitals
electron density

6.5 Quantum Numbers, Orbitals, and Nodes
quantum numbers
principal quantum number
angular momentum quantum number
magnetic quantum number
nodes
subshells

Unit 6   Review and Challenge Problems

7

Electron Configurations and the Properties of Atoms

Unit Outline

In This Unit...

In Electromagnetic Radiation and the Electronic Structure of the Atom (Unit 6) we introduced and explored the concept of orbitals, which define the shapes electrons take around the nucleus of an atom. In this unit we expand this description to atoms that contain more than one electron and compare atoms that differ in their numbers of protons in the nucleus and electrons surrounding that nucleus. Much of what we know and can predict about the properties of an atom, such as its size and the number and types of bonds it will form, can be derived from the number and arrangement of the atom's electrons and the energies of the atom's orbitals.

7.1 Electron Spin and Magnetism

7.1a Electron Spin and the Spin Quantum Number, m_s

Although electrons are too small to observe directly, we can detect the magnetic field that they exert. This magnetic field is generated by **electron spin**, the negatively charged electron spinning on an axis (Interactive Figure 7.1.1).

The magnetic field produced by an electron occurs in only one of two directions, indicating that electron spin is quantized. That is, an electron has only two possible **spin states**. In one spin state, the electron produces a magnetic field with the North pole in one direction. In the other spin state, the North pole is in the opposite direction (Figure 7.1.2).

Spin states are defined by a fourth quantum number, the **spin quantum number, m_s.** Because there are two different spin states, m_s has two possible values: $+\frac{1}{2}$ or $-\frac{1}{2}$. The sign of m_s is used to indicate the fact that the two spin states are in opposite directions and should not be confused with the negative charge on the electron. The electron always has a negative charge, regardless of its spin.

Symbolizing Electron Spin

We saw in Electromagnetic Radiation and the Electronic Structure of the Atom (Unit 6) that lines or boxes can be used to depict orbitals. Electrons in orbitals are shown using arrows, where the direction of the arrow indicates the spin state of the electron. For example, in the diagram that follows, the upward (↑) and downward (↓) arrows indicate electrons in different spin states.

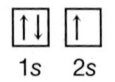

1s 2s

We will arbitrarily assign $m_s = +\frac{1}{2}$ to electrons represented with an upward arrow (also called "spin up" electrons) and $m_s = -\frac{1}{2}$ to electrons represented with a downward arrow (also called "spin down" electrons).

7.1b Types of Magnetic Materials

Magnetic materials derive their magnetic behavior from the magnetic properties of their electrons. Because all electrons produce a magnetic field, you might ask the question, Why aren't all materials magnetic? The answer lies in the fact that the magnetic fields

Relate electron spin and magnetic properties.

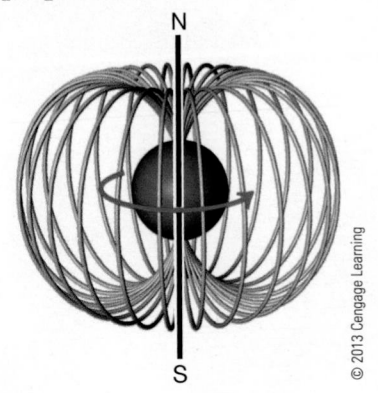

© 2013 Cengage Learning

Electron spin and magnetic field

Interactive Figure 7.1.3

Use spin states to predict magnetic properties.

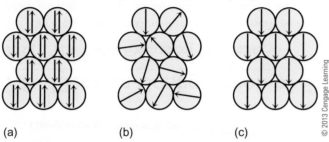

(a)   (b)   (c)

Electron spin representations of a material with (a) diamagnetic, (b) paramagnetic, and (c) ferromagnetic properties

© 2013 Cengage Learning

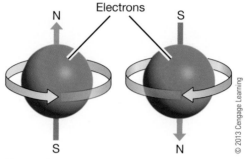

Figure 7.1.2 Electron spin and magnetic fields

© 2013 Cengage Learning

generated by electrons with opposite spin (in a single atom, molecule, or ion) directly counteract and cancel each other. Therefore, any atom or molecule with equal numbers of spin up and spin down electrons will have a net magnetic field of zero. Two electrons with opposite spin are said to be spin paired and produce no net magnetic field. An uneven number of electrons leaves unpaired electrons. Materials with unpaired electrons are magnetic.

The magnetism of most materials can be categorized as being diamagnetic, paramagnetic, or ferromagnetic. In a **diamagnetic** material (Interactive Figure 7.1.3a), all electrons are spin paired and the material does not have a net magnetic field. These materials are slightly repelled by the magnetic field of a strong magnet.

Paramagnetic materials (Interactive Figure 7.1.3b) contain atoms, molecules, or ions with unpaired electrons. In the absence of an external, strong magnetic field, the magnetic fields generated by the individual particles are arranged in random directions and the magnetism produced by each atom or molecule can be cancelled by the magnetic fields around it. This results in a magnetic material, but one with a weak net magnetic field. However, the presence of a strong, external magnet causes the individual spins to align so that the material is attracted to the magnet.

Ferromagnetic materials (Interactive Figure 7.1.3c), like paramagnetic materials, contain particles with unpaired electrons. In these materials, however, the individual magnetic fields align naturally and produce a strong, permanent magnetic field. The common magnets you are familiar with are ferromagnets.

Section 7.1 Mastery

7.2 Orbital Energy

7.2a Orbital Energies in Single- and Multielectron Species

The relationship between the principal quantum number, n, and orbital energy is shown in an orbital energy diagram (Figure 7.2.1). For a single-electron species such as a hydrogen atom, the energy of the atomic orbitals depends only on the value of n. For example, a $2p$ orbital in a hydrogen atom has the same energy as a $2s$ orbital. The energy of an orbital in a single-electron system depends only on the degree of attraction between the electron in that orbital and the nucleus. This is mainly a function of the average distance of the electron from the nucleus, which is controlled by the principal quantum number, n.

In multielectron species such as helium atoms or sodium ions, the orbital energies depend both on the principal quantum number, n, and the type of orbital, given by the angular momentum quantum number, ℓ (Interactive Figure 7.2.2). Multielectron atoms are more complex because the energy of the electron depends on both how close an electron is to the nucleus and the degree to which it experiences repulsive forces with the other electrons present in the atom or ion.

As shown in Interactive Figure 7.2.2, subshell energies increase with increasing ℓ (in a given energy level, $s < p < d < f$). Using Interactive Figure 7.2.2 as a reference, we can make several generalizations about orbital energies in multielectron species.

- As n increases, orbital energy increases for orbitals of the same type.
  A $4s$ orbital is higher in energy than a $3s$ orbital.

- As ℓ increases, orbital energy increases.
  In the $n = 3$ shell, $3s < 3p < 3d$.

- As n increases, the subshell energies become more closely spaced and overlapping occurs.
  The $4f$ orbital is higher in energy than the $5s$ orbital, despite its lower n value.

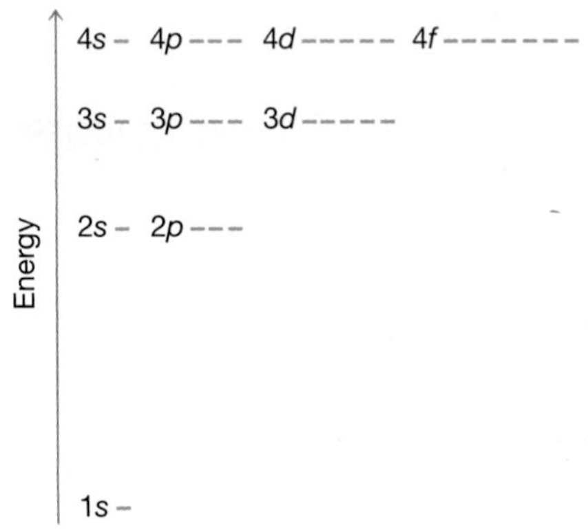

Figure 7.2.1 Orbital energies ($n = 1$ to $n = 4$) in a single-electron species

© 2013 Cengage Learning

Section 7.2 Mastery

7.3 Electron Configuration of Elements

7.3a The Pauli Exclusion Principle

The **electron configuration** of an element shows how electrons are distributed in orbitals—which ones are filled and which ones remain vacant. We can predict the electron configuration of most elements, and we can use electron configurations to predict physical and chemical properties of the elements.

To predict the electron configuration for an atom's **ground state**, the lowest energy state for an atom, electrons are put into the orbitals with the lowest energy possible, placing no more than two electrons in an orbital.

The order of subshell filling is related to n, the principal quantum number, and ℓ, the angular momentum quantum number. In general,

- electrons fill orbitals in order of increasing $(n + \ell)$ and

- when two or more subshells have the same $(n + \ell)$ value, electrons fill the orbital with the lower n value.

These general rules result in the following orbital filling order:

$$1s, 2s, 2p, 3s, 3p, 4s, 3d, 4p, 5s, 4d, 5p, 6s, 4f, 5d, 6p, 7s, 5f, 6d, 7p, 8s, \dots$$

The **Pauli exclusion principle** states that no two electrons within an atom can have the same set of four quantum numbers (n, ℓ, m_ℓ, and m_s). The limits on possible values for the four quantum numbers means that a single orbital can accommodate no more than two electrons, and when an orbital contains two electrons, those electrons must have opposite spins (Figure 7.3.1).

Example Problem 7.3.1 Apply the Pauli exclusion principle.

a. Identify the orbitals in the following list that fill before the $4d$ orbitals: $3d$, $4s$, $5s$, $6p$.
b. What is the maximum number of electrons in the $4f$ subshell?
c. An electron in an orbital has the following quantum numbers: $n = 3$, $\ell = 1$, $m_\ell = +1$, and $m_s = +\frac{1}{2}$. Identify the orbital and the direction of electron spin (up or down).

Solution:

You are asked to answer questions about orbital filling and to apply the Pauli exclusion principle.

You are given information about specific orbitals or electrons.

a. $3d$, $4s$, and $5s$. In general, orbitals fill in order of increasing $(n + \ell)$. When two or more subshells have the same $(n + \ell)$ value, electrons fill the orbital with the lower n value. The order of filling is $1s$, $2s$, $2p$, $3s$, $3p$, $4s$, $3d$, $4p$, $5s$, $4d$, $5p$, $6s$, $4f$, $5d$, $6p$, $7s$, $5f$, $6d$, $7p$, $8s$, ...
b. 14. There are 7 orbitals in the f subshell, and each orbital can hold two electrons.
c. $3p$, up. The set of three quantum numbers $n = 3$, $\ell = 1$, $m_\ell = +1$ together specifies a $3p$ orbital. The electrons in the $3p$ orbital are oriented "spin up" ($m_s = +\frac{1}{2}$) or "spin down" ($m_s = -\frac{1}{2}$), resulting in two unique sets of four quantum numbers for the two electrons in this orbital.

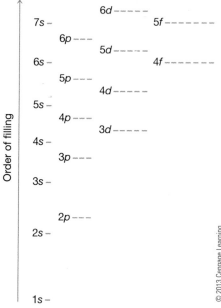

Identify orbital energies in multielectron species.

Orbital energies in a multielectron species

(a)   (b)

Figure 7.3.1 Electron arrangements in an orbital that are (a) allowed and (b) not allowed

Video Solution

Tutored Practice Problem 7.3.1

7.3b Electron Configurations for Elements in Periods 1–3

Hydrogen and Helium

Hydrogen has a single electron that occupies the orbital with the lowest energy, the $1s$ orbital. Two methods are used to represent this electron configuration. The **spdf notation** (also called *spectroscopic notation*) has the general format $n\ell^{\#}$, where subshells are listed in the order in which they are filled and the number of electrons occupying each subshell is shown to the right of the subshell as a superscript. The *spdf* notation for hydrogen is

$$H: 1s^1 \text{ (pronounced "one-ess-one")}$$

Orbital box notation uses boxes or horizontal lines to represent orbitals and arrows to represent electrons. The electron configuration of hydrogen in orbital box notation is

$$H: \boxed{\uparrow}$$
$$1s$$

Helium has two electrons, and both occupy the lowest-energy $1s$ orbital. The electron configuration of helium in *spdf* notation and orbital box notation is therefore

$$n = 1, \ell = 0, m_\ell = 0, m_s = +\tfrac{1}{2}$$
$$He: 1s^2 \text{ (pronounced "one-ess-two")} \quad \boxed{\uparrow\downarrow}$$
$$1s \quad n = 1, \ell = 0, m_\ell = 0, m_s = -\tfrac{1}{2}$$

Each electron in helium has a unique set of four quantum numbers, as required by the Pauli exclusion principle. Notice that hydrogen and helium are in the first row of the periodic table and both elements fill orbitals in the first energy level ($1s$).

Orbital box notations provide information about the number of paired and unpaired electrons in an atom, and that information can be used to determine whether the atoms are paramagnetic or diamagnetic. Hydrogen has one unpaired electron and is a paramagnetic species, whereas helium's electrons are paired and it is diamagnetic.

Lithium to Neon

Lithium has three electrons, two in the $1s$ orbital and one that is in an orbital in the second energy level. As shown previously, the $2s$ orbital is lower in energy than the $2p$ orbitals, so the electron configuration of lithium in *spdf* notation and orbital box notation is

$$Li: 1s^2 2s^1 \quad \boxed{\uparrow\downarrow} \ \boxed{\uparrow}$$
$$1s \quad 2s$$

Notice that it would be more correct to draw the orbital box notation electron configuration of lithium as shown below because the $2s$ orbital is higher in energy than the $1s$ orbital.

$$\boxed{\uparrow}$$
2s

$$\boxed{\uparrow\downarrow}$$
1s

However, it is common to show all orbitals on a horizontal line when writing orbital box notation electron configurations in order to make more efficient use of space.

Beryllium has two electrons in the $1s$ and the $2s$ orbitals.

Be: $1s^2 2s^2$ $\boxed{\uparrow\downarrow}$ $\boxed{\uparrow\downarrow}$
    1s   2s

Boron has five electrons. Four electrons fill the $1s$ and $2s$ orbitals, and the fifth electron is in a $2p$ orbital. Notice that the orbital box diagram shows all three $2p$ orbitals even though only one of the $2p$ orbitals is occupied.

B: $1s^2 2s^2 2p^1$ $\boxed{\uparrow\downarrow}$ $\boxed{\uparrow\downarrow}$ $\boxed{\uparrow}$
    1s   2s   2p

Carbon has six electrons, four in the $1s$ and $2s$ orbitals and two in the $2p$ orbitals. When electrons occupy a subshell with multiple orbitals such as $2p$, **Hund's rule of maximum multiplicity** applies. This rule states that the lowest-energy electron configuration is the one where the maximum number of electrons is unpaired. In the case of carbon, this means that the two $2p$ electrons each occupy a different $2p$ orbital (Interactive Figure 7.3.2).

The electron configurations of nitrogen, oxygen, fluorine, and neon are shown here. Notice that the highest-energy orbital for all second-row elements is in the second energy level ($2s$ or $2p$).

N: $1s^2 2s^2 2p^3$ $\boxed{\uparrow\downarrow}$ $\boxed{\uparrow\downarrow}$ $\boxed{\uparrow|\uparrow|\uparrow}$
    1s   2s   2p

O: $1s^2 2s^2 2p^4$ $\boxed{\uparrow\downarrow}$ $\boxed{\uparrow\downarrow}$ $\boxed{\uparrow\downarrow|\uparrow|\uparrow}$
    1s   2s   2p

Interactive Figure 7.3.2

Apply Hund's rule of maximum multiplicity.

C: $1s^2 2s^2 2p^2$ $\boxed{\uparrow\downarrow}$ $\boxed{\uparrow\downarrow}$ $\boxed{\uparrow|\uparrow|}$
    1s   2s   2p

Carbon's electron configuration has two unpaired electrons in the $2p$ orbitals.

F: $1s^2 2s^2 2p^5$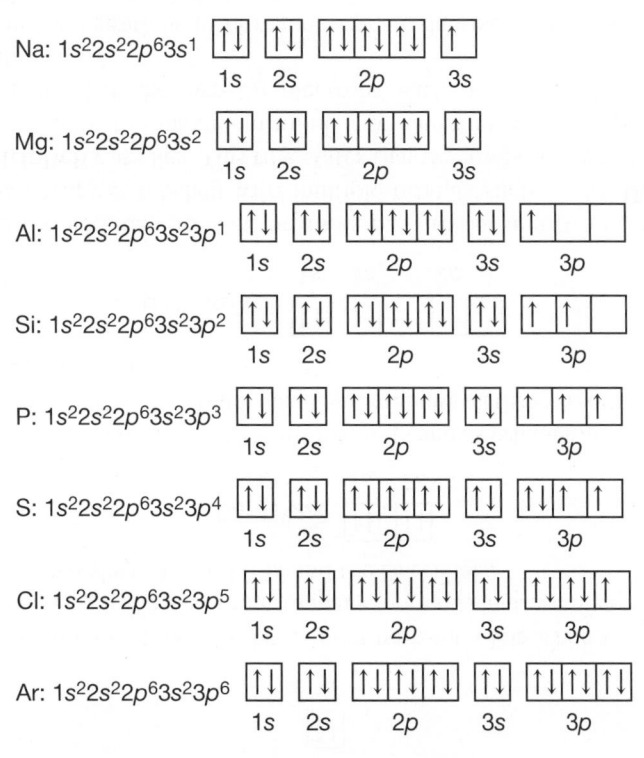
1s   2s   2p

Ne: $1s^2 2s^2 2p^6$
1s   2s   2p

Sodium to Argon

The elements in the third row of the periodic table have electrons that occupy orbitals in the third energy level, as shown.

Na: $1s^2 2s^2 2p^6 3s^1$
1s   2s   2p   3s

Mg: $1s^2 2s^2 2p^6 3s^2$
1s   2s   2p   3s

Al: $1s^2 2s^2 2p^6 3s^2 3p^1$
1s   2s   2p   3s   3p

Si: $1s^2 2s^2 2p^6 3s^2 3p^2$
1s   2s   2p   3s   3p

P: $1s^2 2s^2 2p^6 3s^2 3p^3$
1s   2s   2p   3s   3p

S: $1s^2 2s^2 2p^6 3s^2 3p^4$
1s   2s   2p   3s   3p

Cl: $1s^2 2s^2 2p^6 3s^2 3p^5$
1s   2s   2p   3s   3p

Ar: $1s^2 2s^2 2p^6 3s^2 3p^6$
1s   2s   2p   3s   3p

When electron configurations for multielectron species are written, **noble gas notation** is often used to represent filled shells (these filled shells are also called **core electrons**). In noble gas notation, the symbol for a noble gas is written within square brackets in front of the *spdf* or orbital box notation representing additional, noncore electrons. For example, the electron configuration for chlorine is written using noble gas notation as shown here.

Cl: [Ne]$3s^23p^5$ [Ne] ⇅ | ⇅|⇅|↑
                            3s   3p

The symbol [Ne] represents the 10 lowest-energy electrons ($1s^22s^22p^6$) in the electron configuration.

Example Problem 7.3.2 Write electron configurations for period 1–3 elements.

Write the electron configuration for phosphorus using orbital box notation.

Solution:

You are asked to write the electron configuration for an element in orbital box notation.

You are given the identity of the element.

P: ⇅ | ⇅ | ⇅|⇅|⇅ | ⇅ | ↑|↑|↑
   1s   2s   2p   3s   3p

Video Solution

Tutored Practice
Problem 7.3.2

7.3c Electron Configurations for Elements in Periods 4–7

Both potassium and calcium have electron configurations similar to those of other elements in Groups 1A and 2A.

K: [Ar]$4s^1$ [Ar] ↑
                   4s

Ca: [Ar]$4s^2$ [Ar] ⇅
                     4s

Scandium is the first transition element, and it is the first element to fill the $3d$ orbitals. The electron configurations of the transition elements follow Hund's rule:

Sc: [Ar]$4s^23d^1$ [Ar] ⇅ | ↑ | | | |
                         4s   3d

Ti: [Ar]$4s^23d^2$ [Ar] ⇅ | ↑|↑ | | |
                         4s   3d

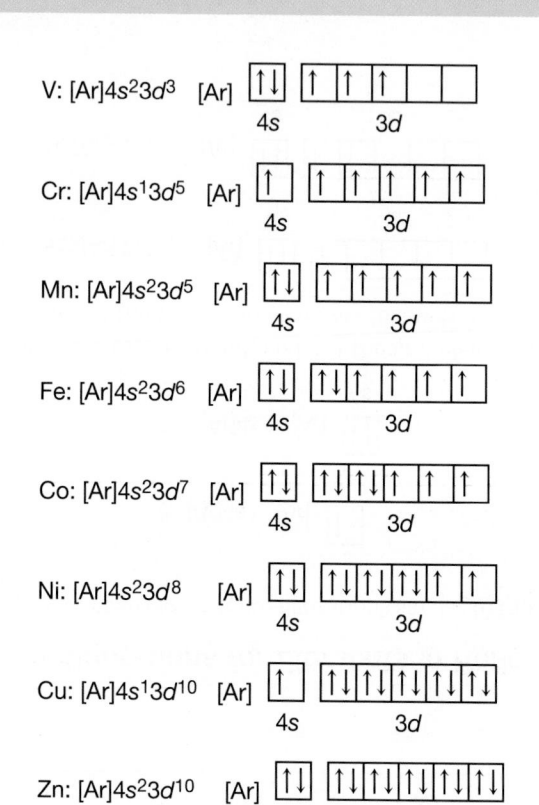

V: [Ar]$4s^2 3d^3$ [Ar] 4s 3d

Cr: [Ar]$4s^1 3d^5$ [Ar] 4s 3d

Mn: [Ar]$4s^2 3d^5$ [Ar] 4s 3d

Fe: [Ar]$4s^2 3d^6$ [Ar] 4s 3d

Co: [Ar]$4s^2 3d^7$ [Ar] 4s 3d

Ni: [Ar]$4s^2 3d^8$ [Ar] 4s 3d

Cu: [Ar]$4s^1 3d^{10}$ [Ar] 4s 3d

Zn: [Ar]$4s^2 3d^{10}$ [Ar] 4s 3d

Both chromium and copper have electron configurations that do not follow the general filling order, for reasons that are complex and related to the similar energies of the $4s$ and $3d$ orbitals in multielectron atoms. The electron configurations for these elements should be memorized. The electron configurations for all elements are shown in Interactive Table 7.3.1. There are a few exceptions to the general filling order in the heavier elements, but most elements follow the general guidelines that we have used to write electron configurations.

Electron Configurations of Atoms in the Ground State

Z	Element	Configuration	Z	Element	Configuration	Z	Element	Configuration
1	H	$1s^1$	41	Nb	$[Kr]5s^14d^4$	81	Tl	$[Xe]6s^25d^{10}4f^{14}6p^1$
2	He	$1s^2$	42	Mo	$[Kr]5s^14d^5$	82	Pb	$[Xe]6s^25d^{10}4f^{14}6p^2$
3	Li	$[He]2s^1$	43	Tc	$[Kr]5s^24d^5$	83	Bi	$[Xe]6s^25d^{10}4f^{14}6p^3$
4	Be	$[He]2s^2$	44	Ru	$[Kr]5s^14d^7$	84	Po	$[Xe]6s^25d^{10}4f^{14}6p^4$
5	B	$[He]2s^22p^1$	45	Rh	$[Kr]5s^14d^8$	85	At	$[Xe]6s^25d^{10}4f^{14}6p^5$
6	C	$[He]2s^22p^2$	46	Pd	$[Kr]4d^{10}$	86	Rn	$[Xe]6s^25d^{10}4f^{14}6p^6$
7	N	$[He]2s^22p^3$	47	Ag	$[Kr]5s^14d^{10}$	87	Fr	$[Rn]7s^1$
8	O	$[He]2s^22p^4$	48	Cd	$[Kr]5s^24d^{10}$	88	Ra	$[Rn]7s^2$
9	F	$[He]2s^22p^5$	49	In	$[Kr]5s^24d^{10}5p^1$	89	Ac	$[Rn]7s^26d^1$
10	Ne	$[He]2s^22p^6$	50	Sn	$[Kr]5s^24d^{10}5p^2$	90	Th	$[Rn]7s^26d^2$
11	Na	$[Ne]3s^1$	51	Sb	$[Kr]5s^24d^{10}5p^3$	91	Pa	$[Rn]7s^26d^15f^2$
12	Mg	$[Ne]3s^2$	52	Te	$[Kr]5s^24d^{10}5p^4$	92	U	$[Rn]7s^26d^15f^3$
13	Al	$[Ne]3s^23p^1$	53	I	$[Kr]5s^24d^{10}5p^5$	93	Np	$[Rn]7s^26d^15f^4$
14	Si	$[Ne]3s^23p^2$	54	Xe	$[Kr]5s^24d^{10}5p^6$	94	Pu	$[Rn]7s^25f^6$
15	P	$[Ne]3s^23p^3$	55	Cs	$[Xe]6s^1$	95	Am	$[Rn]7s^25f^7$
16	S	$[Ne]3s^23p^4$	56	Ba	$[Xe]6s^2$	96	Cm	$[Rn]7s^26d^15f^7$
17	Cl	$[Ne]3s^23p^5$	57	La	$[Xe]6s^25d^1$	97	Bk	$[Rn]7s^25f^9$
18	Ar	$[Ne]3s^23p^6$	58	Ce	$[Xe]6s^25d^14f^1$	98	Cf	$[Rn]7s^25f^{10}$
19	K	$[Ar]4s^1$	59	Pr	$[Xe]6s^24f^3$	99	Es	$[Rn]7s^25f^{11}$
20	Ca	$[Ar]4s^2$	60	Nd	$[Xe]6s^24f^4$	100	Fm	$[Rn]7s^25f^{12}$
21	Sc	$[Ar]4s^23d^1$	61	Pm	$[Xe]6s^24f^5$	101	Md	$[Rn]7s^25f^{13}$
22	Ti	$[Ar]4s^23d^2$	62	Sm	$[Xe]6s^24f^6$	102	No	$[Rn]7s^25f^{14}$
23	V	$[Ar]4s^23d^3$	63	Eu	$[Xe]6s^24f^7$	103	Lr	$[Rn]7s^26d^15f^{14}$
24	Cr	$[Ar]4s^13d^5$	64	Gd	$[Xe]6s^25d^14f^7$	104	Rf	$[Rn]7s^26d^25f^{14}$

Table 7.3.1 (*continued*)

Z	Element	Configuration	Z	Element	Configuration	Z	Element	Configuration
25	Mn	$[Ar]4s^23d^5$	65	Tb	$[Xe]6s^24f^9$	105	Db	$[Rn]7s^26d^35f^{14}$
26	Fe	$[Ar]4s^23d^6$	66	Dy	$[Xe]6s^24f^{10}$	106	Sg	$[Rn]7s^26d^45f^{14}$
27	Co	$[Ar]4s^23d^7$	67	Ho	$[Xe]6s^24f^{11}$	107	Bh	$[Rn]7s^26d^55f^{14}$
28	Ni	$[Ar]4s^23d^8$	68	Er	$[Xe]6s^24f^{12}$	108	Hs	$[Rn]7s^26d^65f^{14}$
29	Cu	$[Ar]4s^13d^{10}$	69	Tm	$[Xe]6s^24f^{13}$	109	Mt	$[Rn]7s^26d^75f^{14}$
30	Zn	$[Ar]4s^23d^{10}$	70	Yb	$[Xe]6s^24f^{14}$	110	Ds	$[Rn]7s^26d^85f^{14}$
31	Ga	$[Ar]4s^23d^{10}4p^1$	71	Lu	$[Xe]6s^25d^14f^{14}$	111	Rg	$[Rn]7s^26d^95f^{14}$
32	Ge	$[Ar]4s^23d^{10}4p^2$	72	Hf	$[Xe]6s^25d^24f^{14}$	112	Cn	$[Rn]7s^26d^{10}5f^{14}$
33	As	$[Ar]4s^23d^{10}4p^3$	73	Ta	$[Xe]6s^25d^34f^{14}$	113	—	$[Rn]7s^26d^{10}5f^{14}7p^1$
34	Se	$[Ar]4s^23d^{10}4p^4$	74	W	$[Xe]6s^25d^44f^{14}$	114	—	$[Rn]7s^26d^{10}5f^{14}7p^2$
35	Br	$[Ar]4s^23d^{10}4p^5$	75	Re	$[Xe]6s^25d^54f^{14}$	115	—	$[Rn]7s^26d^{10}5f^{14}7p^3$
36	Kr	$[Ar]4s^23d^{10}4p^6$	76	Os	$[Xe]6s^25d^64f^{14}$	116	—	$[Rn]7s^26d^{10}5f^{14}7p^4$
37	Rb	$[Kr]5s^1$	77	Ir	$[Xe]6s^25d^74f^{14}$	117	—	$[Rn]7s^26d^{10}5f^{14}7p^5$
38	Sr	$[Kr]5s^2$	78	Pt	$[Xe]6s^15d^94f^{14}$	118	—	$[Rn]7s^26d^{10}5f^{14}7p^6$
39	Y	$[Kr]5s^24d^1$	79	Au	$[Xe]6s^15d^{10}4f^{14}$			
40	Zr	$[Kr]5s^24d^2$	80	Hg	$[Xe]6s^25d^{10}4f^{14}$			

Example Problem 7.3.3 Write electron configurations for period 4–7 elements.

Write electron configurations for the following elements, in *spdf* notation and orbital box notation. Identify the element as paramagnetic or diamagnetic.

a. Zn (Do not use the noble gas notation.)
b. Sm (Use the noble gas notation.)

Solution:

You are asked to write electron configurations in both *spdf* and orbital box notation and to identify the element as paramagnetic or diamagnetic.

You are given the identity of the element.

Example Problem 7.3.3 *(continued)*

a. Zinc is element 30. Use the filling order to write the electron configuration, keeping in mind the maximum number of electrons that can be accommodated in a subshell.

$spdf$ notation: $1s^2 2s^2 2p^6 3s^2 3p^6 4s^2 3d^{10}$

orbital box notation:

           1s   2s   2p   3s   3p   4s   3d

Zinc is diamagnetic because its electron configuration shows no unpaired electrons.

b. Samarium is element 62. Use the symbol [Xe] to represent the first 54 electrons in the electron configuration.

$spdf$ notation: $[Xe]6s^2 4f^6$

orbital box notation: [Xe]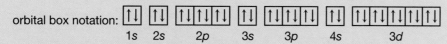

                              6s   4f

Samarium is paramagnetic because its electron configuration shows six unpaired electrons.

Video Solution

Tutored Practice Problem 7.3.3

7.3d Electron Configurations and the Periodic Table

Interactive Figure 7.3.3 shows the electron configurations for the elements in periods 1–3, written using *spdf* notation and noble gas notation.

Interactive Figure 7.3.3

Relate electron configuration and the periodic table.

1A							8A
1 H $1s^1$	2A	3A	4A	5A	6A	7A	2 He $1s^2$
3 Li $[He]2s^1$	4 Be $[He]2s^2$	5 B $[He]2s^2 2p^1$	6 C $[He]2s^2 2p^2$	7 N $[He]2s^2 2p^3$	8 O $[He]2s^2 2p^4$	9 F $[He]2s^2 2p^5$	10 Ne $[He]2s^2 2p^6$
11 Na $[Ne]3s^1$	12 Mg $[Ne]3s^2$	13 Al $[Ne]3s^2 3p^1$	14 Si $[Ne]3s^2 3p^2$	15 P $[Ne]3s^2 3p^3$	16 S $[Ne]3s^2 3p^4$	17 Cl $[Ne]3s^2 3p^5$	18 Ar $[Ne]3s^2 3p^6$

© 2013 Cengage Learning

Electron configurations, periods 1–3.

Notice that elements within a group have the following in common:

- They have the same number of electrons beyond the core electrons (represented by noble gas notation) and similar electron configurations. For example, both Li and Na have one electron in addition to the core electrons, and both elements have the general electron configuration [noble gas]ns^1.

- For the main group elements, the number of electrons beyond the core electrons is equal to the group number (with the exception of He).

The electrons beyond the core electrons are the **valence electrons** for an element. The valence electrons are the highest-energy electrons and are the electrons least strongly attracted to the nucleus. It is these electrons that are involved in chemical reactions and the formation of chemical bonds. As shown in Interactive Figure 7.3.3, for the Group A elements, the number of valence electrons is equal to the group number of the element. The fact that elements within a group have similar electron configurations and the same number of valence electrons suggests that elements within a group have similar properties, something we will investigate later in this unit.

The orbital filling order is related to the structure of the periodic table, as shown in Interactive Figure 7.3.4. The periodic table can be divided into four blocks, indicated

Interactive Figure 7.3.4

Identify electron configuration and valence electrons for elements using the periodic table.

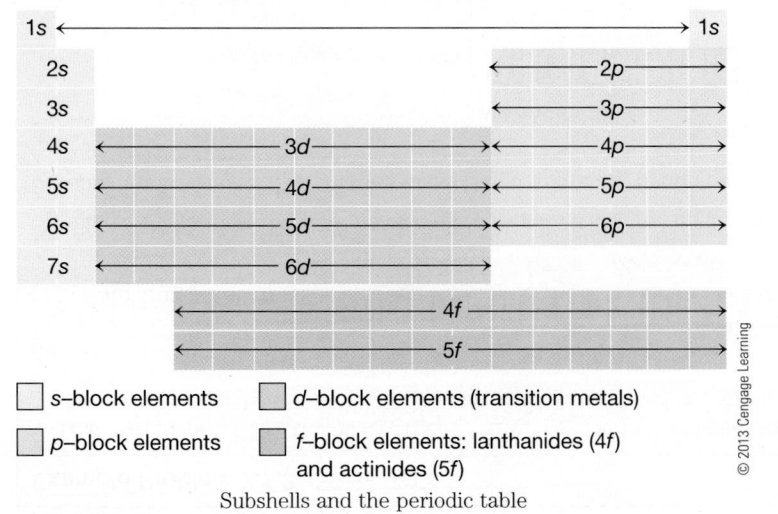

Subshells and the periodic table

© 2013 Cengage Learning

by color in Interactive Figure 7.3.4, each of which represents the type of subshell that is filled with the highest-energy electrons for the elements in that block. Elements in Groups 1A and 2A constitute the ns-block because the highest-energy electron is assigned to the ns orbital. Groups 3A through 8A make up the np-block, where np subshells are filled last. The s-block and p-block elements are the Group A elements in the periodic table and are commonly called the main-group elements. The transition elements sit within the $(n-1)d$-block, and the lanthanide and actinide elements make up the $(n-2)$ f-block.

The number of elements in each horizontal row within a block is related to the number of orbitals within that block and the number of electrons that can fill that subshell. For example, the d-block is 10 elements across because the d subshell contains five d orbitals and can accommodate a maximum of 10 electrons.

The periodic table can be used to generate electron configurations by "counting up" from hydrogen to the desired element. Each element represents the addition of one electron to an orbital. For example, arsenic, As, is a p-block element. To "count up" to As, you begin with the first-row elements, H and He, each of which represents an electron in the $1s$ orbital ($1s^2$). Next, Li and Be represent $2s^2$ and B through Ne represent $2p^6$. Similarly, Na and Mg represent $3s^2$ and Al through Ar represent $3p^6$. In the fourth row of the periodic table, K and Ca represent $4s^2$, Sc through Zn represent $3d^{10}$, and Ga through As represent three electrons in the $4p$ orbital, $4p^3$. The electron configuration of arsenic is

$$\text{As: } 1s^2 2s^2 2p^6 3s^2 3p^6 4s^2 3d^{10} 4p^3 \qquad [\text{Ar}]4s^2 3d^{10} 4p^3$$

Section 7.3 Mastery

7.4   Properties of Atoms

7.4a   Trends in Orbital Energies

As shown in the previous section, the organization of the periodic table is closely related to electron configurations. The energy of atomic orbitals is also related to the structure of the periodic table.

As you move down within a group, the energy of highest-energy occupied orbitals increases. Likewise, the elements have greater numbers of electrons, and they occupy orbitals with higher n. As the value of n increases from element to element, the orbitals are larger, the electrons in these orbitals are farther from the nucleus, and the electrons in these high-energy orbitals experience weaker electron–nucleus attractions.

Explore orbital energies.

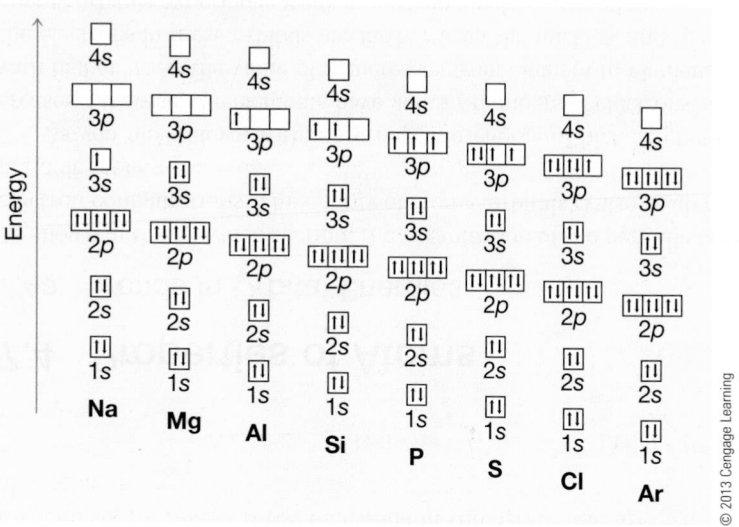

Orbital energies, Na–Ar

© 2013 Cengage Learning

Similarly, as you move left to right across the periodic table within a period, the energy of all atomic orbitals decreases. Consider the elements in the third period, Na through Ar, for example (Interactive Figure 7.4.1). Comparing Na to Mg, two factors affect the energy of the atomic orbitals. First, the Mg nucleus has one more proton than the Na nucleus, which increases the nucleus–electron attractive forces. This will lower the energy of atomic orbitals. Second, Mg has one more electron than Na, which increases the electron–electron repulsions. This will increase the energy of atomic orbitals.

As shown in Interactive Figure 7.4.1, the Mg atomic orbitals are lower in energy than the Na atomic orbitals. This suggests that the attractive forces that result from additional protons in the nucleus are more important than the repulsive forces from the additional electrons. In general, when moving across a period from left to right, the attractive forces between the electrons and the nucleus increase and thus the orbital energies decrease.

Figure 7.4.2 shows the periodic trend for relative energy of atomic orbitals; elements in the upper right of the periodic table have the lowest-energy orbitals and those in the lower left have the highest-energy orbitals.

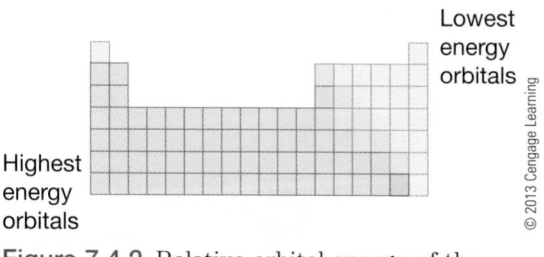

Figure 7.4.2 Relative orbital energy of the elements

© 2013 Cengage Learning

Effective Nuclear Charge

The combination of attractive forces between electrons and the nucleus and electron–electron repulsive forces is called the **effective nuclear charge** and is given the symbol Z^*. The effective nuclear charge for the highest-energy electrons in an atom is the nuclear charge felt by those electrons, taking into account the electron–electron repulsive forces between high-energy electrons and core electrons. A simplified method for calculating Z^* for the highest-energy electrons in an atom is

$$Z^* = Z - [\text{number of core electrons}]$$

Using this simplified method, we can calculate values of Z and Z^* for Na through Ar:

	Na	Mg	Al	Si	P	S	Cl	Ar
Z	11	12	13	14	15	16	17	18
Z^*	1	2	3	4	5	6	7	8

In general, effective nuclear charge increases moving left to right across the periodic table. More sophisticated calculation methods for determining the electron-electron repulsive forces in an atom or ion result in Z^* values that also increase moving left to right across the periodic table, but these values more accurately represent electron energies (Interactive Figure 7.4.3). As Z^* increases, the electrons feel a strong attractive force from the nucleus. This has the effect of decreasing the energy of the atomic orbitals. As the energy of the atomic orbitals decreases, the electrons are held closer to the nucleus and are held more tightly to the nucleus. This general trend in orbital energies controls most of the observed trends we see in atomic properties, as we will explore in the rest of this unit.

7.4b Atomic Size

The size of an atom is controlled by the size of its orbitals, but orbitals are boundless, with no defined outer limit. Chemists use many different definitions for the size of an atom. The **covalent radius** of an element is the distance between the nuclei of two atoms of that element when they are held together by a single bond. For example, the distance between two bonded Cl atoms in Cl_2 is 198 pm (Figure 7.4.4). Therefore, the covalent radius of each Cl atom is 99.0 pm. Other covalent radii are determined from atom distances in molecules. Because distances between bonded atoms vary from molecule to molecule, tables of covalent radii show average values.

The **metallic radius** of an element is the distance between the nuclei of two atoms in a metallic crystal. Tables of atomic radii generally report covalent radii for nonmetals and metallic radii for metals.

Interactive Figure 7.4.3

Explore effective nuclear charge.

Higher Z^*, electrons held more tightly

Relative effective nuclear charge for the elements

© 2013 Cengage Learning

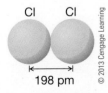

198 pm

© 2013 Cengage Learning

Figure 7.4.4 The distance between two Cl atoms in Cl_2 is 198 pm.

The size of an atom is closely related to its electron configuration. Recall that the lower an orbital's energy, the more tightly an electron in that orbital is held to the nucleus and the smaller the orbital. Therefore, the trend in atomic size follows that of orbital energies, with the smallest atoms in the top right of the periodic table and the largest atoms in the lower left (Interactive Figure 7.4.5).

Interactive Figure 7.4.5

Explore atomic size of the elements.

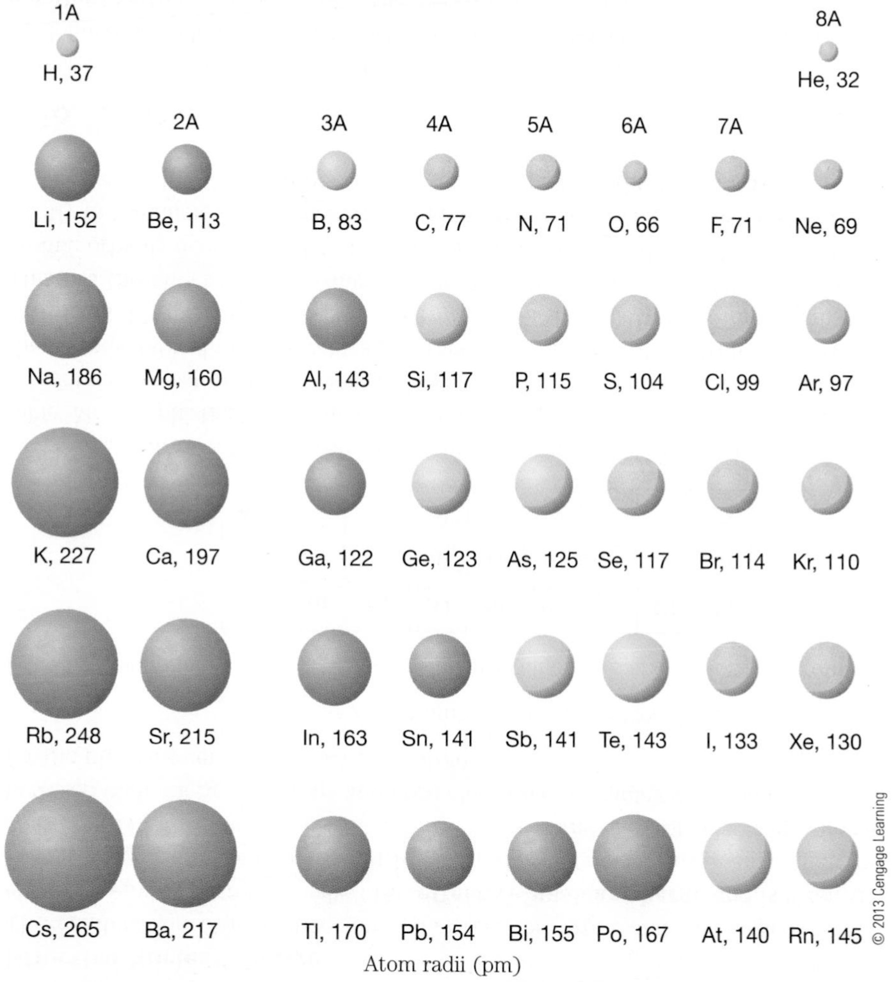

1A							8A
H, 37							He, 32

	2A	3A	4A	5A	6A	7A	
Li, 152	Be, 113	B, 83	C, 77	N, 71	O, 66	F, 71	Ne, 69
Na, 186	Mg, 160	Al, 143	Si, 117	P, 115	S, 104	Cl, 99	Ar, 97
K, 227	Ca, 197	Ga, 122	Ge, 123	As, 125	Se, 117	Br, 114	Kr, 110
Rb, 248	Sr, 215	In, 163	Sn, 141	Sb, 141	Te, 143	I, 133	Xe, 130
Cs, 265	Ba, 217	Tl, 170	Pb, 154	Bi, 155	Po, 167	At, 140	Rn, 145

Atom radii (pm)

© 2013 Cengage Learning

7.4c Ionization Energy

Ionization energy is the amount of energy required to remove an electron from a gaseous atom.

$$X(g) \rightarrow X^+(g) + e^-(g) \qquad \Delta E = \text{ionization energy}$$

All atoms require energy to remove an electron, and therefore ionization energy values are positive. The trends in ionization energies follow those expected based on orbital energies (Interactive Figure 7.4.6).

Notice that the elements farther down within a group, for example, moving from H to Li to Na, have decreased ionization energy. This trend is related to orbital energies. The electron removed from sodium occupies a $3s$ orbital, which is higher in energy than the electron lost by lithium ($2s$), which in turn is higher than the energy of the electron lost by hydrogen ($1s$).

Ionization energy generally increases when moving left to right across a period. This trend is related both to orbital energies and to effective nuclear charge, Z^*. Orbital energies decrease and Z^* increases in the elements as you move left to right across a period. Therefore, the farther right an element is, the more strongly the outermost electrons are attracted to the nucleus and the higher its ionization energy. In general, a large, positive ionization energy value indicates that an element is more stable as a neutral atom than as a cation.

Interactive Figure 7.4.6

Explore ionization energy values.

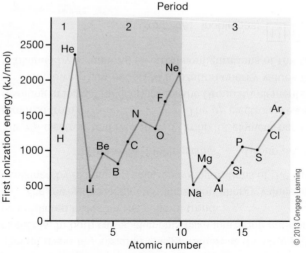

Ionization energy values for elements in periods 1–3

Notice that Interactive Figure 7.4.6 shows two areas where there are exceptions to the general trend in ionization energy values: at the Group 3A elements and at the Group 6A elements. In both cases, elements have ionization energy values that are lower than would be predicted based on the general trend.

The lower-than-expected ionization energy values for the Group 3A elements can be explained by examining electron configurations of the Group 2A and 3A elements.

Group 2A: [noble gas]ns^2 Group 3A: [noble gas]ns^2np^1

The np electron removed from a Group 3A element has higher energy than the ns electron removed from a Group 2A element. The np electron is easier to remove than the ns electron, resulting in ionization energies for the Group 3A elements that are lower than predicted.

The lower-than-expected ionization energy values for the Group 6A elements can be explained by examining electron configurations of the Group 5A and 6A elements.

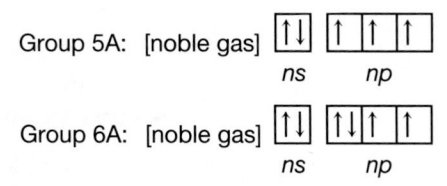

The electron removed from a Group 6A element is in a filled np orbital. Electron–electron repulsive forces make it easier to remove this paired electron than the unpaired np electron removed from a Group 5A element; therefore, Group 6A elements have ionization energies that are lower than predicted.

7.4d Electron Affinity

Electron affinity is the energy change when a gaseous atom gains an electron.

$$X(g) + e^- \rightarrow X^-(g) \qquad \Delta E = \text{electron affinity}$$

Although most elements have negative electron affinity values, some are slightly positive or close to zero. Keep in mind that a negative electron affinity value indicates that energy is released when an electron is added to a gaseous atom and that a large, negative electron affinity value indicates that an element is more stable as an anion than as a neutral atom. Not all atoms have a measurable electron affinity. Despite this, the trends in electron affinities generally follow those of the other periodic properties we have examined thus far (with numerous exceptions) (Interactive Figure 7.4.7).

Elements lower in a group generally have less negative electron affinity values than those higher in a group.

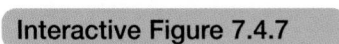

Interactive Figure 7.4.7

Explore electron affinity values.

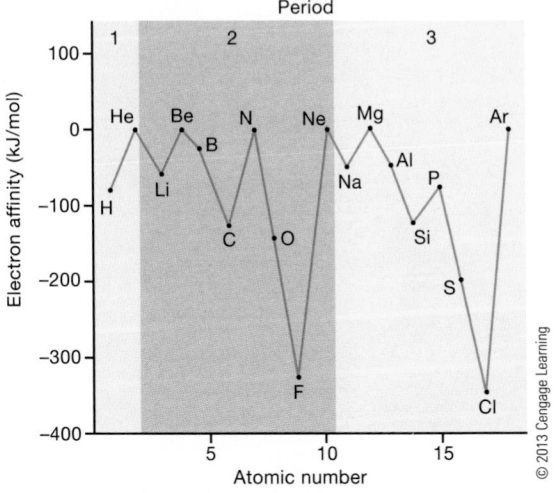

Electron affinity values for elements in periods 1–3

H: −72.8 kJ/mol   Li: −59.6 kJ/mol   Na: −52.9 kJ/mol   K: −48.4 kJ/mol

As atomic size increases down a group, the orbital that is occupied by the new electron increases in energy. As the orbital energy increases, the attractive forces between the new electron and the nucleus decrease, so electron affinity becomes less negative. In general, adding electrons becomes less favorable the farther down a group the element is.

Electron affinity values generally are more negative in elements farther to the right across a period. There are some notable exceptions, however, in the Group 2A and 5A elements.

Group 2A: [noble gas] $\boxed{\uparrow\downarrow}$
                                ns

Group 5A: [noble gas] $\boxed{\uparrow\downarrow}$ $\boxed{\uparrow\,|\,\uparrow\,|\,\uparrow}$
                                ns np

The Group 2A elements have electron affinities that are very close to zero because the added electron occupies the np orbital. Adding an electron to this higher-energy orbital requires energy, so the electron affinity is much lower (less negative) than predicted from general trends. The Group 5A elements have low (less negative) electron affinity values because the added electron fills a half-filled np orbital and introduces new electron–electron repulsion forces.

Notice that the Group 8A elements also have electron affinity values that are close to zero. An electron added to a noble gas element would occupy the $(n + 1)s$ orbital in a higher-energy level. Thus, the noble gases generally do not form anions.

Section 7.4 Mastery

7.5   Formation and Electron Configuration of Ions

7.5a   Cations

A cation forms when an atom loses one or more electrons. Metals have low ionization energy values, so metals generally form cations. For example, sodium, a Group 1A element, loses one electron to form the sodium cation, Na^+. Examination of the Na electron configuration shows that the highest-energy electron occupies the $3s$ orbital. This is the sodium *valence electron*, and it is this electron that is lost upon formation of Na^+.

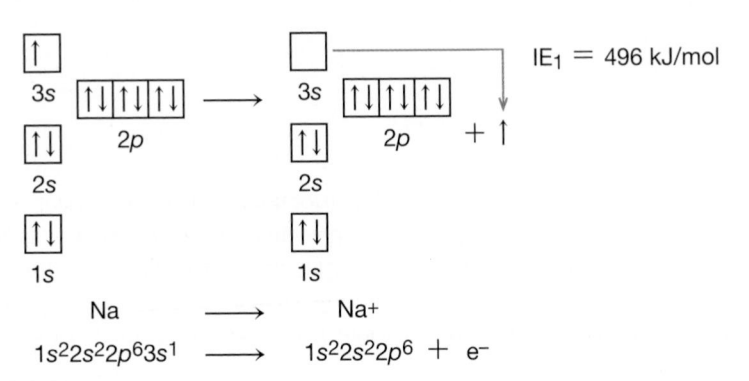

$IE_1 = 496$ kJ/mol

$$Na \longrightarrow Na^+$$

$$1s^2 2s^2 2p^6 3s^1 \longrightarrow 1s^2 2s^2 2p^6 + e^-$$

The energy required to remove a second electron from an atom is the second ionization energy, IE_2. For any atom, $IE_2 > IE_1$. Sodium does not normally form a +2 cation because the second electron is removed from the $2p$ subshell, which is much lower in energy than the $3s$ subshell. Removing an electron from an inner energy shell requires a great deal of energy, in this case almost 10 times as much as is required to remove the $3s$ valence electron.

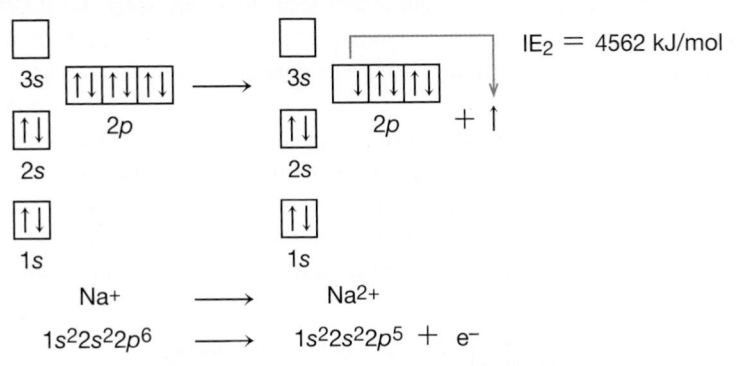

$IE_2 = 4562$ kJ/mol

$$Na^+ \longrightarrow Na^{2+}$$

$$1s^2 2s^2 2p^6 \longrightarrow 1s^2 2s^2 2p^5 + e^-$$

Magnesium, a Group 2A element, loses both of its valence electrons to form a +2 cation.

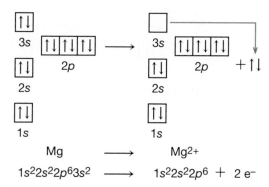

$$Mg \longrightarrow Mg^{2+}$$
$$1s^2 2s^2 2p^6 3s^2 \longrightarrow 1s^2 2s^2 2p^6 + 2\,e^-$$

Like sodium, magnesium is expected to easily lose its valence electrons, but not an electron from a lower-energy shell. As shown in Table 7.5.1, the first and second ionization energies for Mg are relatively low, but the third ionization energy (IE_3) is more than 10 times greater than IE_1.

In general, when metals form cations, they lose electrons from the highest-energy occupied orbitals, those with the highest n value. Transition metals have valence electrons in ns, $(n-1)d$, and in the case of lanthanides and actinides, $(n-2)f$ orbitals. When a transition metal forms a cation, electrons are first lost from the outermost occupied orbital (the orbital with highest n). For example, when iron forms a +2 cation, it loses two $4s$ electrons.

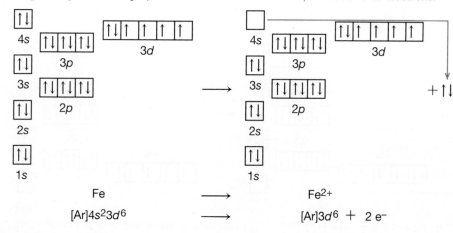

$$Fe \longrightarrow Fe^{2+}$$
$$[Ar]4s^2 3d^6 \longrightarrow [Ar]3d^6 + 2\,e^-$$

Table 7.5.1 First, Second, and Third IE values for Mg	
$Mg(g) \rightarrow Mg^+(g) + e^-$	$IE_1 = 738$ kJ/mol
$Mg^+(g) \rightarrow Mg^{2+}(g) + e^-$	$IE_2 = 1451$ kJ/mol
$Mg^{2+}(g) \rightarrow Mg^{3+}(g) + e^-$	$IE_3 = 7733$ kJ/mol

Recall that the 4s orbital is filled before the 3d orbitals are filled. However, electrons are removed from the highest-energy orbitals because those electrons feel the weakest attractive "pull" from the positively charged nucleus.

Transition metals often form more than one cation. Iron is commonly found in both the +2 and +3 oxidation states. In the formation of Fe^{3+}, two 4s electrons and one 3d electron are lost.

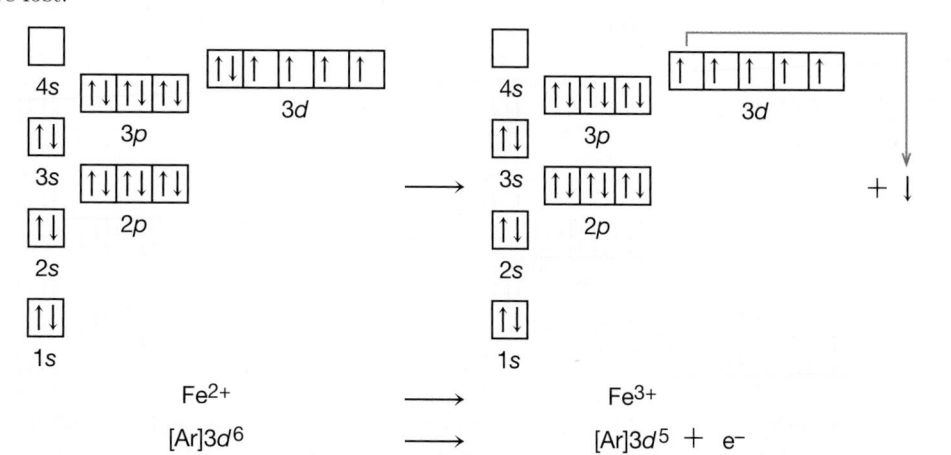

Notice that the electron removed to form Fe^{3+} from Fe^{2+} is taken from a filled 3d orbital, not a half-filled orbital. This minimizes the repulsive forces between electrons.

Example Problem 7.5.1 Write electron configurations for cations.

Write electron configurations for the following ions in *spdf* and orbital box notation.

a. Al^{3+} (Do not use noble gas notation.)
b. Cr^{2+} (Use noble gas notation.)

Solution:

You are asked to write an electron configuration for a cation using *spdf* and orbital box notation.

You are given the identity of the cation.

a. Aluminum is element 13. The element loses three electrons from its highest-energy orbitals to form the Al^{3+} ion.

Al: $1s^2 2s^2 2p^6 3s^2 3p^1$ [box notation] 1s 2s 2p 3s 3p

Example Problem 7.5.1 *(continued)*

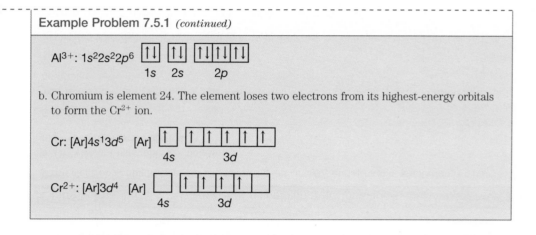

Al^{3+}: $1s^2 2s^2 2p^6$ | ↑↓ | ↑↓ | ↑↓|↑↓|↑↓ |
                  1s   2s   2p

b. Chromium is element 24. The element loses two electrons from its highest-energy orbitals to form the Cr^{2+} ion.

Cr: [Ar]$4s^1 3d^5$ [Ar] | ↑ | ↑|↑|↑|↑|↑ |
                    4s   3d

Cr^{2+}: [Ar]$3d^4$ [Ar] | | ↑|↑|↑|↑| |
                      4s   3d

Video Solution

Tutored Practice
Problem 7.5.1

7.5b Anions

An anion forms when an atom gains one or more electrons. Nonmetals have relatively large, negative electron affinity values, so nonmetals generally form anions. For example, chlorine gains one electron to form the chloride ion, Cl$^-$. Examination of the Cl electron configuration shows that the $3p$ subshell in Cl contains a single vacancy. Cl therefore gains a single electron in a $3p$ orbital to form Cl$^-$.

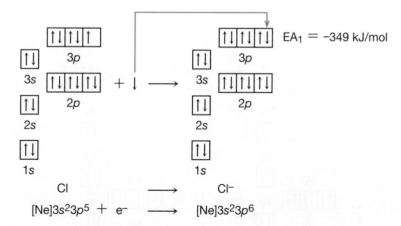

EA$_1$ = −349 kJ/mol

Cl   ⟶   Cl$^-$

[Ne]$3s^2 3p^5$ + e$^-$ ⟶ [Ne]$3s^2 3p^6$

Sulfur has two $3p$ orbital vacancies and can therefore gain two electrons.

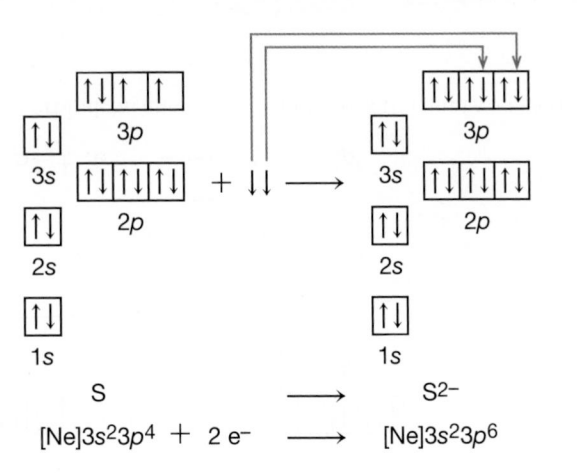

$$S \longrightarrow S^{2-}$$

$$[Ne]3s^23p^4 + 2\ e^- \longrightarrow [Ne]3s^23p^6$$

The total number of electrons gained when an element forms an anion is related to the electron configuration of the element. In the case of Cl, only one electron is gained because a second electron would occupy the higher-energy $4s$ orbital. An electron in this orbital is farther from the nucleus and therefore feels less attractive "pull" from the nucleus. For the same reason, sulfur is often found as a –2 anion and not a –3 anion.

Finally, not all main-group elements commonly form ions. The Group 1A, 2A, and 3A metals lose one, two, and three electrons, respectively, to form cations where the charge on the cation is equal to the group number. The Group 5A, 6A, and 7A nonmetals gain three, two, and one electrons, respectively, to form anions where the charge on the anion is equal to $(8 -$ the group number). The Group 4A elements, however, are not commonly found as $+4$ or -4 ions because the gain or loss of four electrons requires a great deal of energy.

Example Problem 7.5.2 Write electron configurations for anions.

Write electron configurations for the following ions in $spdf$ and orbital box notation.

a. O^{2-} (Do not use noble gas notation.)
b. I^- (Use noble gas notation.)

Solution:

You are asked to write an electron configuration for an anion using $spdf$ and orbital box notation.

You are given the identity of the anion.

a. Oxygen is element 8. The element gains two electrons in its highest-energy orbitals to form the O^{2-} ion.

Example Problem 7.5.2 *(continued)*

O: $1s^2 2s^2 2p^4$ [↑↓] [↑↓] [↑↓|↑|↑]
               1s   2s   2p

O^{2-}: $1s^2 2s^2 2p^6$ [↑↓] [↑↓] [↑↓|↑↓|↑↓]
                 1s   2s   2p

b. Iodine is element 53. The element gains one electron in its highest-energy orbital to form the I^- ion.

I: $[Kr]5s^2 4d^{10} 5p^5$ [Kr] [↑↓] [↑↓|↑↓|↑↓|↑↓|↑↓] [↑↓|↑↓|↑]
                                 5s   4d   5p

I^-: $[Kr]5s^2 4d^{10} 5p^6$ [Kr] [↑↓] [↑↓|↑↓|↑↓|↑↓|↑↓] [↑↓|↑↓|↑↓]
                                 5s   4d   5p

Video Solution

**Tutored Practice
Problem 7.5.2**

7.5c Ion Size

The size of an ion is related both to the size of the atom from which it is formed and the ion charge. In Interactive Figure 7.5.1, notice that

- cations are smaller than the atoms from which they are formed;

- anions are larger than the atoms from which they are formed; and

- anions are generally larger than cations.

Cations are smaller than the atoms from which they are formed primarily because they have fewer electrons. For example, magnesium has 12 electrons and loses 2 of them to form the Mg^{2+} cation.

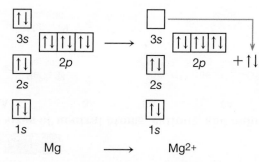

Explore relative sizes of neutral atoms, cations, and anions.

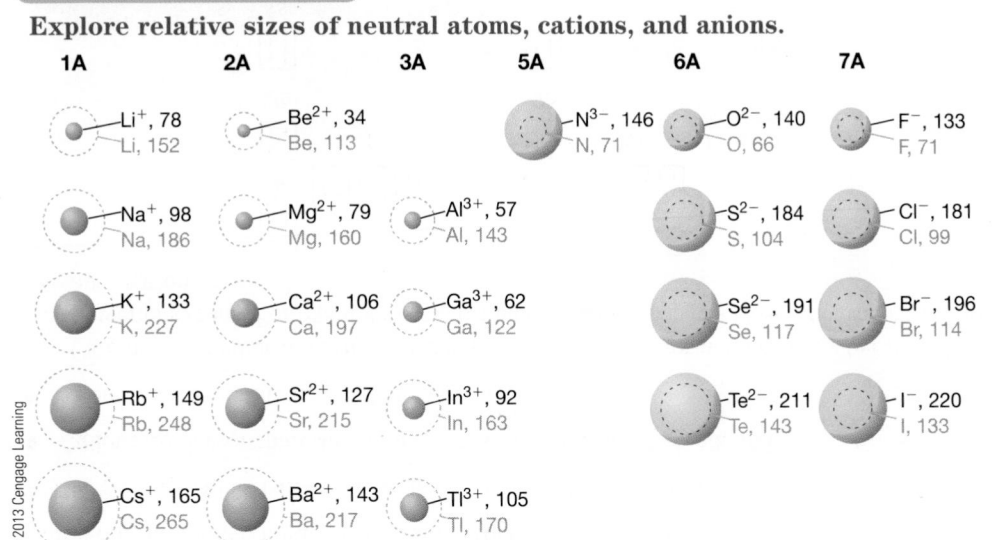

Ionic radii (pm)

A Mg^{2+} ion is smaller than a Mg atom because the highest-energy occupied orbital has changed from $3s$ to $2p$. The $2p$ orbital is smaller than the $3s$ orbital, so the radius of the cation is smaller than the radius of the Mg atom. In addition, the fact that Mg^{2+} has fewer electrons than Mg minimizes electron repulsion forces, also decreasing the size of the cation.

Anions are larger than the atoms from which they are formed primarily because of their added electrons. Consider formation of the sulfide ion, S^{2-}, from a sulfur atom. The sulfide ion has 18 electrons, 2 more than a sulfur atom. The additional electrons occupy the $3p$ subshell, which was already partially occupied in a sulfur atom.

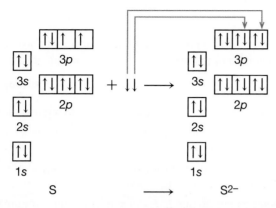

The added electrons increase the existing repulsive forces between electrons in the $3p$ subshell, causing the electrons to move away from one another. This in turn leads to an expansion of the electron cloud and an anion that is larger than the neutral atom.

Consider the following **isoelectronic ions**, species with the same number of electrons but different numbers of protons.

	Protons	Electrons	Electron Configuration	Ion Radius
O^{2-}	8	10	$1s^2 2s^2 2p^6$	140 pm
F^-	9	10	$1s^2 2s^2 2p^6$	133 pm
Na^+	11	10	$1s^2 2s^2 2p^6$	102 pm
Mg^{2+}	12	10	$1s^2 2s^2 2p^6$	66 pm

The ion with the largest radius in this isoelectronic series is O^{2-}, and the smallest ion is Mg^{2+}. The oxide ion has only 8 protons to attract its 10 electrons, whereas the magnesium ion has 12 protons and thus more strongly attracts its 10 electrons.

Section 7.5 Mastery

Unit Recap

Key Concepts

7.1 Electron Spin and Magnetism

- Electrons exert a magnetic field due to electron spin (7.1a).
- An electron has two possible spin states, indicated by the spin quantum number ($m_s = +\frac{1}{2}$ or $-\frac{1}{2}$) (7.1a).
- Materials with unpaired electrons are magnetic (7.1b).
- The magnetic nature of a material can be categorized as diamagnetic, paramagnetic, or ferromagnetic (7.1b).

7.2 Orbital Energy

- In a single-electron atom, the energy of an orbital is directly related to n, the principal quantum number (7.2a).
- In multielectron atoms, orbital energy depends on both n and ℓ (7.2a).

7.3 Electron Configuration of Elements

- The electron configuration of an element shows how electrons are distributed in orbitals (7.3a).
- An atom's ground state is the lowest-energy electron configuration for that atom (7.3a).
- The Pauli exclusion principle states that no two electrons within an atom can have the same set of four quantum numbers (7.3a).
- Electron configurations are written using *spdf* and orbital box notation (7.3b).
- Hund's rule of maximum multiplicity states that the lowest-energy electron configuration is the one where the maximum number of electrons is unpaired (7.3b).
- Noble gas notation is a shortcut used when writing electron configurations, where a noble gas symbol is used to represent core electrons (7.3b).
- Valence electrons are the electrons beyond the core electrons (7.3d).

7.4 Properties of Atoms

- Orbital energy in elements decreases as you move left to right across the periodic table and increases as you move down a periodic group (7.4a).

- The effective nuclear charge for the highest-energy electrons in an atom is the nuclear charge felt by those electrons, taking into account the attractive forces between electrons and the nucleus and the repulsive forces between electrons (7.4a).

- Effective nuclear change generally increases in elements as you move left to right across the periodic table (7.4a).

- The covalent radius of an element is the distance between the nuclei of two atoms of that element when they are held together by a single bond (7.4b).

- Metallic radius is the distance between two atoms in a metallic crystal (7.4b).

- Atom size generally decreases as you move left to right across the periodic table and increases as you move down within a periodic group (7.4b).

- Ionization energy is the amount of energy required to remove an electron from a gaseous atom (7.4c).

- Ionization energy generally increases as you move left to right across the periodic table and decreases as you move down within a periodic group (7.4c).

- Electron affinity is the energy change when an electron is added to a gaseous atom (7.4d).

- Electron affinity values generally become more negative as you move left to right across the periodic table and become less negative (more positive) as you move down within a periodic group (7.4d).

7.5 Formation and Electron Configuration of Ions

- Metals generally form cations by losing their highest-energy valence electrons (7.5a).

- Nonmetals generally form anions by gaining electrons in their highest-energy orbitals (7.5b).

- Cations are generally smaller than the atom from which they are formed, and anions are generally larger than the atoms from which they are formed (7.5c).

- Isoelectronic ions have the same number of electrons but different numbers of protons (7.5c).

Key Terms

7.1 Electron Spin and Magnetism
electron spin
spin state
spin quantum number, m_s
diamagnetic
paramagnetic
ferromagnetic

7.3 Electron Configuration of Elements
electron configuration
ground state
Pauli exclusion principle
spdf notation
orbital box notation
Hund's rule of maximum multiplicity
noble gas notation
core electrons
valence electrons

7.4 Properties of Atoms
effective nuclear charge
covalent radius
metallic radius
ionization energy
electron affinity

7.5 Formation and Electron Configuration of Ions
isoelectronic ions

Unit 7   Review and Challenge Problems

8 Covalent Bonding and Molecular Structure

Unit Outline

In This Unit...

We will examine chemical bonding in detail in this unit and the next. Here we apply what you have learned about atomic structure, electron configurations, and periodic trends to the chemical bonds formed between atoms and ions and the shapes of molecules and ions that contain covalent bonds. This unit and the next primarily address covalent bonding; we examined ionic bonding briefly in Elements and Compounds (Unit 2), and we will do so in more detail in The Solid State (Unit 12). We will also examine the forces that exist between individual particles, called intermolecular forces, in Intermolecular Forces and the Liquid State (Unit 11).

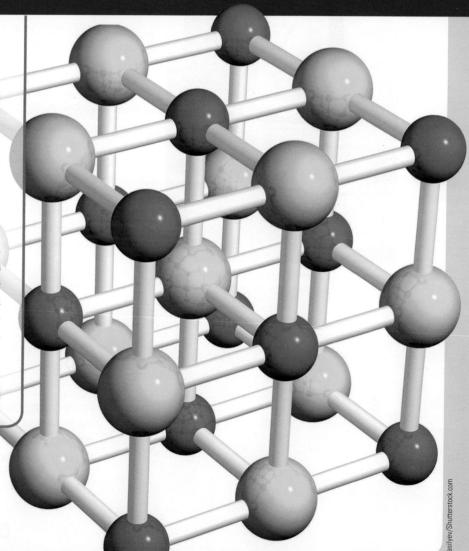

8.1 An Introduction to Covalent Bonding

8.1a Coulomb's Law

Matter is made up of atoms and ions that experience both attractive and repulsive forces. The strength of the force between two charged particles together in any material is described by **Coulomb's law** (Equation 8.1). According to this law, the force of attraction or repulsion between two charged species is directly proportional to the magnitude of the charge on the particles (q_A and q_B in Equation 8.1) and inversely proportional to the square of the distance between the two particles (r in Equation 8.1).

$$\text{Force} \propto \frac{(q_A)(q_B)}{r^2} \qquad \textbf{(8.1)}$$

For electrons, $q = -e$, and for nuclei, $q = +Ze$

where

- e = magnitude of electron charge (1.6022×10^{-19} C)
- Z = nuclear charge (number of protons)
- r = distance between particles A and B

Interactive Figure 8.1.1 shows a demonstration of Coulomb's law using a gold leaf electroscope in which a narrow metal plate and a thin sheet of gold are connected to a conducting rod. The plastic pen held above the electroscope carries a static charge, which is transferred to the metals via inductance. Both metal surfaces pick up the same charge (both become either positively or negatively charged), so they repel and the gold sheet moves away from the fixed metal plate.

Both attractive and repulsive forces exist in matter, and it is the balance of these forces (and the resulting overall decrease in energy) that results in the formation of chemical **bonds**, interactions that hold two or more atoms together. Chemical attractive forces involve opposite charges, such as those between protons in a nucleus and the electrons surrounding that nucleus, and between positive and negative ions. Repulsive forces occur between species with like charges, such as the positively charged nuclei of two bonded atoms.

Chemical bonding can be categorized by the types of attractive forces that exist between particles. For example, in **ionic bonding**, found in ionic solids such as NaCl and CaCO$_3$, there are strong attractive forces between positively and negatively charged ions. In **covalent bonding**, which occurs in compounds such as H$_2$O and NH$_3$, there are attractive forces between electrons and nuclei on adjacent atoms within a molecule. In **metallic bonding**, attractive forces exist between the electrons and nuclei in metal atoms such as Cu and Fe. Attractive forces are responsible for more than chemical bonding. There are forces that exist between molecules, called **intermolecular forces**, which will be discussed in more detail in Intermolecular Forces and the Liquid State (Unit 11).

Interactive Figure 8.1.1

Explore Coulomb's Law

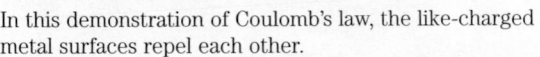

In this demonstration of Coulomb's law, the like-charged metal surfaces repel each other.

8.1b Fundamentals of Covalent Bonding

A covalent bond is characterized by the sharing of valence electrons by two adjacent atoms. This type of bonding occurs most typically between nonmetal elements such as carbon, hydrogen, oxygen, and nitrogen.

For example, consider a simple covalently bonded molecule, H_2 (Interactive Figure 8.1.2). When two isolated H atoms are at a great distance from one another, they feel no attractive or repulsive forces. However, as the atoms approach more closely, the attractive and repulsive forces between the two atoms become important. At very short distances, repulsive forces become more important than attractive forces, and the atoms repel.

Attractive forces between the hydrogen atoms result from the interaction between the positively charged nucleus (represented by the element symbol) on one hydrogen atom and the negatively charged electron (represented by a dot) on the other hydrogen atom. Repulsive forces are the result of both the like-charged nuclei on adjacent atoms and the electrons on the two hydrogen atoms.

Interactive Figure 8.1.2

Explore the relationship between potential energy and interatomic distance.

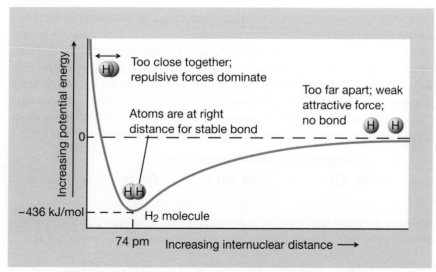

Changing the distance between H atoms affects potential energy.

Nucleus-nucleus repulsion   Electron-electron repulsion   Electron-nucleus attraction

When two atoms approach each other, these repulsive and attractive forces always occur. When the attractive forces are greater than the repulsive forces, a covalent bond forms.

Attraction is greater than repulsion

The balance between the attractive and repulsive forces in H_2 is related to the distance between H atoms, as shown in Interactive Figure 8.1.2. At large distances, neither attractive nor repulsive forces are important and no bond forms between H atoms. At short distances, repulsive forces are stronger than attractive forces and no bond forms. At an internuclear distance where the attractive forces are stronger than the repulsive forces, a bond forms between the H atoms. The two valence electrons "pair up" and are shared between the two hydrogen nuclei in a covalent bond that is represented by a single line connecting the H atoms. (◄Flashback to Section 7.3a The Pauli Exclusion Principle)

$$H\cdot \ + \ H\cdot \ \longrightarrow \ H-H$$

The single line between H atoms in H_2 is a very useful representation of a chemical bond, but it does not give an accurate picture of the distribution of bonding electrons in the H_2 molecule. More sophisticated descriptions of chemical bonding will be discussed in Theories of Chemical Bonding (Unit 9). (► Flashforward to Section 9.1 Valence Bond Theory)

Section 8.1 Mastery

8.2 Lewis Structures

8.2a Lewis Symbols and Lewis Structures

One of the most important tools chemists use to predict the properties of a chemical species is the Lewis structure. A **Lewis structure** (also called a Lewis dot structure or Lewis diagram) shows the arrangement of valence electrons (both bonding and nonbonding) and nuclei in covalently bonded molecules and ions. The simplest Lewis structure is

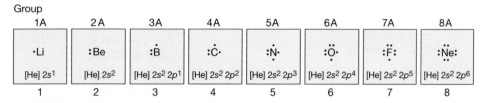

Group

1A	2A	3A	4A	5A	6A	7A	8A
·Li	:Be	:Ḃ	:Ċ·	:N̈·	:Ö·	:F̈:	:N̈e:
[He] $2s^1$	[He] $2s^2$	[He] $2s^2 2p^1$	[He] $2s^2 2p^2$	[He] $2s^2 2p^3$	[He] $2s^2 2p^4$	[He] $2s^2 2p^5$	[He] $2s^2 2p^6$
1	2	3	4	5	6	7	8

Number of valence electrons

Figure 8.2.1 Lewis symbols for the second-period elements

the **Lewis symbol** for an element, where the element symbol represents the nucleus and core (nonvalence) electrons and dots represent valence electrons.

Lewis symbols have the following characteristics, as shown by the examples in Figure 8.2.1:

- in a Lewis symbol, electrons are arranged around the four sides of the element symbol;

- a Lewis symbol shows the total number of valence electrons for an element; and

- Lewis symbols can be drawn to reflect the electron configuration of the element (the number of paired and unpaired electrons in the ground-state electron configuration of the element), but this is not an absolute rule (for example, both :Be and ·Be· are accepted Lewis symbols for beryllium).

The Lewis symbols for the second-period elements in Figure 8.2.1 also show the relationship between the number of valence electrons and group number for the A group (main-group) elements. Recall that the number of valence electrons for any main-group element is equal to the group number for that element.

Example Problem 8.2.1 Interpret Lewis symbols.

a. Draw the Lewis symbol for Te, tellurium.
b. The following Lewis diagram represents the valence electron configuration of a main-group element

X·

Identify the element in the fifth period that has this valence electron configuration.

Example Problem 8.2.1 *(continued)*

Solution:

You are asked to draw a Lewis symbol for an element or to identify an element from a Lewis symbol.

You are given the identity of an element or a Lewis symbol.

a. $:\overset{\cdot\cdot}{\underset{\cdot}{Te}}\cdot$ Tellurium is a Group 6A element and has six valence electrons.

b. Rubidium: This element has one valence electron, so it is a Group 1A element. Rubidium is the Group 1A element in the fifth period.

Video Solution

Tutored Practice
Problem 8.2.1

The representation of H_2 shown previously where the bonding electrons are represented with a line, H—H, is the Lewis structure of H_2. In a Lewis structure, **bonding pairs** of electrons are represented as lines connecting atom symbols, and nonbonding electrons are shown as dots (Interactive Figure 8.2.2). Nuclei and core electrons are represented by element symbols. Notice in Interactive Figure 8.2.2 that there can be more than one bonding pair between atoms and that nonbonding electrons usually appear in pairs (called **lone electron pairs** or *lone pairs*). A bond consisting of two electrons (one line) is called a **single bond**, a bond made up of four electrons (two pairs, two lines) is called a **double bond**, and a bond with six electrons (three pairs, three lines) is called a **triple bond**.

Lewis structures are very useful for visualizing the physical and chemical properties of compounds made up of nonmetal elements. We will begin by learning how to create Lewis structures; later in this unit, we will use Lewis structures to explore bond properties (bond order, bond length, bond energy, and bond polarity), the shapes of molecules, molecular

Interactive Figure 8.2.2

Interpret Lewis structures.

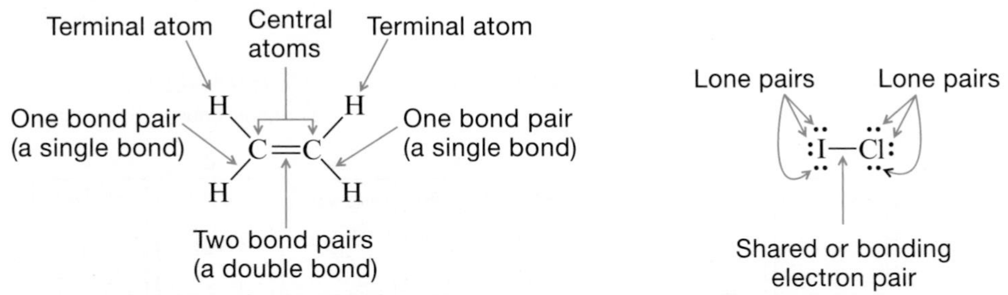

Lewis structures for C_2H_4 and ICl

Unless otherwise noted, all art is © 2013 Cengage Learning.

Unit 8 *Covalent Bonding and Molecular Structure*

230

polarity, and how the shape and polarity of molecules influence chemical properties (Figure 8.2.3).

8.2b Drawing Lewis Structures

An important guideline to follow when drawing Lewis structures is the octet rule. The **octet rule** states that most atoms in a Lewis structure are surrounded by no more than eight electrons (shared bonding electrons and unshared nonbonding electrons). The octet rule is related to the fact that valence shells contain a single s orbital and three p orbitals that can accommodate up to eight electrons, and it is these orbitals that are most often involved in forming covalent bonds between nonmetals in covalent compounds.

When determining whether elements have satisfied octets, sum the number of lone pair electrons and the number of bonding electrons assigned to an atom.

In addition to the octet rule, it is useful to remember that only certain elements (C, N, O, P, and S) form multiple (double and triple) bonds due to their size and electronic properties. Note that this does not mean that no other elements form multiple bonds; however, it is more likely for these elements to do so. Some elements rarely form multiple bonds, notably fluorine because of its high affinity for electrons.

Lewis structures are drawn by following the five steps shown in Table 8.2.1.

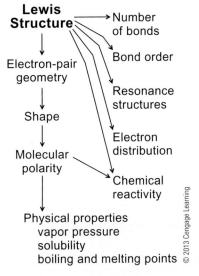

Figure 8.2.3 Relationship between Lewis structure and physical and chemical properties

Table 8.2.1 How to Draw a Lewis Structure

	Example: PF_3	Example: CO_2
Step 1: Count valence electrons. Count the total number of valence electrons in the molecule or ion. Anions have extra electrons, so add 1 electron for each negative charge. Cations have a deficiency of electrons, so subtract 1 electron for each positive charge.	Phosphorus has 5 valence electrons and each fluorine has 7 valence electrons. $5 + (3 \times 7) = 26$ electrons (or 13 electron pairs)	Carbon has 4 valence electrons and each oxygen has 6 valence electrons. $4 + (2 \times 6) = 16$ electrons (or 8 electron pairs)
Step 2: Arrange atoms. The central atom is usually the one with the lowest affinity for electrons (the one farthest from fluorine on the periodic table). Exception: H is never a central atom. Certain elements are found more frequently as central atoms (C, N, S, P) or terminal atoms (halogens, O) but there are, of course, exceptions. Electronegativity (Bond Polarity, 8.4b) can also be used to choose the central atom in a Lewis structure.	F P F F	O C O

Table 8.2.1 (continued)

	Example: PF_3	Example: CO_2
Step 3: Add single bonds. Add one bond (using a line) between each terminal atom and the central atom. Each bond represents 2 electrons.	F—P—F$\quad$ │ $\quad$ F $\quad$ 26 electrons − 6 bonding electrons = 20 electrons remaining	O—C—O $\quad$ 16 electrons − 4 bonding electrons = 12 electrons remaining
Step 4: Add remaining electrons. Assign any remaining electrons to the terminal atoms, in pairs, until the octet rule is satisfied for each terminal atom (except hydrogen). If additional electrons remain, add them to the central atom.	:F̈—P̈—F̈: $\quad$ │ $\quad$:F̈: $\quad$ Each F needs 3 pairs to satisfy the octet rule, and the 1 remaining pair is assigned to P.	:Ö—C—Ö: $\quad$ Each O needs 3 pairs to satisfy the octet rule. No other electrons remain to satisfy the octet rule for C.
Step 5: Check octet rule. Use the octet rule to determine whether multiple bonds are necessary between atoms. If there is an electron deficiency for an element, change a nonbonding electron pair (lone pair) on an adjacent atom into a bonding pair. Continue only until the octet rule is satisfied for all elements (other than the known exceptions described in Exceptions to the Octet Rule, 8.2c).	:F̈—P̈—F̈: $\quad$ │ $\quad$:F̈: $\quad$ All elements have complete octets.	:Ö⤻C⤸Ö: $\quad$ ↓ $\quad$:Ö=C=Ö: $\quad$ The central atom is sharing only 4 electrons. Use one lone pair on each oxygen to make a second bond to carbon, satisfying the octet rule.

The steps in Table 8.2.1 can also be applied to the Lewis structures of ionic species. When the Lewis structure for an ion is drawn, the structure is placed within square brackets and the ion charge is shown outside the brackets. For example, the steps required to draw the Lewis structure for SO_3^{2-} are as follows:

Step 1: Sulfur has 6 valence electrons, each oxygen has 6 valence electrons, and the 2− charge on the ion adds 2 more electrons:

$$6 + (3 \times 6) + 2 = 26 \text{ valence electrons (or 13 pairs)}$$

Steps 2 and 3:

$$O-S-O$$
(with O above S)

Steps 4 and 5:

$$\left[\ddot{:O:} \atop :\ddot{O}-\ddot{S}-\ddot{O}: \right]^{2-}$$

Example Problem 8.2.2 Draw Lewis structures.

Draw the Lewis structure for

a. SCl_2
b. CN^-

Solution:

You are asked to draw the Lewis structure for a molecule or ion.

You are given the chemical formula for a molecule or ion.

a. **Step 1:** $6 + (2 \times 7) = 20$ valence electrons (or 10 pairs)

Steps 2 and 3:

$$Cl-S-Cl$$

Sulfur has a lower affinity for electrons than Cl, so it is the central atom.

Steps 4 and 5:

$$:\ddot{Cl}-\ddot{S}-\ddot{Cl}:$$

All atoms have an octet of electrons, so it is not necessary to create multiple bonds.

b. **Step 1:** $4 + 5 + 1 = 10$ valence electrons (or 5 pairs)

Steps 2 and 3:

$$C-N$$

Step 4:

$$\ddot{C}-\ddot{N}$$

Step 5:

$$[:C\equiv N:]^-$$

A triple bond between carbon and nitrogen results in a satisfied octet for both atoms.

Is your answer reasonable? As a final check, always count the number of valence electrons in your Lewis structure and confirm that it matches the number of valence electrons from Step 1.

Video Solution

Tutored Practice
Problem 8.2.2

8.2c Exceptions to the Octet Rule

Although many Lewis structures follow the octet rule, there are exceptions.

Electron-deficient compounds are compounds in which an element has an incomplete octet. Some elements, notably H, Be, and B, often have fewer than eight electrons in Lewis structures. Hydrogen has a single valence electron in a $1s$ orbital and therefore accommodates only two electrons when it forms covalent bonds. Therefore, it almost always forms only one chemical bond to another atom and does not accommodate lone pairs of electrons. Beryllium (two valence electrons) and boron (three valence electrons) often accommodate only four or six electrons, respectively, in Lewis structures.

For example, BF_3 is an electron-deficient compound.

$$:\ddot{F}-B-\ddot{F}:$$
$$|$$
$$:\ddot{F}:$$

Each fluorine has a complete octet (six nonbonding electrons plus two bonding electrons), but boron has only six electrons (six bonding electrons and no lone pairs). Changing a fluorine lone pair to a bonding pair would alleviate the electron deficiency, but fluorine's high affinity for electrons means this element is unlikely to share its nonbonding electrons with boron. The electron deficiency means that BF_3 is a highly reactive compound. For example, it reacts readily with NH_3.

$$:\ddot{F}: \quad H$$
$$| \quad\quad |$$
$$:\ddot{F}-B \quad + \quad :N-H \quad \longrightarrow \quad :\ddot{F}-B-N-H$$
$$| \quad\quad | \quad\quad\quad\quad\quad\quad | \quad |$$
$$:\ddot{F}: \quad H \quad\quad\quad\quad\quad\quad :\ddot{F}: \quad H$$

In this reaction, the lone pair on N forms a new covalent bond between the compounds. In the new compound, both N and B have full octets and neither compound is electron deficient.

Free radicals are compounds with at least one unpaired electron. Free radicals can also be electron-deficient compounds. Nitrogen monoxide (nitric oxide, NO) is a free-radical compound because it is an odd-electron molecule with 11 valence electrons and 1 unpaired electron.

$$\dot{\ddot{N}}=\ddot{O}$$

Notice that the unpaired electron is placed on the nitrogen atom. In general, the odd electron in free radicals is not located on oxygen because of its high affinity for electrons. Free radicals are highly reactive species because the unpaired electrons react with other molecules. Pure NO, for example, reacts readily with halogens, O_2, and other free radicals.

Elements with an **expanded valence** (also called an expanded octet) have more than 8 electrons (often 10 or 12) in a Lewis structure. Elements in the third period and below, such as phosphorus, sulfur, and bromine, often have an expanded valence because of their larger radii (when compared to the second-row elements) and the availability of empty d orbitals in the valence shell.

Consider the Lewis structure of SF_4.

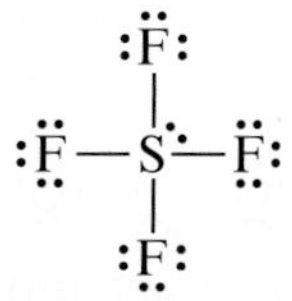

$$:\!\ddot{\text{F}}\!:$$
$$\ddot{\text{F}}\!-\!\dot{\text{S}}\!-\!\ddot{\text{F}}:$$
$$:\!\ddot{\text{F}}\!:$$

Each fluorine has a satisfied octet (6 nonbonding electrons plus 2 bonding electrons), but the central sulfur atom has an expanded octet with 10 electrons around it (8 bonding electrons and 2 nonbonding electrons).

Example Problem 8.2.3 Draw Lewis structures (octet rule exceptions).

Draw the Lewis structure for

a. ClO

b. IBr_3

Solution:

You are asked to draw the Lewis structure for a molecule or ion that contains at least one atom that does not follow the octet rule.

You are given the chemical formula for a molecule or ion.

a. **Step 1:** $7 + 6 = 13$ valence electrons (or 6 pairs and 1 unpaired electron)

Steps 2 and 3:

$$\text{Cl}\!-\!\text{O}$$

Steps 4 and 5:

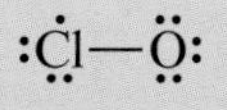

$$:\!\dot{\text{Cl}}\!-\!\ddot{\text{O}}:$$

This is an odd-electron molecule (a free radical). The unpaired electron is placed on chlorine because oxygen has a high affinity for electrons.

b. **Step 1:** $7 + (3 \times 7) = 28$ valence electrons (or 14 pairs)

Steps 2 and 3:

$$Br—I—Br$$
$$\mid$$
$$Br$$

Iodine has a lower affinity for electrons than bromine, so it is the central atom.

Steps 4 and 5:

$$:\ddot{B}r—\overset{\cdot\cdot}{I}—\ddot{B}r:$$
$$\mid$$
$$:\ddot{B}r:$$

Iodine is in the fifth period and therefore can have an expanded valence.

Video Solution

Tutored Practice
Problem 8.2.3

8.2d Resonance Structures

Some molecules have more than one valid Lewis structure. Two or more valid Lewis structures for a species that differ only in the arrangement of electrons, not the arrangement of atoms, are called **resonance structures**. A **resonance hybrid** is the actual electron arrangement for the molecule or ion, and it is intermediate between the resonance structures but not represented by any of the individual resonance structures. There is a great deal of experimental evidence to support resonance, including bond distances and angles that cannot be explained by the existence of a single Lewis structure for a molecule or ion.

Consider the case of ozone, O_3. Completing the first four steps of drawing its Lewis structure results in the following structure for ozone.

$$:\ddot{O}—\ddot{O}—\ddot{O}:$$

To complete the Lewis structure, the central oxygen needs one more pair of electrons. Both terminal oxygens have lone pairs that can be changed to shared bonding pairs, completing the octet for the central oxygen. There are therefore two possible Lewis structures for O_3, and they are drawn separated by a double-headed arrow.

$$\ddot{O}=\ddot{O}—\ddot{O}: \quad\longleftrightarrow\quad :\ddot{O}—\ddot{O}=\ddot{O}$$

The two resonance structures for ozone suggest that the oxygen–oxygen bonds in this molecule are not single bonds or double bonds, but something in between. The resonance hybrid, which is not represented by either resonance structure, has oxygen–oxygen bonds that have equivalent properties such as length and energy.

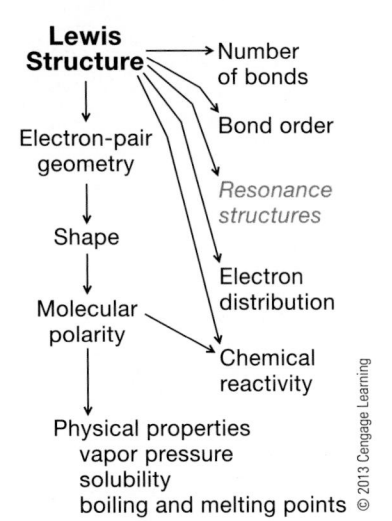

Lewis Structure → Number of bonds

Electron-pair geometry

Bond order

↓

Shape

Resonance structures

↓

Molecular polarity

Electron distribution

↓

Physical properties
vapor pressure
solubility
boiling and melting points

Chemical reactivity

© 2013 Cengage Learning

The ozone resonance structures are equivalent to one another because they contain the same number and type of chemical bonds. Not all resonance structures are equivalent, however. For example, carbon dioxide has three resonance structures, and they are not all equivalent.

$$:O{\equiv}C{-}\ddot{\ddot{O}}: \quad \longleftrightarrow \quad \ddot{\ddot{O}}{=}C{=}\ddot{\ddot{O}} \quad \longleftrightarrow \quad :\ddot{\ddot{O}}{-}C{\equiv}O:$$

The first and third resonance structures are equivalent because each contains one carbon–oxygen triple bond and one carbon–oxygen single bond. The middle resonance structure is unique because it has two carbon–oxygen double bonds. Later in this unit we will look more closely at the electron distribution in molecules and will see how to use that information to determine the most likely resonance structure for a molecule or ion. (▶ Flashforward to Section 8.4c Resonance Structures, Formal Charge, and Electronegativity)

Example Problem 8.2.4 Draw resonance structures.

Draw all resonance structures for the carbonate ion, CO_3^{2-}.

Solution:

You are asked to draw all resonance structures for a molecule or ion.

You are given the chemical formula for the molecule or ion.

The incomplete Lewis structure (octet rule not satisfied for carbon) is

$$\left[\begin{array}{c} :\ddot{O}: \\ | \\ :\ddot{O}{-}C{-}\ddot{O}: \end{array}\right]^{2-}$$

The electron deficiency for carbon is corrected by changing a lone pair on an adjacent oxygen into a bonding pair. There are three oxygen atoms with lone pairs that can correct the carbon electron deficiency, so there are three equivalent resonance structures for the carbonate ion.

$$\left[\begin{array}{c} :\ddot{O}: \\ | \\ \ddot{O}{=}C{-}\ddot{O}: \end{array}\right]^{2-} \longleftrightarrow \left[\begin{array}{c} :O: \\ \| \\ :\ddot{O}{-}C{-}\ddot{O}: \end{array}\right]^{2-} \longleftrightarrow \left[\begin{array}{c} :\ddot{O}: \\ | \\ :\ddot{O}{-}C{=}\ddot{O} \end{array}\right]^{2-}$$

Video Solution

Tutored Practice
Problem 8.2.4

Section 8.2 Mastery

8.3 Bond Properties

8.3a Bond Order, Bond Length, and Bond Energy

Lewis structures show the arrangement of valence electrons in covalently bonded molecules and ions. We can use Lewis structures to predict the properties of the bonds in molecules and ions, such as the properties of bond length and bond energy, and use these bond properties to predict physical and chemical properties of molecules and ions.

Bond Order

One factor that has a great influence on bond properties is **bond order**, the number of bonding electron pairs between two bonded atoms. The bond order of the carbon–carbon single bond in acetone is 1, for example, whereas the carbon–oxygen double bond in acetone has a bond order of 2. The carbon–nitrogen triple bond in the cyanide ion has a bond order of 3.

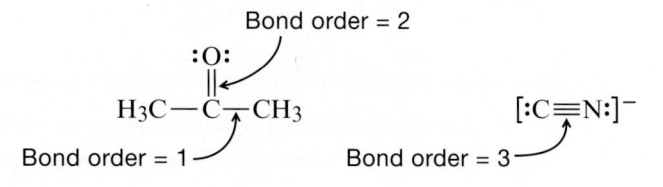

Bond Length

As we saw earlier, chemical bonds form when attractive forces between atoms are stronger than repulsive forces. The distance between the atomic nuclei when energy is minimized is the **bond length**.

Accurate bond distances are determined from careful measurements using techniques such as x-ray crystallography. Bond lengths between two different elements vary slightly between compounds. For that reason, average bond lengths are reported in tables, such as Interactive Table 8.3.1.

The bond lengths in Interactive Table 8.3.1 have units of picometers (1 pm = 10^{-12} m). Other common units for bond lengths are nanometers (1 nm = 10^{-9} nm) and Ångstroms (1 Å = 10^{-10} m). The data in Interactive Table 8.3.1 demonstrate two general trends in bond lengths. First, bond lengths increase with increasing atom size. Consider the trend in H—X bond lengths (where X is a halogen). Atomic radii increase as you move down the periodic table (F < Cl < Br < I), and therefore the H—X bond lengths increase as the halogen radius increases. (◄Flashback to Section 7.4 Properties of Atoms)

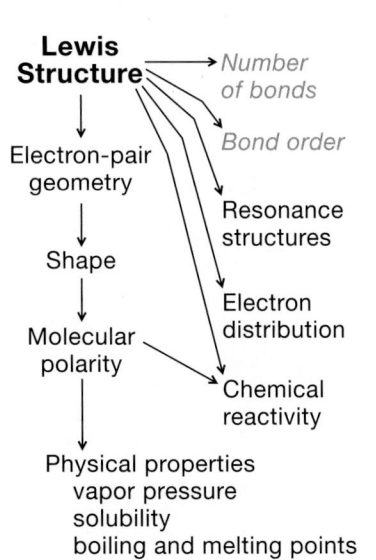

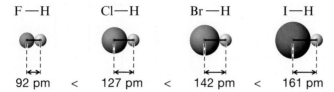

$$F—H \quad Cl—H \quad Br—H \quad I—H$$

$$92 \text{ pm} \quad < \quad 127 \text{ pm} \quad < \quad 142 \text{ pm} \quad < \quad 161 \text{ pm}$$

The second trend demonstrated in Interactive Table 8.3.1 is the relationship between bond length and bond order. As bond order increases, there is an increase in electron density between two nuclei. This results in stronger attractive forces between electrons and nuclei, decreasing the distance between the nuclei. A carbon–carbon single bond has a

Interactive Table 8.3.1

Some Average Bond Lengths (pm)

Single Bonds										
	H	**C**	**N**	**O**	**F**	**Si**	**P**	**S**	**Cl**	**Br**	**I**
H	74	110	98	94	92	145	138	132	127	142	161
C		154	147	143	141	194	187	181	176	191	210
N			140	136	134	187	180	174	169	184	203
O				132	130	183	176	170	165	180	199
F					128	181	174	168	163	178	197
Si						234	227	221	216	231	250
P							220	214	209	224	243
S								208	203	218	237
Cl									200	213	232
Br										228	247
I											226

Multiple Bonds			
C=C	134	O=O	112
C≡C	121	C=O	122
N=N	120	N=O	108
N≡N	110	C=N	127

bond order of 1 and is longer than a carbon–carbon double bond with a bond order of 2. *In general, for a series of bonds that differ only in bond order, an increase in bond order results in a decrease in bond length.*

$$C\!-\!O \ 143 \text{ pm} > C\!=\!O \ 122 \text{ pm} > C\!\equiv\!O \ 113 \text{ pm}$$

Bond Energy

As shown previously in Interactive Figure 8.1.1, a chemical bond forms when the attractive and repulsive forces between atoms result in an energy minimum. **Bond energy** is the energy required to break a chemical bond in a gas-phase molecule. For example, the bond energy of an H—H bond is 436 kJ/mol (at 298 K). This means that 436 kJ of energy is required to break 1 mol of H—H bonds, forming 2 mol of H atoms.

$$H\!-\!H(g) \rightarrow H(g) + H(g) \qquad \Delta H^\circ = +436 \text{ kJ/mol}$$

Breaking bonds is an endothermic process; therefore, bond energies are always positive values. Conversely, bond formation is an exothermic process that always releases energy. Just as bond lengths vary between compounds, so do bond energies. Some average bond energies are shown in Interactive Table 8.3.2.

Interactive Table 8.3.2

Some Average Bond Energies (kJ/mol)

				Single Bonds						
	H	C	N	O	F	Si	P	S	Cl	Br	I
H	436	413	391	463	565	318	322	347	432	366	299
C		346	305	358	485	—	—	272	339	285	213
N			163	201	283	—	—	—	192	243	—
O				146	184	452	335	—	218	201	201
F					155	565	490	284	253	249	278
Si						222	—	293	381	310	234
P							201	—	326	—	184
S								226	255	213	—
Cl									242	216	208
Br										193	175
I											151

Table 8.3.2 *(continued)*			
Multiple Bonds			
C=C	602	O=O	498
C≡C	835	C=O	732
N=N	418	N=O	607
N≡N	945	C≡N	615

A clear trend that can be observed from the data in Interactive Table 8.3.2 is the relationship between bond energy and bond order. *Bond energy increases with increasing bond order*:

C—O 358 kJ/mol < C=O 732 kJ/mol < C≡O 1072 kJ/mol

As bond order increases, the stronger attractive forces between bonding electrons and nuclei mean that more and more energy is required to separate the bonded nuclei.

Example Problem 8.3.1 Relate bond order, bond length, and bond energy.

a. What is the bond order of the boron–oxygen bonds in the borate ion, BO_3^{3-}?
b. Which molecule has the shortest carbon–carbon bond: C_2H_4, C_2H_2, or C_2H_6?
c. Which molecule has the weakest carbon–carbon bond: C_2H_4, C_2H_2, or C_2H_6?

Solution:

You are asked to identify the bond order in a molecule or ion and to use bond order to make predictions about relative bond length or bond energy in a series of molecules or ions.

You are given the chemical formula of a molecule or ion, or a series of molecules or ions.

a. The Lewis structure for the borate ion shows three boron–oxygen single bonds. Each B—O bond has a bond order of 1.

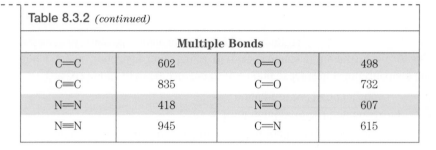

b. Bond length decreases with increasing bond order. C_2H_2 has the highest carbon–carbon bond order and the shortest carbon–carbon bond.

CC bond order: 1   2   3

c. Bond energy decreases with decreasing bond order. C_2H_6 has the lowest carbon–carbon bond order and the weakest carbon–carbon bond.

Video Solution

Tutored Practice Problem 8.3.1

8.3b   Resonance Structures, Bond Order, Bond Length, and Bond Energy

As shown earlier, compounds with resonance structures often have chemical bonds that are not easily described as single, double, or triple bonds. To describe the bond order, bond length, and bond energy of these bonds, the number of resonance structures and bonding pairs must be taken into account. Consider the nitrite ion, NO_2^-, which has two equivalent resonance structures.

$$\left[\ddot{\text{O}} = \ddot{\text{N}} - \ddot{\text{O}} : \right]^- \quad \longleftrightarrow \quad \left[: \ddot{\text{O}} - \ddot{\text{N}} = \ddot{\text{O}} \right]^-$$

Because neither resonance structure represents the actual electron arrangement, the nitrogen–oxygen bonds in this ion are not single bonds (bond order = 1) or double bonds (bond order = 2). Instead, NO_2^- has two equivalent NO bonds where three pairs of bonding electrons are distributed over two equivalent NO bond locations. The best way to describe the bonding in the resonance hybrid is with a fractional bond order:

$$\text{each NO bond order in } NO_2^- = \frac{\text{total number of NO bonding pairs}}{\text{number of NO bond locations}} = \frac{3}{2} = 1.5$$

The NO bonds in the nitrite ion have a bond order that is intermediate between a single and a double bond. The bond length of the NO bonds in the nitrite ion are also intermediate between average NO single and double bond lengths.

Bond	Bond Length (pm)
N—O	136
NO in NO_2^-	125
N=O	115

Example Problem 8.3.2 Relate bond order, bond length, and bond energy (resonance structures).

a. What is the bond order of the nitrogen–oxygen bonds in the nitrate (NO_3^-) ion?
b. In which ion (NO_2^- or NO_3^-) are the nitrogen–oxygen bonds longer?
c. In which ion (NO_2^- or NO_3^-) are the nitrogen–oxygen bonds stronger?

Solution:

You are asked to identify the bond order in a molecule or ion that has resonance structures and to use bond order to make predictions about relative bond length or bond order in a series of molecules or ions that have resonance structures.

You are given the chemical formula of a molecule or ion, or a series of molecules or ions.

a. The nitrate ion has three equivalent resonance structures.

There are four NO bonding pairs distributed over three NO bonding locations.

$$\text{each NO bond order in } NO_3^- = \frac{\text{total number of NO bonding pairs}}{\text{number of NO bond locations}} = \frac{4}{3} = 1.3$$

b. Bond length increases with decreasing bond order. The nitrogen–oxygen bond order in NO_3^- is less than the nitrogen–oxygen bond order in NO_2^- (1.3 versus 1.5, respectively). Therefore, NO_3^- has longer nitrogen–oxygen bonds than does NO_2^-.

c. Bond energy increases with increasing bond order. The nitrogen–oxygen bond order in NO_2^- is greater than the nitrogen–oxygen bond order in NO_3^- (1.5 versus 1.3, respectively). Therefore, NO_2^- has stronger nitrogen–oxygen bonds than does NO_3^-.

Video Solution

Tutored Practice
Problem 8.3.2

Unless otherwise noted, all art is © 2013 Cengage Learning.

Unit 8 *Covalent Bonding and Molecular Structure*   **243**

8.3c Bond Energy and Enthalpy of Reaction

Bond energy values can be used to calculate the enthalpy change for gas-phase reactions. First, assume all chemical bonds are broken in reactant molecules, creating isolated atoms. Next, rearrange the atoms to form products and new chemical bonds. The first step involves breaking bonds and is an endothermic process; the second step involves forming bonds and is an exothermic process. The bond energies are summed as shown in Equation 8.2.

$$\Delta H° = \Sigma(\text{energies of bonds broken}) - \Sigma(\text{energies of bonds formed}) \qquad \textbf{(8.2)}$$

Example Problem 8.3.3 Calculate enthalpy of reaction using bond energies.

Calculate the enthalpy change for the following gas-phase reaction.

$$C_2H_6(g) + Cl_2(g) \rightarrow C_2H_5Cl(g) + HCl(g)$$

Solution:

You are asked to calculate the enthalpy change for a reaction using bond energies.

You are given a chemical equation.

Draw Lewis structures for reactants and products and use them to list all reactant bonds broken and product bonds formed, along with bond energies from Interactive Table 8.3.2.

Reactants:

Products:

Bonds broken		**Bonds formed**	
6 mol C—H	6 × 413 kJ/mol	5 mol C—H	5 × 413 kJ/mol
1 mol C—C	346 kJ/mol	1 mol C—C	346 kJ/mol
1 mol Cl—Cl	242 kJ/mol	1 mol C—Cl	339 kJ/mol
		1 mol H—Cl	432 kJ/mol

Use Equation 8.2 to calculate $\Delta H°$ for the reaction.

$\Delta H° = [6 \text{ mol}(413 \text{ kJ/mol}) + 1 \text{ mol}(346 \text{ kJ/mol}) + 1 \text{ mol}(242 \text{ kJ/mol})]$
$\qquad - [5 \text{ mol}(413 \text{ kJ/mol}) + 1 \text{ mol}(346 \text{ kJ/mol}) + 1 \text{ mol}(339 \text{ kJ/mol}) + 1 \text{ mol}(432 \text{ kJ/mol})]$
$\qquad = -116 \text{ kJ}$

Example Problem 8.3.3 *(continued)*

Notice that five of the C—H bonds and the C—C bond in C_2H_6 are unchanged in the reaction. The enthalpy of reaction can also be calculated using only the energies of the bonds that are broken (1 mol C—H and 1 mol Cl—Cl) and the bonds that are formed (1 mol C—Cl and 1 mol H—Cl).

$$\Delta H^\circ = [1 \text{ mol}(413 \text{ kJ/mol}) + 1 \text{ mol}(242 \text{ kJ/mol})] - [1 \text{ mol}(339 \text{ kJ/mol}) + 1 \text{ mol}(432 \text{ kJ/mol})]$$

$$= -116 \text{ kJ}$$

Video Solution

Tutored Practice Problem 8.3.3

Section 8.3 Mastery

8.4 Electron Distribution in Molecules

8.4a Formal Charge

Lewis structures give chemists one method for visualizing the valence electron arrangement in molecules and ions. However, these structures do not fully represent the way electrons are distributed. Calculated formal charge and electronegativity are two tools chemists use, in addition to a Lewis structure, to more accurately describe the electron distribution in a molecule.

The **formal charge** of an atom in a molecule or ion is the charge it would have if all bonding electrons were shared equally. Formal charge is a "bookkeeping" method of showing electron distribution in a Lewis structure, and although it does not give a completely accurate picture of charge distribution, it is helpful in identifying regions with a large positive or negative charge buildup. Consider the Lewis structure for BCl_3, and in particular the electrons represented by the B—Cl bonds.

$$:\ddot{C}l: \\ | \\ :\ddot{C}l - B - \ddot{C}l:$$

The formal charge on the chlorine atom in the B—Cl bond is a measure of the relationship between the chlorine valence electrons (equal to its group number) and the number of bonding and nonbonding electrons assigned to chlorine in the Lewis structure. In calculating formal charge, we assume that all bonding electrons are shared equally between the atoms in the bond and therefore that each atom gets half of them (Equation 8.3). (◄ Flashback to Section 4.4 Oxidation–Reduction Reactions)

$$\text{formal charge} = \left(\begin{array}{c}\text{number of} \\ \text{valence electrons}\end{array}\right) - \left[\left(\begin{array}{c}\text{number of} \\ \text{nonbonding electrons}\end{array}\right) + \frac{1}{2}\left(\begin{array}{c}\text{number of} \\ \text{bonding electrons}\end{array}\right)\right] \textbf{(8.3)}$$

As shown in Interactive Figure 8.4.1, all of the atoms in BCl_3 have a formal charge of zero and we predict that there is no buildup of positive or negative charge in the molecule. Formal charge, however, assumes all bonding electrons are shared equally. As you will see in the following section, other methods that take into account the unequal sharing of bonding electrons can give a more accurate picture of electron distribution in a molecule.

The sum of the formal charges for all atoms in a molecule or ion is equal to the charge on the molecule or ion, as shown in the following example.

Interactive Figure 8.4.1

Determine formal charge.

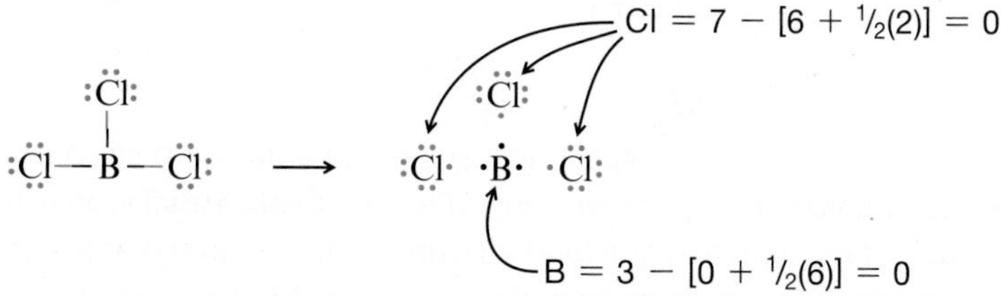

$$Cl = 7 - [6 + \tfrac{1}{2}(2)] = 0$$

$$B = 3 - [0 + \tfrac{1}{2}(6)] = 0$$

Formal charges in BCl_3

Example Problem 8.4.1 Calculate formal charge.

Use the formal charge for each atom in the cyanide ion, CN^-, to predict whether H^+ is more likely to attach to carbon or nitrogen when forming hydrocyanic acid.

Solution:

You are asked to identify the formal charge for every atom in a polyatomic ion and to use formal charge to predict the attachment site of H^+.

You are given the chemical formula for the polyatomic ion.

First draw the Lewis structure of CN^-.

$$[:C \equiv N:]^-$$

Use Equation 8.3 to calculate the formal charge for carbon and nitrogen.

$$C \text{ formal charge} = (4) - [2 + \tfrac{1}{2}(6)] = -1$$
$$N \text{ formal charge} = (5) - [2 + \tfrac{1}{2}(6)] = 0$$

When the positively charged H^+ ion attaches to CN^-, it is more likely to attach to the atom with the negative formal charge, forming $H—C \equiv N$.

Is your answer reasonable? Notice that the sum of the formal charges is equal to the overall charge on the ion (-1).

Video Solution

Tutored Practice
Problem 8.4.1

8.4b Bond Polarity

In a covalent bond, electrons are attracted to two nuclei, but sometimes one nucleus attracts the electrons more strongly than the other. When one nucleus attracts the electrons more strongly, the bonding electrons are located closer to one nucleus than the other. This creates an uneven distribution of bond electron density and a **polar bond** (or **polar covalent bond**). When electrons experience the same attractive force to both nuclei, the bond is **nonpolar**.

The term *polar* refers to the existence of a **dipole**, a separation of partial positive (symbolized $\delta+$) and partial negative (symbolized $\delta-$) charge within a bond or a molecule. The partial charges in a dipole are due to an uneven distribution of electron density, not a transfer of electrons between atoms to form ions.

A dipole is characterized by its **dipole moment**, the measurement of the strength or polarity of the dipole. Dipole moments, which are reported in units of debyes (D), are proportional to the sizes of the charges and their distance apart. Dipoles are commonly indicated by use of an arrow that points at the negative end of the dipole and has a cross to indicate the positive end of the dipole.

$$\overset{\delta+}{X}\longmapsto\overset{\delta-}{Y}$$

Consider a fluorine–fluorine bond and a carbon–fluorine bond.

$$F-F \qquad \overset{\delta+}{C}\longmapsto\overset{\delta-}{F}$$

In the fluorine–fluorine bond, the bonding electrons are attracted equally to both fluorine nuclei. The electron density is therefore evenly distributed between the fluorine atoms, and the bond is nonpolar. The carbon–fluorine bond, in contrast, is polar because the bonding electrons are attracted more strongly to fluorine than to carbon. This results in electron density that is closer to fluorine than to carbon. Recall that fluorine has a higher effective nuclear charge than carbon and its valence atomic orbitals are lower in energy than those of carbon. (◀ Flashback to Section 7.4 Properties of Atoms) A partial negative charge occurs at fluorine because of this uneven electron density distribution, and a corresponding partial positive charge happens at the electron-deficient carbon.

The ability of an atom in a molecule to attract electrons to itself is referred to as its **electronegativity**, χ (Interactive Figure 8.4.2). Electronegativity, a concept first proposed by Linus Pauling in 1932, is a relative scale where the most electronegative element (fluorine) is assigned a value of 4.0. Notice in Interactive Figure 8.4.2 that

- the noble gases (Group 8A) are not assigned electronegativity values (most of these elements do not form covalent bonds);

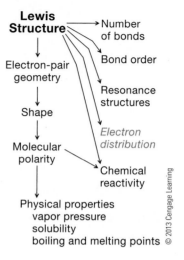

Interactive Figure 8.4.2

Explore electronegativity.

	1A	2A												3A	4A	5A	6A	7A
																	H 2.1	
	Li 1.0	Be 1.5												B 2.0	C 2.5	N 3.0	O 3.5	F 4.0
	Na 1.0	Mg 1.2	3B	4B	5B	6B	7B	8B		1B	2B		Al 1.5	Si 1.8	P 2.1	S 2.5	Cl 3.0

(Transition metal rows)

			3B	4B	5B	6B	7B		8B		1B	2B					
K 0.9	Ca 1.0	Sc 1.3	Ti 1.4	V 1.5	Cr 1.6	Mn 1.6	Fe 1.7	Co 1.7	Ni 1.8	Cu 1.8	Zn 1.6	Ga 1.7	Ge 1.9	As 2.1	Se 2.4	Br 2.8
Rb 0.9	Sr 1.0	Y 1.2	Zr 1.3	Nb 1.6	Mo 1.6	Tc 1.7	Ru 1.8	Rh 1.8	Pd 1.8	Ag 1.6	Cd 1.6	In 1.6	Sn 1.8	Sb 1.9	Te 2.1	I 2.5
Cs 0.8	Ba 1.0	La 1.1	Hf 1.3	Ta 1.4	W 1.5	Re 1.7	Os 1.9	Ir 1.9	Pt 1.8	Au 1.9	Hg 1.7	Tl 1.6	Pb 1.7	Bi 1.8	Po 1.9	At 2.1

Legend:
- <1.0
- 1.0–1.4
- 1.5–1.9
- 2.0–2.4
- 2.5–2.9
- 3.0–3.9
- 4.0

Electronegativity values

- electronegativity values increase as you move left to right across the periodic table and decrease as you move down the periodic table;

- hydrogen does not follow the periodic trend and has an electronegativity similar to that of carbon;

- metals generally have low electronegativity values and nonmetals generally have high electronegativity values; and

- fluorine, oxygen, nitrogen, and chlorine are the most electronegative elements.

The polarity of a chemical bond is related to the difference in electronegativity ($\Delta\chi$) of the two elements that make up the bond. For covalent bonds between nonmetals, a $\Delta\chi$ greater than zero indicates that the bond is polar and a zero $\Delta\chi$ indicates a nonpolar bond. For example, the C—F bond described earlier has $\Delta\chi = (4.0 - 2.5) = 1.5$. It is a polar bond.

When $\Delta\chi$ is very large, the interaction between elements has more ionic character than covalent character. For example, aluminum and fluorine have very different electronegativities

© 2013 Cengage Learning

$[\Delta\chi = F(4.0) - Al(1.5) = 2.5]$, and an Al—F bond has much greater ionic character than a C—F bond.

There is not a defined electronegativity difference that separates ionic from polar bonds. Instead, there is a continuum moving between these two extremes. All polar bonds have some "ionic character" and some "covalent character."

Example Problem 8.4.2 Use electronegativity to determine bond polarity.

a. Which of the following bonds are nonpolar?

$$C—Cl \quad H—Cl \quad H—H \quad F—F \quad P—H \quad S—O \quad B—F$$

b. Which of the following bonds is most polar?

$$C—Cl \quad H—Cl \quad H—H \quad F—F \quad P—H \quad S—O \quad B—F$$

Solution:

You are asked to identify bonds as polar or nonpolar and to identify the most polar bond in a series of bonded atoms.

You are given a series of bonded atoms.

a. The H—H, F—F, and P—H bonds are nonpolar. Any bond between two like atoms is nonpolar because $\Delta\chi = 0$. The P—H bond is nonpolar by the coincidence that both P and H have the same electronegativity value of 2.1. The C—Cl, H—Cl, S—O, and B—F bonds are polar because when two different elements are bonded, the bond is almost always polar.

b. The B—F bond has the greatest $\Delta\chi$ $(4.0 - 2.0 = 2.0)$, so it is the most polar bond.

Video Solution

Tutored Practice
Problem 8.4.2

8.4c Resonance Structures, Formal Charge, and Electronegativity

When nonequivalent resonance structures exist, the most likely resonance structure is the one with formal charges closest to zero. A general rule of chemical stability is that the localization of positive or negative charges within a molecule is destabilizing. Consider the following two possible inequivalent resonance structures for BCl_3 shown here with calculated formal charges.

The resonance structure on the right is less likely than the one on the left because it has localized formal charges of +1 and −1. In addition, chlorine ($\chi = 3.0$) is more electronegative than boron ($\chi = 2.0$), so a resonance structure that has a positive formal charge on the more electronegative element (and a negative formal charge on the less electronegative element) is probably not stable.

The three nonequivalent resonance structures of CO_2 are shown here with calculated formal charges.

$$:O\equiv C-\ddot{O}: \longleftrightarrow \ddot{O}=C=\ddot{O} \longleftrightarrow :\ddot{O}-C\equiv O:$$

$$\begin{array}{ccc} (+1) \quad (0) \quad (-1) & (0) \quad (0) \quad (0) & (-1) \quad (0) \quad (+1) \end{array}$$

The central structure is the most likely resonance structure for CO_2 because it has formal charges of zero on each atom, whereas both of the other resonance structures have formal charges of $+1$ and -1 on the O atoms.

The cyanate ion, OCN^-, has three nonequivalent resonance structures.

$$[:O\equiv C-\ddot{N}:]^- \longleftrightarrow [\ddot{O}=C=\ddot{N}]^- \longleftrightarrow [:\ddot{O}-C\equiv N:]^-$$

$$\begin{array}{ccc} (+1) \quad (0) \quad (-2) & (0) \quad (0) \quad (-1) & (-1) \quad (0) \quad (0) \end{array}$$

None of the three resonance structures have formal charges of zero on all atoms. The first resonance structure is not very likely because of the large formal charges ($+1$ and -2). The other two resonance structures, however, each have small formal charges. In this case, electronegativity can help determine the best resonance structure. The electronegativity of oxygen (3.5) is greater than that of nitrogen (3.0). Therefore, oxygen is more likely to carry a negative formal charge in a Lewis structure, and according to formal charges, the resonance structure on the right is most likely.

Example Problem 8.4.3 Use formal charge and electronegativity to identify best resonance structure.

Three nonequivalent resonance structures for the chlorate ion are shown here. Assign formal charges to all atoms in the resonance structures and identify the more likely resonance structure.

$$\begin{array}{ccc} \mathbf{A} & \mathbf{B} & \mathbf{C} \end{array}$$

Example Problem 8.4.3 *(continued)*

Solution:

You are asked to use formal charge to identify the most likely resonance structure of those given for a polyatomic ion.

You are given three valid resonance structures.

Formal charges:

Structure B is the most likely resonance structure. The formal charges in B are close to zero, and the highly electronegative oxygen atom carries a −1 formal charge. Structure A is unlikely because of the large positive formal charge on chlorine. Structure C is unlikely because the least electronegative element in the ion (chlorine) has a negative formal charge while the most electronegative element in the ion (oxygen) has a formal charge of zero.

Video Solution

Tutored Practice
Problem 8.4.3

When the central atom in a molecule or ion is in the third period or below (for example, P, Se, or Xe), expanding the central atom's octet will often result in lower formal charges and therefore a prediction of more stable resonance structures. However, this can lead to some confusion when calculating bond order or predicting relative bond lengths and bond energies. In general, we use the following guidelines when deciding how to arrange valence electrons in a molecule or ion.

- When drawing a Lewis structure, follow the octet rule and do not expand the octet for a central atom unless it is necessary. Use this structure (or these resonance structures, if appropriate) to predict trends in bond order, bond length and bond strength.

- When using formal charge to predict the best resonance structure, expand the octet for central atoms in the third period or below if it results in calculated formal charges that are closer to zero.

Calculated formal charges and relative electronegativities are useful tools for assigning charge distribution in a Lewis structure because they help identify likely resonance structures and regions with a buildup of positive or negative charge in molecules and ions. They do not, however, give an accurate picture of the actual electron distribution in molecules. Computer modeling programs calculate **partial charges** on atoms that give a more accurate picture of electron distribution. For example, the calculated partial charges on the atoms in carbon dioxide and the cyanate ion are shown in the following figure. Red spheres indicate positive partial charge, and yellow spheres indicate negative partial charge. The size of the sphere is proportional to the magnitude of the partial charge. As was true with formal charge, the sum of partial charges is equal to the charge on the molecule or ion.

$$O=C=O \qquad \left[O=C=N\right]^-$$

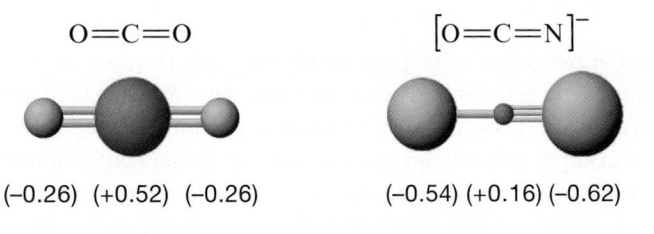

(−0.26) (+0.52) (−0.26)   (−0.54) (+0.16) (−0.62)

Section 8.4 Mastery

In CO_2, each oxygen carries a negative partial charge and the carbon has a positive partial charge. This reflects the fact that oxygen is more electronegative than carbon but does not reflect the calculated formal charges of zero in the most likely resonance structure for CO_2. However, the fact that both oxygen atoms have negative calculated partial charges suggests that the two resonance structures that have a positive formal charge on oxygen are very unlikely. The calculated partial charges also show that the CO bonds are polar, results that are supported by electronegativity values but not by formal charge calculations.

In the cyanate structure, both oxygen and nitrogen carry approximately the same negative partial charge and carbon has a slightly positive partial charge. None of the three OCN^- resonance structures shows a negative formal charge on both nitrogen and oxygen, but two of the resonance structures have a −1 formal charge on either nitrogen or oxygen. The calculated partial charges for OCN^- suggest that both of the resonance structures with −1 formal charges on nitrogen and oxygen are important and both contribute to the resonance hybrid for OCN^-.

8.5 Valence-Shell Electron-Pair Repulsion Theory and Molecular Shape

8.5a VSEPR and Electron-Pair Geometry

Lewis structures show the atom connectivity and number of bonds and lone pairs in a molecule or ion but do not provide information about the three-dimensional shapes of molecules. For example, numerous experiments have shown that water, H_2O, has a bent (nonlinear) shape. The Lewis structure, however, can be drawn to show a linear arrangement of atoms.

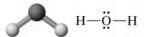

The **valence shell electron-pair repulsion (VSEPR)** theory allows chemists to easily predict the shapes of molecules and ions made up of nonmetals.

According to VSEPR theory,

- positions around a central atom are occupied by **structural electron pairs**, nonbonding or bonding electrons that repel one another and are arranged so as to avoid one another as best as possible;

- structural electron pairs can be nonbonding electrons, where each pair of electrons (or single electrons, in free radicals) is counted as one structural electron pair on a central atom;

- structural electron pairs can be bonding electrons, where each bond (single or multiple) is counted as one structural electron pair on a central atom;

- the **electron-pair geometry** is the arrangement of the structural electron pairs around the central atom; and

- the **shape** (also called **molecular geometry**) is the arrangement of atoms around the central atom.

The electron-pair geometry is defined by the arrangement of structural electron pairs around the central atom. In large molecules with multiple central atoms, we will describe the electron-pair geometry around each central atom. When a central atom is surrounded by two, three, four, five, or six structural pairs of electrons, they are arranged in one of the following ideal electron-pair geometries (Interactive Table 8.5.1).

Interactive Table 8.5.1 also shows the bond angles that are characteristic for each ideal electron-pair geometry. **Bond angle** is the angle formed by the nuclei of two atoms with a central atom at the vertex. The bond angles shown in Interactive Table 8.5.1 are ideal

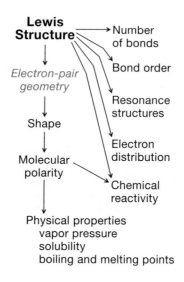

Lewis Structure → Number of bonds

→ Bond order

Electron-pair geometry

↓

Shape → Resonance structures

↓

Molecular polarity → Electron distribution

↓ → Chemical reactivity

Physical properties
vapor pressure
solubility
boiling and melting points

Ideal Electron-Pair Geometries

Number of Structural Pairs	Species Type (A = central atom, X = terminal atom)	Electron-Pair Geometry	Example	X—A—X Bond Angles
2	AX_2	Linear	CO_2	180°
3	AX_3	Trigonal planar	BF_3	120°
4	AX_4	Tetrahedral	CH_4	109.5°
5	AX_5	Trigonal bipyramidal	PCl_5	90° 120°
6	AX_6	Octahedral	SF_6	90°

All images are © 2013 Cengage Learning

angles, and most molecules with these ideal electron-pair geometries have bond angles that are very close to these values.

Notice that of all the electron-pair geometries in Interactive Table 8.5.1, only one has two non-equivalent sites. In a trigonal bipyramidal geometry, terminal atoms can occupy either an axial or an equatorial position. As shown in Figure 8.5.1, equatorial positions are arranged in a trigonal planar geometry, 120° from each other, whereas axial positions are 90° from the equatorial (trigonal planar) plane.

Follow these steps to predict the electron-pair geometry of a molecule with a single central atom:

1. Draw the Lewis structure.
2. Sum the number of structural electron pairs around the central atom.
3. Use the number of structural electron pairs to identify the electron-pair geometry using the ideal geometries shown in Interactive Table 8.5.1.

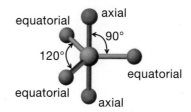

Figure 8.5.1 Equatorial and axial positions in a trigonal bipyramid

---

Example Problem 8.5.1 Identify electron-pair geometry.

Determine the electron-pair geometry of

a. the carbon atoms in acetic acid, CH_3CO_2H
b. SF_4

Solution:

You are asked to determine the electron-pair geometry for a molecule or for specific atoms in a complex molecule.

You are given the chemical formula for the molecule.

Draw the Lewis structure of each species, sum the number of structural electron pairs around the central atom, and use Interactive Table 8.5.1 to determine the electron-pair geometry.

a. The Lewis structure of acetic acid shows that there are four structural electron pairs (four single bonds) around the carbon on the left and three structural electron pairs (two single bonds and one double bond) around the carbon on the right.

$$\begin{array}{ccc} & H & :O: \\ & | & || \\ H- & C-C & -\ddot{O}-H \\ & | & \\ & H & \end{array}$$

The electron-pair geometry is tetrahedral around the carbon on the left (bond angles of 109.5°) and trigonal planar around the carbon on the right (bond angles of 120°).

b. There are five structural electron pairs around the central atom in SF_4: four single bonds and one lone pair of electrons.

Example Problem 8.5.1 (continued)

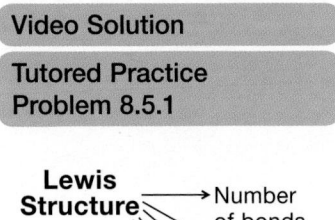

The electron-pair geometry of SF_4 is trigonal bipyramidal.

Video Solution

Tutored Practice
Problem 8.5.1

8.5b Shape (Molecular Geometry)

The shape of a molecule (its molecular geometry) is defined by the atom positions in a molecule. The shape is the same as the electron-pair geometry when none of the structural electron pairs are nonbonding electrons.

For example, boron trichloride, BCl_3, has a trigonal planar electron-pair geometry and a trigonal planar shape. There are three structural electron pairs around the central boron atom, and all are bonding electrons. Notice that a Lewis structure is not always drawn to show the shape of the molecule or the correct bond angles.

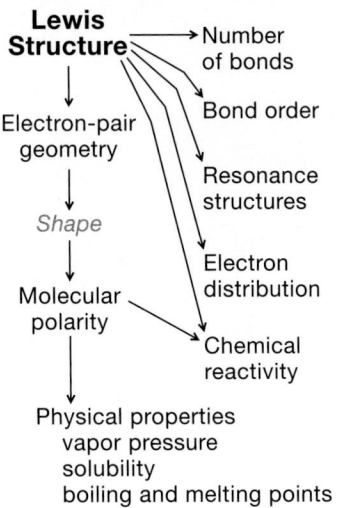

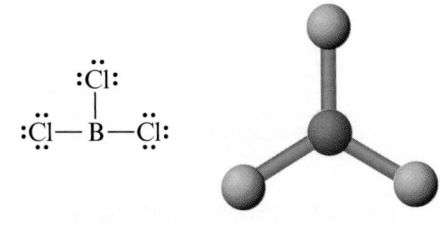

Carbon dioxide, CO_2, has a linear electron-pair geometry and a linear shape. There are two structural electron pairs around the central carbon atom, and both are bonding electrons.

There are many molecules for which the shape is not the same as the electron-pair geometry. For example, ozone, O_3, has a trigonal planar electron-pair geometry and a bent shape. There are three structural electron pairs around the central oxygen atom, but only two are bonding electrons. Because shape is defined by the positions of the atoms in a molecule, not the electrons, ozone has a bent shape rather than a trigonal planar shape.

Interactive Table 8.5.2 shows the shapes for molecules with three to six structural electron pairs and varying numbers of nonbonding electron pairs on the central atom.

Shapes of Molecules with Three to Six Structural Electron Pairs

	Number of Nonbonding Structural Pairs			
	0	**1**	**2**	**3**
3	Trigonal planar BF_3 120°	Bent O_3		
4	Tetrahedral CF_4 109.5°	Trigonal pyramidal NH_3	Bent H_2O	
5	Trigonal bipyramidal PF_5 90° 120°	Seesaw SF_4	T-shaped ClF_3	Linear XeF_2
6	Octahedral SF_6 90°	Square pyramidal BrF_5	Square planar XeF_4	

Number of Structural Electron Pairs

All images are © 2013 Cengage Learning

Example Problem 8.5.2 Identify molecular geometry.

Determine the shape of

a. ICl_5

b. SO_3^{2-}

Solution:

You are asked to determine the shape of a molecule or ion.

You are given the chemical formula for the molecule or ion.

Draw the Lewis structure of each species, sum the number of structural electron pairs around the central atom, and use Interactive Table 8.5.1 to determine the electron-pair geometry. Use the number of bonding and nonbonding structural pairs, along with Interactive Table 8.5.2, to determine the molecular geometry.

a. ICl_5 has six structural electron pairs around the central atom. Five are bonding electrons, and one is nonbonding.

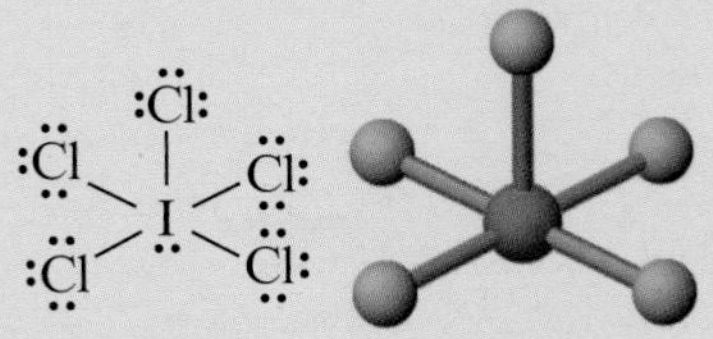

The electron-pair geometry is octahedral, and the molecular geometry is square pyramidal.

b. SO_3^{2-} has four structural electron pairs around the central atom. Three are bonding electrons, and one is nonbonding.

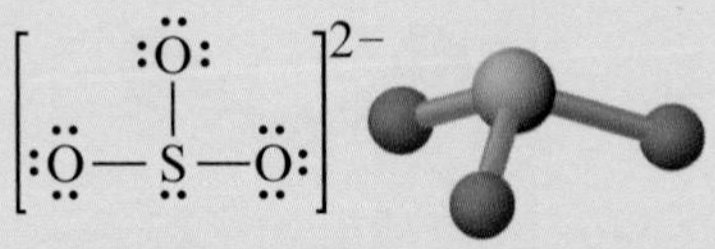

The electron-pair geometry is tetrahedral, and the molecular geometry is trigonal pyramidal.

Video Solution

Tutored Practice
Problem 8.5.2

When determining the shape of a molecule with nonbonding electrons (lone pairs) in structural positions, it is important to note the different steric requirements of lone pairs and bonding pairs. Lone pairs are diffuse because they are attracted to a single nucleus; bonding pairs, in contrast, are localized between and attracted to two nuclei. Lone pairs therefore occupy more space around a central atom than do bonding pairs. Consider SF_4, which has four structural positions occupied by bonding pairs and one occupied by a lone pair. The five structural positions are arranged in a trigonal bipyramid. The lone pair occupies an equatorial position (seesaw shape) because this minimizes repulsive forces between the larger, diffuse lone pair and the bonding pairs (Figure 8.5.2(a)). If the lone pair occupies an axial position (Figure 8.5.2(b)), the close contact (90° angles) between the lone pair and bonding pairs increases repulsive forces, making this configuration less favorable. Thus, lone pairs occupy equatorial positions in trigonal bipyramidal electron-pair geometries. Similarly, lone pairs occupy positions opposite each other in an octahedron, because this also minimizes repulsive forces.

Molecular shapes predicted using the VSEPR ideal electron-pair shapes are usually very close to the shapes measured by experiment, especially when all terminal atoms are identical. However, when lone pairs are present on the central atom, bond angles in particular differ from predicted values. Consider the bond angles in CH_4, NH_3, and H_2O (Figure 8.5.3). All molecules have the same electron-pair geometry (tetrahedral), but they differ in the number of lone pairs on the central atom. As the number of lone pairs on the central atom increases, the H—X—H bond angle decreases due to the steric requirements of the lone pairs.

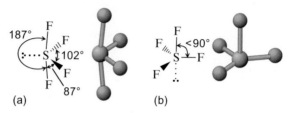

Figure 8.5.2 SF_4 with lone pair in (a) equatorial and (b) axial positions

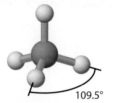 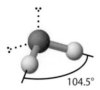

109.5°   107.5°   104.5°

Figure 8.5.3 Effect of lone pairs on bond angles

Section 8.5 Mastery

8.6 Molecular Polarity

8.6a Molecular Polarity

As we saw earlier, covalent bonds are polar when there is an uneven attraction for electrons between the bonded atoms. Polar bonds in a molecule can result in a polar molecule, which affects the physical properties of a compound. For example, polar molecules are often very soluble in water, whereas nonpolar molecules are not.

For a molecule to be polar, it must contain polar bonds and those bonds must be arranged so that there is an uneven charge distribution. The following steps are used to predict whether a molecule is polar.

1. Draw the Lewis structure.
2. Determine the shape.

3. If the molecule has polar bonds, indicate them on the molecule.
4. Use the shape and bond dipoles to determine whether there is an uneven distribution of bond electron density in the molecule.

For example, boron trifluoride, BF_3, is a nonpolar molecule that has polar bonds.

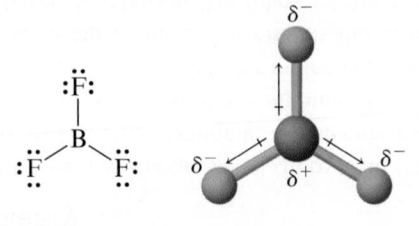

The shape of BF_3 is trigonal planar and the B—F bonds are polar (fluorine is the negative end of the bond dipole). As shown in the preceding illustration, the three bond dipoles are arranged symmetrically in the molecule and do not result in an uneven distribution of electron density. BF_3 is therefore nonpolar.

Trifluoromethane is a polar molecule.

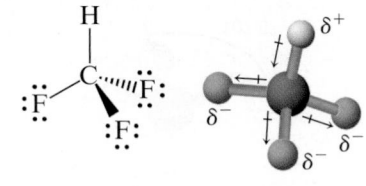

The shape of CHF_3 is tetrahedral, and the C—F and C—H bonds are polar. Carbon is the negative end of the C—H bond dipole, and fluorine is the negative end of each C—F bond dipole. When we draw in the individual bond dipoles, we see that there is a net movement of bond electron density away from hydrogen and toward the fluorine atoms. This uneven electron density distribution means that CHF_3 is polar, and the molecule has a net dipole with the positive end near hydrogen and the negative end near the fluorine atoms.

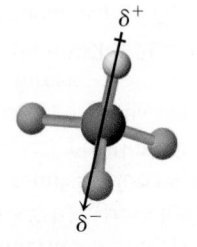

This method is also useful for larger, more complex molecules. For example, consider acetone, CH_3COCH_3.

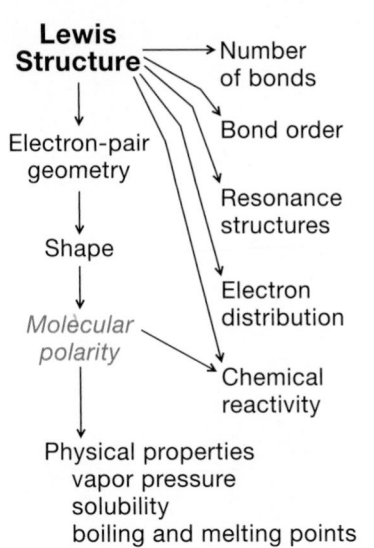

Lewis Structure → Number of bonds

Electron-pair geometry

Bond order

Resonance structures

Shape

Electron distribution

Molecular polarity

Chemical reactivity

Physical properties
vapor pressure
solubility
boiling and melting points

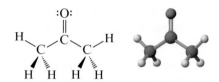

Carbon and hydrogen have very similar electronegativities ($\Delta\chi = 0.4$), so the bonds in the —CH_3 groups are only slightly polar. The carbon–oxygen bond, however, is very polar ($\Delta\chi = 1.0$), and oxygen is the negative end of the C—O bond dipole. There is a net shift of electron density toward oxygen, so the molecule is polar. The net molecular dipole is shown here.

Table 8.6.1 Dipole moments for some common molecules

Molecule	Dipole Moment (debye, D)
BF_3	0
NF_3	0.23
CH_4	0
CH_2Cl_2	1.60
H_2O	1.85
H_2S	0.95
HCl	1.07
HI	0.38

Dipole moments for some common molecules are shown on Table 8.6.1.

Example Problem 8.6.1 Determine polarity of molecules.

Determine whether each of the following is polar.

a. SO_3
b. BrF_3

Solution:

You are asked to determine whether a molecule is polar.

You are given the compound formula.

Draw the Lewis structure for each molecule, determine the shape, and label any polar bonds. Use shape and bond polarity to determine whether the molecule is polar.

a. Nonpolar. SO_3 has a trigonal planar shape, and all three SO bonds are polar. The three bond dipoles are arranged symmetrically in the molecule, so the molecule is not polar.

Unless otherwise noted, all art is © 2013 Cengage Learning.

◄- - - - - - - - -

Example Problem 8.6.1 *(continued)*

b. Polar. BrF$_3$ is T-shaped, and all three BrF bonds are polar. The three bond dipoles are not arranged symmetrically around the molecule, so the molecule is polar. The bromine is the positive end of the net dipole, and one of the fluorine atoms is the negative end of the bond dipole.

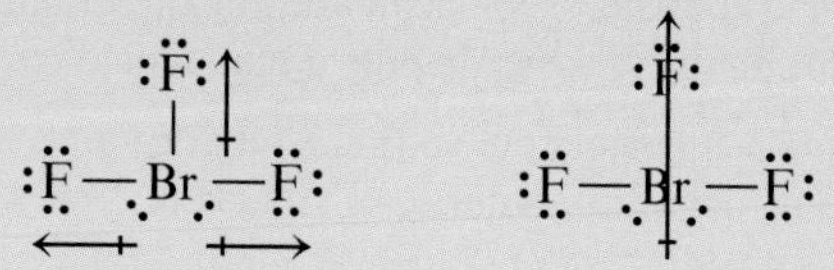

Video Solution

Tutored Practice
Problem 8.6.1

Section 8.6 Mastery

Unit Recap

Key Concepts

8.1 An Introduction to Covalent Bonding

- The force between charged particles is described in Coulomb's law, which shows that the force is directly related to the magnitude of the charges and inversely related to the distance between the particles (8.1a).

- Forces between the particles that make up matter include ionic bonding, covalent bonding, metallic bonding, and intermolecular forces (8.1a).

- Covalent bonds are characterized by the sharing of valence electrons by two adjacent atoms (8.1b).

8.2 Lewis Structures

- A Lewis structure shows the arrangement of valence electrons and nuclei in a covalently bonded molecule or ion (8.2a).

- Lewis structures include nonbonding electrons (lone pairs) and bonding electrons (single, double, or triple bonds) (8.2a).

- The octet rule, an important guideline when drawing Lewis structures, states that most atoms in a Lewis structure are surrounded by no more than eight electrons (8.2b).

- Exceptions to the octet rule include electron-deficient compounds, free radicals, and compounds where an atom has an expanded valence (8.2c).

- Resonance structures are two or more valid Lewis structures for a given species that vary only in the arrangement of electrons. The resonance hybrid, the actual electron arrangement, is intermediate between the resonance structures (8.2d).

8.3 Bond Properties

- Bond order is the number of bonding electron pairs between two bonded atoms (8.3a).
- Bond length is the distance between atomic nuclei when energy is minimized (8.3a).
- Bond energy is the energy required to break a chemical bond in a gas-phase molecule (8.3a).
- For a series of bonds between like atoms, as bond order increases, bond length decreases and bond energy increases (8.3a).
- Bond energy values can be used to calculate the enthalpy change for gas-phase reactions (8.3c).

8.4 Electron Distribution in Molecules

- Formal charge is the charge an atom in a molecule or ion would have if all bonding electrons were shared equally. The sum of formal charges on all atoms in a molecule or ion is equal to the charge on the molecule or ion (8.4a).
- When bonding electrons are unequally shared between atoms, a polar bond results (8.4b).
- A polar bond has a dipole, a separation of partial positive and negative charges, which is characterized by a dipole moment (8.4b).
- Electronegativity is a measure of the ability of an atom in a molecule to attract electrons to itself (8.4b).
- The greater the difference in electronegativity between bonded atoms, the more polar the bond (8.4b).
- Partial charge is a calculated value that gives a more accurate picture of electron distribution in molecules and ions (8.4c).

8.5 Valence-Shell Electron-Pair Repulsion Theory and Molecular Shape

- Shapes of molecules and ions made up of nonmetals are predicted using VSEPR theory, which states that electron pairs (structural and nonbonding) repel one another and are arranged to avoid one another as best as possible. (8.5a).
- Electron-pair geometry is the arrangement of structural electron pairs around the central atom, and molecular geometry is the arrangement of atoms around the central atom (8.5a).

8.6 Molecular Polarity

- For a molecule to be polar, it must contain polar bonds that are arranged so that there is an uneven charge distribution (8.6a).

Key Equations

$$\text{Force} \propto \frac{(q_A)(q_B)}{r^2} \qquad (8.1)$$

$$\Delta H^\circ = \Sigma(\text{energies of bonds broken}) - \Sigma(\text{energies of bonds formed}) \qquad (8.2)$$

$$\text{formal charge} = \begin{pmatrix} \text{number of} \\ \text{valence electrons} \end{pmatrix} - \left[\begin{pmatrix} \text{number of} \\ \text{nonbonding electrons} \end{pmatrix} + \frac{1}{2}\begin{pmatrix} \text{number of} \\ \text{bonding electrons} \end{pmatrix} \right] \qquad (8.3)$$

Key Terms

8.1 An Introduction to Covalent Bonding
Coulomb's law
bond
ionic bonding
covalent bonding
metallic bonding
intermolecular forces

8.2 Lewis Structures
Lewis structure
Lewis symbol
bonding pair
lone (electron) pairs
single bond
double bond

triple bond
octet rule
electron deficient
free radicals
expanded valence
resonance structures
resonance hybrid

8.3 Bond Properties
bond order
bond length
bond energy

8.4 Electron Distribution in Molecules
formal charge
polar bond

polar covalent bond
nonpolar
dipole
dipole moment
electronegativity, χ
partial charges

8.5 Valence-Shell Electron-Pair Repulsion Theory and Molecular Shape
valence-shell electron-pair repulsion (VSEPR)
structural electron pairs
electron-pair geometry
shape (molecular geometry)
bond angle

Unit 8   Review and Challenge Problems

9

Theories of Chemical Bonding

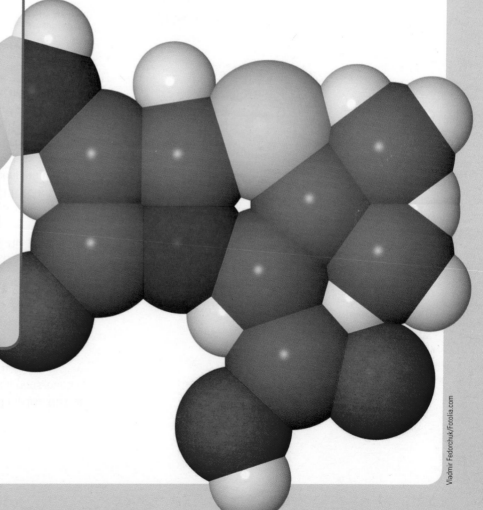

Unit Outline

In This Unit...

In Covalent Bonding and Molecular Structure (Unit 8) we introduced the concept of covalent bonding. In that description, the VSEPR model uses rules for predicting structures that are based on observations of the geometries of many molecules. In this unit we expand this discussion to understand why molecules have these predictable shapes. This deeper understanding involves a model of chemical bonding called valence bond theory and will allow us to predict not only expected structures, but also expected exceptions to the usual rules. In the second major part of this unit, we examine a second theory of chemical bonding, called molecular orbital theory. Molecular orbital theory can be used to explain structures of molecules as well as the energetics of chemical processes, such as what happens when a molecules absorbs a photon of light.

Vladmir Fedorchuk/Fotolia.com

9.1 Valence Bond Theory

9.1a Tenets of Valence Bond Theory

Valence bond theory and **molecular orbital theory** are two theoretical models that describe the chemical bonding in molecules. The two theories share many assumptions, but they also differ in many ways. The two theories are similar in that both assume that

- bonds occur due to the sharing of electrons between atoms;

- the attraction of bonding electrons to the nuclei of the bonded atoms leads to lower energy, and therefore the formation of a bond; and

- two types of bonds can form (sigma and pi).

The two theories differ in how they describe the location of the electrons in bonding orbitals, how they explain the energy of electrons, and how they explain the presence of unpaired electrons in molecules.

Although molecular orbital theory is the more accurate and more broadly useful of the two theories, valence bond theory is easier to use. For example, VSEPR coupled with valence bond theory allows chemists to easily predict shapes of most compounds made up of p-block elements. Molecular orbital theory, on the other hand, predicts molecular shapes but only after a more complicated process. When discussing orbital energies, electronic transitions between orbitals, or bonding, for example, in transition metal compounds, chemists use the more complex molecular orbital theory. At times, the two theories are used in concert to describe different aspects of chemical bonding in a single large molecule.

There are three basic tenets of valence bond theory:

- Valence atomic orbitals on adjacent atoms overlap.

- Each pair of overlapping valence orbitals is occupied by two valence electrons to form a chemical bond.

- Valence electrons are either involved in bonding between two atoms (shared bonding pairs) or reside on a single atom (nonbonding lone pairs).

A covalent bond is the result of **orbital overlap**, the partial occupation of the same region of space by orbitals on adjacent atoms. The bonding region is the location between the atomic nuclei, where electrons occupy the overlapping orbitals. For example, consider the covalent bond in hydrogen, H_2 (Interactive Figure 9.1.1). Each H atom has a single unpaired electron in a 1s orbital (H: $1s^1$). The covalent bond in H_2 is the result of

Interactive Figure 9.1.1

Explore covalent bonding in H_2.

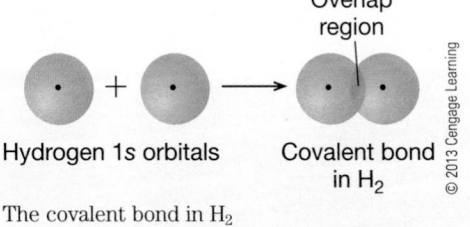

The covalent bond in H_2

the overlap of two $1s$ atomic orbitals on adjacent H atoms, and each H atom contributes one electron to the covalent bond. The covalent bond forms because of the strong attractive forces between the bonding electrons in the overlapping region and the two H nuclei.

Sigma Bonds The covalent bond in H_2 is a **sigma (σ) bond** because the bonding region lies along the internuclear axis, the region of space between the nuclei of the bonded atoms. Sigma bonds are not only formed between two s orbitals. Consider the sigma bond in HF (Interactive Figure 9.1.2). Both hydrogen (H: $1s^1$) and fluorine (F: $[He]2s^22p^5$) have an unpaired electron in an atomic orbital. The sigma bond that forms between H and F is the result of overlap of a $1s$ atomic orbital (on H) with a $2p$ atomic orbital (on F). Each atom contributes one electron to the covalent bond. The H—F bond is a sigma bond because the bonding region lies between the H and F nuclei.

The sigma bond in F_2 is the result of overlap of two $2p$ atomic orbitals, one from each F atom (Figure 9.1.3). Each F atom contributes one electron to the covalent bond.

9.2 Hybrid Orbitals

9.2a sp^3 Hybrid Orbitals

The covalent bonding in methane (CH_4) is not easily explained by the atomic orbital overlap model. Although there are four C—H sigma bonds in methane, the electron configuration of carbon shows two unpaired electrons, each in a $2p$ atomic orbital. This suggests that carbon can form no more than two covalent bonds.

$$C: [He]\ \boxed{\uparrow\downarrow}\ \boxed{\uparrow}\,\boxed{\uparrow}\,\boxed{\ }\qquad H: \boxed{\uparrow}$$
$$\quad\ \ 2s\qquad 2p\qquad\qquad\quad 1s$$

In addition, VSEPR theory predicts that methane has a tetrahedral shape with bond angles of 109.5°, which is not easily explained by the overlap of $2s$ and $2p$ atomic orbitals on carbon with hydrogen $1s$ orbitals. The $2p$ orbitals are arranged at 90° to one another, suggesting that the bond angles in methane should be 90°, not 109.5° (Figure 9.2.1).

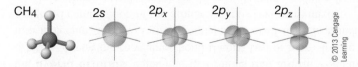

Figure 9.2.1 CH_4 and the orientation of $2s$ and $2p$ atomic orbitals

Explore sigma bonding.

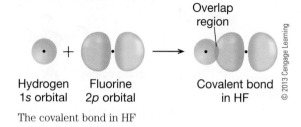

Hydrogen   Fluorine
$1s$ orbital $2p$ orbital

Covalent bond
in HF

© 2013 Cengage Learning

The covalent bond in HF

Section 9.1 Mastery

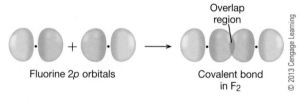

Fluorine $2p$ orbitals

Covalent bond
in F_2

© 2013 Cengage Learning

Figure 9.1.3 The covalent bond in F_2

Valence bond theory explains the bonding in molecules such as methane by introducing **hybrid orbitals**, equal-energy orbitals that are the combination of an atom's atomic orbitals. According to valence bond theory, two or more atomic orbitals on a central atom in a molecule "mix" to form an equal number of hybrid orbitals. Each hybrid orbital is a weighted combination of the atomic orbitals that were mixed.

In methane, one $2s$ and three $2p$ orbitals "mix," or hybridize, forming four equal-energy hybrid orbitals (Interactive Figure 9.2.2). Each carbon hybrid orbital is labeled an **sp^3 hybrid orbital** because it is made up of one part s, one part p_x, one part p_y, and one part p_z atomic orbital. Notice that when four atomic orbitals "mix," or hybridize, four hybrid orbitals result. The four sp^3 hybrid orbitals arrange so that each points to the corner of a tetrahedron; the angle between any two sp^3 hybrid orbitals is 109.5°. Note the connection between valence bond theory and VSEPR theory:

- Four sp^3 hybrid orbitals point to the corners of a tetrahedron.

- Four structural positions are arranged in tetrahedral electron pair geometry.

The C—H sigma bonds in CH_4 result from the overlap of a carbon sp^3 hybrid orbital with a hydrogen $1s$ atomic orbital. The H $1s$ orbitals and C sp^3 hybrid orbitals each have one unpaired electron, and four C—H sigma bonds result from hybrid orbital–atomic orbital overlap (Interactive Figure 9.2.3).

Interactive Figure 9.2.2

Investigate the formation of sp^3 hybrid orbitals.

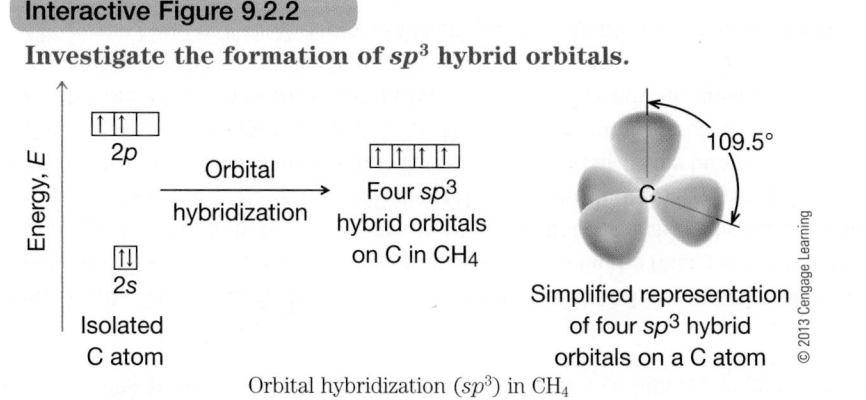

Orbital hybridization (sp^3) in CH_4

Interactive Figure 9.2.3

Explore sigma bonding in CH_4.

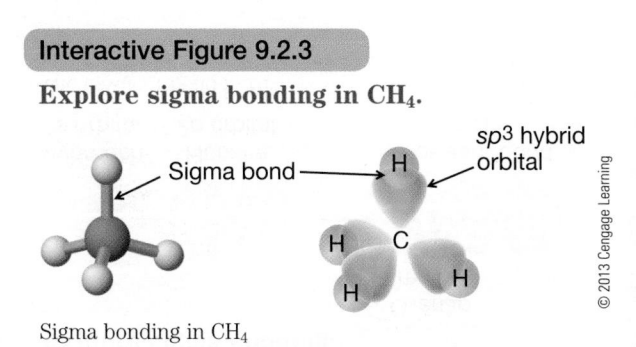

Sigma bonding in CH_4

Hybrid orbitals can also be used to explain the bonding in molecules that have lone pairs on the central atom. Consider NH_3, which has three N—H sigma bonds and a lone pair of electrons on nitrogen (four structural pairs).

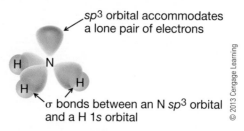

sp^3 orbital accommodates a lone pair of electrons

σ bonds between an N sp^3 orbital and a H 1s orbital

Figure 9.2.4 Sigma bonding in NH_3

The electron-pair geometry of ammonia is tetrahedral, and the H—N—H bond angle in ammonia is close to 109.5°. Both suggest that the nitrogen atom in ammonia is sp^3 hybridized (Figure 9.2.4). Three of the four sp^3 hybrid orbitals overlap with H 1s atomic orbitals to form sigma bonds. The fourth sp^3 hybrid orbital accommodates the nitrogen lone pair.

Methanol, a molecule with two central atoms, also contains sp^3-hybridized atoms.

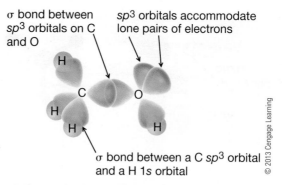

σ bond between sp^3 orbitals on C and O

sp^3 orbitals accommodate lone pairs of electrons

σ bond between a C sp^3 orbital and a H 1s orbital

Figure 9.2.5 Sigma bonding in CH_3OH

Both carbon and oxygen have four structural pairs and tetrahedral electron-pair geometry, and both are sp^3 hybridized (Figure 9.2.5). Three of the four sp^3 hybrid orbitals on C overlap with H 1s atomic orbitals to form C—H sigma bonds. The C—O sigma bond forms from overlap of a carbon sp^3 hybrid orbital and an oxygen sp^3 hybrid orbital. Two oxygen sp^3 hybrid orbitals accommodate lone pairs, and the fourth oxygen sp^3 hybrid orbital overlaps the H 1s orbital to form the O—H sigma bond. Because both carbon and oxygen are sp^3 hybridized, the bond angles around these atoms are approximately 109°.

9.2b sp^2 Hybrid Orbitals

Boron trifluoride, BF_3, has trigonal planar geometry with 120° F—B—F bond angles.

The electron configuration of boron and the F—B—F bond angles suggest that boron hybridizes when it forms covalent bonds with fluorine.

B: [He] $\uparrow\downarrow$ $\uparrow$ □ □   2s 2p   F: [He] $\uparrow\downarrow$ $\uparrow\downarrow$ $\uparrow\downarrow$ $\uparrow$ 2s 2p

Investigate the formation of sp^2 hybrid orbitals.

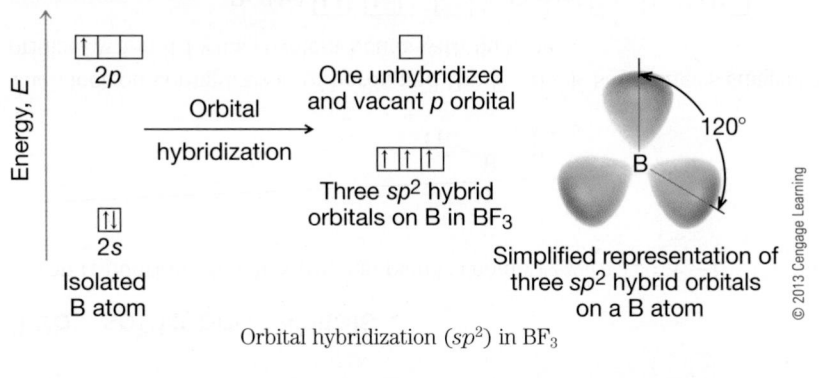

Orbital hybridization (sp^2) in BF_3

Boron forms three equal-energy hybrid orbitals by combining its three lowest-energy valence orbitals, the $2s$ and two $2p$ orbitals (Interactive Figure 9.2.6).

Each boron hybrid orbital is labeled an **sp^2 hybrid orbital** because it is made up of one part s and two parts p atomic orbital. One of the boron $2p$ orbitals does not hybridize; it remains an unhybridized $2p$ atomic orbital. The three sp^2 hybrid orbitals point to the corners of a triangle; the angle between any two sp^2 hybrid orbitals is 120°. Note the connection between valence bond theory and VSEPR theory:

- Three sp^2 hybrid orbitals point to the corners of a triangle.

- Three structural positions are arranged in trigonal planar electron-pair geometry.

Each of the three fluorine atoms has a $2p$ orbital with one unpaired electron, and each $2p$ atomic orbital overlaps a boron sp^2 hybrid orbital, forming three B—F sigma bonds (Figure 9.2.7).

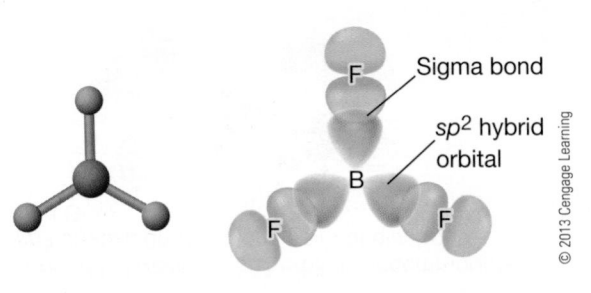

Figure 9.2.7 Sigma bonding in BF_3

9.2c   *sp* Hybrid Orbitals

Beryllium difluoride, BeF_2, has linear geometry with an F—Be—F angle of 180°.

$$:\ddot{F}\text{—}Be\text{—}\ddot{F}:$$

The electron configuration of beryllium and the F—Be—F angle suggest that beryllium hybridizes when it forms covalent bonds with fluorine.

Investigate the formation of *sp* hybrid orbitals.

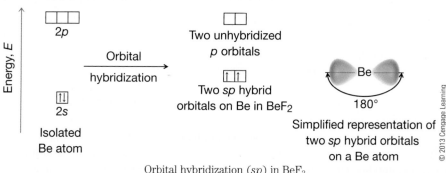

Orbital hybridization (*sp*) in BeF_2

Figure 9.2.9 Sigma bonding in BeF_2

Beryllium forms two equal-energy hybrid orbitals by combining its two lowest-energy valence orbitals, the 2*s* and one 2*p* orbital (Interactive Figure 9.2.8). Each beryllium hybrid orbital is labeled an ***sp* hybrid orbital** because it is made up of one part *s* and one part *p* atomic orbital. Two of the beryllium 2*p* orbitals do not hybridize; they remain unhybridized 2*p* atomic orbitals. The two *sp* hybrid orbitals are arranged in a line; the angle between the two *sp* hybrid orbitals is 180°. Note the connection between valence bond theory and VSEPR theory:

- Two *sp* hybrid orbitals are arranged in a line.

- Two structural positions are arranged in linear electron-pair geometry.

Each of the two fluorine atoms has a 2*p* orbital with one unpaired electron, and each 2*p* atomic orbital overlaps a beryllium *sp* hybrid orbital, forming two Be—F sigma bonds (Figure 9.2.9).

Example Problem 9.2.1 Recognize hybridization.

Determine the hybridization of all nonhydrogen atoms in the following molecule.

$$\underset{H}{\overset{H}{>}}C=C\underset{:NH_2}{\overset{CH_3}{<}}$$

Solution:

You are asked to determine the hybridization of all nonhydrogen atoms in a molecule.

You are given the Lewis structure for the molecule.

The two central carbon atoms are sp^2 hybridized. The carbon on the left has three sigma bonds (two to H atoms and one to the central C atom), and the central carbon has three sigma bonds (one to the left C atom, one to the right C atom, and one to the N atom). Both sp^2-hybridized carbon atoms have trigonal planar electron-pair geometry.

The carbon atom on the right is sp^3 hybridized. This carbon has four sigma bonds (one to the sp^2-hybridized carbon and three to H atoms) and tetrahedral geometry.

The nitrogen atom is sp^3 hybridized. It has three sigma bonds (one to the central C atom and two to H atoms) and one lone pair. The nitrogen atom has tetrahedral electron-pair geometry and a trigonal pyramidal shape.

Video Solution

Tutored Practice
Problem 9.2.1

9.2d sp^3d Hybrid Orbitals

As shown in Covalent Bonding and Molecular Structure (Unit 8), compounds with a central atom that is an element in the third period or below in the periodic table can have an expanded valence where more than eight electrons are associated with the central atom. The hybrid orbitals used to form sigma bonds in these compounds include d orbitals. In PF_5, for example, phosphorus has five structural pairs and therefore forms five hybrid orbitals (Figure 9.2.10).

Each phosphorus hybrid orbital is labeled an **sp^3d hybrid orbital** because it is made up of one part s, three parts p, and one part d atomic orbital. The five sp^3d hybrid orbitals point to the corners of a trigonal bipyramid; the angle between any two sp^3d hybrid orbitals is 90°, 120°, or 180°. Note the connection between valence bond theory and VSEPR theory:

* Five sp^3d hybrid orbitals point to the corners of a trigonal bipyramid.

* Five structural positions are arranged in trigonal bipyramidal electron-pair geometry.

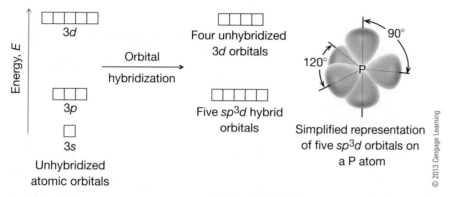

Figure 9.2.10 Orbital hybridization (sp^3d) in PF_5

Each of the five fluorine atoms has a $2p$ orbital with one unpaired electron, and each $2p$ atomic orbital overlaps a phosphorus sp^3d hybrid orbital, forming five P—F sigma bonds (Figure 9.2.11).

Sulfur tetrafluoride, a molecule with five structural pairs, also has an sp^3d-hybridized central atom (Interactive Figure 9.2.12). Four of the five sp^3d hybrid orbitals on S overlap with F $2p$ atomic orbitals to form four S—F sigma bonds. The fifth sulfur sp^3d hybrid orbital accommodates the lone pair of electrons on sulfur. Recall from VESPR theory that in trigonal pyramidal electron-pair geometry, lone pairs occupy equatorial positions.

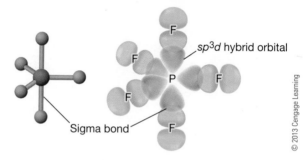

Figure 9.2.11 Sigma bonding in PF_5

Interactive Figure 9.2.12

Investigate sigma bonding in SF_4.

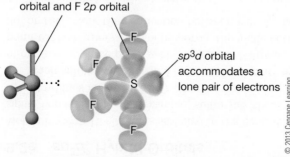

9.2e sp^3d^2 Hybrid Orbitals

Molecules with 12 electrons assigned to the central atom use sp^3d^2 hybrid orbitals to form sigma bonds. In SF_6, for example, sulfur has six structural pairs and therefore forms six hybrid orbitals (Figure 9.2.13).

Each sulfur hybrid orbital is labeled an **sp^3d^2 hybrid orbital** because it is made up of one part s, three parts p, and two parts d atomic orbital. The six sp^3d^2 hybrid orbitals point to the corners of an octahedron; the angle between any two sp^3d^2 hybrid orbitals is 90° or 180°. Note the connection between valence bond theory and VSEPR theory:

- Six sp^3d^2 hybrid orbitals point to the corners of an octahedron.

- Six structural positions are arranged in octahedral electron-pair geometry.

Each of the six fluorine atoms has a $2p$ orbital with one unpaired electron, and each $2p$ atomic orbital overlaps a sulfur sp^3d^2 hybrid orbital, forming six S—F sigma bonds (Figure 9.2.14).

Interactive Table 9.2.1 summarizes the relationship between the hybridization of a central atom and the number of structural pairs on the hybridized atom. To chemists, *hybridization* and *electron-pair geometry* terms are synonymous. For example, the terms sp^2 and *trigonal planar geometry* provide the same information about the geometry and sigma bonding of an atom in a molecule or ion.

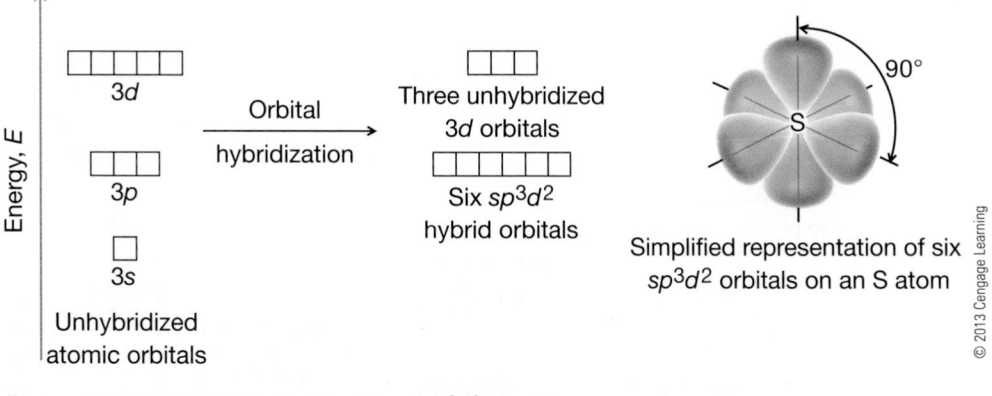

Figure 9.2.13 Orbital hybridization (sp^3d^2) in SF_6

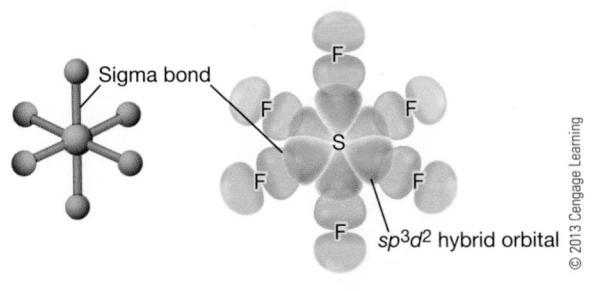

Figure 9.2.14 Sigma bonding in SF_6

Relation Between Structural Pairs and Hybridization

Number of Structural Pairs on the Central Atom	Electron-Pair Geometry	Atomic Orbitals Mixed to Form Hybrid Orbitals	Hybridization
2	Linear	One s, one p	sp
3	Trigonal planar	One s, two p	sp^2
4	Tetrahedral	One s, three p	sp^3
5	Trigonal bipyramidal	One s, three p, one d	sp^3d
6	Octahedral	One s, three p, two d	sp^3d^2

Example Problem 9.2.2 Identify sigma bonding.

Answer the following questions about SO_3.

a. What is the hybridization of the sulfur atom in SO_3?
b. What orbitals make up the sigma bond between S and O in SO_3?
c. What are the approximate bond angles in SO_3?

Solution:

You are asked, for a specific compound, to describe the hybridization of the central atom, describe the orbitals used to form sigma bonds, and give the bond angles.

You are given the formula of the compound.

The Lewis structure of SO_3 is

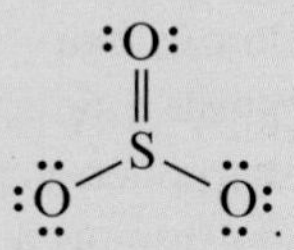

a. The sulfur atom in SO_3 has three sigma bonds and no lone pairs. It is sp^2 hybridized.
b. Oxygen atoms have an unpaired electron in a $2p$ orbital. The S—O sigma bonds result from overlap of a $2p$ orbital on oxygen with one of the sp^2 hybrid orbitals on sulfur.
c. Sulfur is sp^2 hybridized and therefore has trigonal planar electron-pair geometry. The O—S—O bond angles are 120°.

Video Solution

Tutored Practice Problem 9.2.2

Section 9.2 Mastery

9.3 Pi Bonding

9.3a Formation of Pi Bonds

Earlier, we defined a sigma bond as a covalent bond where the bonding region lies along the internuclear axis, between the nuclei of the bonded atoms. A **pi (π) bond** occurs when the two orbitals overlap to form a bond where the bonding region is above and below the internuclear axis.

Pi bonds can form when two unhybridized p orbitals on adjacent atoms overlap (Interactive Figure 9.3.1). The sideways overlap of two p orbitals to form a pi bond is less effective and results in a weaker bond than the sigma bond formed by direct, head-on overlap of atomic or hybrid orbitals. As a result, a single, two-electron bond between two atoms is always a sigma-type bond. When two or more covalent bonds form between two atoms, one is always a sigma bond and the additional bonds are pi bonds.

Interactive Figure 9.3.1

Explore the formation of pi bonds from the overlap of p orbitals.

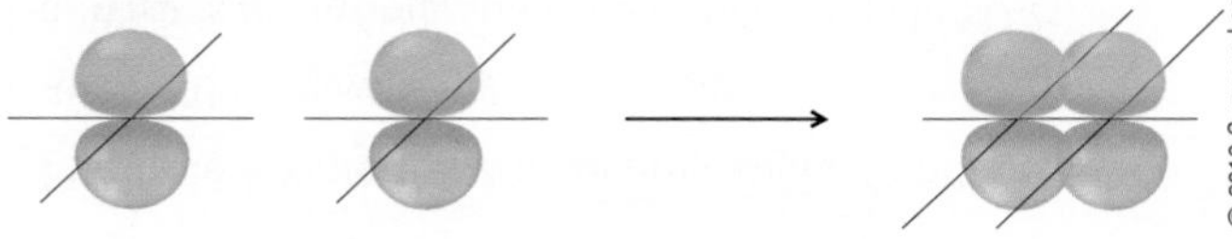

© 2013 Cengage Learning

Pi bond formation from two p orbitals

Example Problem 9.3.1 Identify sigma and pi bonds.

How many sigma and pi bonds are in the following molecule?

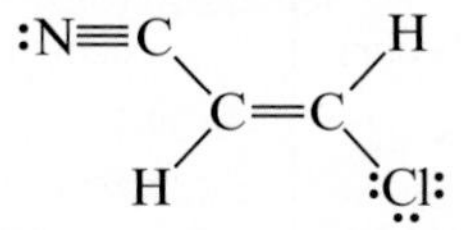

Solution:

You are asked to identify the number of sigma and pi bonds in a molecule.

You are given the Lewis structure of the compound.

Example Problem 9.3.1 (continued)

Each line represents a two-electron bond. A single bond (one line) represents a sigma bond; a double bond (two lines) represents one sigma bond and one pi bond; a triple bond (three lines) represents one sigma bond and two pi bonds.

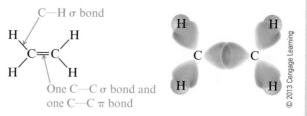

There are six sigma bonds and three pi bonds in the molecule.

Video Solution

Tutored Practice Problem 9.3.1

9.3b Pi Bonding in Ethene, C_2H_4, Acetylene, C_2H_2, and Allene, CH_2CCH_2

Each C atom in ethene is sp^2 hybridized, and each sp^2 hybrid orbital is used to form a sigma bond to another atom (Figure 9.3.2). The sp^2-hybridized carbon atoms in ethene each have an unhybridized p orbital that is not involved in sigma bonding and that contains a single electron (Figure 9.3.3).

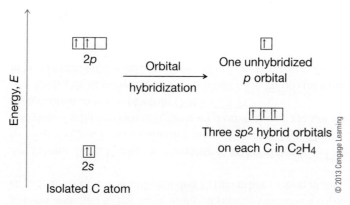

Figure 9.3.3 Formation of hybrid orbitals for the carbon atoms in ethene

The unhybridized $2p$ orbitals are used to form a pi bond between the two carbon atoms (Figure 9.3.4). The pi bond is a two-electron bond, like a sigma bond, and it lies above and

Figure 9.3.2 Sigma bonding in ethene

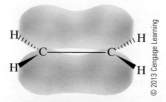

Figure 9.3.4 Pi bonding in ethene

below the plane containing the sigma bonds in ethene. Notice that the sideways alignment of the p orbitals that overlap to form the pi bond result in a flat (planar) molecular shape for ethene. A complete picture of the sigma and pi bonding in ethene is shown in Interactive Figure 9.3.5.

Acetylene, C_2H_2, and allene, CH_2CCH_2, are examples of molecules with more than one pi bond attached to the same atom. Acetylene contains a triple bond, whereas allene has a carbon atom that is double-bonded to two other carbon atoms.

Acetylene Each C atom in acetylene is sp hybridized, and each sp hybrid orbital is used to form a sigma bond to another atom (Figure 9.3.6). The sp-hybridized carbon atoms in acetylene each have two unhybridized p orbitals that are not involved in sigma bonding, and each contains a single electron (Figure 9.3.7).

Each pair of unhybridized $2p$ orbitals (one $2p$ orbital from each carbon atom) forms a pi bond between the carbon atoms. There are two pairs of unhybridized $2p$ orbitals, so two

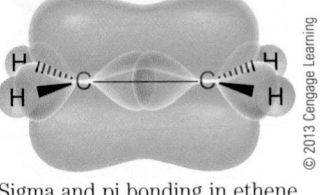

Explore sigma and pi bonding in ethene.

Sigma and pi bonding in ethene

© 2013 Cengage Learning

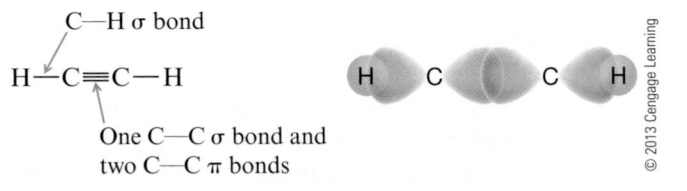

C—H σ bond

H—C≡C—H

One C—C σ bond and
two C—C π bonds

© 2013 Cengage Learning

Figure 9.3.6 Sigma bonding in acetylene

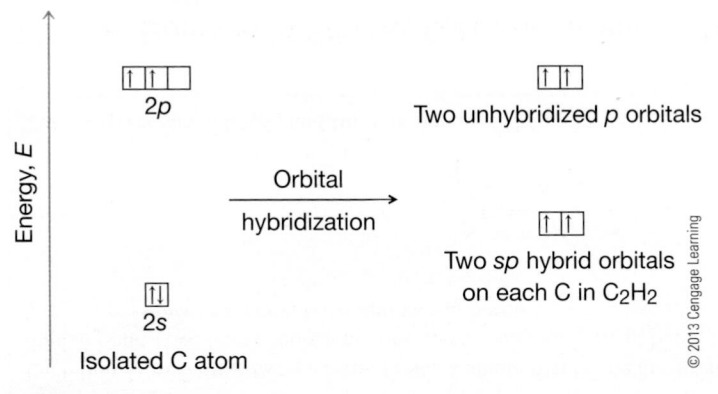

2p

Two unhybridized p orbitals

Energy, E

Orbital
hybridization

Two sp hybrid orbitals
on each C in C_2H_2

2s

Isolated C atom

© 2013 Cengage Learning

Figure 9.3.7 Formation of hybrid orbitals for the carbon atoms in acetylene

pi bonds are formed. The two pi bonds are oriented 90° from each other because the unhybridized $2p$ orbitals on each carbon are at 90° from each other. A complete picture of the sigma and pi bonding in acetylene is shown in Figure 9.3.8.

Allene The outer C atoms in allene are sp^2 hybridized, and the central C atom is sp hybridized. Each hybrid orbital is used to form a sigma bond to another atom (Figure 9.3.9). The three C atoms in allene are also connected by pi bonds. Because each p orbital on the outer C atoms overlaps with a different p orbital on the central C atom, the two outer —CH_2 groups are aligned at 90° to each other (Interactive Figure 9.3.10).

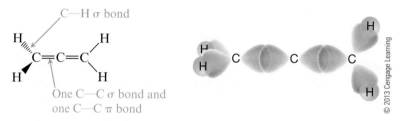

Figure 9.3.9 Sigma bonding in allene

9.3c Pi Bonding in Benzene, C_6H_6

The cyclic compound benzene, C_6H_6, is one of the most important organic molecules. The molecule is composed of six carbon atoms in a ring with alternating single and double bonds, and each carbon also bonds to a single hydrogen atom. Benzene has two equivalent resonance structures.

Each C atom in benzene is sp^2 hybridized, and the sp^2 hybrid orbitals are used to form sigma bonds to two carbon atoms and one hydrogen atom. The trigonal planar electron-pair geometry around each carbon atom results in a planar ring structure (Figure 9.3.11).

Like ethene, each sp^2-hybridized carbon atom in benzene has an unhybridized p orbital that is used to form pi bonds, and the pi bonds lie above and below the plane containing the

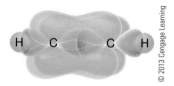

Figure 9.3.8 Sigma and pi bonding in acetylene

Interactive Figure 9.3.10

Explore sigma and pi bonding in allene.

Sigma and pi bonding in allene

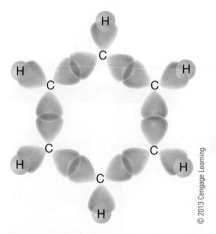

Figure 9.3.11 Sigma bonding in benzene

carbon–carbon and carbon–hydrogen sigma bonds. The pi bonding in benzene is more complex than in ethene, however, because of the benzene resonance structures. The two equivalent resonance structures for benzene indicate that the molecule does not have three pi bonds that are each localized between two carbon atoms. Instead, the six unhybridized p orbitals on carbon form one delocalized pi bonding system that lies above and below the plane of the molecule (Figure 9.3.12).

Although it can be difficult to represent the resonance hybrid for a molecule, chemists often use a ring when drawing the structure of benzene to represent the delocalized pi bonding system.

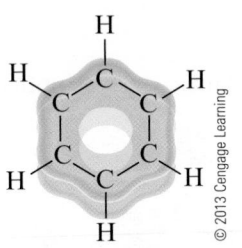

Figure 9.3.12 Pi bonding in benzene

or

Viewing the sigma and pi system in benzene together shows the planar shape of the molecule and the delocalized pi electron density that lies above and below the plane of the molecule (Interactive Figure 9.3.13).

Pi bonds form due to overlap of two or more p orbitals, but they can also form from the overlap of p and d orbitals or two d orbitals. In ethene, acetylene, allene, and benzene, the number of pi bonds formed by a carbon atom is related to the number of unhybridized p orbitals available on that atom. The general relationship between atom hybridization, the number of unhybridized p orbitals, and the number of possible p–p pi bonds is shown in Interactive Table 9.3.1.

Interactive Figure 9.3.13

Explore sigma and pi bonding in benzene.

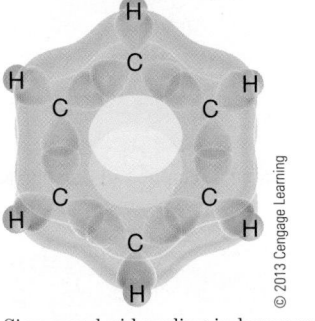

Sigma and pi bonding in benzene

Interactive Table 9.3.1

Relationship Between Hybridization and Number of Possible p–p Pi Bonds

Number of Structural Pairs on the Central Atom	Hybridization	Unhybridized p Orbitals	Number of Possible p–p Pi Bonds
2	sp	Two p	2
3	sp^2	One p	1
4	sp^3	None	0
5	sp^3d	None	0
6	sp^3d^2	None	0

9.3d Conformations and Isomers

The presence or absence of pi bonds in a molecule affects its physical properties. For example, acetylene, C_2H_2, has two pi bonds and reacts with H_2, whereas ethane, C_2H_6, has no pi bonds and does not react with H_2. In addition, pi bonds affect the physical shape of a molecule, which relates to the number of possible conformations and isomers a molecule can adopt.

Conformations are the different three-dimensional arrangements of atoms in a molecule that can be interconverted by rotation around single bonds. Consider the butane conformations shown in Interactive Figure 9.3.14. Each differs only in the orientation of the right side of the molecule. The structures were generated by rotating one half of the molecule with respect to the other half, around the carbon–carbon sigma bond.

Carbon–carbon sigma bond rotation in butane occurs easily because a sigma bond has bonding electron density directly between two bonded atoms. Rotation around a sigma bond does not affect the bonding electrons that lie between the bonded nuclei. There is no limit to the number of possible conformations of a butane molecule, and the ends of the molecule rotate freely at room temperature.

Isomers are two or more substances that have the same chemical formula but have different properties because of the different arrangement of atoms. Molecules with pi bonds are one example of compounds that can exist as more than one isomer. Consider the bond rotation in a molecule containing a pi bond. Because a pi bond has electron density both above and below the internuclear axis, a pi bond cannot rotate freely. Rotation around a double bond (a sigma bond and a pi bond) results in breaking the pi bond, a process that requires a significant amount of energy. Such free rotation therefore does not happen at room temperature (Interactive Figure 9.3.15).

The energy barrier that prevents rotation of pi bonds means that two isomers that differ by rotation about the pi bond will not easily interconvert at room temperature. Consider the two possible isomers of 1,2-dichloroethene (Figure 9.3.16). These two structures differ only in the placement of the Cl and H atoms about the C=C double bond. In the structure labeled *cis*, both Cl atoms are on the same side of the double bond, whereas in the structure

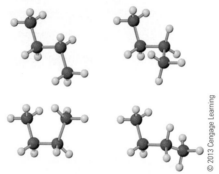
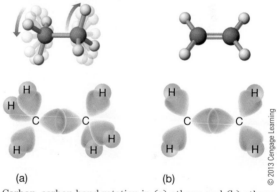

cis-1,2-dichloroethene

trans-1,2-dichloroethene

Figure 9.3.16 Two unique isomers of 1,2-dichloroethene

labeled *trans*, the Cl atoms are on opposite sides of the double bond. These two structures cannot interconvert easily because doing so would require breaking the C—C pi bond; therefore, the two structures represent two unique isomers.

Example Problem 9.3.2 Identify isomers and conformations.

Consider the following set of molecules. Gray spheres represent carbon atoms, white spheres represent hydrogen atoms, and green spheres represent chlorine atoms. Which pairs represent conformations of the same molecule, and which pairs represent isomers?

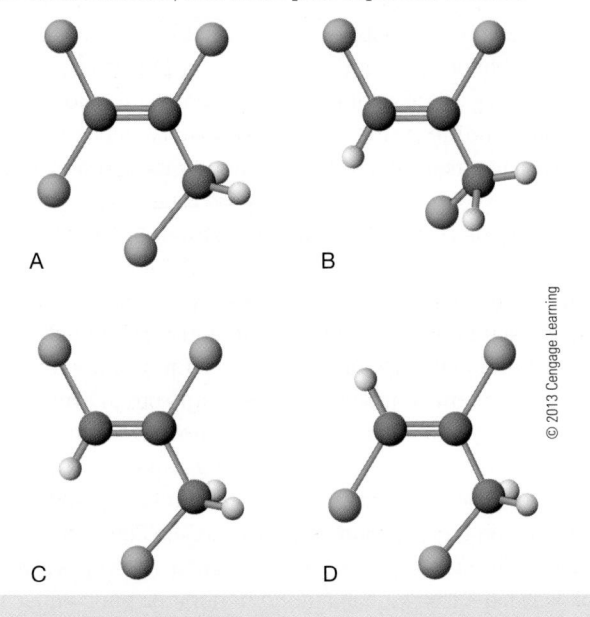

A   B

C   D

© 2013 Cengage Learning

Solution:

You are asked to identify conformations and isomers of a compound.

You are given a set of molecular structures.

B and C are conformations of the same compound. They have the same chemical formula ($C_3H_3Cl_3$) and are related by rotation around a carbon–carbon single bond.

C and D are isomers. They have the same chemical formula ($C_3H_3Cl_3$) and are related by rotation around a carbon–carbon double bond. C is the *cis* isomer, and D is the *trans* isomer.

B and D are also isomers. B is the *cis* isomer, and D is the *trans* isomer. They are related by rotation around a carbon–carbon double bond and rotation around a carbon–carbon single bond.

A is a unique compound ($C_3H_2Cl_4$).

Video Solution

Tutored Practice
Problem 9.3.2

Section 9.3 Mastery

9.4 Molecular Orbital Theory

9.4a Sigma Bonding and Antibonding Molecular Orbitals

Molecular orbital theory is similar in many ways to valence bond theory. Bond formation is viewed in a similar manner, where overlapping orbitals on different atoms increase attractive forces between electrons and nuclei. The two theories differ, however, in how the resulting orbital combinations are viewed. In valence bond theory, orbitals in a molecule are thought to be localized on atoms, with some overlap of the orbitals between bonded nuclei. In molecular orbital theory, orbitals in a molecule are thought to be spread out (delocalized) over many atoms. Valence bond theory is often referred to as a localized bonding theory, whereas molecular orbital theory is referred to as a delocalized bonding theory. One of the most important and unique aspects of molecular orbital theory is its ability to predict the shapes and energies of orbitals that contain no electrons. That is, molecular orbital theory explains not only how electrons are arranged in the ground state but also how they might be arranged in an excited electronic state.

According to molecular orbital theory, when any number of atomic orbitals overlap to form molecular orbitals, an equal number of molecular orbitals are formed. When two s orbitals overlap, for example, they form two new orbitals: one at a lower energy than the original s orbitals and one at a higher energy than the original s orbitals.

Consider the formation of H_2. Each H atom has a single electron in a $1s$ orbital. Adding the two $1s$ orbitals (the $1s$ wave functions) together results in the formation of a **bonding molecular orbital** (Figure 9.4.1) that increases the electron density between the bonded nuclei. This bonding molecular orbital is a sigma (σ) molecular orbital because electron density lies along the internuclear axis. It is labeled σ_{1s}, where the subscript identifies the atomic orbitals that contributed to form the molecular orbital. This molecular orbital is lower in energy than the separated hydrogen $1s$ orbitals because electrons occupying this orbital experience increased attractive forces to the hydrogen nuclei.

Subtracting the two $1s$ orbitals (the $1s$ wave functions) results in the formation of an **antibonding molecular orbital** (Interactive Figure 9.4.2) that decreases the electron density between the bonded nuclei. This bonding molecular orbital is a sigma (σ) molecular orbital because electron density lies along the internuclear axis. It is labeled σ^*_{1s}, where the asterisk (*) indicates its antibonding nature and the subscript identifies the atomic orbitals that contributed to form the molecular orbital. This molecular orbital is higher in energy than the separated hydrogen $1s$ orbitals because electrons occupying this orbital experience decreased attractive forces to the hydrogen nuclei. In addition, the antibonding molecular orbital has a node, a plane on which there is zero probability for finding an electron.

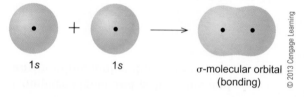

Figure 9.4.1 A bonding (σ_{1s}) molecular orbital

Interactive Figure 9.4.2

Explore bonding and antibonding molecular orbitals.

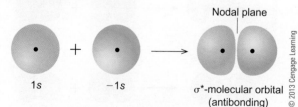

An antibonding (σ^*_{1s}) molecular orbital

Sigma bonding and antibonding molecular orbitals can also form from the interactions of p orbitals. When two $2p$ orbitals are added and subtracted in a head-on alignment along the internuclear axis, one sigma bonding (σ_{2p}) molecular orbital and one sigma antibonding (σ^*_{2p}) molecular orbital results (Figure 9.4.3). Notice the formation of a new planar node in the antibonding (σ^*_{2p}) molecular orbital.

9.4b Pi Bonding and Antibonding Molecular Orbitals

Just as in valence bond theory, pi (π) molecular orbitals result from the sideways overlap of p orbitals. When two $2p$ orbitals are added and subtracted, a pi bonding orbital and a pi antibonding orbital form (Interactive Figure 9.4.4). The π_{2p} molecular orbital is lower in energy than the original $2p$ orbitals; the π^*_{2p} molecular orbital is higher in energy (less stable) than the original $2p$ orbitals.

Because there are two $2p$ orbitals on each atom that can overlap to form pi bonds, a total of four pi molecular orbitals are possible, two pi bonding molecular orbitals (π_{2p}) and two pi antibonding molecular orbitals (π^*_{2p}).

9.4c Molecular Orbital Diagrams (H_2 and He_2)

One of the strengths of molecular orbital theory is its ability to describe the energy of both occupied and unoccupied molecular orbitals for a molecule. A **molecular orbital diagram** shows both the energy of the atomic orbitals (from the atoms that are combining) and the energy of the molecular orbitals.

Consider the molecular orbital diagram for H_2 (Figure 9.4.5). Notice the following features of the H_2 molecular orbital diagram.

- The atomic orbitals are placed on the outside of the diagram, and the molecular orbitals are placed between the atomic orbitals, in the center of the diagram.

- Valence electrons are shown in atomic orbitals, and electrons are assigned to molecular orbitals according to the Pauli exclusion principle and Hund's rule.

- Dashed lines are used to connect molecular orbitals to the atomic orbitals that contribute to their formation.

- Bonding molecular orbitals are lower in energy than the atomic orbitals that contribute to their formation.

- Antibonding molecular orbitals are higher in energy than the atomic orbitals that contribute to their formation.

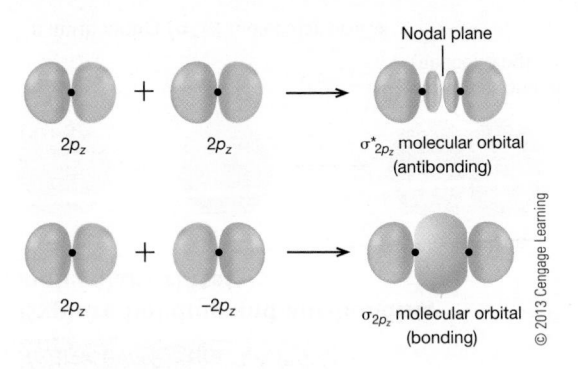

Figure 9.4.3 Bonding (σ_{2p}) and antibonding (σ^*_{2p}) molecular orbitals

© 2013 Cengage Learning

Interactive Figure 9.4.4

Compare sigma and pi bonding and antibonding molecular orbitals.

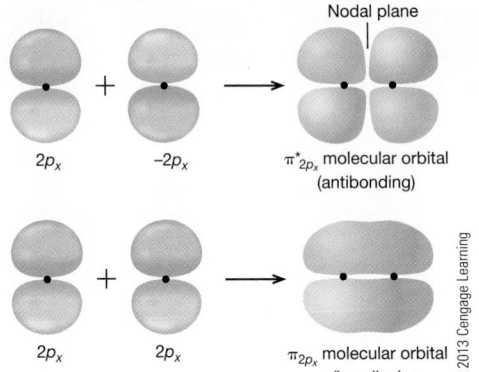

Bonding (π_{2p}) and antibonding (π^*_{2p}) molecular orbitals

© 2013 Cengage Learning

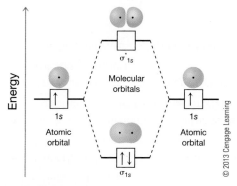

Figure 9.4.5 Molecular orbital diagram for H_2

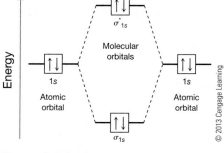

Figure 9.4.6 Molecular orbital diagram for He_2

The electron configuration for H_2 is written $(\sigma_{1s})^2$, showing the presence of two electrons in the σ_{1s} molecular orbital. The molecular orbital diagram for dihelium (He_2) is the same as that of hydrogen, with the addition of two more electrons (Figure 9.4.6). The electron configuration for dihelium is $(\sigma_{1s})^2(\sigma^*_{1s})^2$.

Molecular orbital diagrams provide a method for determining the **bond order** between two atoms in a molecule. (◄ Flashback to Section 8.3a Bond Order, Bond Length, and Bond Energy)

$$\text{bond order} = \tfrac{1}{2} \,[\text{number of electrons in bonding orbitals}$$

$$- \text{ number of electrons in antibonding orbitals}] \qquad \textbf{(9.1)}$$

$$\text{H—H bond order in } H_2 = \tfrac{1}{2}\,[2 - 0] = 1$$

$$\text{He—He bond order in } He_2 = \tfrac{1}{2}\,[2 - 2] = 0$$

The bond order in H_2 is the same as that predicted from its Lewis dot structure. The bond order in He_2 is zero, suggesting that this molecule probably does not exist.

Calculated bond orders for other hydrogen and helium species (Interactive Figure 9.4.7) suggest that H_2^+, H_2^-, and He_2^+ have weak bonds (bond order = 0.5) and are predicted to exist.

9.4d Molecular Orbital Diagrams (Li_2–F_2)

The **homonuclear diatomic molecules** of the second period, Li_2–F_2, have both $2s$ and $2p$ valence atomic orbitals. The molecular orbital diagram for the second-row homonuclear

Interactive Figure 9.4.7 ◄

Predict electron configuration and bond properties for first-row diatomic species.

	H_2^+	H_2	H_2^-	He_2^+	He_2
σ^*	☐	☐	↑	↑	↑↓
σ	↑	↑↓	↑↓	↑↓	↑↓
Species	H_2^+	H_2	H_2^-	He_2^+	He_2
Number of bonding electrons	1	2	2	2	2
Number of antibonding electrons	0	0	1	1	2
Bond order	0.5	1.0	0.5	0.5	0

Molecular orbital diagrams for first-row diatomic species

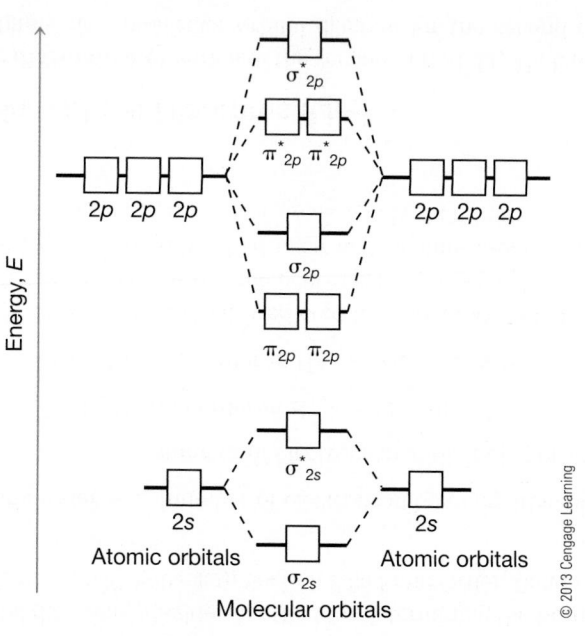

Figure 9.4.8 Molecular orbital diagram for second-row homonuclear diatomic molecules

diatomics (Figure 9.4.8) therefore shows the formation of molecular orbitals from overlap of these valence atomic orbitals.

Notice the following features of the molecular orbital diagram for the second-row homonuclear diatomics.

- Only the valence atomic orbitals and resulting valence molecular orbitals are shown in the molecular orbital diagram.

- Each atom contributes four atomic orbitals ($2s$ and three $2p$ orbitals), resulting in the formation of eight molecular orbitals.

- There are two π_{2p} molecular orbitals (of equal energy) and two π^*_{2p} molecular orbitals (of equal energy).

The homonuclear molecular orbital diagram for oxygen, O_2, is shown in Figure 9.4.9.

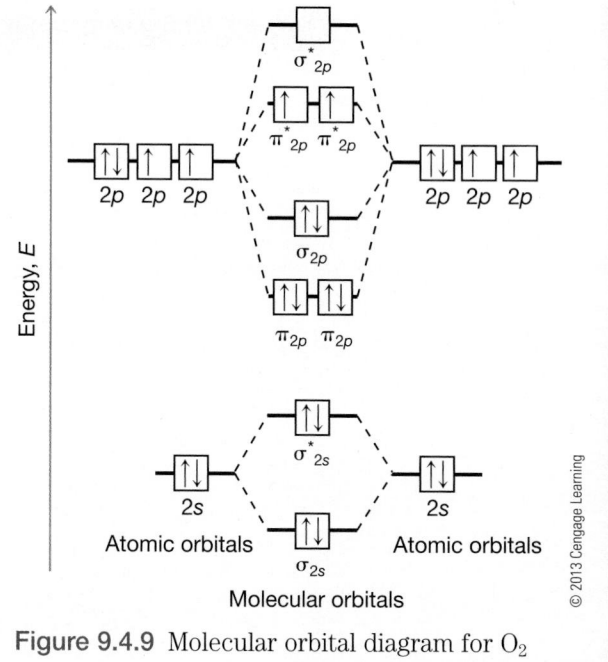

Figure 9.4.9 Molecular orbital diagram for O_2

The molecular orbital diagram shows that the O—O bond order is two and that oxygen is paramagnetic with two unpaired electrons. (◀ Flashback to Section 7.1b Types of Magnetic Materials)

$$O_2: \quad [He](\sigma_{2s})^2(\sigma^*_{2s})^2(\pi_{2p})^4(\sigma_{2p})^2(\pi^*_{2p})^2$$

Paramagnetic (two unpaired electrons)

O—O bond order in $O_2 = \frac{1}{2}[8 - 4] = 2$

As shown in Interactive Figure 9.4.10, in its liquid state, O_2 is attracted to a strong magnet. Notice that valence bond theory and Lewis dot structures do not explain why liquid oxygen is attracted to a magnetic field. This is one example of the more accurate nature of molecular orbital theory.

Abbreviated molecular diagrams for the second-row homonuclear diatomics are shown in Interactive Figure 9.4.11.

Interactive Figure 9.4.10

Explore the paramagnetic nature of O_2.

Liquid oxygen adheres to the poles of a strong magnet.

Interactive Figure 9.4.11

Investigate molecular orbital diagrams for the second-row homonuclear diatomics.

	Li$_2$	Be$_2$	B$_2$	C$_2$	N$_2$	O$_2$	F$_2$
Species	Li$_2$	Be$_2$	B$_2$	C$_2$	N$_2$	O$_2$	F$_2$
Number of bonding electrons	2	2	4	6	8	8	8
Number of antibonding electrons	0	2	2	2	2	4	6
Bond order	1	0	1	2	3	2	1

Molecular orbital diagrams for Li$_2$–F$_2$

© 2013 Cengage Learning

Example Problem 9.4.1 Predict electron configuration and bond properties for homonuclear diatomic molecules.

What is the Ne—Ne bond order in Ne_2 and Ne_2^+?

Solution:

You are asked to identify the bond order in a diatomic molecule or ion.

You are given the formula of the molecule or ion.

To answer this question, the homonuclear diatomic molecular orbital diagram must be filled in with the appropriate number of electrons. The diagram below on the left is filled in for Ne_2, which has a total of eight bonding electrons and eight antibonding electrons. The bond order is therefore 0. In Ne_2^+, however, there are only seven antibonding electrons, and the bond order is 0.5.

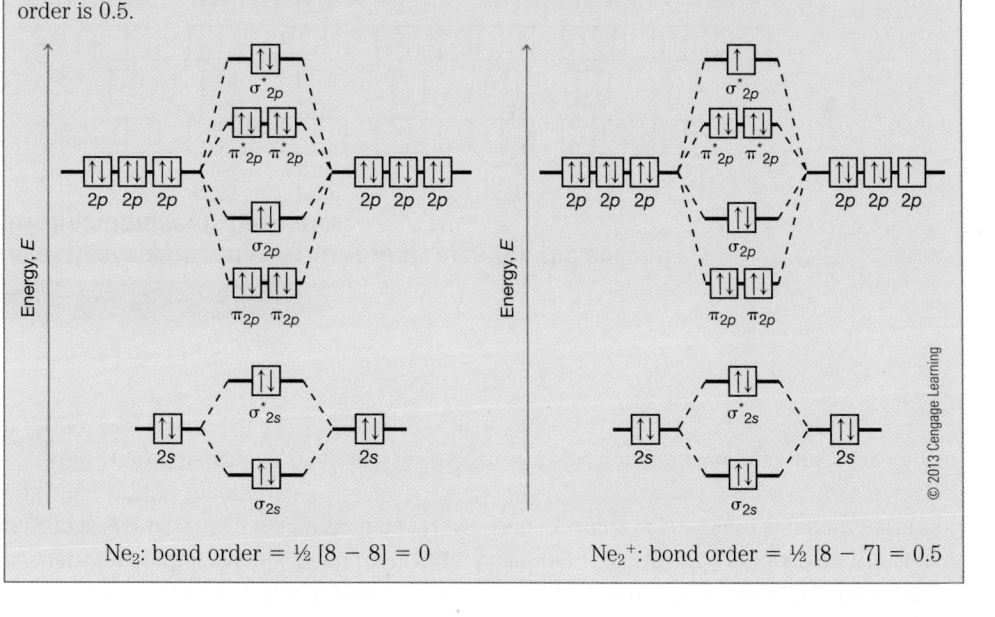

$$Ne_2: \text{ bond order } = \tfrac{1}{2}\,[8 - 8] = 0 \qquad\qquad Ne_2^+: \text{ bond order } = \tfrac{1}{2}\,[8 - 7] = 0.5$$

© 2013 Cengage Learning

Video Solution

Tutored Practice Problem 9.4.1

9.4e Molecular Orbital Diagrams (Heteronuclear Diatomics)

Using what we have learned about homonuclear diatomic molecular orbital diagrams, we can now construct diagrams for **heteronuclear diatomic molecules**, compounds composed of two atoms of different elements. The molecular orbital diagram for the heteronuclear diatomic compound nitrogen monoxide, NO, is shown in Interactive Figure 9.4.12.

$$\text{NO:} \quad [\text{He}](\sigma_{2s})^2(\sigma^*_{2s})^2(\pi_{2p})^4(\sigma_{2p})^2(\pi^*_{2p})^1$$

Paramagnetic (one unpaired electron)

N—O bond order in NO = ½ [8 − 3] = 2.5

Notice that the heteronuclear diatomic diagram is very similar to the homonuclear diagram. For example, the diagrams have the same types of molecular orbitals (sigma and pi, bonding and antibonding). However, the energy of the valence atomic orbitals is not the same. Oxygen is more electronegative than nitrogen, and its atomic orbitals are lower in energy. (◄ Flashback to Section 7.4a Trends in Orbital Energies)

9.4f Molecular Orbital Diagrams (More Complex Molecules)

Most molecules are much more complex than those we have examined here. Molecular orbital theory has proved to be very powerful in interpreting and predicting the bonding in virtually all molecules. The molecular orbital diagram for a slightly more complex compound, methane (CH_4), is shown in Figure 9.4.13, along with the valence bond theory model of bonding in methane.

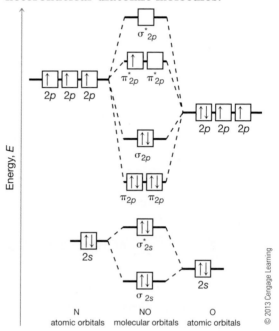

Explore molecular orbital diagrams for heteronuclear diatomic molecules.

Molecular orbital diagram for nitrogen monoxide

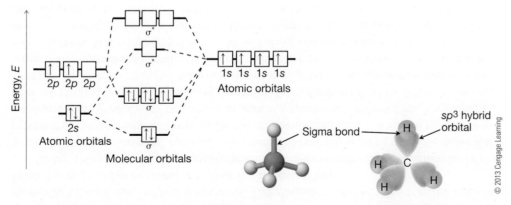

Figure 9.4.13 Both molecular orbital theory and valence bond theory show the presence of four sigma bonds in methane.

Although the molecular orbital diagram in Figure 9.4.13 does not greatly resemble that of the diatomics, it is possible to recognize that the diagram shows four sigma bonding molecular orbitals, each with two electrons. This diagram therefore reinforces the valence bond theory model of methane, with four C—H sigma bonds.

Today, although molecular orbital diagrams for complex molecules can be developed on paper, "molecular modeling" computer programs are typically used to calculate molecular orbital shapes and energies. These programs do not always produce accurate results, however, and their results must be compared against experimental information. For example, photoelectron spectroscopy is one of the more direct methods used to determine orbital energies. In photoelectron spectroscopy, ultraviolet light or x-rays are used to remove electrons from molecules. Assuming the energy levels of an ionized molecule are essentially the same as those in the uncharged molecule, scientists can use photoelectron spectra to understand orbital energies in molecules. Lower-energy peaks correspond to higher-energy orbitals because less energy is required to remove electrons in those orbitals. The photoelectron spectrum for methane (Interactive Figure 9.4.14) shows two distinct energies, which correspond to the two different energies of the occupied sigma bonding molecular orbitals in the molecular orbital diagram for CH_4 shown in Figure 9.4.13.

Explore the MO diagram for a complex molecule.

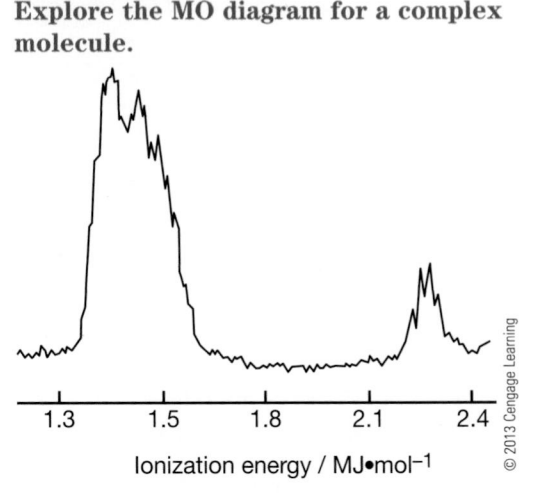

Ionization energy / MJ•mol⁻¹

© 2013 Cengage Learning

Photoelectron spectrum for methane

Unit Recap

Key Concepts

9.1 Valence Bond Theory

- According to valence bond theory, valence atomic orbitals on adjacent atoms overlap, each pair of overlapping valence orbitals is occupied by two valence electrons to form a chemical bond, and valence electrons are either involved in bonding between two atoms (shared bonding pairs) or reside on a single atom (nonbonding lone pairs) (9.1a).

- A sigma (σ) bond is a covalent bond with a bonding region along the internuclear axis (9.1a).

9.2 Hybrid Orbitals

- Hybrid orbitals are equal-energy orbitals that are a combination of atomic orbitals (9.2a).

- Hybrid orbitals are used in valence bond theory to explain bonding in complex molecules (9.2a).

- Four sp^3 hybrid orbitals result from "mixing" an atom's single s and three p valence atomic orbitals. The angle between any two sp^3 hybrid orbitals is 109.5° (9.2a).

- Three sp^2 hybrid orbitals result from "mixing" an atom's single s and two of its three p valence atomic orbitals. The angle between any two sp^2 hybrid orbitals is 120° (9.2b).

- Two sp hybrid orbitals result from "mixing" an atom's single s and one of its three p valence atomic orbitals. The angle between two sp hybrid orbitals is 180° (9.2c).

- Five sp^3d hybrid orbitals result from "mixing" an atom's single s, its three p, and one of its d valence atomic orbitals. The angle between sp^3d hybrid orbitals is 180°, 120°, or 90° (9.2d).

- Six sp^3d^2 hybrid orbitals result from "mixing" an atom's single s, its three p, and two of its d valence atomic orbitals. The angle between sp^3d^2 hybrid orbitals is 180° or 90° (9.2e).

9.3 Pi Bonding

- A pi (π) bond occurs when two orbitals overlap to form a bond where the bonding region is above and below the internuclear axis (9.3a).

- The number of pi bonds an atom forms is related to the number of unhybridized p orbitals available (9.3c).

- Conformations are the different three-dimensional arrangements of atoms in a molecule that can be interconverted by rotation around single bonds (9.3d).

- Isomers are two or more substances that have the same chemical formula but have different properties because of the different arrangement of atoms (9.3d).

9.4 Molecular Orbital Theory

- According to molecular orbital theory, orbitals in a molecule are thought to be spread out (delocalized) over many atoms (9.4a).

- A bonding molecular orbital has increased electron density between bonded nuclei. Bonding molecular orbitals can be either sigma or pi molecular orbitals (9.4a, 9.4b).

- An antibonding molecular orbital has decreased electron density between bonded nuclei. Antibonding molecular orbitals can be either sigma or pi molecular orbitals (9.4a, 9.4b).

- A molecular orbital diagram shows the relative energy of the molecules atomic (valence) orbitals and molecular orbitals (9.4c).

- The bond order between two atoms in a molecule can be calculated from the number of electrons in bonding and antibonding molecular orbitals (9.4c).

Key Equations

bond order = ½ [number of electrons in bonding orbitals

$$- \text{number of electrons in antibonding orbitals}] \qquad \textbf{(9.1)}$$

Key Terms

9.1 Valence Bond Theory
valence bond theory
molecular orbital theory
orbital overlap
sigma (σ) bond

9.2 Hybrid Orbitals
hybrid orbitals
sp^3 hybrid orbital
sp^2 hybrid orbital

sp hybrid orbital
sp^3d hybrid orbital
sp^3d^2 hybrid orbital

9.3 Pi Bonding
pi (π) bond
conformations
isomers

9.4 Molecular Orbital Theory
bonding molecular orbital
antibonding molecular orbital
molecular orbital diagram
bond order
homonuclear diatomic molecule
heteronuclear diatomic molecule

Unit 9   Review and Challenge Problems

10 Gases

Unit Outline

In This Unit...

Matter exists in three main physical states under conditions we encounter in everyday life: gaseous, liquid, and solid. Of these, the most fluid and easily changed is the gaseous state. Gases differ significantly from liquids and solids in that both liquids and solids are condensed states with molecules packed close to one another, whereas gases have molecules spaced far apart. This unit examines the bulk properties of gases and the molecular scale interpretation of those properties.

10.1 Properties of Gases

10.1a Overview of Properties of Gases

Gases are one of the three major states of matter. The physical properties of gases can be manipulated and measured more easily than those of solids or liquids. Because of this, the mathematical relationships between different gas properties were among the first quantitative aspects of chemistry to be studied.

In general, gases differ from liquids and solids more than they differ from each other.

	Solid	Liquid	Gas
Density	High	High	Low
Compressible	No	No	Yes
Fluid	No	Yes	Yes

The most striking property of gases is the simple relationship between the pressure, volume, and temperature of a gas and how a change in one of these properties affects the other properties (Interactive Figure 10.1.1).

Interactive Figure 10.1.1

Explore the properties of gases.

Charles D. Winters

As the gaseous water vapor inside this can condenses to a liquid, the pressure inside the can drops and the can is crushed by the greater external pressure.

These same simple relationships do not exist for solids or liquids. The major properties of gases are given in Table 10.1.1.

Table 10.1.1 Properties of Gases and Their Units

Property	Common Unit	Other Units	Symbol
Mass	grams, g	kg (SI unit), mg	—
Amount	mole, mol (SI unit)	—	n
Volume	liters, L	mL, m^3 (SI unit)	V
Pressure	atmosphere, atm	Pa (SI unit), kPa, bar, mm Hg, psi	P
Temperature	kelvin, K (SI unit)	°C, °F	T

10.1b Pressure

Gases exert **pressure** on surfaces, measured as a force exerted on a given area of surface.

$$\text{pressure} = \frac{\text{force}}{\text{area}}$$

A confined gas exerts pressure on the interior walls of the container holding it and the gases in our atmosphere exert pressure on every surface with which they come in contact.

Gas pressure is commonly measured using a **barometer**. The first barometers consisted of a long, narrow tube that was sealed at one end, filled with liquid mercury, and then inverted into a pool of mercury (Figure 10.1.2a). The gases in the atmosphere push down on the mercury in the pool and balance the weight of the mercury column in the tube. The higher the atmospheric pressure, the higher the column of mercury in the tube. The height of the mercury column, when measured in millimeters, gives the atmospheric pressure in units of **millimeters of mercury** (mm Hg). The **torr** is a unit of pressure (1 torr = 1 mm Hg) named in honor of the inventor of the barometer, Evangelista Torricelli (1608–1647).

Pressure of a gas sample in the laboratory is measured with a manometer, which is shown in Figure 10.1.2b. In this case, mercury is added to a U-shaped tube. One end of the tube is connected to the gas sample under study. The other is open to the atmosphere. If the pressure of the gas sample is equal to atmospheric pressure, the height of the mercury is the same on both sides of the tube. In Figure 10.1.2b, the pressure of the gas is greater than atmospheric pressure by "h" mm Hg.

Gas pressure is expressed in different units. The SI unit for pressure is the **pascal**, Pa, which is equal to the force in newtons exerted on 1 square meter (1 Pa = 1 N/m^2). The

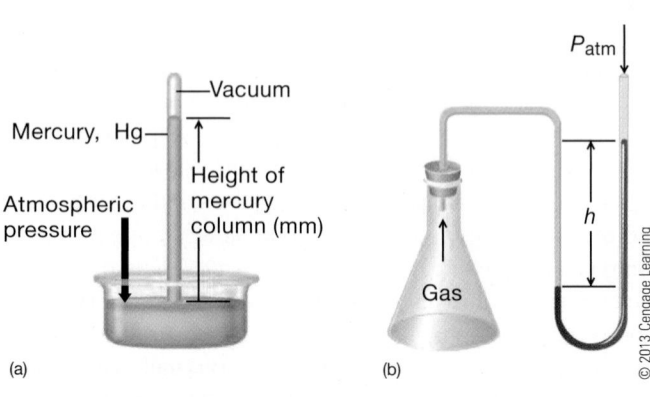

Figure 10.1.2 Measuring pressure using (a) a barometer and (b) a manometer

© 2013 Cengage Learning

English pressure unit, pounds per square inch (psi), is a measure of how many pounds of force a gas exerts on 1 square inch of a surface. Atmospheric pressure at sea level is approximately 14.7 psi. This means that an 8½″ × 11″ piece of paper has a total force on it of more than 1370 pounds. Commonly used pressure units are given in Table 10.1.2. Early pressure units were based on pressure measurements at sea level, where on average the mercury column has a height of 760 mm. This measurement was used to define the **standard atmosphere** (1 atm = 760 mm Hg). Modern scientific studies generally use gas pressure units of atm, kPa, **bar** (1 atm = 1.013 bar), and mm Hg.

Table 10.1.2 Common Units of Gas Pressure		
1 atm	= 1.013 bar	(bar)
	= 101.3 kPa	(kilopascal)
	= 760 mm Hg	(millimeters of mercury)
	= 760 torr	(torr)
	= 14.7 psi	(pounds per square inch)

Example Problem 10.1.1 Convert between pressure units.

A gas sample has a pressure of 722 mm Hg. What is this pressure in atmospheres?

Solution:

You are asked to convert between pressure units.

You are given a pressure value.

Use the relationship 1 atm = 760 mm Hg to convert between these pressure units.

$$722 \text{ mm Hg} \times \frac{1 \text{ atm}}{760 \text{ mm Hg}} = 0.950 \text{ atm}$$

Is your answer reasonable? The pressure, 722 mm Hg, is less than 760 mm Hg, which is the equivalent of 1 atm. Therefore, the pressure expressed in units of atmospheres should be less than 1 atm.

Video Solution

Tutored Practice
Problem 10.1.1

Section 10.1 Mastery

10.2 Historical Gas Laws

10.2a Boyle's Law: $P \times V = k_B$

Boyle's law states that the pressure and volume of a gas sample are inversely related when the amount of gas and temperature are held constant. For example, consider a syringe that is filled with a sample of a gas and attached to a pressure gauge and a thermostat (used to keep the system at a constant temperature). When the plunger is depressed (decreasing the volume of the gas sample), the pressure of the gas increases (Interactive Figure 10.2.1).

When the pressure on the syringe is low, the gas sample has a large volume. When the pressure is high, the gas is compressed and the volume is smaller. The relationship between pressure and volume is therefore an inverse one:

$$\text{volume} \propto \frac{1}{\text{pressure}} \text{ or } P \times V = k_B$$

Because the product of gas pressure and volume is a constant (when temperature and amount of gas are held constant), it is possible to calculate the new pressure or volume of a gas sample when one of the properties is changed.

$$P_1 V_1 = P_2 V_2 \tag{10.1}$$

The subscripts "1" and "2" in Equation 10.1 indicate the different experimental conditions before and after pressure or volume is changed.

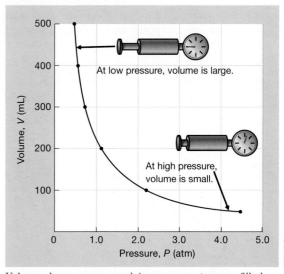

Interactive Figure 10.2.1

Explore Boyle's law.

At low pressure, volume is large.

At high pressure, volume is small.

Volume changes upon applying pressure to a gas-filled syringe. Temperature is constant and no gas escapes from the syringe.

Example Problem 10.2.1 Use Boyle's law to calculate volume.

A sample of gas has a volume of 458 mL at a pressure of 0.970 atm. The gas is compressed and now has a pressure of 3.20 atm. Predict if the new volume is greater or less than the initial volume, and calculate the new volume. Assume temperature is constant and no gas escaped from the container.

Solution:

You are asked to calculate the volume of a gas sample when only the pressure is changed.

You are given the original volume and pressure and the new pressure of the gas sample.

According to Boyle's law, pressure and volume are inversely related when the temperature and amount of gas are held constant. In this case, the pressure increases from 0.970 atm to 3.20 atm, so the new volume should decrease. It will be less than the original volume.

Example Problem 10.2.1 *(continued)*

To calculate the new volume, first make a table of the known and unknown pressure and volume data. In this case, the initial volume and pressure and the new pressure are known and the new volume must be calculated.

$$P_1 = 0.970 \text{ atm} \qquad P_2 = 3.20 \text{ atm}$$
$$V_1 = 458 \text{ mL} \qquad V_2 = ?$$

Rearrange Boyle's law to solve for V_2 and calculate the new volume of the gas sample.

$$P_1V_1 = P_2V_2$$

$$V_2 = \frac{P_1V_1}{P_2} = \frac{(0.970 \text{ atm})(458 \text{ mL})}{3.20 \text{ atm}} = 139 \text{ mL}$$

The pressure units (atm) cancel, leaving volume in units of milliliters.

Is your answer reasonable? The final volume is less than the initial volume, as we predicted.

Video Solution

Tutored Practice
Problem 10.2.1

Interactive Figure 10.2.2

Explore Charles's law.

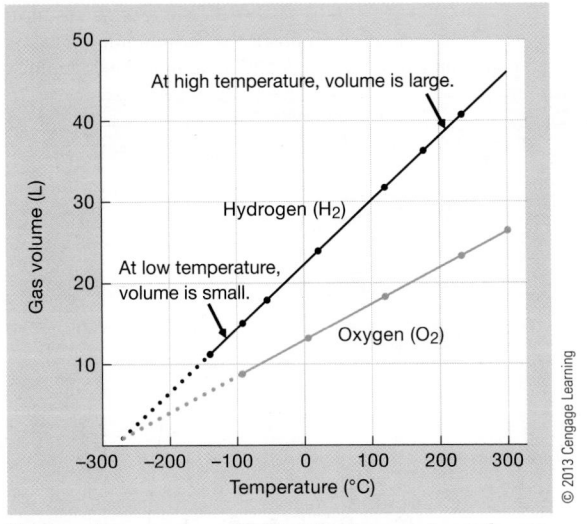

Volume changes upon changing the temperature of a gas sample. The pressure is held constant, and no gas escapes the syringe.

10.2b Charles's Law: $V = k_C \times T$

Charles's law states that the temperature and volume of a gas sample are directly related when the pressure and the amount of gas are held constant. For example, heating the air in a hot air balloon causes it to expand, filling the balloon. Consider a sample of gas held in a syringe attached to a pressure gauge and a temperature control unit (Interactive Figure 10.2.2).

When the pressure and amount of gas are held constant, decreasing the temperature of the gas sample decreases the volume of the gas. The two properties are directly related.

$$V = k_C \times T$$

Because the ratio of gas volume and temperature (in kelvin units) is a constant (when pressure and amount of gas are held constant), it is possible to calculate the new volume or temperature of a gas sample when one of the properties is changed.

$$\frac{V_1}{T_1} = \frac{V_2}{T_2} \qquad \qquad \textbf{(10.2)}$$

The subscripts "1" and "2" in Equation 10.2 indicate the different experimental conditions before and after volume or temperature is changed.

As shown in Interactive Figure 10.2.2, extending a volume–temperature plot for any gas to the point at which the gas volume is equal to zero shows that this occurs at a temperature of −273.15 °C. This temperature is known as absolute zero, or 0 K.

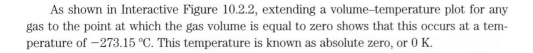

Example Problem 10.2.2 Use Charles's law to calculate volume.

A sample of gas has a volume of 2.48 L at a temperature of 58.0 °C. The gas sample is cooled to a temperature of −5.00 °C (assume pressure and amount of gas are held constant). Predict whether the new volume is greater or less than the original volume, and calculate the new volume.

Solution:

You are asked to calculate the volume of a gas sample when only the temperature is changed.

You are given the original volume and temperature and the new temperature of the gas sample.

According to Charles's law, temperature and volume are directly related when the pressure and amount of gas are held constant. In this case, the temperature decreases from 58.00 °C to −5.00 °C, so the volume should also decrease. It will be less than the original volume.

To calculate the new volume, first make a table of the known and unknown volume and temperature data. In this case, the initial volume and temperature and the new temperature are known and the new volume must be calculated. Note that all temperature data must be in kelvin temperature units.

$$V_1 = 2.48 \text{ L} \qquad\qquad V_2 = ?$$
$$T_1 = 58.00 \text{ °C} + 273.15 = 331.15 \text{ K} \qquad T_2 = -5.00 \text{ °C} + 273.15 = 268.15 \text{ K}$$

Rearrange Charles's law to solve for V_2, and calculate the new volume of the gas sample.

$$\frac{V_1}{T_1} = \frac{V_2}{T_2}$$

$$V_2 = \frac{V_1 T_2}{T_1} = \frac{(2.48 \text{ L})(268.15 \text{ K})}{331.15 \text{ K}} = 2.01 \text{ L}$$

The temperature units (K) cancel, leaving volume in units of liters.

Is your answer reasonable? The final volume is less than the initial volume, as we predicted.

Video Solution

Tutored Practice
Problem 10.2.2

10.2c Avogadro's Law: $V = k_A \times n$

Avogadro's hypothesis states that equal volumes of gases have the same number of particles when they are at the same temperature and pressure. One aspect of the hypothesis is called **Avogadro's law**, which states that the volume and amount (in moles) of a gas are directly related when pressure and temperature are held constant. Consider an experiment where the volume of gas in a syringe is measured as a function of the amount of gas in the syringe (at constant pressure and temperature) (Figure 10.2.3).

As the amount of gas in the syringe is increased (at constant temperature and pressure), the volume of the gas increases.

$$V = k_A \times n$$

Avogadro's law can also be used to calculate the new volume or amount of a gas sample when one of the properties is changed.

$$\frac{V_1}{n_1} = \frac{V_2}{n_2} \qquad (10.3)$$

The subscripts "1" and "2" in Equation 10.3 indicate the different experimental conditions before and after volume or the amount of gas is changed.

Avogadro's law is independent of the identity of the gas, as shown in a plot of volume versus amount of gas (Interactive Figure 10.2.4a). This means that a 1-mol sample of Xe

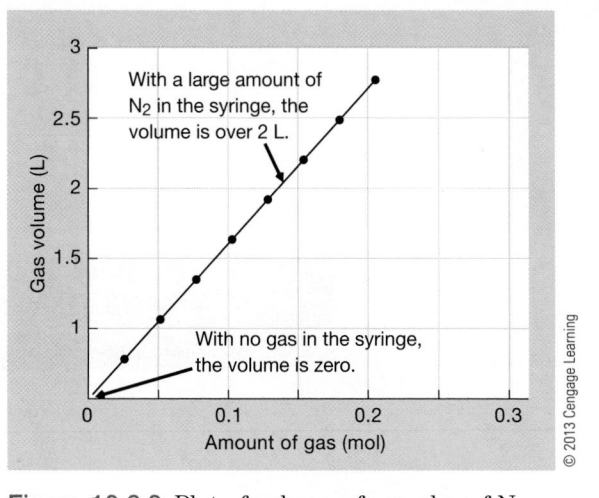

Figure 10.2.3 Plot of volume of samples of N_2 gas with differing amounts of N_2 present

Interactive Figure 10.2.4

Explore Avogadro's law.

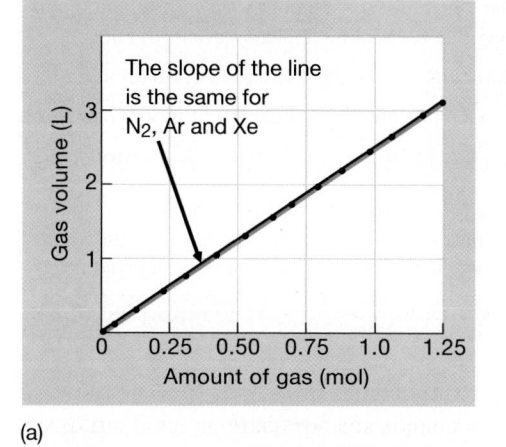 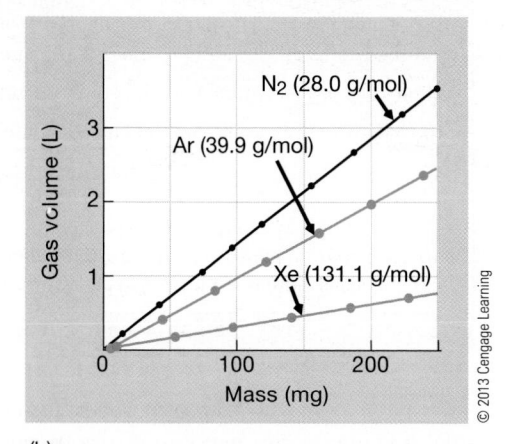

(a)   (b)

(a) Avogadro's law and (b) volume versus mass plots for N_2, Ar, and Xe

has the same volume as a 1-mol sample of N_2 (at the same temperature and pressure), even though the Xe sample has a mass more than 4.5 times as great. The same is not true for gas samples with equal mass, as shown in Interactive Figure 10.2.4b.

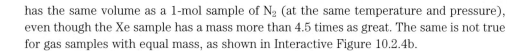

Example Problem 10.2.3 Use Avogadro's law to calculate volume.

A sample of gas contains 2.4 mol of SO_2 and 1.2 mol O_2 and occupies a volume of 17.9 L. The following reaction takes place: $2\ SO_2(g) + O_2(g) \rightarrow 2\ SO_3(g)$

Calculate the volume of the sample after the reaction takes place (assume temperature and pressure are constant).

Solution:

You are asked to calculate the volume of a gas sample when the number of moles of gas changes.

You are given the original volume and amount of the gases in the sample and the balanced equation for the reaction that takes place.

Make a table of the known and unknown volume and temperature data. In this case, the initial volume and amount of reactants and the amounts of product are known and the new volume must be calculated.

$$V_1 = 17.9\text{ L} \qquad\qquad V_2 = ?$$
$$n_1 = 3.6\text{ mol } SO_2 \text{ and } O_2 \qquad n_2 = 2.4\text{ mol } SO_3$$

The reactants are present in a 2:1 stoichiometric ratio, so they are consumed completely upon reaction to form SO_3.

$$\frac{2.4\text{ mol } SO_2}{1.2\text{ mol } O_2} = \frac{2\text{ mol } SO_2}{1\text{ mol } O_2}$$

Use the balanced equation to calculate the amount of SO_3 produced.

$$2.4\text{ mol } SO_2 \times \frac{2\text{ mol } SO_3}{2\text{ mol } SO_2} = 2.4\text{ mol } SO_3$$

Rearrange Avogadro's law to solve for V_2, and calculate the volume of the gas sample after the reaction is complete.

$$\frac{V_1}{n_1} = \frac{V_2}{n_2}$$

$$V_2 = \frac{V_1 n_2}{n_1} = \frac{(17.9\text{ L})(2.4\text{ mol})}{3.6\text{ mol}} = 12\text{ L}$$

Is your answer reasonable? The new volume is smaller than the initial volume, which makes sense because the amount of gas decreased as a result of the chemical reaction.

Video Solution

Tutored Practice
Problem 10.2.3

Section 10.2 Mastery

10.3 The Combined and Ideal Gas Laws

10.3a The Combined Gas Law

We can rewrite the three historical gas laws, solving for volume:

Boyle's Law	Charles's Law	Avogadro's Law
$V = k_B \times \dfrac{1}{P}$	$V = k_C \times T$	$V = k_A \times n$

These three equations can be combined into a single equation that relates volume, pressure, temperature, and the amount of any gas.

$$V = \text{constant} \times \frac{nT}{P} \quad \text{or} \quad \frac{PV}{nT} = \text{constant}$$

Because the ratio involving pressure, volume, amount, and temperature of a gas is a constant, it can be used in the form of the **combined gas law** to calculate the new pressure, volume, amount, or temperature of a gas when one or more of these properties is changed.

$$\frac{P_1 V_1}{n_1 T_1} = \frac{P_2 V_2}{n_2 T_2} \tag{10.4}$$

The combined gas law is most often used to calculate the new pressure, volume, or temperature of a gas sample when two of these properties are changed. Under typical conditions, the amount of the gas is held constant ($n_1 = n_2$).

Example Problem 10.3.1 Use the combined gas law to calculate pressure.

A 2.68-L sample of gas has a pressure of 1.22 atm and a temperature of 29 °C. The sample is compressed to a volume of 1.41 L and cooled to −17 °C. Calculate the new pressure of the gas, assuming that no gas escaped during the experiment.

Solution:

You are asked to calculate the pressure of a gas when only its volume and temperature are changed.

You are given the original volume, pressure, and temperature of the gas and the new volume and temperature of the gas sample.

Make a table of the known and unknown pressure, volume, and temperature data. Note that temperature must be converted to kelvin temperature units and that the amount of gas (n) does not change.

Example Problem 10.3.1 (*continued*)

$P_1 = 1.22$ atm $P_2 = ?$
$V_1 = 2.68$ L $V_2 = 1.41$ L
$T_1 = 29\,°C + 273 = 302$ K $T_2 = -17\,°C + 273 = 256$ K
$n_1 = n_2$

Rearrange the combined gas law to solve for P_2, and calculate the new pressure of the gas sample.

$$\frac{P_1V_1}{n_1T_1} = \frac{P_2V_2}{n_2T_2}$$

$$P_2 = \frac{P_1V_1T_2}{V_2T_1} = \frac{(1.22\text{ atm})(2.68\text{ L})(256\text{ K})}{(1.41\text{ L})(302\text{ K})} = 1.97\text{ atm}$$

Video Solution

Tutored Practice
Problem 10.3.1

10.3b The Ideal Gas Law

The **ideal gas law** incorporates the three historical gas laws and a constant, which is given the symbol R and called the universal gas constant or the **ideal gas constant** ($R = 0.082057$ L·atm/K·mol).

$$PV = nRT \tag{10.5}$$

One property of an **ideal gas** is that it follows the ideal gas law; that is, its variables (P, V, n, and T) vary according to the ideal gas law. The universal gas constant is independent of the identity of the gas. Note that the units of R (L·atm/K·mol) control the units of P, V, T, and n in any equation that includes this constant.

When three of the four properties of a gas sample are known, the ideal gas law can be used to calculate the unknown property.

Example Problem 10.3.2 Use the ideal gas law to calculate an unknown property.

A sample of O_2 gas has a volume of 255 mL, has a pressure of 742 mm Hg, and is at a temperature of 19.6 °C. Calculate the amount of O_2 in the gas sample.

Solution:

You are asked to calculate the amount of gas in a sample.

You are given the pressure, volume, and temperature of the gas sample.

Example Problem 10.3.2 *(continued)*

The ideal gas law contains a constant ($R = 0.082057$ L·atm/K·mol), so all properties must have units that match those in the constant.

$$P = 742 \text{ mm Hg} \times \frac{1 \text{ atm}}{760 \text{ mm Hg}} = 0.976 \text{ atm}$$

$$V = 255 \text{ mL} \times \frac{1 \text{ L}}{1000 \text{ mL}} = 0.255 \text{ L}$$

$$T = (19.6 + 273.15) \text{ K} = 292.8 \text{ K}$$

$$n = ?$$

Rearrange the ideal gas law to solve for n, and calculate the amount of oxygen in the sample.

$$PV = nRT$$

$$n_{O_2} = \frac{PV}{RT} = \frac{(0.976 \text{ atm})(0.255 \text{ L})}{(0.082057 \text{ L} \cdot \text{atm/K} \cdot \text{mol})(292.8 \text{ K})} = 0.0104 \text{ mol}$$

Video Solution

Tutored Practice
Problem 10.3.2

10.3c The Ideal Gas Law, Molar Mass, and Density

If a compound exists in gaseous form at some temperature, it is very easy (and relatively inexpensive) to determine its molar mass from simple laboratory experiments. Molar mass can be calculated from pressure, temperature, and density measurements and the use of the ideal gas law. First, we use the definition of molar mass to incorporate the mass of a gas sample and the molar mass of a gas into the ideal gas law.

$$\text{molar mass } (M) = \frac{m \text{ (mass, in g)}}{n \text{ (amount, in mol)}}$$

$$PV = \left(\frac{m}{M}\right) RT$$

Next, we rearrange this equation to derive a relationship between molar mass, density, temperature, and pressure of a gas.

$$M = \frac{mRT}{PV} = \left(\frac{m}{V}\right) \frac{RT}{P}$$

$$M = \frac{dRT}{P} \qquad\qquad (10.6)$$

In Equation 10.6, M = molar mass (g/mol) and d = gas density (g/L). This form of the ideal gas law can be used to calculate the molar mass or density of a gas, as shown in the following examples.

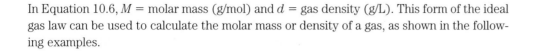

Example Problem 10.3.3 Use the ideal gas law to calculate molar mass.

A 4.07-g sample of an unknown gas has a volume of 876 mL and a pressure of 737 mm Hg at 30.4 °C. Calculate the molar mass of this compound.

Solution:

You are asked to calculate the molar mass of a compound.

You are given the mass, volume, pressure, and temperature of the gas sample.

There are two ways to solve this problem: using Equation 10.6 or using the ideal gas law in its original form. Note that both of these equations contain a constant (R = 0.082057 L·atm/K·mol), so all properties must have units that match those in the constant.

Method 1:

$$P = 737 \text{ mm Hg} \times \frac{1 \text{ atm}}{760 \text{ mm Hg}} = 0.970 \text{ atm}$$

$$T = (30.4 + 273.15) \text{ K} = 303.6 \text{ K}$$

$$V = 876 \text{ mL} \times \frac{1 \text{ L}}{1000 \text{ mL}} = 0.876 \text{ L}$$

$$d = \frac{m}{V} = \frac{4.07 \text{ g}}{0.876 \text{ L}} = 4.65 \text{ g/L}$$

Use Equation 10.6 to calculate the molar mass of the unknown gas.

$$M = \frac{dRT}{P} = \frac{(4.65 \text{ g/L})(0.082057 \text{ L} \cdot \text{atm/K} \cdot \text{mol})(303.6 \text{ K})}{0.970 \text{ atm}} = 119 \text{ g/mol}$$

Method 2:

Rearrange the ideal gas law to calculate the amount (n) of gas present.

$$n = \frac{PV}{RT} = \frac{(0.970 \text{ atm})(0.876 \text{ L})}{(0.082057 \text{ L} \cdot \text{atm/K} \cdot \text{mol})(303.6 \text{ K})} = 0.0341 \text{ mol}$$

Use the mass of gas and the amount to calculate molar mass.

$$M = \frac{m}{n} = \frac{4.07 \text{ g}}{0.0341 \text{ mol}} = 119 \text{ g/mol}$$

Video Solution

Tutored Practice
Problem 10.3.3

Example Problem 10.3.4 Use the ideal gas law to calculate density.

Calculate the density of oxygen gas at 788 mm Hg and 22.5 °C.

Solution:

You are asked to calculate the density of a gas at a given pressure and temperature.

You are given the identity of the gas and its pressure and temperature.

Equation 10.6 contains a constant (R = 0.082057 L·atm/K·mol), so all properties must have units that match those in the constant.

$$P = 788 \text{ mm Hg} \times \frac{1 \text{ atm}}{760 \text{ mm Hg}} = 1.04 \text{ atm}$$

$$T = (22.5 + 273.15) \text{ K} = 295.7 \text{ K}$$

$$M(O_2) = 32.00 \text{ g/mol}$$

Solve Equation 10.6 for density, and use it to calculate the density of oxygen under these conditions.

$$M = \frac{dRT}{P}$$

$$d = \frac{PM}{RT} = \frac{(1.04 \text{ atm})(32.00 \text{ g/mol})}{(0.082057 \text{ L} \cdot \text{atm/K} \cdot \text{mol})(295.7 \text{ K})} = 1.37 \text{ g/L}$$

Gas Densities at STP

Gas densities are often reported under a set of **standard temperature and pressure** conditions (STP) of 1.00 atm and 273.15 K (0 °C). Do not confuse STP with standard state conditions, which typically use a temperature of 25 °C. Under STP conditions, one mole of an ideal gas has a **standard molar volume** of 22.4 L.

$$V = \frac{nRT}{P} = \frac{(1 \text{ mol})(0.082057 \text{ L} \cdot \text{atm/K} \cdot \text{mol})(273.15 \text{ K})}{1.00 \text{ atm}} = 22.4 \text{ L}$$

The relationship between gas density and molar mass is shown in Table 10.3.1, which contains gas densities at STP for some common and industrially important gases. Notice that some flammable gases such as propane and butane are denser than either O_2 or N_2. This means that these gases sink in air and will collect near the ground, so places where these gases might leak (such as in a house or apartment) will have a gas detector mounted near the floor. Carbon monoxide (CO) has a density very similar to that of O_2 and N_2, so it mixes well with these gases. For this reason, CO detectors can be placed at any height on a wall.

Table 10.3.1 Gas Densities at STP

Gas	Molar Mass (g/mol)	Density at STP (g/L)
H_2	2.02	0.0892
He	4.00	0.178
N_2	28.00	1.25
CO	28.01	1.25
O_2	32.00	1.42
CO_2	44.01	1.96
Propane (C_3H_8)	44.09	1.97
Butane (C_4H_{10})	58.12	2.59
UF_6	351.99	15.69

Video Solution

Tutored Practice Problem 10.3.4

Section 10.3 Mastery

10.4 Partial Pressure and Gas Law Stoichiometry

10.4a Introduction to Dalton's Law of Partial Pressures

Our atmosphere is a mixture of many gases, and this mixture changes composition constantly. For example, every time you breathe in and out, you make small changes to the amount of oxygen, carbon dioxide, and water in the air around you. **Dalton's law of partial pressures** states that the pressure of a gas mixture is equal to the sum of the pressures due to the individual gases of the sample, called partial pressures. For a mixture containing the gases A, B, and C, for example,

$$P_{total} = P_A + P_B + P_C \qquad\qquad (10.7)$$

where P_{total} is the pressure of the mixture and P_X is the partial pressure of gas X in the mixture. Each gas in a mixture behaves as an ideal gas and as if it alone occupies the container. This means that although individual partial pressures may differ, all gases in the mixture have the same volume (equal to the container volume) and temperature.

Example Problem 10.4.1 Calculate pressure using Dalton's law of partial pressures.

A gas mixture is made up of O_2 (0.136 g), CO_2 (0.230 g), and Xe (1.35 g). The mixture has a volume of 1.82 L at 22.0 °C. Calculate the partial pressure of each gas in the mixture and the total pressure of the gas mixture.

Solution:

You are asked to calculate the partial pressure of each gas in a mixture of gases and the total pressure of the gas mixture.

You are given the identity and mass of each gas in the sample and the volume and temperature of the gas mixture.

The partial pressure of each gas is calculated from the ideal gas equation.

$$P_{O_2} = \frac{nRT}{V} = \frac{\left(0.136 \text{ g} \times \dfrac{1 \text{ mol } O_2}{32.00 \text{ g}}\right)(0.082057 \text{ L·atm/K·mol})(22.0 + 273.15 \text{ K})}{1.82 \text{ L}} = 0.0566 \text{ atm}$$

$$P_{CO_2} = \frac{nRT}{V} = \frac{\left(0.230 \text{ g} \times \dfrac{1 \text{ mol } CO_2}{44.01 \text{ g}}\right)(0.082057 \text{ L·atm/K·mol})(22.0 + 273.15 \text{ K})}{1.82 \text{ L}} = 0.0695 \text{ atm}$$

Example Problem 10.4.1 (continued)

$$P_{Xe} = \frac{nRT}{V} = \frac{\left(1.35 \text{ g} \times \dfrac{1 \text{ mol Xe}}{131.3 \text{ g}}\right)(0.082057 \text{ L} \cdot \text{atm/K} \cdot \text{mol})(22.0 + 273.15 \text{ K})}{1.82 \text{ L}} = 0.137 \text{ atm}$$

Notice that the three gases have the same volume and temperature but different pressures. The total pressure is the sum of the partial pressures for the gases in the mixture.

$$P_{total} = P_{O_2} + P_{CO_2} + P_{Xe} = 0.0566 \text{ atm} + 0.0695 \text{ atm} + 0.137 \text{ atm} = 0.263 \text{ atm}$$

Video Solution

Tutored Practice
Problem 10.4.1

Collecting a Gas by Water Displacement

A common laboratory experiment involves collecting the gas generated during a chemical reaction by water displacement (Interactive Figure 10.4.1). Because water can exist in gaseous form (as water vapor), the gas that is collected is a mixture of both the gas formed during the chemical reaction and water vapor.

According to Dalton's law of partial pressures, the pressure of the collected gas mixture is equal to the partial pressure of the gas formed during the chemical reaction plus the partial pressure of the water vapor (the **vapor pressure** of the water).

$$P_{total} = P_{gas} + P_{H_2O}$$

Water vapor pressure varies with temperature. Table 10.4.1 shows vapor pressure values for moderate temperatures; a more complete table is found in the reference tools.

Interactive Figure 10.4.1

Apply Dalton's law of partial pressures.

Collecting a gas by water displacement

Charles D. Winters

Example Problem 10.4.2 Calculate the amount of gas produced when collected by water displacement.

Aluminum reacts with strong acids such as HCl to form hydrogen gas.

$$2 \text{ Al}(s) + 6 \text{ HCl}(aq) \rightarrow 3 \text{ H}_2(g) + 2 \text{ AlCl}_3(aq)$$

In one experiment, a sample of Al reacts with excess HCl and the gas produced is collected by water displacement. The gas sample has a temperature of 22.0 °C, a volume of 27.58 mL, and a pressure of 738 mm Hg. Calculate the amount of hydrogen gas produced in the reaction.

Solution:

You are asked to calculate the amount of gas produced in a chemical reaction when it is collected over water at a given pressure and temperature.

You are given the volume, pressure, and temperature of the gas produced in the reaction.

Example Problem 10.4.2 *(continued)*

The gas collected is a mixture containing water vapor, which has a vapor pressure of 19.83 mm Hg at 22.0 °C (Table 10.4.1). Subtract the water vapor pressure from the total pressure of the mixture to calculate the partial pressure of the hydrogen gas in the mixture.

$$P_{H_2} = P_{total} - P_{H_2O} = 739 \text{ mm Hg} - 19.83 \text{ mm Hg} = 718 \text{ mm Hg}$$

Use the ideal gas law to calculate the amount of gas produced in the reaction.

$$P = 718 \text{ mm Hg} \times \frac{1 \text{ atm}}{760 \text{ mm Hg}} = 0.945 \text{ atm}$$

$$V = 27.58 \text{ mL} \times \frac{1 \text{ L}}{1000 \text{ mL}} = 0.02758 \text{ L}$$

$$T = (22.0 + 273.15) \text{ K} = 295.2 \text{ K}$$

$$n_{H_2} = \frac{PV}{RT} = \frac{(0.945 \text{ atm})(0.02758 \text{ L})}{(0.082057 \text{ L} \cdot \text{atm/K} \cdot \text{mol})(295.2 \text{ K})} = 1.08 \times 10^{-3} \text{ mol}$$

Video Solution

Tutored Practice
Problem 10.4.2

10.4b Partial Pressure and Mole Fractions of Gases

Within a gas mixture, the total pressure is the sum of the partial pressures of each of the component gases. The ideal gas law shows that pressure and amount (in moles) of any gas are directly related.

$$P = n\frac{RT}{V}$$

Therefore, the degree to which any one gas contributes to the total pressure is directly related to the amount (in moles) of that gas present in a mixture. In other words, the greater the amount of a gas in a mixture, the greater its partial pressure and the greater amount its partial pressure contributes to the total pressure. Quantitatively, this relationship is shown in Equation 10.8, where P_A and n_A are the partial pressure and amount (in moles), respectively, of gas A in a mixture of gases, and n_{total} is the total amount (in moles) of gas in the mixture:

$$\frac{P_A}{P_{total}} = \frac{n_A}{n_{total}} \tag{10.8}$$

The ratio n_A/n_{total} is the **mole fraction** of gas A in a mixture of gases, and it is given the symbol χ_A. Rearranging Equation 10.8 to solve for the partial pressure of gas A,

$$P_A = \chi_A P_{total} \tag{10.9}$$

Table 10.4.1 Vapor Pressure of Water

Temperature (°C)	Vapor pressure of H_2O (mm Hg)
19	16.48
20	17.54
21	18.65
22	19.83
23	21.07
24	22.38
25	23.76
26	25.21
27	26.74
28	28.35

Note that mole fraction is a unitless quantity. Also, the sum of the mole fractions for all gases in a mixture is equal to 1. For a mixture containing gases A, B, and C, for example,

$$\chi_A + \chi_B + \chi_C = 1$$

Example Problem 10.4.3 Use mole fraction in calculations involving gases.

A gas mixture contains the noble gases Ne, Ar, and Kr. The total pressure of the mixture is 2.46 atm, and the partial pressure of Ar is 1.44 atm. If a total of 18.0 mol of gas is present, what amount of Ar is present?

Solution:

You are asked to calculate the amount of gas present in a mixture.

You are given the total pressure of the mixture, the total amount of gas in the mixture, and the partial pressure of one of the gases in the mixture.

The amount (in moles) of Ar is equal to the total moles of gas in the mixture times its mole fraction. The first step, then, is to determine the mole fraction of Ar in the mixture. Solving Equation 10.9 for mole fraction of Ar,

$$P_{Ar} = \chi_{Ar} P_{total}$$

$$\chi_{Ar} = \frac{P_{Ar}}{P_{total}} = \frac{1.44 \text{ atm}}{2.46 \text{ atm}} = 0.585$$

Use the mole fraction of Ar and the total amount of gases in the mixture to calculate moles of Ar in the mixture.

$$0.585 = \frac{n_{Ar}}{n_{total}} = \frac{n_{Ar}}{18.0 \text{ mol}}$$

$$n_{Ar} = 10.5 \text{ mol}$$

Video Solution

Tutored Practice
Problem 10.4.3

10.4c Gas Laws and Stoichiometry

We investigated stoichiometric relationships for systems involving pure substances and solutions in Stoichiometry (Unit 3) and Chemical Reactions and Solution Stoichiometry (Unit 4), where the amount of a reactant or product was determined from mass data or from volume and concentration data. We now have the tools needed to include the gas properties of pressure, temperature, and volume in the stoichiometric relationships derived in those units.

Interactive Figure 10.4.2 gives a schematic representation of the relationship between gas properties and the amount of a reactant or product.

Interactive Figure 10.4.2

Use gas properties in stoichiometry calculations.

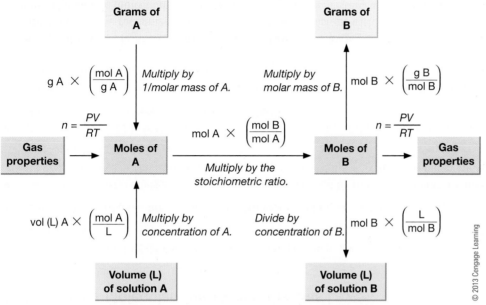

Schematic flow chart of stoichiometric relationships

Example Problem 10.4.4 Use gas laws in stoichiometry calculations.

A sample of O_2 with a pressure of 1.42 atm and a volume of 250. mL is allowed to react with excess SO_2 at 129 °C.

$$2\ SO_2(g) + O_2(g) \rightarrow 2\ SO_3(g)$$

Calculate the pressure of the SO_3 produced in the reaction if it is transferred to a 1.00-L flask and cooled to 35.0 °C.

Solution:

You are asked to calculate the pressure of a gas produced in a chemical reaction.

You are given the balanced equation for the reaction; the pressure, volume, and temperature of a reactant; and the volume and temperature of the gas produced in the reaction.

Step 1. Calculate the amount of reactant (O_2) available using the ideal gas law.

Example Problem 10.4.4 *(continued)*

$$n_{O_2} = \frac{PV}{RT} = \frac{(1.42\text{ atm})(0.250\text{ L})}{(0.082057\text{ L}\cdot\text{atm/K}\cdot\text{mol})(129+273.15\text{ K})} = 0.0108\text{ mol}$$

Step 2. Use the amount of limiting reactant (O_2) and the balanced equation to calculate the amount of SO_3 produced.

$$0.0108\text{ mol }O_2 \times \frac{2\text{ mol }SO_3}{1\text{ mol }O_2} = 0.0216\text{ mol }SO_3$$

Step 3. Use the ideal gas law and the new volume and temperature conditions to calculate the pressure of SO_3.

$$P_{SO_3} = \frac{nRT}{V} = \frac{(0.0216\text{ mol }SO_3)(0.082057\text{ L}\cdot\text{atm/K}\cdot\text{mol})(35.0+273.15\text{ K})}{1.00\text{ L}} = 0.546\text{ atm}$$

Video Solution

Tutored Practice
Problem 10.4.4

Section 10.4 Mastery

10.5 Kinetic Molecular Theory

10.5a Kinetic Molecular Theory and the Gas Laws

According to the kinetic molecular theory,

- gases consist of molecules whose separation is much larger than the molecules themselves;

- the molecules of a gas are in continuous, random, and rapid motion;

- the average kinetic energy of gas molecules is determined by the gas temperature, and all gas molecules at the same temperature, regardless of mass, have the same average kinetic energy; and

- gas molecules collide with one another and with the walls of their container, but they do so without loss of energy in "perfectly elastic" collisions.

Note that when we talk about the behavior of gas molecules, we include monoatomic gaseous atoms in our definition of molecules.

Relating the Kinetic Molecular Theory to the Gas Laws

For any theory to be recognized as useful, it must be consistent with experimental observations. The kinetic molecular theory therefore must be consistent with and help explain the well-known gas laws. At the molecular level, the concept of pressure is considered in terms of collisions between gas molecules and the inside walls of a container. Each collision between a moving gas molecule and the static wall involves the imparting of a force pushing on the inside of the wall. The more collisions there are and the more energetic the collisions on average, the greater the force and the higher the pressure. Changing the number of molecule-wall collisions by, for example, lowering the temperature of a sample of gas (as shown in Interactive Figure 10.5.1) results in a change in the other properties of the gas.

Interactive Figure 10.5.1

Relate kinetic molecular theory to the gas laws.

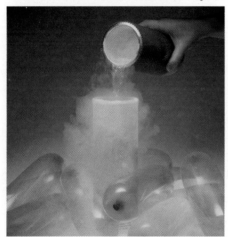 

Charles D. Winters

When air-filled balloons are placed in liquid nitrogen (−196 °C), the gas volume decreases dramatically

P and *n*

If the number of gas molecules inside a container is doubled, the number of molecule–wall collisions will exactly double (assuming volume and temperature are constant). This means that twice as much force will push against the wall and the pressure is twice as great.

P and *T*

As the temperature of a gas in a container increases, the molecules move more rapidly. This does two things: it leads to more frequent collisions between the molecules and the walls of the container, and it also results in more energetic collisions between the molecules and container walls. Therefore, as temperature increases, there are more frequent and energetic collisions on the inside of the walls and a greater pressure (assuming constant amount of gas and volume).

P and V

As the volume of a container is increased, the gas molecules take longer to move across the inside of the container before hitting the wall on the other side. This means that the frequency of collisions decreases, resulting in a lower pressure (assuming constant amount of gas and temperature).

10.5b Molecular Speed, Mass, and Temperature

Gas molecules move through space at very high speeds. As you saw in Thermochemistry (Unit 5), the kinetic energy (KE) of a moving object is directly related to its mass (m) and speed (v).

$$KE = \frac{1}{2}mv^2$$

For a collection of gas molecules, the mean (average) kinetic energy of one mole of gas particles is directly related to the average of the square of the gas velocity (N_A = Avogadro's number).

$$\overline{KE} = N_A \frac{1}{2}m\overline{v^2} \tag{10.10}$$

As described in the third postulate of the kinetic molecular theory, it can also be shown that kinetic energy of one mole of gas particles is directly related to the absolute temperature of the gas [Equation 10.11, where the ideal gas constant has units of J/K·mol (R = 8.314 J/K·mol)].

$$\overline{KE} = \frac{3}{2}RT \tag{10.11}$$

Combining Equations 10.10 and 10.11, we see that the average velocity of molecules in a sample of gas is inversely related to the mass of gas molecules (m) and directly related to the temperature of the gas (T).

$$N_A \frac{1}{2}m\overline{v^2} = \frac{3}{2}RT \tag{10.12}$$

Taking the square root of the average of the square of the gas velocity gives the **root mean square (rms) speed** of a gas. The rms speed is not the same as the average speed of a sample of gas particles, but the two values are similar. Rearranging Equation 10.12 to solve for rms speed gives another important relationship between molecular speed and mass:

$$\overline{v^2} = \frac{3RT}{N_A m}$$

$$v_{rms} = \sqrt{\overline{v^2}} = \sqrt{\frac{3RT}{N_A m}}$$

$$v_{\text{rms}} = \sqrt{\frac{3RT}{M}} \qquad \textbf{(10.13)}$$

In Equation 10.13, R = 8.3145 J/K·mol and M = molar mass of the gas. Notice that the thermodynamic units of R require molar mass to be expressed in units of kilograms per mole when using this equation (1 J = 1 kg·m²/s²).

Example Problem 10.5.1 Calculate root mean square (rms) speed.

Calculate the rms speed of NH_3 molecules at 21.5 °C.

Solution:

You are asked to calculate the rms speed for a gas at a given temperature.

You are given the identity and temperature of the gas.

Use Equation 10.13, with molar mass in units of kilograms per mole.

$$v_{\text{rms}} = \sqrt{\frac{3RT}{M}} = \sqrt{\frac{3(8.3145 \text{ J/K} \cdot \text{mol})(21.5 + 273.15 \text{ K})}{0.01703 \text{ kg/mol}}} = 657 \text{ m/s}$$

Boltzmann Distribution Plots

Not all molecules in a sample of gas move at the same speed. Just like people or cars, a collection of gas molecules shows a range of speeds. The distribution of speeds is called a **Boltzmann distribution**. Interactive Figure 10.5.2 shows a Boltzmann distribution plot for O_2 at 25 °C. Each point in a Boltzmann distribution plot gives the number of gas molecules moving at a particular speed. This Boltzmann distribution for O_2 starts out with low numbers of molecules at low speeds, increases to a maximum at around 400 m/s, and then decreases smoothly to very low numbers at about 1000 m/s. This means that in a sample of O_2 gas at 25 °C, very few O_2 molecules are moving slower than 100 m/s, many molecules are moving at speeds between 300 and 600 m/s, and very few molecules are moving faster than 900 m/s. The peak around 450 m/s indicates the most probable speed at which the molecules are moving at this temperature. It does not mean that most of the molecules are moving at that speed.

Boltzmann distribution plots for a number of different gases, each at the same temperature, are shown in Figure 10.5.3. The heights of the curves in this figure differ because the area under each curve represents the total number of molecules in the sample. In the case of O_2, the gas molecules have a relatively narrow range of speeds and therefore more molecules moving at any particular speed. Helium has a wide range of speeds, and therefore the curve is stretched out, with few molecules moving any particular speed.

Video Solution

Tutored Practice
Problem 10.5.1

Interactive Figure 10.5.2

Explore Boltzmann distribution plots.

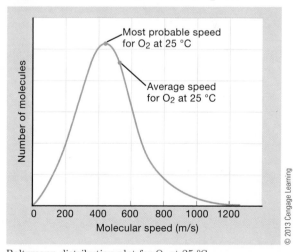

Boltzmann distribution plot for O_2 at 25 °C

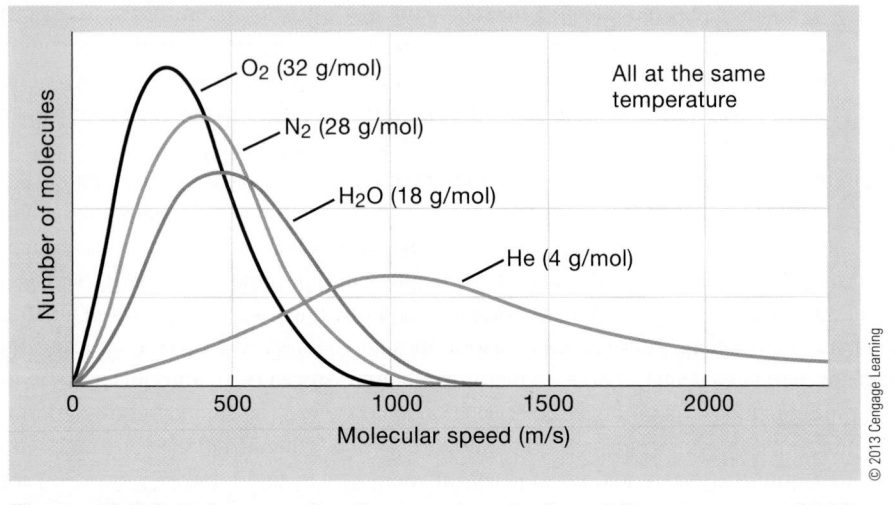

Figure 10.5.3 Boltzmann distribution plots for four different gases at 25 °C

Notice the relationship between the Boltzmann distribution plots and the molar mass of the gases in Figure 10.5.3. The peak in the O_2 curve is farthest to the left, meaning it has the slowest-moving molecules. The peak in the H_2O curve is farther to the right, which means that H_2O gas molecules move, on average, faster than O_2 molecules at the same temperature. The helium curve is shifted well to the right and has very fast-moving molecules. Recall that average molecular speed depends on both the average kinetic energy and on the mass of the moving particles. According to the kinetic molecular theory, all gases at the same temperature have the same kinetic energy. Therefore, if a gas has a smaller mass, it must have a larger average velocity.

Molecular speed changes with temperature, as shown in Interactive Figure 10.5.4. The curve for O_2 at the higher temperature is shifted to the right, indicating that O_2 molecules move faster, on average, at the higher temperature. As the temperature increases, the average kinetic energy increases. Because the mass is constant, as average kinetic energy increases, average molecular speed must increase.

In summary, Boltzmann distributions give us the following information:

1. Gases move with a range of speeds at a given temperature.

2. Gases move faster, on average, at higher temperatures.

3. Heavier gases move more slowly, on average, than lighter gases at the same temperature.

Investigate how temperature affects Boltzmann plots.

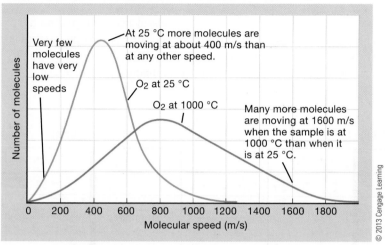

Boltzmann distribution plots for O_2 at 25 °C and 1000 °C

10.5c Gas Diffusion and Effusion

As you saw in a previous example problem, an ammonia molecule moves with an rms speed of about 650 m/s at room temperature. This is the equivalent of almost 1500 miles per hour! If you open a bottle of ammonia, however, it can take minutes for the smell to travel across a room. Why do odors, which are the result of volatile molecules, take so long to travel? The answer lies in gas diffusion. As a gas molecule moves in an air-filled room, it is constantly colliding with other molecules that block its path. It therefore takes much more time for a gas sample to get from one place to another than it would if there were nothing in its way. This gas process is called **diffusion**, the mixing of gases, and is illustrated in Interactive Figure 10.5.5.

A process related to diffusion is **effusion**, the movement of gas molecules through a small opening into a vacuum (Interactive Figure 10.5.6). **Graham's law of effusion** states that the rate of effusion of a gas is inversely related to the square root of its molar mass. That is, lighter gases move faster and effuse more rapidly compared with heavier gases, which move more slowly and effuse more slowly.

$$\text{rate of effusion} \propto \frac{1}{\sqrt{M}}$$

Explore gas diffusion.

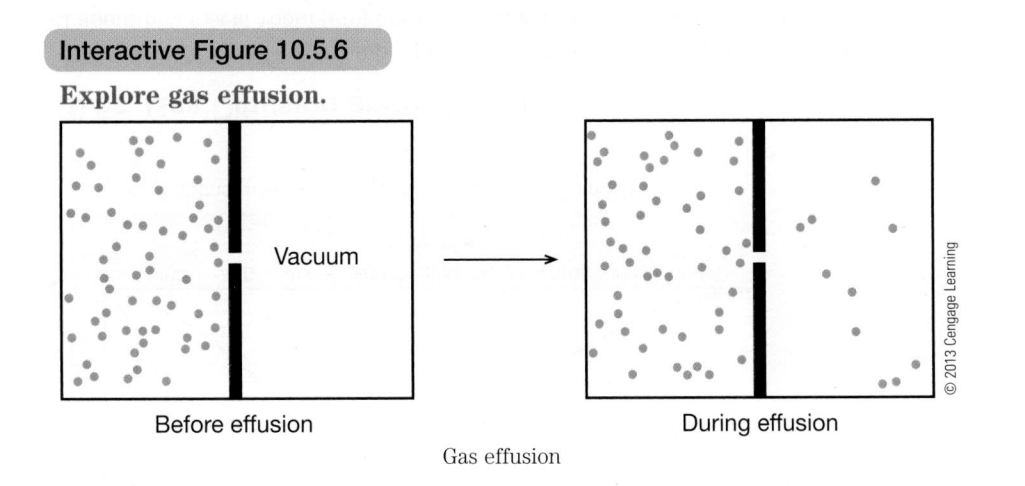

Before mixing   After mixing

© 2013 Cengage Learning

Gas diffusion

Explore gas effusion.

Vacuum

Before effusion   During effusion

© 2013 Cengage Learning

Gas effusion

A useful form of Graham's law is used to determine the molar mass of an unknown gas. The effusion rate of the unknown gas is compared with that of a gas with a known molar mass. The ratio of the effusion rates for the two gases is inversely related to the square root of the ratio of the molar masses.

$$\frac{\text{rate}_1}{\text{rate}_2} = \sqrt{\frac{M_2}{M_1}} \qquad\qquad (10.14)$$

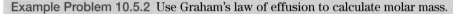

Example Problem 10.5.2 Use Graham's law of effusion to calculate molar mass.

A sample of ethane, C_2H_6, effuses through a small hole at a rate of 3.6×10^{-6} mol/h. An unknown gas, under the same conditions, effuses at a rate of 1.3×10^{-6} mol/hr. Calculate the molar mass of the unknown gas.

Solution:

You are asked to calculate the molar mass of an unknown gas.

You are given the effusion rate and identity of a gas and the effusion rate of the unknown gas measured under the same conditions.

Use Equation 10.14 and the molar mass of ethane to calculate the molar mass of the unknown gas.

$$\frac{\text{rate}_1}{\text{rate}_2} = \sqrt{\frac{M_2}{M_1}}$$

$$\frac{3.6 \times 10^{-6} \text{ mol/h}}{1.3 \times 10^{-6} \text{ mol/h}} = \sqrt{\frac{M_2}{30.07 \text{ g/mol}}}$$

$$M_2 = 230 \text{ g/mol}$$

Is your answer reasonable? The unknown gas effuses at a slower rate than ethane, so its molar mass should be greater than that of ethane.

Video Solution

Tutored Practice
Problem 10.5.2

10.5d Nonideal Gases

According to the kinetic molecular theory, an ideal gas is assumed to experience only perfectly elastic collisions and take up no actual volume. So, although gas molecules have volume, it is assumed that each individual molecule occupies the entire volume of its container and that the other molecules do not take up any of the container volume. At room temperature and pressures at or below 1 atm, most gases behave ideally. However, at high pressures or low temperatures, gases deviate from ideal behavior.

For an ideal gas, $PV/nRT = 1$ at any pressure. Therefore, one way to show deviation from ideal behavior is to plot the ratio PV/nRT as a function of pressure (Interactive Figure 10.5.7). All three of the gases shown in Interactive Figure 10.5.7 deviate from ideal behavior at high pressures.

Two types of deviations occur from ideal behavior.

1. Deviations due to gas volume

At high pressures, the concentration of the gas in a container is very high and as a result, molecules, on average, are closer together than they are at lower pressures. At these high pressures, the gas molecules begin to occupy a significant amount of the container

Interactive Figure 10.5.7

Explore deviations from ideal gas behavior.

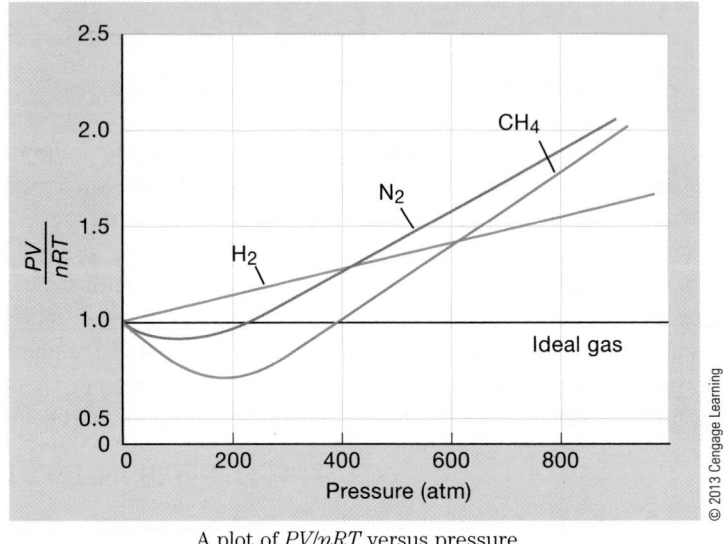

A plot of *PV/nRT* versus pressure

volume; thus, *the predicted volume occupied by the gas is less than the actual (container) volume*. Because the volume occupied by the gas is less than the container volume, the correction is subtracted from the container volume.

The amount of volume occupied by the gas molecules depends on the amount of gas present (n), and a constant (b) that represents how large the gas molecules are. Therefore, under nonideal conditions,

$$V = V_{\text{container}} - nb$$

2. Deviations due to molecular interactions

Under ideal conditions, gas molecules have perfectly elastic conditions. When the temperature decreases, however, gas molecules can interact for a short time after they collide. When this happens, there are fewer effective particles in the container (some molecules form small clusters, decreasing the number of particle-wall collisions) and these molecular interactions decrease the force with which molecules collide with the container walls. Thus, *the predicted pressure of the gas is greater than the actual (measured) pressure*. Because the predicted pressure is greater than measured pressure, the pressure correction is added to the measured pressure.

The strength of these molecular interactions depends greatly on the number of collisions and therefore depends on the amount of gas present (n) and the volume (V) of the container. Gases differ in these interactions, and this is reflected in a constant, a, which is specific to a given gas at a particular temperature. Under nonideal conditions,

$$P = P_{measured} + \frac{n^2a}{V^2_{container}}$$

These two deviations from ideal behavior are combined into a separate, more sophisticated gas law, the **van der Waals equation**. Table 10.5.1 shows some constants for common gases.

$$\left(P_{measured} + \frac{n^2a}{V^2_{measured}}\right)(V - nb) = nRT \qquad \textbf{(10.15)}$$

Notice that the values of b increase roughly with increasing size of the gas molecules. The values of a are related to the tendency of molecules of a given species to interact at the molecular level.

Table 10.5.1 Van der Waals Constants

Gas	a (L²·atm/mol²)	b (L/mol)
H_2	0.244	0.0266
He	0.034	0.0237
N_2	1.39	0.0391
NH_3	4.17	0.0371
CO_2	3.59	0.0427
CH_4	2.25	0.0428

Example Problem 10.5.3 Calculate pressure using the ideal gas law and the van der Waals equation.

A 1.78-mol sample of ammonia gas is maintained in a 2.50-L container at 302 K. Calculate the pressure of the gas using both the ideal gas law and the van der Waals equation (van der Waals constants are listed in Table 10.5.1).

Solution:

You are asked to calculate the pressure of a gas sample assuming ideal and nonideal behavior.

You are given the identity, amount, volume, and temperature of the gas.

First use the ideal gas law to calculate the pressure in the flask.

$$P = \frac{nRT}{V} = \frac{(1.78\ \text{mol})(0.082057\ \text{L} \cdot \text{atm/K} \cdot \text{mol})(302\ \text{K})}{2.50\ \text{L}} = 17.6\ \text{atm}$$

Compare this pressure with that calculated using the van der Waals equation.

$$\left(P_{measured} + \frac{n^2a}{V^2_{measured}}\right)(V - nb) = nRT$$

$$\left[P_{measured} + \frac{(1.78\ \text{mol})^2(4.17\ \text{L}^2 \cdot \text{atm/mol}^2)}{(2.50\ \text{L})^2}\right][2.50\ \text{L} - (1.78\ \text{mol})(0.0371\ \text{L/mol})]$$

$$= (1.78\ \text{mol})(0.082057\ \text{L} \cdot \text{atm/K} \cdot \text{mol})(302\ \text{K})$$

$$P_{measured} = 16.0\ \text{atm}$$

The actual pressure in the container is about 10% less than that calculated using the ideal gas law.

Video Solution

Tutored Practice
Problem 10.5.3

Section 10.5 Mastery

Unit Recap

Key Concepts

10.1 Properties of Gases

- Gas pressure is a measure of the force exerted on a given area of surface (10.1b).
- Gas pressure is measured using a barometer or a manometer (10.1b).
- The SI unit of pressure is the pascal (Pa), but common pressure units are atmosphere, millimeters of mercury, and bar (10.1b).

10.2 Historical Gas Laws

- Boyle's law states that the pressure and volume of a gas sample are inversely related when the amount of gas and temperature are held constant (10.2a).
- Charles's law states that the temperature and volume of a gas sample are directly related when the pressure and the amount of gas are held constant (10.2b).
- Absolute zero (0 K, -273.15 °C) is the temperature at which gas volume is equal to zero (10.2b).
- Avogadro's law states that the volume and amount (in moles) of a gas are directly related when pressure and temperature are held constant (10.2c).

10.3 The Combined and Ideal Gas Laws

- The combined gas law is used to calculate the new pressure, volume, amount, or temperature of a gas when one or more of these properties is changed (10.3a).
- The ideal gas law incorporates the three historical gas laws and a constant, which is given the symbol R and called the universal gas constant or the ideal gas constant (10.3b).
- An ideal gas is a gas whose variables (P, V, n, and T) vary according to the ideal gas law (10.3b).
- Standard temperature and pressure (STP) conditions are defined as 273.15 K (0 °C) and 1 atm (10.3b).
- At STP, one mole of an ideal gas has a standard molar volume of 22.4 L (10.3b).

10.4 Partial Pressure and Gas Law Stoichiometry

- Dalton's law of partial pressures states that the pressure of a gas mixture is equal to the sum of the pressures due to the individual gases of the sample, called partial pressures (10.4a).

10.5 Kinetic Molecular Theory

- According to the kinetic molecular theory, (1) gases consist of molecules whose separation is much larger than the molecules themselves; (2) the molecules of a gas are in continuous, random, and rapid motion; (3) the average kinetic energy of gas molecules is determined by the gas temperature, and all gas molecules at the same temperature, regardless of mass, have the same average kinetic energy; and (4) gas molecules collide with one another and with the walls of their container, but they do so without loss of energy in "perfectly elastic" collisions (10.5a).

- The average velocity of molecules in a sample of gas is inversely related to the mass of gas molecules and directly related to the temperature of the gas (10.5b).

- The root mean square (rms) speed of a gas is directly related to the temperature of the gas and inversely related to the molar mass of the gas (10.5b).

- A Boltzmann distribution plot shows the distribution of speeds in a sample of a gas at a given temperature (10.5b).

- Gases mix in a process called diffusion (10.5c).

- Effusion is the movement of gas molecules through a small opening into a vacuum (10.5c).

- Graham's law of effusion states that the rate of effusion of a gas is inversely related to the square root of its molar mass (10.5c).

- At high pressures or low temperatures, gases deviate from ideal behavior (10.5d).

- The van der Waals equation is used to calculate gas pressure under nonideal conditions (10.5d).

Key Equations

$$P_1V_1 = P_2V_2 \qquad \textbf{(10.1)}$$

$$\frac{V_1}{T_1} = \frac{V_2}{T_2} \qquad \textbf{(10.2)}$$

$$\frac{V_1}{n_1} = \frac{V_2}{n_2} \qquad \textbf{(10.3)}$$

$$\frac{P_1V_1}{n_1T_1} = \frac{P_2V_2}{n_2T_2} \qquad \textbf{(10.4)}$$

$$PV = nRT \qquad \textbf{(10.5)}$$

$$M = \frac{dRT}{P} \qquad \textbf{(10.6)}$$

$$P_{\text{total}} = P_A + P_B + P_C \qquad \textbf{(10.7)}$$

$$\frac{P_A}{P_{\text{total}}} = \frac{n_A}{n_{\text{total}}} \qquad \textbf{(10.8)}$$

$$P_A = \chi_A P_{\text{total}} \qquad \textbf{(10.9)}$$

$$\overline{KE} = N_A \frac{1}{2} m\overline{v^2} \qquad \textbf{(10.10)}$$

$$\overline{KE} = \frac{3}{2} RT \qquad \textbf{(10.11)}$$

$$N_A \frac{1}{2} m\overline{v^2} = \frac{3}{2} RT \qquad \textbf{(10.12)}$$

$$v_{\text{rms}} = \sqrt{\frac{3RT}{M}} \qquad \textbf{(10.13)}$$

$$\frac{\text{rate}_1}{\text{rate}_2} = \sqrt{\frac{M_2}{M_1}} \qquad \textbf{(10.14)}$$

$$\left(P_{\text{measured}} + \frac{n^2a}{V^2_{\text{measured}}}\right)(V - nb) = nRT \qquad \textbf{(10.15)}$$

Key Terms

10.1 Properties of Gases
pressure
barometer
millimeters of mercury (mm Hg)
torr
pascal (Pa)
standard atmosphere (atm)
bar

10.2 Historical Gas Laws
Boyle's law
Charles's law
Avogadro's law

10.3 The Combined and Ideal Gas Laws
combined gas law
ideal gas law
ideal gas constant (R)
ideal gas
standard temperature and pressure (STP)
standard molar volume

10.4 Partial Pressure and Gas Law Stoichiometry
Dalton's law of partial pressures
vapor pressure
mole fraction

10.5 Kinetic Molecular Theory
kinetic molecular theory
root mean square (rms) speed
Boltzmann distribution
diffusion
effusion
Graham's law of effusion
van der Waals equation

Unit 10   Review and Challenge Problems

11 Intermolecular Forces and the Liquid State

Unit Outline

In This Unit...

In previous units, we have described the structures of individual atoms and individual molecules in great detail. For atoms, we have focused on electronic structure, and for molecules, we have studied the arrangement of atoms with respect to one another. We are now ready to move to the next level of complexity in chemical systems: how collections of molecules interact and how those interactions control the physical properties of matter.

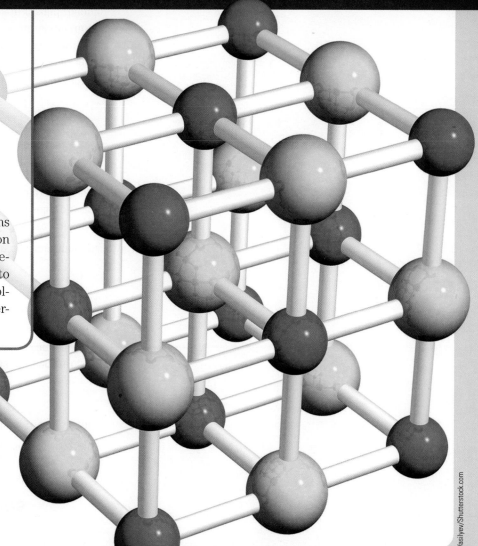

Vasilyev/Shutterstock.com

11.1 Kinetic Molecular Theory, States of Matter, and Phase Changes

11.1a Condensed Phases and Intermolecular Forces

In Gases (Unit 10), we described the kinetic molecular theory as it applies to gases:

- Gases consist of molecules whose separation is much larger than the molecules themselves.

- The molecules of a gas are in continuous, random, and rapid motion.

- The average kinetic energy of gas molecules is determined by the gas temperature, and all gas molecules at the same temperature, regardless of mass, have the same average kinetic energy.

- Gas molecules collide with one another and with the walls of their container, but they do so without loss of energy in "perfectly elastic" collisions.

This theory also applies to liquids and solids.

- The molecules, atoms, and ions that make up liquids and solids are in constant motion.

Liquids and solids are called **condensed phases** because the particles are packed in close proximity to one another. The difference between gaseous, liquid, and solid physical states is shown visually in Interactive Figure 11.1.1 and experimentally by the densities of a substance in the gas, liquid, and solid state.

For water, the densities (measured at 0 °C and 1 atm pressure) are:

$$H_2O(g) \qquad 0.000804 \text{ g/cm}^3$$
$$H_2O(\ell) \qquad 0.9999 \text{ g/cm}^3$$
$$H_2O(s) \qquad 0.9150 \text{ g/cm}^3$$

Liquid and solid water have similar density values, but gaseous water is about 1/1200 as dense as either of the condensed phases. In other words, water molecules in the gaseous state are spaced about 11 times farther apart than in the liquid or solid states. For a real-life analogy, you can think about people attending a basketball game as representing the different states of matter. The fans sitting in a packed stadium are particles in a solid, people standing in slow-moving lines at the concession stands are particles in a liquid, and the basketball players on the court are particles in a gas.

Interactive Figure 11.1.1

Explore the phases of matter.

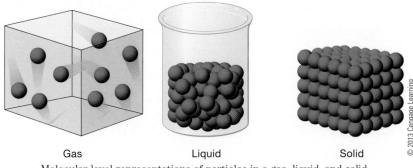

Gas   Liquid   Solid

© 2013 Cengage Learning

Molecular-level representations of particles in a gas, liquid, and solid

The physical state of a substance is controlled by pressure, temperature, and **intermolecular forces** (IMFs), the forces that exist between the individual particles that make up solids, liquids, and gases. Intermolecular forces differ from the chemical bonds in a molecule. Whereas chemical bonds hold bonded atoms together in a single molecule, IMFs are the forces that exist between the molecules, atoms, or ions that make up a particular substance. In other words, chemical bonds explain why atoms in a molecule stay near one another in a solid or liquid, and IMFs explain why molecules stay near one another in a solid or liquid.

The strength of the IMFs in a substance affects many of its properties, such as compressibility, the shape and volume of a substance, and the ability of the particles to flow past one another (Table 11.1.1). Liquids have IMFs that are intermediate between those in solids and gases, and they share some basic properties:

Table 11.1.1 Properties of Solids, Liquids, and Gases

Physical State	IMFs between Particles	Compressibility	Shape and Volume	Ability to Flow
Gas	Generally weak	High	Takes on shape and volume of container	High
Liquid	Generally intermediate	Very low	Takes on shape of container; volume limited by surface area	Moderate
Solid	Generally strong	Almost none	Maintains own shape and volume	Almost none

- Liquids flow when poured out of a container.

- Liquids take on the shape of the container in which they are placed.

- Liquids can evaporate to form vapor.

The IMFs between the molecules that make up liquids vary in strength, and this variation results in a broad range of physical properties among liquids. For example, compared with water, the most common liquid on our planet, ethylene glycol ($HOCH_2CH_2OH$) has a much higher boiling point (100 °C vs. 197 °C, respectively), ethanol (CH_3CH_2OH) evaporates much more easily (Interactive Figure 11.1.2), and honey has a much higher viscosity (resistance to flow).

11.1b Phase Changes

Intermolecular forces play an important role in influencing the physical state of a substance. The stronger the IMFs between particles, the harder it is to separate the particles and the more likely the substance is a solid at typical laboratory temperature and pressure (25 °C and 1 atm). When the IMFs between particles are weak, the substance made up of those particles is most likely a gas at 25 °C and 1 atm. As we will study in the next section, IMFs between particles vary in strength and can have a significant impact on the physical properties of a substance.

Fusion (the physical state change from solid to liquid) and **vaporization** (the physical state change from liquid to gas) are examples of phase changes that require the input of energy. Converting an ordered phase to a less ordered phase requires the addition of energy to overcome the IMFs in the more ordered material. **Condensation** (the physical state change from gas to liquid) is an example of a phase change that results in the release of energy. Energy is released when a less ordered phase is converted to a more ordered phase as IMFs are formed (Table 11.1.2).

Compare the properties of two liquids.

Charles D. Winters

The ethanol in wine evaporates and recondenses on the cool surface of the glass.

Table 11.1.2 Phase Changes

Phase Change	Physical Process	Energy Change
Fusion (melting)	solid → liquid	Energy is absorbed.
Vaporization	liquid → gas	Energy is absorbed.
Sublimation	solid → gas	Energy is absorbed.
Freezing	liquid → solid	Energy is released.
Condensation	gas → liquid	Energy is released.
Deposition	gas → solid	Energy is released.

Example Problem 11.1.1 Investigate phase changes.

In the figure below, N is blue and H is white.

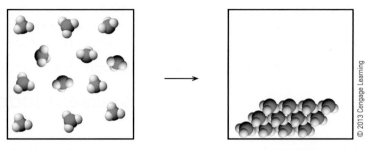

© 2013 Cengage Learning

What kind of change is represented by this illustration? Is energy absorbed or released when this change occurs?

Solution:

You are asked to identify the type of physical change and the energy change that happens during a phase change.

You are given a molecular-level illustration of a phase change.

The figure represents a physical change because the NH_3 molecules are intact in both boxes. The change is from a gas on the left to a solid on the right, a process called deposition. Deposition is a change from a less ordered phase to a more ordered phase, so energy is released.

Video Solution

Tutored Practice
Problem 11.1.1

11.1c Enthalpy of Vaporization

Enthalpy of vaporization, ΔH_{vap}, also called heat of vaporization, is the energy required to convert one mole (or 1 gram) of a liquid to vapor at a given temperature. The enthalpy of vaporization for water at 100 °C is 40.7 kJ/mol. Selected ΔH_{vap} values are shown in Table 11.1.3. Note that enthalpy of vaporization values are nearly independent of temperature. (◄ Flashback to Section 5.3c Energy, Changes of State, and Heating Curves)

Because converting a liquid to vapor involves overcoming the forces between the molecules in the liquid, the magnitude of the enthalpy of vaporization is a direct indication of the strength of the IMFs in the liquid.

> **As the strength of the IMFs in a series of liquids increases, the enthalpy of vaporization values for the liquids increase.**

For example, from the data in Table 11.1.3 we can infer that in the liquid phase, Ar has stronger IMFs than He and that H_2O has stronger IMFs than CH_3OH.

Table 11.1.3 Enthalpy of Vaporization for Some Common Substances

Compound	Enthalpy of Vaporization (kJ/mol)
Helium, He	0.0828
Argon, Ar	6.43
Methane, CH_4	8.17
Ethane, CH_3CH_3	14.7
Methanol, CH_3OH	35.4
Water, H_2O	40.7
Benzene, C_6H_6	34.1

Enthalpy of vaporization values are generally much smaller than the energies of chemical bonds in molecules. As a result, heating a liquid will normally cause it to boil (a physical change where molecules are separated from one another) before it decomposes (a chemical change where bonds between atoms in molecules are broken, resulting in a new compound). For example, the enthalpy of vaporization of H_2O is 40.7 kJ/mol and the energy of an O—H covalent bond is more than ten times stronger, 463 kJ/mol. When water is heated, the added energy overcomes the IMFs in water, vaporizing the liquid to form water vapor where the water molecules have been separated. Breaking the O—H bonds in water requires much more energy (and thus a much higher temperature) than is needed to overcome the IMFs in liquid water. Therefore, when water is heated, the water molecules move away from one another instead of breaking apart into hydrogen and oxygen atoms.

Example Problem 11.1.2 Relate enthalpy of vaporization to IMFs.

The following information is given for two hydrocarbons, propane and hexane, at 1 atm:

$$C_3H_8 \text{ (propane)} \qquad \Delta H_{vap} = 19.0 \text{ kJ/mol}$$

$$C_6H_{14} \text{ (hexane)} \qquad \Delta H_{vap} = 28.9 \text{ kJ/mol}$$

Which hydrocarbon has stronger IMFs?

Solution:

You are asked to determine the relative strength of IMFs in two hydrocarbons.

You are given the enthalpy of vaporization for the two hydrocarbons.

The magnitude of the enthalpy of vaporization is a direct indication of the strength of the IMFs in the liquid. The enthalpy of vaporization for hexane is greater than that of propane, so the IMFs in hexane are stronger.

Video Solution

Tutored Practice
Problem 11.1.2

Section 11.1 Mastery

11.2 Vapor Pressure

11.2a Dynamic Equilibrium and Vapor Pressure

Consider a liquid placed in a sealed, evacuated flask, to which a pressure gauge is attached. Over time, the **vapor pressure** in the flask increases as molecules at the surface of the liquid escape into the gas phase (also called the vapor phase). As molecules escape into the gas phase, some recondense, entering the liquid phase. When the rates of vaporization and condensation become equal, a state of **dynamic equilibrium** is reached (Interactive Figure 11.2.1).

In this equilibrium state, vaporization and condensation both continue to occur, but because they do so at the same rate, the total number of molecules in the liquid and gaseous states is constant. The **equilibrium vapor pressure** for a liquid is the pressure of the vapor over a liquid when a dynamic equilibrium is reached at a given temperature. Vapor pressure values for some common liquids at various temperatures are shown in Table 11.2.1.

Interactive Figure 11.2.1

Investigate dynamic equilibria.

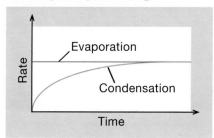

© 2013 Cengage Learning

Example Problem 11.2.1 Use vapor pressure.

A student leaving campus for spring break wants to make sure the air in her dorm room has a high water vapor pressure so that her plants are comfortable. The dorm room measures 4.0 m × 3.0 m × 3.0 m and the student places a large pan containing 8.0 L of water in the room. Assuming that the room is airtight, that there is no water vapor in the air when she closes the door, and that the temperature remains a constant 25 °C, will all of the water in the pan evaporate? The vapor pressure of water at 25 °C is 23.8 mm Hg.

Solution:

You are asked to determine whether a liquid will evaporate in a sealed room.

You are given the size of the room and the volume and temperature of the liquid.

To determine whether or not all the water would evaporate, first calculate the vapor pressure the water would exert if all the water were to evaporate. If this pressure is greater than the vapor pressure of water at 25 °C, it implies that not all of the water can evaporate.

$$n_{H_2O} = 8.0 \text{ L } H_2O \times \frac{10^3 \text{ mL}}{1 \text{ L}} \times \frac{1.00 \text{ g}}{1 \text{ mL}} \times \frac{1 \text{ mol } H_2O}{18.0 \text{ g}} = 440 \text{ mol}$$

$$V_{room} = 4.0 \text{ m} \times 3.0 \text{ m} \times 3.0 \text{ m} \times \left(\frac{10^2 \text{ cm}}{1 \text{ m}}\right)^3 \times \frac{1 \text{ mL}}{1 \text{ cm}^3} \times \frac{1 \text{ L}}{10^3 \text{ mL}} = 3.6 \times 10^4 \text{ L}$$

$$P = \frac{nRT}{V} = \frac{(440 \text{ mol})(0.082057 \text{ L} \cdot \text{atm/K} \cdot \text{mol})(298 \text{ K})}{3.6 \times 10^4 \text{ L}} \times \frac{760 \text{ mm Hg}}{1 \text{ atm}} = 230 \text{ mm Hg}$$

The equilibrium vapor pressure of water at 25 °C (23.8 mm Hg) is much smaller than the vapor pressure in the room if all of the water evaporates (230 mm Hg). Therefore, not all of the water in the pan will evaporate in an airtight room. (In reality, the room is probably not airtight and more water will evaporate than predicted.)

Table 11.2.1 Vapor Pressures (mm Hg) of Some Common Liquids

Liquid	0 °C	25 °C	50 °C	75 °C	100 °C	125 °C
Water	4.6	23.8	92.5	300	760	1741
Benzene	27.1	94.4	271	644	1360	—
Methanol	29.7	122	404	1126	—	—
Diethyl ether	185	470	1325	2680	4859	—

Video Solution

Tutored Practice Problem 11.2.1

11.2b Effect of Temperature and Intermolecular Forces on Vapor Pressure

The vapor pressure of a liquid is directly related to the number of atoms or molecules in the gas phase. Two factors that control how much of a liquid is converted to vapor are temperature and the IMFs in the liquid.

Vapor Pressure and Temperature

Vapor pressure is highly dependent on temperature. The relationship between vapor pressure and temperature for a given liquid is demonstrated in Boltzmann distribution plots for a liquid at two different temperatures (Figure 11.2.2). (◄ Flashback to Section 10.6 Kinetic Molecular Theory) Because molecules in a liquid, on average, have greater kinetic energy at higher temperatures, the high-temperature curve is shifted to the right when compared with the low-temperature curve. The energy threshold for a liquid escaping into the vapor phase is constant because it is a function of the IMFs in the liquid. So at a higher temperature, a greater fraction of molecules exceed the energy threshold needed to overcome the IMFs than at the lower temperature.

Therefore, as temperature increases,

- more molecules have the minimum kinetic energy needed to enter the gas phase,

- the number of molecules in the gas phase increases, and

- vapor pressure increases.

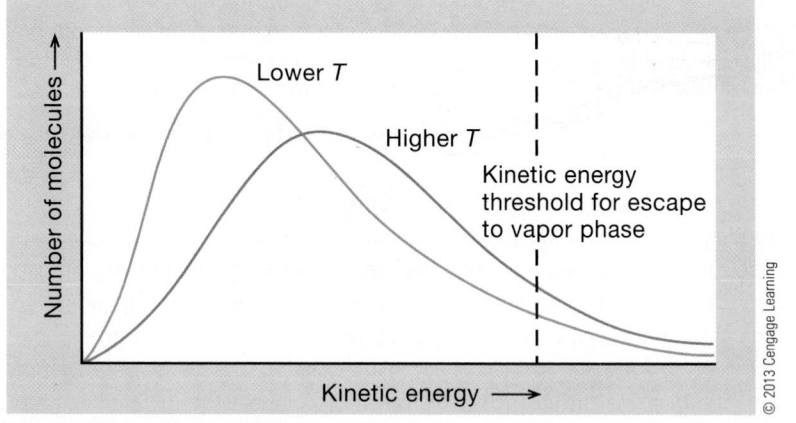

Figure 11.2.2 Boltzmann distributions for a liquid at two different temperatures

Vapor Pressure and Intermolecular Forces

Vapor pressure is also dependent on the IMFs in a liquid. The relationship between IMFs and vapor pressure can be shown using Boltzmann distribution plots for two different liquids at the same temperature (Interactive Figure 11.2.3). The plots have identical shapes because kinetic energy depends only on temperature.

Interactive Figure 11.2.3

Explore the effect of temperature and IMFs on vapor pressure.

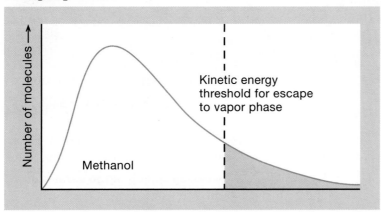

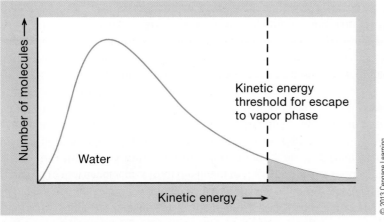

Boltzmann distribution plots for liquid methanol and water at 25 °C

© 2013 Cengage Learning

However, the two liquids do not have the same vapor pressure at a given temperature. Interactive Figure 11.2.3 shows the minimum kinetic energy at a given temperature needed to overcome the IMFs between molecules in the liquid phase, which is the minimum energy required to move molecules from the liquid to the vapor phase. Methanol has a lower kinetic energy threshold than water and therefore a higher number (indicated by shading in the figure) of molecules with the minimum energy needed to escape into the vapor phase. Therefore, more methanol molecules than water molecules occupy the vapor phase at a given temperature and methanol's vapor pressure is higher than the water vapor pressure at a given temperature.

This is in agreement with the enthalpy of vaporization values (Table 11.1.3) for these liquids; methanol has weaker IMFs than water, and therefore methanol molecules are held less tightly to one another in the liquid phase. It is easier to vaporize methanol (lower ΔH_{vap}) than water (higher ΔH_{vap}) at a given temperature, and more methanol escapes into the vapor phase at a given temperature. The liquid with the stronger IMFs, water, has a lower vapor pressure at a given temperature.

As the strength of the IMFs in a series of liquids increases, the vapor pressures of the liquids decrease.

Interactive Figure 11.2.4 shows vapor pressure–temperature curves for carbon disulfide, methanol, ethanol, and heptane.

Interactive Figure 11.2.4

Explore the effect of IMFs on vapor pressure.

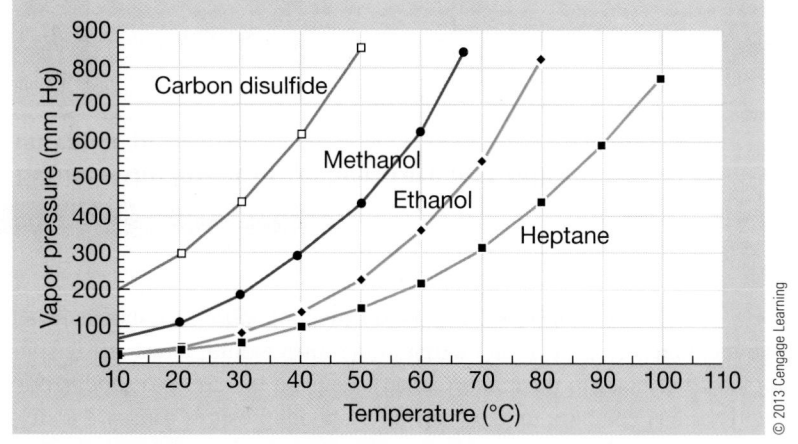

Vapor pressure vs. temperature plots for four compounds

For any liquid, as the temperature increases, the vapor pressure increases because of the increasing fraction of molecules above the escape threshold. At any specific temperature, however, the liquids have different vapor pressures because of the different strengths of the IMFs in the liquids. The compound with the weakest IMFs, carbon disulfide, has the highest vapor pressure at a given temperature. Heptane, the compound with the strongest IMFs in this group, has the lowest vapor pressure at a given temperature.

Example Problem 11.2.2 Relate vapor pressure, enthalpy of vaporization, and IMFs.

Use the plot of vapor pressures versus temperature in Interactive Figure 11.2.4 to answer the following questions.

  a. What is the approximate temperature at which the vapor pressure of heptane is 400 mm Hg?
  b. Which liquid, methanol or heptane, has the higher enthalpy of vaporization?

Solution:

You are asked to determine the temperature at which a liquid has a given vapor pressure and the relative magnitude of the enthalpy of vaporization for two liquids.

You are given a plot of vapor pressure vs. temperature.
  a. Using Interactive Figure 11.2.4, the vapor pressure of heptane is 400 mm Hg at approximately 77 °C.
  b. At any temperature, methanol has a higher vapor pressure than heptane. Therefore, the IMFs in heptane are stronger than the IMFs in methanol. The heat of vaporization of heptane would be expected to be higher than the heat of vaporization of methanol.

Video Solution

Tutored Practice
Problem 11.2.2

11.2c Boiling Point

When a liquid is placed in a container open to the atmosphere and then heated, its vapor pressure increases. If enough heat is added to reach the temperature where the vapor pressure of the liquid equals the pressure of the atmosphere, bubbles containing vapor form in the liquid and the liquid boils. The **boiling point** of a liquid is defined as the temperature at which the vapor pressure of the liquid is equal to the external pressure. The **normal boiling point** is the temperature when the vapor pressure of a liquid is equal to 1 atm.

Because boiling involves breaking IMFs, the boiling point for a liquid is related to the strengths of the IMFs in the liquid. Consider the vapor pressure curves for diethyl ether, ethanol, and water shown in Interactive Figure 11.2.5.

Explore the boiling point of a liquid.

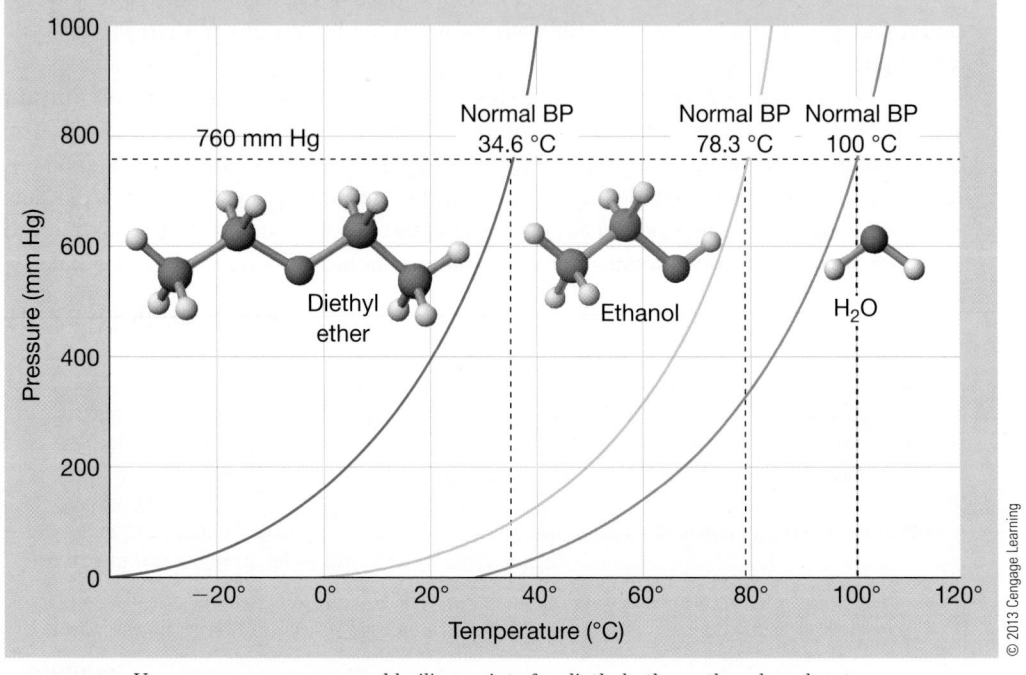

Vapor pressure curves and boiling points for diethyl ether, ethanol, and water

The normal boiling points for these three liquids are shown as the temperature when the vapor pressure of the liquid is equal to 1 atm (760 mm Hg). Of the three liquids shown in the figure, diethyl ether has highest vapor pressure at a given temperature, the lowest boiling point (34.6 °C), and the weakest IMFs of the three liquids. Water has the lowest vapor pressure at a given temperature, the highest boiling point (100 °C), and the strongest IMFs.

As the strength of the IMFs in a series of liquids increases, the boiling points of the liquids increase.

Notice that the vapor pressure curves also allow you to determine the boiling point for a liquid under nonstandard conditions. For example, water boils at a temperature below its normal boiling point when the external pressure is lower than 1 atm. As shown in Interactive Figure 11.2.5, water boils at 93 °C when the external pressure is 600 mm Hg.

You can find an external pressure this low in a typical high-altitude airplane cabin or in a city that sits about 6000 feet above sea level. Because of this variation in boiling point, cooking instructions often indicate longer cooking times for high altitudes.

In summary, IMFs affect the enthalpy of vaporization, vapor pressure, and boiling point of liquids. For a series of liquids, as IMF strength increases,

- the energy needed to vaporize the liquids increases (ΔH_{vap} increases),
- the liquid vapor pressures decrease, and consequently,
- the liquid boiling points increase.

Example Problem 11.2.3 Predict boiling point using a vapor pressure–temperature plot.

Using the figure shown below, estimate the boiling point of heptane when the external pressure is 595 mm Hg.

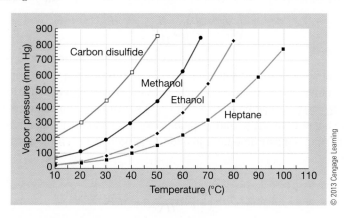

© 2013 Cengage Learning

Solution:

You are asked to estimate the boiling point of a liquid.

You are given a plot of vapor pressure versus temperature for the liquid.

The boiling point is the temperature at which the vapor pressure is equal to the external pressure. From the graph in Interactive Figure 11.2.4, the vapor pressure of heptane is approximately 595 mm Hg at 90 °C.

Video Solution

Tutored Practice
Problem 11.2.3

11.2d Mathematical Relationship between Vapor Pressure and Temperature

The mathematical relationship between vapor pressure (P), temperature (T), and the strength of IMFs (which are related to ΔH_{vap}) is expressed in the Clausius–Clapeyron equation, where R is the ideal gas constant ($R = 8.314 \times 10^{-3}$ kJ/K · mol) and C is a constant of integration.

$$\ln P = \frac{-\Delta H_{vap}}{RT} + C \qquad (11.1)$$

This general form of the Clausius–Clapeyron equation can also be written in the form of a straight-line equation.

$$\ln P = \frac{-\Delta H_{vap}}{R}\left(\frac{1}{T}\right) + C$$

$$y = \qquad m \qquad x \ + b$$

When vapor pressure is measured at a range of temperatures, the data can be used to determine the enthalpy of vaporization for a liquid. A plot of the natural log of vapor pressure ($\ln P$) versus inverse temperatures ($1/T$) has a slope equal to $-\Delta H_{vap}/R$.

A second form of the Clausius–Clapeyron equation is used to calculate the vapor pressure of a liquid at a given temperature if both the enthalpy of vaporization of the liquid and the vapor pressure at another temperature are known.

$$\ln \frac{P_2}{P_1} = \frac{-\Delta H_{vap}}{R}\left(\frac{1}{T_2} - \frac{1}{T_1}\right) \qquad (11.2)$$

The examples that follow demonstrate the use of both forms of the Clausius–Clapeyron equation.

Example Problem 11.2.4 Determine enthalpy of vaporization graphically using the Clausius–Clapeyron equation.

The vapor pressure of sulfur dioxide has been measured over a range of temperatures.

T (K)	P (mm Hg)
220	81.6
230	147.4
240	253.6
250	417.7

Use these data to calculate ΔH_{vap} for SO_2.

Example Problem 11.2.4 *(continued)*

Solution:

You are asked to calculate the enthalpy of vaporization for a liquid.

You are given the vapor pressure of the liquid at different temperatures.

Use the general form of the Clausius–Clapeyron equation and a plot of ln P versus $1/T$ to determine ΔH_{vap}.

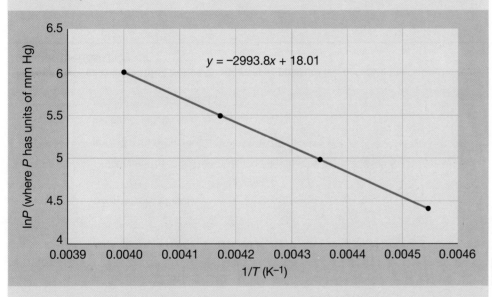

$y = -2993.8x + 18.01$

© 2013 Cengage Learning

According to Equation 11.1, the slope of this linear plot is equal to $-\Delta H_{vap}/R$.

$$\text{slope} = \frac{-\Delta H_{vap}}{R}$$

$$-2993.8\ \text{K}^{-1} = \frac{-\Delta H_{vap}}{8.3145 \times 10^{-3}\ \text{kJ/mol} \cdot \text{K}}$$

$$\Delta H_{vap} = 24.9\ \text{kJ/mol}$$

Video Solution

**Tutored Practice
Problem 11.2.4**

Example Problem 11.2.5 Use the two-point version of the Clausius–Clapeyron equation.

The vapor pressure of liquid aluminum is 400. mm Hg at 2590 K. Assuming that ΔH_{vap} for Al (296 kJ/mol) does not change significantly with temperature, calculate the vapor pressure of liquid Al at 2560 K.

Solution:

You are asked to calculate the vapor pressure of a liquid at a given temperature.

You are given the vapor pressure of the liquid at a different temperature and the enthalpy of vaporization for the liquid.

Use the alternative form of the Clausius–Clapeyron equation (Equation 11.2) along with pressure, temperature, and ΔH_{vap} data to calculate the vapor pressure of aluminum at 2560 K.

$$\ln \frac{P_2}{P_1} = \frac{-\Delta H_{vap}}{R} \left(\frac{1}{T_2} - \frac{1}{T_1} \right)$$

$P_1 = 400.$ mm Hg $T_1 = 2590$ K

$P_2 = ?$ $T_2 = 2560$ K

$\Delta H_{vap} = 296$ kJ/mol

$$\ln \frac{P_2}{400.\text{ mm Hg}} = \frac{-(296 \text{ kJ/mol})}{8.3145 \times 10^{-3} \text{ kJ/K} \cdot \text{mol}} \left(\frac{1}{2560 \text{ K}} - \frac{1}{2590 \text{ K}} \right)$$

$\ln (P_2) - \ln (400.\text{ mm Hg}) = -0.1611$

$\ln (P_2) = 5.830$

$P_2 = e^{5.830} = 340.$ mm Hg

Is your answer reasonable? Notice that the vapor pressure has decreased because the temperature decreased ($T_2 < T_1$).

Video Solution

Tutored Practice
Problem 11.2.5

Section 11.2 Mastery

11.3 Other Properties of Liquids

11.3a Surface Tension

Cohesive forces are attractive forces that exist between molecules at the macroscopic level that affect the physical properties of a liquid. Sometimes called bulk-scale forces, cohesive forces are the result of IMFs, the molecular-level interactions in a liquid.

Surface tension is a measure of force required to "break" the surface of a liquid. The surface tension of water can be seen in Figure 11.3.1a, where an insect rests on the "skin" that results from the high surface tension of water. Surface tension results from the different IMFs experienced by molecules at the surface of a liquid and those in the interior of the liquid. Molecules in the interior of a liquid experience IMFs in all directions. Molecules on the surface, however, are attracted to molecules only to the sides and below; thus, they experience a net inward force (Figure 11.3.1b). This net inward force pulls surface molecules toward the interior, minimizing the surface area and causing droplets of liquid to form a spherical shape (the shape with the smallest surface area for any given volume).

Surface tension can be measured using the angle between a horizontal surface and the tangent to the surface where a liquid droplet meets the horizontal surface, called the *contact angle* (Interactive Figure 11.3.2a). The greater the contact angle when a droplet of liquid sits on a horizontal surface, the stronger the surface tension of the liquid, and the better it resists the force of gravity to attain a spherical shape. Contact angle depends not only on the liquid but also on whether anything is dissolved in it and the material in the horizontal surface. You can see the effect of these factors on contact angles at home or in your dorm room. Try placing droplets of water on glass, ceramic, and plastic, and then repeat the experiment after adding soap to the water. Also, explore the effect of gravity by looking at droplets with different sizes.

Interactive Figure 11.3.2b shows photographs of droplets of pure mercury, water, and acetone. Notice that the mercury forms the most spherical droplets and the acetone, the flattest. This suggests that of these three liquids, acetone has the lowest surface tension and mercury the greatest.

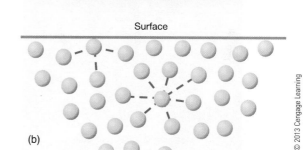

Figure 11.3.1 (a) An insect resting on water using surface tension; (b) surface tension at the molecular level

Interactive Figure 11.3.2

Explore contact angle and surface tension.

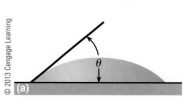 

(a) Measuring contact angle for a liquid; (b) droplets of mercury, water, and acetone

Table 11.3.1 Surface Tension Values and Boiling Points for Some Common Liquids

Substance	Formula	Surface Tension (J/m^2 at 20 °C)	Normal Boiling Point (°C)
Octane	C_8H_{18}	2.16×10^{-2}	125.5
Ethanol	CH_3CH_2OH	2.23×10^{-2}	78.4
Chloroform	$CHCl_3$	2.68×10^{-2}	61.2
Benzene	C_6H_6	2.85×10^{-2}	80.1
Water	H_2O	7.29×10^{-2}	100.0
Mercury	Hg	46×10^{-2}	356.7

In general, liquids with strong IMFs have high surface tension. However, the relationship between surface tension and other liquid properties, such as boiling point, is less direct (Table 11.3.1).

11.3b Viscosity

Viscosity is a measure of a liquid's resistance to flow. We commonly think of viscosity as a measure of the "thickness" of a liquid, where, for example, gasoline is a "thin" liquid and motor oil is a "thick" liquid. Like surface tension, viscosity is not highly predictable based on other properties of liquids, such as boiling point, but it does depend on IMF strength. In general, strong IMFs result in high viscosity for a liquid (Interactive Figure 11.3.3).

Another property that affects the viscosity of a liquid is molecular structure, particularly the length of molecules. Long, flexible molecules tend to become entangled and resist moving past one another as a liquid flows. As a result, some liquids with strong IMFs, such as mercury, have a very low viscosity—the individual mercury atoms move past one another with little effort. Stearic acid, an organic fatty acid with a long molecular shape, has a much higher viscosity than mercury, even though the compound has weaker overall IMFs than mercury.

11.3c Capillary Action

Movement of a liquid up a capillary tube (a glass tube with a narrow diameter that is open at both ends) and the flow of groundwater through soil are examples of capillary action. In **capillary action**, the cohesive forces within a bulk sample of liquid are overcome by **adhesive forces**, the IMFs between a liquid and a surface such as glass or paper.

Interactive Figure 11.3.3

Explore viscosity.

Charles D. Winters

Honey is a viscous liquid because of its strong IMFs.

Consider two capillary tubes, one placed in water and one in a sample of mercury (Interactive Figure 11.3.4). In the tube placed in water, the attractive forces between water and the glass result in the liquid being drawn up in the tube. At the top of the column of water, the *meniscus,* the curved surface of the liquid, draws up at the edges due to the strong adhesive forces between water and glass. In this case the adhesive forces are strong enough to overcome gravitational forces, and the column of water rises in the tube.

In the tube placed in mercury, the meniscus is curved downward and the liquid does not rise in the tube. Here, the cohesive forces in the sample of mercury are stronger than the adhesive forces between mercury and the glass.

11.4 The Nature of Intermolecular Forces

11.4a Dipole–Dipole Intermolecular Forces

We have seen that the properties of liquids depend on the strength of the IMFs in a liquid. The strength of IMFs is controlled by differing charges on adjacent molecules that lead to **electrostatic forces** of attraction. These attractions follow Coulomb's law, which was described in our discussion of chemical bonding in Covalent Bonding and Molecular Structure (Unit 8).

$$\text{force} \propto \frac{(q_A)(q_B)}{r^2}$$

As the magnitude of the charges leading to the attractions between molecules in liquids increases, so does the strength of the attractive force between the molecules. The attractive force also increases as the distance between those charges decreases. Charges that exist on particles in a liquid are primarily partial charges (as compared with a permanent ionic charge).

Dipole–dipole intermolecular forces are the attractive forces that occur between two polar molecules. Polar molecules have a permanent dipole; there is a partial positive charge (δ^+) at one end of the molecule and a negative partial charge (δ^-) at the opposite end. When two polar molecules such as hydrogen chloride approach each other, an electrostatic force of attraction forms between the positive and negative partial charges on the adjacent molecules (Interactive Figure 11.4.1).

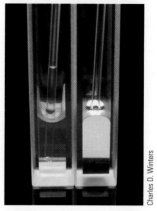
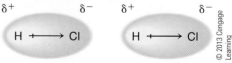

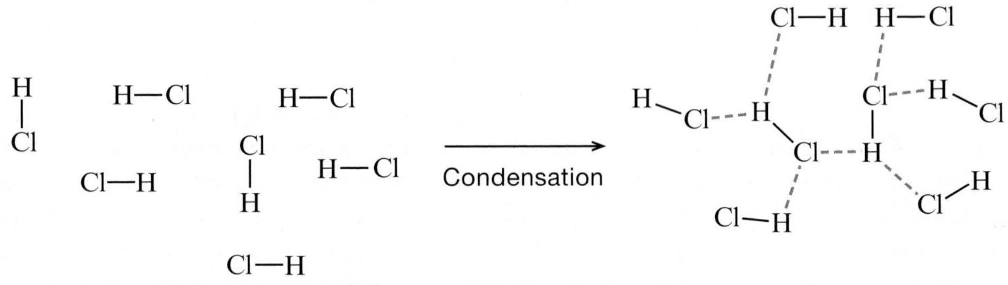

Figure 11.4.2 Condensation of HCl molecules

The dipole–dipole IMFs between HCl molecules allow gaseous HCl to condense to form a liquid at low temperatures. Liquid HCl consists of HCl molecules held together by a network of dipole–dipole IMFs (Figure 11.4.2). Dipole–dipole IMFs are generally weaker than the ionic and covalent forces in ionic solids and covalently bonded compounds, respectively.

Hydrogen Bonding

Hydrogen bonding is an unusually strong type of dipole–dipole IMF that occurs between molecules with H—N, H—O, or H—F bonds. Hydrogen bonding is especially strong for two reasons:

1. The elements bonded to hydrogen are very electronegative, and hydrogen has a relatively low electronegativity, resulting in very polar H—N, H—O, and H—F bonds. The highly polar bonds result in large partial charges and therefore a strong force of attraction between molecules.

2. Hydrogen has a very small size, allowing hydrogen-bonding molecules to approach closely. This decrease in distance between polar molecules results in a strong force of attraction.

Hydrogen bonds are generally stronger than dipole–dipole IMFs.

Example Problem 11.4.1 Identify species that can form hydrogen bonds.

In which of the following pure substances would hydrogen bonding be expected?

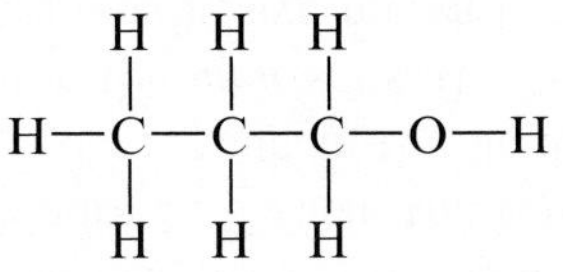

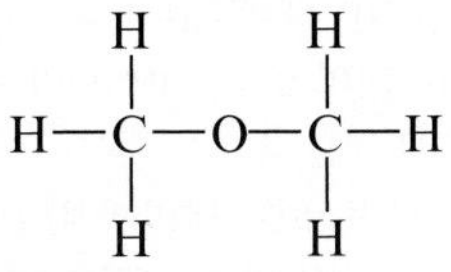

Solution:

You are asked to predict whether hydrogen bonds are present in a sample of a pure substance.

You are given the chemical structure of the substance.

Although both compounds contain oxygen and hydrogen, of the two, only propanol contains an O—H bond. Hydrogen bonding occurs between molecules containing O—H, N—H, or H—F bonds, so we expect there to be hydrogen bonding between propanol molecules.

Video Solution

Tutored Practice
Problem 11.4.1

11.4b Dipole–Induced Dipole Forces

Dipole–induced dipole intermolecular forces are attractive forces that occur between polar and nonpolar molecules. Nonpolar molecules do not have permanent dipoles, but it is possible to create a temporary dipole, called an **induced dipole**, in a nonpolar molecule. Consider the interaction between water, a polar molecule, and oxygen, a nonpolar molecule. When the negative end of an H_2O molecule approaches an O_2 molecule, the negative partial charge repels the O_2 electron cloud, distorting it and creating a temporary dipole on the O_2 molecule (Interactive Figure 11.4.3).

Note that the positive end of a water molecule can also induce a dipole in an O_2 molecule by attracting the electron cloud instead of repelling it. Induced dipoles are temporary; when the H_2O and O_2 molecules separate, the oxygen electron cloud is no longer distorted. The dipole–induced dipole IMF thus occurs between a polar molecule (with a permanent dipole) and a nonpolar molecule (with a temporary or induced dipole).

Although dipole–induced dipole IMFs are generally weaker than dipole–dipole forces, the magnitude of the induced dipole can result in these forces being stronger than dipole–dipole forces. The ease with which the electron cloud in a molecule can be distorted and a dipole can be induced is called its **polarizability**. Polarizability increases as the number of electrons in a molecule increases, and therefore, it increases with increasing molar mass and molecular size.

Interactive Figure 11.4.3

Explore dipole–induced dipole IMFs.

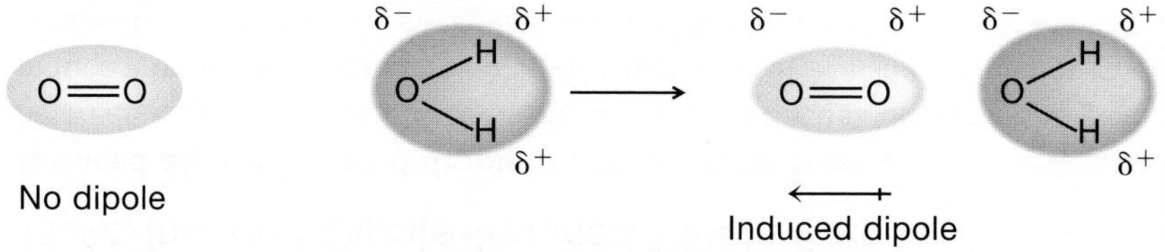

Water induces a temporary dipole in O_2.

© 2013 Cengage Learning

11.4c Induced Dipole–Induced Dipole Forces

Induced dipole–induced dipole intermolecular forces are attractive forces that occur between nonpolar molecules. As you saw in the previous discussion, a polar molecule can induce a temporary dipole in a nonpolar molecule. It is also possible for two nonpolar molecules to induce temporary dipoles in each other. Consider iodine, which consists of nonpolar I_2 molecules. When two I_2 molecules approach each other, the electron clouds repel and temporary dipoles form in each I_2 molecule (Interactive Figure 11.4.4).

The attractive forces between these temporary dipoles, induced dipole–induced dipole IMFs, are also called **London dispersion forces** (or simply *dispersion forces*). Dispersion forces are generally weak, but they can be stronger than dipole–dipole forces when they occur between highly polarizable molecules.

Dispersion forces are important in all liquids, even those consisting of polar molecules, because any molecule's electron cloud can be distorted by the partial charges in surrounding molecules.

Interactive Figure 11.4.4

Explore induced dipole–induced dipole intermolecular forces.

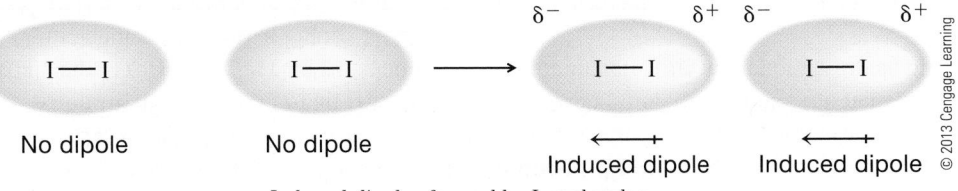

Induced dipoles formed by I_2 molecules

© 2013 Cengage Learning

Example Problem 11.4.2 Identify intermolecular forces in a substance.

Identify all IMFs that exist between XeO_3 molecules.

Solution:

You are asked to identify all IMFs that exist between molecules in a pure substance.

You are given the chemical formula of the substance.

First, draw a Lewis structure and determine the shape of the molecule. XeO_3 has a trigonal pyramidal shape and polar bonds; therefore, it is a polar molecule. Thus, there are dipole–dipole IMFs between XeO_3 molecules. In addition, induced dipole–induced dipole (dispersion) forces are always present between molecules.

Video Solution

Tutored Practice
Problem 11.4.2

Section 11.4 Mastery

11.5 Intermolecular Forces and the Properties of Liquids

11.5a Effect of Polarizability on Physical Properties

Nonpolar species such as hydrocarbons and diatomic elements have weak dispersion (induced dipole–induced dipole) IMFs, so many are gases at room temperature. However, as the molar mass of a compound increases, so does the number of electrons and therefore its polarizability, that is, the ease with which a dipole can be induced. As shown by the data in Table 11.5.1, as the number of electrons increases in a series of nonpolar compounds, so does the enthalpy of vaporization and boiling point.

The effect of dispersion forces on physical properties can also be seen in the physical properties of a series of straight-chain hydrocarbons (Table 11.5.2). Each hydrocarbon is nonpolar, and each has a similar molecular structure, differing only in molar mass and the length of the carbon chain. As shown by the data in the table, as molecular size increases, so does the strength of the IMFs, as indicated by increasing enthalpy of vaporization and boiling point.

Table 11.5.1 Molar Mass, $\Delta H°_{vap}$, and Boiling Point for Some Nonpolar Species

Compound	Molar Mass (g/mol)	$\Delta H°_{vap}$ (kJ/mol)	Boiling Point (°C)
He	4.0	0.08	−268.9
Ne	20.3	1.7	−246.1
N_2	28.0	5.6	−195.8
O_2	32.0	6.8	−183.0
Ar	39.9	6.4	−185.9
Cl_2	70.9	20.4	−34.0
Br_2	159.8	30.0	58.8

Table 11.5.2 Selected Physical Properties of Some Hydrocarbons

Molecular Formula	Name	$\Delta H°_{vap}$ (kJ/mol)	Boiling Point (°C)
CH_4	Methane	8.2	−161
C_2H_6	Ethane	14.7	−88
C_3H_8	Propane	19.0	−42
C_6H_{14}	Hexane	28.9	69
C_8H_{18}	Octane	34.4	126
$C_{10}H_{22}$	Decane	38.8	174
$C_{18}H_{38}$	Octadecane	54.5	317

Example Problem 11.5.1 Predict physical properties using relative polarizability.

Consider the two alcohols methanol (CH_3OH) and 1-hexanol ($C_6H_{13}OH$).

   a. Which alcohol has the higher boiling point?
   b. Which has the lower enthalpy of vaporization?
   c. What type of IMF is primarily responsible for the differences in the physical properties of these alcohols?

Solution:

You are asked to predict the relative boiling point and enthalpy of vaporization for two alcohols and to identify the IMF primarily responsible for the differences in the physical properties of the alcohols.

You are given the chemical formulas of the alcohols.

Both alcohols can form hydrogen bonds, but they differ in their molar masses (CH_3OH, 32.0 g/mol; $C_6H_{13}OH$, 102 g/mol).

   a. The alcohol with the higher molar mass has stronger IMFs and therefore a higher boiling point.

      CH_3OH, 65 °C; $C_6H_{13}OH$, 158 °C

   b. The alcohol with the lower molar mass has weaker IMFs and therefore a lower enthalpy of vaporization.

      CH_3OH, 35.2 kJ/mol; $C_6H_{13}OH$, 44.5 kJ/mol

   c. Dispersion forces (induced dipole–induced dipole) IMFs are primarily responsible for the differences in boiling point and enthalpy of vaporization for these two alcohols.

Video Solution

Tutored Practice
Problem 11.5.1

11.5b Effect of Hydrogen Bonding on Physical Properties

Life as we know it would not exist without water. Most of the planet is covered with water: liquid water in oceans, seas, and rivers, and solid water in the polar ice caps. However, if it were not for hydrogen bonding, most of the water on our planet would not be in liquid form. Instead, based on the boiling points of similar molecules with similar structures, most water would be found as water vapor. Consider the plot of the boiling points of Group 6A hydrides (H_2A, where A = Te, Se, S, and O) shown in Interactive Figure 11.5.1.

Explore the effect of hydrogen bonding on boiling points.

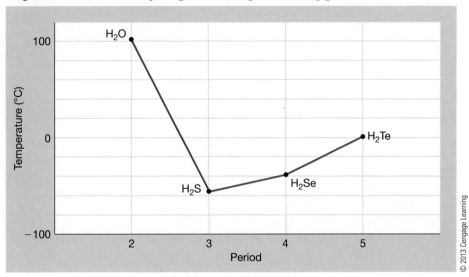

Boiling points of the Group 6A hydrides

The Group 6A hydrides are polar; thus, there are dipole–dipole IMFs in the pure liquids. Boiling point decreases from H_2Te to H_2S as the molar mass of the compounds decrease (polarizability decreases with decreasing molar mass). If the trend in boiling point is extended to H_2O, its boiling point is predicted to be approximately -80 °C. However, we know that water has a boiling point of 100 °C, about 180° higher than predicted based on the trend for the other Group 6A hydrides. This discrepancy is a reflection of the fact that there is extensive hydrogen bonding in water. The hydrogen bonds between water molecules influence many of its physical properties in addition to its high boiling point, including a high enthalpy of vaporization, high surface tension, and high heat capacity.

The influence of hydrogen bonding on physical properties can also be seen in the differences between ethanol (CH_3CH_2OH) and dimethyl ether (CH_3OCH_3), two constitutional isomers (Figure 11.5.2).

Although these two compounds have the same molecular formula, the IMFs in ethanol (hydrogen bonding) are stronger than those in dimethyl ether (dipole–dipole) due to the presence of an O—H bond in the molecular structure of ethanol. This small difference in molecular structure results in a boiling point for ethanol that is about 100° higher than that of dimethyl ether.

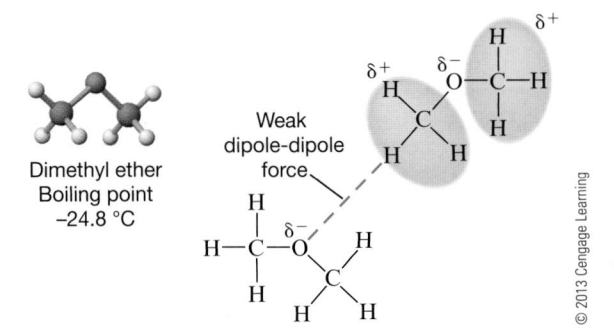

Figure 11.5.2 Ethanol and dimethyl ether have very different physical properties.

Video Solution

Tutored Practice
Problem 11.5.2

Example Problem 11.5.2 Predict physical properties using relative strength of intermolecular forces.

Consider the two compounds C_2H_5OH (ethanol) and C_2H_5SH (ethanethiol).

a. Which compound has the higher molar mass?
b. Which compound has the higher boiling point?

Solution:

You are asked to determine relative molar mass and boiling point for two compounds.

You are given the chemical formulas for the two compounds.

a. Ethanethiol (62.1 g/mol) has a greater molar mass than ethanol (46.1 g/mol).
b. Although ethanethiol has a greater molar mass than ethanol, the IMFs in ethanol are stronger than those in ethanethiol due to the ability of ethanol to form hydrogen bonds. Therefore, we would predict a higher boiling point for ethanol (ethanol, 79 °C; ethanethiol, 35 °C).

11.5c Quantitative Comparison of Intermolecular Forces

Because IMF strength depends on the magnitude of the partial charges present on molecules, it is reasonable to predict that polar molecules will have stronger IMFs than comparably sized nonpolar molecules. Consider the compounds in Table 11.5.3. As expected, in each pair of molecules with similar molar mass, the polar compound has a higher boiling point. To explore the influence of polarity and molar mass on IMF strength, consider a series of compounds that differ only in one element or group of elements. Each compound

Table 11.5.3 Molar Masses and Boiling Points of Some Polar and Nonpolar Substances

	Nonpolar			Polar	
	Molar Mass (g/mol)	Boiling Point (°C)		Molar Mass (g/mol)	Boiling Point (°C)
N_2	28	−196	CO	28	−192
SiH_4	32	−112	PH_3	34	−88
GeH_4	77	−90	AsH_3	78	−62
Br_2	160	59	ICl	162	97

consists of a different hydrocarbon group (represented by "R" in the plots that follow) and a different "end group" (X = H, F, Cl, or OH).

R Group	Compound (X = H, F, Cl, or OH)
Methyl	CH_3—X
Ethyl	CH_3CH_2—X
Propyl	$CH_3CH_2CH_2$—X
Butyl	$CH_3CH_2CH_2CH_2$—X
Pentyl	$CH_3CH_2CH_2CH_2CH_2$—X
Hexyl	$CH_3CH_2CH_2CH_2CH_2CH_2$—X

A plot of boiling point versus the molar mass of each R—X compound (Interactive Figure 11.5.3) shows a fairly linear relationship when X = H, F, or Cl. This is quite interesting because when X = H, the compounds are nonpolar (weak dispersion IMFs), but when X = F or Cl, the compounds are polar and have stronger IMFs (dipole–dipole). The linear plot implies that polarizability, which influences the strength of the dispersion forces in these series, is the controlling factor, whereas polarity plays a relatively unimportant role.

Explore the effect of molar mass and IMFs on boiling point.

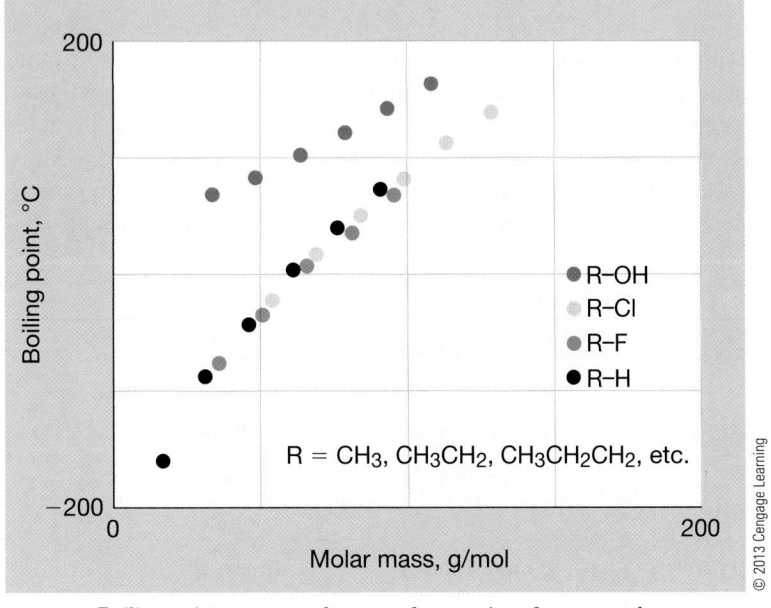

Boiling point versus molar mass for a series of compounds

Interactive Figure 11.5.3 also shows the influence of hydrogen bonding on boiling point. The alcohols, R—OH, have much higher boiling points than the other R—X compounds due to their ability to form hydrogen bonds. It is interesting to compare the effect of hydrogen bonds on boiling point to that of polarizability (increasing molar mass). If greater molar mass leads to stronger dispersion forces but hydrogen bonds have a greater influence on boiling, how do we compare hydrogen-bonding compounds to non–hydrogen-bonding compounds? Using the same data, we can determine the additional molar mass needed by an R—X (X = H, Cl, F) compound for it to have the same boiling point as an R—OH compound (Figure 11.5.4).

For small molecules, a nonpolar molecule needs to have a molar mass about 50 g/mol greater than a corresponding hydrogen-bonding molecule, which is equivalent to a three- to four-carbon-long alkyl group. For larger molecules the difference is smaller, and a two-carbon alkyl group can make up the difference.

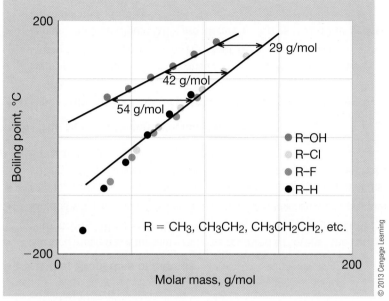

Figure 11.5.4 Comparison of IMF effect on boiling point for R—X compounds versus alcohols

Finally, it is important to understand that these quantitative relationships are not always simple. For example, H_2O and ICl have about the same boiling point (100 °C and 97 °C, respectively), but the molar mass difference between the two is 144 g (18.02 g/mol and 162.4 g/mol, respectively). However, for certain groups of compounds, such as the series of R—X compounds shown earlier, the quantitative relationship is more clear.

Section 11.5 Mastery

Unit Recap

Key Concepts

11.1 Kinetic Molecular Theory, States of Matter, and Phase Changes

- Liquids and solids are condensed phases (11.1a).
- The physical state of a substance is controlled by pressure, temperature, and intermolecular forces (IMFs) (11.1a).

- The strength of the IMFs in a substance affects many of its properties (11.1a).

- Converting an ordered phase to a less ordered phase requires the addition of energy to overcome the IMFs, and energy is released when a less ordered phase is converted to a more ordered phase (11.1b).

- The magnitude of a liquid's enthalpy of vaporization is directly related to the strength of the IMFs in the liquid (11.1c).

11.2 Vapor Pressure

- Equilibrium vapor pressure for a liquid is the pressure of the vapor over a liquid when a dynamic equilibrium is reached at a given temperature (11.2a).

- As temperature increases, the vapor pressure for a liquid increases (11.2b).

- As the strength of the IMFs in a series of liquids increases, the vapor pressures of the liquids decrease (11.2b).

- The temperature at which the vapor pressure for a liquid is equal to the external pressure is the boiling point of the liquid (11.2c).

- As the strength of the IMFs in a series of liquids increases, the boiling points of the liquids increase (11.2c).

- Vapor pressure and temperature are related mathematically by the Clausius–Clapeyron equation (11.2d).

11.3 Other Properties of Liquids

- Surface tension is a measure of force required to "break" the surface of a liquid. Liquids with strong IMFs generally have a strong surface tension (11.3a).

- Liquids with strong IMFs generally have high viscosity, or resistance to flow (11.3b).

- Capillary action occurs when the forces within a bulk sample of liquid are overcome by strong IMFs between the liquid and the surface of the material in contact with the liquid (11.3c).

11.4 The Nature of Intermolecular Forces

- Dipole–dipole IMFs are the attractive forces that occur between two polar molecules (11.4a).

- Hydrogen bonding is an unusually strong type of dipole–dipole IMF that occurs between molecules with H—N, H—O, or H—F bonds (11.4a).

- Dipole–induced dipole IMFs are attractive forces that occur between polar and nonpolar molecules (11.4b).

- Polarizability, which is directly related to molar mass and molecular size, is the ease with which the electron cloud in a molecule can be distorted and a dipole can be induced (11.4b).
- Induced dipole–induced dipole IMFs, also called London dispersion forces, are attractive forces that occur between nonpolar molecules (11.4c).

11.5 Intermolecular Forces and the Properties of Liquids

- As molar mass increases in a series of small nonpolar compounds, enthalpy of vaporization and boiling point generally increase (11.5a).
- As molar mass and the length of the carbon chain in a series of compounds with similar IMFs increases, enthalpy of vaporization and boiling point generally increase (11.5a).
- Hydrogen bonds have a significant impact on the physical properties of compounds (11.5b).
- It is possible to quantify the relationship between IMFs and the physical properties of liquids, but the relationships are not always simple (11.5c).

Key Equations

$$\ln P = \frac{-\Delta H_{vap}}{RT} + C \qquad \textbf{(11.1)}$$

$$\ln \frac{P_2}{P_1} = \frac{-\Delta H_{vap}}{R}\left(\frac{1}{T_2} - \frac{1}{T_1}\right) \qquad \textbf{(11.2)}$$

Key Terms

11.1 Kinetic Molecular Theory, States of Matter, and Phase Changes
condensed phase
intermolecular forces
fusion
vaporization
condensation
enthalpy of vaporization, ΔH_{vap}

11.2 Vapor Pressure
vapor pressure
dynamic equilibrium

equilibrium vapor pressure
boiling point
normal boiling point

11.3 Other Properties of Liquids
cohesive forces
surface tension
viscosity
capillary action
adhesive forces

11.4 The Nature of Intermolecular Forces
electrostatic forces
dipole–dipole intermolecular forces
hydrogen bonding
dipole–induced dipole intermolecular forces
induced dipole
polarizability
induced dipole–induced dipole intermolecular forces
London dispersion forces

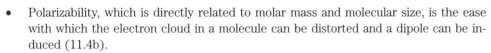

Unit 11   Review and Challenge Problems

12 The Solid State

Unit Outline

In This Unit...

In Intermolecular Forces and the Liquid State (Unit 11), we focused on intermolecular forces, the non-bonding interactions between collections of atoms and molecules, and how these forces manifest themselves in the physical properties of liquids. In this unit we continue this exploration with the study of structure and bonding in solids. We begin by looking at the structural features of some simple types of solids. Then we investigate bonding in different types of solids, and conclude by tying the three states of matter together in what is known as a phase diagram.

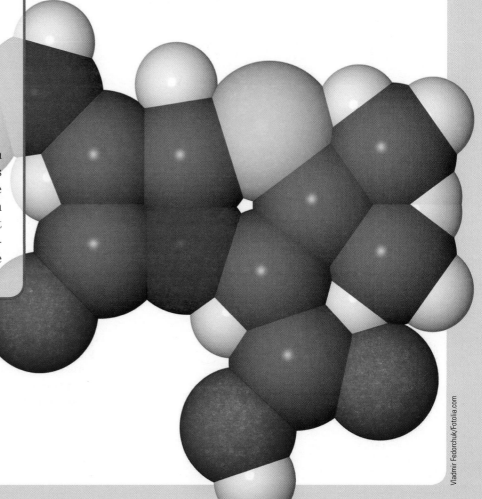

Vladmir Fedorchuk/Fotolia.com

12.1 Introduction to Solids

12.1a Types of Solids

Most solids are best described as either crystalline or amorphous. A **crystalline solid** is one in which the particles in the solid are arranged in a regular way. There is long-range order extending over the entire crystal, which can be described as repeating atomic- or molecular-level building blocks. The atomic-level order in a crystalline solid is often reflected in the well-defined faces of the crystal. Examples of pure substances that are crystalline solids at room temperature and pressure are diamond, table salt (NaCl), and sugar ($C_{12}H_{22}O_{11}$). In an **amorphous solid**, the particles that make up the solid are arranged in an irregular manner and the solid lacks long-range order. Many important solid materials, such as synthetic fibers, plastics, and glasses, are amorphous, but pure solid substances, such as elemental phosphorus or sulfur, may also exist in amorphous forms.

Because the lack of a well-defined repeating structure means that amorphous materials are more difficult to describe systematically, we restrict our discussion here to crystalline solids. Like liquids, solids are condensed phases in which the constituent particles are in contact and the properties are determined by the nature of the interactions holding the particles together. Solids can be broadly classified, based on these interactions, as molecular, ionic, covalent, or metallic (Interactive Table 12.1.1).

Molecular solids (also called **van der Waals solids**) consist of individual molecules held together by relatively weak intermolecular forces (IMFs), such as dipole–dipole IMFs, hydrogen bonds, dipole–induced dipole IMFs, and London dispersion forces. The strengths of the IMFs holding the molecules near one another in the solid (0.05 to 30 kJ/mol) are much weaker than the strengths of the intramolecular covalent bonds (200 to 600 kJ/mol) between atoms within the molecules that make up the solid. The weak, non-ionic IMFs between the molecules result in molecular solids that are generally low melting, do not conduct electricity, and have low hardness (resistance to deformation).

Ionic solids are composed of oppositely charged ions combined to produce a neutral solid. The forces holding the ions together are the coulombic forces between the oppositely charged ions. The attractive forces between ions in ionic solids (generally greater than 700 kJ/mol) are typically much stronger than those between the molecules in molecular solids, resulting in solids that have generally high melting points, conduct electricity when molten, and have high hardness.

Covalent solids (also called **network solids**) consist of a three-dimensional extended network of atoms held together by covalent bonds. For example, in a diamond

Crystalline Solids

	Molecular	Ionic	Covalent	Metallic
Constituent particles	Molecules	Ions	Covalent network	Metal atoms
Melting point	Moderate to low	High to very high	Very high	Variable
Hardness	Soft to brittle	Hard and brittle	Very hard	Variable, malleable
Conductivity	Nonconducting	Nonconducting solid, conducting liquid	Usually nonconducting	Conducting
Attractive forces	Dipole–dipole, hydrogen bonds, London dispersion forces	Ion–ion	Covalent bonds	Metallic bonds
Schematic diagram	© 2013 Cengage Learning	© 2013 Cengage Learning	© 2013 Cengage Learning	© 2013 Cengage Learning
Examples	Carbon dioxide (CO_2), m.p. $-78\ °C$ (sublimes) Sucrose ($C_{12}H_{22}O_{11}$), m.p. 186 °C (decomposes) Water (H_2O), m.p. 0 °C	NaCl, m.p. 801 °C K_2SO_4, m.p. 1689 °C MgO, m.p. 2852 °C	Diamond (C), m.p. 3550 °C SiO_2, m.p. 1650 °C	Na, m.p. 98 °C W, m.p. 3422 °C

crystal, each carbon atom is covalently bonded to its nearest neighbors and the entire crystal can be viewed as one extremely large molecule. The relatively strong covalent bonds between atoms in covalent solids result in generally high-melting solids that have high hardness but typically low or no conductivity. (One notable exception is graphite, which is a network solid that is a good conductor of electricity.)

Metallic solids consist of positively charged metal atom cores held together by attractions to their valence electrons, which are delocalized over the entire crystal. The details of the bonding in metals will be discussed later in this unit.

12.1b The Unit Cell

Crystalline solids consist of a repeating three-dimensional array of particles (atoms, ions, or molecules). The repeating pattern has **translational symmetry**; that is, when it is translated (moved without being rotated) in certain directions, an identical pattern can be seen. In crystalline solids we define **lattice points** as the location of identical points in the repeating pattern. A collection of lattice points is a **crystal lattice**, and the smallest repeating unit of the crystal lattice is the **unit cell**, the building block of a crystal.

Consider the simple two-dimensional representation of a solid shown in Interactive Figure 12.1.1. In Interactive Figure 12.1.1a, the repeating unit is a single foot. Only 12 feet are shown in the array, but imagine that the pattern repeats to cover a plane that contains Avogadro's number of feet. In order to represent the translational symmetry of this array, we chose a lattice point in the repeating unit (the red dot in Interactive Figure 12.1.1b). Notice that the lattice point is in the same place relative to each foot. Connecting the lattice points (Interactive Figure 12.1.1c) forms unit cells. Interactive Figure 12.1.1d shows the unit cells without the repeating unit, and a different choice of the unit cell is outlined in black.

Derive a unit cell from a repeating array.

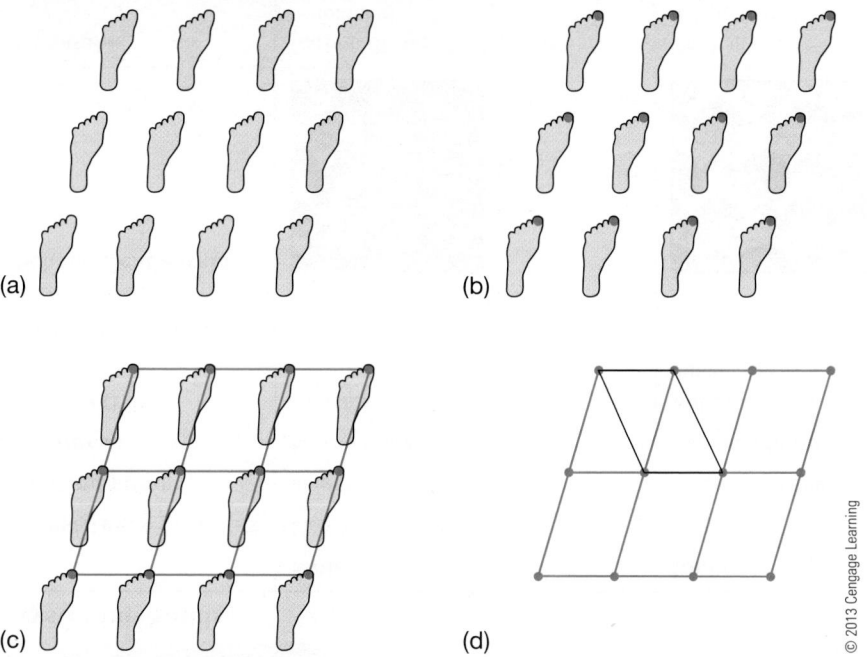

(a)   (b)

(c)   (d)

© 2013 Cengage Learning

(a) A repeating pattern; (b) lattice points; (c) and (d) unit cells

There are some important points to notice in the two-dimensional crystal lattice shown in Interactive Figure 12.1.1.

- Because each lattice point is shared by four cells and each unit cell has a lattice point in each of its four corners, there is one lattice point per unit cell ($4 \times \frac{1}{4} = 1$).

- There is one foot per lattice point and one lattice point per unit cell, so there is one foot per unit cell. (Although to get an entire foot, the pieces of the four feet that are contained in a single unit cell must be combined.)

- Repeatedly moving a single unit cell in any direction a distance equal to its own dimensions could recreate the entire crystal lattice.

- The lattice points did not have to be placed on the toe of the feet. If the lattice points were placed in spaces between the feet, the crystal lattice and unit cell would be the same but each unit cell could contain one whole foot.

There are only 7 crystal systems in three-dimensional crystalline solids, and within these crystal systems, there are only 14 observed repeat patterns for lattice points (called the **Bravais lattices**). With all of the complexity found in crystalline solids, it is amazing that only 14 repeat patterns for lattice points are found in nature! Crystal systems are defined by the relationships between the edge lengths of the unit cell (a, b, and c) and the angles between the unit cell edges (α, β, and γ). The least symmetrical crystal system is called triclinic; in a triclinic unit cell, no two edges have the same length ($a \neq b \neq c$) and all angles between unit cell edges are not equal to one another and are also not equal to 90°. The most symmetrical crystal system is defined by a cubic unit cell, where all cell edges are identical and all angles between unit cell edges are equal to 90° (Figure 12.1.2).

In this unit we study solids that have cubic unit cells, and it is useful to visualize what happens when identical cubic unit cells pack together. Consider a perfect cubic unit cell, which has 12 edges, 6 faces, and 8 corners. As shown in Interactive Figure 12.1.3, when identical cubes are packed together in three dimensions,

- each cube face is shared by two adjacent unit cells;

- each cube edge is shared by four adjacent unit cells; and

- each cube corner is shared by eight adjacent unit cells.

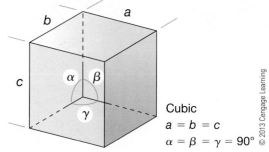

Cubic
$a = b = c$
$\alpha = \beta = \gamma = 90°$

© 2013 Cengage Learning

Figure 12.1.2 Cubic unit cell

Interactive Figure 12.1.3

Explore cubic unit cells in a three-dimensional crystal lattice.

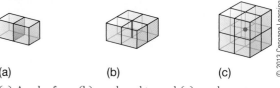

(a)　　　　(b)　　　　(c)

© 2013 Cengage Learning

(a) A cube face, (b) a cube edge, and (c) a cube corner are shared by multiple unit cells.

Section 12.1 Mastery

12.2 Metallic Solids

12.2a Simple Cubic Unit Cell

Metallic solids such as sodium and copper consist of an extended three-dimensional array of metal atoms, and most metals have crystal lattices with cubic unit cells. Unit cells for these metallic solids commonly have one metal atom per lattice point; therefore, it is possible to use the lattice point positions in the unit cell to describe arrangements of the atoms in the metallic crystal.

The **simple cubic** (SC) **unit cell**, also called a **primitive cubic** (PC) **unit cell**, consists of one lattice point on each corner of a cube (Interactive Figure 12.2.1). In the simple cubic lattice, each of the eight corners of the unit cell contains a lattice point (a metal atom) that is shared by the eight unit cells that meet at each corner (only $\frac{1}{8}$ is contained within the unit cell). Therefore, there is one lattice point, or one metal atom, per unit cell.

$$8 \text{ corner atoms} \times \frac{1}{8} = 1 \text{ atom/unit cell}$$

The **coordination number** of an atom in a crystal lattice is defined as the number of nearest neighbors an atom has in the extended lattice. Each atom in the SC unit cell has six nearest neighbors and thus a coordination number of six. Notice that the atoms in an SC unit cell are in contact along the unit cell edge; therefore, the edge length of an SC unit cell is twice the atomic radius of the metal.

Another way to describe a unit cell is the percentage of occupied space in the unit cell.

$$\% \text{ occupied space} = \frac{\text{total volume of atoms in unit cell}}{\text{total volume of unit cell}} \times 100$$

The volume of a cube is calculated by cubing the edge length ($V = a^3$), and the volume of an atom is calculated using the equation for the volume of a sphere, $V = \frac{4}{3}\pi r^3$. In an SC unit cell, the unit cell edge length is equal to twice the sphere radius ($a = 2r$) and there is one atom per unit cell. The percentage of occupied space in an SC unit cell is therefore

$$\% \text{ occupied space} = \frac{\text{volume of 1 atom}}{\text{total volume of unit cell}} \times 100 = \frac{\frac{4}{3}\pi r^3}{(2r)^3} \times 100 = 52.4\%$$

This is a relatively low value, which suggests that only about half of the space available in the solid is occupied when a metal crystallizes in an SC crystal lattice. The low packing efficiency of the SC lattice means that this structure type is rarely found in any metals (polonium is one reported example).

Interactive Figure 12.2.1

Explore a simple cubic unit cell.

Unit cell   Cube face

Representations of a simple cubic unit cell

© 2013 Cengage Learning

12.2b Body-Centered Cubic Structure

The **body-centered cubic** (BCC) **unit cell** consists of one lattice point on each corner of a cube and a single lattice point in the center of the unit cell (Interactive Figure 12.2.2). In the BCC lattice, each of the eight corners of the unit cell is shared by the eight unit cells that meet at each corner (only ⅛ of each lattice point is contained within the unit cell), and the center lattice point is completely within the unit cell. Therefore, there are two lattice points, or two metal atoms, per unit cell.

$$\left(8 \text{ corner atoms} \times \frac{1}{8}\right) + 1 \text{ center atom} = 2 \text{ atoms/unit cell}$$

Each atom in the BCC unit cell has eight nearest neighbors and thus a coordination number of eight. As shown in Interactive Figure 12.2.2, the atoms in a BCC unit cell are not in contact along the unit cell edge; instead, the atoms make contact along the diagonal that connects opposite corners of the cube, called the body diagonal. The body diagonal passes through the diameter of the atom in the center of the cell and halfway through each of the corner atoms, so the entire body diagonal has a length of $4r$.

We can relate atomic radius, r, to the edge length, a, of a BCC unit cell using the Pythagorean theorem, as shown in Figure 12.2.3. The right triangle outlined in red in Figure 12.2.3 is composed of the body diagonal ($4r$), an edge (a), and a face diagonal. The face diagonal of a perfect cube with edge length a is determined using the Pythagorean theorem:

$$(\text{face diagonal})^2 = (a)^2 + (a)^2 = 2a^2$$

$$\text{face diagonal} = a\sqrt{2}$$

Substituting this into the equation for the right triangle that includes the body diagonal,

$$(\text{body diagonal})^2 = (a)^2 + (\text{face diagonal})^2 = (a)^2 + (a\sqrt{2})^2 = 3a^2$$

$$a = \frac{\text{body diagonal}}{\sqrt{3}}$$

Because the body diagonal is equal to $4r$,

$$a = \frac{4r}{\sqrt{3}} \qquad\qquad \textbf{(12.1)}$$

The percentage of occupied space in the BCC unit cell can now be calculated:

$$\% \text{ occupied space} = \frac{\text{volume of 2 atoms}}{\text{total volume of unit cell}} \times 100 = \frac{2 \times \left(\frac{4}{3}\pi r^3\right)}{\left(\frac{4r}{\sqrt{3}}\right)^3} \times 100 = 68.0\%$$

Explore a body-centered cubic unit cell.

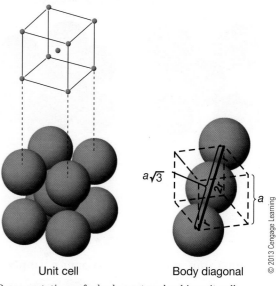

Unit cell   Body diagonal

Representations of a body-centered cubic unit cell

© 2013 Cengage Learning

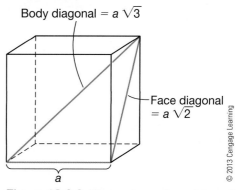

Body diagonal = $a\sqrt{3}$

Face diagonal = $a\sqrt{2}$

Figure 12.2.3 Dimensions of a cubic unit cell

© 2013 Cengage Learning

The percentage of occupied space in a BCC unit cell is higher than that in an SC unit cell. In addition, the coordination number is also higher, suggesting that atoms pack more efficiently in a BCC lattice than in an SC lattice. Many metals adopt the BCC structure, including barium, chromium, iron, lithium, manganese, molybdenum, tungsten, and vanadium.

Example Problem 12.2.1 Relate edge length and atomic radius for a body-centered cubic unit cell.

Metallic uranium crystallizes in a BCC lattice, with one U atom per lattice point. How many atoms are there per unit cell? If the edge length of the unit cell is found to be 343 pm, what is the metallic radius of U in pm?

Solution:

You are asked to determine the number of atoms in a unit cell and the metallic radius of an atom in a unit cell.

You are given the type of unit cell and the edge length of the unit cell.

A BCC lattice has an atom on each corner (each corner shared by eight unit cells) and one atom completely within the unit cell.

$$(8 \text{ atoms on corner}) \frac{1}{8} + 1 \text{ atom in center} = 2 \text{ atoms per unit cell}$$

In this type of lattice the atoms are in contact along the body diagonal.

$$a = \frac{4r}{\sqrt{3}}, \text{ where } r \text{ is the metallic radius and } a \text{ is the edge length of the unit cell.}$$

$$r = \frac{a\sqrt{3}}{4} = \frac{(343 \text{ pm})\sqrt{3}}{4} = 149 \text{ pm}$$

12.2c Closest-Packed Structure

The third cubic unit cell is a **face-centered cubic** (FCC) **unit cell**, in which there are lattice points (metal atoms) on each corner of the cube and a lattice point centered on each cube face (Interactive Figure 12.2.4). In the FCC lattice, each of the eight corners of the unit cell is shared by the eight unit cells that meet at each corner (only ⅛ is contained within the unit cell), and the lattice points on the unit cell faces are shared by two unit cells (only ½ is contained within the unit cell). Therefore, there are four lattice points, or four metal atoms, per unit cell.

$$\left(8 \text{ corner atoms} \times \frac{1}{8}\right) + \left(6 \text{ face atoms} \times \frac{1}{2}\right) = 4 \text{ atoms/unit cell}$$

Video Solution

Tutored Practice
Problem 12.2.1

Interactive Figure 12.2.4

Explore a face-centered cubic unit cell.

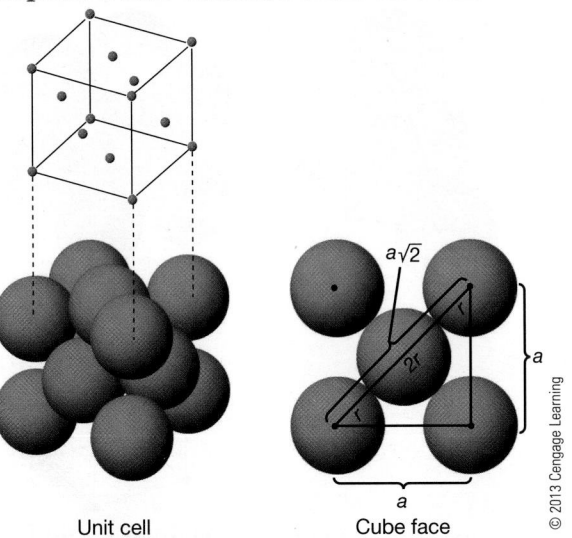

Unit cell   Cube face

© 2013 Cengage Learning

Representations of a face-centered cubic unit cell

The FCC unit cell is the result of a very efficient way of packing spheres called a **closest-packed structure**. In a closest-packed structure, spheres are arranged in layers, and in each layer there are six spheres surrounding every sphere (Interactive Figure 12.2.5). When layers of closest-packed spheres are stacked, there are two possible arrangements of the layers. In the **hexagonal closest-packed** (HCP) **structure**, the spheres in the third layer are aligned directly above the spheres in the first layer (Interactive Figure 12.2.5b). We indicate this type of arrangement as ABABAB…, where A and B represent the different layers in the structure. If the spheres in the third layer are not directly above the spheres in layer A or B, a second kind of structure results, called a **cubic closest-packed** (CCP) **structure**, which is described as an ABCABC… layered packing (Interactive Figure 12.2.5c).

The cubic closest-packed structure has an FCC unit cell, whereas the hexagonal closest-packed structure has a unit cell that is not cubic (it is hexagonal). Figure 12.2.6 shows the ABCABC… packing structure along the body diagonal of the FCC unit cell. In both closest-packed structures, each sphere has a coordination number of 12 (six nearest neighbors in the same layer and three nearest neighbors in the layers above and below) and identical packing efficiency as calculated by percent occupied space.

Using the cubic closest-packed (FCC) unit cell, you saw in Interactive Figure 12.2.4 that the spheres are in contact along a unit cell face, which is equal to $4r$. An FCC unit cell contains four atoms, and as we saw in the previous section, the face diagonal in a perfect cube is equal to (8 atoms on corner) $\frac{1}{8}$ + 1 atom in center = 2 atoms per unit cell.

Interactive Figure 12.2.5

Explore closest-packed structures.

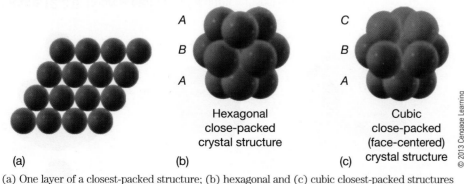

A
B
A

Hexagonal
close-packed
crystal structure

C
B
A

Cubic
close-packed
(face-centered)
crystal structure

© 2013 Cengage Learning

(a)   (b)   (c)

(a) One layer of a closest-packed structure; (b) hexagonal and (c) cubic closest-packed structures

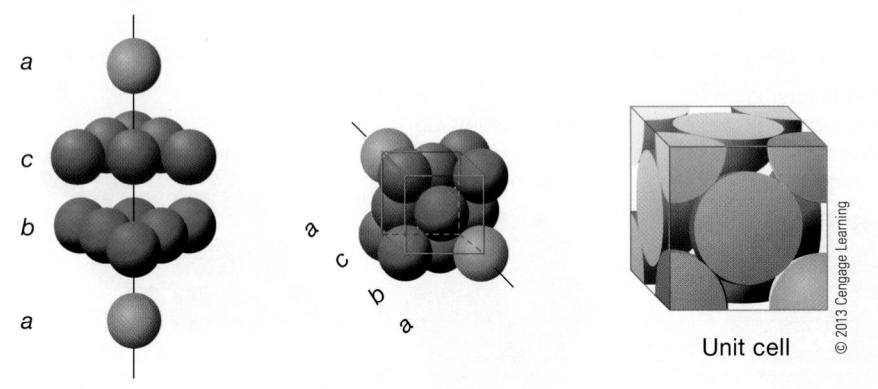

Figure 12.2.6 Representations of a cubic closest-packed unit cell

$$(\text{face diagonal})^2 = (a)^2 + (a)^2 = 2a^2$$

$$\text{face diagonal} = 4r = a\sqrt{2}$$

$$a = \frac{4r}{\sqrt{2}} \tag{12.2}$$

The percentage of occupied space in a closest-packed structure is

$$\% \text{ occupied space} = \frac{\text{volume of 4 atoms}}{\text{total volume of unit cell}} \times 100 = \frac{4 \times \left(\dfrac{4}{3}\pi r^3\right)}{\left(\dfrac{4r}{\sqrt{2}}\right)^3} \times 100 = 74.0\%$$

The closest-packed structures have the highest packing efficiency as measured by both percentage of occupied space and coordination number. It should come as no surprise that most metals crystallize in a closest-packed structure, either CCP (FCC) or HCP. Some of the metals that have FCC lattices are aluminum, calcium, copper, gold, lead, nickel, silver and strontium.

Table 12.2.1 summarizes the properties of the three cubic unit cells, where r is atomic radius and a is unit cell edge length. Keep in mind that these relationships are true only for cubic unit cells with one atom per lattice point.

Table 12.2.1 Summary of Cubic Unit Cells

Lattice	Atoms per Unit Cell	Edge Length–Atomic Radius Relationship	Coordination Number	% Occupied Space
Simple cubic	1	$a = 2r$	6	52.4
Body-centered cubic	2	$a = \dfrac{4r}{\sqrt{3}}$	8	68.0
Face-centered cubic	4	$a = \dfrac{4r}{\sqrt{2}}$	12	74.0

Example Problem 12.2.2 Relate unit cell dimensions, density, and number of atoms per unit cell for a cubic unit cell.

The element copper is found to crystallize in a cubic lattice, with an edge length of 361.5 pm. If the density of solid copper is 8.935 g/cm^3, how many Cu atoms are there per unit cell?

Solution:

You are asked to calculate the number of atoms in one unit cell of a specific element.

You are given the identity of the element, the type of unit cell, the edge length of the unit cell, and the density of the solid.

Step 1. Use the edge length to calculate the volume of the cell in units of cm^3 to match the units used for density.

$$361.5 \text{ pm} \times \frac{10^2 \text{ cm}}{10^{12} \text{ pm}} = 3.615 \times 10^{-8} \text{ cm}$$

$$\text{volume} = (3.615 \times 10^{-8} \text{ cm})^3 = 4.724 \times 10^{-23} \text{ cm}^3$$

Step 2. Use the calculated volume and the density of the solid to calculate the mass of one unit cell.

$$(4.724 \times 10^{-23} \text{ cm}^3)\left(\frac{8.935 \text{ g}}{\text{cm}^3}\right) = 4.221 \times 10^{-22} \text{ g}$$

Step 3. Calculate the mass of a single copper atom from its molar mass and Avogadro's number.

$$\frac{63.55 \text{ g}}{1 \text{ mol Cu}} \times \frac{1 \text{ mol Cu}}{6.022 \times 10^{23} \text{ Cu atoms}} = 1.055 \times 10^{-22} \text{ g/Cu atom}$$

Example Problem 12.2.2 *(continued)*

Step 4. Use the mass of one unit cell and the mass of a single copper atom to calculate the number of atoms per unit cell.

$$\left(\frac{4.221 \times 10^{-22} \text{ g}}{1 \text{ unit cell}}\right)\left(\frac{1 \text{ Cu atom}}{1.055 \times 10^{-22} \text{ g}}\right) = 4 \text{ Cu atoms/unit cell}$$

Video Solution

Tutored Practice Problem 12.2.2

12.2d X-ray Diffraction

Much of our knowledge of the structures of molecules and materials comes from the x-ray diffraction analysis of crystalline solids. In the single-crystal experiment, a beam of monochromatic (single wavelength) x-rays strikes a small single crystal and the positions and the intensities of hundreds to thousands of diffracted beams are measured (Figure 12.2.7). Because it is the electrons in the atoms that scatter x-rays, the experiment gives information about the electron density in the crystal, which in turn shows how the atoms are arranged in the unit cell. An alternative experiment is neutron diffraction, where the neutron beam is scattered by the nuclei of the atoms. X-rays are more commonly used because

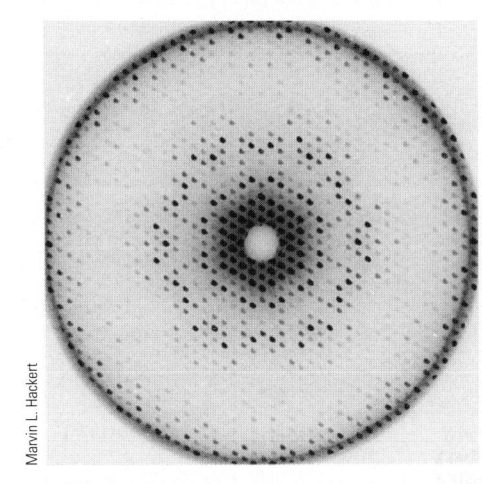

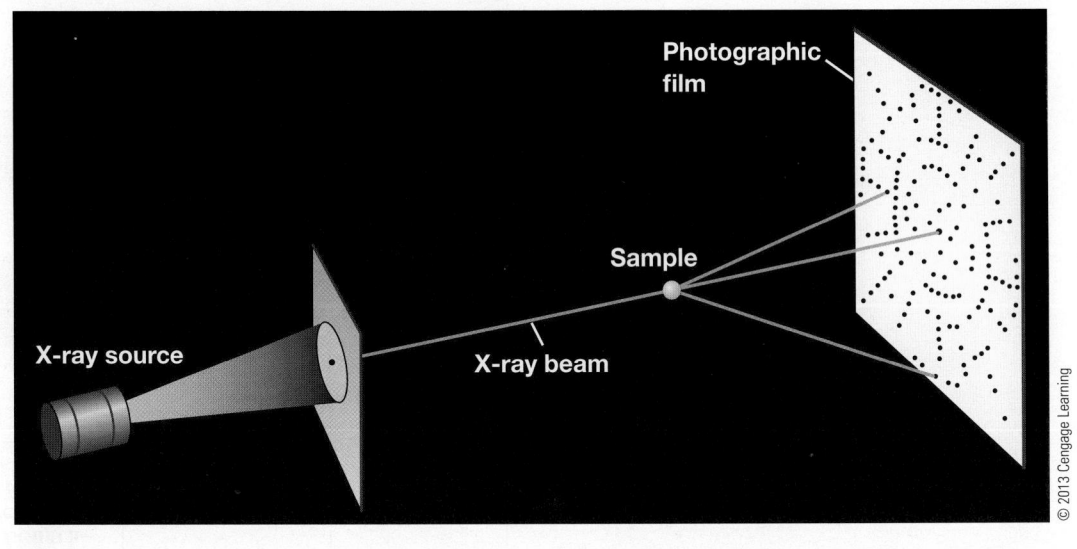

(a)   (b)

Figure 12.2.7 (a) X-ray photograph of a crystalline solid; (b) a diagram of the X-ray diffraction experiment

they are easily generated in sealed tubes in the laboratory by bombarding a metal target with electrons, whereas the production of a neutron beam requires a nuclear reactor.

The size and shape of the unit cell can be determined from the positions of the diffracted beams, whereas the intensities give information about what kinds of atoms are in the unit cell and where they are located. Because x-rays are electromagnetic radiation with wavelengths comparable to atomic dimensions, diffraction can be described in terms of the reflection of the x-ray beam off of sets of parallel planes in the crystal. These planes are drawn through lattice points and therefore reflect the periodicity of the scattering motifs. A diffracted beam is to be expected whenever the scattered radiation from the array of identical motifs is in phase. The condition for observing a reflection by diffraction from a crystal is known as **Bragg's law** and can be stated as:

$$n\lambda = 2d \sin \theta \qquad \textbf{(12.3)}$$

where λ is the wavelength of the x-radiation, n is an integer called the order of the reflection, d is the spacing between the set of planes for the reflection, and θ is the angle that the incident x-ray beam makes with the planes (Interactive Figure 12.2.8). A reflection with $n = 1$ is called a first-order reflection; $n = 2$, a second-order reflection; and so forth.

Because the planes are drawn through the lattice points, they reflect the size and shape of the unit cell. Therefore, the scattering angle from an appropriate set of planes can be used to determine the unit cell edge length, and this in turn can be used to determine the radius of an atom.

Interactive Figure 12.2.8

Explore Bragg's law.

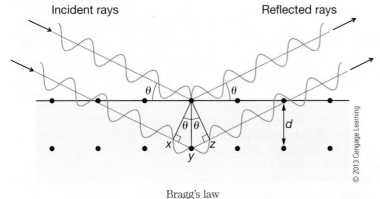

Bragg's law

© 2013 Cengage Learning

Example Problem 12.2.3 Determine d-spacing and radius from a scattering angle.

Silver metal crystallizes in an FCC lattice. Monochromatic x-radiation from a copper target has a wavelength of 154 pm. If this radiation is used in a diffraction experiment with a silver crystal, a second-order diffracted beam is observed at a theta value of 22.18°. If the spacing between these planes corresponds to the unit cell length ($d = a$), what is the d-spacing between the planes that gave rise to this reflection? What is the metallic radius of a silver atom?

Solution:

You are asked to calculate the d-spacing between the planes in a solid lattice and the metallic radius of an atom of a specific element.

You are given the x-ray wavelength, the theta value, and the type of crystal lattice.

Step 1. Calculate the d-spacing using Bragg's law.

$$d = \frac{n\lambda}{2 \sin \theta} = \frac{(2)(154 \text{ pm})}{2 \sin(22.18°)} = 408 \text{ pm}$$

Step 2. Determine the metallic radius of a silver atom.

The d-spacing corresponds to the edge length of the cubic cell. Use the relationship between unit cell edge length and atomic radius for an FCC unit cell to calculate the atomic radius of silver.

$$a = \frac{4r}{\sqrt{2}}$$

$$r = \frac{a\sqrt{2}}{4} = \frac{(408 \text{ pm})\sqrt{2}}{4} = 144 \text{ pm}$$

Video Solution

Tutored Practice
Problem 12.2.3

Section 12.2 Mastery

12.3 Ionic Solids

12.3a Holes in Cubic Unit Cells

The structure of many ionic solids can be visualized as a lattice of anions, typically the larger ion in the solid, with cations, the smaller ions, filling the spaces between the anions. The positions occupied by cations are called the holes in the anionic lattice. The anionic lattice can often be described as either SC or FCC, so we describe the holes within these two cubic lattices.

An SC lattice has a single hole in the center of the unit cell, called a **cubic hole** (Figure 12.3.1). A cation occupying this hole is contained completely within the unit cell (1 cubic hole per unit cell) and has eight nearest anionic neighbors (coordination number = 8).

The FCC unit cell has two different types of holes (Interactive Figure 12.3.2). Within each layer of closest-packed atoms in an FCC (cubic closest-packed) structure, there are triangular depressions formed when three atoms meet. If the next layer results in an atom being placed directly above that triangular depression, a **tetrahedral hole** forms. If the next layer has atoms offset, not sitting above the triangular depression, an **octahedral hole** forms. As you can see in Interactive Figure 12.3.2, an ion occupying a tetrahedral hole has four nearest neighbors (coordination number = 4) and an ion in an octahedral hole has a coordination number of 6.

Tetrahedral and octahedral holes are located in unique positions in the FCC unit cell (Figure 12.3.3). There are eight tetrahedral holes contained completely within the FCC unit cell, each surrounded by four ions. There are 13 octahedral holes in an FCC unit cell in two different locations. One octahedral hole sits at the center of the FCC unit cell and is contained completely within the unit cell. Twelve octahedral holes sit at the center of each edge of the FCC unit cell. Recall that in cubic unit cells, edges are shared by four unit cells, so only ¼ of each ion in an octahedral hole on a cube edge is within the unit cell. Therefore, there are four octahedral holes in an FCC unit cell [(1 hole in center) + (12 holes on cell edges) × ¼) = 4 octahedral holes per unit cell].

Finally, a fourth type of hole is found in anionic lattices. A **trigonal hole** is formed in a closest-packed structure, and it lies at the center of three atoms within a closest-packed layer (coordination number = 3). Experimentally it is very unusual for ions to occupy this type of hole due to its small size, so we do not include it in our discussion of ionic solids. Table 12.3.1 summarizes the most important types of holes found in cubic lattices.

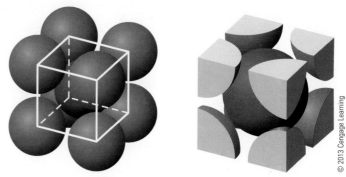

© 2013 Cengage Learning

Figure 12.3.1 The cubic hole in a simple cubic lattice

Interactive Figure 12.3.2

Explore holes in a closest-packed lattice.

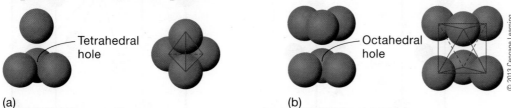

(a)

(b)

© 2013 Cengage Learning

(a) Tetrahedral and (b) octahedral holes in a closest-packed lattice

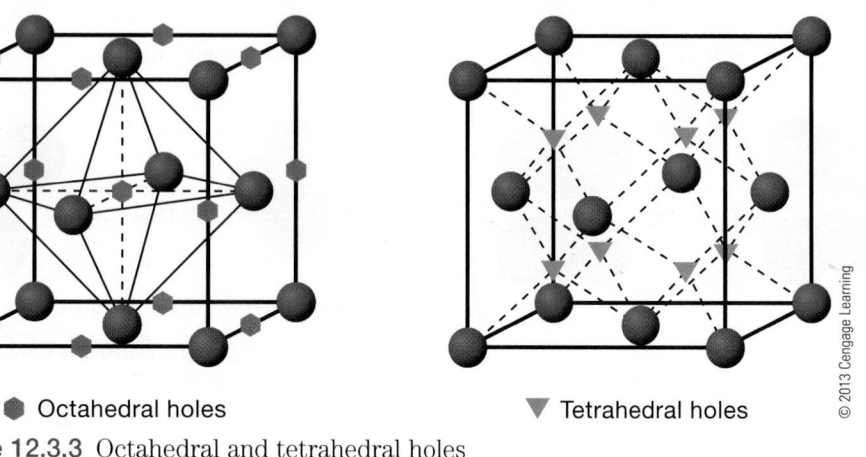

© 2013 Cengage Learning

● Octahedral holes   ▼ Tetrahedral holes

Figure 12.3.3 Octahedral and tetrahedral holes

One thing you will observe as we discuss different structures of ionic solids is that when an ion occupies a hole, it pushes apart the ions that define the hole. For example, when a cation occupies a cubic hole, the anions no longer are in contact along the unit cell edge. The cation has expanded the anionic lattice. Consider the theoretical anion/cation arrangements in Interactive Figure 12.3.4, where cations (red or green) and anions (gray) are arranged in a two-dimensional lattice.

We would expect that, to minimize energy, ions of opposite charge would be in contact as much as possible. At the same time, ions of the same charge, which repel one another, would not be expected to be in close contact. Because holes have different sizes and different coordination numbers (a tetrahedral hole is smaller than an octahedral hole, which in

Table 12.3.1 Important Holes in Cubic Unit Cells

Type of Hole	Holes per Unit Cell	Coordination Number	Location
Tetrahedral	8	4	Contained completely within FCC unit cell of ions
Octahedral	4	6	1 in center and 12 on edges of FCC unit cell of ions
Cubic	1	8	Center of SC unit cell of ions

Interactive Figure 12.3.4

Investigate cation/anion arrangements in ionic solids.

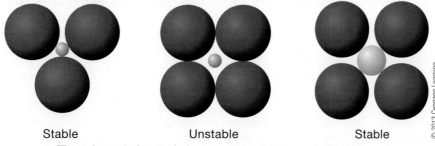

Stable   Unstable   Stable

© 2013 Cengage Learning

Three theoretical cation/anion arrangements in a crystalline lattice

turn is smaller than a cubic hole), relative size can dictate the type of holes that are filled in a lattice; this can, in turn, dictate the type of lattice formed.

An unfavorable situation is depicted in the middle diagram of Interactive Figure 12.3.4. Here the cation (red) is in a hole formed by the anions (gray), but the hole is much larger than the cation. Attractive forces are not maximized, and repulsive forces (between anions) are maximized. A better situation results from placing the small cation into a smaller hole, as shown by the diagram on the left. Here, attractive forces between the cation and anions are maximized, and the repulsive forces are minimized because the anions are no longer in contact with one another. A similar situation is shown on the right, where a large cation is placed in a small hole, which pushes the anions apart.

Regardless of the packing arrangement, ions of opposite charge are assumed to be in contact in ionic solids, and the distance between their centers is taken to be the sum of their ionic radii. Many ionic radii have been obtained from the distances between ion centers obtained from crystallographic data. Because the distance gives a sum rather than an individual ionic radius, the radius of one ion must be known to determine the other value. By looking at values from many different structures, sets of self-consistent experimental ionic radii can be determined.

Example Problem 12.3.1 Analyze an ionic cubic unit cell.

Solid strontium chloride has a crystal structure with the cubic unit cell pictured here, where the Cl^- ions are represented as yellow spheres. Describe the positions of the anions in the lattice. What is the coordination number of the strontium ions in the crystal? How many $SrCl_2$ formula units are there per unit cell?

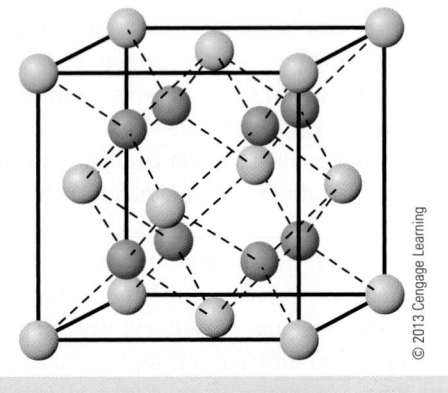

© 2013 Cengage Learning

Solution:

You are asked to describe the position of anions in a crystal lattice, to identify the coordination number of the cations in the lattice, and to determine the number of formula units per unit cell.

You are given a picture of the unit cell for the solid.

The lattice can be viewed as a cubic close-packed array of strontium ions with chloride ions in tetrahedral holes. This gives four strontium ions and eight chloride ions per unit cell. The number of formula units per unit cell is therefore four.

To visualize the coordination number of the strontium ions, focus on the strontium ion on the top face of the cell. This ion has four chloride ions in the unit cell directly beneath it as nearest neighbors. But because of the translational symmetry, the strontium ion on the bottom face of the cell is identical to the one on the top face of the cell. This strontium ion is closest to the four chloride ions directly above it in the cell. Therefore, in the extended structure, each strontium ion lies at the center of a cube of chloride ions and has a coordination number of 8.

Video Solution

Tutored Practice
Problem 12.3.1

12.3b Cesium Chloride and Sodium Chloride Structures

Both cesium chloride and sodium chloride have a 1:1 ratio of cations to anions. However, the unit cell arrangement of ions in these two solids is very different.

CsCl Structure

One of the simplest structural types for ionic compounds with a 1:1 ratio of cation to anions is the cesium chloride structure. In the **cesium chloride structure**, an SC arrangement of cesium ions (the larger ion) surrounds a cubic hole occupied by a chloride ion (the smaller ion). The cation:anion ratio in this structure is 1:1.

SC arrangement of Cs^+: (8 Cs^+ ions on unit cell corners) $\left(\dfrac{1}{8}\right)$ = 1 Cs^+ per unit cell

Cl^- in cubic hole: 1 Cl^- per unit cell

As shown in Interactive Figure 12.3.5, the CsCl unit cell can appear to be a BCC unit cell. However, notice that the ion in the center of the unit cell is not the same as the ions at the corners of the unit cell. Because ionic compounds have expanded lattices, the edge length of a CsCl unit cell is not equal to twice the Cs^+ ionic radius. In the CsCl unit cell, the ions are in contact along the body diagonal of the cube. Using the mathematical relationships developed for metallic unit cells,

$$\text{body diagonal} = (2 \times Cs^+ \text{ radius}) + (2 \times Cl^- \text{ radius}) = a\sqrt{3}$$

The cesium chloride structure is adopted by other ionic compounds such as CsBr and TlCl. The structure is found when the cation and anion have similar radii. Formally, the larger ion defines the SC arrangement of ions and the smaller ion occupies the cubic hole. But as you can see in Interactive Figure 12.3.5, either ion can be used to define the lattice or fill the cubic holes.

NaCl Structure

Sodium chloride has a crystal lattice that is adopted by many other ionic compounds. In the **sodium chloride structure**, the larger ions (Cl^-) adopt an FCC (CCP) arrangement and the smaller ions (Na^+) occupy all of the octahedral holes. The cation:anion ratio in this structure is 1:1.

FCC arrangement of Cl^-:

$$(8\ Cl^- \text{ ions on unit cell corners}) \left(\dfrac{1}{8}\right) + (6\ Cl^- \text{ ions on unit cell faces}) \left(\dfrac{1}{2}\right)$$

$$= 4\ Cl^- \text{ per unit cell}$$

Na^+ in all octahedral holes: (1 Na^+ in center of unit cell) + (12 Na^+ on unit cell edges) (¼)
                $= 4\ Na^+$ per unit cell

In Interactive Figure 12.3.6, sodium ions are represented by white spheres and chloride ions, by yellow spheres. Notice that although there are four NaCl formula units per unit cell, the cation:anion ratio is 1:1, as required by the ionic charges in the compound.

Explore the CsCl structure.

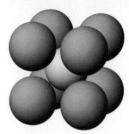

Cl^- lattice and Cs^+ in lattice hole

Cs^+ lattice and Cl^- in lattice hole

CsCl unit cell

© 2013 Cengage Learning

Both the cations and anions have a coordination number of 6, as each is surrounded by an octahedron of counterions. In the NaCl structure, the anions do not touch along the unit cell face (as they would in a metallic solid with an FCC unit cell). Instead, the ions touch along the edge of the unit cell.

$$\text{edge length } (a) = (2 \times Cl^- \text{ radius}) + (2 \times Na^+ \text{ radius})$$

The NaCl structure is very common in ionic compounds. All of the alkali metal halides have this structure, and other examples of solids with this structure include AgCl, CaS, KOH, and MgO.

Interactive Figure 12.3.6

Explore the NaCl structure.

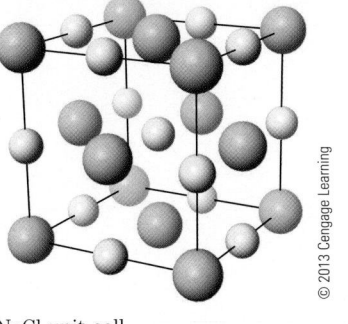

NaCl unit cell

Example Problem 12.3.2 Determine an ionic radius from unit cell dimensions.

Magnesium oxide crystallizes with the sodium chloride structure. If the edge length of the unit cell is determined to be 420 pm and the O^{2-} ion is assigned a radius of 126 pm, what is the radius of the Mg^{2+} ion?

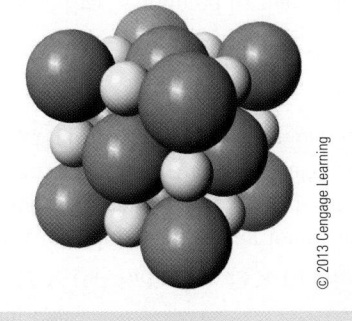

Solution:

You are asked to calculate the radius of a cation in an ionic solid.

You are given the crystal structure of the solid, the edge length of the unit cell, and the ionic radius of the anion.

Step 1. Calculate the sum of the ionic radii from the fact that the edge length of the unit cell is twice the sum of the radii. The sum of the radii is the distance between the ion centers, which is a distance that can be determined from an x-ray diffraction experiment.

$$a = 2(r_{Mg^{2+}} + r_{O^{2-}}) = 420 \text{ pm}$$

$$r_{Mg^{2+}} + r_{O^{2-}} = 210 \text{ pm}$$

Step 2. Use the sum of the ionic radii to calculate the radius of the Mg^{2+} ion, taking the radius of the O^{2-} ion to be 126 pm and assuming that the ions are in contact.

$$r_{Mg^{2+}} = 210 \text{ pm} - r_{O^{2-}} = 210 \text{ pm} - 126 \text{ pm} = 84 \text{ pm}$$

Video Solution

Tutored Practice
Problem 12.3.2

12.3c Zinc Blende (ZnS) Structure

Zinc sulfide, another ionic compound with a 1:1 cation:anion ratio, adopts more than one crystal structure. In the **zinc blende structure**, the larger ions (S^{2-}) adopt an FCC arrangement and the smaller ions (Zn^{2+}) occupy tetrahedral holes in the closest-packed lattice. Recall that there are eight tetrahedral holes in an FCC unit cell, twice as many holes as ions that make up the FCC lattice (four ions per unit cell). To maintain the 1:1 cation:anion ratio, only half of the available tetrahedral holes are occupied in this structure.

FCC arrangement of S^{2-}:   4 S^{2-} per unit cell

Zn^{2+} in ½ of tetrahedral holes: (½)8 = 4 Zn^{2+} per unit cell

The zinc blende unit cell in Interactive Figure 12.3.7 shows sulfide ions as yellow spheres and zinc ions as gray spheres. Notice the tetrahedral environment around each Zn^{2+} ion and the fact that there are two tetrahedral holes occupied in the upper portion of the unit cell and two occupied in the lower portion.

As is the case in the NaCl structure, the anions that make up the FCC array no longer touch along the face diagonal in the zinc blende unit cell. Some compounds that adopt this structure include BeS, AlP, and SiC.

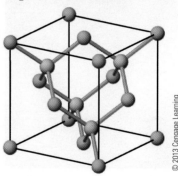

© 2013 Cengage Learning

Interactive Figure 12.3.7

Explore the zinc blende structure.

ZnS unit cell

Example Problem 12.3.3 Calculate the density of a solid using unit cell type and dimensions.

Copper(I) iodide crystallizes with the zinc blende structure. If the edge length of the unit cell is 605 pm, what is the density (in g/cm³) of crystalline CuI?

Solution:

You are asked to calculate the density of a solid.

You are given the identity and crystal structure of the solid and the edge length of the unit cell.

Density is calculated from the mass and the volume of the unit cell.

Step 1. Calculate the mass of the unit cell using the fact that there are four formula units per unit cell in the ZnS structure.

$$\frac{4 \text{ CuI formula units}}{1 \text{ unit cell}} \times \frac{1 \text{ mol CuI}}{6.022 \times 10^{23} \text{ CuI formula units}} \times \frac{190.45 \text{ g}}{1 \text{ mol CuI}}$$

$$= 1.27 \times 10^{-21} \text{ g/unit cell}$$

Example Problem 12.3.3 *(continued)*

Step 2. Calculate the volume of the unit cell (in cm³) from the edge length.

edge length = 605 pm

$$605 \text{ pm} \times \frac{10^2 \text{ cm}}{10^{12} \text{ pm}} = 6.05 \times 10^{-8} \text{ cm}$$

$$\text{volume} = a^3 = (6.05 \times 10^{-8} \text{ cm})^3 = 2.21 \times 10^{-22} \text{ cm}^3$$

Step 3. Calculate density from the mass and volume of the unit cell.

$$\text{density} = \frac{\text{mass}}{\text{volume}} = \frac{1.27 \times 10^{-21} \text{ g}}{2.21 \times 10^{-22} \text{ cm}^3} = 5.71 \text{ g/cm}^3$$

Video Solution

Tutored Practice
Problem 12.3.3

12.3d Complex Solids

Although many ionic compounds have relatively simple crystal structures, it is not true that all ionic solids have simple structures or that only ionic compounds form crystalline solids. Here we look at a few ionic and non-ionic solids with more complex structures.

Ceramics

Ceramics are solid inorganic materials that are typically prepared by heating at elevated temperatures. In general, they are insulators, but in 1986 a lanthanum–barium–copper oxide ceramic was discovered to be a high-temperature **superconductor**, a material that has zero electrical resistance below a given temperature. High temperature in this context was only 30 K, but subsequent investigations have led to materials that are superconducting at temperatures above that of liquid nitrogen (77 K), which is easily attained. These ceramics consist of layers of copper oxide spaced by layers containing other atoms. One such high-temperature superconductor is a yttrium–barium–copper oxide ceramic, with the unit cell shown in Figure 12.3.8. The spheres in the model on the right represent relative ionic radii.

The unit cell in this case is primitive orthorhombic (one lattice point per unit cell, all cell edges different lengths, all angles = 90°). You can see that, in this case, a single lattice point corresponds to a fairly complicated motif. Can you figure out the formula of the compound by looking at the unit cell? This solid is sometimes called a 1-2-3 superconductor, and its formula is $YBa_2Cu_3O_7$.

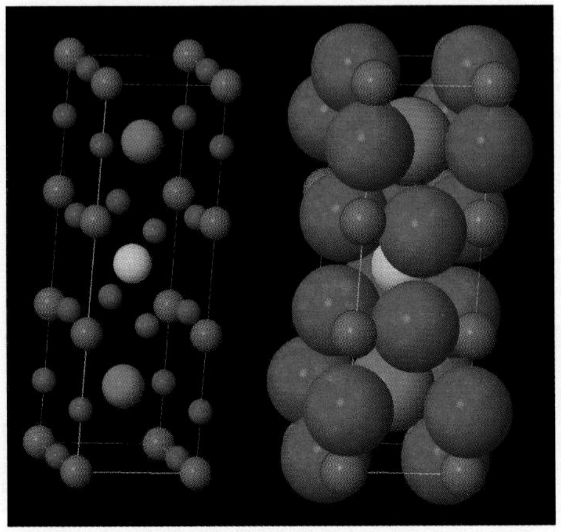

© 2013 Cengage Learning

Figure 12.3.8 The unit cell of $YBa_2Cu_3O_7$ (Y = blue, Ba = green, Cu = brown, O = red)

Example Problem 12.3.4 Determine a formula from a unit cell model.

An oxide of cesium crystallizes in a tetragonal unit cell with a cesium atom on each corner and at the center, as shown to the right in the following figure. There are two oxygen atoms each on four of the edges and two oxygen atoms in the interior of the cell. What is the chemical formula of the compound?

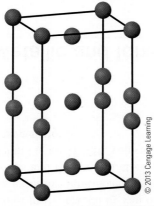

© 2013 Cengage Learning

Solution:

You are asked to determine the chemical formula of a compound.

You are given the arrangement of ions in the unit cell of the compound.

Corner atoms are shared with eight cells and edge atoms with four cells. Interior atoms belong entirely to the cell.

Cs at corners: $8 \times \dfrac{1}{8} = 1$ Cs

Cs at center: $= 1$ Cs

O on edges: $(2 \times 4)(\frac{1}{4}) = 2$ O

O in interior: $= 2$ O

2 Cs $+ 4$ O $= Cs_2O_4$

Formula: CsO_2

Video Solution

Tutored Practice
Problem 12.3.4

Zeolites

Zeolites are solid inorganic materials with an aluminosilicate framework that results in a porous structure. They are network-covalent structures with additional materials housed in the pores. These substances can act as "molecular sieves" because their pore structures have molecular dimensions of variable sizes that can hold some molecules or ions, but not others. For example, zeolites are widely used for ion exchange in water purification and softening and in chemistry as drying agents.

Faujasite is a naturally occurring zeolite with the composition $(Ca,Na_2,Mg)_{3.5}[Al_7Si_{17}O_{48}] \cdot 32 H_2O$. It crystallizes in an FCC lattice, which means that it has four lattice points per unit cell, but it has a complicated structure with a lot of symmetry. The unit cell of a synthetic sodium faujasite is shown in Figure 12.3.9, viewed down a body diagonal of the cube.

Note the silicon–aluminum–oxygen covalent framework that extends throughout the crystal and the large cages and channels that are formed. This synthetic version has been dehydrated, and only sodium ions (purple) are present to maintain electrical neutrality.

Non-ionic Compounds

You might not think that organic molecules will exist as crystalline solids, but they do (take a close look at the crystals in your sugar bowl). Acetamide, CH_3CONH_2, is a small organic molecule that exists as a crystalline solid at room temperature. The unit cell of acetamide is shown in Figure 12.3.10.

Acetamide has a unit cell with symmetry that is difficult to visualize, so instead focus on the facts that the crystal is made up of molecules that are packed together and that the unit cell does repeat to fill space. There are no ions and no covalent networks in the solid, so the molecules are held together in the crystal by van der Waals forces. In addition to these dispersion forces, there is extensive intermolecular hydrogen bonding between the hydrogen atoms of the NH_2 groups and the oxygen atoms of the CO groups. If you look closely at the space-filling model in Figure 12.3.10, you can see places where the H atoms of the NH_2 groups are close to oxygen atoms.

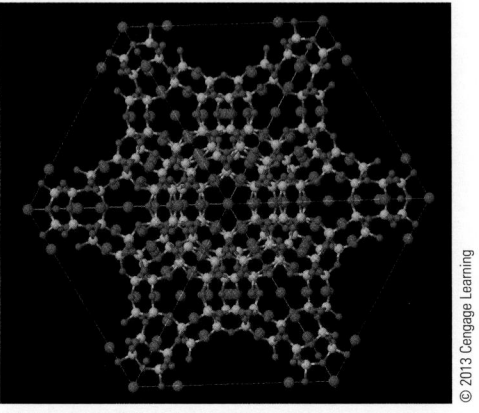

Figure 12.3.9 The unit cell of a synthetic sodium faujasite (Na = purple; Si, Al = brown, O = red)

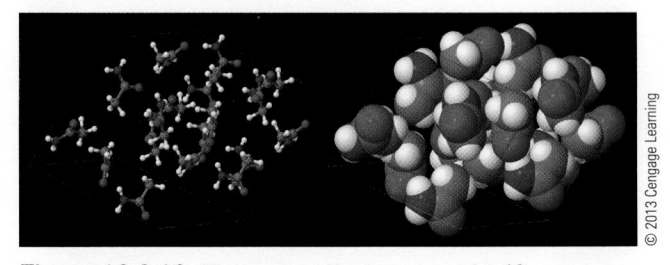

Figure 12.3.10 The unit cell of acetamide (C = gray, H = white, O = red, N = blue)

12.4 Bonding in Metallic and Ionic Solids

Section 12.3 Mastery

12.4a Band Theory

Most metals are solid crystalline materials that are malleable and ductile and good conductors of heat and electricity. In the simplest bonding model, the metal atom cores are imagined to be held together by delocalized valence electrons that are free to move about the crystal to conduct electricity.

Band theory is a more formal approach to the bonding in crystalline metallic solids, similar to the molecular orbital theory that you saw in Theories of Chemical Bonding (Unit 9). In this theory, the valence molecular orbitals in crystals involve all of the atoms in the crystal and therefore extend over the entire array of atoms. Recall from molecular orbital theory that when two identical orbitals from two identical atoms combine, two molecular orbitals, one of lower energy and one of higher energy, are produced. In a crystal that contains a mole of atoms, a mole of such orbitals will combine to produce a mole of molecular orbitals that extend over the crystal and are very closely spaced in energy. Some of the molecular orbitals are of lower energy and thus confer stability, and some are of higher energy and do not (Interactive Figure 12.4.1).

A group of these molecular orbitals is called a *band* because the orbitals are nearly continuous in energy. Orbitals in the valence shells of the atoms combine to give a series of bands, some of which are separated by energy gaps. In the isolated metal crystal, valence electrons occupy the lowest energy orbitals possible in what is called the **valence band**. The electrons in the valence band account for bonding between atoms in the crystal. The next lowest-energy band of orbitals available acts as the **conduction band**. When a metal or other material conducts electricity, electrons in the valence band move into the conduction band and a current flows through the material as electrons move within the conduction band. When an electron moves from the valence band into the conduction band, it leaves behind a "hole" that formally carries a positive charge. As electrons move within the conduction band, they can also move within the valence band by moving into the space occupied by the positive hole.

The conductivity of a material depends on the energy gap between the valence band and the conduction band (Figure 12.4.2). In a **conductor** such as a metal, either the valence band and the conduction band are contiguous or the valence band and the conduction band overlap, so there is no energy gap between the two. In an **intrinsic semiconductor**, the valence band is full and the energy gap is small enough that if the solid is heated, thermal energy becomes sufficient to promote electrons from the valence band into the empty conduction band. Silicon is an example of a semiconductor. In an **insulator**, the valence band is full and the energy gap is so large that motion of electrons from it to the empty conduction band is prohibited. Diamond is an example of an insulator.

The conducting properties of a semiconductor can be improved by adding small amounts of a **dopant**, an impurity with specific properties. For example, in pure silicon the valence band is full and the conduction band is empty. If some arsenic or phosphorus is added as a dopant, for each Si atom replaced an extra electron is introduced into the lattice (Interactive Figure 12.4.3). These electrons occupy levels just below the conduction band and are easily promoted into the conduction band, increasing the conductivity of the

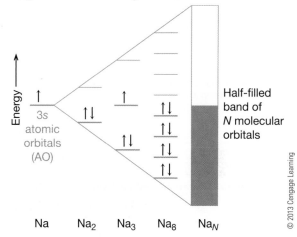

Interactive Figure 12.4.1

Explore band theory.

Energy

3s atomic orbitals (AO)

Half-filled band of *N* molecular orbitals

Na   Na₂   Na₃   Na₈   Na_N

© 2013 Cengage Learning

Formation of bands of molecular orbitals in a solid

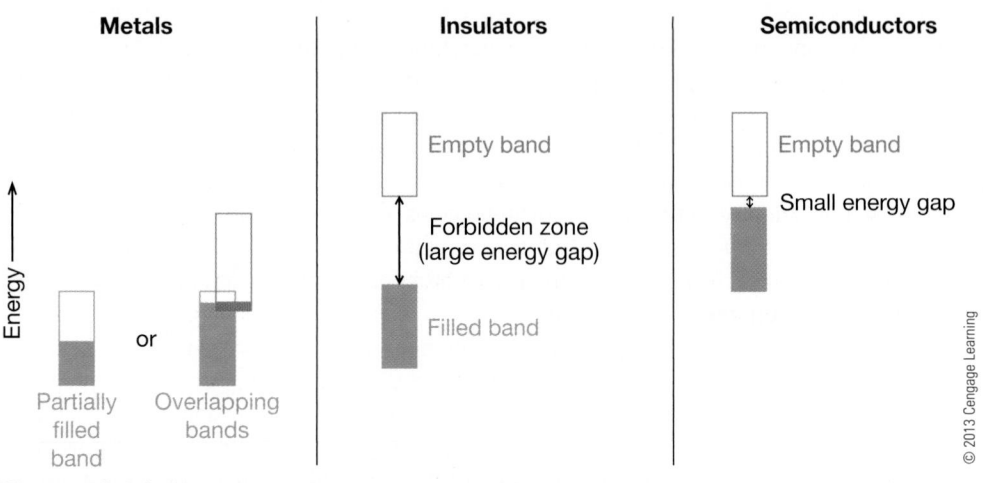

Figure 12.4.2 The effect of band gap on conductivity

silicon. Because the dopant introduces extra electrons, which carry a negative charge, the resulting material is called an **n-type semiconductor**. If, on the other hand, some gallium or indium is added to the silicon, there is one fewer electron in the valence band for each Si atom replaced. This creates positive holes into which electrons can move when a potential difference is applied, again supporting a current. The resulting material is called a **p-type semiconductor** because it is the presence of positive holes that promotes the flow of current.

12.4b Lattice Energy and Born–Haber Cycles

Ionic solids are held together by the strong attractive forces between oppositely charged ions. The energy that holds ions within a crystal lattice is the **lattice energy**, U, and it is defined as the energy *required* to separate one mole of a crystalline ionic solid into its gaseous ions. For the ionic compound sodium chloride,

$$NaCl(s) \rightarrow Na^+(g) + Cl^-(g) \qquad U = \Delta H = \text{lattice energy (kJ/mol)}$$

Alternatively, lattice energy can be defined as the energy *released* when one mole of a crystalline ionic solid forms from its gaseous ions. Note that separating the ions is an endothermic process, whereas the reverse process is exothermic.

Theoretical lattice energy values can be calculated by starting with Coulomb's law and summing all of the attractive and repulsive forces within a crystal lattice. However, it is also

Interactive Figure 12.4.3

Explore n- and p-type semiconductors.

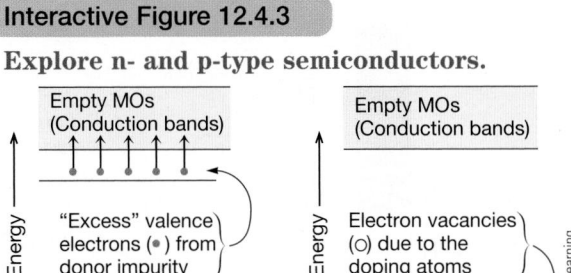
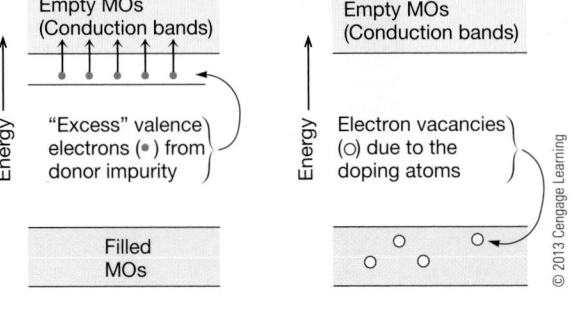

(a) n-Type and (b) p-type semiconductors

possible to calculate lattice energy from experimentally determined thermodynamic values using a Born–Haber cycle.

Recall that Hess's law that says that if a process can be expressed as the sum of other processes, then the enthalpy change for the overall process is the sum of the enthalpy changes for the individual steps. Using sodium chloride as an example, the lattice energy process can be expressed as the sum of the following five steps:

Step 1. Vaporization (sublimation) of solid sodium metal

$$Na(s) \rightarrow Na(g) \qquad \Delta H^\circ_{sublimation} = 107 \text{ kJ}$$

Step 2. First ionization energy of sodium

$$Na(g) \rightarrow Na^+(g) + e^- \qquad IE = 496 \text{ kJ}$$

Step 3. Dissociation of ½ mol of $Cl_2(g)$ (this is equivalent to ½ of the Cl—Cl bond energy)

$$\tfrac{1}{2} Cl_2(g) \rightarrow Cl(g) \qquad \tfrac{1}{2} D_{Cl-Cl} = (\tfrac{1}{2} \text{ mol})(242 \text{ kJ/mol}) = 121 \text{ kJ}$$

Step 4. First electron affinity of chlorine

$$Cl(g) + e^- \rightarrow Cl^-(g) \qquad EA = -349 \text{ kJ}$$

Step 5. Reverse of the enthalpy of formation of NaCl(s)

$$NaCl(s) \rightarrow Na(s) + \tfrac{1}{2} Cl_2(g) \qquad -\Delta H^\circ_f = -(-411 \text{ kJ}) = 411 \text{ kJ}$$

Summing steps 1–5 and canceling terms common to reactants and products,

$$NaCl(s) \rightarrow Na^+(g) + Cl^-(g) \qquad \begin{aligned} U &= 107 \text{ kJ} + 496 \text{ kJ} + 121 \text{ kJ} + (-349 \text{ kJ}) \\ &+ 411 \text{ kJ} = 786 \text{ kJ} \end{aligned}$$

The lattice energy of NaCl(s) is 786 kJ/mol. The energy changes are shown on an energy diagram in Interactive Figure 12.4.4. Because lattice energy is the energy required to separate an ionic compound into is gaseous ions, as the strength of the forces between the ions increases, the lattice energy increases. This is reflected in some of the physical properties of ionic solids, such as hardness and melting point. For example, as lattice energy increases, so generally does melting point (Table 12.4.1).

According to Coulomb's law, the energy of interaction between ions of opposite charge increases with the magnitude of the charges and decreases with the distance between them. You can see these effects in the lattice energies in Table 12.4.1, where all of the compounds included have the sodium chloride structure. Notice that when ion size increases but ion charge remains constant (LiF vs. LiCl), the magnitudes of the lattice energies decrease because the distance between ions has increased. When ion size is relatively constant but ion

Explore lattice energy.

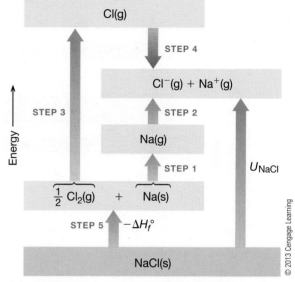

Energy diagram for the lattice energy of NaCl(s)

© 2013 Cengage Learning

Table 12.4.1 Lattice Energies and Melting Points for Some Common Ionic Solids

Compound	U (kJ/mol)	Melting Point (°C)
LiF	1037	870
LiCl	852	605
LiBr	815	552
NaF	926	993
NaCl	786	801
NaBr	752	747
MgO	3850	2852

charge increases (NaF and MgO), lattice energy increases due to the greater attractive force between the ions with higher charges.

Example Problem 12.4.1 Calculate a lattice energy from thermodynamic data.

Calculate the lattice energy of KCl(s) using the following thermodynamic data (all data given is in kJ/mol). Do you expect this value to be larger or smaller than the lattice energy of NaCl(s)?

K(s)	$\Delta H_{sublimation} = 89.6$ kJ/mol	Cl—Cl	$D_{Cl-Cl} = 242$ kJ/mol
KCl(s)	$\Delta H_f^\circ = -437$ kJ/mol	Cl(g)	EA $= -349$ kJ/mol
K(g)	IE $= 419$ kJ/mol		

Solution:

You are asked to calculate the lattice energy for a compound and to predict the relative magnitude of a similar compound's lattice energy.

You are given the identity of the compound and thermodynamic data needed to calculate the lattice energy using a Born–Haber cycle.

The lattice energy of KCl is the enthalpy change for the reaction $KCl(s) \rightarrow K^+(g) + Cl^-(g)$. Use the data given to write five equations that will sum to give this net reaction, and then sum the respective enthalpy changes to calculate $U(KCl(s))$.

1. $K(s) \rightarrow K(g)$ $\qquad\qquad\qquad$ $\Delta H_{sublimation} = 89.6$ kJ

2. $K(g) \rightarrow K^+(g) + e^-$ $\qquad\qquad$ ionization energy $= 419$ kJ

3. $\frac{1}{2} Cl_2(g) \rightarrow Cl(g)$ $\qquad\qquad$ $\frac{1}{2}$ bond energy $= 121$ kJ

4. $Cl(g) + e^- \rightarrow Cl^-(g)$ $\qquad\qquad$ electron affinity $= -349$

5. $KCl(s) \rightarrow K(s) + \frac{1}{2} Cl_2(g)$ $\qquad$ $-\Delta H_f^\circ = -(-437 \text{ kJ}) = 437$ kJ

Sum 1–5: $KCl(s) \rightarrow K^+(g) + Cl^-(g)$ $\qquad$ $U = 89.6 + 419 + 121 - 349 + 437 = 718$ kJ

The lattice energy of KCl is 718 kJ/mol.

This value is smaller than that for NaCl (786 kJ/mol), as expected because of the larger size of K^+ compared with Na^+.

Video Solution

Tutored Practice
Problem 12.4.1

Section 12.4 Mastery

12.5 Phase Diagrams

12.5a Phase Changes Involving Solids

In a solid the particles are held in position relative to one another because they do not have enough energy to overcome the forces that hold them in place. As the temperature of the solid is increased, the average energy of the particles increases. When a temperature is reached where some of the surface particles have enough energy to move away from their neighbors, a dynamic equilibrium is established between the solid and the liquid. For as long as there is some solid present, the addition of energy will not result in an increase in temperature but will convert more solid to liquid. The stronger the attractive forces between the particles, the higher the temperature at which this will occur.

A temperature at which a solid and a liquid are in equilibrium at a given pressure is called a **melting point**. Because solids and liquids are not very compressible and have comparable densities, the melting temperature of a solid does not change much with pressure. However, there is some pressure dependence; therefore, to specify a melting temperature precisely, it is also necessary to specify the pressure. Because a crystal is composed of objects that repeat, the forces holding them together also repeat throughout the lattice. A pure crystalline solid is therefore characterized by a sharp melting point, rather than the gradual softening seen in amorphous materials.

Because the melting point of a solid reflects the strength of the forces holding the particles in the solid together, in general, substances with relatively high melting points also have relatively high boiling points. However, the efficiency with which the particles in a solid pack together varies with their shape and with the geometric details of their interactions. This causes the variation of melting points with strength of attractive forces to be less regular than what is found for boiling points.

As you saw in Thermochemistry (Unit 5), the conversion of a solid to a liquid is called melting or fusion and the energy required to effect the conversion is called the enthalpy of fusion or heat of fusion, ΔH_{fusion}. Fusion is always an endothermic process, because it always takes energy to overcome the forces holding the particles in a solid in place. The reverse process, freezing, is always exothermic. Some heat of fusion values at the **normal melting point** (the melting temperature at a pressure of 1 atm) and heat of vaporization at the normal boiling point are given in Table 12.5.1.

Notice that the heat of fusion is always smaller than the heat of vaporization for a given substance. This reflects the fact that the solid and liquid are both condensed phases in which the particles are fairly close together. To go to the gas phase, all of the attractive forces must be overcome.

Table 12.5.1 Selected Heats of Fusion and Vaporization at the Temperature of the Normal Phase Transition

	ΔH_{fusion} (kJ/mol)	Melting Point (°C)	ΔH_{vap} (kJ/mol)	Boiling Point (°C)
Methane	0.94	−182.5	8.2	−161.6
Ethane	2.86	−182.8	14.7	−88.6
Propane	3.53	−187.6	19.0	−42.1
Methanol	3.16	−97.0	35.3	64.7
Ethanol	5.02	−114.3	38.6	78.4
1-Propanol	5.20	−127	41.4	97.2
Water	6.01	0.0	40.7	100.0
Na	2.60	97.82	97.42	881.4
NaBr	26.11	755	160.7	1390

Many solids are volatile enough to have appreciable vapor pressures, even at temperatures below their melting points. The direct conversion of a solid to a gas is called **sublimation**. Like fusion and vaporization, and for the same reasons, sublimation is an endothermic process. The reverse process, **deposition**, is therefore exothermic. The energy required to sublime a solid is called its **enthalpy of sublimation**, or *heat of sublimation*, $\Delta H_{\text{sublimation}}$. Because melting followed by vaporization results in the same overall physical change as sublimation, according to Hess's law, the heat of sublimation should be equal to the sum of the heat of fusion and the heat of vaporization. Therefore, the heat of sublimation should be the largest of the three (Figure 12.5.1).

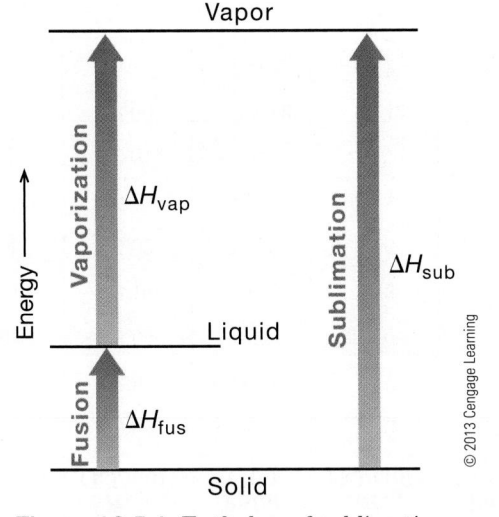

Figure 12.5.1 Enthalpy of sublimation is the sum of enthalpy of vaporization and enthalpy of fusion.

Example Problem 12.5.1 Calculate enthalpy of sublimation from heats of fusion and vaporization.

a. Use the data in Table 12.5.1 to approximate the enthalpy of sublimation for water. Why is this just an approximation?
b. At 298 K, the standard heat of formation of $H_2O(g)$ is -241.8 kJ/mol, whereas the standard heat of formation of $H_2O(\ell)$ is -285.8 kJ/mol. Use this to calculate the heat of vaporization of water and compare your result with the value in Table 12.5.1.

Solution:

You are asked to approximate the enthalpy of sublimation of water and to calculate the heat of vaporization of water.

You are given enthalpies of fusion and vaporization for water and the standard heat of formation for liquid and gaseous water.

a. $\Delta H_{\text{sub}} = \Delta H_{\text{fus}} + \Delta H_{\text{vap}} = 6.01$ kJ/mol $+ 40.7$ kJ/mol $= 46.7$ kJ/mol

For water, the heat of fusion is given at 0 °C and the heat of vaporization is given at 100 °C. Because ΔH does vary somewhat with temperature, the calculated value for ΔH_{sub} is only an approximation.

b. $H_2O(\ell) \rightarrow H_2O(g)$ $\Delta H° = -241.8$ kJ/mol $- (-285.8)$ kJ/mol $= 44.0$ kJ/mol

The value at 25 °C is larger than the value at 100 °C (40.7 kJ/mol). At the higher temperature, the molecules in the liquid have a higher average energy and are more gaslike (farther apart and less short range order), so less energy should be needed for the conversion.

Video Solution

Tutored Practice
Problem 12.5.1

12.5b Phase Diagrams

A **phase diagram** summarizes the conditions at which phases are in equilibrium with each other or at which only a single phase is stable. The independent variables are temperature and pressure, and these are plotted on the x- and y-axes, respectively. The phase diagram for water, not drawn to scale, is shown in Interactive Figure 12.5.2.

Each point on the diagram represents a specific temperature and pressure. Areas on the diagram correspond to regions where only one phase is stable. Note that the gaseous phase is favored at lower pressures and higher temperatures (lower right), whereas the solid phase is favored at higher pressures and lower temperatures (upper left). Also notice that the liquid phase occupies a region between the solid and the gaseous phases and is not stable below a certain pressure, regardless of the temperature.

Separating any two areas are lines where the two phases are in equilibrium. You have already seen the section of the phase diagram shown by the line TC when we studied the

Interactive Figure 12.5.2

Explore a phase diagram.

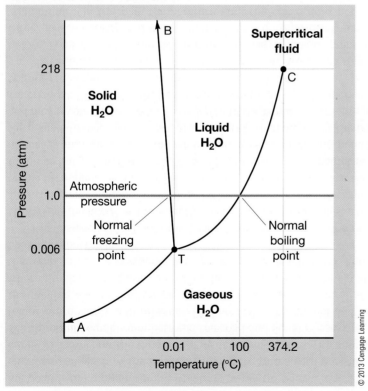

The phase diagram for water

properties of liquids in Intermolecular Forces and the Liquid State (Unit 11). Because boiling point is the temperature at which the vapor pressure is equal to external pressure, this curve also shows how the boiling temperature varies with pressure. Analogous to this, the curve TA shows how the vapor pressure of the solid varies with temperature, whereas the line TB represents equilibrium between the solid and liquid phases and shows how the melting temperature varies with pressure. The normal melting point and the normal boiling point, the melting and boiling temperatures at a pressure of 1 atm, are seen on the diagram where the line $P = 1$ atm intersects the solid–liquid and the liquid–vapor lines. For water the temperatures corresponding to 1 atm are the familiar 0 °C and 100 °C for melting and boiling, respectively.

It is instructive to relate the heating curves that were described in Thermochemistry (Unit 5) to the line $P = 1$ atm on the phase diagram. Imagine starting with the solid at $P = 1$ atm and some temperature below 0 °C. Also imagine that the sample is in a cylinder fitted with a frictionless piston that moves to adjust the volume so that the pressure on both sides of the piston is equal to 1 atm. As you heat the solid at constant pressure, its temperature increases until the solid–liquid equilibrium is reached. At this point, further heating causes the sample to melt at 0 °C. Additional heating then increases the temperature of the liquid and moves you across the liquid region until the liquid–vapor line is reached and the sample vaporizes at 100 °C. Finally, additional heating raises the temperature of the gas and moves you into the region where the gas is the stable phase.

The liquid–vapor curve TC begins and ends in two points that you may not have seen before. Point T is called the **triple point**. This is the unique point on the diagram where all three phases coexist and is the temperature and pressure where all three phases are in equilibrium. On the high temperature end, the curve does not extend beyond point C, which is called the **critical point**. The temperature and pressure at this point are called the **critical temperature** and the **critical pressure**, respectively. Analogous to other points defining equilibrium between gas and liquid, the critical pressure is the pressure required to liquefy the gas at the critical temperature. Beyond the critical temperature, it is not possible to liquefy the gas, regardless of the pressure. Some critical temperatures and pressures are given in Table 12.5.2.

Substances in which the IMFs are relatively strong, such as the polar molecules ammonia, sulfur dioxide and water, have relatively high critical temperatures. Substances in which the IMFs are weak, such as argon, hydrogen, nitrogen, and oxygen, have low critical temperatures. *Permanent gases* are gases whose critical temperatures are below room temperature, so they cannot be liquefied at room temperature, regardless of the pressure.

At temperatures and pressures above the critical temperature and pressure, substances are known as **supercritical fluids**. Supercritical fluids can have properties that make them

Table 12.5.2 Selected Critical Temperatures and Pressures

	Critical Temperature (°C)	Critical Pressure (atm)
NH_3	132.5	111.5
Ar	−122.4	48.1
Br_2	310.8	102
Cl_2	143.8	76.0
F_2	−128.85	51.5
He	−267.96	2.24
H_2	−239.95	12.8
Kr	−63.8	54.3
CH_4	−82.1	45.8
Ne	−228.75	27.2
N_2	−146.9	33.5
O_2	−118.6	49.8
SO_2	157.8	77.7
H_2O	373.936	217.7
Xe	16.6	57.6

of practical importance. For example, supercritical carbon dioxide has been used to decaffeinate coffee beans, replacing potentially carcinogenic solvents that were previously used. It is also used as a solvent for extracting fragrance molecules, permitting this to be done at temperatures that are low enough to avoid decomposition of the compounds.

Because the melting temperature is not very sensitive to pressure, the slope of the solid–liquid line on the phase diagram for water is greatly exaggerated. The negative slope for water is unusual and occurs because for water the density of the solid is less than the density of the liquid, due to the open structure of the hydrogen-bonded ice lattice. According to Le Chatelier's principle, if you increase the pressure on an equilibrium system it will respond to decrease the pressure. A pressure increase will cause the solid-liquid equilibrium to shift to favor the more dense phase because this will alleviate pressure by decreasing volume. For water, the liquid is the more dense phase, and raising the pressure causes the melting point to decrease (negative slope). This means that at a given temperature, increasing the pressure enough will melt the solid. Notice that as you move to higher pressures, the liquid occupies more area on the phase diagram for water.

Contrast this to the behavior of carbon dioxide shown in its phase diagram (Figure 12.5.3). Carbon dioxide is a more typical substance in that its solid phase is more dense than its liquid

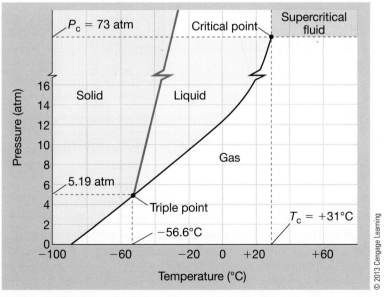

Figure 12.5.3 The phase diagram for carbon dioxide

© 2013 Cengage Learning

phase. Therefore, its melting point increases as the pressure increases, the solid-liquid line has a positive slope, and at higher pressures the solid occupies more area on the diagram. Carbon dioxide has an unusual feature in that it has neither a normal melting point nor a normal boiling point due to the fact that its triple point pressure is above 1 atm. Because the liquid does not exist as a stable phase below the triple point pressure of 5.19 atm, at 1 atm solid CO_2 sublimes to form gaseous CO_2 rather than melting. The low temperature at which this phase change occurs ($-78\ °C$) makes it an ideal substance to use for cooling and gives solid carbon dioxide its common name, "dry ice."

Example Problem 12.5.2 Interpret phase diagrams.

Answer the following questions based on the phase diagram (not drawn to scale) for ammonia, NH_3.

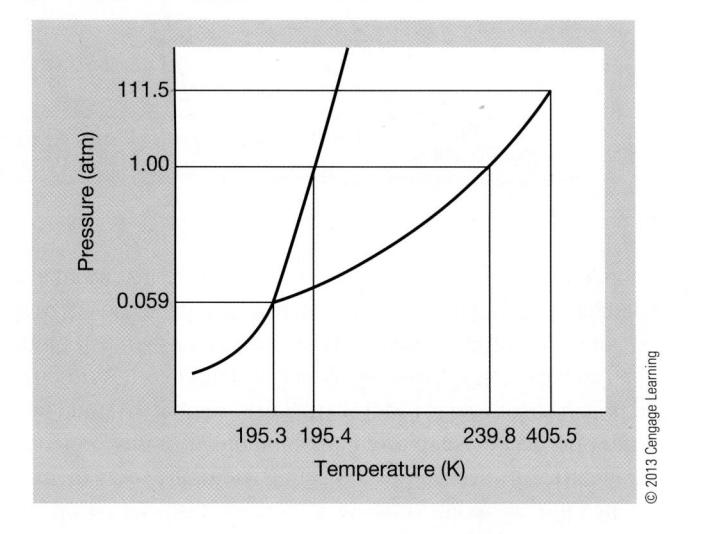

© 2013 Cengage Learning

a. What is the normal melting point? The normal boiling point?
b. At what temperature and pressure do the three phases coexist in equilibrium?
c. Is ammonia a solid, a liquid, or a gas at room temperature and pressure? If a gas, is it a permanent gas?
d. Which is the more dense phase, the liquid or the solid?

Example Problem 12.5.2 *(continued)*

Solution:

You are asked to answer questions about the physical properties and physical state of a compound.

You are given the phase diagram for the compound.

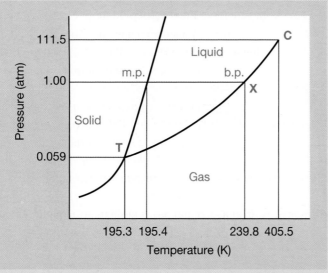

a. Normal refers to a pressure of 1 atm. The point on the liquid–solid line with a pressure of 1 atm has a temperature of 195.4 K. This is the normal melting point. The point on the liquid–vapor curve with a pressure of 1 atm occurs at a temperature of 239.8 K. This is the normal boiling point.

b. Look for the one point that all three phases have in common (T). This is the triple point at 195.3 K and 0.059 atm.

c. Take room temperature and pressure to be 298 K and 1 atm. This corresponds to the point indicated by the letter X on the diagram, which is in the gaseous region. Ammonia is a gas at room temperature, but it is not a permanent gas. Because the temperature is below the critical temperature (405.5 K), it can be liquefied by increasing the pressure to about 11 atm at 298 K.

d. The liquid–solid equilibrium line has a positive slope. Therefore, the solid is more dense than the liquid.

Video Solution

Tutored Practice
Problem 12.5.2

Section 12.5 Mastery

Unit Recap

Key Concepts

12.1 Introduction to Solids

- Most solids are best described as crystalline or amorphous (12.1a).

- Solids are broadly classified based on the nature of the interactions holding the particles that make up the solid together as molecular, ionic, covalent, or metallic (12.1a).

- Crystalline solids consist of a repeating three-dimensional array of particles (12.1b).

- Lattice points are the locations of identical points in the repeating pattern, and a collection of lattice points is a crystal lattice (12.1b).

- The smallest repeating unit in a crystal lattice is the unit cell (12.1b).

- In a cubic unit cell, all edge lengths are the same and all angles between edges are 90° (12.1b).

12.2 Metallic Solids

- The coordination number of an atom is the number of nearest neighbors the atom has in a crystal lattice (12.2a).

- A simple cubic (SC) unit cell has one lattice point on each corner of a cube (12.2a).

- An SC unit cell with one atom per lattice point contains one atom per unit cell, and each atom has a coordination number of 6 (12.2a).

- A body-centered cubic (BCC) unit cell has one lattice point on each corner of a cube and one lattice point in the center of the cube (12.2b).

- A BCC unit cell with one atom per lattice point contains two atoms per unit cell, and each atom has a coordination number of 8 (12.2b).

- A face-centered cubic (FCC) unit cell has one lattice point on each corner of a cube and a lattice point centered on each cube face (12.2c).

- An FCC unit cell with one atom per lattice point contains four atoms per unit cell, and each atom has a coordination number of 12 (12.2c).

- In a closest-packed structure, spheres are arranged in layers, and in each layer there are six spheres surrounding every sphere (12.2c).

- In hexagonal closest-packing (HCP), the closest-packed layers are arranged in an ABAB… repeating pattern, whereas in cubic closest-packing (CCP), the closest-packed layers are arranged in an ABCABC… repeating pattern (12.2c).

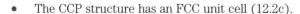

- The CCP structure has an FCC unit cell (12.2c).

- Bragg's law, in combination with an x-ray diffraction experiment, can be used to determine the spacing between the set planes and thus the unit cell edge length (12.2d).

12.3 Ionic Solids

- In an ionic solid, smaller ions (typically the cations) occupy holes in the anionic lattice (12.3a).

- An SC array of ions defines a unit cell with a cubic hole at its center (12.3a).

- An FCC array of anions has two types of holes: tetrahedral and octahedral (12.3a).

- An FCC unit cell with one atom per lattice point has eight tetrahedral holes completely within the unit cell (12.3a).

- An FCC unit cell with one atom per lattice point has 1 octahedral hole completely within the unit cell and 12 that sit at the center of each edge of the unit cell; there are 4 octahedral holes in the FCC unit cell (12.3a).

- In the CsCl structure, an SC arrangement of cesium ions (the larger ion) surrounds a cubic hole occupied by a chloride ion (the smaller ion) (12.3b).

- In the NaCl structure, the larger ions (Cl^-) adopt an FCC arrangement and the smaller ions (Na^+) occupy all of the octahedral holes in the unit cell (12.3b).

- In the ZnS (zinc blende) structure, the larger ions (S^{2-}) adopt an FCC arrangement and the smaller ions (Zn^{2+}) occupy half of the tetrahedral holes in the unit cell (12.3c).

12.4 Bonding in Metallic and Ionic Solids

- In band theory, the valence molecular orbitals in crystals involve all of the atoms in the crystal and therefore extend over the entire array of atoms (12.4a).

- A band of orbitals consists of many molecular orbitals that are very closely spaced in energy (12.4a).

- Valence electrons in an isolated metal crystal occupy the lowest energy band, called the valence band (12.4a).

- The next lowest-energy band is called the conduction band, and electrons move through the conduction band when a metal conducts electricity (12.4a).

- Conductivity is a function of the energy gap between the valence and conduction bands (12.4a).

- In a conductor, either the valence band and the conduction band are contiguous or the valence band and the conduction band overlap (12.4a).

- In an intrinsic semiconductor, the valence band is full and the energy gap is small enough that if the solid is heated, thermal energy becomes sufficient to promote electrons from the valence band into the empty conduction band (12.4a).

- In an insulator, the valence band is full and the energy gap is so large that motion of electrons from it to the empty conduction band is prohibited (12.4a).

- A dopant is an impurity added to a semiconductor to improve conductivity (12.4a).

- Adding a dopant with more electrons than atoms of the pure semiconductor results in an n-type semiconductor. Adding a dopant with fewer electrons than atoms of the pure semiconductor results in a p-type semiconductor (12.4a).

- Lattice energy is the energy that holds ions within a crystal lattice (12.4b).

- Lattice energy can be calculated from thermodynamic values using a Born–Haber cycle (12.4b).

12.5 Phase Diagrams

- Melting point is defined as the temperature at which a solid and liquid are at equilibrium at a given pressure (12.5a).

- The energy required or given off during a phase change is calculated from ΔH_{fusion}, ΔH_{vap}, and $\Delta H_{sublimation}$ values (12.5a).

- Sublimation is the direct conversion of a solid to a gas; the reverse process is called deposition (12.5a).

- A phase diagram summarizes the temperature and pressure conditions at which phases are in equilibrium with each other or at which only a single phase is stable (12.5b).

- The triple point on a phase diagram is the temperature and pressure where all three phases coexist and are at equilibrium (12.5b).

- The end of the liquid–vapor equilibrium line on a phase diagram is called the critical point, and the temperature and pressure at this point are called the critical temperature and critical pressure, respectively (12.5b).

- Critical pressure is the pressure required to liquefy the gas at the critical temperature. Beyond the critical temperature it is not possible to liquefy the gas, regardless of the pressure (12.5b).

- At temperatures and pressures above the critical temperature and pressure, substances are known as supercritical fluids (12.5b).

Key Equations

$$a = \frac{4r}{\sqrt{3}} \text{ for a body-centered cubic unit cell } \textbf{(12.1)} \quad\bigg|\quad a = \frac{4r}{\sqrt{2}} \text{ for a face-centered cubic unit cell } \textbf{(12.2)} \quad\bigg|\quad n\lambda = 2d\sin\theta \qquad\qquad \textbf{(12.3)}$$

Key Terms

12.1 Introduction to Solids
crystalline solid
amorphous solid
molecular solid (van der Waals solid)
ionic solid
covalent solid (network solid)
metallic solid
translational symmetry
lattice point
crystal lattice
unit cell
Bravais lattices

12.2 Metallic Solids
simple cubic (primitive cubic) unit cell
coordination number
body-centered cubic unit cell
face-centered cubic unit cell
closest-packed structure
hexagonal closest-packed structure

cubic closest-packed structure
Bragg's law

12.3 Ionic Solids
cubic hole
tetrahedral hole
octahedral hole
trigonal hole
cesium chloride structure
sodium chloride structure
zinc blende structure
ceramic
superconductor
zeolite

12.4 Bonding in Metallic and Ionic Solids
band theory
valence band
conduction band

conductor
intrinsic semiconductor
insulator
dopant
n-type semiconductor
p-type semiconductor
lattice energy

12.5 Phase Diagrams
melting point
normal melting point
sublimation
deposition
enthalpy of sublimation
phase diagram
triple point
critical point
critical temperature

Unit 12   Review and Challenge Problems

13 Chemical Mixtures: Solutions and Other Mixtures

Unit Outline

In This Unit...

Look around you. How many pure elements or compounds can you identify? Probably not very many. You can probably see some pure aluminum, and maybe some fairly pure water. Most of the materials you see, however, are mixtures of elements and compounds. The air around you, for example, is a mixture of gaseous nitrogen, oxygen, argon, water vapor, and carbon dioxide, and the "lead" in your pencil is a mixture of carbon (graphite), clay, and wax. In this unit we explore mixtures of gases, liquids, and/or solids, and study how to describe their composition and properties.

AlbertSmirnov/iStockphoto.com

13.1 Quantitative Expressions of Concentration

13.1a Review of Solubility

Recall from Chemical Reactions and Solution Stoichiometry (Unit 4) that a solution is a homogeneous mixture in which one substance (the *solute*) is dissolved in another (the *solvent*). In previous units we primarily studied solid–liquid and liquid–liquid solute–solvent combinations, but solutes and solvents can be solids, liquids, or gases.

Solutions are characterized in many different ways. **Solubility** is a quantitative measure of the amount of solute that will dissolve in a solvent at a given temperature. Solubility can be reported in a variety of units, such as g solute/100 mL solvent, g solute/L solvent, or mol solute/L solution (called molar solubility). A solute that dissolves to an appreciable extent in a solvent is said to be **soluble** in that solvent; one that does not dissolve is **insoluble** in that solvent. Recall that in Chemical Reactions and Solution Stoichiometry (Unit 4) we used solubility rules to predict the solubility of an ionic compound in water. These solubility rules are based on quantitative solubility values for ionic compounds, where an insoluble compound is defined as having a solubility less than about 0.01 mol/L.

As you will see later in this unit, solubility varies with temperature. At a given temperature, a solution can be defined as saturated, unsaturated, or supersaturated (Interactive Figure 13.1.1).

- A **saturated** solution is one in which the solute concentration is at the solubility limit at a given temperature. For example, the solubility of sodium acetate in water is 46.4 g/100 mL at 20 °C. This means that at 20 °C you can dissolve 46.4 g of $NaCH_3CO_2$ in 100 mL of water. One method for making a saturated solution is to add solid until no more will dissolve. For example, if 50.0 g of $NaCH_3CO_2$ is added to 100 mL of water at 20 °C, only 46.4 g of the solid will dissolve and the rest will remain as a solid (Interactive Figure 13.1.1, left beaker).

- In an **unsaturated** solution, the solute concentration of a solute is less than the solubility limit at a given temperature. For example, adding 40.0 g of $NaCH_3CO_2$ to 100 mL of water at 20 °C results in an unsaturated solution (Interactive Figure 13.1.1, middle beaker). One way to test whether a solution is unsaturated is to add additional solute and see if it dissolves. If 2.0 g of $NaCH_3CO_2$ is added to the unsaturated solution just described, the additional solute dissolves, indicating that the original solution was unsaturated.

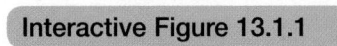

Interactive Figure 13.1.1

Explore saturated, unsaturated, and supersaturated solutions.

Saturated, unsaturated, and supersaturated aqueous sodium acetate solutions

- In a **supersaturated** solution, the solute concentration exceeds the solubility limit at a given temperature. Supersaturated solutions can be prepared only under carefully controlled conditions. For example, a supersaturated sodium acetate solution can be prepared by dissolving a large quantity of the solid in a small amount of warm water and then allowing the solution to cool slowly to room temperature (Interactive Figure 13.1.1, right beaker). A slight disturbance such as the introduction of a small particle of solid or simply bumping the solution container will result in precipitation of solid sodium acetate from the solution and a saturated sodium acetate solution.

Solutions consisting of liquid solutes and solvents can be defined by how well the solute and solvent components mix. At the extreme, two liquids can mix completely or not at all.

- A solute and solvent are **miscible** if the two liquids mix completely, forming a homogeneous solution.

- Liquids are **immiscible** if they do not intermix. Immiscible liquids form a heterogeneous mixture, which can appear as a layered mixture where the more dense liquid sits below the less dense liquid.

Example Problem 13.1.1 Classify solutions.

Answer the following questions about solutions.

a. The solubility of silver sulfate in water is 5.05 g/L. If a silver sulfate solution has a concentration of 0.0474 g/L, is the solution saturated, unsaturated, or supersaturated?
b. The liquids methyl alcohol and water dissolve in each other in all proportions. Are liquid methyl alcohol and water miscible or immiscible?

Solution:

You are asked to describe the nature of a solution.

You are given information about the solubility of a solute in a solvent or the characteristics of a solution made from two components.

a. The amount of solute dissolved in the solution is less than the solubility limit for the solid, so the solution is unsaturated. It is possible to dissolve additional solute in the solution.
b. The two liquids are miscible because they mix completely and form a homogeneous solution.

Video Solution

Tutored Practice
Problem 13.1.1

13.1b Concentration Units

The amount of solute dissolved in a particular amount of solvent or solution is called the solution concentration. Although the general terms *concentrated* or *dilute* can be used to describe solutions with a large or small solute to solvent ratio, respectively, it is more useful to define quantitatively the amount of solute in a solution. Many concentration units are used to express the concentration of a solution. The examples that follow show different ways to represent the concentration of a solution consisting of 26.0 g of NaCl (0.445 mol) and 100.0 g of water (5.55 mol), a solution with a total volume of 121 mL.

Molarity (M)

The concentration unit molarity (M), the amount (in moles) of solute dissolved in 1 L of solution, was introduced earlier. You will often see this concentration unit used in chemical laboratories.

$$\text{molarity} = \frac{\text{amount (mol) solute}}{\text{total volume (L) solution}}$$

$$[\text{NaCl}] = \frac{0.445 \text{ mol NaCl}}{0.121 \text{ L}} = 3.68 \text{ mol/L} = 3.68 \text{ M}$$

Weight Percent (%)

Weight percent is defined as the mass of solute in 100 g of solution.

$$\text{weight percent component A} = \frac{\text{quantity (g) component A}}{\text{total mass (g) of solution}} \times 100\% \qquad \textbf{(13.1)}$$

Many commercial solutions have concentrations reported using weight percent. To avoid confusion about whether the concentration is percent by weight or percent by volume, weight percents are sometimes followed by "w/w" (for weight to weight).

$$\% \text{ NaCl} = \frac{26.0 \text{ g NaCl}}{26.0 \text{ g NaCl} + 100.0 \text{ g H}_2\text{O}} \times 100\% = 20.6\%$$

The sum of the weight percents of all components in a mixture is 100%.

$$\% \text{ H}_2\text{O} = 100.0\% - 20.6\% = 79.4\%$$

Molality (*m*)

Molality is defined as the amount of solute dissolved in 1 kg of solvent.

$$\text{molality} = \frac{\text{amount (mol) solute}}{\text{mass (kg) solvent}} \qquad \textbf{(13.2)}$$

Although this concentration unit is very similar to the molarity concentration unit, note that the denominator here is *mass (kg) of the solvent,* whereas in molarity the denominator is *volume (L) of the solution.* As you will see later in this unit, molality units are required when calculating the boiling point or freezing point of a solution.

$$\text{molality NaCl} = \frac{0.445 \text{ mol NaCl}}{0.100 \text{ kg solvent}} = 4.45 \text{ mol/kg solvent} = 4.45 \; m$$

Mole fraction (χ)

Mole fraction was introduced in Gases (Unit 10), in our study of gases and partial pressures. This concentration unit can be used to describe the composition of any solution, however, not only one that is a mixture of gases.

$$\text{mole fraction component A} = \chi_A = \frac{\text{amount (mol) component A}}{\text{total amount (mol) in solution}} \qquad \textbf{(13.3)}$$

$$\text{mole fraction NaCl} = \chi_{NaCl} = \frac{0.445 \text{ mol NaCl}}{0.445 \text{ mol NaCl} + 5.55 \text{ mol H}_2\text{O}} = 0.0742$$

The sum of the mole fractions of all components of a mixture is equal to 1.

$$\chi_{H_2O} = 1.0000 - 0.0742 = 0.9258$$

Example Problem 13.1.2 Calculate solution concentration.

A solution consists of 20.4 g of silver nitrate, $AgNO_3$, and 113.6 g of water.

a. Calculate the weight percent, the molality, and the mole fraction of $AgNO_3$ in the solution.
b. The solution volume is 117 mL. Calculate the molarity of $AgNO_3$ in the solution.

Solution:

You are asked to calculate the concentration of a solution in weight percent, molality, mole fraction, and molarity units.

You are given the mass of solute and solvent in the solution and the total volume of the solution.

Convert the quantities of solute and solvent to units of moles.

$$20.4 \text{ g AgNO}_3 \times \frac{1 \text{ mol AgNO}_3}{169.9 \text{ g}} = 0.120 \text{ mol AgNO}_3$$

$$113.6 \text{ g H}_2\text{O} \times \frac{1 \text{ mol H}_2\text{O}}{18.015 \text{ g}} = 6.306 \text{ mol H}_2\text{O}$$

a. Use the definition of each concentration unit to calculate the solution concentration.

$$\% \text{ AgNO}_3 = \frac{20.4 \text{ g AgNO}_3}{20.4 \text{ g AgNO}_3 + 113.6 \text{ g H}_2\text{O}} \times 100\% = 15.2\%$$

$$\text{molality AgNO}_3 = \frac{0.120 \text{ mol AgNO}_3}{(113.6 \text{ g H}_2\text{O})\left(\dfrac{1 \text{ kg}}{10^3 \text{ g}}\right)} = 1.06 \; m$$

Example Problem 13.1.2 (continued)

$$\chi_{AgNO_3} = \frac{0.120 \text{ mol AgNO}_3}{0.120 \text{ mol AgNO}_3 + 6.30 \text{ mol H}_2\text{O}} = 0.0187$$

a. Use the definition of molarity to calculate the solution concentration.

$$\text{molarity AgNO}_3 = \frac{0.120 \text{ mol AgNO}_3}{(117 \text{ mL})\left(\dfrac{1 \text{ L}}{10^3 \text{ mL}}\right)} = 1.03 \text{ M}$$

Video Solution

Tutored Practice
Problem 13.1.2

Parts per Million (ppm) and Parts per Billion (ppb)

When a solution contains a very low amount of solute, its concentration is often reported in units of **parts per million** (ppm) or **parts per billion** (ppb). Both of these units are mass ratios; parts per million is the mass of solute (g) in 10^6 g of solution, and parts per billion is the mass of solute (g) in 10^9 g of solution. Concentrations in ppm or ppb are calculated from a solute:solution mass ratio multiplied by 10^6 or 10^9, respectively, or from the ratio of milligrams of solute per kilogram of solution or microgram of solute per kilogram of solution, respectively.

$$\text{parts per million} = \frac{\text{mass (g) solute}}{10^6 \text{ g solution}} = \frac{\text{mass (g) solute}}{\text{mass (g) solution}} \times 10^6 = \frac{\text{mass (mg) solute}}{\text{mass (kg) solution}}$$

$$\textbf{(13.4)}$$

$$\text{parts per billion} = \frac{\text{mass (g) solute}}{10^9 \text{ g solution}} = \frac{\text{mass (g) solute}}{\text{mass (g) solution}} \times 10^9 = \frac{\text{mass } (\mu\text{g}) \text{ solute}}{\text{mass (kg) solution}}$$

$$\textbf{(13.5)}$$

Consider a dilute solution that contains 0.000258 g NaCl and has a total solution mass of 100. g.

$$\text{ppm NaCl} = \frac{0.258 \text{ mg NaCl}}{0.100 \text{ kg solution}} = 2.58 \text{ ppm}$$

$$\text{ppb NaCl} = \frac{258 \ \mu\text{g NaCl}}{0.100 \text{ kg solution}} = 2.58 \times 10^3 \text{ ppb}$$

For dilute aqueous solutions, in which the density of the solution is approximately 1 g/mL, the ppm and ppb concentration units are the equivalent of mg/L and μg/L, respectively.

$$\text{ppm (dilute aqueous solution)} = \frac{\text{mg solute}}{\text{kg solution}} \times \frac{1 \text{ kg}}{10^3 \text{ g}} \times \frac{1 \text{ g}}{1 \text{ mL}} \times \frac{10^3 \text{ mL}}{1 \text{ L}} = \frac{\text{mg solute}}{\text{L solution}}$$

$$\text{ppb (dilute aqueous solution)} = \frac{\mu g \text{ solute}}{\text{kg solution}} \times \frac{1 \text{ kg}}{10^3 \text{ g}} \times \frac{1 \text{ g}}{1 \text{ mL}} \times \frac{10^3 \text{ mL}}{1 \text{ L}} = \frac{\mu g \text{ solute}}{\text{L solution}}$$

Example Problem 13.1.3 Use ppm and ppb concentration units.

The U.S. Environmental Protection Agency's (EPA's) secondary standards for contaminants that may cause cosmetic or aesthetic effects in drinking water suggest an upper limit of 250 mg/L for chloride ion. If 3.88×10^4 L of water in a storage tank contains 4.14 g of Cl, what is the contaminant level in ppm? Is this level acceptable based on EPA guidelines?

Solution:

You are asked to calculate the concentration of a dilute aqueous solution and to compare it with an EPA contaminant concentration.

You are given the mass of solute and solvent in the solution and EPA secondary standard for the contaminant.

Because this is a dilute aqueous solution, we can assume a density of 1 g/mL (ppm = mg solute/L solution).

$$\text{ppm Cl}^- = \frac{(4.14 \text{ g Cl}^-)\left(\dfrac{10^3 \text{ mg}}{1 \text{ g}}\right)}{3.88 \times 10^4 \text{ L solution}} = 0.107 \text{ ppm}$$

The level of the contaminant is acceptable by EPA standards.

Video Solution

Tutored Practice
Problem 13.1.3

Section 13.1 Mastery

13.2 Inherent Control of Solubility

13.2a Entropy and Thermodynamic Control of Chemical Processes

In a chemical system, the overall energy change can be used to predict whether a process is favored to occur at a given temperature. Two different types of energy factors affect whether a process is favored to occur: the change in enthalpy and the change in entropy for a given system. We studied enthalpy change in Thermochemistry (Unit 5) and will describe its connection to the solution process in this section. Entropy and entropy change will be described in detail in Thermodynamics: Entropy and Free Energy (Unit 19), but we will introduce the concept here and use a conceptual definition to describe the influence of this energetic factor on the solution process.

Entropy is a measure of the dissipated energy within a system that is unavailable to do work at a given temperature. It is a measure of the disorder in a system. In a chemical system, entropy is closely linked to the motion (or mobility) of atoms and molecules. The more freedom of motion the particles have, the higher the entropy. Thus, temperature is one factor that is linked to the entropy of a system; as temperature increases, molecular motion increases and so does entropy.

The enthalpy change for a dissolution process is related to the difference between the strength of the intermolecular forces (IMFs) in the separated (pure) substances and the strength of the IMFs between the species in the mixed substances (the solution). If the IMFs in the pure substances are stronger than those formed between species in the solution, the overall enthalpy change for the mixing process is positive (endothermic). However, if the IMFs in the mixture are stronger than those in the separated species, the enthalpy change for the mixing process is negative (exothermic). Recall that exothermic reactions are generally favored.

Both entropy and enthalpy play a role in determining whether a chemical process such as the dissolution of a solute in a solvent will be favored or not at a particular temperature. In general, any chemical process that increases the free motion of molecules tends to be favored, and those that result in the formation of stronger IMFs are generally favored. When a process is favored by both entropy and enthalpy, it is favored overall. If the process is disfavored by both entropy and enthalpy, it will be disfavored overall and not occur to any appreciable extent. When a process is favored by one and disfavored by the other, its overall favorability depends on temperature, as shown in Table 13.2.1.

Table 13.2.1 Entropy, Enthalpy, and Dissolution		
	Positive Enthalpy Change (Reduction in Strength of IMFs)	**Negative Enthalpy Change (Increase in Strength of IMFs)**
Positive Entropy Change (Increase in Mobility of Particles)	Favored at higher temperatures	Favored at all temperatures
Negative Entropy Change (Reduction in Mobility of Particles)	Disfavored at all temperatures	Favored at lower temperatures

Explore evaporation at the molecular level.

Gaseous state: more dispersed, weaker IMFs
(entropy favored, enthalpy disfavored)

Liquid state: less dispersed, stronger IMFs
(entropy disfavored, enthalpy favored)

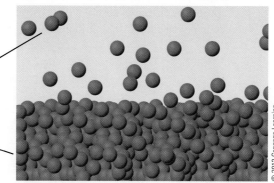

© 2013 Cengage Learning

A molecular-scale view of evaporation

Before considering the effect of entropy and enthalpy on solubility, consider the relationship between entropy, enthalpy, and the simple process of vaporizing liquid water (Interactive Figure 13.2.1). In vaporization, water molecules move from the liquid state, where they are very close to one another (a highly constrained state), to the gaseous state, where water molecules are well separated and more mobile. Vaporization results in an increase in particle mobility and therefore is *entropy favored*. At the same time, water molecules move from a situation where there are strong hydrogen bonds between neighboring water molecules to one where the hydrogen bonds have been broken and any bonding between water molecules is very weak. Thus, vaporization is *enthalpy disfavored*. As shown in Table 13.2.1, processes that are entropy favored and enthalpy disfavored become favored at high temperatures. Thus, water vaporization is favored at higher temperatures and disfavored at lower temperatures.

When making predictions about whether a solute will dissolve in a solvent, it is helpful to think about the effects of enthalpy and entropy separately, as shown in Interactive Figure 13.2.2. This flow chart is often helpful in understanding the factors that control gas–gas, liquid–liquid, and solid–liquid mixtures.

13.2b Gas–Gas Mixtures

When two gases held in separated containers are allowed to mix, they expand to occupy the container completely (Interactive Figure 13.2.3).

You are often exposed to gaseous mixtures. For example, air is a mixture composed mostly of N_2, O_2, Ar, CO_2, and a small amount of H_2O, and natural gas is a mixture made up

Interactive Figure 13.2.2

Use entropy and enthalpy to predict solubility.

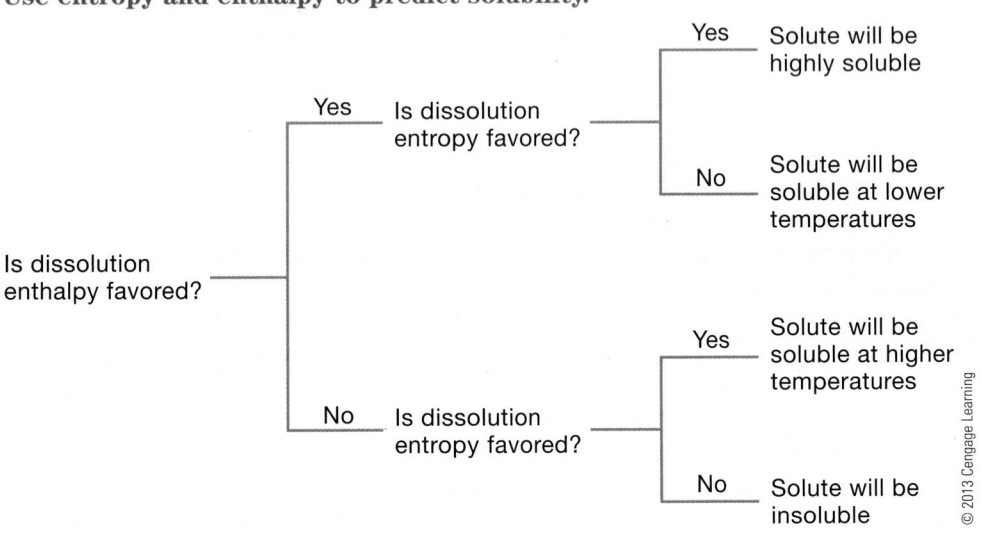

Is dissolution enthalpy favored?

— Yes — Is dissolution entropy favored?
  — Yes — Solute will be highly soluble
  — No — Solute will be soluble at lower temperatures

— No — Is dissolution entropy favored?
  — Yes — Solute will be soluble at higher temperatures
  — No — Solute will be insoluble

© 2013 Cengage Learning

Interactive Figure 13.2.3

Explore gas–gas mixtures.

Stopcock closed

Stopcock open

© 2013 Cengage Learning

Two gases are allowed to mix.

of mostly methane along with some heavier hydrocarbons and other gases such as CO_2 and N_2. Mixing gases increases the disorder of a system, so mixing gases is entropy favored. Enthalpy does not play a role in gas mixing because, as you learned in Intermolecular Forces and the Liquid State (Unit 11), gases have no appreciable IMFs. Therefore, all gases are expected to mix in all proportions; that is, the process is favored at any temperature.

One thing that can prevent the mixing of gases is when a chemical reaction occurs between the gases. For example, when gaseous ammonia and hydrogen chloride are mixed (Figure 13.2.4), a chemical reaction occurs and solid ammonium chloride is produced. The chemical reaction shown in Figure 13.2.4 results in the formation of a mixture called a *smoke*, a solid dispersed in a gas.

13.2c Liquid–Liquid Mixtures

When two liquids containing neutral, covalent molecules are combined, whether they mix (are miscible) or do not mix (are immiscible) depends on both entropy and enthalpy. Consider the mixture of water with methanol at a given temperature (Interactive Figure 13.2.5). In terms

Interactive Figure 13.2.5

Explore miscible liquid–liquid mixtures.

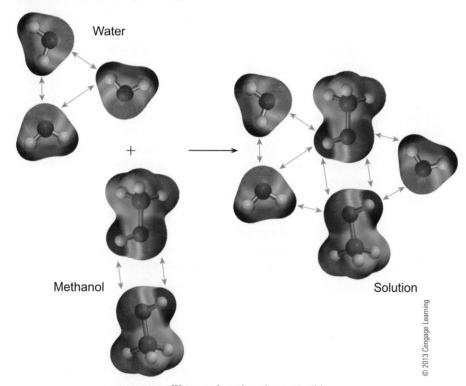

Water and methanol are miscible.

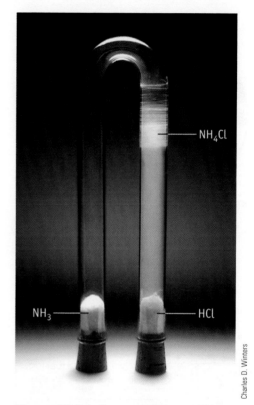

Figure 13.2.4 Gaseous NH_3 and HCl react to form solid NH_4Cl.

of entropy, there is an increase in disorder as the change from these two pure substances to a mixture takes place, and the process is entropy favored. The effect of enthalpy on mixing is more complex, however, as three processes occur:

1. IMFs between H_2O molecules are broken.   Endothermic   Enthalpy disfavored

2. IMFs between CH_3OH molecules are broken.   Endothermic   Enthalpy disfavored

3. New IMFs between H_2O and CH_3OH molecules are formed.   Exothermic   Enthalpy favored

In the specific case of a methanol–water mixture, the net enthalpy change is exothermic because the formation of new IMFs is more exothermic than the sum of the endothermic process, as shown in Figure 13.2.6. The mixing of methanol and water is both entropy favored and enthalpy favored, so water and methanol mix in all proportions—they are miscible.

In contrast, consider what happens when water is mixed with the nonpolar compound carbon tetrachloride (Interactive Figure 13.2.7). When water and a nonpolar compound such as CCl_4 are mixed, the enthalpy gain that would result from forming new dipole–induced dipole forces is minimal. So, instead, when water encounters a nonpolar surface, the water molecules adopt a highly specific, constrained arrangement that allows them to break few if any hydrogen bonds to neighboring water molecules. This results in a more highly ordered system of water molecules, but has little effect on the IMFs present in the mixture. Therefore, the mixing of water and nonpolar compounds is generally found to have little enthalpy change, but is almost always highly entropy disfavored.

In general, polar liquids are miscible in other polar liquids, and nonpolar liquids are miscible in other nonpolar liquids. We use the expression "like dissolves like" to describe this general rule based on the relative strength of the IMFs between different liquids. If the IMFs between molecules in two different liquids are similar, when they mix, the enthalpy effect is not significant and the process is entropy favored and favorable overall. If the IMFs in the two liquids are significantly different, the process is enthalpy disfavored and disfavored overall.

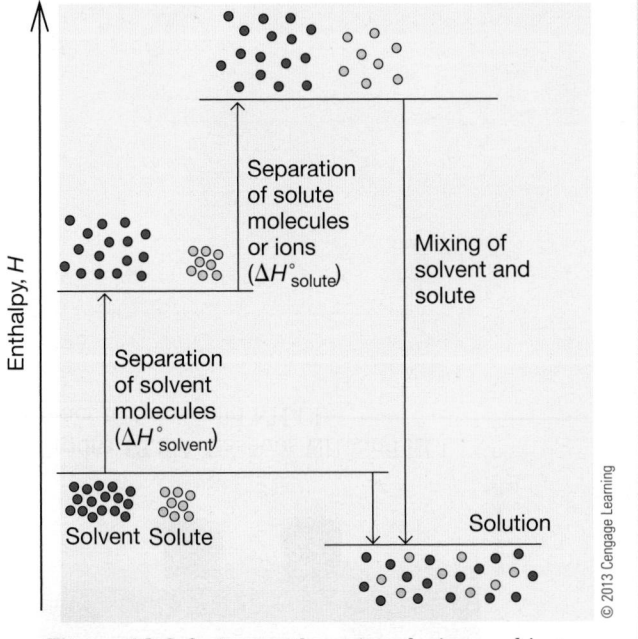

Figure 13.2.6 An exothermic solution-making process

© 2013 Cengage Learning

Example Problem 13.2.1 Predict miscibility of a liquid solute in a liquid solvent.

Predict whether each liquid is likely to be miscible with water.

a. Methylamine, CH_3NH_2
b. Hexane, $CH_3CH_2CH_2CH_2CH_2CH_3$

Solution:

You are asked to predict whether a liquid is miscible with water.

You are given the chemical formula for the compound.

Interactive Figure 13.2.7

Explore immiscible liquid–liquid mixtures.

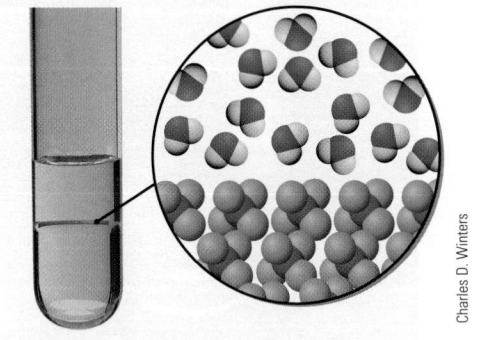

Charles D. Winters

Molecular-scale view of a mixture of H_2O and CCl_4

Water is a polar molecule, and there are strong hydrogen bonds between water molecules in the liquid state.

   a. Methylamine is a polar molecule. Based on the "like dissolves like" general rule, methylamine and water are miscible because they are both polar. In addition, both can form hydrogen bonds, a strong IMF.

   b. Hexane is a nonpolar molecule. Based on the "like dissolves like" general rule, hexane and water are immiscible because they do not have similar IMFs.

13.2d Solid–Liquid Mixtures

The "like dissolves like" rule introduced in our discussion of liquid–liquid mixtures can be used to help explain the solubility of solids in liquids. Consider Figure 13.2.8, which shows a mixture of water, carbon tetrachloride, and iodine. As we saw earlier, water and carbon tetrachloride do not mix because the entropy of water molecules decreases as water contacts a nonpolar surface. Nonpolar iodine, however, dissolves readily in nonpolar carbon tetrachloride (nonpolar species are miscible in other nonpolar species).

    A common type of solid–liquid mixture is an ionic compound dissolved in water. In Chemical Reactions and Solution Stoichiometry (Unit 4), we introduced solubility rules and used them to predict which ionic compounds will dissolve in water. Now we can use the concepts of enthalpy and entropy to understand why some ionic compounds are soluble in water while others are not.

    When an ionic compound dissolves in water, IMFs within the pure substances are broken and new IMFs (between ions and water molecules) are formed. Note that this process involves a physical change, not a chemical change. No covalent bonds are broken in the solid when it dissolves in water; instead, the ion–ion attractive forces in the solid are broken.

    The new IMF between an ion and a polar water molecule is called an **ion–dipole intermolecular force** because it occurs between a permanently charged ion (such as Na^+) and a polar molecule (such as H_2O), as shown in Figure 13.2.9.

    In Figure 13.2.9, the positive ion is attracted to the negative end of the molecular dipole. An anion, such as Cl^-, will be attracted to the positive end of a molecular dipole. For example, when K_2SO_4 dissolves in water (Interactive Figure 13.2.10), the negative end of the water dipoles are attracted to the K^+ ions and the positive end of the water dipoles are attracted to the SO_4^{2-} ions. Note that the ions in the solid separate during the dissolution process; the covalent bonds in the SO_4^{2-} ion are not broken when the solid dissolves.

    The strength of an ion–dipole IMF is directly related to the hydration enthalpy of an ion. The **enthalpy of hydration (ΔH_{hyd})** is the enthalpy change when one mole of

Video Solution

Tutored Practice
Problem 13.2.1

Figure 13.2.8 In this mixture of polar H_2O, nonpolar CCl_4, and purple nonpolar I_2, the two nonpolar species form a solution in the bottom, more dense layer.

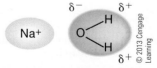

Figure 13.2.9 Sodium ions are attracted to the negative end of the permanent dipole in water.

a gaseous ion dissolves in water, forming a hydrated ion. The enthalpy of hydration is an exothermic process and is thus generally favored. Trends in ΔH_{hyd} are related to Coulomb's law, as shown by the values in Table 13.2.2. As ion size increases (and ion charge is held constant), the two species attracted to each other are farther apart and the enthalpy of hydration decreases. As ion charge increases (and ion size is held constant), the attractive force increases and so does the enthalpy of hydration. Notice that ions with small radii and large charge magnitude have the greatest (most negative) enthalpies of hydration.

Entropy also plays a role in the dissolution of ionic compounds in water. When ions are released into water, water molecules surround the ion, forming a hydrated complex. These water molecules are now constrained from free movement by the presence of the ions, decreasing the entropy of the system. The degree to which this decrease in entropy occurs depends on how many water molecules are constrained, a quantity called the *hydration number* (Interactive Table 13.2.3).

Notice that smaller ions (such as Li^+) have much larger hydration numbers than larger ions with the same charge (such as Cs^+), and ions with large charges also have large hydration numbers. This means that the effect of entropy on dissolution is more significant for ionic compounds containing small, highly charged ions and less significant for ionic compounds containing large ions with lower charges.

To see how both enthalpy and entropy play a role in the dissolution of an ionic solid, consider the dissolution shown in Interactive Figure 13.2.10. When solid K_2SO_4 dissolves, a number of things happen.

Interactive Table 13.2.3

Hydration Numbers of Some Hydrated Ions

Ion	Hydration number	Ion	Hydration number
Li^+	22	Mg^{2+}	36
Na^+	13	Ca^{2+}	29
K^+	7	Sr^{2+}	29
Cs^+	6	Ba^{2+}	28
		Cd^{2+}	39
		Zn^{2+}	44

Interactive Figure 13.2.10

Explore the dissolution of an ionic compound in water.

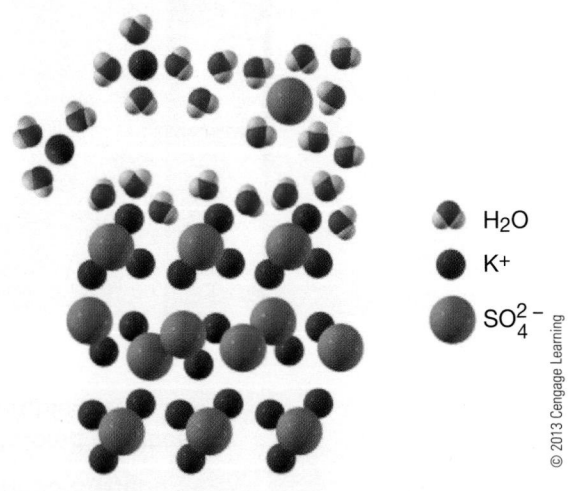

- H_2O
- K^+
- SO_4^{2-}

© 2013 Cengage Learning

Molecular-scale view of the dissolution of K_2SO_4 in water

Table 13.2.2 Enthalpies of Hydration of Selected Metal Cations (kJ/mol)

	+1 Ions			+2 Ions			+3 Ions	
Ion	Radius (pm)	ΔH_{hyd}	Ion	Radius (pm)	ΔH_{hyd}	Ion	Radius (pm)	ΔH_{hyd}
Cs	181	−263	Ra		−1259	La	117	−3283
Rb	166	−296	Ba	149	−1304	Lu	100	−3758
K	152	−321	Sr	132	−1445	Y	104	−3620
Na	116	−405	Ca	114	−1592	Sc	88	−3960
Li	90	−515						
H		−1091						

1. The K^+ and SO_4^{2-} ions are separated from each other.   Endothermic   Enthalpy disfavored

2. The K^+ and SO_4^{2-} ions are hydrated, resulting in new ion–dipole IMFs.   Exothermic   Enthalpy favored

3. The K^+ and SO_4^{2-} ions move from a highly ordered state (an ionic solid) to a more disordered state (hydrated ions).   Entropy favored

4. The water molecules move to a more constrained state in the hydrated ions.   Entropy disfavored

The dissolution of K_2SO_4 in water is both enthalpy favored and disfavored, and it is both entropy favored and disfavored. There are competing factors when ionic compounds dissolve in water, and the overall favorability of a dissolution depends on how significant each factor is in the overall process. This is an area of chemistry where we can *explain* a trend or result, but we cannot necessarily *predict* it.

Because there are competing trends governing solubility of ionic compounds, it is generally difficult to predict ionic compound solubility based on chemical formula alone. However, because of the influence of hydration enthalpy and hydration numbers on solubility, a general trend is that ionic compounds composed of small and/or highly charged ions are usually not water-soluble. For example, recall that according to the solubility rules, salts containing the highly charged anions CO_3^{2-}, PO_4^{3-}, and SO_4^{2-} are generally insoluble unless they are paired with a small ion with a low charge such as Na^+ or K^+.

Example Problem 13.2.2 Predict the relative enthalpy of hydration, hydration number, and solubility.

Which of the following ions will have the enthalpy of hydration with the largest magnitude? Which will have the greatest hydration number? Which will be most favored to dissolve in water in terms of entropy?

     Mg^{2+}, Ca^{2+}, Ba^{2+}, Al^{3+}

Solution:

You are asked to predict the ion with the greatest enthalpy of hydration, the one with the greatest hydration number, and the one that is most favored, in terms of entropy, to dissolve in water.

You are given the formulas of some ions.

Hydration enthalpy increases in magnitude as the ion charge increases and as radius decreases. Of these ions, Al^{3+} has both the highest charge and the smallest radius, so it has the largest magnitude hydration enthalpy. Hydration number also follows the same trends, and Al^{3+} will have the largest hydration number.

◄ -

Example Problem 13.2.2 *(continued)*

In terms of entropy of dissolution, the smaller the number of water molecules bonding to the dissolved ions, the more favored the dissolution, and therefore the ion with the smallest hydration number—that is, the ion with the lowest charge and the largest radius—will be most favored to dissolve in water. Of these, Ba^{2+} will have the smallest hydration number and will be the most favored to dissolve in terms of entropy.

Video Solution

Tutored Practice
Problem 13.2.2

Section 13.2 Mastery

13.3 External Control of Solubility

13.3a Pressure Effects: Solubility of Gases in Liquids

Pressure has a large effect on the solubility of gases in liquids but a negligible effect on the solubility of liquids or solids in a liquid. We therefore restrict our discussion to the effect of pressure on gas–liquid mixtures.

When liquid and gas are combined in a sealed container, an equilibrium is established where the rate at which gas molecules are dissolving in and escaping from the solution are equal and the amount of dissolved gas is constant. At the molecular level, the dissolution involves collision of gas molecules with the surface of the liquid. Those collisions can result in the solvent molecules surrounding the gas molecules, thus dissolving the gas. At the same time, previously dissolved gas solute molecules find their way to the surface, and if they have enough energy, they escape back into the gas phase (Figure 13.3.1a).

If the pressure in the flask is increased, either by introducing more gas into the container or decreasing the volume of the container, the solubility of the gas in the liquid increases (Figure 13.3.1b).

As pressure is increased, there are more collisions between the gas molecules and the solvent; thus the rate of dissolution increases. A new equilibrium is established with a greater amount of dissolved gas. The solubility of a gas in a liquid, therefore, depends on the pressure of the gas above the liquid. This relationship is described by **Henry's law**, which states that *the solubility of a gas is proportional to the pressure of the gas above the solution.*

$$S = k_H \times P \qquad (13.6)$$

where

    S = solubility
    k_H = Henry's law constant for a specific solute, solvent, and temperature
    P = partial pressure of the solute gas

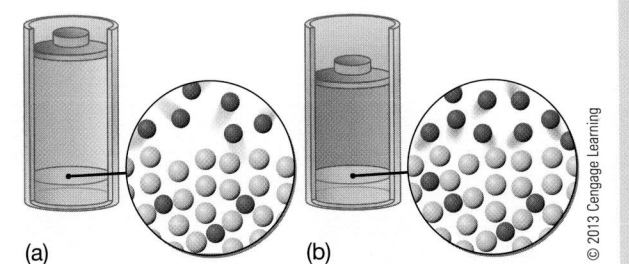

(a)   (b)

Figure 13.3.1 Molecular-scale representations of (a) a gas–liquid mixture in a sealed container and (b) the effect of pressure on the solubility of the gas

© 2013 Cengage Learning

Consider the bottle of carbonated beverage in Figure 13.3.2. The solubility of CO_2 in the beverage is directly related to the partial pressure of the CO_2 above the liquid. Before the bottle is opened, the partial pressure of CO_2 is high, greater than 1 atm. The system is at equilibrium, and a large amount of CO_2 is dissolved. When the bottle is first opened, the "head gas" in the space above the beverage escapes and the partial pressure of CO_2 drops to almost zero because the surrounding atmosphere contains very little CO_2. The solution is now supersaturated in CO_2 because the amount dissolved is greater than its solubility at that temperature, based on the new low pressure of CO_2 in the air above the beverage. The CO_2 therefore escapes out of the soda. After some time, the system reaches equilibrium again with a new, lower solubility of CO_2 in line with its new low atmospheric partial pressure. The soda is flat.

Some Henry's law constants are given in Table 13.3.1. The magnitude of the constants reflects the relative solubility of each gas. For example, CO_2 is much more soluble in water than is O_2 or N_2.

Henry's law predicts that the solubility of a gas is proportional to its partial pressure above the solution. This relationship is shown graphically in Interactive Figure 13.3.3, where gas solubility is plotted as a function of partial pressure of the gas, at a constant temperature. The slope of each line is equal to the Henry's law constant for the gas at that temperature. The steeper the line, the greater the k_H value and the more soluble the gas.

Charles D. Winters

Figure 13.3.2 Bubbles are released when a carbonated beverage is opened.

Table 13.3.1 Some Henry's Law Constants for the Solubility of Gases in Water (25 °C)

Gas	k_H (mol/L · atm) at 25 °C
N_2	6.47×10^{-4}
O_2	1.28×10^{-3}
CO_2	3.36×10^{-2}
He	3.26×10^{-4}

Example Problem 13.3.1 Calculate the solubility of a gas in a solvent.

What is the solubility of oxygen (in units of grams per liter) in water at 25 °C, when the O_2 has a partial pressure of 0.200 atm (typical of oxygen in air under normal conditions)?

Solution:

You are asked to calculate the solubility of a gas in water at a given temperature.

You are given the identity of the gas and the partial pressure of the gas at a given temperature.

The solubility of O_2 is given by Henry's law:

$$S(O_2) = k_H(O_2) \times P(O_2)$$
$$S(O_2) = (1.28 \times 10^{-3} \text{ mol/L} \cdot \text{atm})(0.200 \text{ atm}) = 2.56 \times 10^{-4} \text{ mol/L}$$

$$2.56 \times 10^{-4} \text{ mol/L} \times \frac{32.00 \text{ g}}{1 \text{ mol } O_2} = 8.19 \times 10^{-3} \text{ g/L}$$

Video Solution

Tutored Practice
Problem 13.3.1

13.3b Effect of Temperature on Solubility

Changes in temperature affect both the entropy (particle mobility) and enthalpy of a chemical system. In terms of entropy, as temperature increases, molecular motion increases and so does entropy. Therefore, an increase in temperature always makes the change in entropy larger, making the solution process more favorable.

When considering the effect of temperature on the enthalpy of the solution process, we must take into account the nature of the solute and solvent, the IMFs between the solute and solvent, and the enthalpy change for the dissolution process. Consider a gas–liquid mixture. When a pot of water is placed on a stove and heated, you have probably noticed that bubbles form on the inside of the pot when the water starts to heat up. The bubbles are the result of dissolved gases escaping from solution; as the water temperature increases, the solubility of the dissolved gases in the water decreases. We can explain this by considering the IMFs that are formed when a gas dissolves in a liquid.

Gas dissolution processes are generally exothermic because new IMFs are formed between the gas and solvent molecules but no IMFs are broken between the gas molecules.

1. IMFs between solvent molecules are broken.　　Endothermic　Enthalpy disfavored

2. New IMFs between gas and solvent molecules are formed.　Exothermic　Enthalpy favored

We can write a chemical equation to represent the dissolution of oxygen in water, including heat as a product of the reaction.

$$O_2(g) \rightarrow O_2(aq) + heat$$

Raising the temperature of the mixture adds heat to the system, which favors the reverse process and decreases the solubility of the gas (Figure 13.3.4).

$$heat + O_2(aq) \rightarrow O_2(g)$$

The heat added to the solution both increases molecular motion and overcomes the IMFs between the dissolved gases and the solvent, thus decreasing the solubility of the gas in the solution.

The effect of temperature on the solubility of solids is more complicated. You have probably experienced the fact that sugar is easier to dissolve in hot water than in cold water. What you are observing, however, is an increase in the speed at which the solid dissolves, not necessarily an increase in the solubility of the solid. However, the solubility of solids does vary with temperature, and the solubility of most but not all solids increases with increasing temperature (Interactive Figure 13.3.5).

Explore Henry's law.

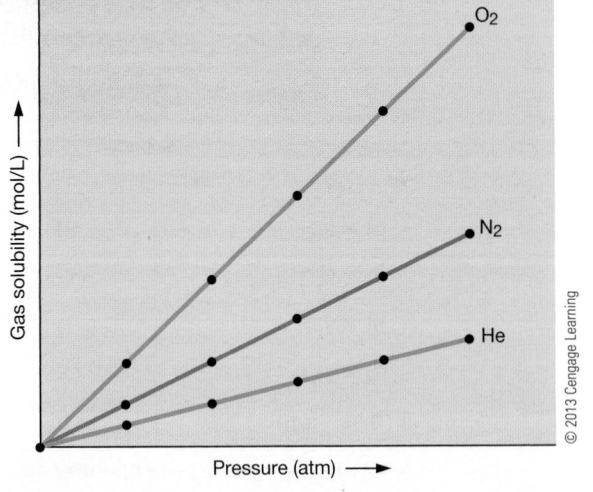

© 2013 Cengage Learning

Plots of solubility vs. pressure for three different gaseous solutes in water at 25 °C

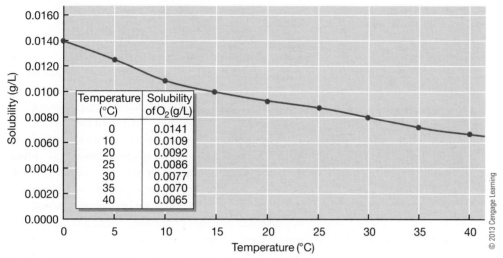

Temperature (°C)	Solubility of O_2 (g/L)
0	0.0141
10	0.0109
20	0.0092
25	0.0086
30	0.0077
35	0.0070
40	0.0065

© 2013 Cengage Learning

Figure 13.3.4 Solubility of oxygen in water at various temperatures

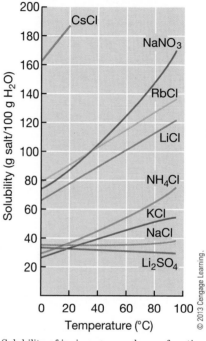

Interactive Figure 13.3.5

Investigate the relationship between temperature and solubility of ionic compounds.

© 2013 Cengage Learning.

Solubility of ionic compounds as a function of temperature

This effect of temperature on solubility is tied to many factors, such as the strength of the IMFs in the solid and the hydration of the ions when the ionic compound dissolves, which are the result of the charge and radius of the ions that make up the solid. In addition, for many ionic compounds, the **enthalpy of dissolution**, or *heat of solution*, the enthalpy change for the process of solution formation, plays a role in the relationship between temperature and solubility. Consider the dissolution of ammonium nitrate, an endothermic process.

$$NH_4NO_3(s) \rightarrow NH_4NO_3(aq) \qquad \Delta H° = +25.7 \text{ kJ/mol}$$

This equation can also be written as

$$\text{heat} + NH_4NO_3(s) \rightarrow NH_4NO_3(aq)$$

As the solid dissolves, it absorbs heat energy from the solvent. If the temperature of the water is increased, some of the additional heat energy is used to dissolve additional solid. Therefore, the solubility of NH_4NO_3 increases with increasing temperature. Compounds with an exothermic heat of solution, such as lithium sulfate, typically show a decrease in solubility with increasing temperature.

$$\text{Li}_2\text{SO}_4(\text{s}) \rightarrow \text{Li}_2\text{SO}_4(\text{aq}) \qquad \Delta H^\circ = -29.8 \text{ kJ/mol}$$

Note that the complexity of the dissolution process means that the enthalpy change for the dissolution of an ionic compound is not always a perfect predictor of the effect of temperature on solubility. For example, lithium chloride has an exothermic enthalpy of dissolution, but its solubility increases with increasing temperature.

$$\text{LiCl}(\text{s}) \rightarrow \text{LiCl}(\text{aq}) \qquad \Delta H^\circ = -37.0 \text{ kJ/mol}$$

Section 13.3 Mastery

13.4 Colligative Properties

13.4a Osmotic Pressure

Up to this point, we have discussed the ability of a solute to dissolve in a solvent. We now turn to how the presence of a dissolved solute affects the properties of the solvent. Properties that are related to the concentration of solute particles, as opposed to the type of solute, are called **colligative properties**. The colligative properties we study here are osmotic pressure, vapor pressure, boiling point, and melting point. In each case, the presence of the solute has the same general effect on the solvent; a solute tends to block solvent molecules from leaving the solution.

Osmosis is the flow of solvent through a **semipermeable membrane** (a thin sheet of material through which only certain chemical species can pass) from a solution of lower solute concentration to a solution of higher solute concentration. Consider the system shown in Figure 13.4.1, where a solution and a sample of pure water are separated by a membrane that allows passage of only water molecules.

Water molecules move through the membrane in both directions. However, the solution contains large molecules and hydrated ions that block some of the openings in the membrane. This decreases the flow of water molecules from the solution into the sample of pure water. As a result, there is an imbalance in the flow of water and a net flow of water from the solution with low solute concentration (the pure water) to the solution with a higher solute concentration (the solution on the left).

Figure 13.4.2 shows the result when osmosis is allowed to take place in a U-shaped tube containing a semipermeable membrane. Water continues to move through the membrane until the height of the solution creates sufficient pressure to stop the net flow of water.

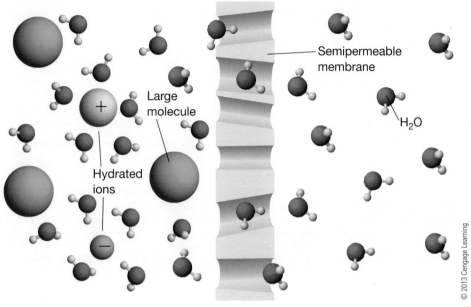

Figure 13.4.1 A solution and a sample of pure water are separated by a semipermeable membrane.

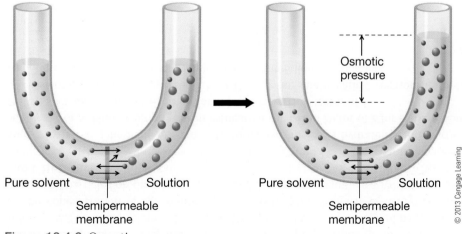

Figure 13.4.2 Osmotic pressure

Osmotic pressure is the amount of pressure required to prevent the flow of a solvent across a semipermeable membrane. This pressure is proportional to the concentration of solute particles in the solution. For a nonelectrolyte, the concentration of particles is equal to the concentration of the solute. For an electrolyte, we must take into account the number of particles produced per solute particle. We use a multiplicative factor, called the **van't Hoff factor**, i, to account for the nature of electrolytes in solution. When a strong electrolyte such as NaCl dissolves in water, the Na^+ and Cl^- ions separate and each becomes hydrated. One mole of solid NaCl forms two moles of hydrated solute particles in solution. For a dilute solution we can estimate a van't Hoff factor of 2 for NaCl. A nonelectrolyte such as sugar, on the other hand, has a van't Hoff factor of 1.

Osmotic pressure is represented by Π, and the mathematical relationship between osmotic pressure and concentration is given by the following equation:

$$\Pi = cRTi \qquad\qquad (13.7)$$

where

c = solution concentration (mol/L)
R = ideal gas constant (0.082057 L · atm/K · mol)
T = temperature (K)
i = van't Hoff factor

Example Problem 13.4.1 Calculate osmotic pressure for a solution containing a nonelectrolyte.

The nonvolatile, nonelectrolyte TNT (trinitrotoluene), $C_7H_5N_3O_6$ (227.10 g/mol), is soluble in benzene (C_6H_6). Calculate the osmotic pressure generated when 14.3 g of TNT is dissolved in 242 mL of a benzene solution at 298 K.

Solution:

You are asked to calculate osmotic pressure for a solution.

You are given the mass of solute in the solution, the volume of the solution, and the temperature of the solution.

$$14.3 \text{ g TNT} \times \frac{1 \text{ mol TNT}}{227.1 \text{ g}} = 0.0630 \text{ mol TNT}$$

$$\frac{0.0630 \text{ mol TNT}}{0.242 \text{ L}} = 0.260 \text{ M}$$

The van't Hoff factor (i) for the solute is 1 because it is a nonelectrolyte.

$$\Pi = cRTi = (0.260 \text{ M})(0.082057 \text{ L} \cdot \text{atm/K} \cdot \text{mol})(298 \text{ K})(1) = 6.36 \text{ atm}$$

Video Solution

Tutored Practice
Problem 13.4.1

van't Hoff factors are determined experimentally and are dependent on the nature of the solute and the interactions between the solute and the solvent. The values also vary with concentration. In solutions with a high concentration of an ionic solute, the formation of *ion pairs*, clusters of oppositely charged ions, reduces the effective number of particles in solution (Figure 13.4.3).

Although we can estimate a van't Hoff factor from the formula of an ionic solute by assuming it dissociates completely in solution, more accurate van't Hoff factors for strong and weak electrolytes come from experiments.

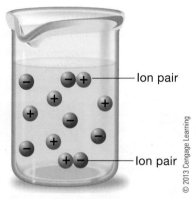

Figure 13.4.3 Formation of ion pairs

> ### Example Problem 13.4.2 Calculate osmotic pressure for a solution containing an electrolyte.
>
> The ionic compound $CaCl_2$ is soluble in water. Calculate the osmotic pressure generated when 0.249 g of $CaCl_2$ is dissolved in 151 mL of an aqueous solution at 298 K. The van't Hoff factor for $CaCl_2$ in this solution is 2.8.
>
> **Solution:**
>
> **You are asked** to calculate osmotic pressure for a solution containing an ionic solute.
>
> **You are given** the mass of solute in the solution, the volume of the solution, the temperature of the solution, and the van't Hoff factor for the ionic solute.
>
> $$0.249 \text{ g } CaCl_2 \times \frac{1 \text{ mol } CaCl_2}{111.0 \text{ g}} = 0.00224 \text{ mol } CaCl_2$$
>
> $$\frac{0.00224 \text{ mol } CaCl_2}{0.151 \text{ L}} = 0.0149 \text{ M}$$
>
> $$\Pi = cRTi = (0.0149 \text{ M})(0.082057 \text{ L} \cdot \text{atm/K} \cdot \text{mol})(298 \text{ K})(2.8) = 1.0 \text{ atm}$$

Video Solution

Tutored Practice Problem 13.4.2

Osmotic pressure has many practical applications in nature, in the home, in commerce, and in the laboratory. For example, drinking water can be purified by the process of reverse osmosis. In **reverse osmosis**, pressure is applied to overcome the osmotic pressure of a solution. This forces a solution with high solute concentration through a semipermeable membrane. When an aqueous salt solution (brine) is desalinated by this method, the result is purified water (Figure 13.4.4).

Osmotic pressures can be very large—the osmotic pressure of seawater, for example, is about 27 atm. A solution with a low solute concentration will have a low osmotic pressure, which is relatively easy to measure in the laboratory. Because solutions containing large

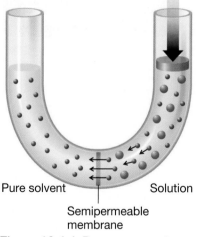

Figure 13.4.4 Reverse osmosis

molecules such as proteins and polymers typically have very low concentrations, scientists can use osmotic pressure to determine the molar mass of these large molecules. Compounds with large molar masses are often difficult to analyze using traditional instrumental methods that are designed for the analysis of small molecules.

Example Problem 13.4.3 Determine the molar mass of a solute from osmotic pressure.

In a laboratory experiment, a student found that a 229.5-mL aqueous solution containing 11.34 g of a compound had an osmotic pressure of 15.4 mm Hg at 298 K. The compound was also found to be nonvolatile and a nonelectrolyte. What is the molar mass of this compound?

Solution:

You are asked to calculate the molar mass of a compound.

You are given the osmotic pressure, temperature, and volume of a solution containing the compound and the mass of compound in the solution.

Use the osmotic pressure, and temperature to calculate the concentration of solute in the solution.

$$15.4 \text{ mm Hg} \times \frac{1 \text{ atm}}{760 \text{ mm Hg}} = 0.0203 \text{ atm}$$

$$\Pi = cRTi$$

$$0.0203 \text{ atm} = c(0.082057 \text{ L} \cdot \text{atm/K} \cdot \text{mol})(298 \text{ K})(1)$$

$$c = 8.30 \times 10^{-4} \text{ mol/L}$$

Use the concentration and the volume of the solution to calculate moles of solute.

$$\frac{8.30 \times 10^{-4} \text{ mol solute}}{1 \text{ L}} \times 0.2295 \text{ L} = 1.90 \times 10^{-4} \text{ mol solute}$$

Use the mass of solute and the moles of solute calculated above to calculate molar mass.

$$\text{molar mass} = \frac{11.34 \text{ g}}{1.90 \times 10^{-4} \text{ mol}} = 5.96 \times 10^{4} \text{ g/mol}$$

Osmosis also occurs regularly in your body as water passes through cell membranes. Figure 13.4.5 shows the effect of placing a red blood cell in three different types of solutions. In the **isotonic** solution, the solute concentrations inside and outside the cell are identical. Here there is no net flow of solvent (water) and the cell appears normal. In the **hypotonic** solution, the solute concentration is lower outside the cell. There is a net flow

Video Solution

Tutored Practice
Problem 13.4.3

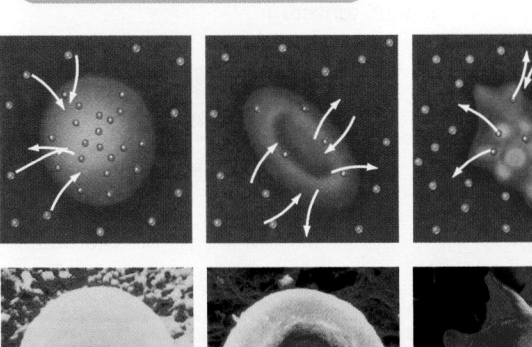

David Phillips/Photo Researchers Inc.

| Cells expand in solution of lower solute concentration (a hypotonic solution) | Normal cells in isotonic solution | Cells shrink in solution of greater solute concentration (a hypertonic solution) |

Figure 13.4.5 Osmosis and red blood cells

of solvent molecules into the cell, causing it to swell. In the **hypertonic** solution, the solute concentration is greater outside the cell. Solvent moves out of the cell, causing it to shrink.

13.4b Vapor Pressure Lowering

As you saw in Intermolecular Forces and the Liquid State (Unit 11), all liquids evaporate to form a vapor, and the pressure of the vapor (when it is in dynamic equilibrium with the liquid) is a function of temperature and the IMFs in the liquid. Adding a solute to a liquid has the effect of lowering the equilibrium vapor pressure. Solvents and solutes can be described as *volatile* or *nonvolatile*, where **volatility** is the tendency for molecules of a substance to escape into the gas phase.

The interface between a liquid and the gas above it is somewhat similar to the semipermeable membrane between two solutions undergoing osmosis. Figure 13.4.6 shows a molecular-scale view of the effect of adding a nonvolatile solute to a volatile solvent.

When a solute is added to a volatile solvent, the solute particles block solvent molecules from escaping to the vapor phase. However, they do not block vapor molecules from

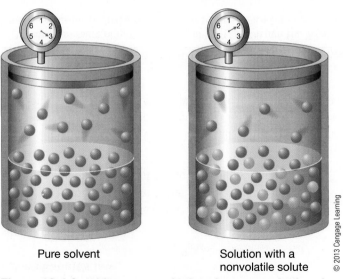

Pure solvent

Solution with a nonvolatile solute

© 2013 Cengage Learning

Figure 13.4.6 Adding a nonvolatile solute to a volatile solvent decreases the vapor pressure of the solvent.

reentering the liquid phase. Because fewer solvent molecules escape into the vapor phase and pressure is directly related to the number of molecules in the gas phase, the result of adding a solute to a volatile solvent is a lowering of the solvent vapor pressure.

The lowering of a liquid's vapor pressure by the presence of a solute mathematically follows **Raoult's law**:

$$P_{solution} = P^o_{solvent}\chi_{solvent}i \qquad \qquad \textbf{(13.8)}$$

where

$P_{solution}$ = vapor pressure of the solvent in the solution
$P^o_{solvent}$ = vapor pressure of the pure solvent
$\chi_{solvent}$ = mole fraction of the solvent in the solution
i = van't Hoff factor

A solution that obeys Raoult's law at all concentrations is said to be an **ideal solution**. In an ideal solution, interactions between the solute and the solvent are minimal. When a solution contains a volatile solute or a solute that interacts with the solvent, deviations from Raoult's law occur.

Example Problem 13.4.4 Calculate the vapor pressure of a solution.

The vapor pressure of hexane (C_6H_{14}) at 50 °C is 399 mm Hg. What is the vapor pressure of a solution consisting of 70.0 g hexane and 0.100 mol of a solute that is a nonvolatile nonelectrolyte?

Solution:

You are asked to calculate the vapor pressure of a solution.

You are given the vapor pressure of a solvent at a given temperature and the quantities of solute and solvent in the solution.

The vapor pressure of the solution is calculated using Raoult's law.

$$P_{solution} = P^o_{hexane}\chi_{hexane}i$$

The van't Hoff factor is 1 because the solute is a nonelectrolyte. First calculate the mole fraction of hexane.

$$\text{mol hexane} = 70.0 \text{ g} \times \frac{1 \text{ mol hexane}}{86.18 \text{ g hexane}} = 0.812 \text{ mol}$$

$$\chi_{hexane} = \frac{\text{mol hexane}}{\text{mol hexane + mol solute}} = \frac{0.812}{0.812 + 0.100} = 0.890$$

Next, calculate the vapor pressure of the solvent in the solution.

$$P_{solution} = (399 \text{ mm Hg})(0.890)(1) = 355 \text{ mm Hg}$$

Is your answer reasonable? A nonvolatile solute was added to the solvent, so the vapor pressure of the solution is less than the vapor pressure of the pure solvent.

Video Solution

Tutored Practice
Problem 13.4.4

An important use of vapor pressure lowering is the process of distillation, where a mixture of two liquids of different volatility are separated by first heating the mixture and then condensing the vapor by exposing it to lower temperatures. Consider a mixture containing equal amounts (moles) of water and ethanol (we will drop the van't Hoff factor in these calculations because both species are nonelectrolytes). First, the mixture is heated to 55 °C. The vapor pressures of the two liquids at this temperature are 126 mm Hg and 290 mm Hg, respectively. The mole fraction of each compound in the gas phase is proportional to the partial pressure of the gas.

$$P_{water} = P°_{water}\chi_{water} = (126 \text{ mm Hg})(0.500) = 63 \text{ mm Hg}$$

$$P_{ethanol} = P°_{ethanol}\chi_{ethanol} = (290 \text{ mm Hg})(0.500) = 145 \text{ mm Hg}$$

Next, the vapor is cooled so that it condenses to form a liquid mixture. Recall that according to Dalton's law of partial pressures, the mole fraction of each component of the mixture is directly related to its partial pressure in the vapor phase.

$$\chi_{water} = \frac{P_{water}}{P_{total}} = \frac{63 \text{ mm Hg}}{63 \text{ mm Hg} + 145 \text{ mm Hg}} = 0.303$$

$$\chi_{ethanol} = 1 - \chi_{water} = 0.697$$

This single distillation step has increased the mole fraction of ethanol from 0.500 to 0.697. Repeating this process will further separate the components of this mixture.

13.4c Boiling Point Elevation

As shown in the previous section, a solution has a vapor pressure that is lower than that of the pure solvent at a given temperature. Interactive Figure 13.4.7 shows vapor pressure–temperature curves for pure water and an aqueous sugar solution.

As defined in Intermolecular Forces and the Liquid State (Unit 11), the normal boiling point for a liquid is the temperature at which the vapor pressure of the liquid is equal to 1 atm (760 mm Hg). Because the vapor pressure of a solution at a given temperature is always lower than the vapor pressure of the pure liquid, a vapor pressure of the solvent in a solution reaches 1 atm at a temperature above the normal boiling point of the solvent. Thus, *the boiling point of a solvent in a solution is always higher than that of the pure solvent.* This effect is called boiling point elevation, and it maintains the following mathematical relationship:

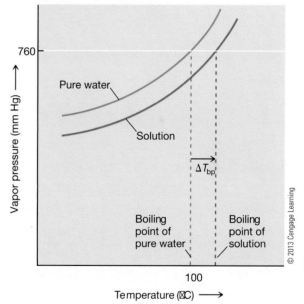

Interactive Figure 13.4.7

Explore the effect of a solute on the boiling point of a solvent.

When sugar, a nonvolatile solute, is added to a solvent, vapor pressure decreases and boiling point increases.

$$\Delta T_{bp} = K_{bp}m_{solute}i \qquad \text{(13.9)}$$

where

ΔT_{bp} = change in boiling point of the solvent

K_{bp} = boiling point elevation constant (unique for each solvent)

m_{solute} = molality of the solute (mol solute/kg solvent)

i = van't Hoff factor

Table 13.4.1 shows boiling points and boiling point elevation constants for some common solvents. Notice that not all solvents are equally susceptible to changes in boiling points. Nitrobenzene, for example, is about 10 times more sensitive to changes in boiling point than water.

Table 13.4.1 Boiling Points and Elevation Constants for Common Solvents

Solvent	T_{bp} (°C)	K_{bp} (°C/m)
Water	100	0.512
Benzene	80.1	2.53
Acetic acid	118.1	3.07
Nitrobenzene	210.9	5.24
Phenol	182	3.56
Camphor	207.4	5.61

Example Problem 13.4.5 Calculate the boiling point of a solution.

What is the boiling point of a solution containing 40.0 g I_2 and 250 g benzene (C_6H_6)?

Solution:

You are asked to calculate the boiling point of a solution.

You are given the mass of solute and solvent in the solution.

The boiling point of the solution is the normal boiling point plus the boiling point elevation. Using data from Table 13.4.1,

$$T_{bp} = 80.1 \text{ °C} + (2.53 \text{ °C}/m)(m_{solute})i$$

The concentration of I_2, a nonelectrolyte, is

$$m_{I_2} = \frac{\text{mol } I_2}{\text{kg benzene}} = \frac{(40.0 \text{ g})\left(\dfrac{1 \text{ mol } I_2}{253.8 \text{ g}}\right)}{0.250 \text{ kg benzene}} = 0.630 \ m$$

Use equation 13.9 to calculate the boiling point of the solution.

$$T_{bp} = 80.1 \text{ °C} + (2.53 \text{ °C}/m)(0.630 \ m)(1) = 81.7 \text{ °C}$$

Is your answer reasonable? A solute was added to the solvent, so the boiling point of the solution is greater than the boiling point of the pure solvent.

Video Solution

Tutored Practice
Problem 13.4.5

Boiling point elevation can be used to determine the molar mass of a solute, as shown in the following example.

Example Problem 13.4.6 Determine the molar mass of a solute from the boiling point of a solution.

In a laboratory experiment, a student found that a solution consisting of 0.315 g of an unknown compound (a nonvolatile nonelectrolyte) and 25 g of $CHCl_3$ has a boiling point of 62.09 °C. The normal boiling point of $CHCl_3$ is 61.70 °C. What is the molar mass of the compound? (K_{bp} = 3.63 °C/m for $CHCl_3$)

Solution:

You are asked to calculate the molar mass of a compound.

You are given the boiling point of the solvent and the solution and the mass of the solute and solvent.

Use the boiling point of the pure liquid and the solution to calculate the molality of the solute.

$$\Delta T_{bp} = K_{bp}m_{solute}i$$
$$(62.09 - 61.70) \text{ °C} = (3.63 \text{ °C/}m)(m_{solute})(1)$$
$$m_{solute} = 0.107 \ m$$

Use the molality and the mass of solvent to calculate moles of solute.

$$\frac{0.107 \text{ mol solute}}{\text{kg } CHCl_3} \times 0.025 \text{ kg } CHCl_3 = 0.00269 \text{ mol solute}$$

Use the mass of solute and the moles of solute calculated above to calculate molar mass.

$$\text{molar mass} = \frac{0.315 \text{ g}}{0.00269 \text{ mol}} = 117 \text{ g/mol}$$

13.4d Freezing Point Depression

As you saw in The Solid State (Unit 12), the normal freezing (melting) point for a substance is the temperature at which the solid and liquid phases are in equilibrium at 1 atm pressure. At the freezing point, solvent particles move from the solid into the liquid phase at the same rate the particles in the liquid phase condense to form a solid (Figure 13.4.8a). In a solution, however, solute particles interfere with this dynamic equilibrium by decreasing the rate at which solvent particles condense from liquid to solid (Figure 13.4.8b).

In order for the rates of these two processes to be equal and for the solvent to freeze, the temperature of a solution must be lowered. Lowering the temperature of the solution

(a)   (b)

Figure 13.4.8 (a) The dynamic equilibrium between solid and liquid in a pure solvent; (b) the solid–liquid equilibrium in a solution

© 2013 Cengage Learning

Video Solution

Tutored Practice Problem 13.4.6

decreases the rate at which solid particles move into the liquid phase, thus allowing the two processes to reach a dynamic equilibrium. Thus, *the freezing point for a solvent in a solution is always lower than the freezing point of the pure solvent.* This effect is called freezing point depression, and it maintains the following mathematical relationship:

$$\Delta T_{fp} = K_{fp} m_{solute} i \qquad \textbf{(13.10)}$$

where

ΔT_{fp} = change in freezing point of the solvent
K_{fp} = freezing point depression constant (unique for each solvent)
m_{solute} = molality of the solute (mol solute/kg solvent)
i = van't Hoff factor

Table 13.4.2 shows freezing points and freezing point depression constants for some common solvents.

Table 13.4.2 Freezing Points and Depression Constants for Common Solvents

Solvent	T_{fp} (°C)	K_{fp} (°C/m)
Water	0	1.86
Benzene	5.5	5.12
Acetic acid	16.6	3.90
Nitrobenzene	5.7	7.00
Phenol	43	7.40
Camphor	178.4	40.0

Example Problem 13.4.7 Calculate the freezing point of a solution.

What is the freezing point of a solution containing 40.0 g I_2, a nonelectrolyte, and 250 g benzene (C_6H_6)?

Solution:

You are asked to calculate the freezing point of a solution.

You are given the quantity of solute and solvent in the solution.

The freezing point is the normal freezing point minus the freezing point depression. Using data from Table 13.4.2,

$$T_{fp} = 5.5 \ °C - (5.12 \ °C/m)(m_{solute})i$$

The molality of I_2 is

$$m_{I_2} = \frac{mol \ I_2}{kg \ benzene} = \frac{(40.0 \ g)\left(\dfrac{1 \ mol \ I_2}{253.8 \ g}\right)}{0.250 \ kg \ benzene} = 0.630 \ m$$

Use equation 13.10 to calculate the freezing point of the solution.

$$T_{fp} = 5.5 \ °C - (5.12 \ °C/m)(0.630 \ m)(1) = 2.3 \ °C$$

Video Solution

Tutored Practice
Problem 13.4.7

An interesting (and useful) application of freezing point depression is when salt is combined with ice in an ice cream maker. To cool an ice cream solution (which consists of sugar, milk, eggs, and flavorings) to form solid ice cream, large amounts of salt are combined with ice and placed in a chamber surrounding the ice cream ingredients. The salt–ice solution has a low melting point (typically around −10 °C), so it cools the ice cream mixture until it freezes.

As shown in Figure 13.4.9, boiling point elevation and freezing point depression combine to result in an increase in the temperature range over which a solvent remains in the liquid phase when a solute is dissolved in the solvent. A common application of this effect is the use of a water–ethylene glycol (or propylene glycol) solution in a car radiator. A 50% (by weight) mixture of water and ethylene glycol results in a solution that freezes at −34 °C and boils at 107 °C (at a pressure of 1 atm), a mixture that eliminates radiator "freeze-up" and "boil-over."

Notice that we have discussed the effect of lowering the temperature on the freezing point of the solvent in a solution. What happens to the solute as the solvent freezes? Under most conditions, only the solvent molecules condense into a solid; the solute particles remain in solution. Figure 13.4.10 shows the result of cooling an aqueous solution containing a colored dye below its freezing point. Notice that the solid, which forms along the test tube walls, is colorless because it is pure ice (solid water). The unfrozen solution, in the center of the test tube, is a concentrated dye solution.

Section 13.4 Mastery

13.5 Other Types of Mixtures

13.5a Alloys

Alloys are mixtures of different metals, almost always in the solid state. Alloys can be heterogeneous mixtures where small crystalline regions of each metal are intermixed to form the solid. This type of structure can often be seen on old brass door handles. Fully homogeneous alloys also exist, where atoms of the two metals are completely intermixed. Some historically important alloys include brass and bronze (Figure 13.5.1). Amalgams are a special type of alloy where mercury acts as the solvent and other metals serve as the solute. Amalgams of mercury and other metals have been used for decades for dental fillings

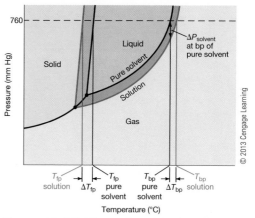

Figure 13.4.9 Phase diagrams for a pure solvent and its solution, showing that the solution has a greater temperature range for the liquid state

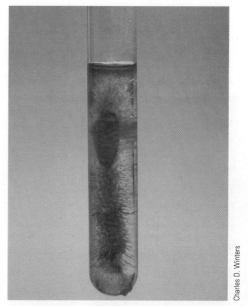

Figure 13.4.10 The result of freezing a solution of a dye in water

because they can be formulated to melt at temperatures not much above body temperature, inserted into cavities in teeth, and allowed to cool and solidify.

Another important class of alloys is steel. Steel is an alloy of iron with carbon and many other trace elements. Inclusion of different elements lends different properties to the steel. Interactive Table 13.5.1 shows some common alloys and their composition.

13.5b Colloids

A **colloid** is a mixture where one substance is distributed uniformly throughout another. The particles that are dispersed in the solvent are larger than the size of a molecule but are invisible to the naked eye (typically 1–1000 nm in diameter). Colloids are classified according to the phases of their components. For example, a **sol** is defined as a colloidal dispersion of a solid in a medium that flows (paint and mud are examples of sols) and a **gel** is a colloidal dispersion with a structure that prevents it from flowing (jam and jelly are examples of gels) (Interactive Table 13.5.2).

Figure 13.5.1 Brass, a metal–metal mixture of copper and zinc.

Interactive Table 13.5.1

Some Common Alloys

Alloy	Composition
Sterling silver	92.5% Ag, 7.5% Cu
18 K "yellow" gold	75% Au, 12.5% Ag, 12.5% Cu
Pewter	91% Sn, 7.5% Sb, 1.5% Cu
Low-alloy steel	98.6% Fe, 1.0% Mn, 0.4% C
Carbon steel	Approximately 99% Fe, 0.2–1.5% C
Stainless steel	72.8% Fe, 17.0% Cr, 7.1% Ni, and approximately 1% each of Al and Mn
Alnico magnets	10% Al, 19% Ni, 12% Co, 6% Cu, remainder Fe
Brass	95–60% Cu, 5–40% Zn
Bronze	90% Cu, 10% Sn

Interactive Table 13.5.2

Types of Colloids

Continuous Phase	Dispersed Phase	Type	Examples
Gas	Liquid	Aerosol	Fog, clouds, aerosol sprays
Gas	Solid	Aerosol	Smoke, airborne viruses, automobile exhaust
Liquid	Gas	Foam	Shaving cream, whipped cream
Liquid	Liquid	Emulsion	Mayonnaise, milk, face cream
Liquid	Solid	Sol	Gold in water, milk of magnesia, mud
Solid	Gas	Foam	Foam rubber, sponge, pumice
Solid	Liquid	Gel	Jelly, cheese, butter
Solid	Solid	Solid sol	Milk glass, many alloys such as steel, some colored gemstones

Emulsions

An **emulsion** is a heterogeneous mixture of two liquids that does not separate into two distinct phases. Consider milk, which consists of a "continuous phase" of water containing "emulsified" globules of nonpolar fat. This is a typical example of an "oil in water" emulsion. Figure 13.5.2 shows a photo of such an emulsion. Emulsions can be formed in the opposite fashion, with a continuous nonpolar oil phase containing globules of water (Figure 13.5.3).

Water and oil do not normally mix, so forming an emulsion requires an external reagent called an emulsifier, a substance that stabilizes the emulsion. One example of an emulsifier is a **surfactant**, a chemical species that has both nonpolar and polar properties.

Figure 13.5.2 An oil-in-water emulsion

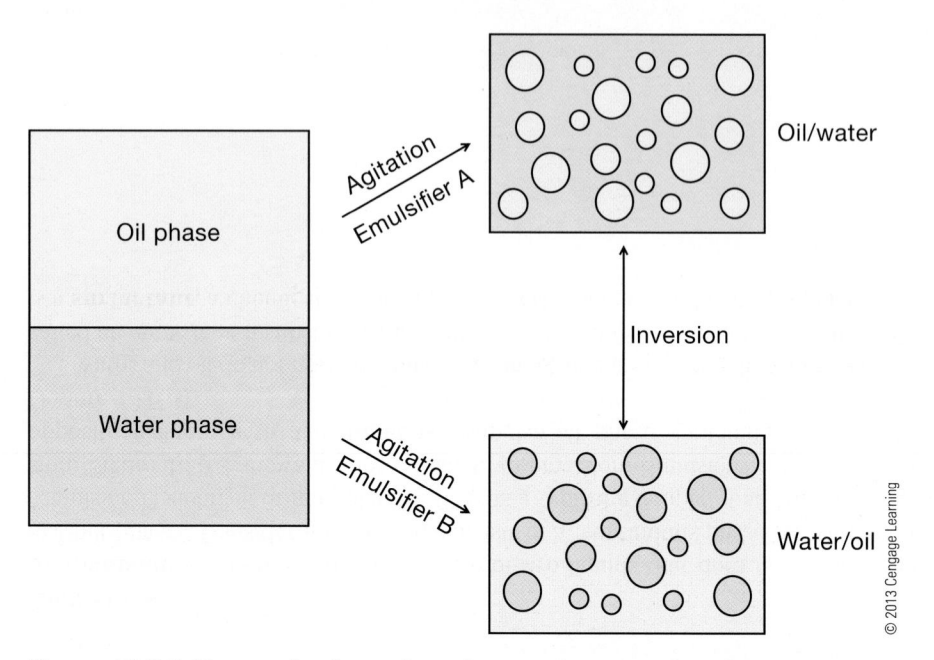

Figure 13.5.3 Process for formation of two main types of emulsions

Emulsions are of great importance in commerce. They are used to form dairy products such as milk and butter, cosmetics such as hand lotion and cold cream, floor and car oils and "waxes," paint and other pigments, and delivery systems for pharmaceuticals.

Aerosols

An **aerosol** is a mixture of small particles of either liquid or solid in a gas. The term *aerosol* is often used when referring to "spray cans" in which a liquid is dispersed into the air in tiny droplets. Other aerosols exist in nature. The haze you see on a hazy day is an aerosol of water in the atmosphere. Solid aerosol particles represent a health risk and are regulated.

Section 13.5 Mastery

Unit Recap

Key Concepts

13.1 Quantitative Expressions of Concentration

- When a solute is combined with a solvent, the resulting solution can be saturated, unsaturated, or supersaturated (13.1a).

- The solute and solvent in a liquid–liquid solution can be described as miscible or immiscible (13.1a).

- The amount of solute dissolved in a solvent can be expressed using concentration units such as molarity, molality, weight percent, mole fraction, parts per million, or parts per billion (13.1b).

13.2 Inherent Control of Solubility

- Both the entropy and enthalpy of a chemical system influence the dissolution process (13.2a).

- When a process is favored by both entropy and enthalpy, it is favored overall. If the process is disfavored by both entropy and enthalpy, it will be disfavored overall and not occur to any appreciable extent. When a process is favored by one and disfavored by the other, its overall favorability depends on temperature (13.2a).

- Gases mix completely in all proportions at any temperature (13.2b).

- The solubility of a liquid solute in a liquid solvent is dependent on the relative strength of the IMFs in both the pure liquids and the solution (13.2c).

- In general, liquids with similar IMFs are miscible; that is, "like dissolves like" (13.2c).

- The solubility of a solid solute in a liquid solvent is dependent on the relative strength of the IMFs in both the pure substances and the solution. In general, "like dissolves like" (13.2d).

- For ionic compounds dissolving in water, solubility is also a function of the strength of the ion–dipole IMFs formed when the compound dissolves, the enthalpy of hydration of the ion, and the hydration number (13.2d).

13.3 External Control of Solubility

- Gas pressure is directly related to the solubility of a gas in a solvent, a relationship described mathematically in Henry's law (13.3a).

- Temperature and the solubility of a gas in a liquid solvent are inversely related (13.3b).

- The effect of temperature on the solubility of a solid in a liquid is tied to many factors, such as the strength of the IMFs in the solid and the hydration of the ions when the ionic compound dissolves (13.3b).

- For most solid–liquid solutions, temperature and solubility are directly related (13.3b).

- For most solids, solubility in water is related to the enthalpy of dissolution for the solid (13.3b).

13.4 Colligative Properties

- Changes in the properties of a solvent are directly related to the concentration of solute particles in the solvent (13.4).

- Osmotic pressure, the result of an imbalance in the flow of solvent between a solution with low solute concentration and a solution with a higher solute concentration, is directly related to the concentration of solute particles and the temperature of the solution (13.4a).

- Application of a pressure greater than osmotic pressure forces a solution with high solute concentration through a semipermeable membrane and can be used to produce pure solvent (13.4a).

- Adding a nonvolatile solute to a volatile solvent decreases the vapor pressure of the solvent, a relationship described mathematically by Raoult's law (13.4b).

- The decrease in vapor pressure for a solvent is directly related to the concentration of solute particles in the solution (13.4b).

- Only ideal solutions follow Raoult's law. When a solution contains a volatile solute or a solute that interacts with the solvent, deviations occur (13.4b).

- The boiling point of a solvent in a solution is always higher than that of the pure solvent (13.4c).

- The increase in boiling point for a solvent is directly related to the concentration of solute particles in the solution (13.4c).

- The freezing point of a solvent in a solution is always lower than that of the pure solvent (13.4d).

- The decrease in freezing point for a solvent is directly related to the concentration of solute particles in the solution (13.4d).

13.5 Other Types of Mixtures

- Some important alloys—mixtures of metals—are steel, brass, and bronze (13.5a).

- Colloids consist of uniform distributions of one substance in another. They are very common and include emulsions and aerosols (13.5b).

Key Equations

$$\text{weight percent component A} = \frac{\text{quantity (g) component A}}{\text{total mass (g) of solution}} \times 100\% \qquad \textbf{(13.1)}$$

$$\text{molality} = \frac{\text{amount (mol) solute}}{\text{mass (kg) solvent}} \qquad \textbf{(13.2)}$$

$$\text{mole fraction component A} = \chi_A = \frac{\text{amount (mol) component A}}{\text{total amount (mol) in solution}} \qquad \textbf{(13.3)}$$

$$\text{parts per million} = \frac{\text{mass (g) solute}}{10^6 \text{ g solution}} = \frac{\text{mass (g) solute}}{\text{mass (g) solution}} \times 10^6 = \frac{\text{mass (mg) solute}}{\text{mass (kg) solution}} \qquad \textbf{(13.4)}$$

$$\text{parts per billion} = \frac{\text{mass (g) solute}}{10^9 \text{ g solution}} = \frac{\text{mass (g) solute}}{\text{mass (g) solution}} \times 10^9 = \frac{\text{mass } (\mu g) \text{ solute}}{\text{mass (kg) solution}} \qquad \textbf{(13.5)}$$

$$S = k_H \times P \qquad \textbf{(13.6)}$$

$$\Pi = cRTi \qquad \textbf{(13.7)}$$

$$P_{\text{solution}} = P^o_{\text{solvent}} \chi_{\text{solvent}} i \qquad \textbf{(13.8)}$$

$$\Delta T_{\text{bp}} = K_{\text{bp}} m_{\text{solute}} i \qquad \textbf{(13.9)}$$

$$\Delta T_{\text{fp}} = K_{\text{fp}} m_{\text{solute}} i \qquad \textbf{(13.10)}$$

Key Terms

13.1 Quantitative Expressions of Concentration
solubility
soluble
insoluble
saturated
unsaturated
supersaturated
miscible
immiscible
molality
parts per million
parts per billion

13.2 Inherent Control of Solubility
entropy
ion–dipole intermolecular force
enthalpy of hydration (ΔH_{hyd})

13.3 External Control of Solubility
Henry's law
enthalpy of dissolution

13.4 Colligative Properties
colligative properties
osmosis
semipermeable membrane
osmotic pressure
van't Hoff factor
reverse osmosis

isotonic
hypotonic
hypertonic
volatility
Raoult's law
ideal solution

13.5 Other Types of Mixtures
alloy
colloid
sol
gel
emulsion
surfactant
aerosol

Unit 13   Review and Challenge Problems

14 Chemical Kinetics

Unit Outline

In This Unit...

This unit investigates the factors that control chemical reactions. Here we study the factors that affect how fast a chemical reaction occurs, such as concentration, temperature, and the presence of a catalyst. In upcoming units we will investigate the factors that affect chemical equilibria and how energy influences chemical reactions.

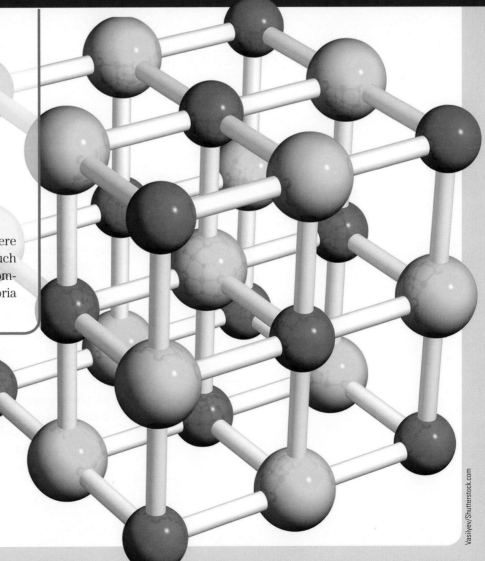

Vasilyev/Shutterstock.com

14.1 Introduction to Kinetics

14.1a Factors that Influence Reactivity

Propane reacts vigorously with oxygen in a Bunsen burner or a gas grill, but left untouched, the two chemicals will not react for years. Meats left at room temperature will invite biochemical reactions that, among other things, generate bad-smelling gases. Put the meat in a refrigerator or freezer, however, and the reactions will take much longer to occur. Burning propane and spoiling meat are just two examples of chemical reactions in which speed, or rate, is influenced by external factors.

Chemical reactivity is controlled by the thermodynamics and kinetics of the reaction. The field of thermodynamics investigates the relative stability of reactants and products. In a way, thermodynamics seeks to answer the question, Should this reaction occur at all? The field of **chemical kinetics**, the study of reaction mechanisms and the rates of chemical reactions under various conditions, seeks to answer the questions, How fast will the reaction occur, and how can the reaction speed be controlled?

For a reaction to occur to an appreciable amount, it must be both thermodynamically and kinetically favored. These are relative terms: the reaction must be thermodynamically favored *enough* to form the amount of product desired, and it must be kinetically favored *enough* to be complete on the timescale required.

The thermodynamics of a reaction are controlled by

- the enthalpy change for the reaction,
- the entropy change for the reaction, and
- the temperature at which the reaction takes place.

The kinetics of a reaction are influenced by

- the manner in which the reaction takes place (its mechanism),
- the energy barrier that must be overcome to convert reactants to products (the activation energy),
- the concentration of the species present, and
- the temperature at which the reaction takes place.

The factors that influence chemical reactivity are summarized visually in Interactive Figure 14.1.1.

Explore control of chemical reactivity.

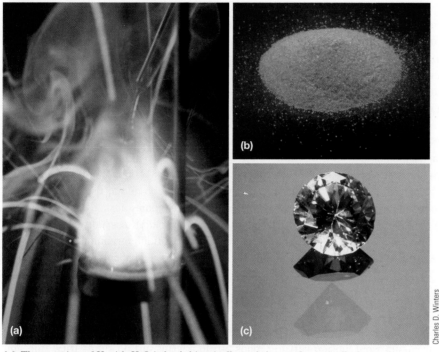

(a) The reaction of K with H_2O is both kinetically and thermodynamically favored.
(b) The conversion of SiO_2 to elemental Si and O_2 is thermodynamically unfavored.
(c) The conversion of C(diamond) to C(graphite) is kinetically unfavored.

14.1b Collision Theory

The speed at which a reaction occurs can be explained by the behavior of reacting species on a molecular level. According to the **collision theory of reaction rates**, a reaction occurs only when three conditions are met:

1. The reacting species come into contact (they collide).

2. The collision has enough energy to overcome the **activation energy**, the energy barrier necessary to initiate the reaction.

3. The reacting species collide in an orientation that allows the necessary bond breaking and bond forming needed to transform reactants to products to take place.

Collision theory tells us that while collisions between reacting species are occurring all of the time, only some fraction—not all—of those collisions will lead to conversion of reactants to products. Using this theory, we can understand how physical state of reactants, temperature, and molecular orientation influence the rate of a reaction.

Number of Collisions

The first condition required for a reaction to occur is for reacting species to come into contact. When more collisions between reacting species occur, the reaction will proceed faster.

For example, compare the reaction between a solid and a gas with the reaction between two gases. The solid–gas reaction (for example, iron and oxygen reacting to form rust) will generally occur at a much slower rate than the gas–gas reaction (for example, oxygen and methane burning in a Bunsen burner) because the gaseous molecules can collide only with atoms on the surface of the solid. For this reason, most chemical reactions are performed either in solution or in the gas phase, where the reacting species can easily come into contact with one another.

Overcoming Activation Energy

For a reaction to occur as a result of a specific collision, the collision must have enough energy to overcome the energy barrier (activation energy).

As you saw in Gases (Unit 10) and Intermolecular Forces and the Liquid State (Unit 11), the molecules in a given sample have a Boltzmann distribution of molecular speeds and thus a corresponding distribution of energies at a given temperature. In a given sample, some molecules will have high energy, some will have low energy, and most will have an intermediate energy. Because an energy barrier (the activation energy) must be overcome for reacting species to form products, only some fraction of reacting species will have enough energy to react at a given temperature.

As the temperature increases, the number of species with enough energy to react will increase (Figure 14.1.2). Because a greater fraction of molecules in the high-temperature sample will have enough energy to provide the activation energy, the high-temperature sample will have more energetic collisions and will experience a faster reaction rate. We will discuss the relationship between temperature and activation energy further later in this unit.

Reactivity also depends on the magnitude of the energy barrier the collision must overcome. As shown in Interactive Figure 14.1.3, a greater fraction of reactant molecules have enough energy to overcome a small activation energy than to overcome a large activation energy. Therefore, more reactant molecules will undergo reaction (the rate of the reaction will be higher) when there is a smaller activation barrier than when the activation barrier is larger.

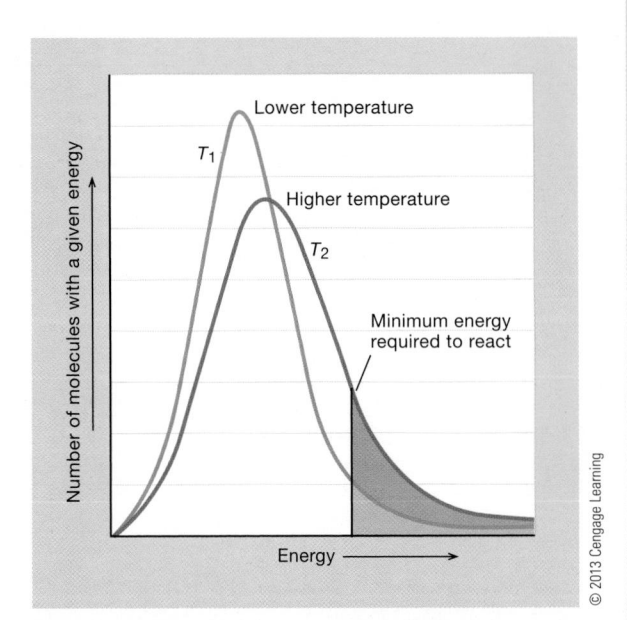

Figure 14.1.2 Boltzmann distributions of molecular energies at different temperatures

Molecular Orientations

The third condition that must be met for a reaction to occur is that the orientation of the collision must allow for the breaking and forming of bonds necessary to form products. For example, chloride ions react with the cationic species $C(CH_3)_3^+$ to form 2-chloro-2-methyl propane. If the chloride ion approaches the central carbon in the cation (Interactive Figure 14.1.4a), product is formed. If, however, the chloride ion approaches one of the terminal carbon atoms (a —CH_3 carbon), as shown in Interactive Figure 14.1.4b, the collision will not result in the formation of product. In reactions involving complex, large molecules with very specific reaction sites, it is not unusual for a small fraction of collisions to lead to products.

Interactive Figure 14.1.4

Explore the effect of molecular orientation on reaction rate.

(a) Effective and (b) ineffective collisions between Cl^- and $C(CH_3)_3^+$

Section 14.1 Mastery

14.2 Expressing the Rate of a Reaction

14.2a Average Rate and Reaction Stoichiometry

How fast does a reaction happen? We can often get a general feel for the speed of a reaction by observing the disappearance of a reactant or how quickly a product is formed. However, we often need to study reaction speed quantitatively. We express **reaction rate**, the speed at which a reaction progresses, as a ratio of change in concentration over change in time.

For the reaction A → 2 B, for example, we can express the rate at which A disappears as

$$\text{Rate} = -\frac{\Delta[A]}{\Delta t} = -\frac{[A]_{t_2} - [A]_{t_1}}{t_2 - t_1}$$

Because [A] decreases over time, the change in its concentration is negative. Reaction rate is defined using the negative of the change in a reactant concentration over the change in time. The negative sign is to follow the convention that reaction rate should have a positive value. If the reaction rate is defined by the increase in concentration of a reactant, the negative sign is not used.

Interactive Figure 14.1.3

Investigate the factors that affect a reaction's energy barrier.

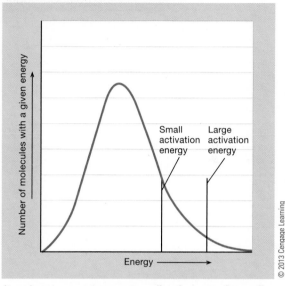

© 2013 Cengage Learning

At a given temperature, more molecules exceed a small activation energy than a large activation energy.

Consider Figure 14.2.1, which shows reactant and product concentration over a period of 8 seconds for the reaction A → 2 B. Reactant A has an initial concentration of 0.50 M, and that concentration drops to 0 M in 8 seconds. Product B has an initial concentration of 0 M, and its concentration increases to 1.0 M in 8 seconds.

An **average reaction rate** is obtained from the change in concentration of a reactant or product over a defined time interval. Using data from Figure 14.2.1, over the entire 8-second reaction time, a rate can be defined using the increase in the concentration of B.

$$\text{Rate} = \frac{\Delta[B]}{\Delta t} = \frac{(1.00 \text{ M} - 0 \text{ M})}{(8 \text{ s} - 0 \text{ s})} = 0.125 \text{ M/s}$$

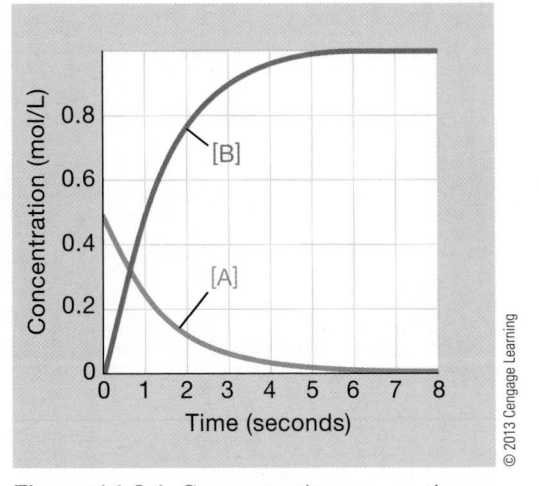

Figure 14.2.1 Concentration versus time for the reaction A → 2 B

© 2013 Cengage Learning

Example Problem 14.2.1 Calculate the average rate of disappearance from concentration–time data.

For the decomposition of hydrogen peroxide in dilute sodium hydroxide at 20 °C,

$$2 \text{ H}_2\text{O}_2(aq) \rightarrow 2 \text{ H}_2\text{O}(\ell) + \text{O}_2(g)$$

the following data have been obtained:

t (min)	$[H_2O_2]$ (mol/L)
0	0.0912
434	0.0566
868	0.0351
1302	0.0218

What is the average rate of disappearance of H_2O_2 over the period from 0 min to 434 min?

Solution:

You are asked to determine the average rate of disappearance of a reactant over a given period of time.

You are given the balanced chemical equation and data showing the reactant concentration as a function of time.

Average reaction rate is equal to the change in concentration over the change in time.

$$\text{rate} = -\frac{\Delta[H_2O_2]}{\Delta t} = -\frac{(0.0566 \text{ M} - 0.0912 \text{ M})}{(434 \text{ min} - 0 \text{ min})} = 7.97 \times 10^{-5} \text{ M/min}$$

Video Solution

Tutored Practice Problem 14.2.1

Reaction Stoichiometry

The relative rate at which reactants are consumed and products are formed is directly related to the reaction stoichiometry. Consider again the reaction $A \rightarrow 2\,B$ and the average rates over the first second of the reaction (Interactive Figure 14.2.2).

$$\text{rate of appearance of B} = \frac{\Delta[\text{B}]}{\Delta t} = \frac{(0.50\ \text{M} - 0\ \text{M})}{(1.0\ \text{s} - 0\ \text{s})} = 0.50\ \text{M/s}$$

$$\text{rate of disappearance of A} = -\frac{\Delta[\text{A}]}{\Delta t} = -\frac{(0.25\ \text{M} - 0.50\ \text{M})}{(1.0\ \text{s} - 0\ \text{s})} = 0.25\ \text{M/s}$$

Notice that the two rates have different values over the same time period. The stoichiometry of the reaction shows that two moles of B are produced for each mole of A that is consumed. Thus, B appears at a rate twice the rate at which A is consumed, or

$$\text{rate (from } t = 0\ \text{s to } t = 1.0\ \text{s)} = -\frac{\Delta[\text{A}]}{\Delta t} = \frac{1}{2}\frac{\Delta[\text{B}]}{\Delta t} = 0.25\ \text{M/s}$$

For the general reaction $a\,\text{A} + b\,\text{B} \rightarrow c\,\text{C} + d\,\text{D}$, a unique reaction rate is defined as:

$$\text{rate} = -\frac{1}{a}\frac{\Delta[\text{A}]}{\Delta t} = -\frac{1}{b}\frac{\Delta[\text{B}]}{\Delta t} = \frac{1}{c}\frac{\Delta[\text{C}]}{\Delta t} = \frac{1}{d}\frac{\Delta[\text{D}]}{\Delta t}$$

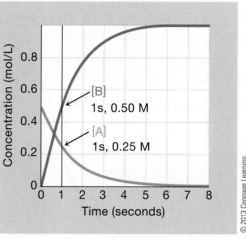
Example Problem 14.2.2 Relate average rates using stoichiometry.

For the decomposition of hydrogen peroxide in dilute sodium hydroxide at 20 °C,

$$2\,\text{H}_2\text{O}_2(\text{aq}) \rightarrow 2\,\text{H}_2\text{O}(\ell) + \text{O}_2(\text{g})$$

the average rate of disappearance of H_2O_2 over the period from $t = 0$ to $t = 516$ min is found to be 8.08×10^{-5} M/min. What is the rate of appearance of O_2 over the same period?

Solution:

You are asked to calculate the rate of appearance of a product in a chemical reaction.

You are given a balanced equation for a chemical reaction and the rate of disappearance of a reactant.

The reaction stoichiometry shows that two moles of H_2O_2 are consumed for every mole of O_2 formed. Therefore, the rate of formation of O_2 is half the rate of H_2O_2 consumption.

$$\frac{\Delta[\text{O}_2]}{\Delta t} = -\frac{1}{2}\frac{\Delta[\text{H}_2\text{O}_2]}{\Delta t} = \frac{1}{2}(8.08 \times 10^{-5}\ \text{M/min}) = 4.04 \times 10^{-5}\ \text{M/min}$$

Video Solution

Tutored Practice
Problem 14.2.2

14.2b Instantaneous and Initial Rates

Instantaneous rate, the rate of a reaction at that point in time, is equal to the slope of a line tangent to the concentration–time curve at a given point in time. One of the most important applications of instantaneous rate is in the study of the mechanism of a chemical reaction, the step-by-step process by which a chemical reaction occurs.

As shown in Figure 14.2.3, the instantaneous rate at time = 2 seconds for the A → 2 B reaction shown previously is determined by drawing tangent lines to the concentration–time curves. When 2 seconds have passed, the instantaneous rate of formation of B is +0.13 M/s and the instantaneous rate of consumption of A is −0.065 M/s.

Initial Rate

A specific type of instantaneous reaction rate is **initial rate**, the rate of the reaction at the beginning of the reaction, when time = 0 (Figure 14.2.4). The initial reaction rate is of interest experimentally because reactant concentrations are easily measured at the start of a chemical reaction.

In the A → 2 B reaction data, the initial rate of disappearance of A is determined by the slope of the line at the beginning of the reaction. The initial rate of formation of B (0.82 M/s) is twice the initial rate of disappearance of A (0.41 M/s).

Change in Rate over Time

The concentration–time plot for the A → 2 B reaction shows that as time increases, the reaction rate slows (Interactive Figure 14.2.5). In this reaction, the reaction rate is directly related to the concentration of A. When [A] is large, the slope of the curve is steep and thus the reaction is fast. When [A] is much smaller, the slope is smaller and the reaction slows.

Recall that according to collision theory, the number of effective collisions affects reaction rate. As the concentration of reactants decreases, reaction rate generally decreases due to fewer collisions with the required energy and correct geometry. Note that although it is typical for reaction rates to slow as the concentration of reactants decreases, this is not true for all chemical reactions.

> Section 14.2 Mastery

14.3 Rate Laws

14.3a Concentration and Reaction Rate

The **rate law** for a chemical reaction shows the quantitative relationship between reaction rate and the concentration of species involved in the chemical reaction. Consider the

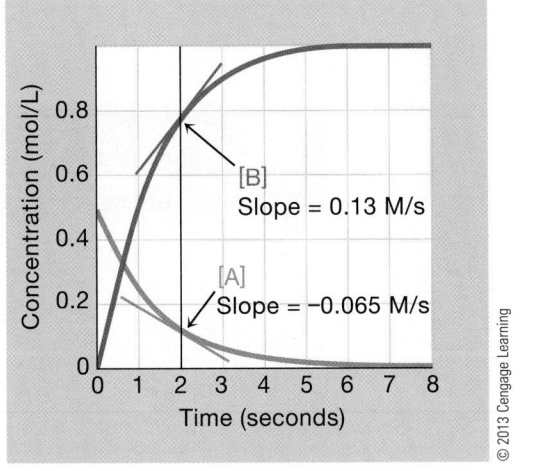

Figure 14.2.3 Instantaneous rate for A → 2 B at time = 2 seconds

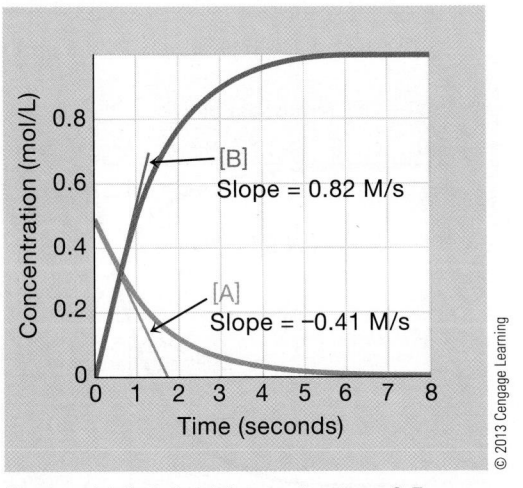

Figure 14.2.4 Initial rate for A → 2 B

following general reaction that includes a **catalyst** (C), a species that speeds up the rate of a chemical reaction but does not undergo any permanent chemical change.

$$a\,A + b\,B \xrightarrow{\ c\,C\ } \text{products}$$

We can write a general rate law for this general reaction that includes all species that participate in the reaction.

$$\text{rate} = k[A]^x[B]^y[C]^z$$

Rate laws consist of three main components:

1. A **rate constant** (k)

2. Concentrations of reacting species (in this example, A, B, and the catalyst, C)

3. The order of the reaction with respect to the reacting species (in this example, x, y, and z)

We can see each component in the rate law for the decomposition of NH_4NCO.

$$NH_4NCO(aq) \rightarrow (NH_2)_2CO(aq)$$

Rate constant for the reaction   Order of the reaction
                                  with respect to NH_4NCO

$$\text{Rate} = k[NH_4NCO]^2$$

Concentration of NH_4NCO

Rate Constant

The rate constant for a reaction is given the symbol k. The value of k, which is determined from experiments, is unique to a given reaction and varies with temperature. The units of the rate constant vary and, as you will see later in this unit, are related to the steps by which reactants are converted to products. The rate constant for the decomposition of NH_4NCO at room temperature is 0.011 L/mol · min.

Concentrations of Reacting Species

The rate law includes the concentrations of reacting species in moles per liter (for species in solution) or pressure units (for gases). Not all species that participate in a reaction will always appear in the rate law, and, as is true for catalysts, it is possible for a species that does not appear in the overall reaction to appear in the rate law.

Use instantaneous rate to describe a chemical reaction.

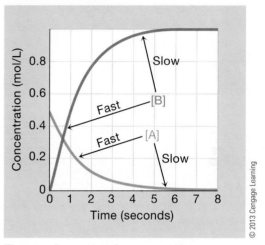

The rate of a reaction changes over time.

Reaction Order

The **reaction order** expresses the order of the reaction with respect to each reactant, and it shows the effect a reacting species has on the rate of the reaction. For example, because the rate law for the decomposition of NH_4NCO is second order with respect to the reactant, doubling the concentration of NH_4NCO will increase the rate of the reaction by a factor of 2^2. Tripling the concentration would increase the rate by a factor of 2^3.

The reaction order is indicated by the exponents x, y, and z in the general rate law given previously. Reaction orders are determined from experiments; they are not taken from the stoichiometric coefficients in the balanced equation.

Three common reaction orders are zero order, first order, and second order (Table 14.3.1). The *overall reaction order* is the sum of the individual reaction orders. Using the general rate law shown previously, the overall reaction order is the sum of the individual reaction orders x, y, and z.

$$\text{rate} = k[A]^x[B]^y[C]^z \qquad \text{overall reaction order} = x + y + z$$

Table 14.3.2 shows some additional examples of rate laws and the relationship between the rate law and reaction order.

Table 14.3.1 Reaction Order

Reaction Order (A → products)	Rate Law
Zero order	rate = $k[A]^0 = k$
First order	rate = $k[A]^1 = k[A]$
Second order	rate = $k[A]^2$

Table 14.3.2 Some Examples of Rate Laws

Reaction	Rate Law	Reaction Order
$2\,N_2O_5(g) \rightarrow 4\,NO_2(g) + O_2(g)$	rate = $k[N_2O_5]$	First order in N_2O_5, first order overall
$NO_2(g) + CO(g) \rightarrow NO(g) + CO_2(g)$	rate = $k[NO_2]^2$	Second order in NO_2, zero order in CO, second order overall
$2\,NO(g) + O_2(g) \rightarrow 2\,NO_2(g)$	rate = $k[NO_2]^2[O_2]$	Second order in NO, first order in O_2, third order overall

Example Problem 14.3.1 Determine orders from the rate law expression.

The reaction of methyl acetate with hydroxide ion

$$CH_3CO_2CH_3 + OH^- \rightarrow CH_3CO_2^- + CH_3OH$$

has the following rate law:

$$rate = k[CH_3CO_2CH_3][OH^-]$$

What is the order of reaction with respect to each reactant? What is the overall reaction order?

Solution:

You are asked to determine the order with respect to each reactant and the overall order for a chemical reaction.

You are given the rate law for a chemical reaction.

Each concentration in the rate law is raised to the first power (the superscript "1" is not shown in the rate law), so the reaction is first order with respect to each reactant, $CH_3CO_2CH_3$ and OH^-. The sum of the reaction orders gives the overall reaction order, so the reaction is second order overall.

Video Solution

Tutored Practice
Problem 14.3.1

14.3b Determining Rate Law Using the Method of Initial Rates

The rate law for an overall reaction is not derived directly from chemical equations. Instead, it is determined from experimental data. Two common methods for determining the order of the reaction with respect to a reactant, and thus the rate law for the reaction, are:

1. The method of initial rates

2. A graphical method (described in the following section)

In the **method of initial rates**, a reaction under investigation is repeated multiple times, and each experiment has a different set of initial concentrations for the reactants. The initial rate is measured for each experiment, and the results are used to determine the order of the reaction with respect to each reacting species.

In a typical experimental setup, the concentration of only one reacting species is changed between sets of experiments while the other concentrations do not change. By comparing the change in reaction rate upon change of only one reactant concentration, the order with respect to that reactant is determined.

For example, consider the following data for the reaction of nitric oxide with hydrogen.

$$2\ NO + 2\ H_2 \rightarrow N_2 + 2\ H_2O \qquad rate = k[NO]^x[H_2]^y$$

Experiment	[NO] (mol/L)	[H$_2$] (mol/L)	Initial Reaction Rate (mol/L · s) at 300 K
1	0.352	0.329	0.0554
2	0.704	0.329	0.222
3	0.352	0.658	0.111

Comparing the data in experiments 1 and 2, the concentration of NO doubles, whereas the concentration of H$_2$ does not change. Therefore, any change in the initial reaction rate is related to the order of the reaction with respect to NO. To determine the reaction order for NO, divide the rate law for experiment 2 by the rate law for experiment 1.

$$\frac{\text{initial rate (experiment 2)}}{\text{initial rate (experiment 1)}} = \frac{k[\text{NO}]^x[\text{H}_2]^y}{k[\text{NO}]^x[\text{H}_2]^y}$$

$$\frac{0.222 \text{ mol/L} \cdot \text{s}}{0.0554 \text{ mol/L} \cdot \text{s}} = \frac{k(0.704 \text{ mol/L})^x(0.329 \text{ mol/L})^y}{k(0.352 \text{ mol/L})^x(0.329 \text{ mol/L})^y} = \left(\frac{0.704 \text{ mol/L}}{0.352 \text{ mol/L}}\right)^x$$

$$4.01 = (2.00)^x$$

$$x = 2$$

When the concentration of NO doubled and the concentration of H$_2$ was held constant, the reaction rate increased by a factor of 4. Thus the reaction is second order with respect to NO.

When data showing the relationship between the initial rate of a reaction and the reactant concentrations are available, we can use the general equation below to determine the order of a reaction with respect to a given reactant. Note that this equation is valid only when using data from two experiments in which only one reactant's concentration has changed.

$$\frac{\text{initial rate (2nd experiment)}}{\text{initial rate (1st experiment)}} = \left(\frac{\text{concentration in 2nd experiment}}{\text{concentration in 1st experiment}}\right)^z$$

where

z = reaction order

Using this equation with the data in experiments 1 and 3, where the concentration of NO is held constant while the concentration of H$_2$ changes,

$$\frac{0.111 \text{ mol/L} \cdot \text{s}}{0.0554 \text{ mol/L} \cdot \text{s}} = \left(\frac{0.658 \text{ mol/L}}{0.329 \text{ mol/L}}\right)^y$$

$$2.00 = (2.00)^y$$

$$y = 1$$

The reaction is first order with respect to H_2. Combining this information into the rate law, we see that the reaction is third order overall.

$$\text{rate} = k[NO]^2[H_2]$$

Notice that, because concentration is raised to the power of reaction order in a rate law, the reaction order with respect to a reactant can sometimes be determined by examining the experiment data instead of using this general equation. For example, if a rate law is second order with respect to a reactant, doubling its concentration (while holding other concentrations constant) will increase the reaction rate by a factor of 4 ($2^2 = 4$) and tripling its concentration will increase the reaction rate by a factor of 9 ($3^2 = 9$). If a reaction is zero order with respect to a reactant, doubling or tripling its concentration (while holding other concentrations constant) will have no effect on reaction rate ($2^0 = 3^0 = 1$).

Finally, the rate constant can be calculated using data from any of the experiments. Using the data in experiment 2,

$$0.222 \text{ mol/L} \cdot \text{s} = k(0.704 \text{ mol/L})^2(0.329 \text{ mol/L})^2$$

$$k = 1.36 \text{ L}^2/\text{mol}^2 \cdot \text{s}$$

Example Problem 14.3.2 Use the method of initial rates to determine rate law.

A reaction is performed between two reactants, A and B.

$$A + B \rightarrow C$$

The following reaction rate data were obtained in four separate experiments.

Experiment	[A] (mol/L)	[B] (mol/L)	Initial Reaction Rate (mol/L · s)
1	0.573	0.252	0.0204
2	1.146	0.252	0.0817
3	0.573	0.504	0.0409
4	0.761	0.630	0.0901

What is the rate law for the reaction, and what is the numerical value of k?

Solution:

You are asked to determine the rate law and value of k for a reaction.

You are given data showing the initial reaction rate as a function of reactant concentration.

▲- -

Example Problem 14.3.2 *(continued)*

First, identify two experiments in which one reactant's concentration changes while the other is held constant. In experiments 1 and 2, [A] changes while [B] is held constant. Next, determine the order of the reaction with respect to A.

$$\frac{0.0817 \text{ mol/L} \cdot \text{min}}{0.0204 \text{ mol/L} \cdot \text{min}} = \left(\frac{1.146 \text{ mol/L}}{0.573 \text{ mol/L}}\right)^z$$

$$4.00 = (2.00)^z$$

$$z = 2$$

The reaction is second order with respect to A.

Next, identify two experiments where [B] changes and [A] is held constant. Using data from experiments 1 and 3,

$$\frac{0.0409 \text{ mol/L} \cdot \text{min}}{0.0204 \text{ mol/L} \cdot \text{min}} = \left(\frac{0.504 \text{ mol/L}}{0.252 \text{ mol/L}}\right)^z$$

$$2.00 = (2.00)^z$$

$$z = 1$$

The reaction is first order with respect to B. The rate law is

$$\text{rate} = k[\text{A}]^2[\text{B}]$$

Finally, use data from any of the experiments to calculate the value of k. Using experiment 4,

$$0.0901 \text{ mol/L} \cdot \text{min} = k(0.761 \text{ mol/L})^2(0.630 \text{ mol/L})$$

$$k = 0.247 \text{ L}^2/\text{mol}^2 \cdot \text{min}$$

Video Solution

Tutored Practice
Problem 14.3.2

Section 14.3 Mastery

14.4 Concentration Change over Time

14.4a Integrated Rate Laws

For reactions that involve a single reactant (A → products), mathematical equations can be derived that show the relationship between the concentration of the reactant and time. The equations are called **integrated rate laws** because they are derived by integration of a rate law equation.

The integrated rate laws for zero-, first-, and second-order reactions where a reactant (A) is converted to products are shown in Table 14.4.1. In the integrated rate laws,

$$t = \text{time}$$

$$[\text{A}]_0 = \text{initial concentration of A (at time} = 0)$$

$$[\text{A}]_t = \text{concentration of A at time } t$$

Table 14.4.1 Integrated Rate Laws for Reactions of Type A → Products

Reaction Order	Rate Law	Integrated Rate Law	
Zero order	rate $= k\,[A]^0 = k$	$[A]_t = [A]_0 - kt$	**(14.1)**
First order	rate $= k[A]$	$\ln\dfrac{[A]_t}{[A]_0} = -kt$	**(14.2)**
Second order	rate $= k[A]^2$	$\dfrac{1}{[A]_t} = \dfrac{1}{[A]_0} + kt$	**(14.3)**

As shown in the example problems that follow, integrated rate laws can be used to calculate the concentration of a reactant after some time has passed, the amount of time required to reduce an initial concentration to some amount, or the rate constant for a reaction.

Example Problem 14.4.1 Use the integrated rate law to calculate reactant concentration at time t.

The decomposition of nitrous oxide at 565 °C

$$N_2O(g) \rightarrow N_2(g) + \tfrac{1}{2}\,O_2(g)$$

is second order in N_2O with a rate constant of 1.10×10^{-3} $M^{-1}s^{-1}$. If an experiment is performed where the initial concentration of N_2O is 0.108 M, what is the N_2O concentration after 1250 seconds?

Solution:

You are asked to calculate the concentration of a reactant after a given period of time.

You are given the balanced equation for the reaction, the order of the reaction with respect to the reactant, the rate constant for the reaction, the initial concentration of the reactant, and the period of time.

Because the reaction is second order, use the second-order integrated rate law.

$$\frac{1}{[N_2O]_t} = \frac{1}{[N_2O]_0} + kt \qquad t = 1250 \text{ seconds}$$

$$[N_2O]_0 = 0.108 \text{ M} \qquad k = 1.10 \times 10^{-3} \text{ M}^{-1}\text{s}^{-1}$$

$$\frac{1}{[N_2O]_t} = \frac{1}{0.108 \text{ M}} + (1.10 \times 10^{-3} \text{ M}^{-1}\text{s}^{-1})(1250 \text{ s})$$

$$\frac{1}{[N_2O]_t} = 10.6 \text{ M}^{-1}$$

$$[N_2O]_t = 0.0940 \text{ M}$$

Video Solution

Tutored Practice
Problem 14.4.1

Example Problem 14.4.2 Use the integrated rate law to calculate time elapsed.

The isomerization of methyl isonitrile to acetonitrile in the gas phase at 250 °C is first order ($k = 3.00 \times 10^{-3}$ s^{-1}).

$$CH_3NC(g) \rightarrow CH_3CN(g)$$

How much time is required for the concentration of CH_3NC to drop to 0.0142 M if its initial concentration was 0.107 M?

Solution:

You are asked to determine the amount of time required for the concentration of a reactant to decrease by a given amount.

You are given the balanced equation for the reaction, the order of the reaction with respect to the reactant, the rate constant for the reaction, the initial concentration of the reactant, and the concentration of the reactant after a period of time.

Use the first order integrated rate equation.

$$\ln \frac{[CH_3NC]_t}{[CH_3NC]_0} = -kt \qquad\qquad [CH_3NC]_t = 0.0142 \text{ M}$$

$$[CH_3NC]_0 = 0.107 \text{ M} \qquad\qquad k = 3.00 \times 10^{-3} \text{ s}^{-1}$$

$$\ln\left(\frac{0.0142}{0.107}\right) = -(3.00 \times 10^{-3} \text{ s}^{-1})t$$

$$-2.02 = -(3.00 \times 10^{-3} \text{ s}^{-1})t$$

$$t = 673 \text{ s}$$

Video Solution

Tutored Practice
Problem 14.4.2

Example Problem 14.4.3 Use the integrated rate law to determine the rate constant.

The gas phase decomposition of dinitrogen pentaoxide at 335 K is first order in N_2O_5.

$$N_2O_5(g) \rightarrow 2 \text{ NO}_2(g) + \frac{1}{2} O_2(g)$$

During one experiment it was found that the N_2O_5 concentration dropped from 0.249 M at the beginning of the experiment to 0.0496 M in 230 seconds. What is the value of the rate constant for the reaction at this temperature?

Solution:

You are asked to determine the rate constant for a reaction.

You are given the balanced equation for the reaction, the order of the reaction with respect to the reactant, the initial concentration of the reactant, and the concentration of the reactant after a known amount of time has passed.

Example Problem 14.4.3 *(continued)*

Use the first order integrated rate equation.

$$\ln\frac{[N_2O_5]_t}{[N_2O_5]_0} = -kt \qquad\qquad [N_2O_5]_t = 0.0496 \text{ M}$$

$$[N_2O_5]_0 = 0.249 \text{ M} \qquad\qquad t = 230 \text{ seconds}$$

$$\ln\left(\frac{0.0496 \text{ M}}{0.249 \text{ M}}\right) = -k(230 \text{ s})$$

$$-1.61 = -k(230 \text{ s})$$

$$k = 7.02 \times 10^{-3} \text{ s}^{-1}$$

Video Solution

Tutored Practice
Problem 14.4.3

In the integrated first-order rate law, the ratio $[A]_t/[A]_0$ is a unitless quantity that is equal to the fraction of reactant that remains unreacted. As shown in Example Problem 14.4.4, you can use this ratio to determine the amount of time needed for some amount of a reactant to be consumed without knowing the amount initially present.

Example Problem 14.4.4 Use the integrated rate law to calculate time for percent reacted.

The isomerization of methyl isonitrile to acetonitrile in the gas phase at 250 °C is first order ($k = 3.00 \times 10^{-3} \text{ s}^{-1}$).

$$CH_3NC(g) \rightarrow CH_3CN(g)$$

How much time is required for 90.0% of the CH_3NC initially present in a reaction flask to be converted to product at 250 °C?

Solution:

You are asked to calculate the amount of time required for a given percentage of a reactant to be converted to product.

You are given the balanced equation for the reaction, the order of the reaction with respect to the reactant, the rate constant for the reaction, and the percentage of reactant to be converted to product.

Example Problem 14.4.4 *(continued)*

Use the first-order integrated rate equation and the fact that if 90.0% of the methyl isonitrile is consumed, only 10.0% remains in the flask.

$$\ln\frac{[CH_3NC]_t}{[CH_3NC]_0} = -kt$$

$$\frac{[CH_3NC]_t}{[CH_3NC]_0} = \frac{10.0}{100.0} = 0.100$$

$$k = 3.00 \times 10^{-3}\ s^{-1}$$

$$\ln(0.100) = -(3.00 \times 10^{-3}\ s^{-1})t$$

$$t = 768\ \text{seconds}$$

Video Solution

Tutored Practice
Problem 14.4.4

14.4b Graphical Determination of Reaction Order

Integrated rate laws for zero-, first-, and second-order reactions can be rewritten in the form of an equation for a straight line, $y = mx + b$, as shown in Interactive Table 14.4.2. We can use these rearranged rate law equations to graphically determine reaction order and the rate constant for a reaction by following the steps below:

1. Collect concentration–time data for a reaction at some temperature.

2. Plot the concentration–time data three different ways:
   a. Concentration versus time
   b. Natural log of concentration versus time
   c. 1/(concentration) versus time

3. If one of the three plots shows that the data fall on a straight line, the reaction is either zero, first, or second order in that reactant.
   a. If the concentration versus time plot is linear, the reaction is zero order with respect to the reactant.
   b. If the natural log of concentration versus time plot is linear, the reaction is first order with respect to the reactant.
   c. If the 1/(concentration) versus time plot is linear, the reaction is second order with respect to the reactant.

4. The slope of the straight-line plot is related to the rate constant for the reaction, as shown in Interactive Table 14.4.2.

Rearranged Integrated Rate Laws for Reactions of Type A → Products

Reaction Order	Integrated Rate Law	Rearranged Rate Law	Straight-Line Plot
Zero order	$[A]_t = [A]_0 - kt$	$\begin{aligned}[A]_t &= -kt + [A]_0 \\ y &= mx + b\end{aligned}$	$\begin{aligned}y &= [A]_t \\ x &= t \\ \text{slope} &= -k\end{aligned}$
First order	$\ln\dfrac{[A]_t}{[A]_0} = -kt$	$\begin{aligned}\ln[A]_t &= -kt + \ln[A]_0 \\ y &= mx + b\end{aligned}$	$\begin{aligned}y &= \ln[A]_t \\ x &= t \\ \text{slope} &= -k\end{aligned}$
Second order	$\dfrac{1}{[A]_t} = \dfrac{1}{[A]_0} + kt$	$\begin{aligned}\dfrac{1}{[A]_t} &= kt + \dfrac{1}{[A]_0} \\ y &= mx + b\end{aligned}$	$\begin{aligned}y &= 1/[A]_t \\ x &= t \\ \text{slope} &= k\end{aligned}$

Example Problem 14.4.5 Use the graphical method to determine reaction order and rate constant.

Concentration versus time data have been collected for the decomposition of H_2O_2 at 300 K.

$$H_2O_2(aq) \rightarrow H_2O(\ell) + \tfrac{1}{2} O_2(g)$$

$[H_2O_2]$ (mol/L)	Time (min)
0.0200	0
0.0118	500.
0.00693	1000.
0.00408	1500.
0.00240	2000.
0.00141	2500.

Use these data to determine the order of the reaction with respect to H_2O_2 and the rate constant for the reaction at this temperature.

Example Problem 14.4.5 *(continued)*

Solution:

You are asked to determine the order of a reaction with respect to a reactant.

You are given the balanced equation for the reaction and data showing reactant concentration as a function of time.

Use the concentration–time data to calculate $\ln[H_2O_2]$ and $1/[H_2O_2]$, and use graphing software to create three plots: $[H_2O_2]$ versus time, $\ln[H_2O_2]$ versus time, and $1/[H_2O_2]$ versus time.

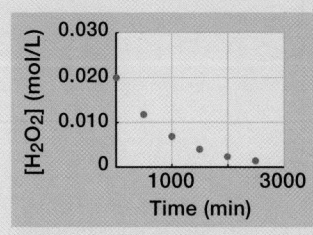 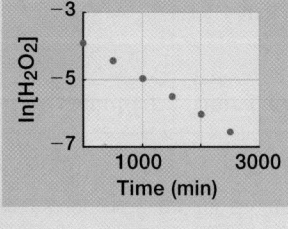 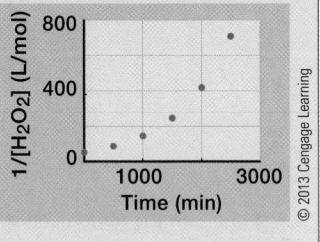

© 2013 Cengage Learning

The data fall on a straight line only in the $\ln[H_2O_2]$ versus time plot. Thus, the reaction is first order with respect to H_2O_2. The rate law for the reaction is

$$\text{rate} = k[H_2O_2]$$

The value of the rate constant is found from the slope of the $\ln[H_2O_2]$ versus time plot. Using the graphing software, the equation for the straight line that passes through the data points in this plot is $y = -0.00106x - 3.912$.

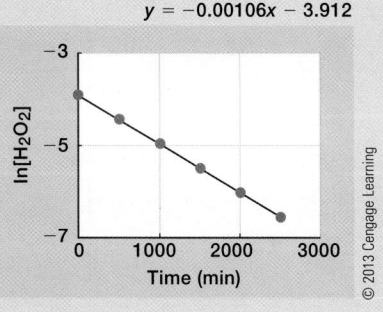

© 2013 Cengage Learning

For a first-order reaction, when the concentration–time data are plotted as a natural log of concentration versus time, the slope of the line is equal to $-k$. For this reaction, $k = -(-0.00106\ \text{min}^{-1}) = 0.00106\ \text{min}^{-1}$ at 300 K.

Video Solution

**Tutored Practice
Problem 14.4.5**

14.4c Reaction Half-Life

The concept of half-life is useful for describing the speed of a reaction. The **half-life ($t_{1/2}$)** for a reaction is the amount of time required for the concentration of a reactant to fall to one half of its initial value. Consider the first-order decomposition of hydrogen peroxide.

$$H_2O_2 \rightarrow H_2O + \tfrac{1}{2} O_2$$

Interactive Figure 14.4.1 shows a concentration–time plot generated from data collected in a typical experiment. In the first 650 minutes, the concentration of H_2O_2 is halved (0.020 M to 0.010 M). After another 650 minutes (a total of 1300 minutes), the concentration is again halved (0.010 M to 0.0050 M). The half-life of this reaction is therefore 650 min.

As you can see from the reaction data, the fraction of H_2O_2 remaining is a function of the number of half-lives that have passed:

Number of half-lives passed	1	2	3	4	x
Fraction of reactant remaining	½	¼	⅛	1⁄16	$(\tfrac{1}{2})^x$

Interactive Figure 14.4.1

Explore half-life.

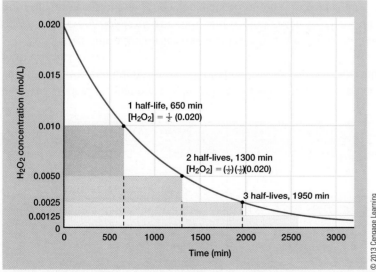

Concentration–time data for the decomposition of H_2O_2

© 2013 Cengage Learning

Table 14.4.3 Half-Life Equations for Reactions of Type A → Products

Reaction Order	Integrated Rate Law	Substitute $t = t_{1/2}$ and $[A]_t = \frac{1}{2}[A]_0$	Half-Life Equation	
Zero order	$[A]_t = [A]_0 - kt$	$\frac{1}{2}[A]_0 = [A]_0 - k(t_{1/2})$	$t_{1/2} = \dfrac{[A]_0}{2k}$	**(14.4)**
First order	$\ln\dfrac{[A]_t}{[A]_0} = -kt$	$\ln(\frac{1}{2}[A]_0) = \ln[A]_0 - kt_{1/2}$	$t_{1/2} = \dfrac{\ln 2}{k} = \dfrac{0.693}{k}$	**(14.5)**
Second order	$\dfrac{1}{[A]_t} = \dfrac{1}{[A]_0} + kt$	$\dfrac{1}{(\frac{1}{2}[A]_0)} = \dfrac{1}{[A]_0} + k(t_{1/2})$	$t_{1/2} = \dfrac{1}{k[A]_0}$	**(14.6)**

We can derive equations for the half-life of zero-, first-, and second-order reactions by substituting the conditions for half-life (when $t = t_{1/2}$, $[A]_t = \frac{1}{2}[A]_0$) into the integrated rate laws (Table 14.4.3).

Notice that only a first-order reaction has a half-life that is *independent of the initial concentration of reactant*. For zero- and second-order reactions, the half-life depends on both the rate constant and the initial concentration of the reactant.

Example Problem 14.4.6 Determine half-life, order, and k from concentration–time data.

The following data were collected for the decomposition of sulfuryl chloride at 383 °C.

$$SO_2Cl_2 \rightarrow SO_2 + Cl_2$$

$[SO_2Cl_2]$ (M)	Time (min)
5.18×10^{-3}	0
2.59×10^{-3}	166
1.30×10^{-3}	332

a. What is the half-life of the reaction when $[SO_2Cl_2] = 5.18 \times 10^{-3}$ M?
b. What is the half-life of the reaction when $[SO_2Cl_2] = 2.59 \times 10^{-3}$ M?
c. Is the reaction first order? If so, what is the rate constant for this reaction?

Solution:

You are asked to calculate the half-life of the reaction at different reactant concentrations, to determine whether the reaction is first order, and to calculate the rate constant for the reaction.

You are given the balanced equation for the reaction and the concentration of the reactant as a function of time.

Example Problem 14.4.6 *(continued)*

a. The concentration of SO_2Cl_2 is halved (5.18×10^{-3} M to 2.59×10^{-3} M) over the first 166 minutes of the reaction. Therefore, the half-life when $[SO_2Cl_2] = 5.18 \times 10^{-3}$ M is 166 minutes.

b. The concentration of SO_2Cl_2 is halved again (2.59×10^{-3} M to 1.30×10^{-3} M) when another 166 minutes pass (166 min to 332 min). The half-life when $[SO_2Cl_2] = 2.59 \times 10^{-3}$ M is 166 minutes.

c. Because the half-life is independent of reactant concentration, the reaction is first order in SO_2Cl_2. The first-order rate constant is calculated using the equation in Table 14.4.3.

$$t_{1/2} = \frac{\ln 2}{k} = \frac{0.693}{k}$$

$$k = \frac{0.693}{t_{1/2}} = \frac{0.693}{166 \text{ min}} = 4.17 \times 10^{-3} \text{ min}^{-1}$$

Video Solution

Tutored Practice
Problem 14.4.6

14.4d Radioactive Decay

All radioactive isotopes decay via first-order reactions. Along with other applications, radioactive decay and half-life are used in radioactive dating, the determination of the age of organic objects, such as those made from plant-based materials, and inorganic objects, such as the rocks and minerals found in meteorites. We will discuss the subject of radioactive dating in detail in Nuclear Chemistry (Unit 24).

When working with radioactive decay, it is common to write concentration–time equations in terms of either the quantity of radioactive particles present or the radioactivity of the sample (N) instead of concentration units.

$$\ln \frac{N_t}{N_0} = -kt$$

Example Problem 14.4.7 Use half-life to calculate mass remaining for radioactive decay.

The radioactive isotope ^{32}P has a half-life of 14.3 days. If a sample contains 0.884 g of ^{32}P, what mass of the isotope will remain after 22 days?

Solution:

You are asked to determine the mass of a radioactive isotope remaining after a given period of time.

You are given the mass of isotope originally present, the half-life of the isotope, and the period of time that passes.

Example Problem 14.4.7 (*continued*)

First, use the half-life to calculate the rate constant for this radioactive decay.

$$t_{1/2} = \frac{0.693}{k}$$

$$k = \frac{0.693}{t_{1/2}} = \frac{0.693}{14.3 \text{ d}} = 0.0485 \text{ d}^{-1}$$

Next, use the integrated first-order rate law to calculate the mass of ^{32}P remaining after 22 days.

$$\ln\frac{N_t}{N_0} = -kt$$

$$\ln\frac{N_t}{0.884 \text{ g}} = -(0.0485 \text{ d}^{-1})(22.0 \text{ d})$$

$$\ln(N_t) - \ln(0.884 \text{ g}) = -1.07$$

$$\ln(N_t) = -1.19$$

$$N_t = e^{-1.19} = 0.304 \text{ g}$$

Video Solution

Tutored Practice
Problem 14.4.7

Section 14.4 Mastery

14.5 Activation Energy and Temperature

14.5a Reaction Coordinate Diagrams

In general, when other reaction conditions are held constant, all reactions proceed more rapidly as temperature increases. Why is this true? All reactions have an *activation barrier*, an energy that must be overcome for reactants to proceed to form products, and changing the temperature affects the ability of reactants to overcome this barrier.

The energetic changes that occur during the progress of a reaction are often displayed using a **reaction coordinate diagram**, a plot that shows energy (on the y-axis) as a function of the progress of the reaction from reactants to products (on the x-axis). Figure 14.5.1 shows reaction coordinate diagrams for two simple reactions, one exothermic and the other endothermic.

As shown in Figure 14.5.1, the conversion of reactants to products passes through a high-energy state where an **activated complex**, or high-energy **transition state**, is formed. The energy required to reach this transition state is the reaction's **activation energy**, E_a. Activation energy is always a positive quantity and, for the reaction where reactants are converted to products, is equal to the energy difference between the reactant energy and the energy of the activated complex. Notice that the overall energy change for the reaction (ΔE) can be either positive or negative.

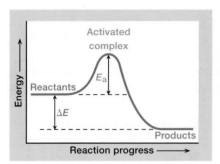
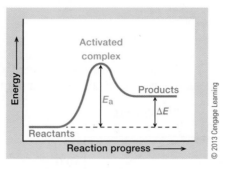

Figure 14.5.1 Reaction coordinate diagrams for an exothermic reaction and an endothermic reaction

In Gases (Unit 10), we saw that any sample of matter will have a Boltzmann distribution of molecular energies. If some minimum amount of energy (the activation energy) is required for the conversion of reactants to products, increasing the temperature will increase the number of molecules that have that minimum energy (Figure 14.5.2a). For any reaction, a high activation energy means that at a given temperature, fewer molecules have the energy required to form products than if the activation energy were lower (Figure 14.5.2b). Recall from our discussion of collision theory that the larger the number of molecules that

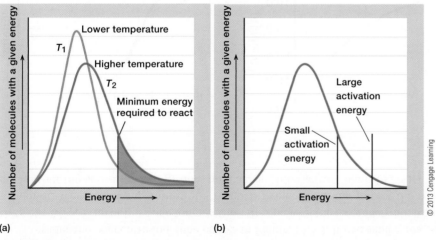

Figure 14.5.2 (a) Boltzmann distributions of molecular energies at different temperatures show that more molecules exceed the activation energy at higher temperature. (b) At a given temperature, more molecules exceed a small E_a than a large E_a.

have the minimum energy needed to form products is, the faster the reaction rate will be. Thus, temperature and activation energy both play an important role in influencing the rate of a chemical reaction.

Consider the concentration–time curves in Figure 14.5.3. If two similar reactions with the same ΔE are run at the same temperature, the one with the larger activation energy will be slower. When the temperature of a reaction is increased, the rate of the reaction increases but the activation energy does not change (Interactive Figure 14.5.4).

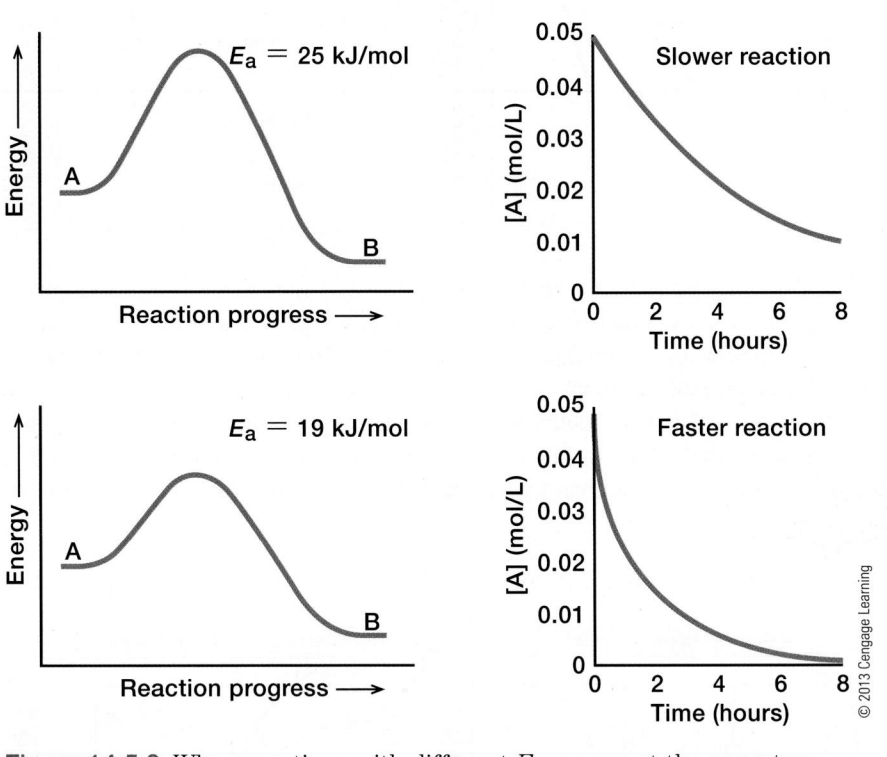

Figure 14.5.3 When reactions with different E_a are run at the same temperature, the reaction with the larger E_a has a slower reaction rate.

Explore the effect of activation energy and temperature on reaction rate.

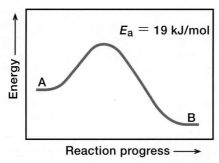

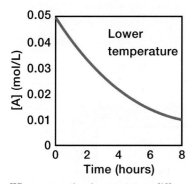

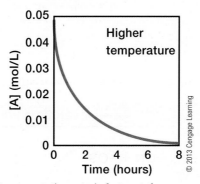

When a reaction is run at two different temperatures, reaction rate is faster at the higher temperature but E_a does not change.

Example Problem 14.5.1 Analyze a reaction profile.

The reaction profile (not drawn to scale) for the reaction $SO_2 + NO_2 \rightarrow SO_3 + NO$ is shown below:

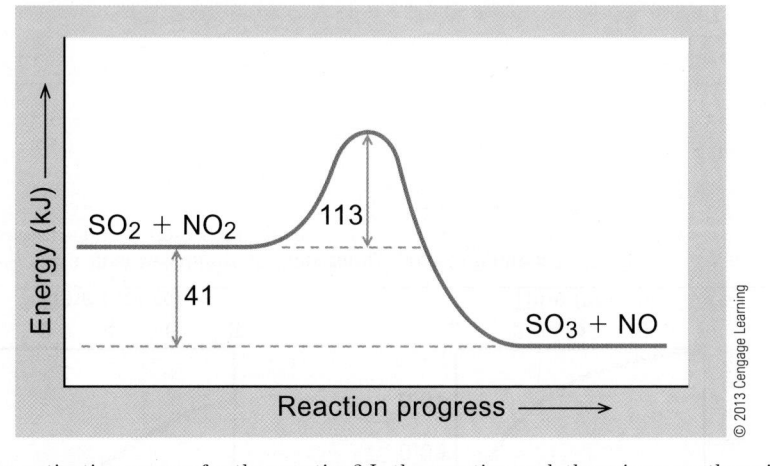

What is the activation energy for the reaction? Is the reaction endothermic or exothermic? What is the value of ΔE for the reaction?

Solution:

You are asked to identify the activation energy for a reaction, to determine whether the reaction is endothermic or exothermic, and to calculate ΔE for the reaction.

You are given an energy diagram for a reaction.

The activation energy, E_a, is the difference in energy between the activated complex and the reactants. This value in the diagram is given as 113 kJ.

The energy of the products is lower than the energy of the reactants, which means that energy is released into the surroundings when the reaction takes place and the reaction is exothermic.

ΔE is the difference in energy between the products and the reactants. The magnitude of this difference is 41 kJ. Because the energy of the products is lower than the energy of the reactants, $\Delta E = E_{products} - E_{reactants} = -41$ kJ. Note that this exothermic reaction has a negative ΔE.

Video Solution

Tutored Practice
Problem 14.5.1

14.5b The Arrhenius Equation

The quantitative relationship among rate, activation energy, and temperature is given by the **Arrhenius equation**.

$$k = Ae^{-E_a/RT} \qquad \textbf{(14.7)}$$

where

k = rate constant
A = frequency factor
E_a = activation energy
R = gas constant (8.3145 J/K · mol)
T = temperature (K)

The frequency factor, A, is a measure of the number of collisions that take place with the correct orientation during a reaction. The other term in the equation, $e^{-E_a/RT}$, is the fraction of collisions that occur with enough energy to overcome the activation energy. As you can see from the equation,

- as E_a increases, $e^{-E_a/RT}$ decreases and the rate constant decreases; and

- as T increases, $e^{-E_a/RT}$ increases and the rate constant increases.

Although the Arrhenius equation in this form is useful, it is more common to use a form of the equation that allows the comparison of two rate constant–temperature data points.

$$\ln\frac{k_2}{k_1} = \frac{-E_a}{R}\left(\frac{1}{T_2} - \frac{1}{T_1}\right) \qquad \textbf{(14.8)}$$

This form of the Arrhenius equation can be used to calculate the rate constant for a reaction at a different temperature when E_a and the rate constant at another temperature are known, or to determine E_a when rate constants at two different temperatures are known.

Example Problem 14.5.2 Use the Arrhenius equation to determine k.

The activation energy for the gas phase decomposition of t-butyl propionate is 164 kJ.

$$C_2H_5COOC(CH_3)_3(g) \rightarrow (CH_3)_2C{=}CH_2(g) + C_2H_5COOH(g)$$

The rate constant for this reaction is $3.80 \times 10^{-4}\,s^{-1}$ at 528 K. What is the rate constant at 569 K?

Solution:

You are asked to calculate the rate constant for a reaction at a specific temperature.

You are given the activation energy for the reaction, the rate constant at a given temperature, and the temperature at which the rate constant is unknown.

Example Problem 14.5.2 (*continued*)

First, create a table of the known and unknown variables.

$$T_1 = 528 \text{ K} \qquad\qquad T_2 = 569 \text{ K}$$
$$k_1 = 3.80 \times 10^{-4}\,\text{s}^{-1} \qquad k_2 = ?$$
$$E_a = 164 \text{ kJ/mol}$$

Next, use the two-point version of the Arrhenius equation (Equation 14.8). The gas constant can be expressed in units of $kJ/K \cdot mol$.

$$\ln\frac{k_2}{k_1} = \frac{-E_a}{R}\left(\frac{1}{T_2} - \frac{1}{T_1}\right)$$

$$\ln\left(\frac{k_2}{3.80 \times 10^{-4}\,\text{s}^{-1}}\right) = \frac{-164\,\text{kJ/mol}}{8.3145 \times 10^{-3}\,\text{kJ/K} \cdot \text{mol}}\left(\frac{1}{569\,\text{K}} - \frac{1}{528\,\text{K}}\right)$$

$$\ln\left(\frac{k_2}{3.80 \times 10^{-4}\,\text{s}^{-1}}\right) = 2.69$$

$$\frac{k_2}{3.80 \times 10^{-4}\,\text{s}^{-1}} = e^{2.69} = 14.8$$

$$k_2 = 5.61 \times 10^{-3}\,\text{s}^{-1}$$

Is your answer reasonable? At a higher temperature, the reaction is faster and the rate constant has increased.

Video Solution

Tutored Practice
Problem 14.5.2

Example Problem 14.5.3 Use the Arrhenius equation to determine E_a.

For the gas phase decomposition of ethyl chloroformate, the rate constant is $1.05 \times 10^{-3}\,\text{s}^{-1}$ at 470 K and $1.11 \times 10^{-2}\,\text{s}^{-1}$ at 508 K.

$$\text{ClCOOC}_2\text{H}_5(g) \rightarrow \text{C}_2\text{H}_5\text{Cl}(g) + \text{CO}_2(g)$$

Calculate the activation energy for this reaction.

Solution:

You are asked to determine the activation energy for a reaction.

You are given the rate constants for the reaction at two different temperatures.

First, create a table of the known and unknown variables.

$$T_1 = 470 \text{ K} \qquad\qquad T_2 = 508 \text{ K}$$
$$k_1 = 1.05 \times 10^{-3}\,\text{s}^{-1} \qquad k_2 = 1.11 \times 10^{-2}\,\text{s}^{-1}$$
$$E_a = ?$$

Example Problem 14.5.3 *(continued)*

Next, use the two-point version of the Arrhenius equation (Equation 14.8). The gas constant can be expressed in units of kJ/K · mol.

$$\ln \frac{k_2}{k_1} = \frac{-E_a}{R}\left(\frac{1}{T_2} - \frac{1}{T_1}\right)$$

$$\ln\left(\frac{1.11 \times 10^{-2}\,\text{s}^{-1}}{1.05 \times 10^{-3}\,\text{s}^{-1}}\right) = \frac{-E_a}{8.3145 \times 10^{-3}\,\text{kJ/K} \cdot \text{mol}}\left(\frac{1}{508\,\text{K}} - \frac{1}{470\,\text{K}}\right)$$

$$2.358 = \frac{-E_a}{8.3145 \times 10^{-3}\,\text{kJ/K·mol}}(-1.59 \times 10^{-4}\,\text{K}^{-1})$$

$$E_a = 123\,\text{kJ/mol}$$

Video Solution

Tutored Practice
Problem 14.5.3

14.5c Graphical Determination of E_a

The Arrhenius equation can also be used to graphically determine the activation energy for a reaction using multiple temperature–rate constant data points. Rewriting the equation in the form of a straight-line equation,

$$\ln k = \frac{-E_a}{R}\left(\frac{1}{T}\right) + \ln A$$

$$y = m\quad x\ +\ b$$

A plot of the natural log of the rate constant versus the inverse of the Kelvin temperature produces a straight line with a slope that is equal to the quantity $(-E_a/R)$.

Example Problem 14.5.4 Use Arrhenius plot information to determine E_a.

The rate of the following reaction was measured at different temperatures and rate constants were determined.

$$N_2O_5(g) \rightarrow 2\,NO_2(g) + \tfrac{1}{2}\,O_2(g)$$

Temperature (K)	k (s^{-1})
298	3.46×10^{-5}
328	1.50×10^{-3}
358	3.34×10^{-2}
378	0.210

Use these data to determine the activation energy for the reaction.

◄- - - - -

Example Problem 14.5.4 *(continued)*

Solution:

You are asked to determine the activation energy for a reaction.

You are given data that show the rate constant for a reaction as a function of temperature.

First, create a data table of the natural log of the rate constant and the inverse of temperature. Plot this data ($\ln k$ on the y-axis and $1/T$ on the x-axis) using a plotting program, and determine the equation for the straight line.

$$y = -12249x + 30.834$$

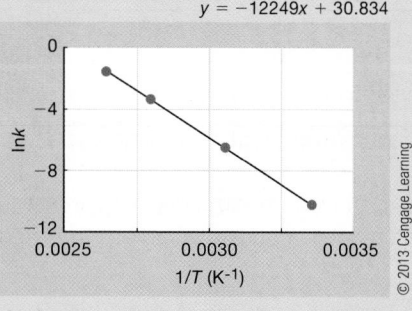

© 2013 Cengage Learning

Next, use the slope of the line to calculate activation energy.

$$\text{slope} = \frac{-E_a}{R}$$

$$-12{,}249 \text{ K} = \frac{-E_a}{8.3145 \times 10^{-3} \text{ kJ/K} \cdot \text{mol}}$$

$$E_a = 102 \text{ kJ/mol}$$

Video Solution

Tutored Practice
Problem 14.5.4

Section 14.5 Mastery

14.6 Reaction Mechanisms and Catalysis

14.6a The Components of a Reaction Mechanism

Most chemical reactions involve the rearrangement of atoms; atoms are transferred between molecules, molecules are broken down into smaller molecules, or molecules are built up into larger ones. A **reaction mechanism** is a detailed description at the molecular level of steps by which reactants are converted to products. Reaction mechanisms cannot be calculated or predicted with any certainty. Instead, reaction mechanisms must be studied experimentally.

A reaction mechanism consists of a series of individual steps that happen in sequence, each of which is called an **elementary step**. The sum of the elementary steps in a mechanism gives the overall equation for the chemical reaction.

Elementary steps are categorized by the number of particles that must collide to make the desired reaction occur. The **molecularity** of an elementary step tells us how many species react in the step. The most common types of elementary steps are **unimolecular**, involving a single reacting species, and **bimolecular**, involving the collision of two reacting species. A **termolecular** step, in which three species collide simultaneously, does not occur very often. Recall that according to collision theory, in a termolecular step all three species would have to collide with sufficient energy and the correct orientation for the reaction to occur. Only a small handful of reaction mechanisms have termolecular steps, so we will focus on unimolecular and bimolecular elementary steps.

From a chemical perspective, three common chemical events happen during a reaction:

- Bond breaking

- Bond forming

- A concerted process where more than one event happens simultaneously (such as simultaneous bond breaking and bond forming)

Bond-Breaking Step

A bond-breaking step involves a single species breaking apart into two or more species. In the elementary step shown in Figure 14.6.1, a C—O bond breaks and a single reactant is converted to two products. Because only one reactant molecule is involved, this is a unimolecular elementary step.

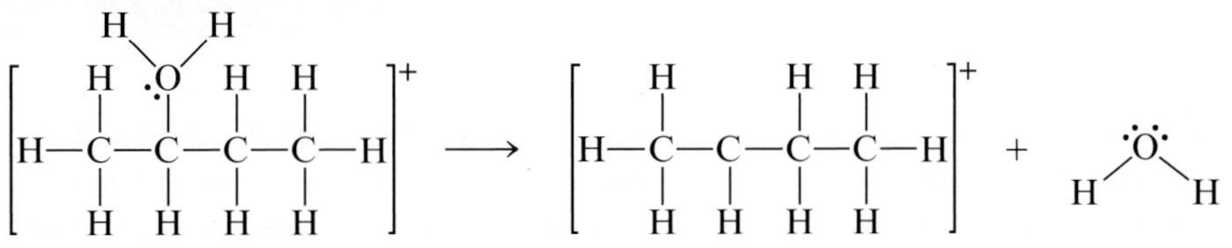

Figure 14.6.1 A unimolecular elementary step

Bond-Forming Step

A bond-forming step involves the combination of two or more species to form a single product. In the elementary step shown in Figure 14.6.2, a C—O bond is formed when the two reactants form a single product. Because two reactant molecules are involved, this is a bimolecular elementary step.

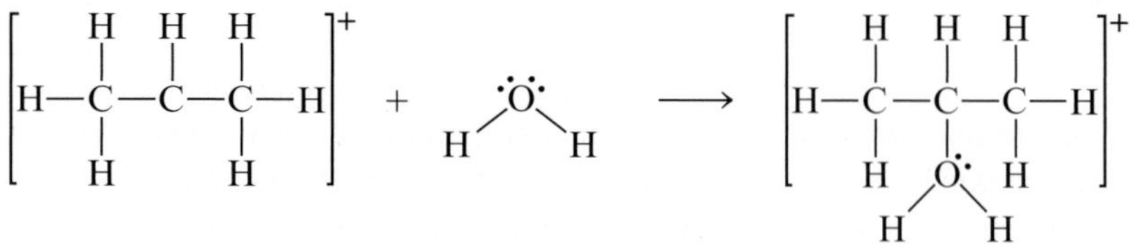

Figure 14.6.2 A bimolecular elementary step

Concerted Process

A **concerted process** involves more than one chemical process happening simultaneously. For example, as shown in the elementary step in Interactive Figure 14.6.3, a C—F bond is formed while at the same time, a C—Cl bond breaks. This concerted elementary step is bimolecular.

Interactive Figure 14.6.3

Investigate different types of elementary steps.

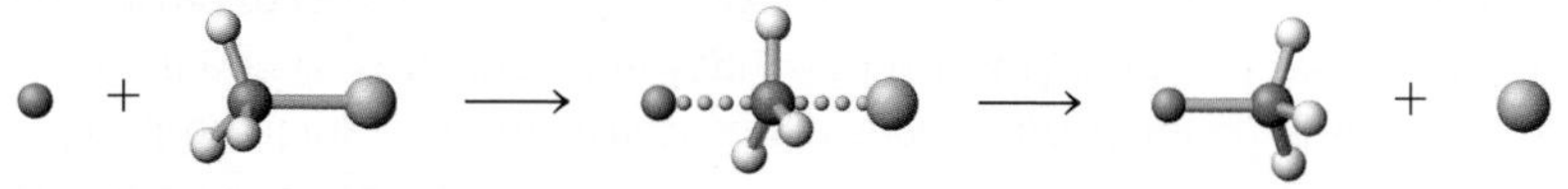

F⁻   CH₃Cl   [F•••CH₃•••Cl]⁻   CH₃F   Cl⁻

A concerted elementary step

Example Problem 14.6.1 Identify the molecularity of elementary steps in a mechanism.

The thermal decomposition of nitryl chloride is proposed to occur by the following mechanism:

Step 1. $NO_2Cl \rightarrow NO_2 + Cl$
Step 2. $Cl + NO_2Cl \rightarrow NO_2 + Cl_2$

Identify the molecularity of each step in the mechanism.

Solution:

You are asked to identify the molecularity of the elementary steps in a mechanism.

You are given a multistep mechanism.

In the first step, one mole of reactant forms product, so the step is unimolecular. In the second step, there are two moles of reactants, so the step is bimolecular.

Video Solution

Tutored Practice
Problem 14.6.1

14.6b Multistep Mechanisms

A chemical reaction that occurs in a single step has only one elementary step, which is the same as the balanced equation for the chemical reaction. For example, many acid–base reactions occur in a single step. The protonation of ammonia by hydronium ion involves the transfer of an H^+ ion from H_3O^+ to NH_3. This reaction happens in a single concerted step where the O—H bond breaking and the H—N bond formation happen at the same time.

$$H_3O^+(aq) + NH_3(aq) \rightarrow H_2O(aq) + NH_4^+(aq)$$

When a chemical reaction occurs in more than one step, the mechanism can involve species that do not appear in the overall net reaction. An **intermediate** is a chemical species that is formed in one step in the mechanism and then consumed in a later step. A catalyst, a species that speeds up a chemical reaction, is consumed in one step and then produced, in its original form, in a later step in the mechanism.

For example, consider the decomposition of ozone, a process that can occur in a two-step mechanism.

Step 1. $O_3(g) \rightarrow O_2(g) + O(g)$ *Unimolecular; bonds are broken*

Step 2. $O_3(g) + O(g) \rightarrow 2\,O_2(g)$ *Bimolecular; bonds are broken and formed*

The net reaction is the sum of all the steps in the mechanism. Add the mechanism steps and cancel any species that are common to the reactants and products.

Step 1. $O_3(g) \rightarrow O_2(g) + O(g)$

Step 2. $O_3(g) + O(g) \rightarrow 2\,O_2(g)$

$\overline{\qquad\qquad 2\,O_3(g) + \cancel{O(g)} \longrightarrow 3\,O_2(g) + \cancel{O(g)} \qquad}$

Net reaction: $2\,O_3(g) \rightarrow 3\,O_2(g)$

While the species $O(g)$ does not appear in the net reaction, it does exist (and play an important role) in the decomposition of ozone. It is an intermediate in the mechanism because it is produced in one step and consumed in a later step. Intermediates are often short-lived chemical species that are very difficult to detect, but chemists have developed special techniques that can be used to trap reaction intermediates in order to better understand a reaction mechanism.

Catalysts also appear in a mechanism but not in the net reaction. Consider the decomposition of hydrogen peroxide in the presence of I^-, a reaction that occurs in two steps.

Step 1. $\quad H_2O_2(aq) + I^-(aq) \rightarrow IO^-(aq) + H_2O(\ell)$ *Bimolecular, concerted*

Step 2. $\quad IO^-(aq) + H_2O_2(aq) \rightarrow I^-(aq) + H_2O(\ell) + O_2(g)$ *Bimolecular, concerted*

Net reaction: $2\,H_2O_2(aq) \rightarrow 2\,H_2O(\ell) + O_2(g)$

In this mechanism, the IO^- ion is an intermediate (it is formed in the first step and then consumed in the second step) and the I^- ion is a catalyst (it is consumed in the first step and then reformed in the second step in its original form).

Example Problem 14.6.2 Analyze a mechanism, including molecularities; write a net reaction; and identify any catalysts or intermediates.

Chlorofluorocarbons break down in the upper atmosphere to form, among other products, chlorine atoms. The chlorine atoms are involved in the breakdown of ozone via the following mechanism.

Step 1. $Cl(g) + O_3(g) \rightarrow ClO(g) + O_2(g)$

Step 2. $ClO(g) + O_3(g) \rightarrow Cl(g) + 2\,O_2(g)$

a. Identify the molecularity of each step in the mechanism.
b. Write the net reaction.
c. Identify any intermediates and/or catalysts in this mechanism.

Solution:

You are asked to identify the molecularity of the steps in a mechanism, to write a net reaction, and to identify any catalysts or intermediates in the mechanism.

You are given a multistep mechanism.

a. Both steps in the mechanism involve two reactants, so both are bimolecular elementary steps.
b. The net reaction is found by adding the mechanism steps together and cancelling any species common to both reactants and products.

Step 1. $\quad Cl(g) + O_3(g) \rightarrow ClO(g) + O_2(g)$

Step 2. $\quad ClO(g) + O_3(g) \rightarrow Cl(g) + 2\,O_2(g)$

$$\overline{Cl(g) + ClO(g) + 2\,O_3(g) \longrightarrow Cl(g) + ClO(g) + 3\,O_2(g)}$$

Net reaction: $2\,O_3(g) \rightarrow 3\,O_2(g)$

c. $Cl(g)$ is consumed in the first step and then produced in its original form in the second step; it is a catalyst in this mechanism. The species ClO is produced in the first step and then consumed in the second step; it is an intermediate in this mechanism. Note that neither the catalyst nor intermediate appear in the net reaction.

Video Solution

Tutored Practice
Problem 14.6.2

Interactive Figure 14.6.4

Investigate multistep mechanisms.

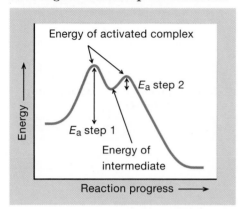
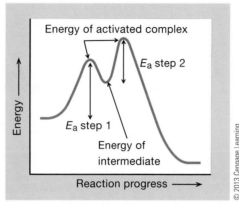

Reaction coordinate diagrams for two-step mechanisms

Activation Energy and Intermediates

The reaction coordinate diagram for a multistep mechanism shows how activation energy can affect how long an intermediate persists during a chemical reaction. Consider the two reaction coordinate diagrams shown in Interactive Figure 14.6.4. In the first reaction, the second activation energy (for the conversion of the intermediate to products) is relatively small. It is likely that the intermediate is converted to product as fast as it is formed, and the intermediate is probably not detectable. In the second reaction, the second activation energy is relatively large. In this case, the intermediate probably exists for an amount of time that allows it to be detected during the reaction.

Mechanisms Involving Multiple Physical States

Reactions that involve more than one phase are inherently complex, and mechanisms are generally not written for these types of reactions. Take, for example, the formation of solid silver chloride from aqueous solutions of silver nitrate and sodium chloride (Interactive Figure 14.6.5). The overall reaction and net ionic equations are both relatively simple:

Net reaction: $\quad AgNO_3(aq) + NaCl(aq) \rightarrow AgCl(s) + NaNO_3(aq)$

Net ionic equation: $\quad Ag^+(aq) + Cl^-(aq) \rightarrow AgCl(s)$

Although the equations for this reaction appear simple, the fact that a solid is involved in the reaction makes the mechanism very complex. The solid AgCl that forms during the

Interactive Figure 14.6.5

Explore a reaction mechanism involving multiple physical states.

Formation of solid silver chloride

reaction is very large on the molecular scale, containing billions of Ag^+ and Cl^- ions. Each crystal forms by the addition of more and more Ag^+ and Cl^- ions.

There is no clear way to write a mechanism for this process. It involves billions of steps, and although the general process for each step is always the same, the intermediate crystals involved are different for each step in the mechanism.

14.6c Reaction Mechanisms and the Rate Law

In the generalized rate law we wrote earlier for a chemical reaction, the reaction order (the exponents in the rate law) were *unrelated* to the reaction stoichiometry. However, the rate law for any elementary step in a mechanism is *directly related* to the stoichiometry of the elementary step because elementary steps describe the molecular-level collisions that are occurring during the reaction. For example, consider the two-step mechanism for the decomposition of hydrogen peroxide in the presence of I^-.

Step 1. $H_2O_2(aq) + I^-(aq) \rightarrow IO^-(aq) + H_2O(\ell)$ rate $= k_1[H_2O_2][I^-]$

Step 2. $IO^-(aq) + H_2O_2(aq) \rightarrow I^-(aq) + H_2O(\ell) + O_2(g)$ rate $= k_2[IO^-][H_2O_2]$

The rate law for each elementary step is equal to the rate constant for the elementary step multiplied by the concentration of each reactant raised to the power of its stoichiometric constant (Table 14.6.1).

The rate laws for each step in this mechanism are different, so how is it possible to determine the rate law for the net reaction? The rate of a reaction can be no greater than the rate of the slowest step in the mechanism, the **rate-determining step** (sometimes called the *rate-limiting step*). Thus, it makes sense that a reaction's rate law directly relates to the

Table 14.6.1 Rate Law and Elementary Steps

Molecularity	Order of Elementary Step	Example	Rate Law
Unimolecular	First order	A $\rightarrow$ products	rate $= k[A]$
Bimolecular	Second order	A + B $\rightarrow$ products 2 A $\rightarrow$ products	rate $= k[A][B]$ rate $= k[A]^2$
Termolecular	Third order	A + B + C $\rightarrow$ products 2 A + B $\rightarrow$ products 3 A $\rightarrow$ products	rate $= k[A][B][C]$ rate $= k[A]^2[B]$ rate $= k[A]^3$

stoichiometry of the rate-determining step in the mechanism. For the H_2O_2 decomposition reaction shown previously, the experimentally determined rate law is rate = $k[H_2O_2][I^-]$. This rate law matches the rate law for the first step in the mechanism. This suggests that the first step in the decomposition reaction is the rate-determining step in the mechanism.

The experimentally determined rate law is one of the most useful pieces of information to have when trying to determine a reaction mechanism. The process of proving or disproving a mechanism usually involves three different steps:

1. Proposing a mechanism, including identifying the rate-determining step.

2. Predicting the rate law using the stoichiometry of the rate-determining step.

3. Determining the experimental rate law for the reaction using one of the methods covered earlier in this unit. If the predicted and experimental rate laws do not match, then the proposed mechanism is incorrect. If they match, then the proposed mechanism might be correct. However, the proposed mechanism might not be correct because different mechanisms can lead to the same predicted rate law. Further experimentation is required to confirm whether the mechanism is correct or not.

Example Problem 14.6.3 Match mechanisms to rate law.

An oxygen atom transfer reaction occurs between NO_2 and CO in the gas phase.

$$NO_2(g) + CO(g) \rightarrow NO(g) + CO_2(g)$$

There are two proposed mechanisms:

Mechanism A

Step 1. $NO_2(g) + CO(g) \rightarrow NO(g) + CO_2(g)$ *Rate-determining slow step*

Mechanism B

Step 1. $2 NO_2(g) \rightarrow NO_3(g) + NO(g)$ *Rate-determining slow step*

Step 2. $NO_3(g) + CO(g) \rightarrow NO_2(g) + CO_2(g)$ *Fast second step*

The experimentally determined rate law is rate = $k[NO_2]^2$.

Which of these two mechanisms is supported by the experimental evidence?

Solution:

You are asked to identify the mechanism that is supported by experimental evidence.

You are given two possible mechanisms and the experimentally determined rate law.

Mechanism A is a single-step mechanism, so the predicted rate law is given by the stoichiometry of the overall reaction.

Example Problem 14.6.3 *(continued)*

$$\text{rate} = k[\text{NO}_2][\text{CO}]$$

This rate law does not match the experimental rate law, so this mechanism is not correct.

Mechanism B is a two-step mechanism with a slow first step. This rate law for this mechanism is directly related to the stoichiometry of the rate-determining step.

$$\text{rate} = k[\text{NO}_2]^2$$

This rate law agrees with the experimental rate law, so this mechanism could be correct. Further experiments are needed to prove that this is the correct mechanism for the reaction.

Video Solution

Tutored Practice
Problem 14.6.3

14.6d More Complex Mechanisms

Some elementary steps can proceed both in the forward and reverse directions; that is, they are reversible. If both the forward and reverse reactions are fast and occur at the same rate, a state of dynamic equilibrium is formed in which both reactants and products are present in the reaction flask.

For example, consider the reaction of bromine and hydrogen to form hydrogen bromide:

$$\text{Br}_2(g) + \text{H}_2(g) \rightarrow 2\,\text{HBr}(g)$$

A proposed mechanism is

Step 1. $\text{Br}_2(g) \underset{k_{-1}}{\overset{k_1}{\rightleftarrows}} 2\,\text{Br}(g)$ Fast in both directions

Step 2. $\text{Br}(g) + \text{H}_2(g) \xrightarrow{k_2} \text{HBr}(g) + \text{H}(g)$ Slow

Step 3. $\text{H}(g) + \text{Br}_2(g) \xrightarrow{k_3} \text{HBr}(g) + \text{Br}(g)$ Fast

In this mechanism the rate constant for each step is written above (or below) its reaction arrow. The number of the step is the subscript for each rate constant. The reversible first step is written with a double arrow, with the rate constant for the reverse step denoted with a minus sign.

The rate law for the overall reaction is the rate law for the slow rate-determining step.

$$\text{rate} = k_2[\text{Br}][\text{H}_2]$$

However, Br is an intermediate in the mechanism and its concentration is not easily measured during the reaction. To write a useful rate law for the reaction, the concentration of Br must be expressed in terms of species whose concentration can be measured (such as the reactants).

Recognizing that the first step is a fast equilibrium, where the rate of the forward reaction is equal to the rate of the reverse reaction, we can write an equation that relates [Br] to [Br$_2$].

rate of Step 1 forward reaction = rate of Step 1 reverse reaction

$$k_1[Br_2] = k_{-1}[Br]^2$$

$$[Br] = \sqrt{\frac{k_1[Br_2]}{k_{-1}}} = \left(\frac{k_1}{k_{-1}}\right)^{1/2}[Br_2]^{1/2}$$

Substituting this relationship into the rate law for the rate-determining step,

$$rate = k_2[Br][H_2] = k_2\left(\frac{k_1}{k_{-1}}\right)^{1/2}[Br_2]^{1/2}[H_2]$$

Experimentally, we express the combination of rate constants as a single value.

$$rate = k[Br_2]^{1/2}[H_2]$$

Example Problem 14.6.4 Write the rate law for a mechanism with a slow second step.

A kinetic study is done to investigate the following reaction:

$$2\,H_2(g) + 2\,NO(g) \rightarrow N_2(g) + 2\,H_2O(g)$$

A proposed mechanism is:

Step 1. $2\,NO(g) \rightleftarrows N_2O_2(g)$ Fast in both directions

Step 2. $N_2O_2(g) + H_2(g) \rightarrow N_2O(g) + H_2O(g)$ Slow

Step 3. $N_2O(g) + H_2(g) \rightarrow N_2(g) + H_2O(g)$ Fast

a. Does this mechanism account for the overall reaction?
b. What experimental rate law would be observed if this mechanism is correct?

Solution:

You are asked to determine whether a given mechanism is possible for a reaction and to predict the experimental rate law for the reaction.

You are given a balanced chemical equation and a proposed mechanism for the reaction.

a. The equation for the overall reaction may be obtained by adding the equations for the elementary steps and simplifying.

$$2\,NO(g) \rightleftarrows N_2O_2(g)$$

$$N_2O_2(g) + H_2(g) \rightarrow N_2O(g) + H_2O(g)$$

$$N_2O(g) + H_2(g) \rightarrow N_2(g) + H_2O(g)$$

$$\overline{2\,NO(g) + \cancel{N_2O_2(g)} + \cancel{N_2O(g)} + 2\,H_2(g) \longrightarrow \cancel{N_2O_2(g)} + \cancel{N_2O(g)} + N_2(g) + 2\,H_2O(g)}$$

$$2\,NO(g) + 2\,H_2(g) \rightarrow N_2(g) + 2\,H_2O(g)$$

Example Problem 14.6.4 *(continued)*

The sum of the proposed mechanism steps is the same as the equation for the reaction being studied.

b. First write the rate law from the slow rate-determining Step 2, where k_2 is the rate constant for Step 2.

$$\text{rate} = k_2[N_2O_2][H_2]$$

To replace the intermediate (N_2O_2) in the rate law, assume that the first step comes to equilibrium, equate the forward (k_1) and backward (k_{-1}) rates, and solve for [N_2O_2] in terms of [NO].

$$k_1[NO]^2 = k_{-1}[N_2O_2]$$

$$[N_2O_2] = (k_1/k_{-1})[NO]^2$$

Finally, replace [N_2O_2] in the rate equation.

$$\text{Rate} = k_2(k_1/k_{-1})[NO]^2[H_2] = k[NO]^2[H_2], \text{ where } k = k_2(k_1/k_{-1})$$

Video Solution

**Tutored Practice
Problem 14.6.4**

Advanced Techniques for Determining Reaction Mechanisms

It is not always a simple process to determine a reaction mechanism. Often, advanced techniques are required, some of which are described here.

1. **Detection of an intermediate using spectroscopy.** If an intermediate can be detected, it can give a clue to a reaction mechanism. Detection can be done using spectroscopic methods if the intermediate absorbs light in the infrared, ultraviolet, or visible region of the electromagnetic spectrum. The main difficulty in detecting intermediates is that they are unstable and tend to react quickly.

2. **Trapping experiments.** These experiments are used to chemically detect intermediates by having the intermediate react with a secondary reagent to produce an identifiable product. That is, you design an experiment to trap an intermediate and stop the rest of the steps in the mechanism from taking place (Figure 14.6.6).
   In the reaction diagrams, the proposed mechanism involves three steps and involves two intermediates, B and C. In the trapped mechanism, a reagent, R, that is known to react with intermediate C is added. If the addition of R leads to formation of E and less D is formed, this supports the proposed mechanism. If no E is observed and D is formed, this means the proposed mechanism is probably incorrect. (This assumes that the trapping reaction occurs faster than conversion of C $\rightarrow$ D.)

3. **Isotopic labeling.** In this type of experiment, one of the reactants is synthesized containing a particular isotope of one of the reacting atoms. The location of the labeled

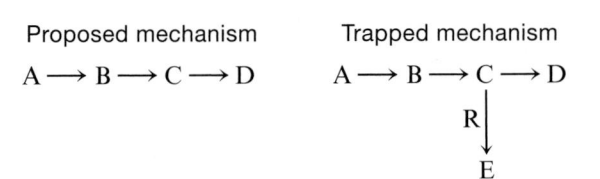

Figure 14.6.6 An example of a trapping experiment

atom can be tracked using a variety of detection methods. For example, in aqueous oxidation reactions involving metal oxides, it is possible to track whether the oxygen transferred in the reaction came from the metal oxide or from the water. The metal oxide can be synthesized using the isotope ^{18}O in a solution in which the water contains almost all ^{16}O. If the product contains ^{18}O, the metal oxide was the source of oxygen. If the product contains only ^{16}O, the oxygen atoms came from water. To be sure this experiment is valid, the experimental conditions need to be carefully controlled to make sure the metal oxide and water do not exchange O atoms.

14.6e Catalysis

Catalysis is the technique of using a catalyst to influence the rate of a reaction. Catalysts serve one of two functions: making reactions faster or making them more "selective." In either case, the catalyst provides an alternative mechanistic pathway for the reaction, which changes the activation energy and changes the reaction rate.

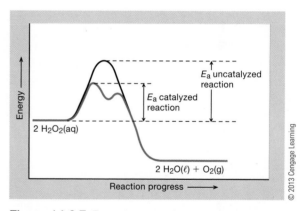

Figure 14.6.7 Reaction coordinate diagrams for the uncatalyzed and the catalyzed decomposition of hydrogen peroxide

Catalysis and Reaction Speed

Earlier, we saw the mechanism of the iodide ion–catalyzed decomposition of H_2O_2. The reaction is speeded up by the presence of I^-. However, H_2O_2 can also decompose on its own, without any iodide ion. If you have an unopened bottle of hydrogen peroxide that has been on the shelf for a few months, notice what happens when you first open it—there will often be pressure built up in the bottle due to the buildup of O_2 gas from the slow decomposition of H_2O_2 over time.

Figure 14.6.7 shows a reaction coordinate diagram for this reaction, both with a catalyst present and without one. The lower activation energy in the catalyzed case leads to a faster reaction.

Catalysis and Selectivity

Catalysts can also affect the **selectivity** of a reaction, which is a measure of the tendency of a reaction to form one set of products over another. Until now we have been assuming all reactions are 100% selective. However, often a chemical reaction can proceed by many different pathways, often with similar activation energies, which results in the formation of many different products.

A catalyst can increase selectivity by changing the pathway of one reaction path and not another. When this happens, the reaction will favor the formation of one set of products over another and the reaction becomes more selective (Interactive Figure 14.6.8).

Investigate selectivity in chemical reactions.

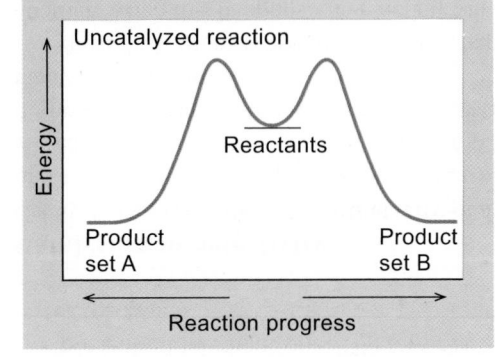
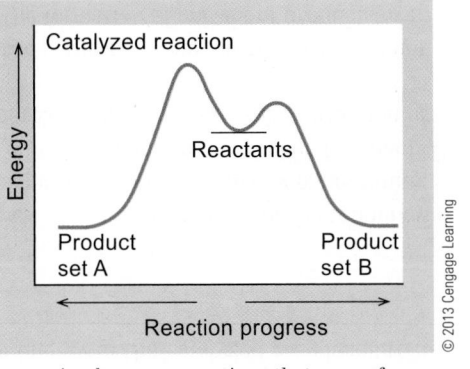

Reaction coordinate diagrams showing the progress for two simultaneous reactions that occur from a single set of reactants

In the uncatalyzed system, the two reactions have similar activation energies and proceed at close to the same rate, forming significant quantities of products A and B. Addition of the catalyst decreases the activation energy for formation of product B, but not product A. In this case, formation of B is much faster than formation of A, so the reaction is selective for formation of B.

Homogeneous and Heterogeneous Catalysis

Catalysts can be homogeneous or heterogeneous. **Homogeneous catalysts** are in the same phase as the compounds undergoing reaction. An example of a homogeneous catalyst is the aqueous iodide ions that catalyze the decomposition of aqueous hydrogen peroxide. These types of catalysts have the advantage of being able to be chemically modified to a great degree. Homogeneous catalysts are good at modifying reactions in a highly selective way. The main disadvantage of homogeneous catalysts is that they must be removed from the reaction mixture when the reaction is complete. That is, once the products are formed, they are still mixed with the catalyst and the two must be separated. This can be an expensive process.

Heterogeneous catalysts are not in the same phase as the compounds undergoing reaction. An example of a heterogeneous catalyst is catalytic hydrogenation, the use of solid platinum or nickel to catalyze the reaction of a hydrocarbon containing a carbon–carbon double bond (such as ethene, C_2H_4) with hydrogen (Interactive Figure 14.6.9). When the heterogeneous catalyst is a solid, it often acts as a reactive surface upon which reactants in

Interactive Figure 14.6.9

Explore a reaction mechanism that includes a heterogeneous catalyst.

Diagrams showing the mechanism for the platinum-catalyzed hydrogenation of ethylene

either the gas or solution phase can react. These catalysts have advantages and disadvantages that are the opposite of those for homogeneous catalysts. For example, they are easy to separate from the reaction products. They are, however, much more difficult to modify. Because of these differences, homogeneous catalysts are more often used for reactions in which great selectivity is needed; heterogeneous catalysts are used where selectivity is not an issue.

Section 14.6 Mastery

Unit Recap

Key Concepts

14.1 Introduction to Kinetics

- Chemical reactivity is controlled by both the thermodynamics and kinetics of the reaction (14.1a).

- According to collision theory, a reaction occurs only when the reacting species collide, the collision has enough energy to overcome the energy barrier necessary to initiate the reaction, and the reacting species collide with the correct orientation (14.1b).

- In a given reaction, not all collisions overcome the activation energy, but the number of effective collisions increases with increasing temperature (14.1b).

14.2 Expressing the Rate of a Reaction

- Reaction rate is expressed as the change in concentration of a reactant or product over time (14.2a).

- The relative rate at which reactants are consumed and products are formed is directly related to the reaction stoichiometry (14.2a).

- The rate of a reaction can be expressed as an average rate or an instantaneous rate (14.2a, 14.2b).

14.3 Rate Laws

- A rate law shows the relationship among reaction rate, the rate constant, the concentrations of reacting species, and the reaction order with respect to the reacting species (14.3a).

- The rate law for a reaction can be determined by using the method of initial rates; the experiments are repeated multiple times, and each experiment has a different set of initial concentrations for the reactants (14.3b).

14.4 Concentration Changes over Time

- Integrated rate laws show the relationship among the initial concentration of a reactant, the concentration of a reactant at a specific time, the rate constant, and time (14.4a).

- For a first-order reaction, the ratio $[A]_t/[A]_0$, the fraction of reactant that remains unreacted, can be used to determine the amount of time needed for some amount of a reactant to be consumed without knowing the amount initially present (14.4a).

- The rate law for a reaction can be determined graphically by using rearranged forms of the integrated rate laws (written in the form of a straight-line equation) and graphing concentration–time data (14.4b).

- The half-life ($t_{1/2}$) for a reaction is the amount of time required for the concentration of a reactant to fall to one half of its initial value (14.4c).

- For a first-order reaction, a half-life is independent of the initial concentration of reactant; for zero- and second-order reactions, half-life depends on both the rate constant and the initial concentration of the reactant (14.4c).

- All radioactive isotopes decay via first-order reactions (14.4d).

14.5 Activation Energy and Temperature

- A reaction coordinate diagram is a plot that shows energy as a function of the progress of the reaction from reactants to products (14.5a).

- Reaction coordinate diagrams show the relative energy of reactants and products, the energy of high-energy activated complexes, activation energy, and the energy of any intermediates (14.5a).

- The magnitude of the activation energy and the temperature of a reaction both affect reaction rate (14.5a).

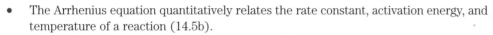

- The Arrhenius equation quantitatively relates the rate constant, activation energy, and temperature of a reaction (14.5b).

- Activation energy can be determined graphically using the Arrhenius equation written in the form of an equation for a straight line (14.5c).

14.6 Reaction Mechanisms and Catalysis

- A reaction mechanism is a detailed description at the molecular level of steps by which reactants are converted to products (14.6).

- Reaction mechanisms consist of a series of elementary steps that, when summed, give the overall equation for the chemical reaction (14.6a).

- Multistep mechanisms involve the presence of an intermediate, a chemical species that is formed in one step in the mechanism and then consumed in a later step (14.6b).

- When a catalyst appears in a reaction mechanism, it is consumed in one step and then produced, in its original form, in a later step (14.6b).

- The rate law for any elementary step is directly related to the stoichiometry of the elementary step (14.6c).

- The overall rate law for a reaction is the rate law for the rate-determining step, the slowest step in the reaction mechanism (14.6c).

- When the rate-determining step involves an intermediate as a reactant, the rate law is more complicated and might include species not participating in the rate-determining step (14.6d).

- Catalysts increase reaction rate by changing the reaction mechanism and lowering the activation energy (14.6e).

- Catalysts can affect reaction selectivity and can be homogeneous or heterogeneous (14.6e).

Key Equations

$$[A]_t = [A]_0 - kt \qquad (14.1)$$

$$\ln\frac{[A]_t}{[A]_0} = -kt \qquad (14.2)$$

$$\frac{1}{[A]_t} = \frac{1}{[A]_0} + kt \qquad (14.3)$$

$$t_{1/2} = \frac{[A]_0}{2k} \qquad (14.4)$$

$$t_{1/2} = \frac{\ln 2}{k} = \frac{0.693}{k} \qquad (14.5)$$

$$t_{1/2} = \frac{1}{k[A]_0} \qquad (14.6)$$

$$k = Ae^{-E_a/RT} \qquad (14.7)$$

$$\ln\frac{k_2}{k_1} = \frac{-E_a}{R}\left(\frac{1}{T_2} - \frac{1}{T_1}\right) \qquad (14.8)$$

Key Terms

14.1 Introduction to Kinetics
chemical kinetics
collision theory of reaction rates
activation energy

14.2 Expressing the Rate of a Reaction
reaction rate
average reaction rate
instantaneous rate
initial rate

14.3 Rate Laws
rate law
catalyst
rate constant
reaction order
method of initial rates

14.4 Concentration Changes over Time
integrated rate laws
half-life ($t_{1/2}$)

14.5 Activation Energy and Temperature
reaction coordinate diagram
activated complex
transition state
activation energy, E_a
Arrhenius equation

14.6 Reaction Mechanisms and Catalysis
reaction mechanism
elementary step
molecularity
unimolecular
bimolecular
termolecular
concerted process
intermediate
rate-determining step
catalysis
selectivity
homogeneous catalysts
heterogeneous catalysts

Unit 14   Review and Challenge Problems

15

Chemical Equilibrium

Unit Outline

15.1 The Nature of the Equilibrium State

15.2 The Equilibrium Constant, *K*

15.3 Using Equilibrium Constants in Calculations

15.4 Disturbing a Chemical Equilibrium: Le Chatelier's Principle

In This Unit...

We now begin our coverage of chemical equilibria, and in other units we will apply the concepts and tools learned in this unit to acid–base equilibria and the chemistry of insoluble ionic compounds. Chemical equilibrium and the reversibility of chemical reactions plays an important role in one of the most important chemical reactions in the human body: the binding of oxygen molecules to heme groups in the protein hemoglobin. When hemoglobin is exposed to oxygen in the lungs, oxygen molecules attach to the heme groups. When the oxygenated hemoglobin reaches oxygen-depleted cells, the oxygen is released. The process is reversible, and as you will see in this unit, it is the reversible nature of chemical reactions that is the basis of chemical equilibrium.

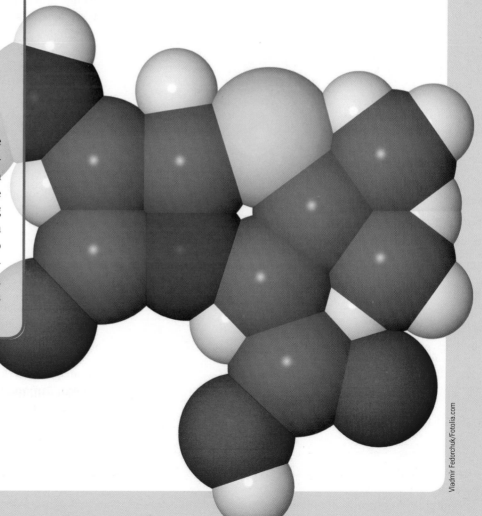

15.1 The Nature of the Equilibrium State

15.1a Principle of Microscopic Reversibility

In Chemical Kinetics (Unit 14) we learned that chemical reactions proceed in a series of steps called a reaction mechanism. The **principle of microscopic reversibility** tells us that the elementary steps in a reaction mechanism are reversible; that is, the mechanism of the reverse net reaction is reverse of the forward reaction mechanism (when the reaction conditions are the same). We define a **reversible process** as one in which it is possible to return to the starting conditions along the exact same path without altering the surroundings. Consider the two-step mechanism for the reaction of nitrogen dioxide with carbon monoxide.

Step 1 $2\ NO_2(g) \rightarrow NO(g) + NO_3(g)$

Step 2 $NO_3(g) + CO(g) \rightarrow NO_2(g) + CO_2(g)$

Net reaction: $NO_2(g) + CO(g) \rightarrow NO(g) + CO_2(g)$

The reverse reaction has a two-step mechanism that is the reverse of the forward reaction mechanism.

Step 1 $NO_2(g) + CO_2(g) \rightarrow NO_3(g) + CO(g)$ *(Step 2 reversed)*

Step 2 $NO_3(g) + NO(g) \rightarrow 2\ NO_2(g)$ *(Step 1 reversed)*

Net reaction: $NO(g) + CO_2(g) \rightarrow NO_2(g) + CO(g)$

In chemical equations, an equilibrium arrow ($\rightleftarrows$) is used to indicate that a reaction is reversible.

$$NO_2(g) + CO(g) \rightleftarrows NO(g) + CO_2(g)$$

The reversibility of chemical reactions can be very useful in chemical synthesis, such as the conversion between alkenes and alcohols (Interactive Figure 15.1.1). Although all chemical reactions are reversible, there are occasions when the use of an equilibrium arrow is not appropriate. For example, when hydrogen and oxygen react to form water vapor, essentially all reactants are converted to product and no noticeable amounts of reactants are formed by the reverse reaction. The chemical equation representing this reaction therefore uses a single reaction arrow ($\rightarrow$).

$$2\ H_2(g) + O_2(g) \rightarrow 2\ H_2O(g)$$

Interactive Figure 15.1.1

Investigate the reversibility of chemical reactions.

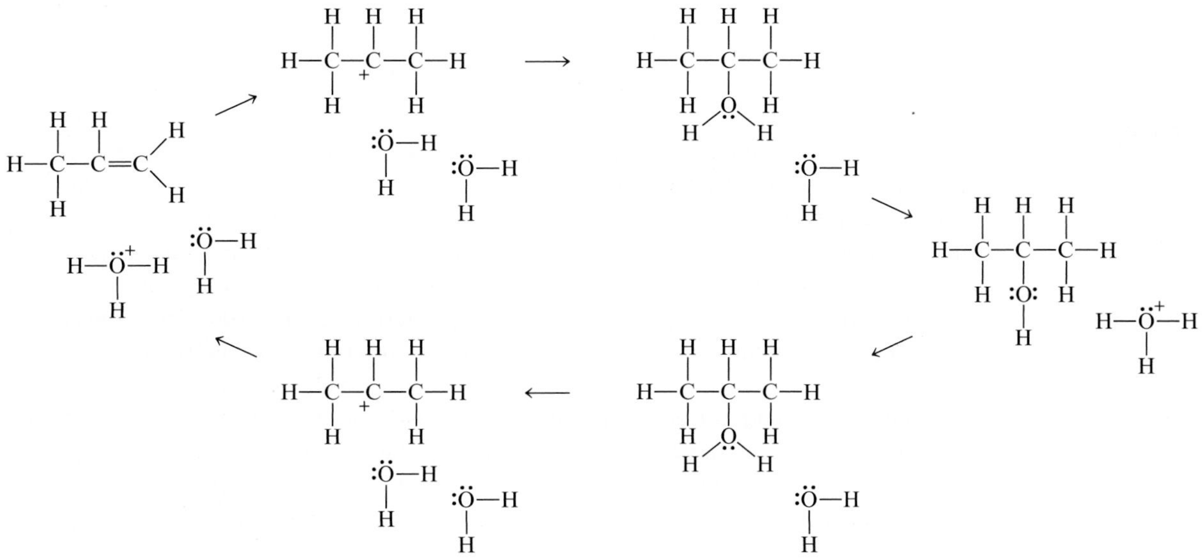

Alkenes are hydrolyzed to form alcohols in a reversible reaction.

Also, if the reverse reaction is unlikely to occur, such as when one of the reaction products is physically separated from a reaction mixture as a gas (the reaction of a metal carbonate with acid, for example), a single reaction arrow is used.

$$CuCO_3(s) + 2\ HCl(aq) \rightarrow CuCl_2(aq) + H_2O(\ell) + CO_2(g)$$

15.1b The Equilibrium State

In Intermolecular Forces and the Liquid State (Unit 11) we described the nature of a dynamic equilibrium between a liquid and its vapor in a sealed flask. (◄ Flashback to Section 11.2 Vapor Pressure) When equilibrium was reached in the flask, the amount of liquid did not change, but the vaporization and condensation processes continued. When the rate of vaporization is equal to the rate of condensation, a state of dynamic equilibrium between the two phases is reached. We can apply this concept to reversible chemical reactions and use it to describe the nature of a dynamic chemical equilibrium.

Consider the reversible reaction between Fe^{3+} and the thiocyanate ion.

$$Fe^{3+}(aq) + SCN^-(aq) \rightleftarrows FeSCN^{2+}(aq)$$

When the two reactants are mixed, they react in the forward direction to form $FeSCN^{2+}$ in a second-order process.

$$Fe^{3+}(aq) + SCN^-(aq) \rightarrow FeSCN^{2+}(aq) \qquad \text{Rate (forward)} = k_{forward}[Fe^{3+}][SCN^-]$$

As the reaction proceeds, the product concentration increases and the reverse reaction begins to take place.

$$FeSCN^{2+}(aq) \rightarrow Fe^{3+}(aq) + SCN^-(aq) \qquad \text{Rate (reverse)} = k_{reverse}[FeSCN^{2+}]$$

As the reaction continues, the rate of the forward reaction decreases (because $[Fe^{3+}]$ and $[SCN^-]$ decrease) and the rate of the reverse reaction increases (because $[FeSCN^{2+}]$ increases). Eventually, a state of **chemical equilibrium** is reached where the rate of the forward reaction is equal to the rate of the reverse reaction:

$$\text{Rate (forward)} = \text{Rate (reverse)}$$

$$k_{forward}[Fe^{3+}][SCN^-] = k_{reverse}[FeSCN^{2+}]$$

When this equilibrium state is achieved, the concentrations of all the species in solution are constant, even though the forward and reverse reactions continue to take place. Note that when a system is at equilibrium, while the rates of the forward and reverse reactions are equal, the rate constants for those reactions and the concentrations of reactant and products are generally not equal.

The equilibrium state can be represented graphically by plotting the concentration of reactants and products over time (Interactive Figure 15.1.2). Notice that during the first stage of the reaction, the forward reaction is faster than the reverse reaction and reactant concentrations decrease while the product concentration increases. When the system reaches equilibrium, the forward and reverse rates are equal and concentrations of reactants and product do not change. Also notice that significant amounts of both reactants and products (Fe^{3+}, SCN^-, and $FeSCN^{2+}$) are present in the equilibrium system.

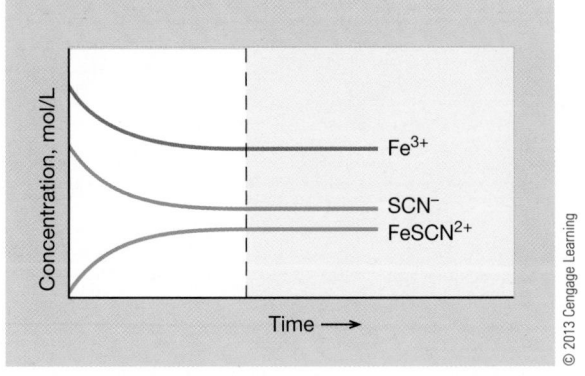

15.2 The Equilibrium Constant, K

15.2a Equilibrium Constants

The relationship between forward and reverse rate constants for an equilibrium system is shown in an **equilibrium constant expression** and quantified by an **equilibrium constant (K)**. An equilibrium constant expression is written by rearranging the equation relating forward and reverse reaction rates. When the forward and reverse reaction rates are equal, the chemical system is at equilibrium and the ratio of forward and reverse rate constants is equal to the equilibrium constant, K. For the $Fe(SCN)^{2+}$ equilibrium system, for example,

$$\text{Rate (forward)} = \text{Rate (reverse)}$$

$$k_{forward}[Fe^{3+}][SCN^-] = k_{reverse}[FeSCN^{2+}]$$

$$\frac{k_{forward}}{k_{reverse}} = K = \frac{[FeSCN^{2+}]}{[Fe^{3+}][SCN^-]} = 142 \text{ at } 25\ °C$$

Note that because rate constants change with temperature, the equilibrium constant will also vary with temperature.

The magnitude of the equilibrium constant provides information about the relative rate constants of the forward and reverse reactions and the relative amounts of reactants and products at equilibrium.

- $K \gg 1$

  A large value of K ($K \gg 1$) means that at equilibrium, the concentration of products is much larger than the concentration of reactants. The rate constant for the forward reaction is much larger than the reverse reaction rate constant ($k_{forward} \gg k_{reverse}$), and at equilibrium the system contains mostly products. This is called a **product-favored reaction**. For example, consider the product-favored reaction between gaseous iodine and chlorine:

$$I_2(g) + Cl_2(g) \rightleftarrows 2\ ICl(g) \qquad K = 2.1 \times 10^5 \text{ at } 25\ °C$$

  This is a product-favored reaction, and at equilibrium there will be very little I_2 or Cl_2 in the reaction flask.

- $K \ll 1$

  A small value of K ($K \ll 1$) means that at equilibrium, the concentration of reactants is much larger than the concentration of products. The rate constant for the reverse reaction is much larger than the forward reaction rate constant ($k_{reverse} \gg k_{forward}$), and at equilibrium the system contains mostly reactants. This is called a **reactant-favored reaction**. For example, consider the reaction of acetic acid with water:

$$CH_3CO_2H(aq) + H_2O(\ell) \rightleftharpoons CH_3CO_2^-(aq) + H_3O^+(aq) \qquad K = 1.8 \times 10^{-5} \text{ at } 25 \text{ °C}$$

This is a reactant-favored reaction, and at equilibrium there will be very little $CH_3CO_2^-$ or H_3O^+ in the reaction flask.

- $K \approx 1$

  A value of K very close to 1 ($K \approx 1$) means that at equilibrium, significant amounts of both reactants and product are found in the reaction vessel. The rate constants for the forward and reverse reactions are comparable ($k_{forward} \approx k_{reverse}$), and at equilibrium the system contains a mixture of both reactants and products. For example, consider the dimerization of nitrogen dioxide to form dinitrogen tetraoxide:

$$2\,NO_2(g) \rightleftharpoons N_2O_4(g) \qquad K = 1.4 \text{ at } 50 \text{ °C}$$

  The equilibrium constant has a value close to 1, so at equilibrium significant amounts of both NO_2 and N_2O_4 will be found in the reaction flask.

  The difference between a product-favored reaction and a reactant-favored reaction can be shown graphically, as in Interactive Figure 15.2.1.

Interactive Figure 15.2.1

Recognize the significance of the magnitude of the equilibrium constant.

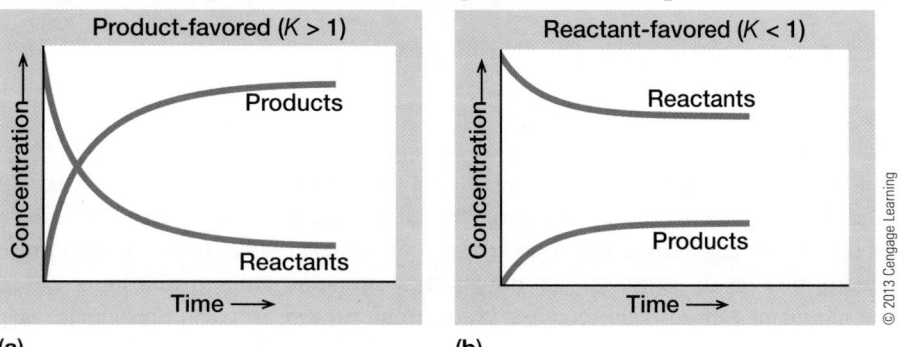

(a)   (b)

Graphical representations of (a) product-favored and (b) reactant-favored chemical reactions

© 2013 Cengage Learning

15.2b Writing Equilibrium Constant Expressions

For the general equilibrium system

$$a\,A + b\,B \rightleftharpoons c\,C + d\,D$$

the equilibrium constant expression is written

$$K = \frac{[C]^c[D]^d}{[A]^a[B]^b} \qquad\qquad (15.1)$$

In the equilibrium constant expressions,

- product concentrations are multiplied in the numerator, and each is raised to the power of the stoichiometric coefficient from the balanced equation;

- reactant concentrations are multiplied in the denominator, and each is raised to the power of the stoichiometric coefficient from the balanced equation; and

- the concentrations of A, B, C, and D are those measured when the reaction is at equilibrium.

For example, the equilibrium constant expression for the dimerization of nitrogen dioxide has the concentration of the product, N_2O_4, in the numerator and the concentration of the reactant, NO_2, in the denominator. Each concentration is raised to the power of the corresponding stoichiometric coefficient from the balanced equation.

$$\text{implied "1"}$$
$$2\,NO_2(g) \rightleftharpoons N_2O_4(g) \quad K = \frac{[N_2O_4]}{[NO_2]^2}$$

© 2013 Cengage Learning

There are two cases when a reactant or product does not appear in the equilibrium expression.

1. *Pure solids do not appear in equilibrium constant expressions.* When an equilibrium system includes a pure solid as a separate phase, it is observed that the equilibrium concentrations of the species in solution or in the gas phase are independent of the amount of solid present. For example, consider the equilibrium between solid ammonium chloride and gaseous ammonia and hydrogen chloride:

$$NH_4Cl(s) \rightleftharpoons NH_3(g) + HCl(g) \qquad K = [NH_3][HCl]$$

The equilibrium concentrations of ammonia and hydrogen chloride do not depend on the amount of solid present as long as there is some solid present for the system to reach equilibrium. The concentration of a pure solid depends only on the density of the substance, a constant that can be incorporated into the equilibrium constant.

2. *Pure liquids and solvents do not appear in equilibrium constant expressions.* When an equilibrium system involves a pure liquid, often a solvent, its concentration is typically very large compared with the concentrations of the other species present. For example, at 25 °C, the concentration of H_2O in a dilute aqueous solution can be as high as 56 mol/L! As the system reaches equilibrium, the solvent concentration remains essentially constant, equal to that of the pure liquid. For example, consider the reaction that takes place in a solution of aqueous ammonia:

$$NH_3(aq) + H_2O(\ell) \rightleftharpoons NH_4^+(aq) + OH^-(aq) \qquad K = \frac{[NH_4^+][OH^-]}{[NH_3]}$$

Water is the solvent for this reaction and it also participates in the chemical equilibrium. Its concentration remains essentially constant during the equilibrium process, so it is incorporated into the equilibrium constant and does not appear in the equilibrium constant expression.

Example Problem 15.2.1 Write equilibrium constant expressions.

Write equilibrium constant expressions for the following reactions:

a. $H_2(g) + Cl_2(g) \rightleftharpoons 2\ HCl(g)$
b. $C(s) + H_2O(g) \rightleftharpoons H_2(g) + CO(g)$
c. $CH_3CO_2H(aq) + H_2O(\ell) \rightleftharpoons H_3O^+(aq) + CH_3CO_2^-(aq)$

Solution:

You are asked to write an equilibrium expression for a given reaction.

You are given a chemical equation for a reaction.

a. $K = \dfrac{[HCl]^2}{[H_2][Cl_2]}$

Note that the HCl concentration is squared because of its stoichiometric coefficient (2) in the balanced equation.

Example Problem 15.2.1 *(continued)*

b. $K = \dfrac{[H_2][CO]}{[H_2O]}$

Note that C(s) does not appear in the equilibrium constant expression. Also, notice that in this example, H_2O appears in the equilibrium constant expression because it is not a solvent.

c. $K = \dfrac{[H_3O^+][CH_3CO_2^-]}{[CH_3CO_2H]}$

Note that water, the solvent in this aqueous equilibrium system, is treated as a pure liquid. Its concentration does not change significantly during the reaction, so it does not appear in the equilibrium constant expression.

Video Solution

Tutored Practice Problem 15.2.1

All of the equilibrium constant expression examples we have seen to this point express reactant and product concentrations in molarity units (mol/L). For gas-phase equilibria, it is also possible to express concentrations in terms of partial pressures of reactants and products. To differentiate between these two expressions, we use the symbol K_c to indicate a constant calculated using molarity units and K_p to indicate a constant calculated using partial pressure units. In this text, all equilibrium constants labeled K are K_c values.

For equilibria involving gases, K_c and K_p values are related by the equation

$$K_p = K_c(RT)^{\Delta n} \qquad\qquad (15.2)$$

where R is the ideal gas constant, T is temperature (in kelvin units), and Δn is the change in number of moles of gas in the reaction (Δn = moles gaseous products − moles gaseous reactants).

Example Problem 15.2.2 Interconvert K_p and K_c values.

Calculate K_p for the following reactions at the indicated temperature.

a. $2\ NOBr(g) \rightleftarrows 2\ NO(g) + Br_2(g)$ $K_c = 6.50 \times 10^{-3}$ at 298 K
b. $NH_4I(s) \rightleftarrows NH_3(g) + HI(g)$ $K_c = 7.00 \times 10^{-5}$ at 673 K

Solution:

You are asked to calculate K_p for an equilibrium reaction at a given temperature.

You are given a K_c value for an equilibrium reaction at a given temperature.

First determine the change in number of moles of gas in the reaction, then use Equation 15.2 to calculate K_p.

Example Problem 15.2.2 *(continued)*

a. $\Delta n = $ (3 mol gaseous products) $-$ (2 mol gaseous reactants) $= 1$

$K_p = K_c(RT)^{\Delta n} = (6.50 \times 10^{-3})[(0.082057 \text{ L} \cdot \text{atm/K} \cdot \text{mol})(298 \text{ K})]^1 = 0.159$

b. $\Delta n = $ (2 mol gaseous products $-$ 0 mol gaseous reactants) $= 2$

$K_p = K_c(RT)^{\Delta n} = (7.00 \times 10^{-5})[(0.082057 \text{ L} \cdot \text{atm/K} \cdot \text{mol})(673 \text{ K})]^2 = 0.213$

Video Solution

Tutored Practice
Problem 15.2.2

15.2c Manipulating Equilibrium Constant Expressions

The value of the equilibrium constant for a chemical system is a function of how the equilibrium expression is written. Consider, for example, the reaction between $Cu^{2+}(aq)$ and $NH_3(aq)$. The chemical equation that describes this reaction can be written as:

$$Cu^{2+}(aq) + 4 NH_3(aq) \rightleftarrows Cu(NH_3)_4^{2+}(aq)$$

$$K = \frac{[Cu(NH_3)_4^{2+}]}{[Cu^{2+}][NH_3]^4} = 6.8 \times 10^{12} \text{ at 25 °C}$$

This equation can be manipulated by (1) multiplying the equation by a constant, (2) writing the reaction in the reverse direction, and (3) combining the equation with another chemical equation to describe a different equilibrium system. Each manipulation results in a new equilibrium constant, each of which is a correct value for the reaction as described by the chemical equation. The equation describing a chemical system can be written in many ways, and each equation has its own K value. This means that it is important to be specific when reporting an equilibrium constant by including the chemical equation along with the value for K.

1. **Multiply the equation by a constant.**
   If the chemical equation for the copper-ammonia equilibrium is manipulated by multiplying the stoichiometric coefficients by ¼, the equilibrium constant expression is written as:

   $$¼ Cu^{2+}(aq) + NH_3(aq) \rightleftarrows ¼ Cu(NH_3)_4^{2+}(aq)$$

   $$K_{new} = \frac{[Cu(NH_3)_4^{2+}]^{1/4}}{[Cu^{2+}]^{1/4}[NH_3]}$$

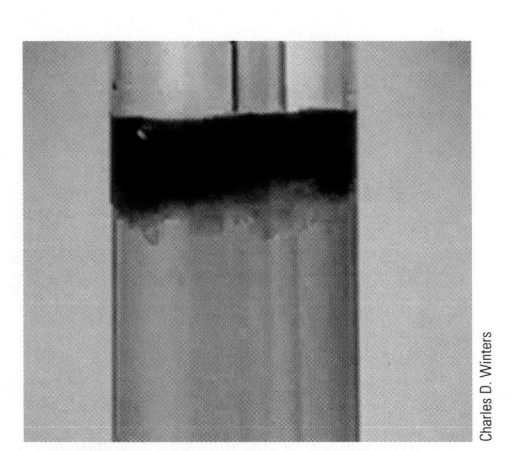

Charles D. Winters

Aqueous Cu^{2+} reacts with ammonia to form deep blue $Cu(NH_3)_4^{2+}$.

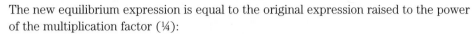

The new equilibrium expression is equal to the original expression raised to the power of the multiplication factor (¼):

$$K_{new} = \frac{[Cu(NH_3)_4^{2+}]^{1/4}}{[Cu^{2+}]^{1/4}[NH_3]} = \left(\frac{[Cu(NH_3)_4^{2+}]}{[Cu^{2+}][NH_3]^4}\right)^{1/4} = (K_{old})^{1/4} = (6.8 \times 10^{12})^{1/4} = 1.6 \times 10^3$$

In general, when an equilibrium equation is multiplied by a constant, n,

$$K_{new} = (K_{old})^n \qquad \text{(15.3)}$$

2. **Reverse the reaction direction.**

If the chemical equation for the copper-ammonia equilibrium is manipulated by writing it in the reverse direction, the equilibrium constant expression is written as:

$$Cu(NH_3)_4{}^{2+}(aq) \rightleftharpoons Cu^{2+}(aq) + 4\,NH_3(aq)$$

$$K_{new} = \frac{[Cu^{2+}][NH_3]^4}{[Cu(NH_3)_4^{2+}]}$$

The new equilibrium expression is equal to the inverse of the original expression:

$$K_{new} = \frac{[Cu^{2+}][NH_3]^4}{[Cu(NH_3)_4^{2+}]} = \left(\frac{[Cu(NH_3)_4^{2+}]}{[Cu^{2+}][NH_3]^4}\right)^{-1} = \frac{1}{K_{old}} = \frac{1}{6.8 \times 10^{12}} = 1.5 \times 10^{-13}$$

In general, when an equilibrium equation is written in the reverse direction,

$$K_{new} = \frac{1}{K_{old}} \qquad \text{(15.4)}$$

3. **Combine reactions.**

If the chemical equation for the copper-ammonia equilibrium is manipulated by combining it with the copper(II) hydroxide dissolution equilibrium ($K = 1.6 \times 10^{-19}$), the new equilibrium is written as:

(1) $Cu^{2+}(aq) + 4\,NH_3(aq) \rightleftharpoons Cu(NH_3)_4{}^{2+}(aq)$ $\qquad K_1 = \dfrac{[Cu(NH_3)_4^{2+}]}{[Cu^{2+}][NH_3]^4}$

(2) $Cu(OH)_2(s) \rightleftharpoons Cu^{2+}(aq) + 2\,OH^-(aq)$ $\qquad K_2 = [Cu^{2+}][OH^-]^2$

Net reaction:

$Cu(OH)_2(s) + 4\,NH_3(aq) \rightleftharpoons Cu(NH_3)_4{}^{2+}(aq) + 2\,OH^-(aq)$ $\quad K_{net} = \dfrac{[Cu(NH_3)_4^{2+}][OH^-]^2}{[NH_3]^4}$

The new equilibrium expression is equal to the product of the two original equilibrium expressions:

$$K_{net} = \frac{[Cu(NH_3)_4^{2+}][OH^-]^2}{[NH_3]^4} = \frac{[Cu(NH_3)_4^{2+}]}{[Cu^{2+}][NH_3]^4} \times [Cu^{2+}][OH^-]^2$$

$$= K_1 \times K_2 = (6.8 \times 10^{12})(1.6 \times 10^{-19})$$

$$= 1.1 \times 10^{-6}$$

In general, when two equations representing equilibrium systems are added together,

$$K_{new} = K_1 \times K_2 \qquad \qquad \textbf{(15.5)}$$

Example Problem 15.2.3 Manipulate equilibrium constant expressions.

The equilibrium constant for the reaction of hydrogen with oxygen to form water is 5.7×10^{40} at 25 °C.

$$H_2(g) + \tfrac{1}{2} O_2(g) \rightleftarrows H_2O(g) \qquad K = 5.7 \times 10^{40} \text{ at } 25 \text{ °C}$$

Calculate the equilibrium constant for the following reactions.

a. $2 H_2(g) + O_2(g) \rightleftarrows 2 H_2O(g)$
b. $H_2O(g) \rightleftarrows H_2(g) + \tfrac{1}{2} O_2(g)$

Solution:

You are asked to calculate an equilibrium constant for a reaction.

You are given a chemical equation, with an equilibrium constant, that is related to the equation for the given reaction.

a. Multiplying the original equation by a factor of 2 results in the new equation, so the new equilibrium constant is equal to the original constant squared (raised to the power of 2).

$$K_{original} = \frac{[H_2O]}{[H_2][O_2]^{1/2}} \qquad K_{new} = \frac{[H_2O]^2}{[H_2]^2[O_2]} = K_{original}^2$$

$$K_{new} = K_{original}^2 = (5.7 \times 10^{40})^2 = 3.2 \times 10^{81}$$

b. Reversing the original equation results in the new equation, so the new equilibrium constant is equal to the inverse of the original constant.

$$K_{original} = \frac{[H_2O]}{[H_2][O_2]^{1/2}} \qquad K_{new} = \frac{[H_2][O_2]^{1/2}}{[H_2O]} = \frac{1}{K_{original}}$$

$$K_{new} = \frac{1}{K_{original}} = \frac{1}{5.7 \times 10^{40}} = 1.8 \times 10^{-41}$$

Video Solution

Tutored Practice
Problem 15.2.3

Section 15.2 Mastery

15.3 Using Equilibrium Constants in Calculations

15.3a Determining an Equilibrium Constant Using Experimental Data

The value of an equilibrium constant can be determined using the equilibrium concentrations of the chemical species involved in the equilibrium process. The concentrations are substituted into the equilibrium expression and the constant is calculated. For any equilibrium system there are an infinite number of combinations of equilibrium concentrations, all of which result in the same value of K.

Example Problem 15.3.1 Use equilibrium concentrations to calculate K.

Some sulfur trioxide is placed in a flask and heated to 1400 K. When equilibrium is reached, the flask is found to contain SO_3 (0.152 M), SO_2 (0.0247 M), and O_2 (0.0330 M). What is the value of the equilibrium constant for the following at reaction at 1400 K?

$$2\ SO_3(g) \rightleftarrows 2\ SO_2(g) + O_2(g)$$

Solution:

You are asked to calculate the equilibrium constant for a reaction at a given temperature.

You are given equilibrium concentrations of reactants and products at a given temperature.

Step 1. Write the equilibrium constant expression.

$$K = \frac{[SO_2]^2[O_2]}{[SO_3]^2}$$

Step 2. Substitute the equilibrium concentrations into the equilibrium expression and calculate K.

$$K = \frac{[SO_2]^2[O_2]}{[SO_3]^2} = \frac{(0.0247)^2(0.0330)}{(0.152)^2} = 8.71 \times 10^{-4}$$

Notice that we do not assign units to the equilibrium constant, even though the concentrations have units of mol/L.

Video Solution

Tutored Practice
Problem 15.3.1

It is often difficult to accurately measure the concentration of all species in solution. When this is the case, the equilibrium concentration of only one species is measured and stoichiometric relationships are used to determine the concentrations of the other species at equilibrium. We will use the *ICE* method to determine the concentration of species at equilibrium. The *ICE* method involves the use of a table showing the **I**nitial concentrations, the **C**hange in concentrations as the reaction approaches equilibrium, and the **E**quilibrium concentrations. The *ICE* method is demonstrated in the following example problem.

Example Problem 15.3.2 Use an *ICE* table to calculate *K*.

A mixture of nitrogen and hydrogen is allowed to react in the presence of a catalyst.

$$N_2(g) + 3 H_2(g) \rightleftarrows 2 NH_3(g)$$

The initial concentrations of the reactants are $[N_2] = 0.1000$ M and $[H_2] = 0.2200$ M. After the system reaches equilibrium, it is found that the nitrogen concentration has decreased to 0.0271 M. Determine the value of the equilibrium constant, *K*, for this reaction.

Solution:

You are asked to calculate the equilibrium constant for a reaction.

You are given the initial concentrations of reactants and the equilibrium concentration of one of the reactants.

Step 1. Write the equilibrium constant expression.

$$K = \frac{[NH_3]^2}{[N_2][H_2]^3}$$

Step 2. Set up an *ICE* table (*initial* concentrations, the *change* in concentrations as the reaction progresses, and the *equilibrium* concentrations) and fill in the initial concentrations.

	$N_2(g)$	+	$3 H_2(g)$	$\rightleftarrows$	$2 NH_3(g)$
Initial (M)	0.1000		0.2200		0
Change (M)					
Equilibrium (M)					

Step 3. Use the stoichiometry of the balanced chemical equation to define the change in concentration for each species as the reaction proceeds toward equilibrium in terms of the unknown quantity, x. Thus, the concentration of N_2 decreases by x, the concentration of H_2 decreases by $3x$, and the concentration of NH_3 increases by $2x$.

	$N_2(g)$	+	$3 H_2(g)$	$\rightleftarrows$	$2 NH_3(g)$
Initial (M)	0.1000		0.2200		0
Change (M)	$-x$		$-3x$		$+2x$
Equilibrium (M)					

Notice that the reactant concentrations decrease and the product concentration increases as the reaction proceeds toward equilibrium.

Example Problem 15.3.2 *(continued)*

Step 4. Add the initial and change in concentrations to represent the equilibrium concentrations in terms of x.

	$N_2(g)$	$+$	$3 H_2(g)$	$\rightleftarrows$	$2 NH_3(g)$
Initial (M)	0.1000		0.2200		0
Change (M)	$-x$		$-3x$		$+2x$
Equilibrium (M)	$0.1000 - x$		$0.2200 - 3x$		$2x$

Step 5. Determine the value of x using the actual equilibrium concentration of N_2 (given in the problem, 0.0271 M) and the N_2 equilibrium concentration described in terms of x.

$$0.1000 - x = 0.0271$$
$$x = 0.0729 \text{ M}$$

Step 6. Use the value of x to determine the equilibrium concentrations of H_2 and NH_3.

$$[H_2]_{\text{equilibrium}} = 0.2200 - 3x = 0.2200 - 3(0.0729) = 0.0013 \text{ M}$$

$$[NH_3]_{\text{equilibrium}} = 2x = 2(0.0729) = 0.146 \text{ M}$$

Step 7. Substitute the equilibrium concentrations into the equilibrium constant expression and calculate K.

$$K = \frac{[NH_3]^2}{[N_2][H_2]^3} = \frac{(0.146)^2}{(0.0271)(0.0013)^3} = 3.6 \times 10^8$$

Video Solution

Tutored Practice
Problem 15.3.2

15.3b Determining Whether a System Is at Equilibrium

Many chemical systems, particularly those found in nature, are not at equilibrium. We can determine whether a system is at equilibrium by comparing the ratio of concentrations at a specific point in the reaction process (the **reaction quotient**, Q) to the ratio found at equilibrium (the equilibrium constant). A ratio of reactant and product concentrations, called the **reaction quotient expression**, can be written for any reaction that may or may not be at equilibrium. For the reaction

$$a \text{ A} + b \text{ B} \rightleftarrows c \text{ C} + d \text{ D}$$

the reaction quotient expression is written

$$Q = \frac{[C]^c[D]^d}{[A]^a[B]^b} \qquad \qquad \textbf{(15.6)}$$

Notice that the reaction quotient expression has the same form as the equilibrium constant expression, with the exception *that the concentrations of A, B, C, and D may or may not be equilibrium concentrations.*

As was true for equilibrium constant expressions, pure liquids and pure solids do not appear in reaction quotient expressions.

Comparing Q to K for a specific reaction allows us to determine whether a system is at equilibrium (Interactive Figure 15.3.1). There are three possible relationships between the two values:

- $Q < K$

  The system is not at equilibrium. Reactants will be consumed and product concentration will increase until $Q = K$. The product concentration (in the numerator of the Q expression) is too small, so the reaction will proceed in the forward direction (to the right) to reach equilibrium.

- $Q > K$

  The system is not at equilibrium. Products will be consumed and reactant concentration will increase until $Q = K$. The product concentration (in the numerator of the

Interactive Figure 15.3.1

Use Q and K to determine if a system is at equilibrium.

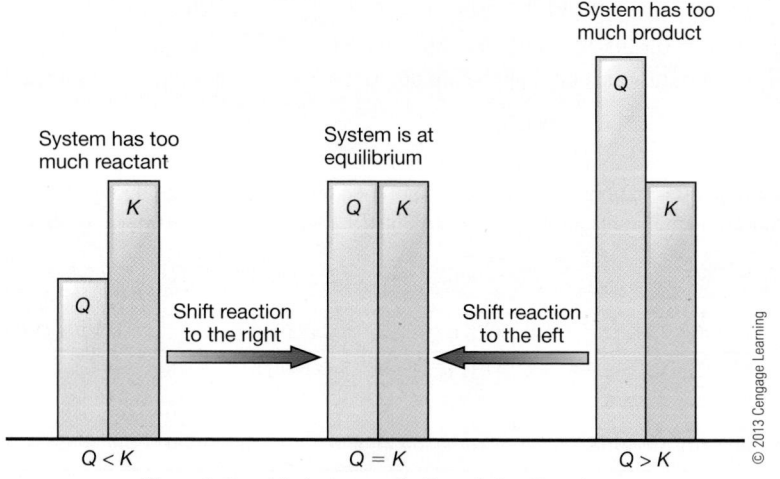

The relationship between Q, K, and the direction
a reaction proceeds to reach equilibrium

© 2013 Cengage Learning

Q expression) is too large, so the reaction will proceed in the reverse direction (to the left) to reach equilibrium.

- $Q = K$

The system is at equilibrium and no further change in reactant or product concentration will occur.

Example Problem 15.3.3 Use the reaction quotient, Q, to determine whether a system is at equilibrium.

Consider the following system ($K = 22.3$ at a given temperature):

$$Cl_2(g) + F_2(g) \rightleftarrows 2\ ClF(g)$$

The concentrations of reactants and products are measured at a particular time and found to be:

$$[Cl_2] = 0.300\ M \qquad [F_2] = 0.620\ M \qquad [ClF] = 0.120\ M$$

Is the system at equilibrium? If not, in which direction will it proceed to reach equilibrium?

Solution:

You are asked to determine if a system is at equilibrium and, if it is not, the direction it will proceed to reach equilibrium.

You are given the chemical equation and equilibrium constant for a reaction, and concentrations of reactants and products.

Substitute the concentrations into the reaction quotient expression and compare the calculated value of Q to the equilibrium constant.

$$Q = \frac{[ClF]^2}{[Cl_2][F_2]} = \frac{(0.120)^2}{(0.300)(0.620)} = 0.0774$$

$Q < K$, so the system is not at equilibrium. The reaction will proceed in the forward direction (to the right), consuming chlorine and fluorine and forming additional ClF until equilibrium is reached.

Video Solution

Tutored Practice
Problem 15.3.3

15.3c Calculating Equilibrium Concentrations

If a chemical system is not at equilibrium, the equilibrium constant can be used to determine the equilibrium concentrations of reactants and products. The *ICE* method is used to determine equilibrium concentrations.

Example Problem 15.3.4 Use K and initial concentrations to calculate equilibrium concentrations.

Consider the equilibrium system involving two isomers of butane.

$$H_3C-CH_2-CH_2-CH_3 \rightleftharpoons H_3C-\overset{\overset{\displaystyle CH_3}{|}}{\underset{\underset{\displaystyle H}{|}}{C}}-CH_3 \qquad K = \frac{[\text{isobutane}]}{[\text{butane}]} = 2.5 \text{ at } 20\ ^\circ C$$

Butane   Isobutane

A flask originally contains 0.200 M butane. Calculate the equilibrium concentrations of butane and isobutane.

Solution:

You are asked to calculate equilibrium concentrations of reactants and products in a reaction.

You are given a chemical equation and equilibrium constant and the initial concentration of a reactant.

Step 1. Use the *ICE* method to determine equilibrium concentrations. Use the unknown quantity x to define the equilibrium concentrations of butane and isobutane.

	[butane]	$\rightleftharpoons$	[isobutane]
Initial (M)	0.200		0
Change (M)	$-x$		$+x$
Equilibrium (M)	$0.200 - x$		x

Step 2. Substitute the equilibrium concentrations into the equilibrium constant expression and determine the value of x, the change in butane and isobutane concentrations as the system approaches equilibrium.

$$K = 2.5 = \frac{x}{0.200 - x}$$

$$x = 0.143$$

Step 3. Use the value of x to calculate the equilibrium concentrations of butane and isobutane.

$$[\text{butane}]_{\text{equilibrium}} = 0.200 - x = 0.057 \text{ M} \qquad [\text{isobutane}]_{\text{equilibrium}} = x = 0.143 \text{ M}$$

Is your answer reasonable? As a final check of your answer, confirm that these are equilibrium concentrations by substituting them into the equilibrium constant expression and calculating K.

$$K = \frac{[\text{isobutane}]}{[\text{butane}]} = \frac{0.143}{0.057} = 2.5$$

Video Solution

Tutored Practice
Problem 15.3.4

Section 15.3 Mastery

15.4 Disturbing a Chemical Equilibrium: Le Chatelier's Principle

15.4a Addition or Removal of a Reactant or Product

According to **Le Chatelier's principle**, if a chemical system at equilibrium is disturbed so that it is no longer at equilibrium, the system will respond by reacting in either the forward or reverse direction so as to counteract the disturbance, resulting in a new equilibrium composition. We often refer to this as a reaction "shifting" to the right (forming additional product) or to the left (forming additional reactant) as a result of an equilibrium disturbance. Typical disturbances include (1) the addition or removal of a reactant or product, (2) a change in the volume of a system involving gases, and (3) a change in temperature.

Consider the reversible reaction between Fe^{3+} and SCN^-.

$$Fe^{3+}(aq) + SCN^-(aq) \rightleftarrows FeSCN^{2+}(aq) \qquad K = \frac{[FeSCN^{2+}]}{[Fe^{3+}][SCN^-]}$$

If the concentration of Fe^{3+} is increased by the addition of some solid $Fe(NO_3)_3$, the system is no longer at equilibrium. The increased Fe^{3+} concentration results in a situation where $Q < K$.

$$[Fe^{3+}]_{new} > [Fe^{3+}]_{equilibrium}$$

$$Q = \frac{[FeSCN^{2+}]}{[Fe^{3+}]_{new}[SCN^-]} < K$$

The reaction will shift to the right, consuming SCN^- and some of the additional Fe^{3+} and forming additional $FeSCN^{2+}$. In terms of Le Chatelier's principle, the system shifts to the right, away from the disturbance of added reactant (Interactive Figure 15.4.1).

A common practice in the chemical industry involves the use of Le Chatelier's principle to drive a chemical reaction to completion. If one of the products is continually removed as it is formed, a chemical reaction can be shifted to the right, forming more and more products until all reactants are consumed. The result of adding or removing reactants or products on the equilibrium composition is summarized in Table 15.4.1.

After a change in reactant or product concentration, new equilibrium concentrations can be calculated using the *ICE* method, as shown in the following example.

Predict effect of concentration change on an equilibrium system.

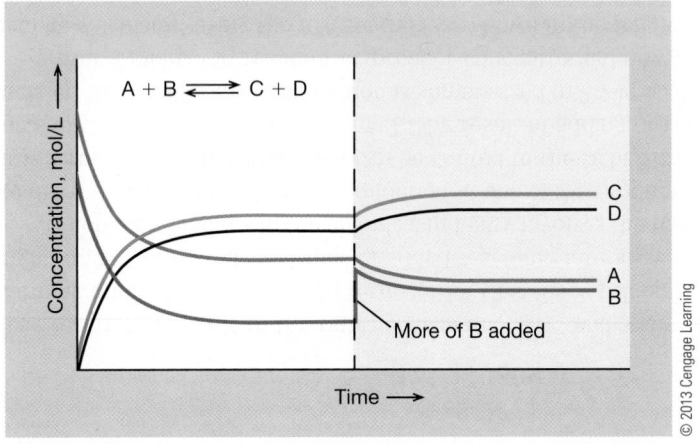

Graphical representation of the addition of a reactant to an equilibrium system

Table 15.4.1 Effect of Concentration Change on an Equilibrium System

Change	System Response		Effect on K
Add reactant	Shifts to the right	Reactants → Products	No change
Add product	Shifts to the left	Reactants ← Products	No change
Remove reactant	Shifts to the left	Reactants ← Products	No change
Remove product	Shifts to the right	Reactants → Products	No change

Example Problem 15.4.1 Predict and calculate the effect of concentration changes on an equilibrium system.

Some FeSCN^{2+} is allowed to dissociate into Fe^{3+} and SCN$^-$ at 25 °C. At equilibrium, [FeSCN^{2+}] = 0.0768 M, and [Fe^{3+}] = [SCN$^-$] = 0.0232 M. Additional Fe^{3+} is added so that [Fe^{3+}]$_{new}$ = 0.0300 M and the system is allowed to once again reach equilibrium.

$$Fe^{3+}(aq) + SCN^-(aq) \rightleftharpoons FeSCN^{2+}(aq) \qquad K = 142 \text{ at } 25 °C$$

a. In which direction will the reaction proceed to reach equilibrium?
b. What are the new concentrations of reactants and products after the system reaches equilibrium?

Example Problem 15.4.1 *(continued)*

Solution:

You are asked to predict the direction a reaction will proceed when additional reactant is added and the new equilibrium concentrations of reactants and products.

You are given a chemical equation and equilibrium constant for a reaction, equilibrium concentrations of reactants and products, and a new (non-equilibrium) concentration of reactant.

a. Additional reactant has been added to the reaction flask, so the system will respond by consuming the excess reactant. The reaction proceeds to the right, consuming reactants (Fe^{3+} and SCN^-) and forming additional product ($FeSCN^{2+}$).

b. **Step 1.** Predict the direction the reaction will proceed to reach equilibrium. In this case, additional reactant has been added to the system, so the reaction will proceed to the right, forming additional product and consuming reactants.

Step 2. Use an *ICE* table to define equilibrium concentrations in terms of x. The initial concentration of Fe^{3+} is equal to the new iron(III) concentration, and the initial concentrations of the other species are equal to the first equilibrium concentrations.

	$Fe^{3+}(aq)$	+	$SCN^-(aq)$	$\rightleftarrows$	$FeSCN^{2+}(aq)$
Initial (M)	0.0300		0.0232		0.0768
Change (M)	$-x$		$-x$		$+x$
Equilibrium (M)	$0.0300 - x$		$0.0232 - x$		$0.0768 + x$

Step 3. Substitute the equilibrium concentrations into the equilibrium constant expression and solve for x. (Note that you must use the quadratic equation to solve this problem.)

$$K = \frac{[FeSCN^{2+}]}{[Fe^{3+}][SCN^-]} = \frac{0.0768 + x}{(0.0300 - x)(0.0232 - x)} = 142$$

$$0 = 142x^2 - 8.55x + 0.0220$$

$x = 0.00267$ or 0.0575 (not a physically possible answer
because it results in negative equilibrium concentrations for Fe^{3+} and SCN^-)

Step 4. Use the value of x to calculate the new equilibrium concentrations.

$$[Fe^{3+}] = 0.0300 - x = 0.0273 \text{ M}$$

$$[SCN^-] = 0.0232 - x = 0.0205 \text{ M}$$

$$[FeSCN^{2+}] = 0.0768 + x = 0.0795 \text{ M}$$

Is your answer reasonable? As a final check of your answer, substitute the new equilibrium concentrations into the equilibrium constant expression and calculate K.

$$K = \frac{[FeSCN^{2+}]}{[Fe^{3+}][SCN^-]} = \frac{0.0795}{(0.0273)(0.0205)} = 142$$

Video Solution

Tutored Practice
Problem 15.4.1

15.4b Change in the Volume of the System

A change in system volume is particularly important for gas-phase equilibria. A change in system volume does not affect systems involving only liquids or aqueous solutions because liquids are not easily compressed or expanded.

Consider the equilibrium system involving solid carbon, carbon dioxide, and carbon monoxide.

$$C(s) + CO_2(g) \rightleftharpoons 2\ CO(g)$$

If this system is at equilibrium and the volume of the reaction container is decreased at constant temperature, the pressure inside the container increases (recall that pressure and volume are inversely proportional). When this occurs, the system shifts in the direction that decreases the pressure inside the container—the direction that decreases the amount of gas present in the container. In this case, the reaction will shift to the left, forming 1 mol of CO_2 for every 2 mol of CO consumed. The direction the reaction shifts is related to the stoichiometry of the gaseous reactants and products, as demonstrated by the effect of volume change on the NO_2/N_2O_4 equilibrium system shown in Interactive Figure 15.4.2.

Interactive Figure 15.4.2

Explore the effect of volume change on an equilibrium system.

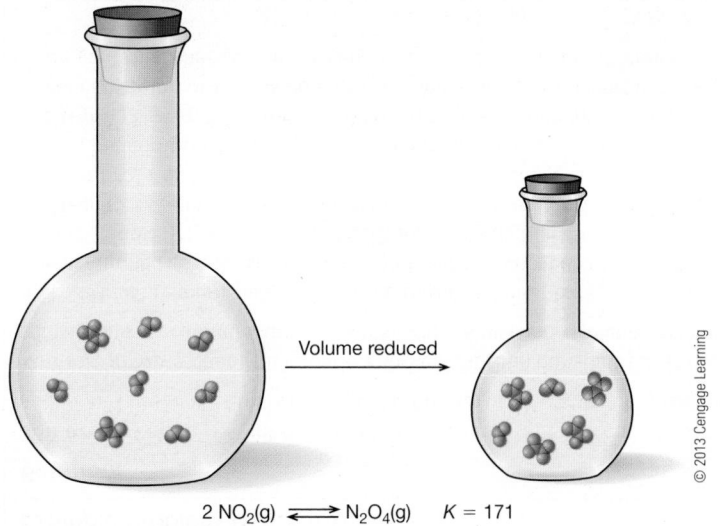

$$2\ NO_2(g) \rightleftharpoons N_2O_4(g) \qquad K = 171$$

$[N_2O_4]$ increases when the reaction flask volume is reduced.

The result of changing the volume on an equilibrium system involving gases is summarized in Table 15.4.2.

Changing the volume of a system involving gases effectively changes the concentrations of species in the reaction vessel. The *ICE* method can be used to calculate new equilibrium concentrations, as shown in the following example.

Table 15.4.2 Effect of Volume Change on an Equilibrium System

Change	System Response	Effect on K
Decrease volume	Shifts to form fewer moles of gas	No change
Increase volume	Shifts to form more moles of gas	No change

Example Problem 15.4.2 Predict and calculate the effect of volume change on an equilibrium system.

Consider the equilibrium between NO_2 and N_2O_4.

$$N_2O_4(g) \rightleftarrows 2\ NO_2(g) \qquad K = 0.690 \text{ at } 50\ °C$$

The reaction is allowed to reach equilibrium in a 2.00-L flask. At equilibrium, $[NO_2] = 0.314$ M and $[N_2O_4] = 0.143$ M.

a. Predict the change in NO_2 concentration when the equilibrium mixture is transferred to a 1.00-L flask.
b. Calculate the new equilibrium concentrations that result when the equilibrium mixture is transferred to a 1.00-L flask.

Solution:

You are asked to predict the change in product concentration when the reaction flask volume is changed and to calculate the new equilibrium concentrations in the new reaction flask.

You are given a chemical equation and equilibrium constant for a reaction, equilibrium concentrations and reaction flask volume, and the volume of the new reaction flask.

a. The reaction flask volume decreases from 2.00 L to 1.00 L. The equilibrium will shift to form fewer moles of gas. In this reaction, there is 1 mol of gaseous reactants and 2 mol of gaseous product, so the reaction shifts to the left, decreasing the NO_2 concentration and increasing the N_2O_4 concentration.
b. **Step 1.** Determine the new initial concentrations of NO_2 and N_2O_4 after the volume is decreased but prior to reequilibration. Because the volume of the reaction vessel is halved, the gas concentrations double.
New initial $[NO_2] = 2(0.314 \text{ M}) = 0.628$ M
New initial $[N_2O_4] = 2(0.143 \text{ M}) = 0.286$ M
Step 2. Predict the direction the reaction will proceed to reach equilibrium. In this case, the volume of the reaction flask decreased, so the reaction will proceed to the left, forming fewer moles of gas.

Example Problem 15.4.2 *(continued)*

Step 3. Use the *ICE* method to define equilibrium concentrations in terms of x.

	$N_2O_4(g)$	$\rightleftarrows$	$2\ NO_2(g)$
Initial (M)	0.286		0.628
Change (M)	$+x$		$-2x$
Equilibrium (M)	$0.286 + x$		$0.628 - 2x$

Step 4. Substitute the equilibrium concentrations into the equilibrium constant expression and solve for x. (Note that you must use the quadratic equation to solve this problem.)

$$K = \frac{[NO_2]^2}{[N_2O_4]} = \frac{(0.628 - 2x)^2}{0.286 + x} = 0.690$$

$$0 = 4x^2 - 3.20x + 0.197$$

$$x = 0.0672 \text{ and } 0.733 \text{ (not a possible answer because it results}$$
$$\text{in a negative NO}_2 \text{ equilibrium concentration)}$$

Step 5. Use the value of x to calculate the new equilibrium concentrations.

$$[NO_2] = 0.628 - 2x = 0.494 \text{ M}$$

$$[N_2O_4] = 0.286 + x = 0.353 \text{ M}$$

Is your answer reasonable? As a final check of your answer, substitute the new equilibrium concentrations into the equilibrium constant expression and calculate K.

$$K = \frac{[NO_2]^2}{[N_2O_4]} = \frac{(0.494)^2}{0.353} = 0.691$$

Video Solution

Tutored Practice
Problem 15.4.2

15.4c Change in Temperature

Only in the case of a temperature change does the value of the equilibrium constant for a reaction change. To understand the effect of a temperature change on an equilibrium system, consider heat as a reactant or product of a reaction. An exothermic reaction releases energy in the form of heat, so heat can be considered a product of an exothermic reaction. In the same way, heat can be considered a reactant in an endothermic reaction.

Exothermic reaction: reactants $\rightleftarrows$ products + *heat*

Endothermic reaction: reactants + *heat* $\rightleftarrows$ products

When heat is added or removed from an equilibrium system to change the temperature, the system responds by shifting to offset the addition or loss of heat. When the temperature increases, the equilibrium system shifts to consume the added energy.

Exothermic reaction shifts to the left: reactants ← products + *heat*

Endothermic reaction shifts to the right: reactants + *heat* → products

When the temperature decreases, the equilibrium system shifts to produce additional energy.

Exothermic reaction shifts to the right: reactants → products + *heat*

Endothermic reaction shifts to the left: reactants + *heat* ← products

Consider the equilibrium shown in Interactive Figure 15.4.3.

$$2\ NO_2(g) \rightleftharpoons N_2O_4(g) \qquad K = 170\ \text{at}\ 298K \qquad \Delta H^\circ = -57.1\ \text{kJ/mol}$$

Because this reaction is exothermic, decreasing the temperature of this equilibrium system results in a decrease in $[NO_2]$, as shown by the loss of color in the reaction vessel (NO_2 is dark red gas and N_2O_4 is colorless).

The result of temperature changes on an equilibrium system is summarized in Table 15.4.3.

Explore the effect of temperature changes on an equilibrium system.

Photos: Charles D. Winters; art: © 2013 Cengage Learning

$[NO_2]$ decreases when the reaction flask temperature is reduced.

Table 15.4.3 Effect of Temperature Change on an Equilibrium System

Change	Reaction Type	System Response	Effect on K
Increase temperature	Exothermic	Shifts to the left ($\leftarrow$)	Decreases
Increase temperature	Endothermic	Shifts to the right ($\rightarrow$)	Increases
Decrease temperature	Exothermic	Shifts to the right ($\rightarrow$)	Increases
Decrease temperature	Endothermic	Shifts to the left ($\leftarrow$)	Decreases

It is possible to estimate the new value for the equilibrium constant when there is a change in temperature by using the **van't Hoff equation**,

$$\ln\left(\frac{K_2}{K_1}\right) = -\frac{\Delta H°}{R}\left(\frac{1}{T_2} - \frac{1}{T_1}\right) \qquad (15.7)$$

where K_1 and K_2 are the equilibrium constants at temperatures T_1 and T_2, respectively, $\Delta H°$ is the enthalpy change for the reaction, and R is the ideal gas constant in thermodynamic units (8.3145×10^{-3} kJ/K · mol). This equation is similar to the Arrhenius equation and the Clausius–Clapeyron equation.

Example Problem 15.4.3 Predict and calculate the effect of temperature change on an equilibrium system.

Sulfur dioxide reacts with oxygen to form sulfur trioxide. The equilibrium constant, K_p, for this reaction is 0.365 at 1150 K.

$$2\,SO_2(g) + O_2(g) \rightleftarrows 2\,SO_3(g)$$

The standard enthalpy change for this reaction ($\Delta H°$) is −198 kJ/mol.

a. Predict the effect on the O_2 concentration when the temperature of the equilibrium system is increased.
b. Use the van't Hoff equation to estimate the equilibrium constant for this reaction at 1260 K.

Solution:

You are asked to predict the effect on reactant concentration when the temperature of an equilibrium system changes and to calculate the equilibrium constant for a reaction at a different temperature.

Example Problem 15.4.3 *(continued)*

You are given the chemical equation and equilibrium constant at a given temperature for a reaction, the standard enthalpy change for the reaction, and a new temperature.

a. This is an exothermic reaction ($\Delta H^\circ = -198$ kJ/mol). Increasing the temperature of the system will cause the equilibrium to shift to the left, consuming SO_3 and forming additional SO_2 and O_2. The O_2 concentration will increase.

b. In this problem, $T_1 = 1150$ K, $T_2 = 1260$ K, and $K_1 = 0.365$. Substitute these values into the van't Hoff equation and calculate the equilibrium constant at the new temperature.

$$\ln\left(\frac{K_2}{K_1}\right) = -\frac{\Delta H^\circ}{R}\left(\frac{1}{T_2} - \frac{1}{T_1}\right)$$

$$\ln\left(\frac{K_2}{0.365}\right) = -\frac{-198 \text{ kJ/mol}}{8.3145 \times 10^{-3} \text{ kJ/K} \cdot \text{mol}}\left(\frac{1}{1260 \text{ K}} - \frac{1}{1150 \text{ K}}\right)$$

$$K_2 = 0.0599$$

Is your answer reasonable? Notice that the equilibrium constant decreases as predicted for an exothermic reaction at a higher temperature.

Video Solution

Tutored Practice
Problem 15.4.3

Section 15.4 Mastery

Unit Recap

Key Concepts

15.1 The Nature of the Equilibrium State

- The principle of microscopic reversibility tells us that the elementary steps in a reaction mechanism are reversible (15.1a).

- An equilibrium arrow ($\rightleftarrows$) is used to indicate a reversible chemical reaction (15.1a).

- When a chemical reaction reaches a state of chemical equilibrium the rate of the forward reaction is equal to the rate of the reverse reaction (15.1b).

15.2 The Equilibrium Constant, *K*

- The relationship between forward and reverse rate constants for an equilibrium system is shown in an equilibrium constant expression and quantified by an equilibrium constant (K) (15.2a).

- The magnitude of the equilibrium constant provides information about the relative amounts of reactants and products at equilibrium (15.2a).

- A reactant-favored reaction has a small equilibrium constant ($K \ll 1$) and a product-favored reaction has a large equilibrium constant ($K \gg 1$) (15.2a).

- The equilibrium constant expression is related to the stoichiometry of the equilibrium reaction and does not include concentrations of pure solids, pure liquids, or solvents (15.2b).

- There are two forms of equilibrium constant expressions: K_c, where reactant and product concentrations are expressed in mol/L, and K_p, where reactant and product concentrations are expressed in partial pressure units (15.2b).

- Equilibrium constant expressions and the corresponding equilibrium constants can be manipulated by multiplying by a constant, reversing the reaction direction, or combining with other equilibrium expressions (15.2c).

15.3 Using Equilibrium Constants in Calculations

- Equilibrium constants are determined from experimental data, and *ICE* tables are often used to keep track of changes to reactant and product concentrations as the system approaches equilibrium (15.3a).

- The reaction quotient, Q, is used to determine whether a system is at equilibrium and in which direction the reaction proceeds to reach equilibrium (15.3b).

15.4 Disturbing a Chemical Equilibrium: Le Chatelier's Principle

- Le Chatelier's principle states that if a chemical system at equilibrium is disturbed so that it is no longer at equilibrium, the system will respond by reacting in either the forward or reverse direction so as to counteract the disturbance, resulting in a new equilibrium composition (15.4).

- Typical disturbances include the addition or removal of a reactant or product (15.4a), a change in the volume of a system involving gases (15.4b), and a change in temperature (15.4c).

- The van't Hoff equation is used to quantify the relationship between the equilibrium constant, reaction temperature, and the enthalpy change for the reaction (15.4c).

Key Equations

$$K = \frac{[C]^c[D]^d}{[A]^a[B]^b} \qquad \textbf{(15.1)}$$

$$K_p = K_c(RT)^{\Delta n} \qquad \textbf{(15.2)}$$

When an equilibrium equation is multiplied by a constant, n,

$$K_{new} = (K_{old})^n \qquad \textbf{(15.3)}$$

When an equilibrium equation is written in the reverse direction,

$$K_{new} = \frac{1}{K_{old}} \qquad \textbf{(15.4)}$$

When two equations representing equilibrium systems are added together,

$$K_{new} = K_1 \times K_2 \qquad \textbf{(15.5)}$$

$$Q = \frac{[C]^c[D]^d}{[A]^a[B]^b} \qquad \textbf{(15.6)}$$

$$\ln\left(\frac{K_2}{K_1}\right) = -\frac{\Delta H^\circ}{R}\left(\frac{1}{T_2} - \frac{1}{T_1}\right) \qquad \textbf{(15.7)}$$

Key Terms

15.1 The Nature of the Equilibrium State
principle of microscopic reversibility
reversible process
chemical equilibrium

15.2 The Equilibrium Constant, *K*
equilibrium constant expression
equilibrium constant (K)
product-favored reaction
reactant-favored reaction

15.3 Using Equilibrium Constants in Calculations
reaction quotient (Q)
reaction quotient expression

15.4 Disturbing a Chemical Equilibrium: Le Chatelier's Principle
Le Chatelier's principle
van't Hoff equation

Unit 15   Review and Challenge Problems

16 Acids and Bases

Unit Outline

In This Unit...

We now continue our discussion of chemical equilibria, applying the concepts and techniques developed in Chemical Equilibrium (Unit 15) to the chemistry of acids and bases. In Advanced Acid-Base Equilibria (Unit 17) and Precipitation and Lewis Acid-Base Equilibria (Unit 18) we will continue to study chemical equilibria as it applies to acid–base reactions, buffers, and the chemistry of sparingly soluble compounds.

16.1 Introduction to Acids and Bases

16.1a Acid and Base Definitions

We begin our study of acids and bases with the acid and base definitions we introduced in Chemical Reactions and Solution Stoichiometry (Unit 4), the Arrhenius definitions.

Arrhenius acid: A substance containing hydrogen that, when dissolved in water, increases the concentration of H^+ ions.
Arrhenius base: A substance containing the hydroxide group that, when dissolved in water, increases the concentration of OH^- ions.

The Brønsted–Lowry definition is a broader description of the nature of acids and bases.

Brønsted–Lowry acid: A substance that can donate a proton (H^+ ion).
Brønsted–Lowry base: A substance that can accept a proton (H^+ ion).

This definition allows us to define a larger number of compounds as acids or bases and to describe acid–base reactions that take place in solvents other than water (such as ethanol or benzene). Ammonia, NH_3, for example, is not an Arrhenius base (its formula does not contain a hydroxide group). However, it acts as a Brønsted–Lowry base when it accepts a proton from an acid such as HCl.

Brønsted–Lowry acid–base reactions are called proton transfer reactions because they involve the transfer of a proton from an acid (a proton donor) to a base (a proton acceptor), as shown in Interactive Figure 16.1.1. Notice that, as shown in Interactive Figure 16.1.1, protons (H^+) do not exist as isolated species in water. In aqueous solutions the hydronium ion, $H_3O^+(aq)$, is a more accurate representation of a hydrated proton.

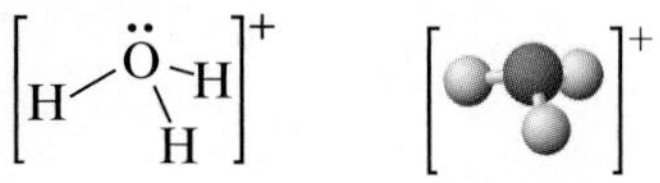

Even this is a simplification, however, because protons are likely found at the center of large clusters containing multiple water molecules.

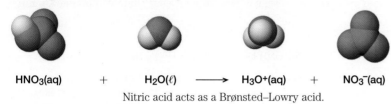

Interactive Figure 16.1.1

Investigate Brønsted–Lowry acids and bases.

$HNO_3(aq)$ + $H_2O(\ell)$ ⟶ $H_3O^+(aq)$ + $NO_3^-(aq)$

Nitric acid acts as a Brønsted–Lowry acid.

The Lewis acid–base definitions are broader still and are often used to describe reactions that take place in the gas phase. For example, ammonia, NH_3, is acting as a Lewis base when it donates a lone pair to a Lewis acid such as borane, BH_3.

Lewis acid: A substance that can accept an electron pair.
Lewis base: A substance that can donate an electron pair.

Most of the acid–base reactions we will study take place in aqueous solution, so we will use the Brønsted–Lowry definitions when referring to acids and bases. The chemistry of Lewis acids and bases will be discussed in Precipitation and Lewis Acid-Base Equilibria (Unit 18).

16.1b Simple Brønsted–Lowry Acids and Bases

A Brønsted–Lowry acid–base reaction involves the transfer of a proton from an acid to a base. For example, in the following reaction,

H^+ transfer

$HF(aq)$ + $NH_3(aq)$ ⟶ $F^-(aq)$ + $NH_4^+(aq)$
acid   base

a proton (H^+) is transferred from the acid HF (the proton donor) to the base NH_3 (the proton acceptor). When viewed from the reverse direction,

H^+ transfer

$HF(aq)$ + $NH_3(aq)$ ⟵ $F^-(aq)$ + $NH_4^+(aq)$
base   acid

a proton is transferred from the acid (NH_4^+) to the base (F^-). The overall equilibrium is represented as

$$HF(aq) \quad + \quad NH_3(aq) \quad \rightleftharpoons \quad F^-(aq) \quad + \quad NH_4^+(aq)$$

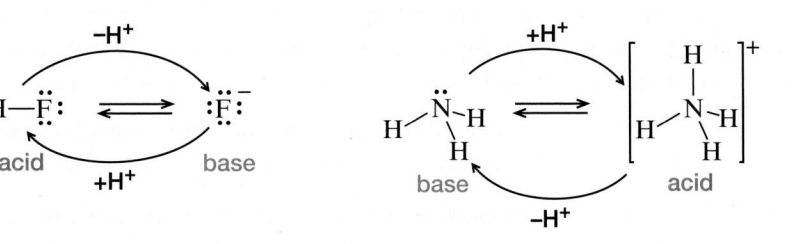

| ACID donates H^+ to NH_3 | base accepts H^+ from HF | BASE accepts H^+ from NH_4^+ | acid donates H^+ to F^- |

The acid in the forward reaction (HF) and the base in the reverse reaction (F^-) differ only by the presence or absence of H^+ and are called a **conjugate acid–base pair**. The other conjugate acid–base pair in this reaction is NH_4^+/NH_3. Because the Brønsted–Lowry definitions are based on the donating or accepting a proton, every Brønsted–Lowry acid has a conjugate base, every Brønsted–Lowry base has a conjugate acid, and every Brønsted–Lowry acid–base reaction involves two conjugate acid–base pairs.

Example Problem 16.1.1 Identify acid–base conjugate pairs.

a. What is the conjugate acid of the iodate ion, IO_3^-?

b. What is the conjugate base of formic acid, HCO_2H?

c. Identify the acid, base, conjugate acid, and conjugate base in the following reaction:

$$HCN(aq) + NO_2^-(aq) \rightleftharpoons HNO_2(aq) + CN^-(aq)$$

Solution:

You are asked to identify the conjugate acid of a base, the conjugate base of an acid, or the conjugate acid–base pairs in an acid–base reaction.

You are given the identity of an acid or a base, or an acid–base reaction.

Example Problem 16.1.1 *(continued)*

a. IO_3^- accepts a proton to form its conjugate acid, HIO_3:

$$IO_3^-(aq) + H^+(aq) \rightarrow HIO_3(aq)$$

b. HCO_2H donates a proton to form its conjugate base, HCO_2^-:

$$HCO_2H\ (aq) \rightarrow H^+(aq) + HCO_2^-(aq)$$

c. In this reaction, the acid (HCN) donates a proton to the base (NO_2^-), resulting in the formation of the conjugate base CN^- and the conjugate acid HNO_2.

$$\underset{\text{acid}}{HCN(aq)} + \underset{\text{base}}{NO_2^-\ (aq)} \rightleftarrows \underset{\text{conj. acid}}{HNO_2(aq)} + \underset{\text{conj. base}}{CN^-(aq)}$$

Video Solution

Tutored Practice
Problem 16.1.1

16.1c More Complex Acids

The Brønsted–Lowry acids we have seen so far are capable of donating only one proton and are called **monoprotic acids**. **Polyprotic acids** can donate more than one proton. Carbonic acid, H_2CO_3, is an example of a **diprotic acid**, a polyprotic acid that can donate two protons.

Step 1:

$$H_2CO_3(aq) + H_2O(\ell) \rightleftarrows HCO_3^-(aq) + H_3O^+(aq)$$

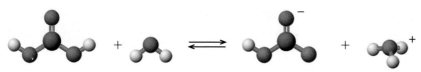

Step 2:

$$HCO_3^-(aq) + H_2O(\ell) \rightleftarrows CO_3^{2-}(aq) + H_3O^+(aq)$$

Interactive Figure 16.1.2

Identify acidic hydrogen atoms.

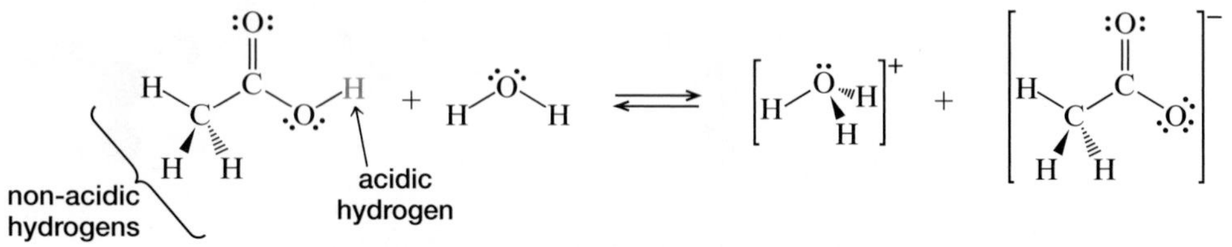

In acetic acid, only the hydrogen atom attached to oxygen is acidic.

Notice that the bicarbonate ion, HCO_3^-, can act as a base (accepting a proton to form H_2CO_3) or as an acid (donating a proton to form CO_3^{2-}). We call such species **amphiprotic**. An amphiprotic species is formed when any polyprotic acid loses a proton.

HCO_3^- as an acid: $HCO_3^-(aq) + H_2O(aq) \rightleftharpoons CO_3^{2-}(aq) + H_3O^+(aq)$ (H⁺ transfer)

acid, base, conj. base, conj. acid

HCO_3^- as a base: $HCO_3^-(aq) + H_2O(aq) \rightleftharpoons H_2CO_3(aq) + OH^-(aq)$ (H⁺ transfer)

base, acid, conj. acid, conj. base

Polyprotic acids are not the only acids that contain more than one hydrogen atom. Many acids contain both acidic and nonacidic hydrogen atoms. As shown in Interactive Figure 16.1.2, acidic hydrogens in most acids are those attached to highly electronegative atoms such as oxygen.

Section 16.1 Mastery

16.2 Water and the pH Scale

16.2a Autoionization

Because some of the most important acid–base chemistry (including biologically related reactions) occurs in aqueous solution, it is crucial to understand the acid–base nature of water itself. Water is an example of an amphiprotic substance, one that can sometimes act as an acid and at other times as a base in acid–base reactions. For example, you may have noticed that in the bicarbonate ion reactions shown earlier, water acted as a base in one reaction and as an acid in a different reaction.

Water acts as an acid, a proton donor, to form the hydroxide ion when it reacts with a base:

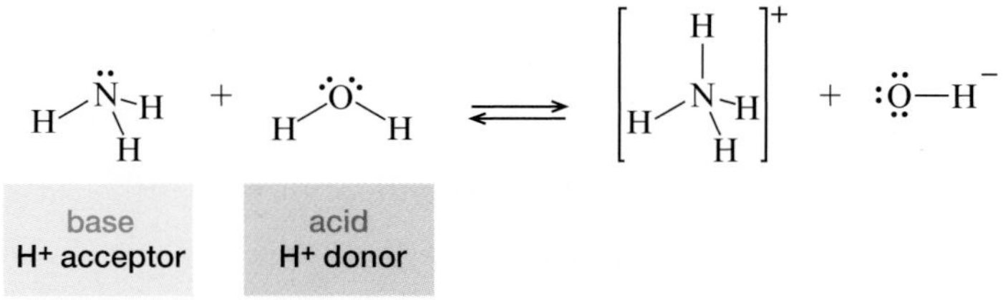

Water acts as a base, a proton acceptor, to form the hydronium ion (H_3O^+, a hydrated proton) when it reacts with an acid:

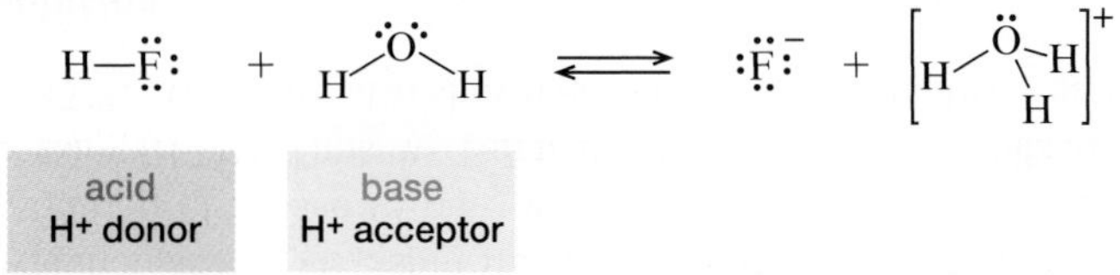

Species such as water that can act either as an acid or as a base can undergo **autoionization**, the reaction between two molecules of a chemical substance to produce ions. Water autoionizes to produce hydronium and hydroxide ions by a proton transfer reaction.

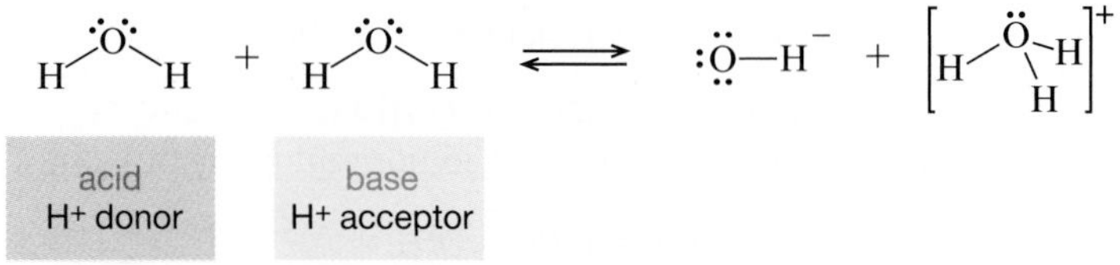

The autoionization of water is a reactant-favored process. Water autoionizes to a very small extent (approximately two out of every billion water molecules in a sample of pure water undergo autoionization), and it is a very weak electrolyte.

$$2\ H_2O(\ell) \rightleftarrows H_3O^+(aq) + OH^-(aq) \qquad K = \frac{[H_3O^+][OH^-]}{[H_2O]^2} << 1$$

Equilibrium expressions do not include the concentration of pure liquids or solvents, so this expression can be simplified by including the essentially constant $[H_2O]$ in the equilibrium constant:

$$K[H_2O]^2 = K_w = [H_3O^+][OH^-]$$

This process is so important that the equilibrium constant is given a special designation and is known as the **ionization constant for water, K_w**. As is true for all equilibrium constants, K_w varies with temperature.

At 25 °C,

$$K_w = [H_3O^+][OH^-] = 1.0 \times 10^{-14} \qquad \textbf{(16.1)}$$

In pure water the autoionization of two water molecules produces one H_3O^+ ion and one OH^- ion. Thus, in pure water $[H_3O^+] = [OH^-]$, and at 25 °C,

$$K_w = [H_3O^+][OH^-] = 1.0 \times 10^{-14}$$

$$[H_3O^+] = [OH^-] = \sqrt{1.0 \times 10^{-14}} = 1.0 \times 10^{-7}\,M$$

A solution in which $[H_3O^+] = [OH^-]$ is a **neutral solution**. In an **acidic** or **basic solution**, $[H_3O^+]$ is not equal to $[OH^-]$. According to Le Chatelier's principle, addition of an acid to pure water will increase $[H_3O^+]$, shift the autoionization equilibrium to the left, and decrease $[OH^-]$. Addition of a base to pure water will have the opposite effect, increasing $[OH^-]$ and decreasing $[H_3O^+]$. Thus, we can make the following generalizations about aqueous solutions at 25 °C, which are also shown graphically in Interactive Figure 16.2.1:

In a neutral solution,

$$[H_3O^+] = [OH^-] = 1.0 \times 10^{-7}\,M$$

In an acidic solution,

$$[H_3O^+] > [OH^-]$$

$$[H_3O^+] > 1.0 \times 10^{-7}\,M \text{ and } [OH^-] < 1.0 \times 10^{-7}\,M$$

In a basic solution,

$$[OH^-] > [H_3O^+]$$

$$[H_3O^+] < 1.0 \times 10^{-7}\,M \text{ and } [OH^-] > 1.0 \times 10^{-7}\,M$$

Because K_w is a constant, knowing either $[H_3O^+]$ or $[OH^-]$ for any aqueous solution allows the other to be calculated.

Interactive Figure 16.2.1

Explore the relationship between [OH⁻] and [H₃O⁺] in aqueous solutions.

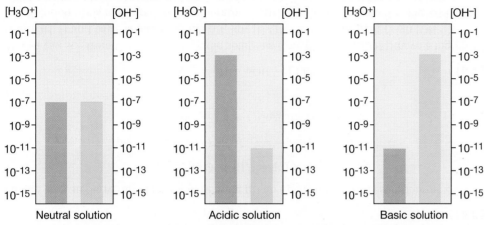

Neutral, acidic, and basic aqueous solutions contain H_3O^+ and OH^- ions.

Example Problem 16.2.1 Use K_w to calculate [H₃O⁺] and [OH⁻].

A solution at 25 °C has a hydronium ion concentration of 4.5×10^{-4} M.

a. What is the hydroxide ion concentration in this solution?
b. Is the solution acidic or basic?

Solution:

You are asked to calculate the hydroxide ion concentration in a solution and to determine whether the solution is acidic or basic.

You are given the hydronium ion concentration in the solution.

a. Use the equilibrium constant expression for K_w to calculate the hydroxide ion concentration.

$$[H_3O^+][OH^-] = 1.0 \times 10^{-14}$$

$$[OH^-] = \frac{1.0 \times 10^{-14}}{[H_3O^+]} = \frac{1.0 \times 10^{-14}}{4.5 \times 10^{-4} \, M} = 2.2 \times 10^{-11} \, M$$

b. In this solution, $[H_3O^+] > [OH^-]$, so the solution is acidic.

Video Solution

Tutored Practice
Problem 16.2.1

16.2b pH and pOH Calculations

Hydronium ion and hydroxide ion concentrations can vary over a wide range in common solutions. In chlorine bleach, $[H_3O^+]$ is around 10^{-12} M, whereas in vinegar, it is closer to 10^{-3} M. To more easily work with these values, we use a logarithmic scale called the p scale:

$$\text{In general, } pX = -\log X \qquad \textbf{(16.2)}$$

The **pH** and **pOH** scales are used to represent the acidity or basicity of an aqueous solution.

$$pH = -\log[H_3O^+] \qquad \textbf{(16.3)}$$

$$pOH = -\log[OH^-] \qquad \textbf{(16.4)}$$

In a neutral aqueous solution at 25 °C, for example, pH = 7.00.

$$pH = -\log[H_3O^+] = -\log(1.0 \times 10^{-7} \text{ M}) = 7.00$$

Notice that when doing calculations that involve logarithms, the number of significant figures in the result of the log operation is indicated by the number of digits following the decimal point. For example, $\log(29) = 1.46$, not 1.5.

Solving the pH and pOH expressions for hydronium ion and hydroxide ion concentration, respectively, allows calculation of $[H_3O^+]$ and $[OH^-]$ from pH and pOH.

$$[H_3O^+] = 10^{-pH} \qquad \textbf{(16.5)}$$

$$[OH^-] = 10^{-pOH} \qquad \textbf{(16.6)}$$

Finally, we can derive an expression relating pH and pOH for aqueous solutions at 25 °C.

$$K_w = [H_3O^+][OH^-] = 1.0 \times 10^{-14}$$

$$pK_w = -\log([H_3O^+][OH^-]) = -\log(1.0 \times 10^{-14})$$

$$pK_w = -\log[H_3O^+] + (-\log[OH^-]) = 14.00$$

Thus,

$$pK_w = pH + pOH = 14.00 \qquad \textbf{(16.7)}$$

As you can see from this equation, acidic solutions (low pH, high $[H_3O^+]$) have a high pOH (low $[OH^-]$) and basic solutions (high pH, low $[H_3O^+]$) have a low pOH (high $[OH^-]$).

Solution pH is measured in the laboratory with a pH meter (Interactive Figure 16.2.2), which uses an electrode whose electrical potential is sensitive to H_3O^+ ions in solution. A pH meter must be carefully calibrated before each use by using solutions of known pH, called buffer solutions.

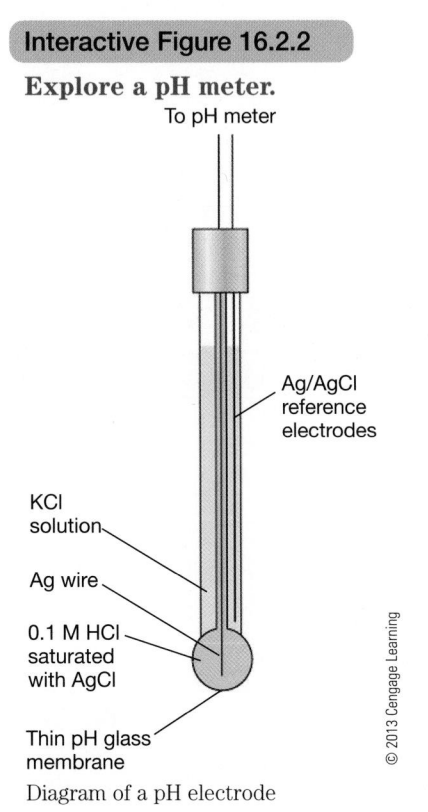

Interactive Figure 16.2.2

Explore a pH meter.

To pH meter

Ag/AgCl reference electrodes

KCl solution

Ag wire

0.1 M HCl saturated with AgCl

Thin pH glass membrane

© 2013 Cengage Learning

Diagram of a pH electrode

Interactive Figure 16.2.3

Relate pH, pOH, $[H_3O^+]$, and $[OH^-]$.

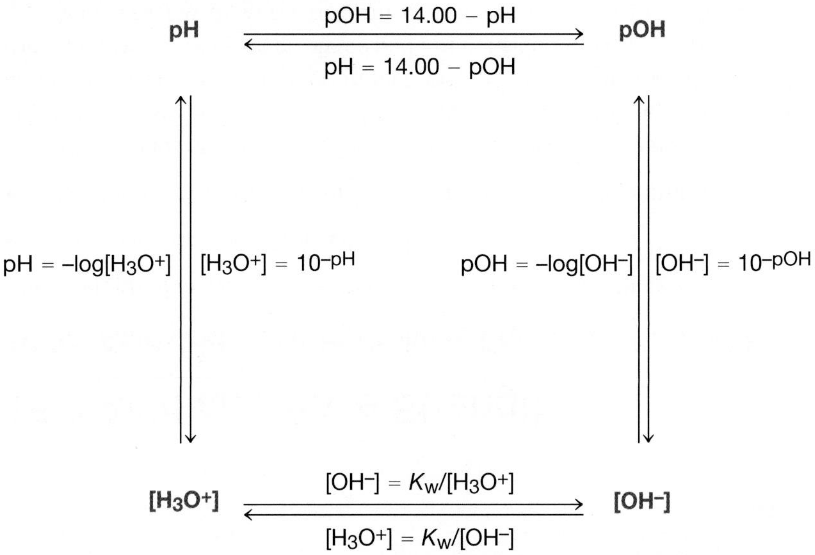

pH $\xrightarrow{\text{pOH} = 14.00 - \text{pH}}$ pOH
$\xleftarrow{\text{pH} = 14.00 - \text{pOH}}$

$\text{pH} = -\log[H_3O^+]$ $[H_3O^+] = 10^{-\text{pH}}$ $\text{pOH} = -\log[OH^-]$ $[OH^-] = 10^{-\text{pOH}}$

$[H_3O^+]$ $\xrightarrow{[OH^-] = K_w/[H_3O^+]}$ $[OH^-]$
$\xleftarrow{[H_3O^+] = K_w/[OH^-]}$

Calculation relationships between pH, pOH, $[H_3O^+]$, and $[OH^-]$

Interactive Figure 16.2.3 shows a combined set of calculation relationships between pH, pOH, $[H_3O^+]$, and $[OH^-]$. Note that it is not simple to directly convert between pH and $[OH^-]$, or directly between pOH and $[H_3O^+]$.

Example Problem 16.2.2 Use pH, pOH, $[H_{3O}^+]$, and $[OH^-]$ relationships.

a. The hydronium concentration in an aqueous solution of HCl is 4.4×10^{-2} M. Calculate $[OH^-]$, pH, and pOH for this solution.
b. The pH of an aqueous solution of NaOH is 10.73. Calculate $[H_3O^+]$, $[OH^-]$, and pOH for this solution.

Solution:

You are asked to calculate hydroxide ion and hydronium ion concentrations and the pH and pOH for a solution.

You are given the hydronium ion concentration or the solution pH.

a. Rearrange the K_w expression, solving for $[OH^-]$.

$$[\text{OH}^-] = \frac{K_\text{w}}{[\text{H}_3\text{O}^+]} = \frac{1.0 \times 10^{-14}}{4.4 \times 10^{-2}\,\text{M}} = 2.3 \times 10^{-13}\,\text{M}$$

Use $[\text{H}_3\text{O}^+]$ and $[\text{OH}^-]$ to calculate pH and pOH.

$$\text{pH} = -\log[\text{H}_3\text{O}^+] = -\log(4.4 \times 10^{-2}\,\text{M}) = 1.36$$

$$\text{pOH} = -\log[\text{OH}^-] = -\log(2.3 \times 10^{-13}\,\text{M}) = 12.64$$

As a final check, verify that pH + pOH = 14.00.

$$\text{pH} + \text{pOH} = 1.36 + 12.64 = 14.00$$

b. Use the rearranged pH equation to calculate $[\text{H}_3\text{O}^+]$ in this solution.

$$[\text{H}_3\text{O}^+] = 10^{-\text{pH}} = 10^{-10.73} = 1.9 \times 10^{-11}\,\text{M}$$

Rearrange the K_w expression, solving for $[\text{OH}^-]$.

$$[\text{OH}^-] = \frac{K_\text{w}}{[\text{H}_3\text{O}^+]} = \frac{1.0 \times 10^{-14}}{1.9 \times 10^{-11}\,\text{M}} = 5.3 \times 10^{-4}\,\text{M}$$

Because pH + pOH = 14.00,

$$\text{pOH} = 14.00 - \text{pH} = 14.00 - 10.73 = 3.27.$$

Video Solution

Tutored Practice
Problem 16.2.2

Section 16.2 Mastery

16.3 Acid and Base Strength

16.3a Acid and Base Hydrolysis Equilibria, K_a, and K_b

The strength of an acid or base solution depends on two things:

- The concentration of the acid or base

- The tendency of the acid or base to donate or accept a proton, respectively

Strong acids and bases, as described in Chemical Reactions and Solution Stoichiometry (Unit 4), are strong electrolytes that ionize completely in aqueous solution. Not all acids and bases are strong, however. In fact, only a small number of acids and bases are strong; the majority are weak electrolytes and are considered weak acids and weak bases.

Acids react with water to form hydronium ions and the conjugate base of the acid. **Hydrolysis** is the general term used to describe the reaction of a substance with water. You will also see the term *ionization* used to describe the reaction between an acid and water or a base and water. (Remember that K_w is called the ionization constant for water.)

Strong acids (Table 16.3.1) such as HCl ionize completely when they react with water.

$$HCl(aq) + H_2O(\ell) \rightarrow H_3O^+(aq) + Cl^-(aq)$$

Although this reaction is theoretically reversible, strong acids are strong electrolytes and thus the reactants are essentially 100% converted to products (Figure 16.3.1a). (Sulfuric acid is an exception; only in its first ionization is sulfuric acid a strong acid.)

None of the acids in Table 16.3.1 exists in its nonionized, molecular form in water, and aqueous solutions of these acids are essentially solutions containing the hydronium ion, H_3O^+. Thus, the hydronium ion is the strongest acid that can exist in aqueous solution, and the acidity of the strong acids is leveled by the solvent, water.

Not all acids are strong, however. In fact, the majority are weak electrolytes (weak acids). Only a small fraction of weak acid molecules are ionized in solution at any moment (Figure 16.3.1b). For example, less than 1% of acetic acid molecules ionize in a 0.2 M aqueous solution of the acid.

When weak acids react with water, an equilibrium system occurs where both reactants and products are found in solution in measurable concentrations. The relative strength of an acid is indicated by the equilibrium constant for the reaction of the acid with water.

Table 16.3.1	Strong Acids		
HCl	Hydrochloric acid	HNO_3	Nitric acid
HBr	Hydrobromic acid	H_2SO_4	Sulfuric acid
HI	Hydroiodic acid	$HClO_4$	Perchloric acid

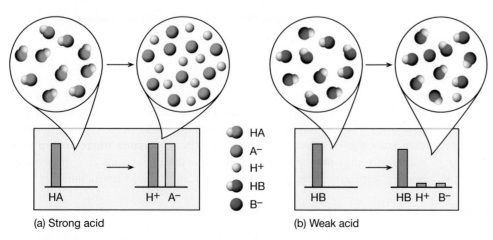

| HA |
| A⁻ |
| H⁺ |
| HB |
| B⁻ |

(a) Strong acid   (b) Weak acid

Figure 16.3.1 Molecular view of (a) strong and (b) weak acids

For the weak acid HA,

$$HA(aq) + H_2O(\ell) \rightleftharpoons H_3O^+(aq) + A^-(aq) \qquad K_a = \frac{[H_3O^+][A^-]}{[HA]} \qquad \textbf{(16.8)}$$

The equilibrium constant for weak acid hydrolysis is given the subscript "a" to indicate that it is an equilibrium constant for a weak acid. The magnitude of $\boldsymbol{K_a}$, the **acid dissociation constant**, indicates the relative strength of a weak acid. Acids with larger K_a values are stronger acids that ionize to a greater extent in solution than do acids with smaller K_a values. For example, chlorous acid ($HClO_2$, $K_a = 1.1. \times 10^{-2}$) is a stronger acid than hydrocyanic acid (HCN, $K_a = 6.2 \times 10^{-10}$). Strong acids have very large K_a values ($K_a \gg 1$), which are determined by performing experiments in nonaqueous solvents.

Example Problem 16.3.1 Write acid hydrolysis reactions.

The value of K_a for acetylsalicylic acid (aspirin), $HC_9H_7O_4$, is 3.00×10^{-4}. Write the equation for the reaction that goes with this equilibrium constant.

Solution:

You are asked to write the chemical equation for the hydrolysis of a weak acid.

You are given the formula of the weak acid.

Acetylsalicylic acid is a weak acid ($K_a \ll 1$). It donates a proton to water, forming the hydronium ion and the salicylate ion.

$$HC_9H_7O_4(aq) + H_2O(\ell) \rightleftharpoons H_3O^+(aq) + C_9H_7O_4^-(aq)$$

Video Solution

Tutored Practice
Problem 16.3.1

Bases react with water to form hydroxide ions and the conjugate acid of the base. Strong bases (Table 16.3.2) such as NaOH ionize completely when they react with water. Hydrolysis reactions for strong bases are typically written without including water as a reactant.

$$NaOH(aq) \rightarrow Na^+(aq) + OH^-(aq)$$

As we saw for strong acids, strong bases are strong electrolytes and thus the reactants are essentially 100% converted to products. The hydroxide ion, OH^-, is the strongest base that can exist in aqueous solution, so strong bases are essentially solutions containing the hydroxide ion.

There are very few strong bases in aqueous solution; most are weak, as we saw for acids. When weak bases react with water, an equilibrium system forms in which both reactants and products are found in solution in measurable concentrations. The relative strength of a weak base can be expressed with an equilibrium constant for the reaction of the base with water. For the weak base B,

$$B(aq) + H_2O(\ell) \rightleftharpoons HB^+(aq) + OH^-(aq) \qquad K_b = \frac{[HB^+][OH^-]}{[B]} \qquad \textbf{(16.9)}$$

Notice that K_a expressions have $[H_3O^+]$ in the numerator, whereas K_b expressions have $[OH^-]$ in the numerator. The equilibrium constant for weak base hydrolysis is given the subscript "b" to indicate that it is an equilibrium constant for a weak base. Like the weak acid K_a values, the magnitude of $\boldsymbol{K_b}$, the **base dissociation constant**, indicates the relative strength of a weak base. For example, ammonia (NH_3, $K_b = 1.8 \times 10^{-5}$) is a stronger base than pyridine (C_6H_5N, $K_b = 1.5 \times 10^{-9}$).

Table 16.3.2 Strong Bases			
LiOH	Lithium hydroxide	$Ca(OH)_2$	Calcium hydroxide (sparingly soluble)
NaOH	Sodium hydroxide	$Ba(OH)_2$	Barium hydroxide
KOH	Potassium hydroxide		

Example Problem 16.3.2 Write base hydrolysis reactions.

The value of K_b for pyridine, C_5H_5N, is 1.5×10^{-9}. Write the equation for the reaction that goes with this equilibrium constant.

Solution:

You are asked to write the chemical equation for the hydrolysis of a weak base.

You are given the formula of the weak base.

Pyridine is a weak base ($K_b \ll 1$). It accepts a proton from water, forming the hydroxide ion and the pyridinium ion.

$$C_5H_5N(aq) + H_2O(\ell) \rightleftharpoons OH^-(aq) + C_5H_5NH^+(aq)$$

Video Solution

Tutored Practice
Problem 16.3.2

16.3b K_a and K_b Values and the Relationship Between K_a and K_b

Interactive Table 16.3.3 shows K_a and K_b values for some common acids and bases. More extensive tables can be found in chemistry handbooks. Notice the following in the table:

- The strongest acids have large K_a values and small $\boldsymbol{pK_a}$ values [$pK_a = -\log(K_a)$]. As acid strength increases, pK_a decreases.

- The strongest bases have large K_b values and small pK_b values [$pK_b = -\log(K_b)$]. As base strength increases, pK_b decreases.

- The conjugate base of a strong acid has a very small K_b value. In general, as acid strength (and K_a) increases, the strength of the acid's conjugate base decreases.
- The conjugate acid of a strong base has a very small K_a value. In general, as base strength (and K_b) increases, the strength of the base's conjugate acid decreases.

The K_a and K_b values in Interactive Table 16.3 show that as acid strength increases, conjugate base strength decreases; as base strength increases, conjugate acid strength decreases. This relationship can be illustrated mathematically using the K_a and K_b expressions for a conjugate acid–base pair.

Weak acid: $HA(aq) + H_2O(\ell) \rightleftarrows A^-(aq) + H_3O^+(aq)$

$$K_a = \frac{[H_3O^+][A^-]}{[HA]}$$

Conjugate base: $A^-(aq) + H_2O(\ell) \rightleftarrows HA(aq) + OH^-(aq)$

$$K_b = \frac{[HA][OH^-]}{[A^-]}$$

Multiplying the two equilibrium expressions and simplifying gives

$$K_a \times K_b = \frac{[H_3O^+][A^-]}{[HA]} \times \frac{[HA]}{[A^-]}[OH^-]$$

$$K_a \times K_b = [H_3O^+][OH^-]$$

Therefore,

$$K_a \times K_b = K_w = 1.0 \times 10^{-14} \text{ at } 25 \text{ }^\circ C \qquad (16.10)$$

Notice that as K_a increases, acid strength increases and K_b for the conjugate base must decrease. Thus, stronger acids have weak conjugate bases, and stronger bases have weak conjugate acids. This relationship can be used to determine, for example, the K_b for a weak base if the K_a for the conjugate acid is known.

Interactive Table 16.3.3

Selected K_a and K_b Values

Acid	K_a	pK_a	Base	K_b	pK_b
Perchloric acid $HClO_4$	large	large negative	Perchlorate ion ClO_4^-	very small	large
Sulfuric acid H_2SO_4	large	large negative	Hydrogen sulfate ion HSO_4^-	very small	large
Hydrochloric acid HCl	large	large negative	Chloride ion Cl^-	very small	large
Nitric acid HNO_3	large	large negative	Nitrate ion NO_3^-	very small	large
Hydronium ion H_3O^+	1.0	0	Water H_2O	1.0×10^{-14}	14.00
Sulfurous acid H_2SO_3	1.7×10^{-2}	1.77	Hydrogen sulfite ion HSO_3^-	5.9×10^{-13}	12.23
Hydrogen sulfate ion HSO_4^-	1.2×10^{-2}	1.92	Sulfate ion SO_4^{2-}	8.3×10^{-13}	12.08
Phosphoric acid H_3PO_4	7.5×10^{-3}	2.12	Dihydrogen phosphate ion $H_2PO_4^-$	1.3×10^{-12}	11.89
Hexaaquairon(III) ion $[Fe(H_2O)_6]^{3+}$	4.0×10^{-3}	2.40	Pentaaquahydroxoiron(III) ion $[Fe(H_2O)_5OH]^{2+}$	2.5×10^{-12}	11.60
Hydrofluoric acid HF	7.2×10^{-4}	3.14	Fluoride ion F^-	1.4×10^{-11}	10.85
Nitrous acid HNO_2	4.5×10^{-4}	3.35	Nitrite ion NO_2^-	2.2×10^{-11}	10.66
Formic acid HCO_2H	1.8×10^{-4}	3.74	Formate ion HCO_2^-	5.6×10^{-11}	10.25
Benzoic acid $C_6H_5CO_2H$	6.3×10^{-5}	4.20	Benzoate ion $C_6H_5CO_2^-$	1.6×10^{-10}	9.80
Acetic acid CH_3CO_2H	1.8×10^{-5}	4.74	Acetate ion $CH_3CO_2^-$	5.6×10^{-10}	9.25
Carbonic acid H_2CO_3	4.2×10^{-7}	6.38	Hydrogen carbonate ion (bicarbonate ion) HCO_3^-	2.4×10^{-8}	7.62
Hydrogen sulfide (hydrosulfuric acid) H_2S	1.0×10^{-7}	7.00	Hydrogen sulfide ion HS^-	1.0×10^{-7}	7.00
Dihydrogen phosphate ion $H_2PO_4^-$	6.2×10^{-8}	7.21	Hydrogen phosphate ion HPO_4^{2-}	1.6×10^{-7}	6.80
Hydrogen sulfite ion HSO_3^-	6.4×10^{-8}	7.19	Sulfite ion SO_3^{2-}	1.6×10^{-7}	6.80

Continued

Selected K_a and K_b Values *(continued)*

Acid	K_a	pK_a	Base	K_b	pK_b
Hypochlorous acid HClO	3.5×10^{-8}	7.46	Hypochlorite ion ClO$^-$	2.9×10^{-7}	6.54
Ammonium ion NH$_4{}^+$	5.6×10^{-10}	9.25	Ammonia NH$_3$	1.8×10^{-5}	4.74
Hydrogen cyanide (hydrocyanic acid) HCN	4.0×10^{-10}	9.40	Cyanide ion CN$^-$	2.5×10^{-5}	4.60
Hydrogen carbonate ion (bicarbonate ion) HCO$_3{}^-$	4.8×10^{-11}	10.32	Carbonate ion CO$_3{}^{2-}$	2.1×10^{-4}	3.68
Hydrogen phosphate ion HPO$_4{}^{2-}$	3.6×10^{-13}	12.44	Phosphate ion PO$_4{}^{3-}$	2.8×10^{-2}	1.55
Water H$_2$O	1.0×10^{-14}	14.00	Hydroxide ion OH$^-$	1.0	0
Hydrogen sulfide ion HS$^-$	1.0×10^{-19}	19.00	Sulfide ion S^{2-}	1.0×10^5	-5.00
Ethanol CH$_3$CH$_2$OH	very small	large	Ethoxide ion CH$_3$CH$_2$O$^-$	large	large negative
Ammonia NH$_3$	very small	large	Amide ion NH$_2{}^-$	large	large negative
Hydrogen H$_2$	very small	large	Hydride ion H$^-$	large	large negative

---

Example Problem 16.3.3 Use the relationship between K_a and K_b.

Chloroacetic acid, CH$_2$ClCO$_2$H, is a weak acid ($K_a = 1.3 \times 10^{-3}$). What is the value of K_b for its conjugate base, the weak base chloroacetate ion, CH$_2$ClCO$_2{}^-$?

Solution:

You are asked to calculate K_b for the conjugate base of a weak acid.

You are given the identity of the weak acid and K_a for the weak acid.

Use the relationship $K_a \times K_b = K_w$.

$$K_b = \frac{K_w}{K_a} = \frac{1.0 \times 10^{-14}}{1.3 \times 10^{-3}} = 7.7 \times 10^{-12}$$

Is your answer reasonable? Chloroacetic acid is a relatively strong weak acid ($K_a \approx 10^{-3}$), so its conjugate base is very weak ($K_b \approx 10^{-11}$).

Video Solution

Tutored Practice
Problem 16.3.3

16.3c Determining K_a and K_b Values in the Laboratory

The K_a and K_b values found in reference tables are determined experimentally. One method for determining these values involves measuring the pH of a solution containing a known concentration of a weak acid or a weak base. The information about $[H_3O^+]$ obtained is then used to determine the equilibrium concentrations of all species and the equilibrium constant for the acid or base hydrolysis. (◄Flashback to Section 15.3 Using Equilibrium Constant in Calculations)

Example Problem 16.3.4 Calculate K_a or K_b from experimental data.

The pH of a 0.086-M solution of nitrous acid (HNO_2) is 2.22. Use this information to determine the value of K_a for nitrous acid.

$$HNO_2(aq) + H_2O(\ell) \rightleftarrows NO_2^-(aq) + H_3O^+(aq)$$

Solution:

You are asked to calculate K_a for a weak acid.

You are given the identity and concentration of the weak acid in the solution and the pH of the solution.

Step 1. Determine the equilibrium concentration of H_3O^+ from the solution pH.

$$[H_3O^+] = 10^{-pH} = 10^{-2.22} = 0.0060 \text{ M}$$

Step 2. Set up an *ICE* table using the variable x to represent the degree to which the weak acid ionizes, in units of mol/L. Based on the stoichiometry of the hydrolysis reaction, the change in concentration upon weak acid ionization is therefore $-x$ for the acid (HNO_2) and $+x$ for both the conjugate base (NO_2^-) and H_3O^+. (The initial concentration of H_3O^+ is actually 1.0×10^{-7} M due to the autoionization of water, but this is so small an amount in an acidic solution as to be negligible.)

	$HNO_2(aq)$	+	$H_2O(\ell)$	$\rightleftarrows$	$NO_2^-(aq)$	+	$H_3O^+(aq)$
Initial (M)	0.086				0		0
Change (M)	$-x$				$+x$		$+x$
Equilibrium (M)	$0.086 - x$				x		x

Step 3. The pH measurement tells us the equilibrium concentration of H_3O^+, which is equal to x. We can replace x in the table with that numerical value.

	$HNO_2(aq)$	+	$H_2O(\ell)$	$\rightleftarrows$	$NO_2^-(aq)$	+	$H_3O^+(aq)$
Initial (M)	0.086				0		0
Change (M)	-0.0060				$+0.0060$		$+0.0600$
Equilibrium (M)	$0.086 - 0.0060$				0.0060		0.0060

◄- - - - - - - - - - - -

Example Problem 16.3.4 *(continued)*

Step 4. The concentrations of all the species are now known at equilibrium, and K_a can be calculated by substituting these in the equilibrium constant expression.

$$K_a = \frac{[NO_2^-][H_3O^+]}{[HNO_2]}$$

$$= \frac{(x)(x)}{0.086 - x} = \frac{(0.0060)(0.0060)}{0.086 - 0.0060} = 4.5 \times 10^{-4}$$

Video Solution

Tutored Practice
Problem 16.3.4

Section 16.3 Mastery

16.4 Estimating the pH of Acid and Base Solutions

16.4a Strong Acid and Strong Base Solutions

Strong acids such as HCl and HNO_3 have large K_a values and are strong electrolytes. Therefore, $[H_3O^+]$ in a solution containing a monoprotic strong acid is equal to the concentration of the acid itself. Note that the diprotic acid H_2SO_4 is both a strong acid and a diprotic acid. However, it acts as a strong acid in its first ionization only ($K_{a1} \gg K_{a2}$).

The pH of a strong acid solution can be calculated if the strong acid concentration is known. An example of this type of calculation follows.

Example Problem 16.4.1 Calculate the pH of a strong acid solution.

A solution of nitric acid has $[HNO_3] = 0.028$ M. What is the pH of this solution?

Solution:

You are asked to calculate the pH of a solution containing a strong acid.

You are given the identity and concentration of the strong acid.

Nitric acid is a strong acid (assume 100% ionization), so the $[H_3O^+]$ is equal to the initial acid concentration.

$$HNO_3(aq) + H_2O(\ell) \rightarrow H_3O^+(aq) + NO_3^-(aq)$$

$$[H_3O^+] = [HNO_3]_0 = 0.028 \text{ M}$$

$$pH = -\log[H_3O^+] = -\log(0.028 \text{ M}) = 1.55$$

Notice that K_a is not used in the calculation of a strong acid solution pH.

Video Solution

Tutored Practice
Problem 16.4.1

Strong bases are strong electrolytes. Therefore, $[OH^-]$ in a solution containing a strong base is directly related to the concentration of the base itself. The pH of this solution can be calculated if the strong base concentration is known.

Example Problem 16.4.2 Calculate the pH of a strong base solution.

What is the pH of a 9.5×10^{-3} M solution of $Ba(OH)_2$?

Solution:

You are asked to calculate the pH of a solution containing a strong base.

You are given the identity and concentration of the strong base.

Barium hydroxide is a strong base (assume 100% ionization), and due to the compound stoichiometry, $[OH^-]$ is equal to twice the initial base concentration.

$$Ba(OH)_2(aq) \rightarrow Ba^{2+}(aq) + 2\,OH^-(aq)$$

$$[OH^-] = [Ba(OH)_2]_0 \times \frac{2\text{ mol }OH^-}{1\text{ mol }Ba(OH)_2} = 0.019\text{ M}$$

$$[H_3O^+] = \frac{K_w}{[OH^-]} = \frac{1.0 \times 10^{-14}}{0.019\text{ M}} = 5.3 \times 10^{-13}\text{ M}$$

$$pH = -\log[H_3O^+] = -\log(5.3 \times 10^{-13}\text{ M}) = 12.28$$

Notice that K_b is not used in the calculation of a strong base pH.

Video Solution

Tutored Practice
Problem 16.4.2

16.4b Solutions Containing Weak Acids

The pH of a solution containing a weak acid can be calculated using the equilibrium constant for the weak acid hydrolysis, K_a. Many weak acids (and weak bases) have very small equilibrium constants, allowing us to simplify the calculation of equilibrium concentrations of reactants and products according to the following general rule.

• If the value of K_a is small when compared to the initial acid concentration ($[HA]_0 > 100 \cdot K_a$), then the amount of weak acid ionized is very small when compared to $[HA]_0$.

Under these conditions, we can assume that the equilibrium concentration of the weak acid is equal to its initial concentration. An example follows.

Example Problem 16.4.3 Calculate the pH of a weak acid solution ($[HA]_0 > 100 \cdot K_a$).

Calculate the pH of a 0.250-M solution of hypochlorous acid (HClO, $K_a = 3.5 \times 10^{-8}$) and the equilibrium concentrations of the weak acid and its conjugate base.

Solution:

You are asked to calculate the pH of a solution containing a weak acid and the equilibrium concentrations of the weak acid and its conjugate base.

Example Problem 16.4.3 (continued)

You are given the identity and concentration of the weak acid.

Step 1. One of the most important steps in any weak acid or weak base equilibrium problem is the identification of the correct equilibrium. In this case, the correct equilibrium is the weak acid hydrolysis reaction.

$$HClO(aq) + H_2O(\ell) \rightleftharpoons ClO^-(aq) + H_3O^+(aq)$$

Step 2. Determine the initial concentrations of the equilibrium species, assuming the acid has not ionized. The initial acid concentration is equal to the solution concentration, 0.250 M. The initial concentrations of ClO^- and H_3O^+ are zero because no ionization has occurred. (The initial concentration of H_3O^+ is actually 1.0×10^{-7} M due to the autoionization of water, but this is such a small amount in an acidic solution that it is considered negligible.)

$HClO(aq) + H_2O(\ell) \rightleftharpoons$	$ClO^-(aq)$	$+ \; H_3O^+(aq)$
Initial (M)	0.250	0	0
Change (M)			
Equilibrium (M)			

Step 3. Use the variable x to represent the degree to which the weak acid ionizes, in units of mol/L. Based on the stoichiometry of the hydrolysis reaction, the change in concentration upon weak acid ionization is therefore $-x$ for the acid (HClO) and $+x$ for both the conjugate base (ClO^-) and for H_3O^+.

$HClO(aq) + H_2O(\ell) \rightleftharpoons$	$ClO^-(aq)$	$+ \; H_3O^+(aq)$
Initial (M)	0.250	0	0
Change (M)	$-x$	$+x$	$+x$
Equilibrium (M)			

Step 4. Complete the table by summing the initial concentration and change in concentration to arrive at equilibrium expressions for each species. For the acid, it is equal to the initial concentration of the acid minus the amount that ionized $(0.250 - x)$. For the conjugate base and hydronium ion, it is equal to the amount of weak acid ionized (x).

$HClO(aq) + H_2O(\ell) \rightleftharpoons$	$ClO^-(aq)$	$+ \; H_3O^+(aq)$
Initial (M)	0.250	0	0
Change (M)	$-x$	$+x$	$+x$
Equilibrium (M)	$0.250 - x$	x	x

In the completed ICE table,

$x =$ amount of weak acid ionized

$x = [ClO^-]$ at equilibrium

$x = [H_3O^+]$ at equilibrium

Step 5. Substitute these equilibrium concentrations into the K_a expression.

$$K_a = \frac{[\text{ClO}^-][\text{H}_3\text{O}^+]}{[\text{HClO}]}$$

$$= 3.5 \times 10^{-8} = \frac{(x)(x)}{0.250-x}$$

Because the value of K_a is small when compared to the initial acid concentration ($[\text{HClO}]_0 > 100 \cdot K_a$), the amount of weak acid ionized (x) is very small when compared to $[\text{HClO}]_0$ ($0.250 - x \approx 0.250$), and the expression can be simplified to

$$K_a = 3.5 \times 10^{-8} = \frac{x^2}{0.250}$$

Solving for x, the amount of weak acid ionized and the equilibrium hydronium ion concentration,

$$x = \sqrt{K_a \times 0.250} = 9.4 \times 10^{-5} \text{ M}$$

$$x = [\text{H}_3\text{O}^+] = 9.4 \times 10^{-5} \text{ M}$$

Step 6. Use x, the equilibrium concentration of H_3O^+ and OCl^-, to calculate the pH of the solution and the equilibrium concentrations of HClO and OCl^-.

$$\text{pH} = -\log[\text{H}_3\text{O}^+] = -\log(9.4 \times 10^{-5} \text{ M}) = 4.02$$

$$[\text{OCl}^-] = x = 9.4 \times 10^{-5} \text{ M}$$

$$[\text{HClO}] = 0.250 - x = 0.250 \text{ M}$$

Notice that our assumption (that the amount of weak acid ionized is insignificant when compared to the initial acid concentration) is valid in this case.

Is your answer reasonable? This is a solution containing an acid, so the pH should be less than 7.

Video Solution

Tutored Practice
Problem 16.4.3

It is not always possible to simplify the K_a expression by assuming that x is small when compared to the initial acid concentration. In these cases, the weak acid ionization occurs to such an extent that it has a significant effect on the equilibrium concentration and $[\text{HA}]_{eq} \neq [\text{HA}]_0$. Instead of simplifying the K_a expression, you must solve for x using the quadratic equation (or a quadratic-solving routine on a calculator).

Example Problem 16.4.4 Calculate the pH of a weak acid solution (quadratic equation).

Calculate the pH of a 0.055-M solution of hydrofluoric acid (HF, $K_a = 7.2 \times 10^{-4}$) and the equilibrium concentrations of the weak acid and its conjugate base.

Solution:

You are asked to calculate the pH of a solution containing a weak acid and the equilibrium concentrations of the weak acid and its conjugate base.

You are given the identity and concentration of the weak acid.

Step 1. Write the balanced equation for the hydrolysis reaction.

$$HF(aq) + H_2O(\ell) \rightleftarrows F^-(aq) + H_3O^+(aq)$$

Step 2. Set up an *ICE* table for the weak acid hydrolysis reaction. (The initial concentration of H_3O^+ is actually 1.0×10^{-7} M due to the autoionization of water, but this is such a small amount in an acidic solution that it is considered negligible.)

	$HF(aq)$ + $H_2O(\ell)$	$\rightleftarrows$	$F^-(aq)$	+	$H_3O^+(aq)$
Initial (M)	0.055		0		0
Change (M)	$-x$		$+x$		$+x$
Equilibrium (M)	$0.055 - x$		x		x

In the completed *ICE* table,

$$x = \text{amount of weak acid ionized}$$

$$x = [F^-] \text{ at equilibrium}$$

$$x = [H_3O^+] \text{ at equilibrium}$$

Step 3. Substitute these equilibrium concentrations into the K_a expression.

$$K_a = \frac{[F^-][H_3O^+]}{[HF]}$$

$$= 7.2 \times 10^{-4} = \frac{(x)(x)}{0.055 - x}$$

In this case the value of K_a is significant when compared to the initial acid concentration ($[HF]_0 < 100 \cdot K_a$), and the expression cannot be simplified. Solve for x using the quadratic equation.

$$7.2 \times 10^{-4} = \frac{(x)(x)}{0.055 - x}$$

$$(7.2 \times 10^{-4})(0.055 - x) = x^2$$

$$0 = x^2 + 7.2 \times 10^{-4}x - 4.0 \times 10^{-5}$$

Example Problem 16.4.4 *(continued)*

$$x = \frac{-7.2 \times 10^{-4} \pm \sqrt{(7.2 \times 10^{-4})^2 - 4(1)(-4.0 \times 10^{-5})}}{2(1)}$$

$x = [H_3O^+] = 6.0 \times 10^{-3}$ M; -6.7×10^{-3} M (not a physically possible answer; concentrations cannot be negative)

Step 4. Use the positive root, the equilibrium concentration of H_3O^+, to calculate the pH of the solution and the equilibrium concentrations of HF and F^-.

$$pH = -\log[H_3O^+] = -\log(6.0 \times 10^{-3} \text{ M}) = 2.22$$

$$[F^-] = x = 6.0 \times 10^{-3} \text{ M}$$

$$[HF] = 0.055 - x = 0.049 \text{ M}$$

Notice that in this case, the amount of weak acid ionized is significant when compared to the initial weak acid concentration.

Is your answer reasonable? This is a solution containing an acid, so the pH should be less than 7.

Video Solution

Tutored Practice
Problem 16.4.4

In Example Problem 16.4.3, the amount of HClO ionized was seen to be negligible compared to the amount originally present ($[HClO]_{eq} = [HClO]_{initial}$), while in Example Problem 16.4.4, the amount of HF ionized was significant when compared to the amount originally present ($[HF]_{eq} < [HF]_{initial}$). We can quantify the amount of weak acid ionized with *percent ionization*, a quantity that compares the amount of acid ionized to the amount originally present (Equation 16.11, where x is the amount of weak acid ionized). For the monoprotic acid HA, $x = [H_3O^+] = [A^-]$ at equilibrium.

$$\text{percent ionization} = \left(\frac{x}{[HA]_{initial}}\right) \times 100\% \qquad \textbf{(16.11)}$$

Using the solutions in Example Problems 16.4.3 and 16.4.4, the acid is 0.038% ionized in a 0.250 M HClO solution ($x = 9.4 \times 10^{-5}$ M), while the acid is 11% ionized in a 0.055 M HF solution ($x = 6.0 \times 10^{-3}$ M). Interactive Table 16.4.1 shows the relationship between K_a, initial acid concentration, and percent ionization for some weak acids.

K_a, $[HA]_{initial}$, and Percent Ionization

	CH_3CO_2H ($K_a = 1.8 \times 10^{-5}$)			HF ($K_a = 7.2 \times 10^{-4}$)		
$[HA]_{initial}$ (M)	0.30	0.010	0.0010	0.30	0.010	0.0010
x (M)	0.0023	4.2×10^{-4}	1.3×10^{-4}	0.014	0.0023	5.6×10^{-4}
Percent Ionization	0.77%	4.2%	13%	4.7%	23%	56%

16.4c Solutions Containing Weak Bases

The pH of a weak base solution is calculated using the same method used for weak acid pH calculations. We can also apply the same general rule we used to simplify the pH calculations for a weak acid solution.

- If the value of K_b is small when compared to the initial base concentration ($[B]_0 > 100 \cdot K_b$), then the amount of weak base ionized is very small when compared to $[B]_0$.

Under these conditions, we can assume that the equilibrium concentration of the weak base is equal to its initial concentration.

Example Problem 16.4.5 Calculate the pH of a weak base solution ($[B]_0 > 100 \cdot K_b$).

Calculate the pH of a 0.0177-M solution of pyridine (C_5H_5N, $K_b = 1.5 \times 10^{-9}$) and the equilibrium concentrations of the weak base and its conjugate acid.

Solution:

You are asked to calculate the pH of a solution containing a weak base and the equilibrium concentrations of the weak base and its conjugate acid.

You are given the identity and concentration of the weak base.

Step 1. One of the most important steps in any weak acid or weak base equilibrium problem is the identification of the correct equilibrium. In this case, the correct equilibrium is the weak base hydrolysis reaction.

$$C_5H_5N(aq) + H_2O(\ell) \rightleftarrows C_5H_5NH^+(aq) + OH^-(aq)$$

Step 2. Determine the initial concentrations of the equilibrium species, assuming the base has not reacted with water. The initial base concentration is equal to the solution concentration, 0.0177 M. The concentrations of $C_5H_5NH^+$ and OH^- are zero because no hydrolysis has

Example Problem 16.4.5 *(continued)*

occurred. (The initial concentration of OH⁻ is actually 1.0×10^{-7} M due to the autoionization of water, but this is such a small amount in a basic solution that it is considered negligible.)

	$C_5H_5N(aq) + H_2O(\ell)$	$\rightleftarrows$	$C_5H_5NH^+(aq)$	$+$	$OH^-(aq)$
Initial (M)	0.0177		0		0
Change (M)					
Equilibrium (M)					

Step 3. Use the variable x to represent the degree to which the weak base hydrolyzes, in units of mol/L. Based on the stoichiometry of the hydrolysis reaction, the change in concentration upon weak base hydrolysis is $-x$ for the base (C_5H_5N) and $+x$ for both the conjugate acid ($C_5H_5NH^+$) and for OH^-.

	$C_5H_5N(aq) + H_2O(\ell)$	$\rightleftarrows$	$C_5H_5NH^+(aq)$	$+$	$OH^-(aq)$
Initial (M)	0.0177		0		0
Change (M)	$-x$		$+x$		$+x$
Equilibrium (M)					

Step 4. Complete the table by summing the initial concentration and change in concentration to arrive at equilibrium expressions for each species. For the base, it is equal to the initial concentration of the base minus the amount that hydrolyzed ($0.0177 - x$). For the conjugate acid and hydroxide ion, it is equal to the amount of weak base hydrolyzed (x).

	$C_5H_5N(aq) + H_2O(\ell)$	$\rightleftarrows$	$C_5H_5NH^+(aq)$	$+$	$OH^-(aq)$
Initial (M)	0.0177		0		0
Change (M)	$-x$		$+x$		$+x$
Equilibrium (M)	$0.0177 - x$		x		x

In the completed *ICE* table,

$$x = \text{amount of weak base hydrolyzed}$$

$$x = [C_5H_5NH^+] \text{ at equilibrium}$$

$$x = [OH^-] \text{ at equilibrium}$$

Step 5. Substitute these equilibrium concentrations into the K_b expression.

$$K_b = \frac{[C_5H_5NH^+][OH^-]}{[C_5H_5N]}$$

$$= 1.5 \times 10^{-9} = \frac{(x)(x)}{0.0177 - x}$$

Because the value of K_b is small when compared to the initial base concentration ($[C_5H_5N]_0 > 100 \cdot K_b$), the amount of weak base hydrolyzed (x) is very small when compared to $[C_5H_5N]_0$, and the expression can be simplified to

Example Problem 16.4.5 (continued)

$$K_b = 1.5 \times 10^{-9} = \frac{x^2}{0.0177}$$

Solving for x, the amount of weak base hydrolyzed and the equilibrium hydroxide ion concentration,

$$x = \sqrt{K_b \times 0.0177} = 5.2 \times 10^{-6} \text{ M}$$

$$x = [OH^-] = 5.2 \times 10^{-6} \text{ M}$$

Step 6. Use x, the equilibrium concentration of OH^-, to calculate the equilibrium H_3O^+ concentration, the pH of the solution, and the equilibrium concentrations of and $C_5H_5NH^+$ and C_5H_5N.

$$[H_3O^+] = \frac{K_w}{[OH^-]} = \frac{1.0 \times 10^{-14}}{5.2 \times 10^{-6}} = 1.9 \times 10^{-9} \text{ M}$$

$$pH = -\log[H_3O^+] = -\log(1.9 \times 10^{-9} \text{ M}) = 8.71$$

$$[C_5H_5NH^+] = x = 5.2 \times 10^{-6} \text{ M}$$

$$[C_5H_5N] = 0.0177 - x = 0.0177 \text{ M}$$

The assumption that the amount of weak base hydrolyzed is insignificant when compared to the initial base concentration is correct.

Is your answer reasonable? This is a solution containing a base, so the pH should be greater than 7.

Video Solution

Tutored Practice
Problem 16.4.5

It is not always possible to simplify the K_b expression by assuming that x is small when compared to the initial base concentration. In these cases, the weak base reaction with water occurs to such an extent that it has a significant effect on the equilibrium concentration and $[B]_{eq} \neq [B]_0$. Instead of simplifying the K_b expression you must solve for x using the quadratic equation (or a quadratic-solving routine on a calculator), as shown in the following example.

Example Problem 16.4.6 Calculate the pH of a weak base solution (quadratic equation).

Calculate the pH of a 0.0246-M solution of ethylamine ($C_2H_5NH_2$, $K_b = 4.3 \times 10^{-4}$) and the equilibrium concentrations of the weak base and its conjugate acid.

Solution:

You are asked to calculate the pH of a solution containing a weak base and the equilibrium concentrations of the weak base and its conjugate acid.

Example Problem 16.4.6 *(continued)*

You are given the identity and concentration of the weak base.

Step 1. Write the balanced equation for the hydrolysis reaction.

$$C_2H_5NH_2(aq) + H_2O(\ell) \rightleftarrows C_2H_5NH_3^+(aq) + OH^-(aq)$$

Step 2. Set up an *ICE* table for the weak base hydrolysis reaction. (The initial concentration of OH^- is actually 1.0×10^{-7} M due to the autoionization of water, but this is such a small amount in a basic solution that it is considered negligible.)

	$C_2H_5NH_2(aq) + H_2O(\ell)$	$\rightleftarrows$	$C_2H_5NH_3^+(aq)$	$+$	$OH^-(aq)$
Initial (M)	0.0246		0		0
Change (M)	$-x$		$+x$		$+x$
Equilibrium (M)	$0.0246 - x$		x		x

In the completed *ICE* table,

$$x = \text{amount of weak base ionized}$$

$$x = [C_2H_5NH_3^+] \text{ at equilibrium}$$

$$x = [OH^-] \text{ at equilibrium}$$

Step 3. Substitute these equilibrium concentrations into the K_b expression.

$$K_b = \frac{[C_2H_5NH_3^+][OH^-]}{[C_2H_5NH_2]}$$

$$= 4.3 \times 10^{-4} = \frac{(x)(x)}{0.0246 - x}$$

In this case the value of K_b is significant when compared to the initial acid concentration ($[C_2H_5NH_2]_0 < 100 \cdot K_b$), and the expression cannot be simplified. Solve for x using the quadratic equation.

$$4.3 \times 10^{-4} = \frac{(x)(x)}{0.0246 - x}$$

$$4.3 \times 10^{-4}(0.0246 - x) = x^2$$

$$0 = x^2 + 4.3 \times 10^{-4}x - 1.06 \times 10^{-5}$$

$$x = \frac{-4.3 \times 10^{-4} \pm \sqrt{(4.3 \times 10^{-4})^2 - 4(1)(-1.06 \times 10^{-5})}}{2(1)}$$

$$x = [OH^-] = 3.0 \times 10^{-3} \text{ M}; -3.5 \times 10^{-3} \text{ M (not a physically possible answer; concentrations cannot be negative)}$$

Step 4. Use the positive root, the equilibrium concentration of OH^-, to calculate the pH of the solution and the equilibrium concentrations of $C_2H_5NH_3^+$ and $C_2H_5NH_2$.

◄ -

Example Problem 16.4.6 *(continued)*

$$pOH = -\log[OH^-] = -\log(3.0 \times 10^{-3} \text{ M}) = 2.52$$

$$pH = 14.00 - pOH = 11.48$$

$$[C_2H_5NH_3^+] = x = 3.0 \times 10^{-3} \text{ M}$$

$$[C_2H_5NH_2] = 0.0246 - x = 0.0216 \text{ M}$$

Notice that in this case, the amount of weak base ionized is significant when compared to the initial weak base concentration.

Is your answer reasonable? This is a solution containing a base, so the pH should be greater than 7.

Video Solution

Tutored Practice Problem 16.4.6

Section 16.4 Mastery

16.5 Acid–Base Properties of Salts

16.5a Acid–Base Properties of Salts: Hydrolysis

As you can see in Table 16.5.1, which contains species found in Interactive Table 16.3.3, many common acids and bases are ionic species. Notice, for example, that the ammonium ion, NH_4^+, is a weak acid, as is the hydrated iron(III) ion, $Fe(H_2O)_6^{3+}$. Similarly, many anions are weak bases, such as the fluoride ion, F^-, and the hypochlorite ion, ClO^-. Salts that contain these ions will mimic these acid–base properties, and it is important to be able to predict this behavior. Three general rules can be used to predict acid–base behavior of ions and their salts (Table 16.5.2).

If the acid–base properties of each ion are known, then the acid–base nature of the salt that contains those ions can be predicted using the following guidelines:

- If both the cation and anion are neutral, the salt is neutral.

- If only one ion is neutral, the other ion controls the acid–base nature of the salt.

  If either the cation or anion is acidic and the other is neutral, the solution is acidic.

  If either the cation or anion is basic and the other is neutral, the solution is basic.

- If neither ion is neutral, the acid–base nature of the salt can be determined by comparing the relative values of K_a and K_b for the ions. In this situation, normally the cation is acidic and the anion is basic.

  If K_a (cation) > K_b (anion), then the solution is mildly acidic.

  If K_b (anion) > K_a (cation), then the solution is mildly basic.

Table 16.5.1 K_a and K_b Values for Ionic Species

Acid		K_a
Hydrogen sulfate ion	HSO_4^-	1.2×10^{-2}
Hexaaquairon(III) ion	$[Fe(H_2O)_6]^{3+}$	4.0×10^{-3}
Dihydrogen phosphate ion	$H_2PO_4^-$	6.2×10^{-8}
Ammonium ion	NH_4^+	5.6×10^{-10}
Base		K_b
Phosphate ion	PO_4^{3-}	2.8×10^{-2}
Carbonate ion	CO_3^{2-}	2.1×10^{-4}
Cyanide ion	CN^-	2.5×10^{-5}
Hypochlorite ion	ClO^-	2.9×10^{-7}
Hydrogen carbonate ion (bicarbonate ion)	HCO_3^-	2.4×10^{-8}

Table 16.5.2 Determining the Acid–Base Properties of Salts

1. Anions associated with strong acids and cations associated with strong bases are acid–base neutral.

   - Na^+ is the cation associated with the strong base NaOH, and it is acid–base neutral. Cl^- is the anion associated with the strong acid HCl, and it is acid–base neutral. Neither ion reacts with water.

   $$Na^+(aq) + H_2O(\ell) \rightarrow no\ reaction$$
   $$Cl^-(aq) + H_2O(\ell) \rightarrow no\ reaction$$

2. The conjugate base of a weak acid is itself a weak base, and the conjugate acid of a weak base is itself a weak acid.

   - The $CH_3CO_2^-$ anion is the conjugate base of acetic acid and is a basic anion.

   $$CH_3CO_2^-(aq) + H_2O(\ell) \rightleftarrows CH_3CO_2H(aq) + OH^-(aq)$$

   - The NH_4^+ cation is the conjugate acid of ammonia and is an acidic cation.

   $$NH_4^+(aq) + H_2O(\ell) \rightleftarrows NH_3(aq) + H_3O^+(aq)$$

3. Small, highly charged (+2 or +3) metal cations are hydrated in water and act as acidic cations.

   - The hydrated iron(III) ion is an acidic cation.

   $$[Fe(H_2O)_6]^{3+}(aq) + H_2O(\ell) \rightleftarrows [Fe(H_2O)_5(OH)]^{2+}(aq) + H_3O^+(aq)$$

Example Problem 16.5.1 Predict acid–base properties of salts.

What is the acid–base nature of the following salts?

a. Potassium nitrite, KNO_2
b. Ammonium carbonate, $(NH_4)_2CO_3$

Solution:

You are asked to identify the acid–base nature of a compound.

You are given the identity of the compound.

a. KNO_2

Step 1. Consider the acid–base properties of the cation.

The potassium ion, K^+, is associated with the strong base KOH. This tells us that potassium ion does not contribute to the acid–base properties of a solution. It is acid–base neutral.

Example Problem 16.5.1 (continued)

Step 2. Consider the acid–base properties of the anion.

The nitrite ion is the conjugate base of nitrous acid, HNO_2, a weak acid. The nitrite ion is therefore a weak base. $K_b(NO_2^-) = 2.2 \times 10^{-11}$.

Step 3. Combine to determine the acid–base properties of the salt.

Potassium nitrite contains an acid–base neutral cation and a basic anion. This combination means that KNO_2 is a basic salt.

b. $(NH_4)_2CO_3$

Step 1. Consider the acid–base properties of the cation.

The ammonium ion, NH_4^+, is the conjugate acid of ammonia, NH_3, a weak base. The ammonium ion is therefore a weak acid. $K_a(NH_4^+) = 5.6 \times 10^{-10}$.

Step 2. Consider the acid–base properties of the anion.

The carbonate ion is the conjugate base of the bicarbonate ion, HCO_3^-, a weak acid (and an amphiprotic species). The carbonate ion is therefore a weak base. $K_b(CO_3^{2-}) = 2.1 \times 10^{-4}$.

Step 3. Combine to determine the acid–base properties of the salt.

Ammonium carbonate contains an acidic cation and a basic anion. Here, $K_b(CO_3^{2-}) > K_a(NH_4^+)$. This combination means that NH_4CO_3 is a basic salt.

> **Video Solution**
>
> **Tutored Practice Problem 16.5.1**

16.5b Determining pH of a Salt Solution

The pH of a solution containing a salt can be calculated by first considering the acid–base nature of the ions of the salt and then using the procedures outlined earlier for calculating the pH of weak acid and weak base solutions.

Example Problem 16.5.2 Calculate pH of a salt solution.

What is the pH of a 0.25-M solution of sodium hypochlorite, NaClO?

Solution:

You are asked to calculate the pH of a solution containing a salt.

You are given the identity and concentration of the salt.

Step 1. Determine the acid–base nature of the salt.

The Na^+ ion is an acid–base neutral cation, as are all Group 1A cations. The ClO^- ion is the conjugate base of the weak acid HClO. It is therefore a weak base ($K_b = 2.9 \times 10^{-7}$). Sodium hypochlorite is a basic salt.

Example Problem 16.5.2 *(continued)*

Step 2. Write the equation for the hydrolysis of the basic anion. (Na^+ is a spectator ion and is not shown in the reaction.)

$$ClO^-(aq) + H_2O(\ell) \rightleftarrows HClO(aq) + OH^-(aq)$$

Step 3. Use the method shown earlier to set up an *ICE* table for a weak base hydrolysis reaction.

	$ClO^-(aq) + H_2O(\ell)$ $\rightleftarrows$	$HClO(aq)$ +	$OH^-(aq)$
Initial (M)	0.25	0	0
Change (M)	$-x$	$+x$	$+x$
Equilibrium (M)	$0.25 - x$	x	x

Substitute the equilibrium concentrations into the K_b expression.

$$K_b = \frac{[HClO][OH^-]}{[ClO^-]}$$

$$= 2.9 \times 10^{-7} = \frac{(x)(x)}{0.25-x}$$

Because the value of K_b is small when compared to the initial base concentration ($[ClO^-]_0 > 100 \cdot K_b$), the amount of weak base hydrolyzed (x) is very small when compared to $[ClO^-]_0$, and the expression can be simplified to

$$K_b = 2.9 \times 10^{-7} = \frac{x^2}{0.25}$$

Solve for x, the amount of weak base hydrolyzed and the equilibrium hydroxide ion concentration,

$$x = \sqrt{K_b \times 0.25} = 2.7 \times 10^{-4}\, M$$

$$x = [OH^-] = 2.7 \times 10^{-4}\, M$$

Step 4. Use x, the equilibrium concentration of OH^-, to calculate the equilibrium H_3O^+ concentration and the pH of the solution.

$$[H_3O^+] = \frac{K_w}{[OH^-]} = \frac{1.0 \times 10^{-14}}{2.7 \times 10^{-4}} = 3.7 \times 10^{-11}\, M$$

$$pH = -\log[H_3O^+] = -\log(3.7 \times 10^{-11}\, M) = 10.43$$

Is your answer reasonable? Sodium hypochlorite is a basic salt, so the pH of the solution should be greater than 7.

Video Solution

Tutored Practice
Problem 16.5.2

Section 16.5 Mastery

16.6 Molecular Structure and Control of Acid–Base Strength

16.6a Molecular Structure and Control of Acid–Base Strength

We have described the properties of acids and bases and how to determine the pH of an aqueous solution containing an acid or base. We now turn to an explanation of what makes one compound more acidic or basic than another. We will restrict this analysis to the control of acid strength, but a similar analysis can be performed to describe base strength.

In a Brønsted–Lowry acid–base reaction, an acid donates a proton to a base:

$$H—A + Base \rightarrow [H—Base]^+ + A^-$$

When this type of reaction takes place, two different events occur. First, the H—A bond is broken, and second, the group directly attached to the acidic hydrogen in the acid accepts both electrons from the H—A bond, gaining a negative charge in the process. This stepwise process allows us to identify two factors that can control acid strength:

1. Molecules with weaker H—A bonds tend to be stronger acids.
2. Molecules in which the group of atoms (—A) attached to the acidic hydrogen has a high electron-accepting ability tend to be stronger acids.

Two types of acids can be used to demonstrate the factors that control acid strength: the HX acids (X = halogen atom) and a series of closely related oxoacids (acids in which the acidic hydrogen is attached to an oxygen atom). The effect of H—X bond energy on the acid strength of the HX acids is shown in Table 16.6.1.

The order of acid strength for this series,

$$HF << HCl < HBr < HI$$

illustrates the relationship between H—A bond energy and relative acid strength. The compound with the weakest H—A bond is the strongest acid.

The best illustration of the effect of the electron-accepting ability of a group of atoms on acid strength is seen in a series of oxoacids. For example, the chlorine oxoacids shown here differ only in the number of oxygen atoms attached to the chlorine atom.

Table 16.6.1 Bond Energies and K_a Values for the HX Acids

H—X	Bond Energy (kJ/mol)	K_a
H—F	569	$\sim 7 \times 10^{-4}$
H—Cl	431	$\sim 10^7$
H—Br	368	$\sim 10^9$
H—I	297	$\sim 10^{10}$

Hypochlorous acid **Chlorous acid** **Chloric acid** **Perchloric acid**

Each additional electronegative oxygen atom increases the electron-accepting ability of the group, drawing electron density away from the Cl atom. This electron-withdrawing effect, also known as the inductive effect, draws electron density away from the oxygen atom bonded to the acidic hydrogen and makes it easier to remove H (as H^+) from the O—H group. As the number of electronegative, electron-withdrawing atoms increases in a series of oxoacids, it becomes easier to lose H^+ from the O—H group and the acid strength increases.

Hypochlorous acid **Chlorous acid** **Chloric acid** **Perchloric acid**

This trend in oxoacid strength can be explained from another perspective where the relative stability of the conjugate base of the weak acid is considered. In general, chemical species that can distribute charge over multiple atoms are more stable than similar species where the charge is concentrated on a single atom or only a few atoms.

In the chlorine oxoacid series, each conjugate base carries a −1 charge. In the case of the hypochlorite ion, OCl^-, the negative charge resides primarily on the more electronegative oxygen atom.

As the number of oxygen atoms increases, the −1 overall charge on the conjugate base is distributed over more and more electronegative oxygen atoms. This distribution of charge means, for example, that the ClO_4^- ion is better able to accommodate the negative charge than the ClO^- ion. As a result, $HClO_4$ is a stronger acid than $HClO$.

Example Problem 16.6.1 Explore the relationship between molecular structure and acid strength.

Use the relationship between molecular structure and acid strength to explain why HNO_3 is a much stronger acid than HNO_2.

Solution:

You are asked to explain the relative strength of two acids.

You are given the identity of the two acids.

The two possible molecular structure explanations are (1) relative strength of the O—H bonds and (2) the electron-accepting ability of NO_3^- versus NO_2^-.

Possibility 1:

The O–H bonds in the two acids have essentially the same bond energy, so this is not a major factor influencing relative acid strength for these two acids.

Possibility 2:

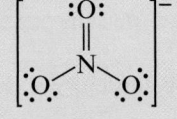

Nitrite ion   Nitrate ion

The nitrate ion has more oxygen atoms than the nitrite ion. As the number of oxygen atoms increases, the negative charge on the conjugate base is distributed over a greater number of highly electronegative atoms. Thus, the nitrate ion is better able to accommodate the negative charge than the nitrite ion.

Video Solution

Tutored Practice
Problem 16.6.1

Section 16.6 Mastery

Unit Recap

Key Concepts

16.1 Introduction to Acids and Bases

- Brønsted–Lowry acids and bases are proton acceptors and proton donors, respectively (16.1a).

- Lewis acids and bases are electron pair acceptors and donors, respectively (16.1a).

- Brønsted–Lowry acid–base reactions involve proton transfer and the formation of a conjugate acid and a conjugate base (16.1b).

- A conjugate acid–base pair differ only by the presence or absence of H^+ (16.1b).

- While monoprotic acids are capable of donating one proton, polyprotic acids such as diprotic acids can donate more than one proton (16.1c).

- An amphiprotic species can both donate and accept protons (16.1c).

16.2 Water and the pH Scale

- Water undergoes autoionization, where two water molecules react to form the hydronium and hydroxide ions (16.2a).

- The ionization constant for water, K_w, is the name given to the equilibrium constant for the autoionization of water (16.2a).

- An acidic solution is one where the hydronium ion concentration is greater than the hydroxide ion concentration; a basic solution has a hydroxide ion concentration greater than the hydronium ion concentration; a neutral solution has equal hydroxide and hydronium ion concentrations (16.2a).

- The p scale is a logarithmic scale used for very large or small numbers. For example, pH is the negative log of the hydronium ion concentration (16.2b).

16.3 Acid and Base Strength

- The strength of an acid or base depends on the concentration of the acid or base and the tendency of the acid or base to donate or accept a proton, respectively (16.3).

- Strong acids and bases ionize 100% in water (16.3a).

- The equilibrium constant for the hydrolysis of a weak acid is given the symbol K_b and the equilibrium constant for the hydrolysis of a weak base is given the symbol K_b (16.3a).

- Stronger acids have larger K_a values and stronger bases have larger K_b values (16.3b).
- As acid strength increases, the strength of the acid's conjugate base decreases and as base strength increases, the strength of the base's conjugate acid decreases (16.3b).

16.4 Estimating the pH of Acid and Base Solutions

- The pH of a solution containing a strong acid or strong base is calculated directly from the strong acid or strong base concentration (16.4a).
- The pH of a solution containing a weak acid or weak base is calculated from the weak acid or weak base concentration and the equilibrium constant for the hydrolysis reaction of the weak acid or weak base (16.4b, 16.4c).

16.5 Acid–Base Properties of Salts

- The acid–base properties of a salt are determined by the hydrolysis reactions of the cation and anion that make up the salt (16.5a).

16.6 Molecular Structure and Control of Acid–Base Strength

- Relative acid–base strength can be predicted or explained by the strength of the chemical bonds in the acid or base and the number of highly electronegative atoms in the acid or base (16.6a).

Key Equations

$K_w = [H_3O^+][OH^-] = 1.0 \times 10^{-14}$ at 25 °C   **(16.1)**

In general, $pX = -\log X$ **(16.2)**

$pH = -\log[H_3O^+]$ **(16.3)**

$pOH = -\log[OH^-]$ **(16.4)**

$[H_3O^+] = 10^{-pH}$ **(16.5)**

$[OH^-] = 10^{-pOH}$ **(16.6)**

$pK_w = pH + pOH = 14.00$ at 25 °C   **(16.7)**

$HA(aq) + H_2O(\ell) \rightleftarrows H_3O^+(aq) + A^-(aq)$ $K_a = \dfrac{[H_3O^+][A^-]}{[HA]}$ **(16.8)**

$B(aq) + H_2O(\ell) \rightleftarrows HB^+(aq) + OH^-(aq)$ $K_b = \dfrac{[HB^+][OH^-]}{[B]}$ **(16.9)**

$K_a \times K_b = K_w = 1.0 \times 10^{-14}$ at 25 °C   **(16.10)**

$\text{percent ionization} = \left(\dfrac{x}{[HA]_{\text{initial}}}\right) \times 100\%$ **(16.11)**

Key Terms

16.1 Introduction to Acids and Bases
Arrhenius acid
Arrhenius base
Brønsted–Lowry acid
Brønsted–Lowry base
Lewis acid
Lewis base
conjugate acid–base pair
monoprotic acid
polyprotic acid
diprotic acid
amphiprotic

16.2 Water and the pH Scale
autoionization
ionization constant for water, K_w
neutral solution
acidic solution
basic solution
pH
pOH

16.3 Acid and Base Strength
hydrolysis reaction
acid dissociation constant, K_a
base dissociation constant, K_b
pK_a

Unit 16   Review and Challenge Problems

17 Advanced Acid–Base Equilibria

Unit Outline

In This Unit

We will now expand the introductory coverage of acid–base equilibria and explore the chemistry of more complicated aqueous solutions containing acids and bases. First, we will address the different types of acid–base reactions, and then we will move on to study buffer solutions, acid–base titrations, and polyprotic acids. One of the more important types of acid–base solutions in terms of commercial and biological applications is buffers because they allow us to control the pH of a solution. So that you can get a feeling for the importance of buffers in your world, we will also briefly discuss the chemistry of two important buffers in biological systems. In Precipitation and Lewis Acid–Base Equilibria (Unit 18), we will conclude our coverage of chemical equilibria with Lewis acids and bases and the equilibria of sparingly soluble compounds.

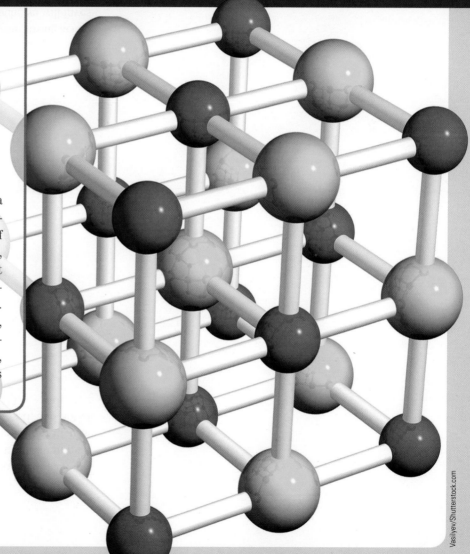

Vasilyev/Shutterstock.com

17.1 Acid–Base Reactions

17.1a Strong Acid/Strong Base Reactions

In Chemical Reactions and Solution Stoichiometry (Unit 4) you learned that acids and bases react to form water and a salt and that these reactions are called neutralization reactions because, on completion of the reaction, the solution is neutral. However, acid–base reactions do not always result in the formation of a solution with a neutral pH, and not all acid-base reactions proceed to 100% completion (Interactive Figure 17.1.1).

Interactive Figure 17.1.1

Investigate the extent of acid-base reactions.

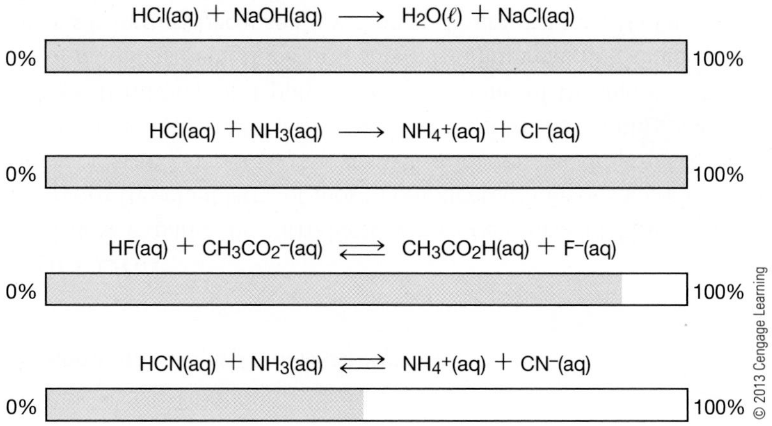

$$HCl(aq) + NaOH(aq) \longrightarrow H_2O(\ell) + NaCl(aq)$$

0% [██████████████████████████] 100%

$$HCl(aq) + NH_3(aq) \longrightarrow NH_4^+(aq) + Cl^-(aq)$$

0% [██████████████████████████] 100%

$$HF(aq) + CH_3CO_2^-(aq) \rightleftharpoons CH_3CO_2H(aq) + F^-(aq)$$

0% [████████████████████░░░] 100%

$$HCN(aq) + NH_3(aq) \rightleftharpoons NH_4^+(aq) + CN^-(aq)$$

0% [█████████░░░░░░░░░░░░] 100%

© 2013 Cengage Learning

Figure 17.1.1 Not all acid-base reactions go to 100% completion

There are four classes of acid–base reactions: strong acid + strong base, strong acid + weak base, weak acid + strong base, and weak acid + weak base. For each, we will investigate the extent of the reaction and the pH of the resulting solution when equimolar amounts of reactants are combined.

Reaction	Example	pH at Equilibrium
Strong acid + strong base	$HCl(aq) + NaOH(aq) \rightarrow H_2O(\ell) + NaCl(aq)$	7
Strong acid + weak base	$HCl(aq) + NH_3(aq) \rightarrow NH_4Cl(aq)$	< 7
Strong base + weak acid	$NaOH(aq) + HClO(aq) \rightarrow H_2O(\ell) + NaClO(aq)$	> 7
Weak acid + weak base	$HClO(aq) + NH_3(aq) \rightleftarrows NH_4ClO(aq)$	Depends on K_a and K_b values

The reaction between a strong acid and a strong base produces water and a salt (an ionic compound consisting of the cation from the strong base and the anion from the strong acid):

$$HCl(aq) + NaOH(aq) \rightarrow H_2O(\ell) + NaCl(aq)$$
$$\text{acid} \qquad \text{base} \qquad \text{water} \qquad \text{salt}$$

The net ionic equation for any reaction between a strong acid and a strong base is the reverse of the K_w reaction.

Complete ionic equation:

$$H_3O^+(aq) + Cl^-(aq) + Na^+(aq) + OH^-(aq) \rightarrow 2\,H_2O(\ell) + Na^+(aq) + Cl^-(aq)$$

Net ionic equation:

$$H_3O^+(aq) + OH^-(aq) \rightarrow 2\,H_2O(\ell) \qquad K = 1/K_w = 1.0 \times 10^{14}$$

The large value of K for this reaction indicates that in a strong acid + strong base reaction, the reactants are completely consumed to form products. The resulting solution is pH neutral (pH = 7). Notice that a single arrow ($\rightarrow$) is used to indicate that the reaction goes essentially to completion.

17.1b Strong Acid/Weak Base and Strong Base/ Weak Acid Reactions

A strong acid/weak base or strong base/weak acid reaction has a large equilibrium constant ($K \gg 1$) and goes essentially to completion. That is, when combined in stoichiometric amounts, all reactants are consumed to form products. For example, consider the reaction between the strong acid HCl and the weak base NH_3.

$$HCl(aq) + NH_3(aq) \rightarrow NH_4Cl(aq)$$

The net ionic equation for this reaction is the reverse of the hydrolysis reaction for the weak acid NH_4^+.

$$H_3O^+(aq) + NH_3(aq) \rightarrow H_2O(\ell) + NH_4^+(aq)$$

$$K = 1/K_a(NH_4^+) = 1/(5.6 \times 10^{-10}) = 1.8 \times 10^9$$

We can predict the large magnitude of the equilibrium constant for this reaction by recognizing that of the two Brønsted acids in this reaction (H_3O^+ and NH_4^+), the hydronium ion is a much better proton donor (a much stronger acid) than the ammonium ion. Therefore, the acid–base reaction favors the formation of the weaker acid, NH_4^+. *In general, all acid–base reactions favor the direction where a stronger acid and base react to form a weaker acid and base.*

When the reaction between hydrochloric acid and ammonia is complete (all reactants are consumed), the solution contains the chloride ion (a pH-neutral anion) and the ammonium ion (an acidic cation). As a result, the solution is acidic (pH < 7).

$$NH_4^+(aq) + H_2O(\ell) \rightleftharpoons NH_3(aq) + H_3O^+(aq) \qquad K_a = 5.6 \times 10^{-10}$$

In general, the pH of a solution resulting from the reaction of a strong acid with weak base is less than 7 (acidic) because of the presence of the conjugate acid of the weak base.

Similarly, the reaction between a weak acid and a strong base has a large equilibrium constant ($K \gg 1$) and therefore goes essentially to completion. For example, consider the reaction of the strong base NaOH with the weak acid HClO.

$$NaOH(aq) + HClO(aq) \rightarrow H_2O(\ell) + NaClO(aq)$$

The net ionic equation for this reaction is the reverse of the hydrolysis reaction for the weak base ClO^-.

$$OH^-(aq) + HClO(aq) \rightarrow H_2O(\ell) + ClO^-(aq)$$

$$K = 1/K_b(ClO^-) = 1/(2.8 \times 10^{-7}) = 3.5 \times 10^6$$

In this reaction, HClO is a stronger acid than H_2O (and OH^- is a stronger base than ClO^-); therefore, the reaction favors the formation of products, the weaker acid (H_2O) and base (ClO^-).

When the reaction is complete (all reactants are consumed), the solution contains the sodium ion (a pH-neutral cation) and the hypochlorite ion (a basic anion). As a result, the solution is basic (pH > 7).

$$OCl^-(aq) + H_2O(\ell) \rightleftharpoons HOCl(aq) + OH^-(aq) \qquad K_b = 2.8 \times 10^{-7}$$

In general, the pH of a solution resulting from the reaction of a strong base with a weak acid is greater than 7 (basic) because of the presence of the conjugate base of the weak acid.

Example Problem 17.1.1 Describe strong acid/weak base and strong base/weak acid reactions.

Write the net ionic equation for the reaction between nitrous acid and potassium hydroxide. Is the reaction reactant- or product-favored? Is the solution acidic or basic after all reactants are consumed?

Solution:

You are asked to write a net ionic equation for an acid–base reaction and to predict whether the reaction is reactant- or product-favored and whether the solution is acidic or basic after all reactants are consumed.

You are given the identity of the acid and the base.

$$HNO_2(aq) + OH^-(aq) \rightarrow H_2O(\ell) + NO_2^-(aq)$$

HNO_2 is a much stronger acid than H_2O (and OH^- is a much stronger base than NO_2^-). The reaction will favor the formation of the weaker acid and base, so the reaction is product-favored. Reactions between weak acids and strong bases are assumed to be 100% complete. The pH of the solution is greater than 7 (basic) after all reactants are consumed because of the presence of the weak base NO_2^-.

Video Solution

Tutored Practice
Problem 17.1.1

17.1c Weak Acid/Weak Base Reactions

In the reaction between a weak acid and a weak base, the magnitude of the equilibrium constant and therefore the extent of reaction depend on the relative strength of the acids and bases in the reaction. For example, consider the reaction between the weak acid HClO and the weak base NH_3.

$$HClO(aq) + NH_3(aq) \rightleftarrows NH_4^+(aq) + ClO^-(aq)$$

In this example, HClO ($K_a = 3.5 \times 10^{-8}$) is a stronger acid than NH_4^+ ($K_a = 5.6 \times 10^{-10}$), so the reaction is product-favored. However, the equilibrium constant for the reaction is not large, so the reaction does not go essentially to completion. A significant concentration of all four species (along with H_3O^+ and OH^-) is found at equilibrium.

$HClO(aq) + H_2O(\ell) \rightleftarrows H_3O^+(aq) + ClO^-(aq)$	$K_a = 3.5 \times 10^{-8}$
$NH_3(aq) + H_2O(\ell) \rightleftarrows NH_4^+(aq) + OH^-(aq)$	$K_b = 1.8 \times 10^{-5}$
$H_3O^+(aq) + OH^-(aq) \rightleftarrows 2\,H_2O(\ell)$	$K = 1/K_w = 1/(1.0 \times 10^{-14})$
$HClO(aq) + NH_3(aq) \rightleftarrows NH_4^+(aq) + ClO^-(aq)$	$K_{net} = K_a K_b / K_w = 63$

The equilibrium pH of the solution depends on the acid–base strength of the predominant species in solution, NH_4^+ and ClO^-. In this example, because HOCl is a weaker acid than NH_3 is a base [K_a(HOCl) $< K_b$(NH_3)], the hypochlorite ion is a stronger base than the ammonium ion is an acid [K_b(ClO^-) $> K_a$(NH_4^+)] and the solution is basic (pH > 7).

Consider the reaction between the weak acid HClO and the weak base HCO_2^-.

$$HClO(aq) + HCO_2^-(aq) \rightleftarrows HCO_2H(aq) + ClO^-(aq)$$

We can predict the extent of reaction by comparing the relative strength of the bases in this reaction. Here, ClO^- ($K_b = 2.9 \times 10^{-7}$) is a stronger base than HCO_2^- ($K_b = 5.6 \times 10^{-11}$), so the reaction is reactant-favored. The pH of the solution at equilibrium is predicted by comparing the relative acid and base strength of the two predominant species at equilibrium. Hypochlorous acid is a stronger acid than the formate ion is a base [K_a(HClO) $> K_b$(HCO_2^-)], so the solution is acidic (pH < 7).

Example Problem 17.1.2 Describe weak acid/weak base reactions.

When equimolar amounts of hydrocyanic acid and the acetate ion react, is the reaction reactant- or product-favored? Predict the pH of the solution after all reactants are consumed.

Solution:

You are asked to predict whether an acid–base reaction is reactant- or product-favored and what the pH of the solution is after the reactants are consumed.

You are given the identity of the acid and the base.

$$HCN(aq) + CH_3CO_2^-(aq) \rightleftarrows CH_3CO_2H(aq) + CN^-(aq)$$

Acetic acid is a stronger acid than hydrocyanic acid (and the cyanide ion is a stronger base than the acetate ion). The equilibrium will favor the weaker acid and base (reactant-favored). The pH of the solution at equilibrium is controlled by the predominant species in solution, HCN and $CH_3CO_2^-$. The solution is basic (pH > 7) because K_b($CH_3CO_2^-$) $> K_a$(HCN).

Video Solution

Tutored Practice
Problem 17.1.2

Section 17.1 Mastery

17.2 Buffers

17.2a Identifying Buffers

Buffers are one of the more important types of acid–base solutions in terms of commercial and biological applications because they allow us to control the pH of a solution. For example, humans are composed of molecules that depend on hydrogen bonding for their structure and function and are therefore highly sensitive to pH. Most of the reactions in a human body occur in aqueous solutions containing buffering agents. It is not surprising that human blood is highly buffered, for if blood is not maintained at a pH near 7.4, death can occur.

A **buffer solution** contains a mixture of a weak acid and a weak base, typically the conjugate base of the weak acid. For example, a buffer solution commonly used in chemistry laboratories contains both acetic acid (CH_3CO_2H, a weak acid) and sodium acetate ($NaCH_3CO_2$, the sodium salt of the conjugate base of acetic acid). Some other examples of buffers are KH_2PO_4/K_2HPO_4 ($H_2PO_4^-$ is a weak acid and HPO_4^{2-} is a weak base) and NH_4Cl/NH_3 (NH_4^+ is a weak acid and NH_3 is a weak base).

The principle property of a buffer solution is that it experiences a relatively small change in pH when a strong acid or a strong base is added. It is the weak acid and weak base components of a buffer that make it possible for buffer solutions to absorb strong acid or strong base without a significant pH change.

- When a strong acid is added to a buffer, the acid reacts with the conjugate base and is completely consumed. Despite the addition of a strong acid, the pH of the buffer solution decreases only slightly.
  Example: When H_3O^+ is added to a $CH_3CO_2H/NaCH_3CO_2$ buffer, it consumes some of the conjugate base, forming additional acetic acid.

$$H_3O^+(aq) + CH_3CO_2^-(aq) \rightarrow H_2O(\ell) + CH_3CO_2H(aq)$$

- When a strong base is added to a buffer, the base reacts with the weak acid and is completely consumed. Despite the addition of a strong base, the pH of the buffer solution increases only slightly.
  Example: When OH^- is added to a $CH_3CO_2H/NaCH_3CO_2$ buffer, it consumes some of the weak acid, forming additional acetate ion.

$$OH^-(aq) + CH_3CO_2H(aq) \rightarrow H_2O(\ell) + CH_3CO_2^-(aq)$$

Example Problem 17.2.1 Identify buffer solutions.

Identify buffer solutions from the following list.

a. 0.13 M sodium hydroxide + 0.27 M sodium bromide
b. 0.13 M nitrous acid + 0.14 M sodium nitrite
c. 0.24 M nitric acid + 0.17 M sodium nitrate
d. 0.31 M calcium chloride + 0.25 M calcium bromide
e. 0.34 M ammonia + 0.38 M ammonium bromide

◄- - - - - -

Example Problem 17.2.1 *(continued)*

Solution:

You are asked to determine whether a given combination of species results in a buffer solution.

You are given the identify of the species in the solution.

a. This is not a buffer. Sodium hydroxide is a strong base, and sodium bromide is a neutral salt.

b. This is a buffer solution. Nitrous acid is a weak acid, and sodium nitrate is a source of its conjugate base, the nitrite ion.

c. This is not a buffer. Nitric acid is a strong acid, and sodium nitrate is a neutral salt.

d. This is not a buffer. Both are neutral salts.

e. This is a buffer solution. The ammonium ion (present as ammonium bromide) is a weak acid, and ammonia is its conjugate base.

Video Solution

Tutored Practice
Problem 17.2.1

17.2b Buffer pH

The **common ion effect** is the shift in an equilibrium that results from adding to a solution a chemical species that is common to an existing equilibrium; we can use this to help us understand the pH of buffer solutions. Consider a solution containing acetic acid, a weak acid:

$$CH_3CO_2H(aq) + H_2O(\ell) \rightleftharpoons H_3O^+(aq) + CH_3CO_2^-(aq)$$

The addition of sodium acetate, a source of the weak base $CH_3CO_2^-$, shifts the equilibrium to the left, suppressing the acid hydrolysis (the forward reaction) and increasing the pH. The following example problem demonstrates the common ion effect in an acetic acid/sodium acetate buffer solution.

Example Problem 17.2.2 Calculate pH of a weak acid/conjugate base buffer solution.

Calculate the pH of 125 mL of a 0.15-M solution of acetic acid before and after the addition of 0.015 mol of sodium acetate.

Solution:

You are asked to calculate the pH of a weak acid solution before and after the addition of its conjugate base.

You are given the identity of the weak acid and its conjugate base, the volume and concentration of the weak acid, and the amount of conjugate base.

Example Problem 17.2.2 *(continued)*

Step 1. Write the balanced equation for the acid hydrolysis reaction.

$$CH_3CO_2H(aq) + H_2O(\ell) \rightleftarrows H_3O^+(aq) + CH_3CO_2^-(aq)$$

When solving problems involving buffers, it is important to first write the weak acid hydrolysis reaction before considering the effect of added conjugate base.

Step 2. Set up an *ICE* table and calculate the pH of the weak acid solution.

	$CH_3CO_2H(aq)$	+	$H_2O(\ell)$	$\rightleftarrows$	$H_3O^+(aq)$	+	$CH_3CO_2^-(aq)$
Initial (M)	0.15				0		0
Change (M)	$-x$				$+x$		$+x$
Equilibrium (M)	$0.15 - x$				x		x

$$K_a = 1.8 \times 10^{-5} = \frac{[CH_3CO_2^-][H_3O^+]}{[CH_3CO_2H]} = \frac{(x)(x)}{0.15-x} \approx \frac{x^2}{0.15}$$

$$x = [H_3O^+] = 1.6 \times 10^{-3} \text{ M}$$

$$pH = 2.78$$

Step 3. Set up a new *ICE* table for the buffer that now includes the concentration of the common ion, the acetate ion.

$$[CH_3CO_2^-]_{initial} = \frac{0.015 \text{ mol } CH_3CO_2^-}{0.125 \text{ L}} = 0.12 \text{ M}$$

	$CH_3CO_2H(aq)$	+	$H_2O(\ell)$	$\rightleftarrows$	$H_3O^+(aq)$	+	$CH_3CO_2^-(aq)$
Initial (M)	0.15				0		0.12
Change (M)	$-x$				$+x$		$+x$
Equilibrium (M)	$0.15 - x$				x		$0.12 + x$

Step 4. Substitute these equilibrium concentrations into the K_a expression.

$$K_a = 1.8 \times 10^{-5} = \frac{[CH_3CO_2^-][H_3O^+]}{[CH_3CO_2H]} = \frac{(0.12 + x)(x)}{0.15-x} \approx \frac{(0.12)(x)}{0.15}$$

We can make the assumption that x is small compared with the initial acid and conjugate base concentrations because (1) K_a is small and (2) the presence of a common ion suppresses the weak acid hydrolysis.

Step 5. Solve the expression for $[H_3O^+]$ and calculate pH.

$$x = [H_3O^+] = K_a\frac{[CH_3CO_2H]}{[CH_3CO_2^-]} = 1.8 \times 10^{-5}\left(\frac{0.15}{0.12}\right) = 2.3 \times 10^{-5} \text{ M}$$

$$pH = 4.65$$

Is your answer reasonable? The pH of the solution has increased (the solution is more basic) because a weak base, the acetate ion, was added to the solution to form a buffer.

Video Solution

Tutored Practice
Problem 17.2.2

A buffer solution can consist of either a weak acid and its conjugate base or a weak base and its conjugate acid. To be consistent, we will treat all buffers as weak acid systems as shown in the following example.

Example Problem 17.2.3 Calculate pH of a weak base/conjugate acid buffer solution.

A 0.30-M aqueous solution of NH_3 has a pH of 11.37. Calculate the pH of a buffer solution that is 0.30 M in NH_3 and 0.23 M in ammonium bromide.

Solution:

You are asked to calculate the pH of a buffer solution.

You are given the identity of the weak acid and weak base in the buffer and the concentration of the species in the buffer solution.

Step 1. Write the balanced equation for the acid hydrolysis reaction. In this example the weak acid is the ammonium ion.

$$NH_4^+(aq) + H_2O(\ell) \rightleftharpoons H_3O^+(aq) + NH_3(aq)$$

Step 2. Set up an *ICE* table for the buffer solution.

	$NH_4^+(aq)$	$+$	$H_2O(\ell)$	$\rightleftharpoons$	$H_3O^+(aq)$	$+$	$NH_3(aq)$
Initial (M)	0.23				0		0.30
Change (M)	$-x$				$+x$		$+x$
Equilibrium (M)	$0.23 - x$				x		$0.30 + x$

Step 3. Substitute these equilibrium concentrations into the K_a expression.

$$K_a = 5.6 \times 10^{-10} = \frac{[NH_3][H_3O^+]}{[NH_4^+]} = \frac{(0.30 + x)(x)}{0.23 - x} \approx \frac{(0.30)(x)}{0.23}$$

Once again, we can make the assumption that x is small compared with the initial acid and conjugate base concentrations.

Step 4. Solve the expression for $[H_3O^+]$ and calculate pH.

$$x = [H_3O^+] = K_a\frac{[NH_4^+]}{[NH_3]} = 5.6 \times 10^{-10}\left(\frac{0.23}{0.30}\right) = 4.3 \times 10^{-10} \text{ M}$$

$$pH = 9.37$$

Is your answer reasonable? Addition of a weak acid, ammonium ion, decreases the pH of the solution (the solution is more acidic than the solution containing only ammonia).

Video Solution

Tutored Practice
Problem 17.2.3

The Henderson–Hasselbalch Equation

The rearranged K_a expression used in previous example problems can be used for calculating the pH of any buffer solution. In general, for a weak acid/conjugate base buffer,

$$[H_3O^+] = K_a \frac{[\text{weak acid}]}{[\text{conjugate base}]}$$

We can rewrite this equation in terms of pH and pK_a by taking the negative logarithm of both sides.

$$-\log[H_3O^+] = -\log\left(K_a \frac{[\text{weak acid}]}{[\text{conjugate base}]}\right)$$

$$-\log[H_3O^+] = -\log(K_a) + \left(-\log \frac{[\text{weak acid}]}{[\text{conjugate base}]}\right)$$

$$pH = pK_a - \log \frac{[\text{weak acid}]}{[\text{conjugate base}]}$$

Because $\left(-\log \dfrac{[\text{weak acid}]}{[\text{conjugate base}]}\right) = \left(+\log \dfrac{[\text{conjugate base}]}{[\text{weak acid}]}\right)$,

$$pH = pK_a + \log \frac{[\text{conjugate base}]}{[\text{weak acid}]} \qquad \textbf{(17.1)}$$

Equation 17.1 is the **Henderson–Hasselbalch equation**, a very useful form of the K_a expression that is often used to calculate the pH of buffer solutions. It is important to note that the Henderson–Hasselbalch equation is used only for calculations involving buffer solutions. It is not used if a solution contains only a weak acid or only a weak base.

The Henderson–Hasselbalch equation shows the mathematical relationship between the weak acid pK_a and the pH of a buffer, and that buffer pH can be manipulated by changing the ratio of [conjugate base] to [weak acid]. Notice that in the special case where [weak acid] = [conjugate base], the ratio of concentrations is equal to 1 and the pH of the buffer solution is equal to the weak acid pK_a.

When [weak acid] = [conjugate base], pH = pK_a + log(1) = pK_a + 0 = pK_a

Buffer components are chosen based on the relationship between weak acid pK_a and the target pH for the buffer. For the buffer to be effective, it should contain significant amounts of both weak acid and conjugate base. Effective buffers, those that can best resist pH change upon addition of a strong acid or base, have a [conjugate base] to [weak acid] ratio between 1:10 and 10:1. Because log(10/1) = 1, this results in a buffer pH that is approximately equal to the weak acid pK_a ± 1. This type of buffer solution has a high

buffer capacity, which is defined as the amount of strong acid or base that can be added to a buffer without a drastic change in pH.

[conjugate base]/ [weak acid]	pH of buffer solution
1	pH = pK_a
10/1	pH = $pK_a + 1$
1/10	pH = $pK_a - 1$

Example Problem 17.2.4 Use the Henderson–Hasselbalch equation to calculate pH of a buffer solution.

Using the Henderson–Hasselbalch equation, calculate the pH of a buffer solution that is 0.18 M in $H_2PO_4^-$ and 0.21 M in HPO_4^{2-}.

Solution:

You are asked to calculate the pH of a buffer solution using the Henderson–Hasselbalch equation.

You are given the concentration and identity of the species in the buffer solution.

Step 1. Write the balanced equation for the acid hydrolysis reaction.

$$H_2PO_4^-(aq) + H_2O(\ell) \rightleftharpoons H_3O^+(aq) + HPO_4^{2-}(aq)$$

Step 2. Set up an *ICE* table for the buffer solution.

	$H_2PO_4^-(aq)$	+	$H_2O(\ell)$	$\rightleftharpoons$	$H_3O^+(aq)$	+	$HPO_4^{2-}(aq)$
Initial (M)	0.18				0		0.21
Change (M)	$-x$				$+x$		$+x$
Equilibrium (M)	$0.18 - x$				x		$0.21 + x$

Step 3. Substitute these equilibrium concentrations into the Henderson–Hasselbalch equation and calculate pH.

$$pH = pK_a + \log\frac{[HPO_4^-]}{[H_2PO_4^-]} = -\log(6.2 \times 10^{-8}) + \log\frac{(0.21 + x)}{(0.18 - x)} \approx 7.21 + \log\left(\frac{0.21}{0.18}\right) = 7.27$$

Once again, we can make the assumption that x is small compared with the initial acid and conjugate base concentrations.

Is your answer reasonable? Notice that the pH of the buffer solution is greater than pK_a for the weak acid because [conjugate base]/[weak acid] > 1. If [conjugate base]/[weak acid] < 1, the pH is less than pK_a.

Video Solution

Tutored Practice
Problem 17.2.4

Adding Acid or Base to a Buffer

It is a common misconception that buffer pH remains constant when some strong acid or base is added. This is not the case. As shown in the following example problem, a buffer minimizes the pH change upon addition of a strong acid or base because only the weak acid/conjugate base ratio of the buffer is affected. The pH changes, but only by a small amount.

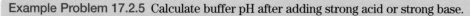

Example Problem 17.2.5 Calculate buffer pH after adding strong acid or strong base.

Determine the pH change when 0.020 mol HCl is added to 1.00 L of a buffer solution that is 0.10 M in CH_3CO_2H and 0.25 M in $CH_3CO_2^-$.

Solution:

You are asked to calculate the pH change when a strong acid is added to a buffer solution.

You are given the amount of strong acid, the concentration of the species in the buffer solution, and the volume of the buffer solution.

Step 1. Write the balanced equation for the acid hydrolysis reaction.

$$CH_3CO_2H(aq) + H_2O(\ell) \rightleftarrows H_3O^+(aq) + CH_3CO_2^-(aq)$$

Step 2. Use the Henderson–Hasselbalch equation to calculate the pH of the buffer solution before the addition of HCl.

$$pH = pK_a + \log\frac{[CH_3CO_2^-]}{[CH_3CO_2H]} = -\log(1.8 \times 10^{-5}) + \log\left(\frac{0.25}{0.10}\right) = 5.14$$

Step 3. Assume that the strong acid reacts completely with the conjugate base. Set up a table that shows the amount (in moles) of species initially in the solution, the change in amounts of reactants and products (based on the amount of limiting reactant), and the amounts of reactants and products present after the acid–base reaction is complete.

	$H_3O^+(aq)$	+	$CH_3CO_2^-(aq)$	$\rightarrow$	$H_2O(\ell)$	+	$CH_3CO_2H(aq)$
Initial (mol)	0.020		0.25				0.10
Change (mol)	−0.020		−0.020				+0.020
After reaction (mol)	0		0.23				0.12

Step 4. Use the new weak acid and conjugate base concentrations to calculate the buffer pH after adding strong acid.

$$[CH_3CO_2H] = \frac{0.12 \text{ mol}}{1.00 \text{ L}} = 0.12 \text{ M} \qquad [CH_3CO_2^-] = \frac{0.23 \text{ mol}}{1.00 \text{ L}} = 0.23 \text{ M}$$

$$pH = pK_a + \log\frac{[CH_3CO_2^-]}{[CH_3CO_2H]} = -\log(1.8 \times 10^{-5}) + \log\left(\frac{0.23}{0.12}\right) = 5.03$$

$$\Delta pH = 5.03 - 5.14 = -0.11$$

Is your answer reasonable? Addition of 0.020 mol HCl to the buffer decreases the pH only slightly, by 0.11 pH units. If the same amount of HCl is added to 1.00 L of water, the pH decreases by 5.30 pH units, from a pH of 7.00 to a pH of 1.70.

Video Solution

Tutored Practice
Problem 17.2.5

17.2c Making Buffer Solutions

The preparation of a buffer solution with a known pH is a two-step process.

1. A weak acid/conjugate base pair is chosen. The weak acid pK_a must be within about 1 pH unit of the desired pH. This guarantees that the [weak acid]/[conjugate base] ratio is between 10:1 and 1:10, ensuring that the solution will contain significant amounts of weak acid and conjugate base and will be able to buffer against the addition of strong acid or base.

2. The desired pH and the weak acid pK_a are used to determine the relative concentrations of weak acid and conjugate base needed to give the desired pH.

Once the desired weak acid and conjugate base concentrations are known, the solution is prepared in one of two ways:

Direct addition: The correct amounts of the weak acid and conjugate base are added to water.

Acid–base reaction: For example, a conjugate base is created by reacting a weak acid with enough strong base to produce a solution containing the correct weak acid and conjugate base concentrations.

Example Problem 17.2.6 Prepare a buffer by direct addition.

Describe how to prepare 500. mL of a buffer solution with pH = 9.85 using one of the weak acid/conjugate base systems shown here.

Weak Acid	Conjugate Base	K_a	pK_a
CH_3CO_2H	$CH_3CO_2^-$	1.8×10^{-5}	4.74
$H_2PO_4^-$	HPO_4^{2-}	6.2×10^{-8}	7.21
HCO_3^-	CO_3^{2-}	4.8×10^{-11}	10.32

Solution:

You are asked to describe how to prepare a buffer solution with a known pH using the direct addition method.

You are given the identity and K_a values for three possible weak acid/conjugate base pairs and the volume and target pH of the buffer solution.

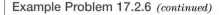

Example Problem 17.2.6 *(continued)*

Step 1. Choose a weak acid/conjugate base pair. The bicarbonate ion/carbonate ion buffer system is the best choice here because the desired pH is close to the pK_a of the weak acid. Write the balanced equation for the acid hydrolysis reaction.

$$HCO_3^-(aq) + H_2O(\ell) \rightleftharpoons H_3O^+(aq) + CO_3^{2-}(aq)$$

Step 2. Determine the necessary weak acid/conjugate base ratio using the rearranged K_a expression for the weak acid.

(The Henderson–Hasselbalch equation can also be used to determine the weak acid/conjugate base ratio.)

$$[H_3O^+] = 10^{-pH} = 10^{-9.85} = 1.4 \times 10^{-10} \text{ M}$$

$$[H_3O^+] = 1.4 \times 10^{-10} = K_a\frac{[HCO_3^-]}{[CO_3^{2-}]} = (4.8 \times 10^{-11})\frac{[HCO_3^-]}{[CO_3^{2-}]}$$

$$\frac{[HCO_3^-]}{[CO_3^{2-}]} = \frac{2.9 \text{ mol/L}}{1.0 \text{ mol/L}} = \frac{2.9 \text{ mol HCO}_3^-}{1.0 \text{ mol CO}_3^{2-}}$$

Notice that the volume of buffer is cancelled in the ratio. The required amounts of weak acid and conjugate base are independent of the solution volume, so *the volume of a buffer has no effect on the buffer pH*.

Step 3. Determine the amount of weak acid and conjugate base that must be combined to produce the buffer solution. Mixing 2.9 mol HCO_3^- and 1.0 mol CO_3^{2-} (or any multiple of this ratio) will result in a buffer with a pH of 9.85. Alternately, assuming that each is present in the form of a sodium salt, combine 240 g $NaHCO_3$ and 110 g Na_2CO_3 in a flask and add water (to a total solution volume of 500 mL) to make the buffer solution.

$$2.9 \text{ mol NaHCO}_3 \times \frac{84.0 \text{ g}}{1 \text{ mol NaHCO}_3} = 240 \text{ g NaHCO}_3$$

$$1.0 \text{ mol Na}_2CO_3 \times \frac{106 \text{ g}}{1 \text{ mol Na}_2CO_3} = 110 \text{ g Na}_2CO_3$$

Video Solution

Tutored Practice Problem 17.2.6

When performing a calculation to determine how to make a buffer solution using an acid–base reaction, it is helpful to set up a stoichiometry table to assist in keeping track of the initial amounts of reactants and the amounts of reactants and products in the solution after a reaction takes place. Note that a stoichiometry table is different from an *ICE* table, which is used to keep track of changes that occur in an equilibrium system. A stoichiometry table shows the amount (in moles) of species initially in the solution, the change in amounts of reactants and products (based on the amount of limiting reactant), and the amounts of reactants and products present after the reaction is complete.

Example Problem 17.2.7 Prepare a buffer by acid–base reactions.

Describe how to prepare a buffer solution with pH = 5.25 (using one of the weak acid/conjugate base systems shown here) by combining a 0.50-M solution of weak acid with any necessary amount of 1.00 M NaOH.

Weak Acid	Conjugate Base	K_a	pK_a
CH_3CO_2H	$CH_3CO_2^-$	1.8×10^{-5}	4.74
$H_2PO_4^-$	HPO_4^{2-}	6.2×10^{-8}	7.21
HCO_3^-	CO_3^{2-}	4.8×10^{-11}	10.32

Solution:

You are asked to describe how to prepare a buffer solution with a known pH using an acid–base reaction.

You are given the identity and K_a values for three possible weak acid/conjugate base pairs, the concentration of the weak acid and the strong base used to make the buffer, and the target pH of the buffer solution.

Step 1. Choose a weak acid/conjugate base pair. The acetic acid/acetate ion buffer system is the best choice here because the desired pH is close to the pK_a of the weak acid. Write the balanced equation for the acid hydrolysis reaction.

$$CH_3CO_2H(aq) + H_2O(\ell) \rightleftarrows H_3O^+(aq) + CH_3CO_2^-(aq)$$

Step 2. Determine the necessary weak acid/conjugate base ratio using the rearranged K_a expression for the weak acid.

(The Henderson–Hasselbalch equation can also be used to determine the weak acid/conjugate base ratio.)

$$[H_3O^+] = 10^{-5.25} = 5.6 \times 10^{-6} \text{ M}$$

$$[H_3O^+] = 5.6 \times 10^{-6} = K_a \frac{[CH_3CO_2H]}{[CH_3CO_2^-]} = (1.8 \times 10^{-5}) \frac{[CH_3CO_2H]}{[CH_3CO_2^-]}$$

$$\frac{[CH_3CO_2H]}{[CH_3CO_2^-]} = \frac{0.31 \text{ mol/L}}{1.0 \text{ mol/L}} = \frac{0.31 \text{ mol } CH_3CO_2H}{1.0 \text{ mol } CH_3CO_2^-}$$

Notice that volume units (L) are cancelled in the $\frac{[CH_3CO_2H]}{[CH_3CO_2^-]}$ ratio. The required amounts of weak acid and conjugate base are independent of the total solution volume, so the volume of a buffer has no effect on the buffer pH.

Step 3. Determine the amount of weak acid and strong base that must be combined to produce the buffer. Recall that a strong base will react completely with a weak acid forming water and the conjugate base of the weak acid. In this case, the weak acid and strong base react in a 1:1 stoichiometric ratio, so

initial amount of weak acid required = amount of weak acid in buffer
+ amount of conjugate base in buffer

initial amount of weak acid required = 0.31 mol + 1.0 mol = 1.31 mol CH_3CO_2H

The amount of strong base required is determined by the reaction stoichiometry. Set up a stoichiometry table that shows the amount (in moles) of species initially in the solution, the change in amounts of reactants and products (based on the amount of limiting reactant), and the amounts of reactants and products present after the acid–base reaction is complete. In this case, the stoichiometry table is used to determine the initial amount of reactants needed to produce a buffer containing 0.31 mol acetic acid and 1.00 mol acetate ion.

	$CH_3CO_2H(aq)$ +	$OH^-(aq)$ →	$H_2O(\ell)$ +	$CH_3CO_2^-(aq)$
Initial (mol)	1.31	1.00		0
Change (mol)	−1.00	−1.00		+1.00
After reaction (mol)	0.31	0		1.00

The combination of 1.31 mol CH_3CO_2H with 1.00 mol OH^- (as NaOH) will produce the buffer solution.

Step 4. Determine the volume of weak acid and strong base solutions that must be combined to produce the buffer solution.

$$1.31 \text{ mol } CH_3CO_2H \times \frac{1.0 \text{ L}}{0.50 \text{ mol } CH_3CO_2H} = 2.6 \text{ L } CH_3CO_2H \text{ solution}$$

$$1.00 \text{ mol NaOH} \times \frac{1.0 \text{ L}}{1.00 \text{ mol NaOH}} = 1.00 \text{ L NaOH solution}$$

Mix 2.6 L of 0.50 M CH_3CO_2H with 1.00 L of 1.00 M NaOH to produce a buffer with a pH of 5.25. Note that any ratio of these volumes will produce the buffer with a pH of 5.25. For example, combing 1.0 L of 0.50 M CH_3CO_2H with 0.38 L of 1.00 M NaOH also produces a buffer with a pH of 5.25.

Video Solution

Tutored Practice
Problem 17.2.7

Alpha Plots

As shown in the preceding example problems, the pH of a buffer solution is controlled by the relative amounts of weak acid and conjugate base present and by the weak acid pK_a. That is, buffer pH is independent of solution volume. If a buffer solution is diluted, the pH does not change. Regardless of the amounts of weak acid and conjugate base that are present in the solution, however, the relative amounts can change if the pH is changed by external agents such as a strong acid or a strong base. Consider the acetic acid/acetate ion buffer system:

$$CH_3CO_2H(aq) + H_2O(\ell) \rightleftharpoons H_3O^+(aq) + CH_3CO_2^-(aq)$$

When the solution is highly acidic, the acid form predominates and very little of the conjugate base is present ($[CH_3CO_2H] \gg [CH_3CO_2^-]$). When the solution is highly basic, the base form predominates and very little of the acid form is present ($[CH_3CO_2^-] \gg [CH_3CO_2H]$). We quantify this using an alpha (α) value, which defines the mole fraction of the acid–base pair present as either the acid or the base. For the acetic acid/acetate ion buffer system,

$$\alpha_{CH_3CO_2H} = \frac{mol\ CH_3CO_2H}{mol\ CH_3CO_2H\ +\ mol\ CH_3CO_2^-}$$

$$\alpha_{CH_3CO_2^-} = \frac{mol\ CH_3CO_2^-}{mol\ CH_3CO_2H\ +\ mol\ CH_3CO_2^-}$$

The relationship between pH and solution composition is shown graphically in an **alpha (α) plot**, a plot of solution composition (α) versus pH. The alpha plot for the acetic acid/acetate ion buffer system shown in Interactive Figure 17.2.1 has the following features:

- When the solution pH is more than about 2 pH units below pK_a (4.74), the solution consists of almost all weak acid and almost no conjugate base.

- When the solution pH is more than about 2 pH units above pK_a (4.74), the solution consists of almost all conjugate base and almost no weak acid.

- When the solution pH is near pK_a, the solution contains a significant concentration of both weak acid and conjugate base.

- When the solution pH is equal to pK_a, the solution is composed of equal parts weak acid and weak base ($\alpha = 0.5$).

The ammonia/ammonium ion alpha plot (Figure 17.2.2) shows that equal amounts of weak acid and conjugate base are found when the pH is equal to the ammonium ion pK_a (9.26).

Explore alpha plots.

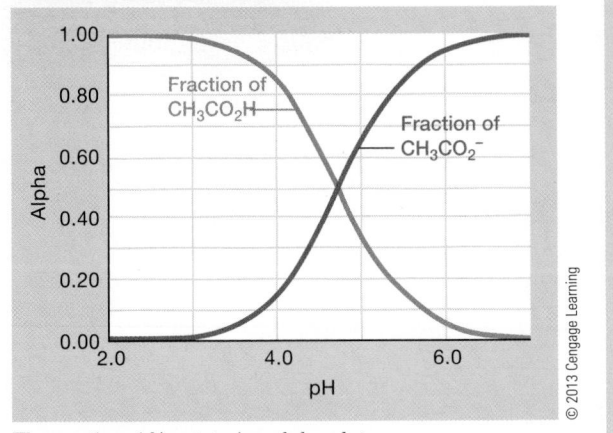

The acetic acid/acetate ion alpha plot

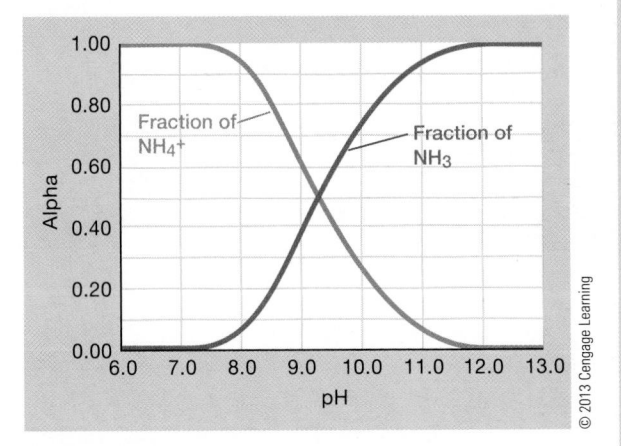

Figure 17.2.2 The ammonia/ammonium ion alpha plot

Section 17.2 Mastery

17.3 Acid–Base Titrations

17.3a Strong Acid/Strong Base Titrations

As you learned in Chemical Reactions and Solution Stoichiometry (Unit 4) an acid–base titration is when one reactant with a known concentration is placed in a buret and is added to the other reactant (of unknown concentration). The progress of the reaction is monitored externally using an acid–base indicator or pH meter, and the titration end point is used to calculate the unknown concentration. We will now consider how pH changes during acid–base titrations involving strong and weak acids and bases. We will construct pH titration plots (a plot of solution pH versus volume of added **titrant**, the substance being added during the titration) that reveal information about the nature of the acid or base under study, the equivalence point in the reaction, and, for weak acids and bases, the weak acid K_a (or weak base K_b).

The plot of pH versus volume of NaOH solution for the titration of 50.0 mL of 0.100 M HCl with 0.100 M NaOH is shown in Interactive Figure 17.3.1. The strong acid/strong base titration plot has four distinct regions:

1. The initial pH is less than 7 because of the presence of an acid.

2. The pH rises slowly with the addition of strong base as the base is completely consumed and a pH-neutral salt is formed.

3. Near the equivalence point, where the acid is completely consumed by added base, the pH increases rapidly. The midpoint of this vertical section of the plot (where the pH is 7.00) is the equivalence point of the titration.

4. After the equivalence point, the pH is high and increases slowly as excess base is added.

The pH plot for a strong base/strong acid titration (Figure 17.3.2) has a similar shape. Again, there are four distinct regions in the pH plot when a strong base is titrated with a strong acid:

1. The initial pH is greater than 7 because of the presence of a base.

2. The pH decreases slowly with the addition of strong acid as the acid is completely consumed and a pH-neutral salt is formed.

3. Near the equivalence point, where the base is completely consumed by added acid, the pH decreases rapidly. The midpoint of this vertical section of the plot (where the pH is 7.00) is the equivalence point of the titration.

4. After the equivalence point, the pH is low and decreases slowly as excess acid is added.

As shown in the following example problem, we can use solution stoichiometry to calculate the pH at four different points of a strong acid/strong base titration. In some of these calculations we will again use a stoichiometry table to assist us in keeping track of the initial amounts of reactants and the amounts of reactants and products in the solution after a reaction takes place. Note that a stoichiometry table is different from an *ICE* table, which is used to keep track of changes that occur in an equilibrium system.

Interactive Figure 17.3.1

Explore a strong acid/strong base pH titration plot.

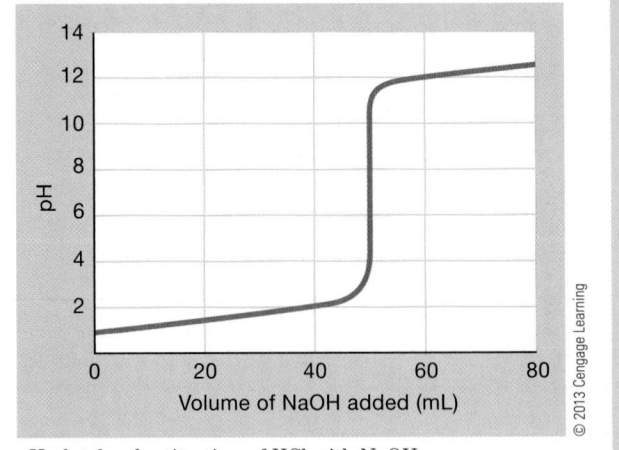

pH plot for the titration of HCl with NaOH

Example Problem 17.3.1 Calculate pH for a strong acid/strong base titration.

Determine the pH during the titration of 75.0 mL of 0.100 M HCl by 0.100 M NaOH at the following points:

a. Before the addition of any NaOH
b. After the addition of 20.0 mL of NaOH
c. At the equivalence point
d. After the addition of 100.0 mL of NaOH

Solution:

You are asked to calculate the pH at four different points in a strong acid/strong base titration.

You are given the volume and concentration of the strong acid and the concentration of the strong base.

Write the balanced net ionic equation for the acid–base reaction.

$$H_3O^+(aq) + OH^-(aq) \rightarrow 2\ H_2O(\ell)$$

a. HCl is a strong acid and is 100% dissociated in solution, so $[H_3O^+] = [HCl] = 0.100$ M.

$$pH = -\log(0.100) = 1.000$$

b. Before the equivalence point, all NaOH is consumed by excess HCl. Set up a stoichiometry table to track the amounts of reactants and products present in solution before and after the reaction is complete.

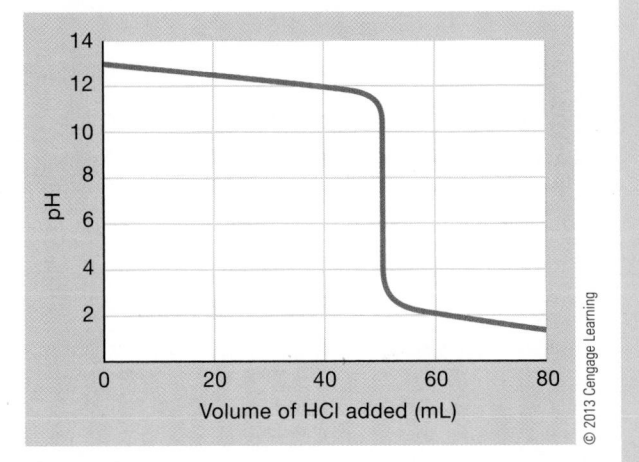

Figure 17.3.2 pH plot for the titration of NaOH with HCl

Example Problem 17.3.1 *(continued)*

$$\text{mol } H_3O^+ = (0.0750 \text{ L})(0.100 \text{ mol/L}) = 0.00750 \text{ mol } H_3O^+$$
$$\text{mol } OH^- = (0.0200 \text{ L})(0.100 \text{ mol/L}) = 0.00200 \text{ mol } OH^-$$

total volume of solution = 75.0 mL + 20.0 mL = 95.0 mL

	$H_3O^+(aq)$	+	$OH^-(aq)$	→	$2 H_2O(\ell)$
Initial (mol)	0.00750		0.00200		
Change (mol)	−0.00200		−0.00200		
After reaction (mol)	0.00550		0		

Concentration after reaction (M) $\quad [H_3O^+] = \dfrac{0.00550 \text{ mol}}{0.0950 \text{ L}} = 0.0579 \text{ M}$

$$pH = -\log(0.0579) = 1.237$$

c. Because this acid and base react in a 1:1 stoichiometric ratio and the two solutions have the same concentration, the equivalence point is reached when an equal volume (75.0 mL) of NaOH is added to the HCl. All base and acid is consumed, forming the pH-neutral salt NaCl and water. At this point, $[H_3O^+] = 1.0 \times 10^{-7}$ M and pH = 7.00.

d. After the equivalence point, all HCl has been consumed and the solution contains excess NaOH. Set up a stoichiometry table.

$$\text{mol } H_3O^+ = (0.0750 \text{ L})(0.100 \text{ mol/L}) = 0.00750 \text{ mol } H_3O^+$$
$$\text{mol } OH^- = (0.1000 \text{ L})(0.100 \text{ mol/L}) = 0.0100 \text{ mol } OH^-$$

total volume of solution = 75.0 mL + 100.0 mL = 175.0 mL

	$H_3O^+(aq)$	+	$OH^-(aq)$	→	$2 H_2O(\ell)$
Initial (mol)	0.00750		0.0100		
Change (mol)	−0.00750		−0.00750		
After reaction (mol)	0		0.0025		

Concentration after reaction (M) $\quad [OH^-] = \dfrac{0.0025 \text{ mol}}{0.1750 \text{ L}} = 0.014 \text{ M}$

$$pOH = -\log(0.014) = 1.85$$
$$pH = 14.000 - 1.85 = 12.15$$

17.3b Weak Acid/Strong Base and Weak Base/Strong Acid Titrations

The shape of the pH titration plot for an acetic acid/sodium hydroxide titration (Interactive Figure 17.3.3) is somewhat different than a strong acid/strong base pH titration plot. There are four regions of interest in the weak acid/strong base pH plot:

Video Solution

Tutored Practice
Problem 17.3.1

Interactive Figure 17.3.3

Explore a weak acid/strong base pH titration plot.

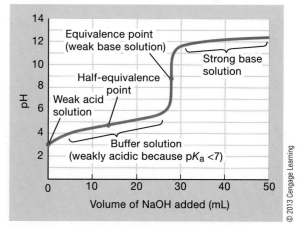

pH plot for the titration of CH_3CO_2H with NaOH

1. The initial pH is less than 7 because the solution contains an acid. The initial pH is greater than the initial pH in the strong acid/strong base titration because this solution contains a weak acid.

2. As the strong base is added, the pH rises sharply and then increases gradually until the equivalence point is reached. This is the buffer region of the titration, where the solution contains a weak acid and its conjugate base. At the midpoint of this region, half of the acid originally present in the flask has been consumed and the **half-equivalence point**, also called the titration midpoint, is reached. As shown in the example problem that follows, at this point in the titration the solution pH is equal to the weak acid pK_a.

3. Near the equivalence point, where the acid is completely consumed by added base, the pH increases rapidly. The equivalence point of the titration has a pH greater than 7 because of the presence of the acetate ion, a basic anion.

4. After the equivalence point, the pH is high because the solution contains excess strong base.

The pH of the solution is calculated at five points of the titration in the following example problem.

Example Problem 17.3.2 Calculate pH for a weak acid/strong base titration.

Determine the pH during the titration of 75.0 mL of 0.100 M benzoic acid ($K_a = 6.3 \times 10^{-5}$) by 0.100 M NaOH at the following points.

a. Before the addition of any NaOH
b. After the addition of 20.0 mL of NaOH
c. At the half-equivalence point (the titration midpoint)
d. At the equivalence point
e. After the addition of 100.0 mL of NaOH

Solution:

You are asked to calculate the pH at five different points of a weak acid/strong base titration.

You are given the volume and concentration of the weak acid and the concentration of the strong base.

Write the balanced net ionic equation for the acid–base reaction.

$$C_6H_5CO_2H(aq) + OH^-(aq) \rightarrow H_2O(\ell) + C_6H_5CO_2^-(aq)$$

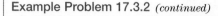

Example Problem 17.3.2 *(continued)*

a. Benzoic acid is a weak acid and the pH of the solution is calculated using methods introduced in Acids and Bases (Unit 16):

$$[H_3O^+] = \sqrt{K_a \times [C_6H_5CO_2H]_0} = \sqrt{(6.3 \times 10^{-5})(0.100)} = 0.00251 \text{ M}$$

$$pH = -\log(0.00251) = 2.600$$

b. Before the equivalence point, all NaOH is consumed by excess weak acid. First consider the stoichiometry of the acid–base reaction; then, because the solution contains a buffer, use the rearranged K_a expression or the Henderson–Hasselbalch equation to calculate pH.

$$\text{mol } CH_3CO_2H = (0.0750 \text{ L})(0.100 \text{ mol/L}) = 0.00750 \text{ mol } CH_3CO_2H$$

$$\text{mol } OH^- = (0.0200 \text{ L})(0.100 \text{ mol/L}) = 0.00200 \text{ mol } OH^-$$

total volume of solution = 75.0 mL + 20.0 mL = 95.0 mL

	$C_6H_5CO_2H(aq)$ +	$OH^-(aq)$ →	$H_2O(\ell)$ +	$C_6H_5CO_2^-(aq)$
Initial (mol)	0.00750	0.00200		0
Change (mol)	−0.00200	−0.00200		+0.00200
After reaction (mol)	0.00550	0		0.00200

Concentration after reaction (M)
$$[C_6H_5CO_2H] = \frac{0.00550 \text{ mol}}{0.0950 \text{ L}} = 0.0579 \text{ M}$$

$$[C_6H_5CO_2^-] = \frac{0.00200 \text{ mol}}{0.0950 \text{ L}} = 0.0211 \text{ M}$$

Use the rearranged K_a expression (or the Henderson–Hasselbalch equation) to calculate pH.

$$[H_3O^+] = K_a \frac{[C_6H_5CO_2H]}{[C_6H_5CO_2^-]} = 6.3 \times 10^{-5}\left(\frac{0.0579}{0.0211}\right) = 1.7 \times 10^{-4} \text{ M}$$

$$pH = 3.76$$

c. The half-equivalence point, the titration midpoint, is reached when half the amount of strong base required to reach the equivalence point has been added to the weak acid solution and, as a result, half of the weak acid originally in the flask has been consumed. In this example the midpoint in the titration is when 37.5 mL of NaOH is added to the solution.

$$\text{mol } CH_3CO_2H = (0.0750 \text{ L})(0.100 \text{ mol/L}) = 0.00750 \text{ mol } CH_3CO_2H$$

$$\text{mol } OH^- = (0.0375 \text{ L})(0.100 \text{ mol/L}) = 0.00375 \text{ mol } OH^-$$

total volume of solution = 75.0 mL + 37.5 mL = 112.5 mL

	$C_6H_5CO_2H(aq)$	$+$	$OH^-(aq)$	$\rightarrow$	$H_2O(\ell)$	$+$	$C_6H_5CO_2^-(aq)$
Initial (mol)	0.00750		0.00375				0
Change (mol)	−0.00375		−0.00375				+0.00375
After reaction (mol)	0.00375		0				0.00375

Concentration after reaction (M) $\quad [C_6H_5CO_2H] = \dfrac{0.00375\ \text{mol}}{0.1125\ \text{L}} = 0.0333\ \text{M}$

$$[C_6H_5CO_2^-] = \dfrac{0.00375\ \text{mol}}{0.1125\ \text{L}} = 0.0333\ \text{M}$$

Notice that at the half-equivalence point, [weak acid] = [conjugate base]. As a result, $[H_3O^+] = K_a$ and $pH = pK_a$.

Use the rearranged K_a expression (or the Henderson–Hasselbalch equation) to calculate pH.

$$[H_3O^+] = K_a \frac{[C_6H_5CO_2H]}{[C_6H_5CO_2^-]} = 6.3 \times 10^{-5}\left(\frac{0.0333}{0.0333}\right) = 6.3 \times 10^{-5}\ \text{M}$$

$$pH = 4.20$$

d. Because of the 1:1 reaction stoichiometry and the equal concentrations of acid and base, the equivalence point is reached when an equal volume (75.0 mL) of NaOH is added to the weak acid solution. All of the acid in the flask and the added base are consumed, forming the water and the conjugate base, $C_6H_5CO_2^-$. The solution now contains a weak base and the pH can be calculated using the methods introduced in Acids and Bases (Unit 16):

$$\text{mol } CH_3CO_2H = \text{mol } OH^- = (0.0750\ \text{L})(0.100\ \text{mol/L}) = 0.00750\ \text{mol } CH_3CO_2H$$

total volume of solution = 75.0 mL + 75.0 mL = 150.0 mL

	$C_6H_5CO_2H(aq)$	$+$	$OH^-(aq)$	$\rightarrow$	$H_2O(\ell)$	$+$	$C_6H_5CO_2^-(aq)$
Initial (mol)	0.00750		0.00750				0
Change (mol)	−0.00750		−0.00750				+0.00750
After reaction (mol)	0		0				0.00750

Concentration after reaction (M) $\quad [C_6H_5CO_2^-] = \dfrac{0.00750\ \text{mol}}{0.1500\ \text{L}} = 0.0500\ \text{M}$

$$C_6H_5CO_2^-(aq) + H_2O(\ell) \rightleftarrows C_6H_5CO_2H(aq) + OH^-(aq)$$

$$K_b(C_6H_5CO_2^-) = \frac{K_w}{K_a} = \frac{1.00 \times 10^{-14}}{6.3 \times 10^{-5}} = 1.6 \times 10^{-10}$$

$$[OH^-] = \sqrt{K_b \times [C_6H_5CO_2^-]_0} = \sqrt{(1.6 \times 10^{-10})(0.0500)} = 2.8 \times 10^{-6}\ \text{M}$$

$$pOH = -\log(2.8 \times 10^{-6}) = 5.55$$

$$pH = 14.00 - pOH = 8.45$$

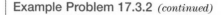

Example Problem 17.3.2 (continued)

e. After the equivalence point, all of the weak acid has been consumed and the solution contains the conjugate base and excess NaOH. The benzoate ion is a much weaker base than NaOH, so the pH is controlled by the hydroxide ion concentration in solution.

$$\text{mol } CH_3CO_2H = (0.0750 \text{ L})(0.100 \text{ mol/L}) = 0.00750 \text{ mol } CH_3CO_2H$$

$$\text{mol } OH^- = (0.1000 \text{ L})(0.100 \text{ mol/L}) = 0.01000 \text{ mol } OH^-$$

$$\text{total volume of solution} = 75.0 \text{ mL} + 100.0 \text{ mL} = 175.0 \text{ mL}$$

	$C_6H_5CO_2H(aq)$	+	$OH^-(aq)$	$\rightarrow$	$H_2O(\ell)$	+	$C_6H_5CO_2^-(aq)$
Initial (mol)	0.00750		0.01000				0
Change (mol)	−0.00750		−0.00750				+0.00750
After reaction (mol)	0		0.00250				0.00750

$$\textit{Concentration after reaction} \text{ (M)} \quad [OH^-] = \frac{0.00250 \text{ mol}}{0.1750 \text{ L}} = 0.0143 \text{ M}$$

$$pOH = -\log(0.0143) = 1.845$$

$$pH = 14.000 - 1.845 = 12.155$$

Video Solution

Tutored Practice Problem 17.3.2

The shape of the titration plot for an ammonia/hydrochloric acid titration (Interactive Figure 17.3.4) is very similar to that of a weak acid/strong base titration plot. The pH plot for the titration of a weak base with a strong acid has four regions of interest:

1. The initial pH is greater than 7 because the solution contains a base.

2. As the strong acid is added, the pH drops sharply and then decreases gradually until the equivalence point is reached. This is the buffer region of the titration, where the solution contains a weak acid and its conjugate base. At the midpoint of this region, half of the base originally present in the flask has been consumed and the half-equivalence point, also called the titration midpoint, is reached. As shown in the example problem that follows, at this point in the titration, the solution pOH is equal to the weak base pK_b.

3. Near the equivalence point, where the base is completely consumed by added acid, the pH decreases rapidly. The pH at the equivalence point of the titration is less than 7 because of the presence of the ammonium ion, an acidic cation.

4. After the equivalence point, the pH is low because the solution contains excess strong acid.

Interactive Figure 17.3.4

Explore a weak base/strong acid pH titration plot.

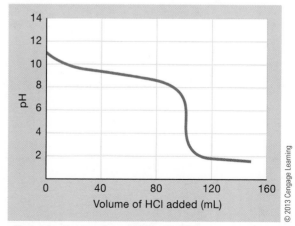

pH plot for the titration of NH_3 with HCl

The calculation of solution pH for a weak base/strong acid titration is very similar to the calculations used in the weak acid/strong base titration, as shown in the following example problem.

Example Problem 17.3.3 Calculate pH for a weak base/strong acid titration.

Determine the pH during the titration of 75.0 mL of 0.100 M ammonia ($K_b = 1.8 \times 10^{-5}$) by 0.100 M HCl at the following points.

a. Before the addition of any HCl
b. After the addition of 20.0 mL of HCl
c. At the titration midpoint
d. At the equivalence point
e. After the addition of 100.0 mL of HCl

Solution:

You are asked to calculate the pH at five different points in a weak base/strong acid titration.

You are given the concentration and volume of the weak base and the concentration of the strong acid.

Write the balanced net ionic equation for the acid–base reaction.

$$NH_3(aq) + H_3O^+(aq) \rightarrow H_2O(\ell) + NH_4^+(aq)$$

a. Ammonia is a weak base, and the pH of the solution is calculated using methods introduced in Acids and Bases (Unit 16):

$$[OH^-] = \sqrt{K_b \times [NH_3]_0} = \sqrt{(1.8 \times 10^{-5})(0.100)} = 0.0013 \text{ M}$$
$$pOH = -\log(0.0013) = 2.87$$
$$pH = 14.00 - pOH = 11.13$$

b. Before the equivalence point, all HCl is consumed by excess weak base. First consider the stoichiometry of the acid–base reaction; then, because the solution contains a buffer, calculate pH using the rearranged K_a expression or the Henderson–Hasselbalch equation.

$$\text{mol } NH_3 = (0.0750 \text{ L})(0.100 \text{ mol/L}) = 0.00750 \text{ mol } NH_3$$
$$\text{mol } H_3O^+ = (0.0200 \text{ L})(0.100 \text{ mol/L}) = 0.00200 \text{ mol } H_3O^+$$
$$\text{total volume of solution} = 75.0 \text{ mL} + 20.0 \text{ mL} = 95.0 \text{ mL}$$

	$NH_3(aq)$	$+$	$H_3O^+(aq)$	$\rightarrow$	$H_2O(\ell)$	$+$	$NH_4^+(aq)$
Initial (mol)	0.00750		0.00200				0
Change (mol)	−0.00200		−0.00200				+0.00200
After reaction (mol)	0.00550		0				0.00200

Example Problem 17.3.3 *(continued)*

Concentration after reaction (M)　　$[NH_3] = \dfrac{0.00550 \text{ mol}}{0.0950 \text{ L}} = 0.0579 \text{ M}$

$$[NH_4^+] = \dfrac{0.00200 \text{ mol}}{0.0950 \text{ L}} = 0.0211 \text{ M}$$

Use the rearranged K_b expression (or the Henderson–Hasselbalch equation) to calculate pH.

$$[OH^-] = K_b\dfrac{[NH_3]}{[NH_4^+]} = 1.8 \times 10^{-5}\left(\dfrac{0.0579}{0.0211}\right) = 4.9 \times 10^{-5} \text{ M}$$

$$pOH = -\log(4.9 \times 10^{-5}) = 4.31$$

$$pH = 14.00 - pOH = 9.69$$

c. The half-equivalence point, the titration midpoint, is reached when half the amount of strong acid required to reach the equivalence point has been added to the weak base solution and, as a result, half of the weak base originally in the flask has been consumed. As shown in the weak acid/strong base titration calculations, at the midpoint in the titration, [weak base] = [conjugate acid]. As a result, $[OH^-] = K_b$ and $pOH = pK_b$.

$$[OH^-] = K_b\dfrac{[NH_3]}{[NH_4^+]} = 1.8 \times 10^{-5} \text{ M}$$

$$pOH = -\log(1.8 \times 10^{-5}) = 4.74$$

$$pH = 14.00 - pOH = 9.26$$

　　Alternately, the Henderson–Hasselbalch equation shows that at the half-equivalence point in a weak base/strong acid titration, the solution pH is equal to the conjugate acid pK_a $[K_a(NH_4^+) = 5.6 \times 10^{-10}]$.

$$pH = pK_a + \log\dfrac{[NH_3]}{[NH_4^+]} = -\log(5.6 \times 10^{-10}) + \log(1) = 9.26$$

d. Because of the 1:1 reaction stoichiometry and the equal concentrations of acid and base, the equivalence point is reached when an equal volume (75.0 mL) of HCl is added to the weak base solution. All acid and base are consumed, forming the water and the conjugate acid, NH_4^+. The solution now contains a weak acid and the pH can be calculated using the methods introduced in Acids and Bases (Unit 16):

$$\text{mol } NH_3 = \text{mol } H_3O^+ = (0.0750 \text{ L})(0.100 \text{ mol/L}) = 0.00750 \text{ mol } NH_3$$

total volume of solution = 75.0 mL + 75.0 mL = 150.0 mL

Example Problem 17.3.3 *(continued)*

	$NH_3(aq)$	$+$	$H_3O^+(aq)$	$\rightarrow$	$H_2O(\ell)$	$+$	$NH_4^+(aq)$
Initial (mol)	0.00750		0.00750				0
Change (mol)	−0.00750		−0.00750				+0.00750
After reaction (mol)	0		0				0.00750

Concentration after reaction (M) $[NH_4^+] = \dfrac{0.00750 \text{ mol}}{0.1500 \text{ L}} = 0.0500 \text{ M}$

$$NH_4^+(aq) + H_2O(\ell) \rightleftarrows NH_3(aq) + H_3O^+(aq)$$

$$K_a(NH_4^+) = \frac{K_w}{K_b} = \frac{1.00 \times 10^{-14}}{1.8 \times 10^{-5}} = 5.6 \times 10^{-10}$$

$$[H_3O^+] = \sqrt{K_a \times [NH_4^+]_0} = \sqrt{(5.6 \times 10^{-10})(0.0500)} = 5.3 \times 10^{-6} \text{ M}$$

$$pH = -\log(5.3 \times 10^{-6}) = 5.28$$

e. After the equivalence point, all of the weak base has been consumed and the solution contains the conjugate acid and excess HCl. The ammonium ion is a much weaker acid than HCl, so the pH is controlled by the concentration of H_3O^+ in solution.

$$\text{mol } NH_3 = (0.0750 \text{ L})(0.100 \text{ mol/L}) = 0.00750 \text{ mol } NH_3$$

$$\text{mol } H_3O^+ = (0.1000 \text{ L})(0.100 \text{ mol/L}) = 0.01000 \text{ mol } H_3O^+$$

$$\text{total volume of solution} = 75.0 \text{ mL} + 100.0 \text{ mL} = 175.0 \text{ mL}$$

	$NH_3(aq)$	$+$	$H_3O^+(aq)$	$\rightarrow$	$H_2O(\ell)$	$+$	$NH_4^+(aq)$
Initial (mol)	0.00750		0.01000				0
Change (mol)	−0.00750		−0.00750				+0.00750
After reaction (mol)	0		0.00250				0.00750

Concentration after reaction (M) $[H_3O^+] = \dfrac{0.00250 \text{ mol}}{0.1750 \text{ L}} = 0.0143 \text{ M}$

$$pH = -\log(0.0143) = 1.845$$

Video Solution

**Tutored Practice
Problem 17.3.3**

17.3c pH Titration Plots as an Indicator of Acid or Base Strength

We have examined and explained the important features of pH titration plots. Titration plots can be used to determine the species being titrated, the relative strength of the acid (or base) being titrated, and the K_a value for the weak acid (or K_b value for the weak base) being titrated.

1. The initial pH is an indicator of the species being titrated.

- A strong or weak acid solution will have a pH < 7 before the addition of base.

- A strong or weak base solution will have a pH > 7 before the addition of acid.

2. The relative strength of the acid or base being titrated can be determined by the pH at the equivalence point of the titration:

- If pH = 7 at the equivalence point, the acid (or base) being titrated is strong. At the equivalence point in a strong acid (or strong base) titration, the solution contains water and a pH-neutral salt.

- If pH > 7 at the equivalence point, the acid being titrated is weak. The pH is greater than 7 because at the equivalence point in a weak acid/strong base titration, the weak acid is consumed, forming the conjugate base. The relative strength of the weak acid cannot be determined from a pH titration plot because it also depends on the initial acid concentration and the concentration of the titrant.

- If pH < 7 at the equivalence point, the base being titrated is weak. The pH is less than 7 because at the equivalence point in a weak base/strong acid titration, the weak base is consumed, forming the conjugate acid. The relative strength of the weak base cannot be determined from a pH titration plot because it also depends on the initial base concentration and the concentration of the titrant.

3. The K_a for a weak acid (or K_b for a weak base) can be determined from the pH at the half-equivalence point in an acid–base titration.

- In a weak acid/strong base titration, [weak acid] = [conjugate base] at the half-equivalence point (titration midpoint) and pH = pK_a for the weak acid. Note that if the weak acid is relatively strong ($K_a > 10^{-3}$) or if the solution is very dilute ([weak acid] < 10^{-3} M), the pH at the titration midpoint will vary slightly from the acid pK_a.

- In a weak base/strong acid titration, the pH at the half-equivalence point (titration midpoint) is equal to the pK_a of the conjugate acid. Because $K_aK_b = K_w$ for any acid–base conjugate pair, the K_b for the weak base can then be calculated.

Example Problem 17.3.4 Interpret pH titration plots.

Answer the following about the titration plot shown below.

a. Is the species being titrated an acid or a base?
b. Is the species being titrated strong or weak?
c. What is the value of K_a or K_b for the species being titrated?

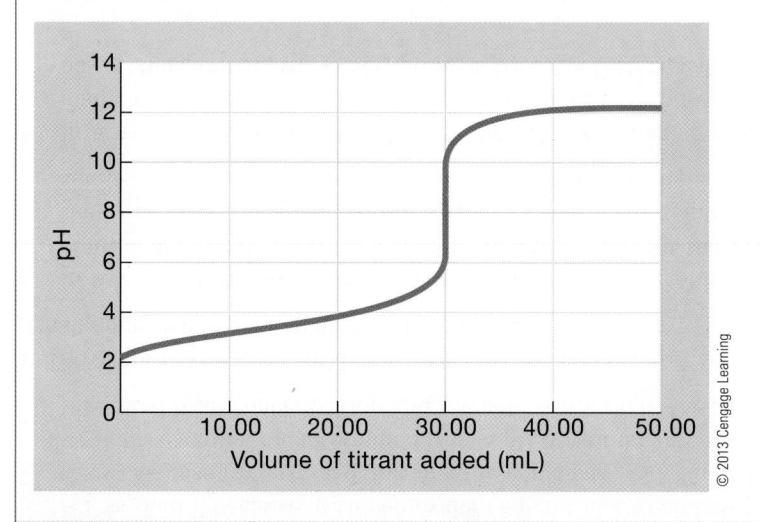

© 2013 Cengage Learning

Solution:

You are asked to use a titration plot to determine information about an acid–base titration.

You are given a pH plot for an acid–base titration.

a. The species titrated is an acid because the initial pH is less than 7.
b. The species titrated is a weak acid. The pH at the equivalence point in the titration is greater than 7 (basic) because of the presence of the conjugate base of the weak acid.
c. The equivalence point in the titration occurs when 30 mL of base has been added. At the half-equivalence point (after 15 mL of base is added), the pH (which is equal to pK_a) is approximately 3.7. The K_a for the weak acid is $10^{-3.7}$ or 2×10^{-4}.

Video Solution

Tutored Practice
Problem 17.3.4

17.3d pH Indicators

The pH of an acidic or basic solution can be determined using an acid–base indicator or a pH meter. An **acid–base indicator** is a weak organic acid that can be used to indicate the pH of a solution because the acid form of the indicator has a different color than its conjugate base and because the color change takes place over a relatively narrow pH range. Some common acid–base indicators are shown in Interactive Figure 17.3.5.

Indicators are chosen for acid–base titrations based on the pH at the titration equivalence point, which is determined by the substance being titrated. The indicator color change is visible over a pH range given approximately by indicator $pK_a \pm 1$. A color change in the indicator is intended to signal that the equivalence point in the titration has been reached. To ensure that this will happen, the pK_a of the indicator should be as close as possible to the pH at the equivalence point of the titration.

In the titration of a strong acid with a strong base, the pH change is so large in the immediate vicinity of the equivalence point that a variety of indicators can be used successfully. A common indicator chosen for strong acid/strong base titrations (pH = 7.00 at the equivalence point) is phenolphthalein, but as shown in Figure 17.3.6, bromothymol blue or methyl red could also be used.

When weak acids and/or bases are involved, the pH range in the immediate vicinity of the equivalence point is smaller and the choice of indicator is more critical. The benzoic acid/sodium hydroxide titration described earlier has a pH of 8.45 at the equivalence point. The indicator cresol red could be used to detect the equivalence point of this titration (Figure 17.3.7).

Interactive Figure 17.3.5

Explore acid–base indicators.

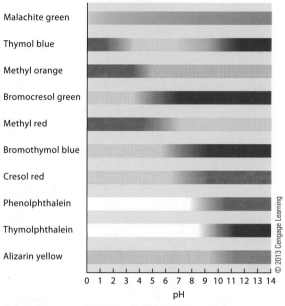

Color range as a function of pH for some acid–base indicators

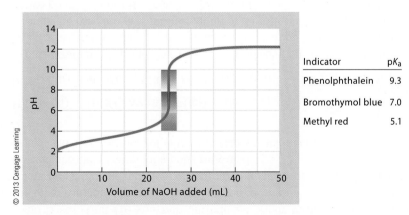

Indicator	pK_a
Phenolphthalein	9.3
Bromothymol blue	7.0
Methyl red	5.1

Figure 17.3.6 Indicator choices for a strong acid/ strong base titration

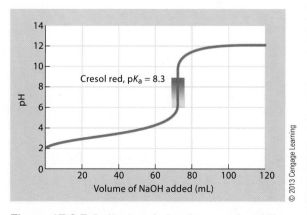

Figure 17.3.7 Indicator choice for a weak acid/ weak base titration

17.3e Polyprotic Acid Titrations

In Acids and Bases (Unit 16) you were introduced briefly to polyprotic acids, Brønsted–Lowry acids that can donate more than one proton to a base. We now consider this class of acids in more detail by studying the concentration of all species in a solution containing a polyprotic acid and the titration of a polyprotic acid.

Carbonic acid is an example of a diprotic acid, an acid that can donate two protons to a base.

$$H_2CO_3(aq) + H_2O(\ell) \rightleftarrows HCO_3^-(aq) + H_3O^+(aq) \qquad K_{a1} = 4.2 \times 10^{-7}$$

$$HCO_3^-(aq) + H_2O(\ell) \rightleftarrows CO_3^{2-}(aq) + H_3O^+(aq) \qquad K_{a2} = 4.8 \times 10^{-11}$$

The concentrations of species in a polyprotic acid solution can be shown using an alpha plot. Recall that when the solution pH is equal to the weak acid pK_a, equal amounts of weak acid and conjugate base are present in solution. A diprotic acid has two K_a values and thus there are two points where pH is equal to pK_a (at pK_{a1} and pK_{a2}). In the alpha plot for the carbonic acid/bicarbonate ion/carbonate ion system (Figure 17.3.8), there are two points where $\alpha = 0.5$ and pH $= pK_a$. In addition, notice that

- at pH values below pK_{a1}, the system contains mostly H_2CO_3;

- at intermediate pH (between pK_{a1} and pK_{a2}), the system contains mostly HCO_3^-; and

- at high pH, above pK_{a2}, the system contains mostly CO_3^{2-}.

The concentrations of species in a polyprotic acid solution can be calculated using the same techniques and assumptions used when considering buffer solutions. The approach involves identifying the species present in significant quantities in the solution, then determining the reaction that represents the predominant equilibrium in solution, and finally solving the equilibrium system using an *ICE* table.

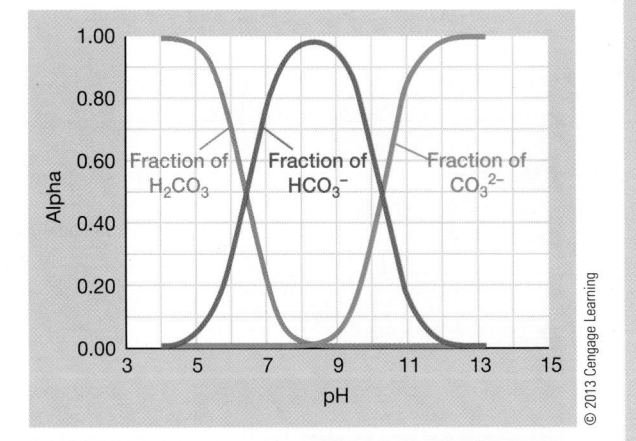

Figure 17.3.8 The alpha plot for carbonic acid

Example Problem 17.3.5 Calculate the pH and concentration of species present in a polyprotic acid solution.

For a 0.20 M solution of H_2CO_3, calculate both the pH and the carbonate ion concentration.

$$H_2CO_3(aq) + H_2O(\ell) \rightleftarrows HCO_3^-(aq) + H_3O^+(aq) \qquad K_{a1} = 4.2 \times 10^{-7}$$

$$HCO_3^-(aq) + H_2O(\ell) \rightleftarrows CO_3^{2-}(aq) + H_3O^+(aq) \qquad K_{a2} = 4.8 \times 10^{-11}$$

Example Problem 17.3.5 *(continued)*

Solution:

You are asked to calculate the pH and carbonate ion concentration in a solution of carbonic acid.

You are given the concentration of the carbonic acid solution.

Step 1. When calculating pH, recognize that because K_{a1} is much greater than K_{a2} the predominant species in solution are H_2CO_3, HCO_3^-, and H_3O^+.

$$H_2CO_3(aq) + H_2O(\ell) \rightleftarrows HCO_3^-(aq) + H_3O^+(aq) \qquad K_{a1} = 4.2 \times 10^{-7}$$

Step 2. Set up an *ICE* table for the first acid hydrolysis.

	$H_2CO_3(aq)$	+	$H_2O(\ell)$	$\rightleftarrows$	$HCO_3^-(aq)$	+	$H_3O^+(aq)$
Initial (M)	0.20				0		0
Change (M)	$-x$				$+x$		$+x$
Equilibrium (M)	$0.20 - x$				x		x

Step 3. Substitute these equilibrium concentrations into the K_{a1} expression and calculate the pH of the solution.

$$K_{a1} = 4.2 \times 10^{-7} = \frac{[HCO_3^-][H_3O^+]}{[H_2CO_3]} = \frac{(x)(x)}{0.20-x} = \frac{x^2}{0.20}$$

Because the value of K_{a1} is small compared with the initial acid concentration, it is reasonable to assume that the amount of weak acid ionized (x) is very small compared with $[H_2CO_3]_0$.

$$x = [H_3O^+] = \sqrt{K_{a1} \times 0.20} = 2.9 \times 10^{-4} \, M$$

$$pH = -\log(2.9 \times 10^{-4}) = 3.54$$

Notice that in the carbonic acid alpha plot (Figure 17.3.8), at a pH of about 3.5 the predominant species in solution is H_2CO_3. It is reasonable to assume that the second ionization is not important in determining the pH of this solution.

Step 4. Use the equation for the second ionization and the pH of the solution to determine the carbonate ion concentration.

$$HCO_3^-(aq) + H_2O(\ell) \rightleftarrows CO_3^{2-}(aq) + H_3O^+(aq) \qquad K_{a2} = 4.8 \times 10^{-11}$$

Step 5. Set up an *ICE* table for the second acid hydrolysis.

	$HCO_3^-(aq)$	+	$H_2O(\ell)$	$\rightleftarrows$	$CO_3^{2-}(aq)$	+	$H_3O^+(aq)$
Initial (M)	2.9×10^{-4}				0		2.9×10^{-4}
Change (M)	$-x$				$+x$		$+x$
Equilibrium (M)	$2.9 \times 10^{-4} - x$				x		$2.9 \times 10^{-4} + x$

Example Problem 17.3.5 *(continued)*

Step 6. Substitute these equilibrium concentrations into the K_{a2} expression and solve for x, the carbonate ion concentration.

$$K_{a2} = 4.8 \times 10^{-11} = \frac{[CO_3^{2-}][H_3O^+]}{[HCO_3^-]} = \frac{(x)(2.9 \times 10^{-4} + x)}{2.9 \times 10^{-4} - x} = \frac{x(2.9 \times 10^{-4})}{2.9 \times 10^{-4}}$$

The value of K_{a2} is very small compared with $[H_3O^+]$ and $[HCO_3^-]$, so it is reasonable to assume that x is very small compared with $[HCO_3^-]_0$.

$$x = [CO_3^{2-}] = K_{a2} = 4.8 \times 10^{-11} \text{ M}$$

Notice that the carbonate ion concentration is equal to the K_{a2} value for this weak acid.

Video Solution

**Tutored Practice
Problem 17.3.5**

Polyprotic Acid Titration Plots

The pH plot for the titration of the diprotic acid maleic acid, $HO_2C(CH)_2CO_2H$, with a strong base is shown in Figure 17.3.9. The plot shows two distinct equivalence points where the two protons are removed by reaction with OH^-.

$$HO_2C(CH)_2CO_2H(aq) + OH^-(aq) \rightarrow H_2O(\ell) + HO_2C(CH)_2CO_2^-(aq)$$

$$HO_2C(CH)_2CO_2^-(aq) + OH^-(aq) \rightarrow H_2O(\ell) + {}^-O_2C(CH)_2CO_2^-(aq)$$

Although the presence of two equivalence points in Figure 17.3.9 clearly indicates that this is the titration of a diprotic acid, there are cases when a polyprotic acid titration plot does not show all of the possible equivalence points. Interactive Figure 17.3.10 shows a series of calculated pH plots for the titration of two weak diprotic acids that have the same pK_{a1} value

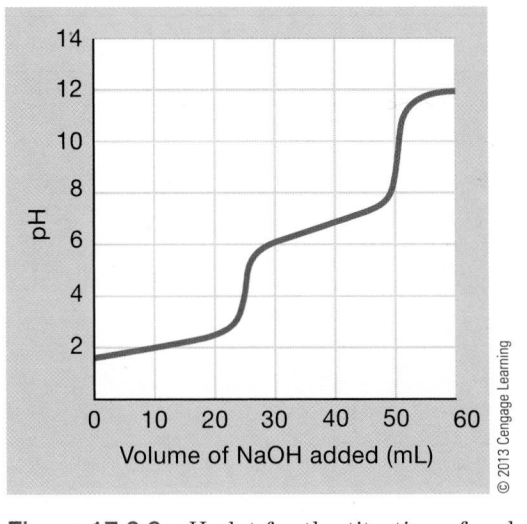

Figure 17.3.9 pH plot for the titration of maleic acid with NaOH

Interactive Figure 17.3.10

Explore alpha plots and pH plots for diprotic acids.

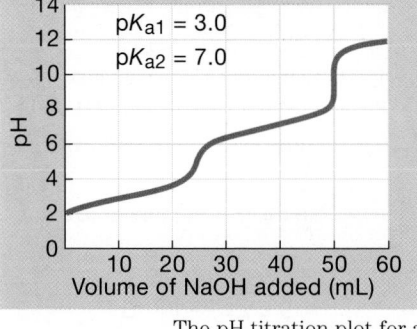 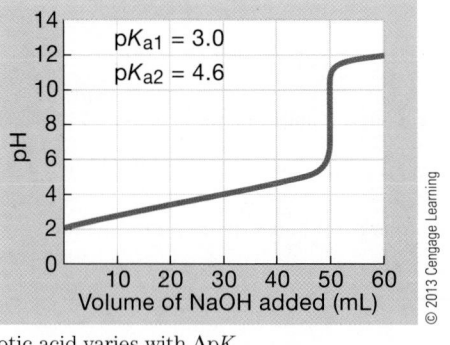

The pH titration plot for a diprotic acid varies with ΔpK_a.

($pK_{a1} = 3$) and different pK_{a2} values. When the difference between pK_{a1} and pK_{a2} is greater than or equal to 3, two distinct equivalence points are observed in the titration plot. However, when the two values are relatively similar ($\Delta pK_a < 3$), only a single equivalence point is observed. Thus, it is important to remember that the presence of a single equivalence point in a titration plot does not necessarily indicate the presence of a monoprotic acid. It could be due to the titration of a diprotic acid where K_{a1} and K_{a2} values are similar.

Section 17.3 Mastery

17.4 Some Important Acid–Base Systems

17.4a The Carbonate System: $H_2CO_3/HCO_3^-/CO_3^{2-}$

The carbonic acid/bicarbonate ion/carbonate ion system is the principal buffer system in our blood. The alpha plot for this system is shown in Interactive Figure 17.4.1. The pH of blood is typically about 7.4. The alpha plot shows that, at this pH, the concentration of carbonate ion, CO_3^{2-}, is very low in the blood and that the concentrations of both H_2CO_3 and HCO_3^- are significant ($[HCO_3^-] >> [H_2CO_3]$).

Blood pH is regulated in part by CO_2 respiration. In the body, CO_2 reacts with water to form carbonic acid, a diprotic acid.

$$CO_2(g) + H_2O(\ell) \rightleftharpoons H_2CO_3(aq)$$

$$H_2CO_3(aq) + H_2O(\ell) \rightleftharpoons HCO_3^-(aq) + H_3O^+(aq) \qquad K_{a1} = 4.2 \times 10^{-7}$$

$$HCO_3^-(aq) + H_2O(\ell) \rightleftharpoons CO_3^{2-}(aq) + H_3O^+(aq) \qquad K_{a2} = 4.8 \times 10^{-11}$$

Under conditions of respiratory acidosis, there is an excess of acid in body fluids. The equilibrium system lies to the left on the alpha plot. Respiratory acidosis can be treated by exhaling large amounts of CO_2. This will decrease the concentration of H_2CO_3 in the blood, shifting the buffer position to the right on the alpha plot and increasing blood pH.

When CO_2 is lost from the body (when you exhale or hyperventilate), the concentration of H_2CO_3 in the blood decreases. This represents a shift to the right on the alpha plot, to higher pH. Excess hyperventilation can cause respiratory alkalosis, an excess of base in body fluids. Breathing into a paper bag can increase the CO_2 levels in the body, increasing $[H_2CO_3]$ and decreasing the pH of body fluids such as blood.

Interactive Figure 17.4.1

Explore the carbonate system alpha plot.

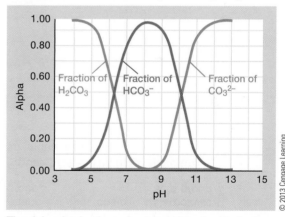

The alpha plot for the carbonic acid/bicarbonate ion/carbonate ion buffer system

17.4b Amino Acids

Amino acids are the building blocks of proteins in our bodies. There are a large number of amino acids known, and 21 are essential for human life. All have the basic structure shown in Figure 17.4.2.

Amino acids contain both a carboxylic acid group ($-CO_2H$) and an amine group ($-NH_2$). The central, sp^3-hybridized carbon, called the alpha carbon, is bonded to the carboxylic acid group, the amino group, a hydrogen atom, and a fourth group, a side chain molecular fragment commonly labeled R. The R group is different for each different amino acid and can be pH neutral, acidic, or basic. Two of the simplest amino acids are glycine (R = H) and alanine (R = CH_3).

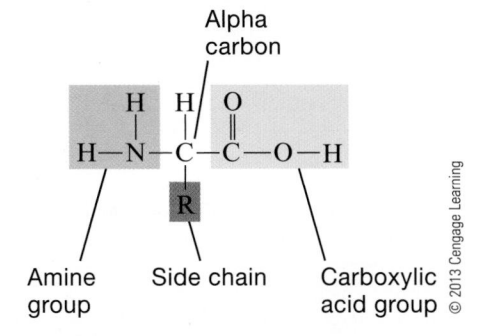

Figure 17.4.2 The basic structure of an amino acid

Glycine

Alanine

Although amino acids are often written as shown above, with $-CO_2H$ and $-NH_2$ groups, this is not a completely accurate representation of their molecular structure. Amino acids exist as **zwitterions**, compounds containing both a positive and negative charge, when dissolved in water, in bodily fluids, and in the solid state. Increasing the pH of a solution containing an amino acid deprotonates the zwitterion, and decreasing the pH protonates the zwitterion.

pH decreases   pH increases

Acidic form   Zwitterion   Basic form

The pH at which an amino acid has equal numbers of positive and negative charges is called the **isoelectric point** (pI). Each different amino acid has a unique isoelectric point. Most of the amino acids have pI values near 6, with the exception of the amino acids with acidic side groups (lower pI values) and the amino acids with basic side groups (higher pI values).

Zwitterionic amino acids can act as buffers, absorbing acid or base to produce the fully protonated or fully deprotonated form, respectively. Under physiological pH conditions, amino acids exist almost exclusively in the zwitterionic form, as shown by the alpha plot for glycine (Interactive Figure 17.4.3).

Unit Recap

Key Concepts

17.1 Acid–Base Reactions

- The reaction between a strong acid and a strong base has a large equilibrium constant and produces water and a pH-neutral salt (17.1a).

- The reaction between a strong acid and a weak base has a large equilibrium constant and produces water and an acidic solution because of the presence of the conjugate acid of the weak base (17.1b).

- The reaction between a strong base and a weak acid has a large equilibrium constant and produces water and a basic solution because of the presence of the conjugate base of the weak acid (17.1b).

- In general, all acid–base reactions favor the direction where a stronger acid and base react to form a weaker acid and base (17.1b).

- The reaction between a weak acid and a weak base has a small equilibrium constant and produces water and a solution whose pH is dependent on the nature of the predominant species in solution (17.1b).

17.2 Buffers

- A buffer solution contains a mixture of a weak acid and a weak base, typically the conjugate base of the weak acid (17.2a).

- The Henderson–Hasselbalch equation is one method used to calculate the pH of a buffer solution (17.2b).

- In a buffer solution, when [weak acid] = [conjugate base], the pH is equal to the weak acid pK_a (17.2b).

- The most effective buffers contain significant amounts of weak acid and conjugate base and have a pH equal to the weak acid $pK_a \pm 1$ (17.2b).

- Buffer capacity is the amount of strong acid or base that can be added to a buffer without a drastic change in pH (17.2b).

- Two methods for preparing a buffer are the direct addition method and the acid–base reaction method (17.2c).

Explore the relationship between amino acid structure and pH.

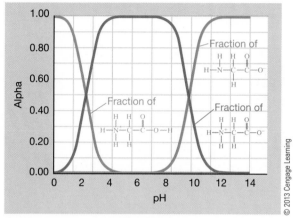

The alpha plot for the amino acid glycine

Section 17.4 Mastery

- The composition of a buffer solution as a function of pH is shown graphically in an alpha (α) plot (17.2c).

17.3 Acid–Base Titrations

- The pH at the equivalence point of a strong acid/strong base titration is equal to 7 (17.3a).
- The pH at the equivalence point of a weak acid/strong base titration is greater than 7 because of the presence of the conjugate base of the weak acid (17.3b).
- After the addition of titrant but before the equivalence point in a weak acid/strong base or weak base/strong acid titration, the solution contains a buffer (17.3b).
- The half-equivalence point in a titration is when half of the species being titrated has been consumed (17.3b).
- In a weak acid/strong base titration, pH = pK_a at the half-equivalence point (17.3b).
- The pH at the equivalence point of a weak base/strong acid titration is less than 7 because of the presence of the conjugate acid of the weak base (17.3b).
- In a weak base/strong acid titration, pOH = pK_b at the half-equivalence point (17.3b).
- Titration plots can be used to determine the species being titrated, the relative strength of the acid (or base) being titrated, and the K_a value for the weak acid (or K_b value for the weak base) being titrated (17.3c).
- The pK_a of an acid–base indicator should be as close as possible to the pH at the equivalence point of the titration (17.3d).
- Polyprotic acid titration plots show more than one equivalence point if there is a large difference between the pK_a values for the acid ($\Delta pK_a > 3$) (17.3e).

17.4 Some Important Acid–Base Systems

- The carbonic acid/bicarbonate ion/carbonate ion system is the principal buffer system in our blood (17.4a).
- Amino acids, compounds containing both a carboxylic acid group ($-CO_2H$) and an amine group ($-NH_2$), exist as zwitterions, compounds containing both a positive and negative charge, when dissolved in water, in bodily fluids, and in the solid state (17.4b).
- The pH at which an amino acid has equal numbers of positive and negative charges is called the isoelectric point (17.4b).

Key Equations

$$\text{pH} = \text{p}K_a + \log \frac{[\text{conjugate base}]}{[\text{weak acid}]} \qquad \textbf{(17.1)}$$

Key Terms

17.2 Buffers
buffer solution
common ion effect
Henderson–Hasselbalch equation
buffer capacity
alpha (α) plot

17.3 Acid–Base Titrations
titrant
half-equivalence point
acid–base indicator

17.4 Some Important Acid–Base Systems
amino acid
zwitterion
isoelectric point

Unit 17   Review and Challenge Problems

18 Precipitation and Lewis Acid–Base Equilibria

Unit Outline

18.1 Solubility Equilibria and K_{sp}
18.2 Using K_{sp} in Calculations
18.3 Lewis Acid–Base Complexes and Complex Ion Equilibria
18.4 Simultaneous Equilibria

In This Unit...

This is the third unit in our study of chemical equilibria. After studying acid–base equilibria in some depth, we now turn to equilibria involving sparingly soluble compounds and the equilibria of Lewis acid–base complexes. You were first introduced to sparingly soluble compounds in Chemical Reactions and Solution Stoichiometry (Unit 4), when we covered precipitation reactions and the solubility of ionic compounds. We will discover in this unit that even insoluble ionic compounds dissolve in water to a small extent and that this solubility can be affected by a variety of chemical species, including Lewis bases. Lewis acids and bases were briefly introduced in Acids and Bases (Unit 16). In this unit we take a closer look at equilibria involving Lewis acid–base complexes and how they can be used to influence the solubility of ionic compounds.

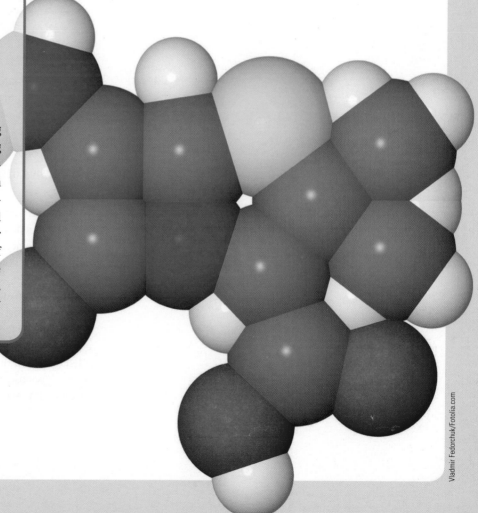

18.1 Solubility Equilibria and K_{sp}

18.1a Solubility Units

This unit is about the chemistry of binding ions and molecules together. The tendency of different species to bind together and our ability to control these binding processes is important. For example, biological systems are rife with examples of precipitation chemistry. The formation of seashells and coral involve precipitation of calcium carbonate, $CaCO_3$. The formation of caves and the interesting structures within them is the result of a combination of precipitation reactions coupled with Lewis acid–base reactions and Brønsted–Lowry acid–base reactions. We begin our study by introducing methods that allow us to quantitatively study the solubility of sparingly soluble compounds.

Ionic compounds we labeled as "insoluble in water" in Chemical Reactions and Solution Stoichiometry (Unit 4) actually dissolve in water to a small extent. The solubility of an ionic compound is determined by measuring the amount of a solid that dissolves in a quantity of water. Solubility values are reported in g/100 mL, g/L, or mol/L (also called molar solubility).

$$\text{Solubility of AgCl} = 1.9 \times 10^{-4} \text{ g/100 mL} = 1.9 \times 10^{-3} \text{ g/L} = 1.3 \times 10^{-5} \text{ mol/L}$$

In Chemical Reactions and Solution Stoichiometry (Unit 4) you learned how to predict the solubility of an ionic compound based on a set of solubility rules. (◄Flashback to Section 4.2b Solubility of Ionic Compounds) These solubility rules are based on the measured solubility of ionic compounds, where an insoluble compound is defined as having a solubility less than about 0.01 mol/L. In this unit we will work with experimental solubility values to more accurately describe the solubility of ionic compounds.

Example Problem 18.1.1 Interconvert solubility units.

The molar solubility of silver sulfate is 0.0144 mol/L. Express the solubility in units of g/L, and calculate the concentration of Ag^+ in a saturated silver sulfate solution.

Solution:

You are asked to convert between solubility units.

You are given the solubility of an ionic compound.

Use the molar mass of silver sulfate, Ag_2SO_4, to convert between solubility units.

$$\frac{0.0144 \text{ mol Ag}_2\text{SO}_4}{1 \text{L}} \times \frac{311.8 \text{ g}}{1 \text{ mol Ag}_2\text{SO}_4} = 4.49 \text{ g/L}$$

Example Problem 18.1.1 *(continued)*

Use the molar solubility to calculate the Ag^+ concentration in a saturated Ag_2SO_4 solution.

$$\frac{0.0144 \text{ mol } Ag_2SO_4}{1L} \times \frac{2 \text{ mol } Ag^+}{1 \text{ mol } Ag_2SO_4} = 0.0288 \text{ M } Ag^+$$

Video Solution

Tutored Practice
Problem 18.1.1

18.1b The Solubility Product Constant

In Chemical Mixtures: Solutions and Other Mixtures (Unit 13) we described a solution as saturated when no additional solid could be dissolved in a solvent. In such a solution, a dynamic equilibrium occurs between the hydrated ions and the undissolved solid. For $PbCl_2$, for example, the equilibrium process is represented as

$$PbCl_2(s) \rightleftarrows Pb^{2+}(aq) + 2\ Cl^-(aq) \qquad K = [Pb^{2+}][Cl^-]^2$$

Notice that the equilibrium is written as a dissolution process (solid as a reactant and aqueous ions as products) and that the pure solid does not appear in the equilibrium constant expression. Because the equilibrium constant expression for dissolution reactions is always expressed as the product of the ion concentrations, the equilibrium constant is given the special name **solubility product constant** and the symbol K_{sp}.

$$K_{sp}(PbCl_2) = [Pb^{2+}][Cl^-]^2 = 1.7 \times 10^{-5}$$

The K_{sp} values for some ionic compounds are shown in Interactive Table 18.1.1. Notice that the values range from relatively large (around 10^{-5}) to very small (around 10^{-40}).

Example Problem 18.1.2 Write solubility product constant expressions.

Write the K_{sp} expression for both of the following sparingly soluble compounds:

a. $PbSO_4$
b. $Zn_3(PO_4)_2$

Solution:

You are asked to write the K_{sp} expression for a sparingly soluble compound.

You are given the chemical formula of the compound.

a. **Step 1.** Write the balanced equation for the dissolution of the ionic compound.

$$PbSO_4(s) \rightleftarrows Pb^{2+}(aq) + SO_4^{2-}(aq)$$

Interactive Table 18.1.1

K_{sp} Values for Some Ionic Compounds

Compound	K_{sp} at 25 °C
$PbBr_2$	6.3×10^{-6}
$AgBr$	3.3×10^{-13}
$CaCO_3$	3.8×10^{-9}
$CuCO_3$	2.5×10^{-10}
$NiCO_3$	6.6×10^{-9}
Ag_2CO_3	8.1×10^{-12}
$PbCl_2$	1.7×10^{-5}
$AgCl$	1.8×10^{-10}
BaF_2	1.7×10^{-6}
CaF_2	3.9×10^{-11}
$Cu(OH)_2$	1.6×10^{-19}
$Fe(OH)_3$	6.3×10^{-38}
$Ni(OH)_2$	2.8×10^{-16}
$Zn(OH)_2$	4.5×10^{-17}
$Ca_3(PO_4)_2$	1.0×10^{-25}
$CaSO_4$	2.4×10^{-5}
$PbSO_4$	1.8×10^{-8}

Step 2. Write the equilibrium constant expression. Remember that the solid, $PbSO_4$, does not appear in the equilibrium constant expression.

$$K_{sp} = [Pb^{2+}][SO_4{}^{2-}]$$

b. **Step 1.** Write the balanced equation for the dissolution of the ionic compound.

$$Zn_3(PO_4)_2(s) \rightleftharpoons 3\ Zn^{2+}(aq) + 2\ PO_4{}^{3-}(aq)$$

Step 2. Write the equilibrium constant expression. Remember that the solid, $Zn_3(PO_4)_2$, does not appear in the equilibrium expression and that each ion concentration is raised to the power of the stoichiometric coefficient in the balanced equation.

$$K_{sp} = [Zn^{2+}]^3[PO_4{}^{3-}]^2$$

Video Solution

Tutored Practice
Problem 18.1.2

18.1c Determining K_{sp} Values

Solubility product equilibrium constants are determined from measured equilibrium ion concentrations or directly from the solubility of an ionic compound, as shown in the following examples. In both cases, you must use the stoichiometry of the compound to determine the concentration of both the cation and anion before calculating K_{sp}.

Example Problem 18.1.3 Use equilibrium ion concentration to calculate K_{sp}.

The Pb^{2+} concentration in a saturated solution of lead chloride is measured and found to be 0.016 M. Use this information to calculate the K_{sp} for lead chloride.

Solution:

You are asked to calculate the K_{sp} of a sparingly soluble compound.

You are given the identity of the compound and one ion's equilibrium concentration.

Step 1. Write the balanced equation for the equilibrium and the K_{sp} expression.

$$PbCl_2(s) \rightleftharpoons Pb^{2+}(aq) + 2\ Cl^-(aq) \qquad K_{sp} = [Pb^{2+}][Cl^-]^2$$

Step 2. Use the lead concentration to determine the chloride ion concentration at equilibrium. Notice that for this salt, the anion concentration is twice the cation concentration ($[Cl^-] = 2 \times [Pb^{2+}]$).

$$[Cl^-] = \frac{0.016\ \text{mol}\ Pb^{2+}}{1\ L} \times \frac{2\ \text{mol}\ Cl^-}{1 \text{mol}\ Pb^{2+}} = 0.032\ M$$

Step 3. Use the equilibrium concentrations to calculate K_{sp}.

$$K_{sp} = [Pb^{2+}][Cl^-]^2 = (0.016\ M)(0.032\ M)^2 = 1.6 \times 10^{-5}$$

Example Problem 18.1.3 *(continued)*

Alternatively, you can use the relationship between the anion and cation concentrations for this salt ($[Cl^-] = 2 \times [Pb^{2+}]$) to calculate K_{sp}:

$$K_{sp} = [Pb^{2+}][Cl^-]^2 = [Pb^{2+}](2 \times [Pb^{2+}])^2 = 4 \times [Pb^{2+}]^3 = 4(0.016\ M)^3 = 1.6 \times 10^{-5}$$

Video Solution

Tutored Practice
Problem 18.1.3

The solubility of a compound can also be used to calculate K_{sp}. Recall that solubility is the amount of solid (in grams or moles) that will dissolve in a given volume of water.

Example Problem 18.1.4 Use solubility to calculate K_{sp}.

The solubility of calcium fluoride, CaF_2, is 0.0167 g/L. Use this information to calculate K_{sp} for calcium fluoride.

Solution:

You are asked to calculate the K_{sp} of a sparingly soluble compound.

You are given the identity of the compound and its solubility.

Step 1. Write the balanced equation for the equilibrium and the K_{sp} expression.

$$CaF_2(s) \rightleftharpoons Ca^{2+}(aq) + 2\ F^-(aq) \qquad K_{sp} = [Ca^{2+}][F^-]^2$$

Step 2. Use solubility to calculate the ion concentrations at equilibrium.

Calcium fluoride dissolves to an extent of 0.0167 g per L of solution. In terms of calcium ion concentration,

$$[Ca^{2+}] = \frac{0.0167\ g\ CaF_2}{1\ L} \times \frac{1\ mol\ CaF_2}{78.08\ g\ CaF_2} \times \frac{1\ mol\ Ca^{2+}}{1\ mol\ CaF_2} = 2.14 \times 10^{-4}\ M$$

The compound stoichiometry tells us that the fluoride ion concentration is twice the calcium ion concentration.

$$[F^-] = \frac{2.14 \times 10^{-4}\ mol\ Ca^{2+}}{1\ L} \times \frac{2\ mol\ F^-}{1\ mol\ Ca^{2+}} = 4.28 \times 10^{-4}\ M$$

Step 3. Use the equilibrium concentrations to calculate K_{sp}.

$$K_{sp} = [Ca^{2+}][F^-]^2 = (2.14 \times 10^{-4})(4.28 \times 10^{-4})^2 = 3.92 \times 10^{-11}$$

Alternatively, you can use the relationship between the anion and cation concentrations for this salt ($[F^-] = 2 \times [Ca^{2+}]$) to calculate K_{sp}.

$$K_{sp} = [Ca^{2+}][F^-]^2 = [Ca^{2+}](2 \times [Ca^{2+}])^2 = 4 \times [Ca^{2+}]^3 = 4(2.14 \times 10^{-4}\ M)^3 = 3.92 \times 10^{-11}$$

Video Solution

Tutored Practice
Problem 18.1.4

Section 18.1 Mastery

18.2 Using K_{sp} in Calculations

18.2a Estimating Solubility

The solubility of a salt in pure water is defined as the amount of solid that will dissolve per liter of solution (g/L or mol/L). As we will see later, many secondary reactions can influence the solubility of an ionic compound. Because of this, when we use K_{sp} values to estimate the solubility of an ionic compound, we assume that none of these secondary reactions are taking place. Solubility can be estimated from K_{sp} using the same techniques we have applied to other equilibrium systems, as shown in the following example.

Example Problem 18.2.1 Estimate solubility from K_{sp}.

Calculate the solubility of mercury(II) iodide, HgI_2, in units of grams per liter.

$$K_{sp}(HgI_2) = 4.0 \times 10^{-29}$$

Solution:

You are asked to calculate the solubility of an ionic salt.

You are given the formula of the salt and the K_{sp} for the salt.

Step 1. Write the balanced equation for the equilibrium and the K_{sp} expression.

$$HgI_2(s) \rightleftharpoons Hg^{2+}(aq) + 2\,I^-(aq) \qquad K_{sp} = [Hg^{2+}][I^-]^2$$

Step 2. Set up an *ICE* table where x = amount of HgI_2 that dissolves in solution (x = molar solubility of HgI_2). We will assume that the small amount of HgI_2 that dissolves dissociates completely in water and that initially no ions are present. Note that the amount of solid present is not indicated in the *ICE* table but that solid must be present in order for equilibrium to be established.

	$HgI_2(s)$	$\rightleftharpoons$	$Hg^{2+}(aq)$	+	$2\,I^-(aq)$
Initial (M)			0		0
Change (M)			$+x$		$+2x$
Equilibrium (M)			x		$2x$

Step 3. Substitute the equilibrium concentrations into the K_{sp} expression and solve for x.

$$K_{sp} = [Hg^{2+}][I^-]^2 = (x)(2x)^2 = 4x^3$$

$$x = \text{solubility} = \sqrt[3]{\frac{K_{sp}}{4}} = \sqrt[3]{\frac{4.0 \times 10^{-29}}{4}} = 2.2 \times 10^{-10}\,M$$

Video Solution

Tutored Practice
Problem 18.2.1

Example Problem 18.2.1 *(continued)*

Step 4. Use x, the molar solubility of HgI_2, to calculate solubility in units of g/L.

$$\frac{2.2 \times 10^{-10} \text{ mol HgI}_2}{1 \text{ L}} \times \frac{454 \text{ g HgI}_2}{1 \text{ mol HgI}_2} = 9.8 \times 10^{-8} \text{ g/L}$$

Mercury(II) iodide is considered an insoluble salt (molar solubility is less than 0.01 mol/L).

The relationship between x, the molar solubility of the salt, and K_{sp} is a function of the stoichiometry of the ionic compound (the relative number of cations and anions per formula unit). These relationships are summarized for ionic compounds with a 1:1, 1:2, 1:3, and 2:3 cation:anion ratio in Table 18.2.1. Notice that each relationship applies to two different types of ionic compounds. The 1:2 relationship, for example, is correct for a salt with twice as many anions as cations (such as $PbCl_2$) or one with twice as many cations as anions (such as Ag_2S).

Relative Solubility

When the solubility of two or more ionic compounds are being compared, both K_{sp} and the stoichiometry of the ionic compounds must be considered. For example, both

Table 18.2.1 Relationship Between Molar Solubility and K_{sp}

General Formula	Example	K_{sp} **Expression**	K_{sp} **as a Function of Molar Solubility (x)**	**Solubility (x) as a Function of K_{sp}**
MY	AgCl	$K_{sp} = [M^+][Y^-]$	$K_{sp} = (x)(x) = x^2$	$x = \sqrt{K_{sp}}$
MY$_2$	HgI$_2$	$K_{sp} = [M^{2+}][Y^-]^2$	$K_{sp} = (x)(2x)^2 = 4x^3$	$x = \sqrt[3]{\dfrac{K_{sp}}{4}}$
MY$_3$	BiI$_3$	$K_{sp} = [M^{3+}][Y^-]^3$	$K_{sp} = (x)(3x)^3 = 27x^4$	$x = \sqrt[4]{\dfrac{K_{sp}}{27}}$
M$_2$Y$_3$	Fe$_2$(SO$_4$)$_3$	$K_{sp} = [M^{3+}]^2[Y^{2-}]^3$	$K_{sp} = (2x)^2(3x)^3 = 108x^5$	$x = \sqrt[5]{\dfrac{K_{sp}}{108}}$

silver chloride (AgCl, $K_{sp} = 1.8 \times 10^{-10}$) and calcium carbonate (CaCO$_3$, $K_{sp} = 3.8 \times 10^{-9}$) have a 1:1 cation:anion ratio. Because they have the same stoichiometry, K_{sp} alone can be used to determine the relative solubility of these compounds in water. Calcium carbonate is more soluble in water than silver chloride because K_{sp}(CaCO$_3$) is greater than K_{sp}(AgCl).

When the relative solubility of silver chloride (AgCl, $K_{sp} = 1.8 \times 10^{-10}$) and silver dichromate (Ag$_2$CrO$_4$, $K_{sp} = 9.0 \times 10^{-12}$) are being compared, it is not possible to use only K_{sp} values. The two salts do not have the same stoichiometry, and thus the molar solubility must be calculated for each salt. Although silver chloride has the greater K_{sp} value, the calculated solubility shows that silver dichromate is more soluble in water. The solubility of Ag$_2$CrO$_4$ in water (1.3×10^{-4} mol/L) is about 10 times greater than the solubility of AgCl (1.3×10^{-5} mol/L).

Example Problem 18.2.2 Determine relative solubility.

Determine the relative solubility of the following calcium compounds: CaSO$_4$ ($K_{sp} = 2.4 \times 10^{-5}$), Ca(OH)$_2$ ($K_{sp} = 7.9 \times 10^{-6}$), and CaF$_2$ ($K_{sp} = 3.9 \times 10^{-11}$).

Solution:

You are asked to determine the relative solubility of three ionic compounds.

You are given the compound formulas and their respective K_{sp} values.

Because the three compounds do not have the same stoichiometry, the solubility must be calculated for each one. Using the relationships shown in Table 18.2.1,

$$\text{CaSO}_4 \text{ solubility} = x = \sqrt{K_{sp}} = \sqrt{2.4 \times 10^{-5}} = 4.9 \times 10^{-3} \text{ mol/L}$$

$$\text{Ca(OH)}_2 \text{ solubility} = x = \sqrt[3]{\frac{K_{sp}}{4}} = \sqrt[3]{\frac{7.9 \times 10^{-6}}{4}} = 0.013 \text{ mol/L}$$

$$\text{CaF}_2 \text{ solubilty} = x = \sqrt[3]{\frac{K_{sp}}{4}} = \sqrt[3]{\frac{3.9 \times 10^{-11}}{4}} = 2.1 \times 10^{-4} \text{ mol/L}$$

Ca(OH)$_2$ is the most soluble of the three compounds, and CaF$_2$ is the least soluble.

Video Solution

Tutored Practice
Problem 18.2.2

18.2b Predicting Whether a Solid Will Precipitate or Dissolve

In Chemical Equilibrium (Unit 15) we used Q, the reaction quotient, to determine whether or not a system is at equilibrium. ($\blacktriangleleft$ Flashback to Section 15.3b Determining Whether a System Is at Equilibrium) The reaction quotient can also be used with precipitation equilibria to determine whether a solution is at equilibrium and to answer questions such as, if 3 g of solid silver sulfate is added to 250 mL of water, will the solid dissolve completely?

Recall that the reaction quotient has the same form as the equilibrium expression, but it differs in that the concentrations may or may not be equilibrium concentrations. For example, the reaction quotient for the silver chloride equilibrium is written as

$$AgCl(s) \rightleftarrows Ag^+(aq) + Cl^-(aq) \qquad Q = [Ag^+][Cl^-]$$

Comparing Q with K_{sp} for a specific solubility equilibrium allows us to determine whether a system is at equilibrium (Interactive Figure 18.2.1). There are three possible relationships between the two values:

Interactive Figure 18.2.1

Explore the relationship between Q and K_{sp}.

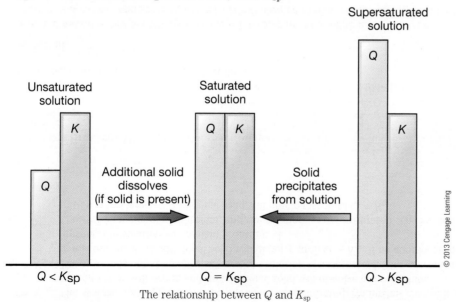

The relationship between Q and K_{sp}

© 2013 Cengage Learning

$Q = K_{sp}$ The system is at equilibrium, and the solution is saturated. No further change in ion concentration will occur, and no additional solid will dissolve or precipitate.

$Q < K_{sp}$ The system is not at equilibrium, and the solution is unsaturated. The ion concentration is too small, so additional solid will dissolve until $Q = K_{sp}$. If no additional solid is present, the solution will remain unsaturated.

$Q > K_{sp}$ The system is not at equilibrium, and the solution is supersaturated. The ion concentration is too large, so additional solid will precipitate until $Q = K_{sp}$.

Example Problem 18.2.3 Use the reaction quotient to predict whether a precipitate will form.

If 5.0 mL of 1.0×10^{-3} M NaCl is added to 1.0 mL of 1.0×10^{-3} M Pb(NO$_3$)$_2$, will solid PbCl$_2$ ($K_{sp} = 1.7 \times 10^{-5}$) precipitate? If a precipitate will not form, what chloride ion concentration will cause a precipitate of lead chloride to form?

Solution:

You are asked to predict whether a solid will form when two solutions are combined. If a solid does not form, you are asked to calculate the concentration of one ion required for the formation of a precipitate.

You are given the volume and concentration of each solution and K_{sp} for the insoluble compound formed in the reaction.

Step 1. Write the balanced net ionic equation for the equilibrium and the Q expression.

$$PbCl_2(s) \rightleftarrows Pb^{2+}(aq) + 2\,Cl^-(aq) \qquad Q = [Pb^{2+}][Cl^-]^2$$

Step 2. Calculate the concentration of Pb^{2+} and Cl$^-$. The total volume of the solution is 6.0 mL.

$$[Pb^{2+}] = \frac{(0.0010\ \text{L})(1.0 \times 10^{-3}\ \text{mol Pb}^{2+}/\text{L})}{0.0060\ \text{L}} = 1.7 \times 10^{-4}\ \text{M}$$

$$[Cl^-] = \frac{(0.0050\ \text{L})(1.0 \times 10^{-3}\ \text{mol Cl}^-/\text{L})}{0.0060\ \text{L}} = 8.3 \times 10^{-4}\ \text{M}$$

Step 3. Substitute the ion concentrations into the equilibrium expression and calculate Q.

$$Q = [Pb^{2+}][Cl^-]^2 = (1.7 \times 10^{-4})(8.3 \times 10^{-4})^2 = 1.2 \times 10^{-10}$$

Step 4. Compare Q and K_{sp}. In this case, Q (1.2×10^{-10}) is less than K_{sp} (1.7×10^{-5}) and the solution is unsaturated. Lead chloride will not precipitate.

Step 5. Determine the chloride ion concentration required for lead chloride precipitation. Substitute the lead ion concentration into the K_{sp} expression to calculate the chloride ion concentration in a saturated solution of lead chloride.

Example Problem 18.2.3 *(continued)*

$$K_{sp} = [Pb^{2+}][Cl^-]^2$$
$$1.7 \times 10^{-5} = (1.7 \times 10^{-4})[Cl^-]^2$$
$$[Cl^-] = 0.32 \text{ M}$$

The chloride ion concentration in a saturated solution of lead chloride (where $[Pb^{2+}]$ = 1.7×10^{-4}) is equal to 0.32 M. If $[Cl^-] > 0.32$ M, the solution will be supersaturated ($Q > K_{sp}$) and a precipitate of lead chloride will form.

Video Solution

Tutored Practice Problem 18.2.3

18.2c The Common Ion Effect

As we saw in Advanced Acid-Base Equilibria (Unit 17), adding a chemical species that is common to an existing equilibrium (a common ion) will cause the system to shift, forming additional reactant or product. (◄ Flashback to Section 17.2b Buffer pH) Because solubility equilibria are always written so that the aqueous ions are reaction products, adding a common ion causes the system to shift to the left (toward the formation of additional solid), decreasing the solubility of the ionic compound. For example, consider the sparingly soluble salt lead chloride, $PbCl_2$.

$$PbCl_2(s) \rightleftharpoons Pb^{2+}(aq) + 2\,Cl^-(aq) \qquad K_{sp} = [Pb^{2+}][Cl^-]^2 = 1.7 \times 10^{-5}$$

When an ion common to the equilibrium is added, Pb^{2+} in the form of $Pb(NO_3)_2$ or Cl^- in the form of NaCl, for example, the system shifts to the left, the solubility of $PbCl_2$ decreases, and additional solid lead chloride precipitates from solution (Interactive Figure 18.2.2).

The common ion effect plays a role in the solubility of ionic compounds in natural systems and even in most laboratory settings. For example, in the examination of how much $PbCl_2$ will dissolve in a natural (nonpurified) water system, additional Cl^- could be present from dissolved NaCl. The effect of a common ion on the solubility of an ionic compound is demonstrated in the following example problem.

Interactive Figure 18.2.2

Explore the effect of a common ion on solubility.

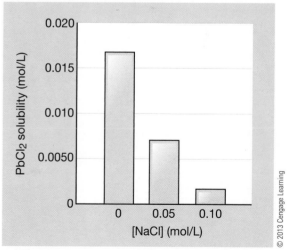

As chloride ion concentration increases, the solubility of $PbCl_2$ decreases.

Example Problem 18.2.4 Calculate solubility in the presence of a common ion.

Calculate the solubility of PbI_2

a. In pure water

b. In a solution in which $[I^-] = 0.15$ M

Solution:

You are asked to calculate the solubility of an ionic compound in pure water and in the presence of an ion that is common to the equilibrium.

You are given the identity of the ionic compound and the concentration of the common ion.

a. Using the relationship between solubility and K_{sp} from Table 18.2.1, calculate the solubility of PbI_2 in pure water.

$$x = \text{solubility of } PbI_2 \text{ in pure water} = \sqrt[3]{\frac{K_{sp}}{4}} = \sqrt[3]{\frac{8.7 \times 10^{-9}}{4}} = 1.3 \times 10^{-3} \text{ mol/L}$$

b. **Step 1.** Write the balanced equation for the dissolution equilibrium and the K_{sp} expression for PbI_2.

$$PbI_2(aq) \rightleftarrows Pb^{2+}(aq) + 2\,I^-(aq) \qquad K_{sp} = [Pb^{2+}][I^-]^2$$

Step 2. Set up an *ICE* table, where the variable y represents the amount of PbI_2 that dissociates in a solution with $[I^-] = 0.15$ M. The variable y also represents the molar solubility of PbI_2 in the presence of the common ion.

	$PbI_2(aq)$	$\rightleftarrows$	$Pb^{2+}(aq)$	+	$2\,I^-(aq)$
Initial (M)			0		0.15
Change (M)			$+y$		$+2y$
Equilibrium (M)			y		$0.15 + 2y$

Step 3. Substitute the equilibrium concentrations into the K_{sp} expression and solve for y. Because the addition of a common ion will shift the system to the left and decrease the solubility of PbI_2, it is reasonable to assume that $2y \ll 0.15$.

$$K_{sp} = [Pb^{2+}][I^-]^2 = (y)(0.15 + 2y)^2 \approx (y)(0.15)^2$$

$$y = \text{solubility of } PbI_2 \text{ in } 0.15 \text{ M } I^- = \frac{K_{sp}}{(0.15)^2} = \frac{8.7 \times 10^{-9}}{(0.15)^2} = 3.9 \times 10^{-7} \text{ mol/L}$$

Note that our assumption $(2y \ll 0.15)$ was valid.

Is your answer reasonable? The solubility of PbI_2 has decreased from 1.3×10^{-3} M in pure water to 3.9×10^{-7} M in the presence of a common ion, the iodide ion. The presence of a common ion will always decrease the solubility of a sparingly soluble salt.

Video Solution

Tutored Practice
Problem 18.2.4

Section 18.2 Mastery

18.3 Lewis Acid–Base Complexes and Complex Ion Equilibria

18.3a Lewis Acids and Bases

The two most commonly used acid–base definitions, the Brønsted and Lewis models, are described in Interactive Table 18.3.1. In Lewis acid–base chemistry, a **Lewis base** is a species that donates a lone pair of electrons to a **Lewis acid**, which is an electron-pair acceptor. The product of a Lewis acid–base reaction is a **Lewis acid–base adduct** or *acid–base complex* (Interactive Figure 18.3.1).

The new bond formed between the Lewis acid and Lewis base is called a **coordinate–covalent bond** because both bonding electrons come from a single species, the Lewis base. The coordinate–covalent bond is often represented using an arrow that points from the Lewis base (the electron-pair donor) to the Lewis acid (the electron-pair acceptor).

Lewis acids and bases can be neutral or ionic. Neutral Lewis acids include compounds containing Group 3A elements ($AlCl_3$, BF_3), organic compounds containing multiple bonds, and CO_2, whereas neutral Lewis bases are compounds that have at least one lone pair of electrons (NH_3, H_2O). Ionic Lewis acids are typically cations (Fe^{2+}, Na^+), whereas ionic Lewis bases are usually anions (OH^-, Cl^-).

Interactive Figure 18.3.1

Investigate the formation of a Lewis acid–base adduct.

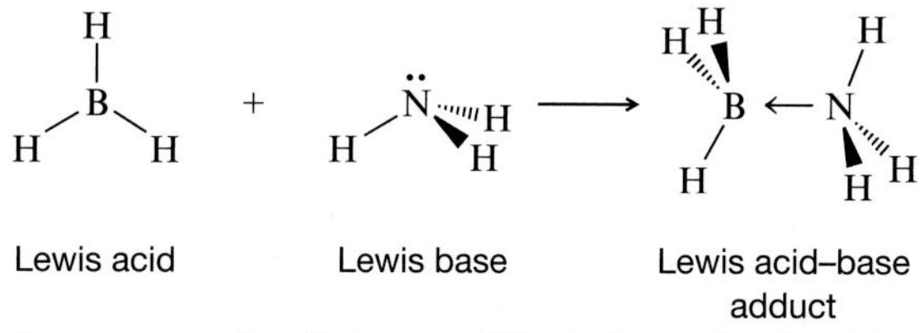

| Lewis acid | Lewis base | Lewis acid–base adduct |

Ammonia reacts with borane, BH_3, to form a Lewis acid–base adduct.

Interactive Table 18.3.1

Brønsted–Lowry and Lewis Acid–Base Models

Acid–Base Model	Acid Definition	Base Definition	Reaction Product(s)
Brønsted–Lowry	Proton donor	Proton acceptor	Conjugate base and conjugate acid
Lewis	Electron-pair acceptor	Electron-pair donor	Lewis acid–base complex

Example Problem 18.3.1 Identify Lewis acids and bases.

Identify the Lewis acid and Lewis base in the reaction between dimethyl ether, CH_3OCH_3, and borane, BH_3.

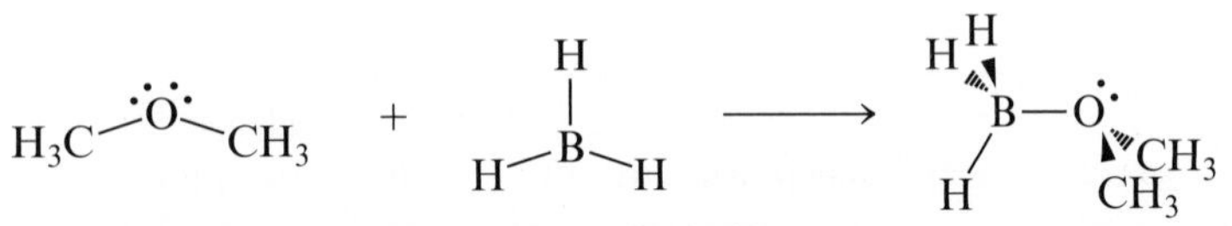

Solution:

You are asked to identify the Lewis acid and Lewis base in a reaction.

You are given a chemical reaction.

Dimethyl ether can act as a Lewis base because it can donate an electron pair to a Lewis acid. Borane accepts an electron pair from a Lewis base, so it is a Lewis acid. Note that the boron atom in BH_3 is sp^2-hybridized and therefore can accept an electron pair because it has an empty, unhybridized $2p$ orbital. In the Lewis acid–base complex, the boron atom is sp^3-hybridized.

Video Solution

Tutored Practice
Problem 18.3.1

When a Lewis acid–base complex carries an overall charge, it is called a **complex ion**. You have encountered complex ions in many chemical reactions that take place in water. For example, when an iron(III) ion is written as $Fe^{3+}(aq)$, this represents the aqueous ion surrounded by up to six water molecules. The water molecules act as Lewis bases, and the iron(III) ion is a Lewis acid. Notice that in a complex ion formula, the Lewis acid and bases are traditionally written inside square brackets and the charge on the complex ion is written outside the brackets. However, this format is not always followed.

$[Fe(H_2O)_6]^{3+}$

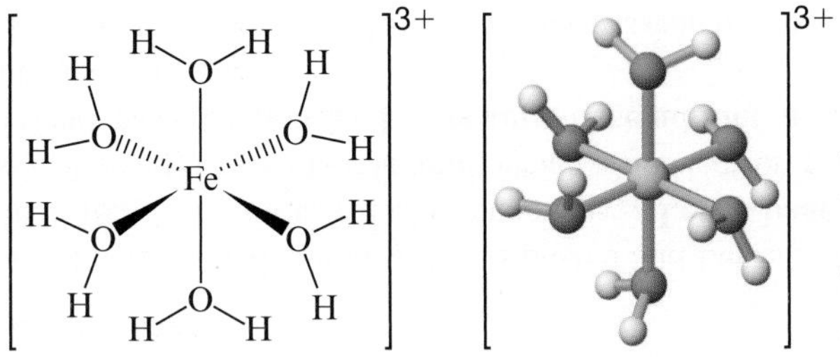

© 2013 Cengage Learning

It is very common to find Lewis acid–base complexes in which one Lewis acid species (such as a metal cation) is coordinated to multiple Lewis base species. It is not simple, however, to predict the number of Lewis bases that coordinate to a given Lewis acid. For example, consider the Lewis acid–base reaction between aqueous Cu^{2+} ions and NH_3.

$$Cu^{2+}(aq) + 4\ NH_3(aq) \rightleftharpoons \left[\begin{array}{c} NH_3 \\ \downarrow \\ H_3N \rightarrow Cu \leftarrow NH_3 \\ \uparrow \\ NH_3 \end{array} \right]^{2+}$$

Here the central Cu^{2+} ion (a Lewis acid) is bonded to four NH_3 molecules (Lewis bases) in the copper–ammonia complex ion.

18.3b Complex Ion Equilibria

Chemical equilibria involving complex ions can be written to show the formation or dissociation of a complex ion. When written to show the formation of a complex ion from the reaction of a Lewis acid with a Lewis base, the equilibrium constant is given the special name **formation constant** and the symbol K_f. For the complex ion $Cu(NH_3)_4{}^{2+}$,

$$Cu^{2+}(aq) + 4\ NH_3(aq) \rightleftharpoons Cu(NH_3)_4{}^{2+}(aq)$$

$$K_f = \frac{[Cu(NH_3)_4^{2+}]}{[Cu^{2+}][NH_3]^4} = 6.8 \times 10^{12}$$

When written in the reverse direction, the equilibrium constant is called a **dissociation constant**, K_d. The dissociation constant is equal to the inverse of K_f ($K_d = 1/K_f$).

$$Cu(NH_3)_4{}^{2+}(aq) \rightleftharpoons Cu^{2+}(aq) + 4\ NH_3(aq)$$

$$K_d = \frac{[Cu^{2+}][NH_3]^4}{[Cu(NH_3)_4^{2+}]} = \frac{1}{K_f} = 1.5 \times 10^{-13}$$

Many formation reactions are strongly product favored and have large K_f values. Table 18.3.2 lists the values of K_f for a number of common complex ions.

Table 18.3.2 Formation Constants for Some Complex Ions

Formation Equilibrium	K_f
$Ag^+ + 2\ Br^- \rightleftharpoons [AgBr_2]^-$	2.1×10^7
$Ag^+ + 2\ Cl^- \rightleftharpoons [AgCl_2]^-$	1.1×10^5
$Ag^+ + 2\ CN^- \rightleftharpoons [Ag(CN)_2]^-$	1.3×10^{21}
$Ag^+ + 2\ S_2O_3{}^{2-} \rightleftharpoons [Ag(S_2O_3)_2]^{3-}$	2.9×10^{13}
$Ag^+ + 2\ NH_3 \rightleftharpoons [Ag(NH_3)_2]^+$	1.1×10^7
$Al^{3+} + 6\ F^- \rightleftharpoons [AlF_6]^{3-}$	6.9×10^{19}
$Al^{3+} + 4\ OH^- \rightleftharpoons [Al(OH)_4]^-$	1.1×10^{33}
$Au^+ + 2\ CN^- \rightleftharpoons [Au(CN)_2]^-$	2.0×10^{38}
$Cd^{2+} + 4\ CN^- \rightleftharpoons [Cd(CN)_4]^{2-}$	6.0×10^{18}
$Cd^{2+} + 4\ NH_3 \rightleftharpoons [Cd(NH_3)_4]^{2+}$	1.3×10^7
$Co^{2+} + 6\ NH_3 \rightleftharpoons [Co(NH_3)_6]^{2+}$	1.3×10^5
$Cu^+ + 2\ CN^- \rightleftharpoons [Cu(CN)_2]^-$	1.0×10^{24}
$Cu^+ + 2\ Cl^- \rightleftharpoons [CuCl_2]^-$	3.2×10^5
$Cu^{2+} + 4\ NH_3 \rightleftharpoons [Cu(NH_3)_4]^{2+}$	2.1×10^{13}
$Fe^{2+} + 6\ CN^- \rightleftharpoons [Fe(CN)_6]^{4-}$	1.0×10^{35}
$Hg^{2+} + 4\ Cl^- \rightleftharpoons [HgCl_4]^{2-}$	1.2×10^{15}
$Ni^{2+} + 4\ CN^- \rightleftharpoons [Ni(CN)_4]^{2-}$	2.0×10^{31}
$Ni^{2+} + 6\ NH_3 \rightleftharpoons [Ni(NH_3)_6]^{2+}$	5.5×10^8
$Zn^{2+} + 4\ OH^- \rightleftharpoons [Zn(OH)_4]^{2-}$	4.6×10^{17}
$Zn^{2+} + 4\ NH_3 \rightleftharpoons [Zn(NH_3)_4]^{2+}$	2.9×10^9

Using Formation Constants

The principles and techniques we have used for treating other equilibrium systems are also applied to predicting the concentrations of species in complex ion equilibria. In general, the mathematical treatment of these systems is more complex because of the large stoichiometric coefficients and the resulting need to solve equations of high order, as shown in the following example.

Example Problem 18.3.2 Calculate concentration of species in a solution containing a complex ion.

In the presence of excess OH^-, the Zn^{2+}(aq) ion forms a hydroxide complex ion. Calculate the concentration of free Zn^{2+} ion when 0.010 mol $Zn(NO_3)_2$(s) is added to 1.00 L of solution in which $[OH^-]$ is held constant (buffered at pH 12.00).

Solution:

You are asked to calculate the concentration of free Lewis acid in a solution containing a complex ion.

You are given the initial concentrations of the Lewis acid and base.

Step 1. First, assume that the $Zn(NO_3)_2$ dissociates completely (it is a strong electrolyte) to form Zn^{2+} and NO_3^-. Because K_f is large, we will approach this problem by first assuming that all of the Zn^{2+} is converted to the $Zn(OH)_4^{2-}$ complex ion. We will use the dissociation equilibrium for $[Zn(OH)_4]^{2-}$ and K_d to calculate the free Zn^{2+} concentration:

$$[Zn(OH)_4]^{2-}(aq) \rightleftharpoons Zn^{2+}(aq) + 4\ OH^-(aq)$$

$$K_d = \frac{[Zn^{2+}][OH^-]^4}{[Zn(OH)_4^{2-}]} = \frac{1}{K_f} = 2.2 \times 10^{-18}$$

Step 2. Calculate the solution $[OH^-]$ from the solution pH:

$$pOH = 14.00 - pH = 14.00 - 12.00 = 2.00$$

$$[OH^-] = 10^{-pOH} = 10^{-2.00} = 1.0 \times 10^{-2}\ M$$

Step 3. Set up an ICF table, where the unknown variable x represents the amount of complex ion that dissociates in solution and is also equal to the concentration of free Zn^{2+} in solution. The initial concentration of $Zn(OH)_4^{2-}$ is equal to the concentration of the limiting reactant, Zn^{2+}. (Recall that we assumed that all of the Zn^{2+} was converted to complex ion when combined with OH^- in solution.)

	$Zn(OH)_4^{2-}$(aq)	$\rightleftharpoons$	Zn^{2+}(aq)	+	$4\ OH^-$(aq)
Initial (M)	0.010		0		1.0×10^{-2}
Change (M)	$-x$		$+x$		
Equilibrium (M)	$0.010 - x$		x		1.0×10^{-2}

Example Problem 18.3.2 (*continued*)

Step 4. Substitute the equilibrium concentrations into the K_d expression and solve for x. We will assume that x is very small compared with the initial complex ion concentration. We assume this for two reasons. First, K_d is very small (2.2×10^{-18}), and second, the presence of a common ion (OH^-) will shift the equilibrium to the left, decreasing the amount of dissociated complex ion.

$$K_d = 2.2 \times 10^{-18} = \frac{[Zn^{2+}][OH^-]^4}{[Zn(OH)_4^{2-}]} = \frac{(x)(1.0 \times 10^{-2})^4}{0.010 - x} \approx \frac{(x)(1.0 \times 10^{-2})^4}{0.010}$$

$$x = [Zn^{2+}] = 2.2 \times 10^{-12} \text{ M}$$

Is your answer reasonable? The large formation constant for the complex ion suggests that there will be a small amount of free Zn^{2+} in solution.

Video Solution

Tutored Practice
Problem 18.3.2

Section 18.3 Mastery

18.4 Simultaneous Equilibria

18.4a Solubility and pH

Many factors affect the solubility of ionic compounds. As we have seen, adding an ion common to the equilibrium decreases the solubility of a sparingly soluble salt. Here we address two additional factors that affect the solubility of ionic compounds—pH and the formation of complex ions—and demonstrate how these factors can be used to isolate one metal ion from a mixture. These are examples of simultaneous equilibria, where more than one equilibrium system is present in a solution.

Consider the dissolution of nickel(II) carbonate.

$$NiCO_3(s) \rightleftharpoons Ni^{2+}(aq) + CO_3^{2-}(aq) \qquad K_{sp} = [Ni^{2+}][CO_3^{2-}] = 1.3 \times 10^{-7}$$

In a saturated solution, Ni^{2+} and CO_3^{2-} are in equilibrium with solid $NiCO_3$. If a strong acid is added to the solution, the carbonate ion can react to form the bicarbonate ion and water.

$$CO_3^{2-}(aq) + H_3O^+(aq) \rightleftharpoons HCO_3^-(aq) + H_2O(\ell) \qquad K = 1/K_a(HCO_3^-) = 1.8 \times 10^{10}$$

Adding these two equilibrium reactions shows the effect of adding a strong acid to a saturated $NiCO_3$ solution.

$NiCO_3(s) \rightleftharpoons Ni^{2+}(aq) + CO_3^{2-}(aq)$ $\qquad\qquad\qquad\qquad K_{sp} = 1.3 \times 10^{-7}$

$CO_3^{2-}(aq) + H_3O^+(aq) \rightleftharpoons HCO_3^-(aq) + H_2O(\ell)$ $\qquad K = 1/K_a(HCO_3^-) = 1.8 \times 10^{10}$

$\overline{NiCO_3(s) + H_3O^+(aq) \rightleftharpoons Ni^{2+}(aq) + HCO_3^-(aq) + H_2O(\ell) \quad K_{net} = K_{sp}/K_a = 2.3 \times 10^3}$

The reaction between $NiCO_3$ and acid is product favored.

In general, the solubility of any ionic compound containing a basic anion such as CO_3^{2-}, S^{2-}, or PO_4^{3-} is increased in the presence of acid. For example, consider Interactive Figure 18.4.1, which shows the effect of adding strong acid to a mixture of AgCl and Ag_3PO_4. The Ag_3PO_4 dissolves because PO_4^{3-} is a basic anion, and AgCl, which does not contain a basic anion, does not dissolve.

Interactive Figure 18.4.1

Investigate the effect of pH on solubility.

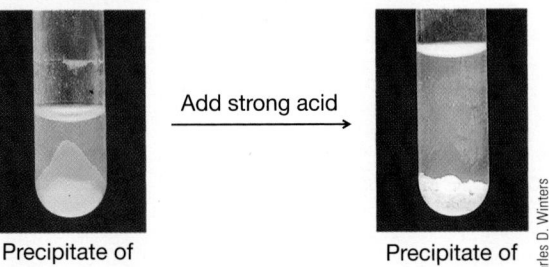

Add strong acid

Precipitate of AgCl and Ag_3PO_4 Precipitate of AgCl

Ag_3PO_4 (yellow solid) dissolves in the presence of strong acid but AgCl (white solid) does not.

Example Problem 18.4.1 Show the relationship between solubility and pH.

Write an equation to show why the solubility of $Fe(OH)_2$ increases in the presence of a strong acid, and calculate the equilibrium constant for the reaction of this sparingly soluble salt with acid.

Solution:

You are asked to write an equation for the reaction between a sparingly soluble salt and a strong acid and to calculate the equilibrium constant for the reaction.

You are given the identity of the sparingly soluble salt.

Iron(II) hydroxide contains a basic anion, so its solubility increases in the presence of acid. Use $K_{sp}[Fe(OH)_2]$ and K_w to calculate the overall equilibrium constant for this reaction.

$Fe(OH)_2(s) \rightleftarrows Fe^{2+}(aq) + 2\ OH^-(aq)$	$K_{sp} = 7.9 \times 10^{-15}$
$2\ OH^-(aq) + 2\ H_3O^+(aq) \rightleftarrows 4\ H_2O(\ell)$	$K = (1/K_w)^2 = 1.0 \times 10^{28}$
$Fe(OH)_2(s) + 2\ H_3O^+(aq) \rightleftarrows Fe^{2+}(aq) + 4\ H_2O(\ell)$	$K_{net} = K_{sp}/(K_w)^2 = 7.9 \times 10^{13}$

Video Solution

Tutored Practice Problem 18.4.1

18.4b Solubility and Complex Ions

The formation of a complex ion can be used to increase the solubility of sparingly soluble ionic compounds. For example, as shown in Interactive Figure 18.4.2, adding the Lewis base NH_3 to a solution containing solid AgCl dissolves the salt due to the formation of the complex ion $Ag(NH_3)_2^+$.

Interactive Figure 18.4.2

Explore the effect of complex ion formation on solubility.

$NH_3(aq)$

AgCl(s) $[Ag(NH_3)_2]^+(aq)$

Addition of ammonia causes silver chloride to dissolve, forming a silver–ammonia complex ion.

Example Problem 18.4.2 Show the effect of complex ion formation on solubility.

Consider the insoluble compound zinc cyanide, $Zn(CN)_2$. The zinc ion also forms a complex ion with hydroxide ions. Write an equation to show why the solubility of $Zn(CN)_2$ increases in the presence of hydroxide ions and calculate the equilibrium constant for this reaction.

Example Problem 18.4.2 *(continued)*

Solution:

You are asked to write an equation for the reaction between an insoluble compound and a Lewis base and to calculate the equilibrium constant for the reaction.

You are given the identity of the insoluble compound and the Lewis base.

Use the equations representing the dissolution of $Zn(CN)_2$ and the formation of the $[Zn(OH)_4]^{2-}$ complex ion to write the new equation showing why the solubility of $Zn(CN)_2$ increases in the presence of hydroxide ions.

$$Zn(CN)_2(s) \rightleftarrows Zn^{2+}(aq) + 2\ CN^-(aq) \qquad\qquad K_{sp} = 8.0 \times 10^{-12}$$

$$Zn^{2+}(aq) + 4\ OH^-(aq) \rightleftarrows [Zn(OH)_4]^{2-}(aq) \qquad\qquad K_f = 4.6 \times 10^{17}$$

$$\overline{Zn(CN)_2(s) + 4\ OH^-(aq) \rightleftarrows [Zn(OH)_4]^{2-}(aq) + 2\ CN^-(aq) \quad K_{net} = K_{sp}K_f = 3.7 \times 10^6}$$

The reaction of $Zn(CN)_2(s)$ with hydroxide is a product-favored reaction.

18.4c Solubility, Ion Separation, and Qualitative Analysis

The principles of simultaneous equilibria as applied to sparingly soluble compounds and complex ions can be used to separate mixtures of ions and to identify the components of a mixture. This process is called **qualitative analysis**. For example, consider a solution that contains a mixture of Ni^{2+} and Zn^{2+} (Interactive Figure 18.4.3a). Is it possible to physically separate the two ions by precipitating one of them while the other remains dissolved in solution? Examination of Tables 18.1.1 and 18.3.2 shows that although both ions form insoluble hydroxide salts and complex ions with ammonia, only zinc forms a complex ion with the hydroxide ion.

Ni^{2+} and Zn^{2+} hydroxide salts:

$$Ni(OH)_2(s) \rightleftarrows Ni^{2+}(aq) + 2\ OH^-(aq) \qquad K_{sp} = 2.8 \times 10^{-16}$$

$$Zn(OH)_2(s) \rightleftarrows Zn^{2+}(aq) + 2\ OH^-(aq) \qquad K_{sp} = 4.5 \times 10^{-17}$$

Ni^{2+} and Zn^{2+} ammonia and hydroxide complex ions:

$$Ni^{2+}(aq) + 6\ NH_3(aq) \rightleftarrows Ni(NH_3)_6^{2+}(aq) \qquad K_f = 5.5 \times 10^8$$

$$Zn^{2+}(aq) + 4\ NH_3(aq) \rightleftarrows Zn(NH_3)_4^{2+}(aq) \qquad K_f = 2.9 \times 10^9$$

$$Zn^{2+}(aq) + 4\ OH^-(aq) \rightleftarrows Zn(OH)_4^{2-}(aq) \qquad K_f = 4.6 \times 10^{17}$$

Video Solution

Tutored Practice
Problem 18.4.2

Interactive Figure 18.4.3

Explore the separation of a mixture of cations.

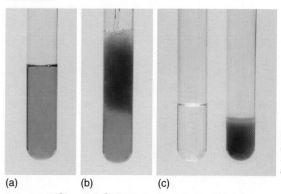

(a)   (b)   (c)

Aqueous Ni^{2+} and Zn^{2+} ions can be separated by the addition of excess sodium hydroxide.

Charles D. Winters

Using this information, a possible separation scheme could involve reacting the mixture of Ni^{2+} and Zn^{2+} with excess hydroxide ion. The Ni^{2+} present in solution will precipitate as $Ni(OH)_2$, and the Zn^{2+} will remain dissolved in solution as the $Zn(OH)_4{}^{2-}$ complex ion (Interactive Figure 18.4.3b). Carefully pouring the solution containing $Zn(OH)_4{}^{2-}$ into a separate test tube separates the two ions (Interactive Figure 18.4.3c).

As shown in the following example problems, it is possible to address the selective precipitation and separation of ions from a mixture both qualitatively and quantitatively.

Example Problem 18.4.3 Determine precipitation sequence (qualitative).

A solution contains 0.10 M sodium sulfate and 0.10 M potassium sulfite. Solid calcium acetate is added slowly to this mixture. What ionic compound precipitates first from the solution?

Solution:

You are asked to determine which of two salts precipitates first when a solid is added to a mixture of soluble salts.

You are given the concentration and identity of the salts in the mixture and the identity of the solid added to the mixture.

Two possible precipitates of calcium can form when solid calcium acetate is added to the mixture: calcium sulfate, $CaSO_4$, and calcium sulfite, $CaSO_3$. The compound that requires the lowest concentration of Ca^{2+} to initiate precipitation will precipitate first.

$$CaSO_4(s) \rightleftarrows Ca^{2+}(aq) + SO_4{}^{2-}(aq) \qquad K_{sp} = [Ca^{2+}][SO_4{}^{2-}] = 2.4 \times 10^{-5}$$
$$CaSO_3(s) \rightleftarrows Ca^{2+}(aq) + SO_3{}^{2-}(aq) \qquad K_{sp} = [Ca^{2+}][SO_3{}^{2-}] = 1.3 \times 10^{-8}$$

Because the two compounds have a 1:1 cation:anion ratio and the cations are both present in the same concentration (0.10 M), the K_{sp} values for the salts can be used to determine relative solubility. In this case, $K_{sp}(CaSO_3) < K_{sp}(CaSO_4)$, so $CaSO_3$ is less soluble than $CaSO_4$. Thus, $CaSO_3$ will precipitate first from the solution.

It is also possible to solve this problem quantitatively.

$$[Ca^{2+}] = \frac{K_{sp}(CaSO_4)}{[SO_4^{2-}]} = \frac{2.4 \times 10^{-5}}{0.10} = 2.4 \times 10^{-6} \text{ M for } CaSO_4 \text{ precipitation}$$

$$[Ca^{2+}] = \frac{K_{sp}(CaSO_3)}{[SO_3^{2-}]} = \frac{1.3 \times 10^{-8}}{0.10} = 1.3 \times 10^{-9} \text{ M for } CaSO_3 \text{ precipitation}$$

The concentration of Ca^{2+} at which $CaSO_3$ begins to precipitate is lower than that required for precipitation of $CaSO_4$. Therefore, $CaSO_3$ precipitates first.

Video Solution

Tutored Practice
Problem 18.4.3

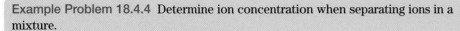

Example Problem 18.4.4 Determine ion concentration when separating ions in a mixture.

A solution contains 1.27×10^{-2} M calcium nitrate and 1.25×10^{-2} M barium acetate. Solid ammonium fluoride is added slowly to this mixture. What is the concentration of calcium ion when barium fluoride begins to precipitate?

Solution:

You are asked to determine the concentration of one ion in a mixture when the other ion begins to precipitate.

You are given the identity and concentration of ions in the mixture and the identity of the solid added to the mixture.

When ammonium fluoride is added to a solution of calcium nitrate and barium acetate, two possible precipitates can form: calcium fluoride, CaF_2, and barium fluoride, BaF_2. These compounds have the fluoride ion in common and the same general stoichiometry (1:2 cation:anion ratio). BaF_2 is more soluble than CaF_2 [$K_{sp}(BaF_2) > K_{sp}(CaF_2)$], so CaF_2 will precipitate first as fluoride ion is added to the cation mixture.

Step 1. Use the Ba^{2+} concentration and the K_{sp} of BaF_2 to calculate the concentration of F^- required to precipitate BaF_2.

$$BaF_2(s) \rightleftarrows Ba^{2+}(aq) + 2\,F^-(aq) \qquad\qquad K_{sp} = [Ba^{2+}][F^-]^2 = 1.7 \times 10^{-6}$$

$$[F^-] = \sqrt{\frac{K_{sp}}{[Ba^{2+}]}} = \sqrt{\frac{1.7 \times 10^{-6}}{1.25 \times 10^{-2}}} = 0.0117 \text{ M}$$

If $[F^-]$ is greater than 0.0117 M, BaF_2 will precipitate.

Step 2. Use the minimum $[F^-]$ required to precipitate BaF_2 to calculate $[Ca^{2+}]$ when BaF_2 begins to precipitate.

$$CaF_2(s) \rightleftarrows Ca^{2+}(aq) + 2\,F^-(aq) \qquad\qquad K_{sp} = [Ca^{2+}][F^-]^2 = 3.9 \times 10^{-11}$$

$$[Ca^{2+}] = \frac{K_{sp}}{[F^-]^2} = \frac{3.9 \times 10^{-11}}{(0.0117)^2} = 2.87 \times 10^{-7} \text{ M}$$

When barium begins to precipitate as BaF_2, the calcium ion concentration has decreased from 1.27×10^{-2} M to 2.87×10^{-7} M. The calcium ion is effectively removed from solution (in the form of insoluble CaF_2) before BaF_2 begins to precipitate.

Video Solution

Tutored Practice Problem 18.4.4

Section 18.4 Mastery

Unit Recap

Key Concepts

18.1 Solubility Equilibria and K_{sp}

- The solubility product constant, K_{sp}, is the equilibrium constant for the dissolution of a sparingly soluble compound (18.1b).

- K_{sp} values are determined from experimental equilibrium ion concentrations or calculated from solubility values (18.1c).

18.2 Using K_{sp} in Calculations

- The solubility of a salt in pure water is defined as the amount of solid that will dissolve per liter of solution (g/L or mol/L) and can be calculated from K_{sp} values (18.2a).

- Solubility and K_{sp} are related by compound stoichiometry (18.2a).

- The reaction quotient, Q, can be used to determine whether a solution is at equilibrium (saturated solution), whether additional solid can dissolve (unsaturated solution), or whether a precipitate will form (supersaturated solution) (18.2b).

- The presence of a common ion always decreases the solubility of an ionic compound (18.2c).

18.3 Lewis Acid–Base Complexes and Complex Ion Equilibria

- A Lewis base, an electron-pair donor, donates a lone pair of electrons to a Lewis acid, an electron-pair acceptor, to form a Lewis acid–base adduct (or Lewis acid–base complex) (18.3a).

- The bond between a Lewis acid and a Lewis base is called a coordinate–covalent bond (18.3a).

- A Lewis acid–base complex with an overall charge is called a complex ion (18.3a).

- The formation constant, K_f, is the equilibrium constant for the formation of a complex ion and the dissociation constant, K_d, is the equilibrium constant for the dissociation of a complex ion (18.3b).

18.4 Simultaneous Equilibria

- Addition of acid to a sparingly soluble salt containing a basic anion will increase the solubility of the salt (18.4a).

- Addition of a Lewis base to a sparingly soluble salt can increase the solubility of the salt (18.4b).
- Qualitative analysis is an experimental method used to separate mixtures of ions and to identify the components of a mixture (18.4c).

Key Terms

18.1 Solubility Equilibria and K_{sp}
solubility product constant, K_{sp}

18.3 Lewis Acid–Base Complexes and Complex Ion Equilibria
Lewis base
Lewis acid

Lewis acid–base adduct
coordinate–covalent bond
complex ion
formation constant, K_f
dissociation constant, K_d

18.4 Simultaneous Equilibria
qualitative analysis

Unit 18   Review and Challenge Problems

19 Thermodynamics: Entropy and Free Energy

Unit Outline

In This Unit...

As we saw in our studies of chemical equilibria, reactions proceed to equilibrium at which point no net change occurs. The position of equilibrium may be very close to reactants, very close to products, or somewhere in between. The spontaneous direction of change toward equilibrium depends upon the reaction conditions. In this unit, we will examine the second law of thermodynamics, which tells us that the direction of spontaneous change is the direction in which the total entropy of the universe increases. This can be expressed in the form of the free energy of the system, which we will use to tell us whether a reaction is thermodynamically favored.

19.1 Entropy and the Three Laws of Thermodynamics

19.1a The First and Second Laws of Thermodynamics

The word *thermodynamics* includes the ideas of heat *(thermo-)* and change *(dynamics)*.

There are many ways of stating the **first law of thermodynamics**:

- The internal energy of an isolated system is constant (E_{isolated} = constant).

- The total energy of the universe is constant (E_{universe} = constant).

- The change in the energy of the universe is zero ($\Delta E_{\text{universe}}$ = 0).

- For a closed system, $\Delta E = q + w$.

Recall that **internal energy** is the sum of the kinetic and potential energies of the particles of a system. It does not include the translation or rotation of the entire system. We cannot measure internal energy, E, directly; we can measure only the change in internal energy, ΔE.

The first law of thermodynamics proposes the concept of internal energy, E, and tells us that although energy may change forms, it is not created or destroyed. However, the first law of thermodynamics does not tell us *what* changes are allowed or *in which direction* change will proceed. To illustrate this, consider the sequence of events shown in Interactive Figure 19.1.1.

Based on our experience, we know that the sequence of events in Interactive Figure 19.1.1 will always begin with Step 3 and proceed to Step 1, where the hydrogen balloon is ignited and explodes, producing water and energy. We never observe the reverse happening, where the pieces of balloon spontaneously gather together and the water turns back into hydrogen + oxygen gas. Energy is conserved whether the process runs backward or forward, but there is an *allowed direction* in which these events always occur.

In fact, most chemical and physical changes naturally occur in one direction and can occur in the opposite direction only with assistance. For example, hydrogen and oxygen react spontaneously to create water. Although a spark is needed to initiate the reaction, once begun, it continues without any further input of energy. A **spontaneous process** is one that occurs without a continual external source of energy.

The opposite reaction, where water decomposes into hydrogen and oxygen gas, can be induced by passing an electric current through a salt solution, but this requires a continuous input of electrical energy, as shown in Figure 19.1.2. This is a nonspontaneous process.

Explore directionality in chemical processes.

1   2   3

Hydrogen balloon explosion sequence

19.1b Entropy and the Second Law of Thermodynamics

The first law of thermodynamics tells us how to measure the change in terms of heat and work. The **second law of thermodynamics** tells us the *allowed direction* of change.

According to the second law of thermodynamics, all physical and chemical changes occur such that

- at least some energy *disperses,*

- the total concentrated or organized energy of the universe *decreases,* and

- the total diffuse or *dis*organized energy of the universe *increases.*

The first law is about the amount of energy. The second law is about the quality of the energy. The first law introduces the thermodynamic variable E, the internal energy. The second law introduces a new thermodynamic variable S, the **entropy**, which is a measure of the dissipated energy within a system at a given temperature. Like the internal energy, entropy is a state function and it has units of energy per kelvin, J/K.

Many different terms are used to describe entropy: *disorganized, chaotic, random, dispersed, diffuse, unconcentrated, dissipated, disordered,* and *delocalized.* In all

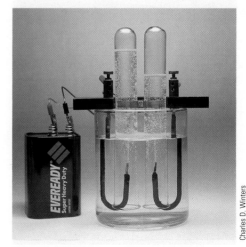

Figure 19.1.2 A nonspontaneous process: splitting water into hydrogen and oxygen

cases, these terms refer to the random molecular energy within a system in the form of translational, rotational, and vibrational motion of the particles. Increasing the temperature increases the random motion of the particles within a substance and, thus, increases its entropy.

Spontaneous change always occurs in the direction of increasing total entropy. An isolated system, one that cannot exchange matter or energy with the surroundings (imagine a thermos bottle), will never spontaneously decrease in entropy. It will remain unchanged or move to higher entropy. Therefore, entropy change is about the *dispersal* of energy, in the form of matter or thermal energy, as shown in Interactive Figure 19.1.3.

Here are some statements of the second law using the concept of entropy:

- All physical and chemical changes occur such that the total entropy of the universe increases.

$$\Delta S_{universe} = \Delta S_{system} + \Delta S_{surroundings} > 0 \text{ for a spontaneous process} \quad \textbf{(19.1)}$$

- The entropy of an isolated system never spontaneously decreases.

$$\Delta S_{isolated} \geq 0$$

Note that the entropy of a *system* could go down, provided that the entropy of the *surroundings* goes up more, or visa versa, so that the sum of the changes is greater than zero. The underlying message is that for an exchange of energy to occur spontaneously, some of the energy must become more diffuse. The consequence is that some energy must always be discarded as heat. Other ways to say this include:

- It is impossible to completely turn heat (diffuse energy) into work (concentrated energy).

- It is impossible to extract thermal energy from a source at one temperature and turn it completely into work. Some energy must be simultaneously discarded as heat to a body at a lower temperature.

- Heat flows from hot to cold.

19.1c Entropy and Microstates

At a molecular level we can relate entropy to the number of ways the total energy can be distributed among all of the particles. Each possible distribution of the energy at the molecular level characterizes a **microstate** of the system. Entropy is a measure of the number of microstates that are accessible at a given total energy. The entropy of a system measures

Investigate dispersal of energy and matter.

Serhiy Kobyakov/Shutterstock.com

Dispersal of energy

the number of ways the system can be different microscopically and still have the same macroscopic state. A high number of microstates (such as the system where the balloon has exploded and energy is dispersed in the enormous number of nitrogen, oxygen, and water molecules in the air) corresponds to a high entropy, and the natural direction of change is toward higher entropy (Figure 19.1.4).

This interpretation was developed very elegantly by Austrian physicist Ludwig Boltzmann and is summarized in Equation 19.2.

$$S = k_B \ln W \qquad\qquad (19.2)$$

where

k_B = Boltzmann's constant = 1.381×10^{-23} J/K
W = number of microstates

As the number of microstates, W, increases, the entropy, S, increases logarithmically. Boltzmann's constant, k_B, is a proportionality constant in this equation and has the same units as entropy, energy/kelvin (J/K).

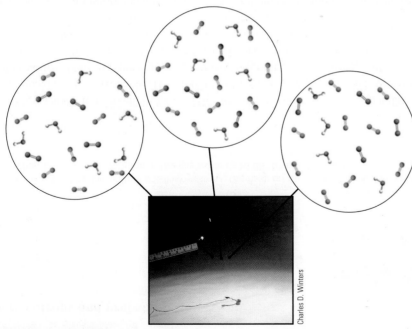

Figure 19.1.4 Entropy and microstates

Explore entropy and probability.

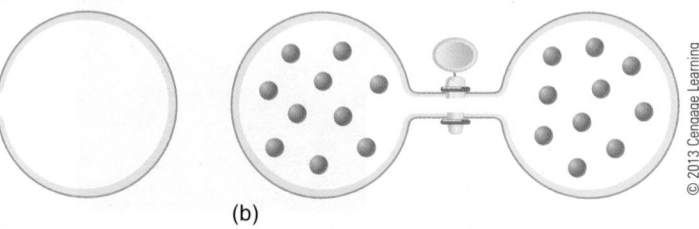

(a)   (b)

(a) A gas is held in one bulb of a 2-flask system (the flask on the right is empty, a vacuum).
(b) When the stopcock is opened, the gas expands to fill the empty space.

A system can "bump" itself into any of the accessible W microstates, depending on how much thermal energy is available. The amount of thermal energy available is determined by the temperature. Thus, temperature is a critical factor when considering entropy. Entropy can be considered to be the amount of diffused energy within a substance at a given temperature. (We will come back to this when we discuss the third law of thermodynamics.)

Thus, we can view the direction of spontaneous change from a macroscopic perspective as a dispersal of energy from more concentrated energy to more diffuse energy. From a molecular perspective we can view the direction of spontaneous change as moving from a limited number of ways to distribute the energy among all of the particles of the system to a larger number of ways (Interactive Figure 19.1.5).

19.1d Trends in Entropy

A number of factors contribute to the entropy of a substance, such as the physical state, temperature, molecular size, intermolecular forces, and mixing (Interactive Figure 19.1.6).

Physical State
Gases tend to have the highest entropies because the motion of gas particles is highly random. Liquids tend to have higher entropies than solids, which are much more restricted in their motions. At room temperature, one mole of $CO_2(g)$ has a much higher entropy than one mole of liquid water, which has a higher entropy than one mole of solid copper metal.

Interactive Figure 19.1.6

Examine trends in entropy.

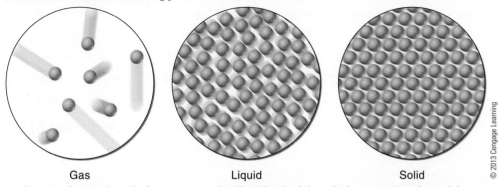

Gas   Liquid   Solid

© 2013 Cengage Learning

In general, gases have higher entropies than liquids, which have higher entropies than solids.

Temperature

The entropy of a substance increases with temperature because the translational and rotational motions of the molecules increase with temperature. Water at 50 °C is higher in entropy than water at 25 °C.

Molecular Size

Larger molecules have more ways to randomly distribute energy at a given temperature and therefore have higher entropies. The entropy of propane, $C_3H_8(g)$, is higher than the entropy of methane, $CH_4(g)$, at room temperature.

Forces Between Particles in a Solid

The entropy of a solid is higher if there are weak forces between the particles. Solids with strong forces have lower entropies because the motion within the solid is more restricted. Diamond, a highly ordered crystal with covalent bonds between all of the carbon atoms, has very low entropy. For ionic solids, both charge and ionic radius impact entropy. Sodium fluoride, an ionic crystal with charges of $+1$ and -1 on the ions, has higher entropy than magnesium oxide with charges of $+2$ and -2 because the electrostatic forces between the ions are higher between the more highly charged Mg^{2+} and O^{2-} ions. Cesium iodide and sodium fluoride both contain $+1$ and -1 ions, but CsI has higher entropy than NaF because it is made up of ions with much larger ionic radii.

Dissolution and Mixing

The entropy of a solid increases significantly when dissolved in solution. NaCl(aq) has a much higher entropy than NaCl(s). When two or more substances mix together, the entropy increases.

Example Problem 19.1.1 Predict which substances have higher entropy.

Predict which substance in each pair has the higher entropy. Assume there is one mole of each substance at 25 °C and 1 bar.

a. $Hg(\ell)$ or $CO(g)$
b. $CH_3OH(\ell)$ or $CH_3CH_2OH(\ell)$
c. $KI(s)$ or $CaS(s)$

Solution:

You are asked to choose the substance with the higher entropy at 25 °C.

You are given two substances and their physical states.

a. At 25 °C, carbon monoxide gas has a higher entropy than liquid mercury because the motion of the gas molecules is much more random than the motion of the mercury atoms in the liquid state.
b. Liquid ethanol, $CH_3CH_2OH(\ell)$, has a higher entropy than liquid methanol, $CH_3OH(\ell)$, because it is a larger, more complex molecule and therefore has a larger number of ways to distribute energy at a given temperature.
c. Potassium iodide, composed of K^+ and I^- ions, has a higher entropy than calcium sulfide, composed of Ca^{2+} and S^{2-} ions. In addition to the different ionic charges, KI contains ions with larger ionic radii. The entropy of ionic solids tends to increase as the attractive forces between the ions decrease because vibrational motion between the ions is easier.

Video Solution

Tutored Practice
Problem 19.1.1

19.1e Spontaneous Processes

A spontaneous process is

- any process in which the total entropy of the universe increases, or

- any process that is able to occur without being continuously *driven* by an external source of energy.

Time is not a part of the thermodynamic definition of a spontaneous process. A spontaneous process may or may not happen immediately, or at all. For example, the conversion of diamond to graphite is a spontaneous process at 25 °C and a pressure of 1 bar, even though this process is so slow that it cannot be observed in a human lifetime (Figure 19.1.7).

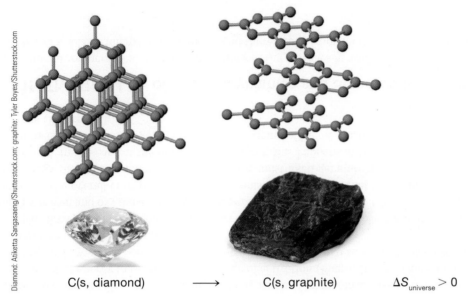

C(s, diamond) $\longrightarrow$ C(s, graphite) $\Delta S_{universe} > 0$

Figure 19.1.7 The conversion of diamond to graphite is a spontaneous process at 25 °C and 1 bar.

Example Problem 19.1.2 Determine the spontaneity of a process.

a. A candle is burning in air. Is a reaction occurring? Is this a spontaneous system?
b. A candle is sitting in air but has not been lit. Is a reaction occurring? Is this a spontaneous system?

Solution:

You are asked to decide if a chemical system is spontaneous or nonspontaneous, and whether a reaction is occurring.

You are given a description of the chemical system.

a. A reaction is occuring. When a candle burns, a combustion reaction is taking place. Any reaction that is actually occurring is a spontaneous system.
b. The unlit candle in the presence of oxygen in the air is not, at least visibly, reacting. However, this still represents a spontaneous system, one that is stopped from reacting due to the high kinetic barrier for the oxidation process. Spontaneity is a function of the nature of the reactants and products of an event, not the rate at which the transformation can occur.

Video Solution

Tutored Practice
Problem 19.1.2

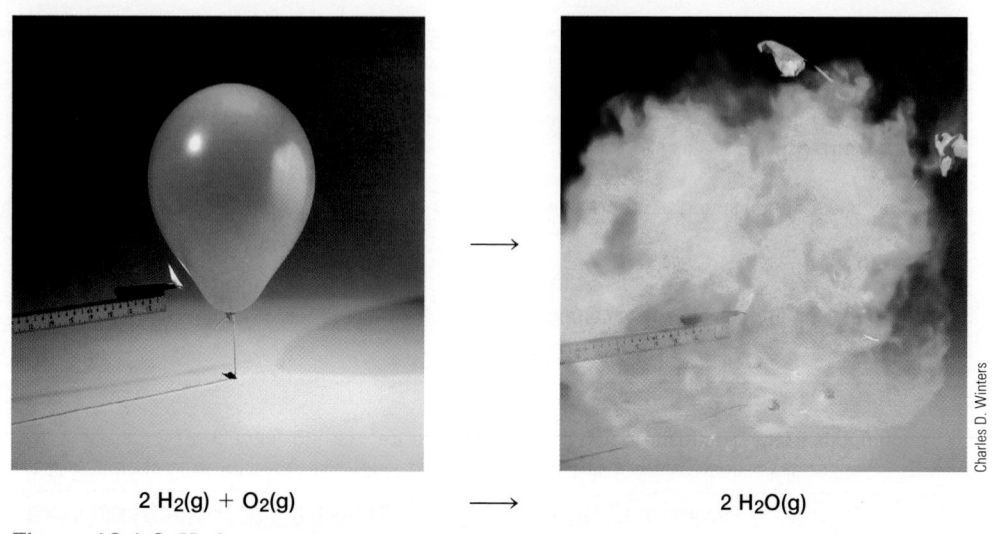

$2 H_2(g) + O_2(g)$ $\longrightarrow$ $2 H_2O(g)$

Figure 19.1.8 Hydrogen spontaneously reacts with oxygen after energy is added to overcome the energy of activation.

Figure 19.1.9 Entropy generated by production of electricity

Another point about spontaneous processes is that they may require an initial input of energy to begin the process. For example, the reaction of hydrogen and oxygen requires an initial input of energy to overcome the energy of activation (Figure 19.1.8). Once the reaction begins, however, it continues without being driven by an external source of energy and is therefore a spontaneous process.

One final point is that nonspontaneous reactions are not impossible. Water *can* be split into hydrogen and oxygen gas if a *constant input of energy* is provided. For this to happen, however, the overall entropy of the universe must still increase. That is, the production of the energy driving the nonspontaneous process must itself generate more entropy than is consumed. In the case of the electrolysis of water, this happens at the power plant where fossil fuels are burned to generate electricity, giving off some wasted heat energy in the process (Figure 19.1.9).

19.1f The Third Law of Thermodynamics and Standard Entropies

As temperature increases, the random motions of the particles in a substance increase, the number of available microstates increase, and therefore entropy increases. As temperature decreases, random motions and entropy decrease. The **third law of thermodynamics** addresses the point at which the entropy of a substance becomes zero (Figure 19.1.10):

Third law: The entropy of a pure crystalline substance is zero at absolute zero.

$$\text{As } T \rightarrow 0 \text{ K, } S \text{ (pure crystal)} \rightarrow 0 \text{ J/K}$$

A pure crystalline substance is one in which the molecules are in perfect alignment with one another and every molecule is identical. If the molecules do not align themselves perfectly, a substance will have some residual entropy at absolute zero, which is not a violation of the third law. Although we cannot actually reach absolute zero in the laboratory, the third law of thermodynamics tells us that there is a zero of entropy. As the temperature is increased, the entropy of a substance increases; therefore, entropy values for all substances are always greater than zero.

Standard Molar Entropy

The third law of thermodynamics allows us to determine the entropy of a substance at a given temperature, or its **standard molar entropy ($S°$)**. The standard molar entropy of a pure substance can be determined by careful measurements of its molar heat capacity as a function of temperature and of the heat absorbed for each phase change, $\Delta H°_{fusion}$, $\Delta H°_{vaporization}$, and $\Delta H°_{sublimation}$. Recall that heat capacity is the energy required to raise the temperature of a substance by one degree Celsius or one kelvin. It has the same units as standard molar entropy, J/mol · K. Table 19.1.1 shows standard molar entropy values for water at various temperatures. Interactive Figure 19.1.11 shows a generalized plot of the entropy of a substance at each temperature from 0 K to above its boiling point.

Note in Table 19.1.1 and Interactive Figure 19.1.11 that at each phase change there is a dramatic rise in entropy while the temperature remains constant. As we noted earlier, the

Table 19.1.1 Entropy of H$_2$O from 2 K to 373 K

Phase	Temperature (K)	$S°$ (J/mol K)
Solid	2	0.00049
	12	0.14
	38	3.1
Liquid	273	64
	300	70
	373	87
Gas	373	197

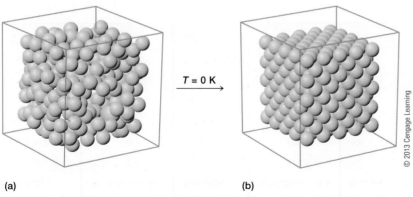

© 2013 Cengage Learning

(a)   (b)

Figure 19.1.10 (a) A pure crystalline substance at $T > 0$ K ($S > 0$). (b) A pure crystalline substance at $T = 0$ K ($S = 0$).

Interactive Figure 19.1.11

Relate entropy and temperature.

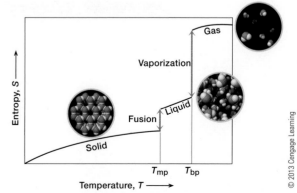

© 2013 Cengage Learning

Entropy as a function of temperature

Table 19.1.2 Standard Molar Entropies for Some Common Substances at 25 °C

Solids	S° (J/mol K)	Liquids	S° (J/mol K)	Gases	S° (J/mol K)
C(diamond)	2.4	Hg	75.9	He	126.2
C(graphite)	5.7	H_2O	69.9	Ne	146.3
Fe	27.8	CH_3OH	126.8	H_2	130.7
Al	28.3	Br_2	152.2	O_2	205.1
Cu	33.2	H_2SO_4	156.9	H_2O	188.8
CaS	56.5	CH_3CH_2OH	160.7	CO	197.7
KI	106.3	C_6H_6	172.8	CO_2	213.7

Section 19.1 Mastery

entropy of a substance depends on its physical state. The liquid phase of a substance has higher entropy than the solid form at the same temperature, and the gas phase has much higher entropy than the liquid phase at the same temperature. Thus, the increase in entropy when water vaporizes is much greater than when ice melts.

Table 19.1.2 shows standard molar entropy values at 25 °C for some common solids, liquids, and gases. Note that, in general, solids have lower entropies than liquids and gases have the highest entropies, but there can be overlap of these regions as compounds get heavier or more complex. In Table 19.1.2, entropy values are reported as an absolute value (S°), not as a change in entropy. This is because there is an absolute reference point for entropy ($S = 0$ J/K at 0 K). Contrast this with energy, where there is no temperature at which the internal energy or enthalpy of a substance goes to zero. We can talk about the enthalpy of a substance only *relative to* other substances. Because of this, thermodynamic tables list the enthalpy of formation of a substance, ΔH_f°, its enthalpy *relative to* the enthalpy of its elements. Recall also that for an element in its standard state, $\Delta H_f^\circ = 0$ (there is no energy change to form it from itself), whereas all substances, including elements, have a positive value for S°.

19.2 Calculating Entropy Change

19.2a Standard Entropy Change for a Phase Change

When a solid melts, it undergoes an increase in entropy. The temperature remains constant during this process, and at constant pressure, the energy change is equal to ΔH_{fusion}. The entropy change for this process is $\Delta H_{fusion}/T_{melting\ point}$. Similarly, when a liquid boils, the entropy change for this process is $\Delta H_{vaporization}/T_{boiling\ point}$. A general expression for the change in entropy for any phase change is given by Equation 19.3, where the temperature must be in kelvins:

Table 19.2.1 Standard Enthalpy and Entropy Changes for Fusion and Vaporization

Substance	Melting Point (K)	$\Delta H°_{fusion}$ (kJ/mol)	$\Delta S°_{fusion}$ (J/mol K)	Boiling Point (K)	$\Delta H°_{vaporization}$ (kJ/mol)	$\Delta S°_{vaporization}$ (J/mol K)
Hg	234	2.30	9.83	630	59.1	93.8
H_2O	273	6.01	22.0	373	40.7	109
NH_3	195	5.66	29.0	240	23.3	97.1
CH_3OH	176	3.22	18.3	338	35.2	104
CH_3CH_2OH	159	5.02	31.6	352	38.6	110
C_6H_6	279	9.95	35.7	353	30.7	87.0

$$\Delta S_{\text{phase change}} = \frac{\Delta H_{\text{phase change}}}{T_{\text{phase change}}} \qquad (19.3)$$

Table 19.2.1 lists the phase change data for a number of substances. Note that for every substance, ΔH_{fusion} and ΔS_{fusion} are much smaller than $\Delta H_{vaporization}$ and $\Delta S_{vaporization}$, as discussed previously.

If the process is reversed, a liquid freezes, or a gas condenses, the sign of ΔH changes. In this case the change in entropy is negative, so the system becomes more ordered.

Example Problem 19.2.1 Calculate the entropy change when a gas condenses at its boiling point.

Calculate the entropy change when two moles of carbon monoxide condense at its boiling point of -191.4 °C. The heat of vaporization of CO is 6.04 kJ/mol.

Solution:

You are asked to calculate ΔS for the condensation of CO(g) at its boiling point.

You are given:

$$n = 2 \text{ mol CO}$$
$$T_{\text{boiling point}} = (-191.4 \text{ °C} + 273.15) = 81.75 \text{ K}$$
$$\Delta H_{\text{vaporization}} = 6.04 \text{ kJ/mol}$$

Use Equation 19.3 to calculate ΔS for this process. Remember that the gas condenses—the opposite of vaporization—so the sign of $\Delta H_{\text{vaporization}}$ is reversed.

$$\Delta S_{\text{condense}} = \frac{-\Delta H_{\text{vaporization}}}{T_{\text{boiling point}}} = \frac{-(6.04 \text{ kJ/mol})(2 \text{ mol})}{81.75 \text{ K}} = -(0.1478 \text{ kJ/K})(1000 \text{ J/kJ}) = -148 \text{ J/K}$$

Video Solution

Tutored Practice
Problem 19.2.1

19.2b Standard Entropy Change for a Chemical Reaction

Similar to the standard enthalpy change for a reaction, $\Delta H°_{rxn}$, the **standard entropy change** for a reaction, $\Delta S°_{rxn}$, can be calculated from standard molar entropy values (Equations 19.4a and 19.4b), where each standard molar entropy value is multiplied by the stoichiometric coefficient in the balanced chemical equation.

$$\Delta H°_{rxn} = \Sigma \Delta H_f° \text{ (products)} - \Sigma \Delta H_f° \text{(reactants)} \qquad \textbf{(19.4a)}$$

$$\Delta S°_{rxn} = \Sigma S° \text{(products)} - \Sigma S° \text{(reactants)} \qquad \textbf{(19.4b)}$$

$\Delta S°_{rxn}$ is the entropy change when pure unmixed reactants in their standard states are converted to pure unmixed products in their standard states. The standard state is the most stable state of an element or compound at a pressure of 1 bar, a fixed temperature, typically 25 °C, and concentrations of 1 M for any solutions.

The sign of $\Delta S°_{rxn}$ can often be estimated by taking into account the stoichiometry of a reaction and the physical states of reactants and products. If the total number of moles of gas increases going from reactants to products, we can predict that the sign of $\Delta S°_{rxn}$ is positive. The products are higher in entropy than are the reactants. Conversely, if the number of moles of gaseous products is less than the number of moles of gaseous reactants, the sign of $\Delta S°_{rxn}$ is negative.

Example Problem 19.2.2 Predict the sign and calculate $\Delta S°$ for a reaction.

Hydrogen and oxygen react to form water vapor in a spontaneous reaction.

$$2\,H_2(g) + O_2(g) \rightarrow 2\,H_2O(g)$$

Predict the sign of the entropy change for this reaction, and using Table 19.2.2, calculate $\Delta S°_{rxn}$ at 25 °C.

Solution:

You are asked to predict the sign of $\Delta S°_{rxn}$ and then to calculate this value.

You are given the balanced chemical equation and the standard molar entropies for water vapor and for hydrogen and oxygen gas.

When hydrogen and oxygen gas react to form water vapor, three moles of gas are converted into two moles of gas, for every mole of reaction. The total number of moles of gas decreases, so $\Delta S°_{rxn}$ for this reaction should be negative.

Use standard molar entropy values and Equation 19.4b to calculate the standard entropy change.

Example Problem 19.2.2 *(continued)*

$\Delta S^\circ_{\text{rxn}} = \Sigma S^\circ(\text{products}) - \Sigma S^\circ(\text{reactants})$

$\quad = (2 \text{ mol})(S^\circ[H_2O(g)]) - \{(2 \text{ mol})(S^\circ[H_2(g)]) + (1 \text{ mol})(S^\circ[O_2(g)])\}$

$\quad = (2 \text{ mol})(188.8 \text{ J/K} \cdot \text{mol}) - \{(2 \text{ mol})(130.7 \text{ J/K} \cdot \text{mol}) + (1 \text{ mol})(205.1 \text{ J/K} \cdot \text{mol})\}$

$\quad = -88.9 \text{ J/K}$

The entropy change for this reaction is negative, as predicted based on the reaction stoichiometry.

> **Video Solution**
>
> **Tutored Practice Problem 19.2.2**

19.2c Entropy Change in the Surroundings

The reaction between hydrogen and oxygen to form water has a negative entropy change even though the reaction is spontaneous. This is possible because the reaction is highly exothermic. As the heat of the reaction expands outward, it increases the thermal energy of the molecules in the air surrounding the chemical reaction. As a result, the increased entropy of the surroundings more than compensates for the fact that the entropy of the system has decreased.

$$\Delta S_{\text{surr}} > - \Delta S_{\text{system}}$$

The change in entropy of the surroundings is related to q, the amount of heat transferred during a chemical or physical change, and the temperature at which the change takes place.

$$\Delta S_{\text{surr}} = \frac{-q_{\text{system}}}{T}$$

For a process that takes place at constant pressure, q_{system} is the enthalpy change for the reaction, ΔH_{rxn}.

$$\Delta S_{\text{surr}} = \frac{-\Delta H_{\text{rxn}}}{T} \text{ at constant pressure} \qquad \textbf{(19.5)}$$

For an exothermic reaction, ΔS_{surr} is positive and the entropy of the surroundings increases. Exothermic reactions create entropy in the surroundings, which helps drive a reaction forward.

For an endothermic reaction, ΔS_{surr} is negative and the entropy of the surroundings decreases. Endothermic reactions decrease entropy in the surroundings. For an endothermic reaction to be spontaneous, the entropy change of the system must be a positive value, large enough to compensate for the decrease in entropy the endothermic reaction causes to the surroundings.

Example Problem 19.2.3 Calculate ΔS_{surr} and $\Delta S_{universe}$ for a reaction.

Hydrogen and oxygen react to form water vapor in a spontaneous reaction with a negative standard entropy change.

$$2 \, H_2(g) + O_2(g) \rightarrow 2 \, H_2O(g) \qquad \Delta S^\circ_{rxn} = -88.9 \, J/K$$

Calculate the entropy change of the surroundings and the universe at 25 °C.

Solution:

You are asked to calculate $\Delta S_{surroundings}$ and $\Delta S_{universe}$ at 25 °C for a reaction.

You are given the balanced chemical equation and ΔS°_{rxn}.

$$\Delta S_{surr} = \frac{-\Delta H_{rxn}}{T}$$

Use standard heats of formation to calculate the enthalpy change for this reaction under standard state conditions. Recall that the standard heat of formation for an element in its standard state is equal to zero.

$$\begin{aligned}
\Delta H^\circ_{rxn} &= \Sigma \Delta H_f^\circ(\text{products}) - \Sigma \Delta H_f^\circ(\text{reactants}) \\
&= (2 \, mol)(\Delta H_f^\circ[H_2O(g)]) - (2 \, mol)(\Delta H_f^\circ[H_2(g)]) - \Delta H_f^\circ[O_2(g)] \\
&= (2 \, mol)(-241.8 \, kJ/mol) - 0 - 0 \\
&= -483.6 \, kJ
\end{aligned}$$

At 25 °C

$$\Delta S_{surr} = \frac{-\Delta H_{rxn}}{T} = \frac{-(-483.6 \, kJ)}{298 \, K} = 1.62 \, kJ/K$$

The reaction is exothermic, so ΔS_{surr} is positive (the entropy of the surroundings increases).

To calculate the entropy change of the universe, add the entropy changes for the system and surroundings, converting the system entropy change units to kJ/K.

$$\Delta S_{universe} = \Delta S_{system} + \Delta S_{surr} = (-88.9 \times 10^{-3} \, kJ/K) + 1.62 \, kJ/K = 1.53 \, kJ/K$$

The reaction is spontaneous ($\Delta S_{universe} > 0$).

Video Solution

Tutored Practice
Problem 19.2.3

Section 19.2 Mastery

19.3 Gibbs Free Energy

19.3a Gibbs Free Energy and Spontaneity

As we have seen, the spontaneity of a reaction is determined by both the entropy change of the system and the entropy change of the surroundings. The **Gibbs free energy (G)**, or simply free energy, is a state function that combines enthalpy and entropy, where T is the absolute temperature. G is defined as follows:

$$G = H - TS \qquad \textbf{(19.6)}$$

For a process occurring at constant temperature, the change in free energy is

$$\Delta G = \Delta H - T\Delta S \qquad \textbf{(19.7)}$$

For a closed system at constant temperature and pressure:

$$\Delta S_{universe} = \Delta S_{system} + \frac{-\Delta H_{system}}{T} > 0 \text{ for a spontaneous process, constant } T \text{ and } P \qquad \textbf{(19.8)}$$

We can set up the reaction conditions so that the system and surroundings are at the same initial temperature, allow the reaction to proceed, wait until the temperature of the system returns to the initial temperature, and then measure the total amount of heat absorbed or released in this process. Constant temperature means a net temperature change of zero. The actual temperature can, and usually does, change during the course of the reaction, but the final and initial temperatures are always the same.

Multiplying both sides of Equation 19.8 by $-T$ and rearranging,

$$-T\Delta S_{universe} = \Delta H_{system} - T\Delta S_{system} < 0 \qquad \textbf{(19.9)}$$

Remember that an inequality reverses when multiplied by a negative number. Examining Equations 19.7 and 19.9, we see that the change in free energy (ΔG) for a closed system at constant temperature and pressure can be expressed in terms of the total entropy of the universe.

$$\Delta G = -T\Delta S_{universe} \text{ at constant } T \text{ and } P \qquad \textbf{(19.10)}$$

Because $\Delta S_{universe} > 0$ for a spontaneous process, we can now relate reaction spontaneity to free energy.

- A reaction that is spontaneous in the forward direction has $\Delta G < 0$.

- A reaction that is not spontaneous in the forward direction has $\Delta G > 0$. This reaction can occur only with the application of energy from an external source, and it is spontaneous in the reverse direction.

- A reaction that is at equilibrium has $\Delta G = 0$. There will be no net change in either the forward or reverse direction.

These points are summarized in Interactive Table 19.3.1.

Three Thermodynamic Energies

We have now defined three different kinds of thermodynamic energies, E, H, and G. Because it can be confusing to know when to use the different energies and to what situations they apply, their definitions are summarized here.

1. Internal energy, E, is the sum of all the submicroscopic kinetic and potential energies of all the particles that make up a system. The change in internal energy, ΔE, is calculated from the first law of thermodynamics, $\Delta E = q + w$.
2. Enthalpy, H, is a defined energy that is based on the internal energy, $H = E + PV$. It is convenient to use enthalpy in the constant-pressure processes that are common in chemistry because $\Delta H = q$, the heat absorbed at constant pressure. You can think about enthalpy as the internal energy measured at constant pressure.
3. Gibbs free energy, G, like enthalpy is a defined energy ($G = H - TS$). Free energy is related to reaction spontaneity by the second law, $\Delta S_{universe} > 0$. At constant temperature and pressure, the change in the entropy of the universe is equal to $-\Delta G/T$, and under these conditions, ΔG points the direction of chemical change.

Internal energy, E, and enthalpy, H, are first law energies. They are associated with the *amount* of energy exchanged. Gibbs free energy is a second law energy. It has to do with the *direction* of change and the *quality* of the energy, the maximum amount of energy that is available to do work.

Interactive Table 19.3.1

Relationship Between ΔG and Reaction Spontaneity

$\Delta G < 0$	Spontaneous in the forward direction
$\Delta G = 0$	At equilibrium; no net change will occur
$\Delta G > 0$	Nonspontaneous in the forward direction; spontaneous in the reverse direction

19.3b Standard Gibbs Free Energy

The **standard free energy of reaction**, ΔG°_{rxn}, is the change in the Gibbs free energy when the reaction is run under standard state conditions. It can be calculated from Equation 19.7 using standard enthalpy and entropy changes for the reaction.

$$\Delta G^{\circ}_{rxn} = \Delta H^{\circ}_{rxn} - T\Delta S^{\circ}_{rxn} \qquad (19.11)$$

We commonly drop the "rxn" subscript from the terms in Equation 19.11. The standard free energy of reaction can be calculated using ΔH_f° and S° values from thermodynamic tables to first calculate ΔH° and ΔS°.

Example Problem 19.3.1 Calculate standard free energy change using ΔH_f° and S° values.

Carbon monoxide and oxygen gas react to form carbon dioxide.

$$CO(g) + \tfrac{1}{2} O_2(g) \rightarrow CO_2(g)$$

Calculate the standard free energy change for this reaction at 25 °C from ΔH°_{rxn} and ΔS°_{rxn}.

Solution:

You are asked to calculate ΔG°_{rxn} from ΔH°_{rxn} and ΔS°_{rxn} at 25 °C for a reaction.

You are given the balanced chemical equation and ΔH_f° and S° data for each substance.

Step 1. Use standard heats of formation to calculate the enthalpy change for this reaction under standard state conditions.

$$
\begin{aligned}
\Delta H^{\circ}_{rxn} &= \Sigma \Delta H_f^{\circ}(\text{products}) - \Sigma \Delta H_f^{\circ}(\text{reactants}) \\
&= \Delta H_f^{\circ}[CO_2(g)] - \Delta H_f^{\circ}[CO(g)] - \tfrac{1}{2}\, \Delta H_f^{\circ}[O_2(g)] \\
&= -393.5 \text{ kJ/mol} - (-110.5 \text{ kJ/mol}) - 0 \\
&= -283.0 \text{ kJ/mol}
\end{aligned}
$$

Step 2. Use standard entropy values to calculate the entropy change for this reaction under standard state conditions.

$$
\begin{aligned}
\Delta S^{\circ}_{rxn} &= \Sigma S^{\circ}(\text{products}) - \Sigma S^{\circ}(\text{reactants}) \\
&= S^{\circ}[CO_2(g)] - \Delta S^{\circ}[CO(g)] - \tfrac{1}{2}\, S^{\circ}[O_2(g)] \\
&= 213.7 \text{ J/mol} \cdot \text{K} - (197.7 \text{ J/mol} \cdot \text{K}) - (\tfrac{1}{2})(205.1 \text{ J/mol} \cdot \text{K}) \\
&= -86.6 \text{ J/mol} \cdot \text{K}
\end{aligned}
$$

Note that ΔS°_{rxn} is negative, as predicted by the stoichiometry: 1.5 moles of gas forms 1 mole of gas.

Step 3. Calculate ΔG°_{rxn} using Equation 19.11.

$$\Delta G^\circ_{rxn} = \Delta H^\circ_{rxn} - T\Delta S^\circ_{rxn}$$
$$= -283.0 \text{ kJ/mol} - (298 \text{ K})(-86.6 \text{ J/mol} \cdot \text{K})(1 \text{ kJ/1000 J})$$
$$= -257 \text{ kJ/mol}$$

Video Solution

Tutored Practice
Problem 19.3.1

There is a second way to use thermodynamic data to calculate ΔG°_{rxn}. Similar to the standard enthalpy of a reaction (Equation 19.4a), the standard free energy change for a reaction can also be calculated from tabulated values of **standard free energy of formation**, ΔG_f°, the change in the Gibbs free energy when one mole of the substance is formed from its elements in their standard states.

$$\Delta G^\circ = \Sigma \Delta G_f^\circ(\text{products}) - \Sigma \Delta G_f^\circ(\text{reactants}) \qquad \textbf{(19.12)}$$

Some standard free energies of formation for pure substances at 25 °C are shown in Table 19.3.2. Notice that for an element in its standard state at 298.15 K, ΔG_f° is zero, similar to ΔH_f° for elements. This simply means that there is no energy change to form an element in its standard state from itself in its standard state. For most compounds, ΔG_f° is negative. This means that under standard state conditions, compounds are generally more stable than the elements from which they are composed. That is, they are *lower* in free energy and could form spontaneously from their constituent elements. This should make sense, because only a few elements are found uncombined in nature. Another way to say this is that elements are generally higher in free energy than compounds, and energy is dissipated when they combine chemically under standard state conditions.

Table 19.3.2 Selected Standard Free Energies of Formation for Pure Substances at 25 °C

Substance	ΔG_f° (kJ/mol)	Substance	ΔG_f° (kJ/mol)
C(diamond)	2.9	Hg(ℓ)	0
C(graphite)	0	HCl(g)	−95.3
CO(g)	−137.2	HCl(aq)	−131.2
CO$_2$(g)	−394.4	FeCl$_2$(s)	−302.3
CH$_3$OH(ℓ)	−166.3	Cu(s)	0
CH$_3$CH$_2$OH(ℓ)	−174.8	H$_2$O(g)	−228.6
C$_6$H$_6$(ℓ)	124.5	H$_2$O(ℓ)	−273.1
H$_2$(g)	0		

Example Problem 19.3.2 Calculate standard free energy change using ΔG_f° values.

Use standard free energies of formation to calculate the standard free energy change for the formation of carbon dioxide from carbon monoxide and oxygen gas at 25 °C.

$$CO(g) + \tfrac{1}{2} O_2(g) \rightarrow CO_2(g)$$

Solution:

You are asked to calculate ΔG°_{rxn} using ΔG_f° values.

You are given the balanced chemical equation and ΔG_f° data for each substance.

▲ - - - - - - - - - -

Example Problem 19.3.2 *(continued)*

Use Equation 19.12 and ΔG_f° data to calculate the standard free energy change for this reaction.

$$\Delta G^\circ_{rxn} = \Sigma \Delta G_f^\circ (\text{products}) - \Sigma \Delta G_f^\circ (\text{reactants})$$

$$= \Delta G_f^\circ [CO_2(g)] - \Delta G_f^\circ [CO(g)] - \tfrac{1}{2} \Delta G_f^\circ [O_2(g)]$$

$$= -394.4 \text{ kJ/mol} - (-137.2 \text{ kJ/mol}) - 0$$

$$= -257.2 \text{ kJ/mol}$$

Is your answer reasonable? Note that this agrees with the result obtained in the previous example using ΔH_f° and S° data.

Video Solution

Tutored Practice
Problem 19.3.2

19.3c Free Energy, Standard Free Energy, and the Reaction Quotient

Recall from Chemical Equilibrium (Unit 15) that by comparing the reaction quotient, Q, with the value of the equilibrium constant, K, we can predict whether a reaction will proceed in the forward or reverse direction to reach equilibrium. For the reaction a A + b B $\rightleftarrows$ c C + d D, the reaction quotient expression is written as

$$Q = \frac{[C]^c [D]^d}{[A]^a [B]^b} \qquad \textbf{(19.13)}$$

There are three possibilities:

$Q < K$ The system is not at equilibrium. Reactants will be consumed, and product concentration will increase until $Q = K$. The reaction will proceed in the forward direction as written (reactants → products).

$Q = K$ The system is at equilibrium, and no further change in reactant or product concentration will occur.

$Q > K$ The system is not at equilibrium. Products will be consumed, and reactant concentration will increase until $Q = K$. The reaction proceeds to the left as written (reactants ← products).

From the second law of thermodynamics, we know that at constant T and P,

- $\Delta G < 0$ for a reaction that is spontaneous in the forward direction (reactants → products).

- $\Delta G > 0$ for a reaction that is spontaneous in the reverse direction (reactants ← products).

- $\Delta G = 0$ for a reaction at equilibrium.

The mathematical relationship that ties together Q and ΔG is given by the following equation, where R is the ideal gas constant (8.3145 J/K · mol):

$$\Delta G = \Delta G^\circ + RT\ln Q \qquad \qquad \textbf{(19.14)}$$

From this equation, you can see that the free energy change for a reaction is equal to the free energy change under standard state conditions plus a correction factor. The value of Q tells us how far away from standard state the chemical or physical change is at a given temperature.

If a reaction occurs under standard state conditions, the pressure of all gases = 1 bar and the concentrations of all solutions = 1 M. In this case, $Q = 1$ and $\ln(1) = 0$. Equation 19.14 becomes $\Delta G = \Delta G^\circ + 0 = \Delta G^\circ$.

If the reaction does not occur under standard state conditions, we can use Equation 19.14 to calculate the free energy change, as shown in the following example.

Example Problem 19.3.3 Calculate the free energy change under nonstandard state conditions.

Use standard thermodynamic data to calculate ΔG at 298.15 K for the following reaction, assuming that all gases have a pressure of 22.51 mm Hg.

$$4\ HCl(g) + O_2(g) \rightarrow 2\ H_2O(g) + 2\ Cl_2(g)$$

Solution:

You are asked to calculate ΔG_{rxn} at 298.15 K.

You are given the balanced chemical equation, the pressure of each gas, and the ΔG_f° data for each substance.

Use Equation 19.14 to calculate the free energy change under nonstandard state conditions. First, calculate ΔG° using ΔG_f° values. Recall that the standard free energy of formation for elements is zero.

$$\Delta G^\circ = \Sigma\Delta G_f^\circ(\text{products}) - \Sigma\Delta G_f^\circ(\text{reactants})$$
$$= (2\ \text{mol})(\Delta G_f^\circ[\text{H}_2\text{O(g)}]) - (4\ \text{mol})(\Delta G_f^\circ[\text{HCl(g)}])$$
$$= (2\ \text{mol})(-228.6\ \text{kJ/mol}) - (4\ \text{mol})(-95.3\ \text{kJ/mol})$$
$$= -76.0\ \text{kJ}$$

For a gas phase reaction, Q is expressed in terms of partial pressures (atm).

Example Problem 19.3.3 (*continued*)

$$22.51 \text{ mm Hg} \times \frac{1 \text{ atm}}{760 \text{ mm Hg}} = 0.02962 \text{ atm}$$

$$Q = \frac{P_{H_2O}^2 \cdot P_{Cl_2}^2}{P_{HCl}^4 \cdot P_{O_2}} = \frac{(0.02962)^2(0.02962)^2}{(0.02962)^4(0.02962)} = 33.76$$

Finally, substitute these values, along with temperature and the ideal gas constant, into Equation 19.14 and calculate ΔG. We will use units of kJ/mol for the calculated $\Delta G°$ value to indicate the standard free energy change is for the reaction as written above.

$$\Delta G = \Delta G° + RT\ln Q = (-76.0 \text{ kJ/mol}) + (8.3145 \times 10^{-3} \text{ kJ/K} \cdot \text{mol})(298.15 \text{ K})[\ln(33.76)]$$

$$= -76.0 \text{ kJ/mol} + 8.72 \text{ kJ/mol}$$

$$= -67.3 \text{ kJ/mol}$$

The negative ΔG value tells us that the reaction is not at equilibrium and must proceed in the forward direction to reach equilibrium.

Video Solution

Tutored Practice
Problem 19.3.3

19.3d Standard Free Energy and the Equilibrium Constant

Now consider how Equation 19.14 applies to a system at equilibrium. At equilibrium, $Q = K$ and $\Delta G = 0$. Substituting into Equation 19.14 we can derive an important relationship between $\Delta G°$ and K:

$$0 = \Delta G° + RT\ln K$$

or

$$\Delta G° = -RT\ln K \qquad\qquad \textbf{(19.15)}$$

Because R and K are constants, this equation is valid whether the system is at equilibrium or not. The standard free energy change for a reaction can be used to determine the equilibrium constant, or the equilibrium constant can be used to determine the standard free energy change.

If $K > 1$, then $\ln K > 0$ and $\Delta G°$ is negative. A large negative $\Delta G°$ means the equilibrium position lies very close to products, and we say that the reaction is product favored at equilibrium.

If $K < 1$, then $\ln K < 0$ and $\Delta G°$ is positive. A large positive $\Delta G°$ means the equilibrium position lies very close to reactants, and we say that the reaction is reactant favored at equilibrium.

Example Problem 19.3.4 Calculate the equilibrium constant from the standard free energy change.

Using standard thermodynamic data, calculate the equilibrium constant at 298.15 K for the reaction shown here.

$$Fe(s) + 2\,HCl(aq) \rightarrow FeCl_2(s) + H_2(g)$$

Solution:

You are asked to calculate the equilibrium constant at 298.15 K.

You are given the balanced chemical equation and the ΔG_f° data for each substance (in the thermodynamic tables).

Use Equation 19.15 to calculate the equilibrium constant at this temperature. First, calculate ΔG° using ΔG_f° values.

$$\begin{aligned}
\Delta G^\circ &= \Sigma \Delta G_f^\circ(\text{products}) - \Sigma \Delta G_f^\circ(\text{reactants}) \\
&= (1\text{ mol})(\Delta G_f^\circ[FeCl_2(s)]) - (2\text{ mol})(\Delta G_f^\circ[HCl(aq)]) \\
&= (1\text{ mol})(-302.3\text{ kJ/mol}) - (2\text{ mol})(-131.2\text{ kJ/mol}) \\
&= -39.9\text{ kJ/mol}
\end{aligned}$$

Rearrange Equation 19.15 to solve for $\ln K$ and substitute this value, along with temperature and the ideal gas constant, to calculate K.

$$\Delta G^\circ = -RT\ln K$$

$$\ln K = \frac{-\Delta G^\circ}{RT} = \frac{-(-39.9\text{ kJ/mol})}{(8.3145 \times 10^{-3}\text{ kJ/K}\cdot\text{mol})(298.15\text{ K})} = 16.1$$

$$K = e^{16.1} = 9.78 \times 10^6$$

Is your answer reasonable? The large negative ΔG° value (-39.9 kJ/mol) tells us that the position of equilibrium for this reaction lies very far toward products, $K \gg 1$.

Video Solution

Tutored Practice
Problem 19.3.4

Interactive Figure 19.3.1 shows the relationship between free energy and the equilibrium position. Chemical reactions proceed until the composition of the reaction mixture corresponds to a minimum free energy, the equilibrium position. At this point, the composition of the reaction mixture ceases to change and its composition is characterized by the equilibrium constant, K.

Both Q and ΔG tell us the direction the reaction must run in order to reach equilibrium. As the reaction proceeds, the concentrations of reactants and products change and thus Q and ΔG change throughout a reaction. When starting with only reactants, $Q < K$ and ΔG is negative. The reaction runs in the forward direction to reach equilibrium. As the number of products increase, Q increases, approaching the value of K, and ΔG becomes less negative,

Interactive Figure 19.3.1

Relate product-favored reactions and spontaneity.

(a)
Reaction is product-favored at equilibrium
$\Delta G°$ is negative, $K > 1$

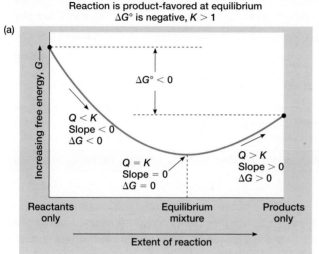

(b)
Reaction is reactant-favored at equilibrium
$\Delta G°$ is positive, $K < 1$

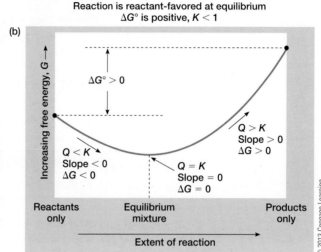

© 2013 Cengage Learning

Free energy and reaction progress for (a) product-favored and (b) reactant-favored reactions

eventually reaching zero. At this point the reaction has reached equilibrium and no further change in concentrations of reactants or products is observed. To summarize:

- Before equilibrium, $\Delta G < 0$ and $Q < K$. The reaction is spontaneous in the forward direction.

- At equilibrium, $\Delta G = 0$ and $Q = K$.

- Beyond the equilibrium position, $\Delta G > 0$ and $Q > K$. The reaction is spontaneous in the reverse direction.

The equilibrium position, or how close to products or how close to reactants the reaction will go, is defined by the equilibrium constant and $\Delta G°$.

- If $K > 1$, $\Delta G° < 0$ and products are favored at equilibrium.

- If $K \approx 1$, $\Delta G° \approx 0$ and neither reactants nor products are favored at equilibrium.

- If $K < 1$, $\Delta G° > 0$ and reactants are favored at equilibrium.

Both K and $\Delta G°$ tell us how far toward products the reaction will run. K and $\Delta G°$ are constants that characterize the reaction at a given temperature. They do not change through the course of the reaction.

Equations 19.14 and 19.15 can be combined into a single equation that shows how Q and K relate to ΔG:

$$\Delta G = -RT\ln K + RT\ln Q = RT\ln(Q/K) \qquad \textbf{(19.16)}$$

When $Q < K$, ΔG is negative and the reaction runs in the forward direction to reach equilibrium. When $Q > K$, ΔG is positive and the reaction runs in the reverse direction to reach equilibrium.

19.3e Gibbs Free Energy and Temperature

Equations 19.7 and 19.11 show that the free energy change for a reaction is related not only to enthalpy and entropy change but also to the temperature at which a reaction occurs.

$$\Delta G = \Delta H - T\Delta S$$

$$\Delta G°_{rxn} = \Delta H°_{rxn} - T\Delta S°_{rxn}$$

Experimentally, we find that enthalpy and entropy changes are relatively constant over a reasonable temperature range. However, as temperature increases, the $(-T\Delta S)$ term plays a larger and larger role in the magnitude of ΔG and thus the spontaneity of a reaction, or the magnitude of the equilibrium constant.

Example Problem 19.3.5 Calculate temperature at which $\Delta G°$ changes sign.

Calculate the temperature (in kelvins) at which the sign of $\Delta G°$ changes from positive to negative. This corresponds to the temperature at which $K < 1$ changes to $K > 1$. Assume that $\Delta H°$ and $\Delta S°$ are constant and do not change with temperature.

$$CO_2(g) + H_2(g) \rightarrow CO(g) + H_2O(g) \qquad \Delta H° = 41.2 \text{ kJ and } \Delta S° = 42.1 \text{ J/K}$$

Solution:

You are asked to calculate the temperature at which $\Delta G°$ changes sign.

You are given the balanced chemical equation and $\Delta H°$ and $\Delta S°$ for the reaction.

Use Equation 19.11 to calculate the temperature at which $\Delta G°$ changes from a positive to a negative value. At this crossover point, $\Delta G°$ is equal to zero. Note that entropy change is converted to units of kJ/K in the calculation.

Example Problem 19.3.5 (continued)

$$\Delta G^\circ = \Delta H^\circ - T\Delta S^\circ = 0$$

$$\Delta H^\circ = T\Delta S^\circ \text{ and } T = \frac{\Delta H^\circ}{\Delta S^\circ} \text{ at the crossover point}$$

$$T = \frac{41.2 \text{ kJ}}{42.1 \times 10^{-3} \text{ kJ/K}} = 979 \text{ K}$$

At temperatures above 979 K, the $(-T\Delta S^\circ)$ term is greater than ΔH° and the standard free energy change for this reaction will be negative. The reaction becomes product favored above 979 K.

Video Solution

Tutored Practice Problem 19.3.5

Changes in enthalpy and entropy can be positive or negative, so the magnitude of K for any chemical or physical change is related to the sign of ΔH° and ΔS°. As shown in Interactive Table 19.3.3, there are four possible situations, two of which depend on the temperature at which the change takes place.

Interactive Table 19.3.3

Effect of Temperature on ΔG° and K

	ΔH°	ΔS°	$\Delta G^\circ = \Delta H^\circ - T\Delta S^\circ$		
1.	−	+	< 0 at all T	Enthalpy and entropy *favorable*	$K > 1$
2.	−	−	< 0 at low T	Enthalpy controlled, *favorable*	$K > 1$
			> 0 at high T	Entropy controlled, *unfavorable*	$K < 1$
3.	+	+	> 0 at low T	Enthalpy controlled, *unfavorable*	$K < 1$
			< 0 at high T	Entropy controlled, *favorable*	$K > 1$
4.	+	−	> 0 at all T	Enthalpy and entropy *unfavorable*	$K < 1$

Here are examples of the four possible combinations of enthalpy and entropy change:

1. Combustion of sugar, $C_6H_{12}O_6(s)$

The combustion of any hydrocarbon is an exothermic reaction ($\Delta H° < 0$).

$$C_6H_{12}O_6(s) + 6\ O_2(g) \rightarrow 6\ CO_2(g) + 6\ H_2O(g)$$

We can predict that the entropy change for this reaction is positive because a total of 12 mol of gaseous products are produced from 1 mol of solid and 6 mol of O_2 gas ($\Delta S° > 0$). Because the kelvin temperature is always positive, the ($-T\Delta S°$) term will be negative and the standard free energy change for this reaction will be negative at all temperatures.

$$\Delta G° = \Delta H° - T\Delta S°$$
$$< 0 \quad (-) \quad (+)(+)$$

Exothermic reactions that increase in entropy are product favored at any temperature.

2. Freezing of a liquid

Freezing a liquid such as water is an exothermic process ($\Delta H° < 0$).

$$H_2O(\ell) \rightarrow H_2O(s)$$

At a fixed temperature, the entropy of a solid is less than that of a liquid, so entropy decreases during this change ($\Delta S° < 0$). Because the kelvin temperature is always positive, the ($-T\Delta S°$) term will be positive.

$$\Delta G° = \Delta H° - T\Delta S°$$
$$? \quad (-) \quad (+)(-)$$

In this case we see that the negative $\Delta H°$ decreases the overall free energy, but the entropy change of the system increases it. Here, temperature plays an important role in whether the reaction is product favored. At high temperatures, the entropy term ($-T\Delta S°$) has a large, positive value; the reaction is reactant favored ($\Delta G° > 0$). At low temperatures, however, the entropy term ($-T\Delta S°$) has a small, positive value. At low temperatures, the reaction becomes product favored ($\Delta G° < 0$) and the liquid will freeze.

We refer to reactions that have favorable enthalpy but unfavorable entropy as *enthalpy-driven reactions*. These reactions must be run at a temperature low enough to keep the unfavorable entropy term from overcoming the favorable enthalpy term.

3. Boiling of a liquid

Boiling a liquid such as water is an endothermic process ($\Delta H° > 0$).

$$H_2O(\ell) \rightarrow H_2O(g)$$

At a fixed temperature, the entropy of a liquid is less than that of a gas, so entropy increases during this change ($\Delta S° > 0$). Because the kelvin temperature is always positive, the ($-T\Delta S°$) term will be negative.

$$\Delta G° = \Delta H° - T\Delta S°$$
$$?\quad (+)\quad (+)(+)$$

In this case we see that the positive $\Delta S°$ decreases the overall free energy, but the enthalpy change of the system increases it. Here, temperature also plays an important role. At low temperatures, the entropy term ($-T\Delta S°$) has a small, negative value; the reaction is reactant favored ($\Delta G° > 0$). At high temperatures, however, the entropy term ($-T\Delta S°$) has a large, negative value. At high temperatures, the reaction is product favored ($\Delta G° < 0$) and the liquid will boil.

We refer to reactions that have favorable entropy but unfavorable enthalpy as *entropy-driven reactions*. These reactions must be run at a temperature high enough to allow the favorable entropy term to overcome the unfavorable enthalpy term.

4. Synthesis of sugar from CO_2 and H_2O

Consider the reverse of the sugar combustion reaction (case 1 above).

$$6\ CO_2(g) + 6\ H_2O(\ell) \rightarrow C_6H_{12}O_6(s) + 6\ O_2(g)$$

We know this reaction is endothermic ($\Delta H° > 0$) and has a negative entropy change ($\Delta S° < 0$). If the burning of sugar is exothermic, the reverse of this reaction is endothermic, $\Delta H° > 0$. Because the kelvin temperature is always positive, the ($-T\Delta S°$) term will be positive and the standard free energy change for this reaction will be positive at all temperatures.

$$\Delta G° = \Delta H° - T\Delta S°$$
$$> 0\quad (+)\quad (+)(-)$$

The reaction is reactant favored at any temperature. This reaction happens in plants during photosynthesis, but only because there is a constant input of energy from an external source, the sun (Figure 19.3.2).

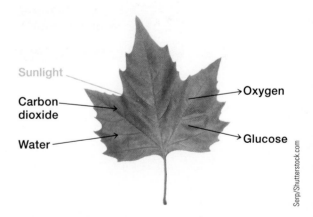

Sunlight

Carbon dioxide

Water

→Oxygen

→Glucose

Figure 19.3.2 Photosynthesis

Example Problem 19.3.6 Predict the temperature dependence for a chemical or physical change.

Predict the temperature conditions under which the following reaction will be product favored.

$$Fe_2O_3(s) + 3 H_2(g) \rightarrow 2 Fe(s) + 3 H_2O(g) \qquad \Delta H° = 98.8 \text{ kJ and } \Delta S° = 143 \text{ J/K}$$

Solution:

You are asked to predict the temperature conditions under which the reaction will be product favored, $K > 1$.

You are given the balanced chemical equation and $\Delta H°$ and $\Delta S°$ for the reaction.

For a reaction to be product favored, $\Delta G°$ must be less than zero. The dependence of $\Delta G°$ on temperature is given by the following equation, assuming that $\Delta H°$ and $\Delta S°$ are independent of temperature.

$$\Delta G° = \Delta H° - T\Delta S°$$

For this reaction, $\Delta H° = 98.8$ kJ and $\Delta S° = 143$ J/K. Because both $\Delta H°$ and $\Delta S°$ are positive, temperature will control the position of the equilibrium. At high temperatures, the term $(-T\Delta S°)$ has a large, negative value, resulting in a negative free energy change ($\Delta G° < 0$).

$$\Delta G° = \Delta H° - T\Delta S°$$
$$? \qquad (+) \quad (+)(+)$$

The reaction is product favored at high temperatures and reactant favored at low temperatures. It is an entropy-driven reaction.

Video Solution

Tutored Practice
Problem 19.3.6

Section 19.3 Mastery

Unit Recap

Key Concepts

19.1 Entropy and the Three Laws of Thermodynamics

- The first law of thermodynamics states that the total energy of the universe is constant but it does not tell the direction in which energy changes naturally occur (19.1a).

- The second law of thermodynamics tells us that energy changes occur in the direction in which at least some energy disperses (19.1a).

- Entropy, S, is a thermodynamic state function that is a measure of the dispersed energy within a system at a given temperature (19.1b).

- All spontaneous changes in nature occur in the direction in which the total entropy of the universe increases (19.1b).

- Entropy is a measure of the number of microstates accessible at a given total energy (19.1c).

- The entropy of a substance depends on its phase, temperature, molecular size, and the forces between its particles (19.1d).

- The third law of thermodynamics states that the entropy of a pure crystalline substance is zero at absolute zero (19.1f).

- The standard molar entropy of a substance, $S°$, is the entropy of the substance at a given temperature and a pressure of 1 bar and is always positive at temperatures above 0 K (19.1f).

19.2 Calculating Entropy Change

- The entropy change for a phase change can be calculated from ΔH for the phase change divided by the kelvin temperature (19.2a).

- The standard entropy of a reaction, $\Delta S°_{rxn}$, can be calculated from the standard molar entropy of products minus reactants (19.2b).

- It is often possible to estimate the sign of $\Delta S°_{rxn}$ by looking at the states of the reactants and products (19.2b).

- The entropy change for the surroundings is equal to $-\Delta H_{rxn}$ divided by the kelvin temperature (19.2c).

- Exothermic reactions increase the entropy of the surroundings and endothermic reactions decrease the entropy of the surroundings (19.2c).

19.3 Gibbs Free Energy

- At constant temperature and pressure, the change in the Gibbs free energy is equal to $\Delta H - T\Delta S$; it is also equal to $T\Delta S_{universe}$ under these conditions (19.3a).

- At constant temperature and pressure, the sign of ΔG tells the direction a reaction must run to reach equilibrium (19.3a).

- The three thermodynamic energies used in general chemistry are the internal energy, E; the enthalpy, H; and the Gibbs free energy, G (19.3a).

- The standard Gibbs free energy, $\Delta G°$, is the change in the free energy of a reaction run under standard state conditions (19.3b).

- ΔG° can be calculated from standard enthalpy and entrophy data and from standard Gibbs free energy of formation data (19.3b).

- The standard free energy of formation, ΔG_f°, for an element in its standard state is equal to zero (19.3b).

- Both ΔG and the reaction quotient, Q, tell the direction of change for a chemical reaction; they are related by the equation $\Delta G = \Delta G^\circ - RT\ln Q$ (19.3c).

- Both ΔG° and the equilibrium constant, K, tell the position of equilibrium; they are related by the equation $\Delta G^\circ = -RT\ln K$ (19.3d).

- Both ΔG and Q depend on the amount of products and reactants present at any given instant for a reaction and therefore change throughout the progress of a reaction (19.3d).

- ΔG° and K are constants characteristic of a reaction; they do not change throughout the progress of a reaction.

- ΔG and ΔG° depend on temperature (19.3d).

- It is possible to predict whether a reaction is product or reactant favored by examining the sign of ΔH° and ΔS° for the reaction (19.3e).

Key Equations

$$\Delta S_{\text{universe}} = \Delta S_{\text{system}} + \Delta S_{\text{surroundings}} > 0 \text{ for a spontaneous process} \qquad \textbf{(19.1)}$$

$$S = k_{\text{B}} \ln W \qquad \textbf{(19.2)}$$

$$\Delta S_{\text{phase change}} = \frac{\Delta H_{\text{phase change}}}{T_{\text{phase change}}} \qquad \textbf{(19.3)}$$

$$\Delta H^{\circ}_{\text{rxn}} = \Sigma \Delta H_f^{\circ}(\text{products}) - \Sigma \Delta H_f^{\circ}(\text{reactants}) \qquad \textbf{(19.4a)}$$

$$\Delta S^{\circ}_{\text{rxn}} = \Sigma S^{\circ}(\text{products}) - \Sigma S^{\circ}(\text{reactants}) \qquad \textbf{(19.4b)}$$

$$\Delta S_{\text{surr}} = \frac{-\Delta H_{\text{rxn}}}{T} \text{ at constant pressure} \qquad \textbf{(19.5)}$$

$$G = H - TS \qquad \textbf{(19.6)}$$

$$\Delta G = \Delta H - T\Delta S \qquad \textbf{(19.7)}$$

$$\Delta S_{\text{universe}} = \Delta S_{\text{system}} + \frac{-\Delta H_{\text{system}}}{T} > 0 \text{ for a spontaneous process, constant } T \text{ and } P \qquad \textbf{(19.8)}$$

$$-T\Delta S_{\text{universe}} = \Delta H_{\text{system}} - T\Delta S_{\text{system}} < 0 \qquad \textbf{(19.9)}$$

$$\Delta G = -T\Delta S_{\text{universe}} \text{ at constant } T \text{ and } P \qquad \textbf{(19.10)}$$

$$\Delta G^{\circ}_{\text{rxn}} = \Delta H^{\circ}_{\text{rxn}} - T\Delta S^{\circ}_{\text{rxn}} \qquad \textbf{(19.11)}$$

$$\Delta G^{\circ} = \Sigma \Delta G_f^{\circ}(\text{products}) - \Sigma \Delta G_f^{\circ}(\text{reactants}) \qquad \textbf{(19.12)}$$

$$Q = \frac{[\text{C}]^{\text{c}}[\text{D}]^{\text{d}}}{[\text{A}]^{\text{a}}[\text{B}]^{\text{b}}} \qquad \textbf{(19.13)}$$

$$\Delta G = \Delta G^{\circ} + RT\ln Q \qquad \textbf{(19.14)}$$

$$\Delta G^{\circ} = -RT\ln K \qquad \textbf{(19.15)}$$

$$\Delta G = -RT\ln K + RT\ln Q = RT\ln(Q/K) \qquad \textbf{(19.16)}$$

Key Terms

19.1 Entropy and the Three Laws of Thermodynamics
first law of thermodynamics
internal energy
spontaneous process
second law of thermodynamics

entropy (S)
microstate
third law of thermodynamics
standard molar entropy (S°)

19.2 Calculating Entropy Change
standard entropy change (ΔS°_{rxn})

19.3 Gibbs Free Energy
Gibbs free energy (G)
standard free energy of reaction (ΔG°_{rxn})
standard free energy of formation (ΔG_f°)

Unit 19   Review and Challenge Problems

20 Electrochemistry

Unit Outline

In This Unit...

Why aren't there more electric cars? Why do we still use corded power tools instead of only using battery-powered tools? Why do the batteries in our portable electronic devices run down so quickly? The answer to all these questions lies in our ability to make good batteries. In this unit we explore the chemistry of batteries, where spontaneous reactions take place by the indirect transfer of electrons from one reactant to another. We will also investigate electrolysis, the process where we use external power supplies such as batteries to force nonspontaneous reactions to form products.

Vasilyev/Shutterstock.com

20.1 Oxidation–Reduction Reactions and Electrochemical Cells

20.1a Overview of Oxidation–Reduction Reactions

In Chemical Reactions and Solution Stoichiometry (Unit 4) we first encountered oxidation–reduction reactions in our study of chemical reactions. **Electrochemistry** is the area of chemistry that studies oxidation–reduction reactions, also called *redox reactions,* which involve electron transfer between two or more species. Recall that in a redox reaction,

- the species that loses electrons has been *oxidized* and is the *reducing agent* in the reaction, and

- the species that gains electrons has been *reduced* and is the *oxidizing agent* in the reaction. (◄ Flashback to Section 4.4 Oxidation–Reduction Reactions)

Oxidizing and reducing agents are identified in a chemical reaction by using the oxidation number (or oxidation state) of the species in the reaction. Recall that the oxidation number of an oxidized species increases during the reaction, whereas the oxidation number of a reduced species decreases during the reaction. Table 20.1.1 summarizes the rules for assigning oxidation numbers.

Consider the net ionic equation for the single displacement redox reaction involving zinc metal and a solution of copper(II) ions (Interactive Figure 20.1.1):

$$Zn(s) + Cu^{2+}(aq) \rightarrow Zn^{2+}(aq) + Cu(s)$$

In this redox reaction, zinc gives up two electrons (the Zn oxidation state increases from 0 to +2) and the copper(II) ions gain two electrons (the Cu oxidation state decreases from +2 to 0). Thus, Zn is oxidized and is the reducing agent, and Cu^{2+} is reduced and is the oxidizing agent.

	$Zn(s)$	+	$Cu^{2+}(aq)$	→	$Zn^{2+}(aq)$	+	$Cu(s)$
Oxidation state:	0		+2		+2		0
Electron gain/loss:	Loses 2 e⁻		Gains 2 e⁻				
Oxidized or reduced?	Oxidized		Reduced				
Reducing or oxidizing agent?	Reducing agent		Oxidizing agent				

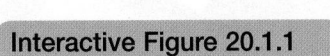

Explore an oxidation–reduction reaction.

Copper(II) nitrate reacts with zinc.

Table 20.1.1 Rules for Assigning Oxidation Numbers

Rule	Example
1. Each atom in a pure element has an oxidation number of zero.	Fe in $Fe(s)$ oxidation number = 0 Each O in $O_2(g)$ oxidation number = 0
2. A monoatomic ion has an oxidation number equal to the ion charge.	Cl in Cl^- oxidation number = -1 Mg in Mg^{2+} oxidation number = $+2$
3. In compounds, halogens (F, Cl, Br, I) have an oxidation number of -1. *Exception:* When Cl, Br, and I are combined with oxygen or fluorine, their oxidation number is not -1.	Each F in CF_4 oxidation number = -1 Cl in ClF_3 oxidation number = $+3$
4. In compounds, oxygen has an oxidation number of -2. *Exception:* In compounds containing the peroxide ion (O_2^{2-}) or superoxide ion (O_2^-), oxygen has an oxidation number of -1 or $-\frac{1}{2}$, respectively.	Each O in CO_2 oxidation number = -2 Each O in H_2O_2 oxidation number = -1
5. When combined with nonmetals, hydrogen is assigned an oxidation number of $+1$. With metals, hydrogen has an oxidation number of -1.	Each H in CH_4 oxidation number = $+1$ H in LiH oxidation number = -1
6. The sum of the oxidation numbers for all atoms in a species is equal to the overall charge on the species.	CO_2 (C oxidation number) + 2(O oxidation number) = 0 (C oxidation number) + 2(-2) = 0 C oxidation number = $+4$ ClO_4^- (Cl oxidation number) + 4(O oxidation number) = -1 (Cl oxidation number) + 4(-2) = -1 Cl oxidation number = $+7$

Example Problem 20.1.1 Use oxidation numbers to identify the species oxidized and reduced in an oxidation–reduction reaction.

Assign oxidation states to all of the species in the following redox reaction. For the reactants, identify electron loss or gain, the species oxidized, the species reduced, the oxidizing agent, and the reducing agent.

$$Hg^{2+}(aq) + Cd(s) \rightarrow Hg(\ell) + Cd^{2+}(aq)$$

Solution:

You are asked to assign oxidation states and to use those oxidation states to identify electron loss or gain, the species oxidized, the species reduced, the oxidizing agent, and the reducing agent.

You are given an oxidation–reduction reaction.

These are all monatomic ions or uncombined elements. For a monatomic ion, the oxidation state is equal to the charge on the ion. For a pure element, the oxidation state is zero.

$$Hg^{2+}(aq) + Cd(s) \rightarrow Hg(\ell) + Cd^{2+}(aq)$$

Oxidation state: +2   0   0   +2

Reduction is defined as a gain of electrons or a decrease in oxidation number. The oxidation state of mercury decreases from +2 to 0, so Hg^{2+} is reduced and is the oxidizing agent. The reduction half-reaction is $Hg^{2+} + 2\,e^- \rightarrow Hg$.

Oxidation is defined as a loss of electrons or an increase in oxidation number. The oxidation state of cadmium increases from 0 to +2 in the reaction, so Cd is oxidized and is the reducing agent. The oxidation half-reaction is $Cd \rightarrow Cd^{2+} + 2\,e^-$.

Video Solution

Tutored Practice
Problem 20.1.1

20.1b Balancing Redox Reactions: Half-Reactions

One way to balance redox reactions is the **half-reaction method**. In the half-reaction method, an oxidation–reduction reaction is represented as the sum of two half-reactions, one that shows the species involved in oxidation and another that shows the species involved in reduction. Recall that half-reactions include the electrons lost or gained by each species and that adding balanced and adjusted oxidation and reduction half-reactions results in the net ionic equation, with no leftover electrons.

Because oxidation–reduction reactions involve the transfer of electrons, a balanced redox reaction has both mass balance and charge balance; that is, the number of electrons lost by the oxidized species must be equal to the number of electrons gained by the reduced species. Consider the reaction between copper metal and silver ions.

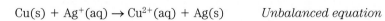

$$Cu(s) + Ag^+(aq) \rightarrow Cu^{2+}(aq) + Ag(s) \qquad \textit{Unbalanced equation}$$

This chemical equation is balanced for mass, but not for charge. We can balance the equation for charge and mass by using the half-reaction method:

Step 1. Assign oxidation numbers to all species in the reaction and identify the oxidized and reduced species.

$$\overset{0}{Cu(s)} + \overset{+1}{Ag^+(aq)} \rightarrow \overset{+2}{Cu^{2+}(aq)} + \overset{0}{Ag(s)}$$

oxidized   reduced

Step 2. Divide the reaction into oxidation and reduction half-reactions.

$$Cu(s) \rightarrow Cu^{2+}(aq) \qquad \textit{Oxidation}$$

$$Ag^+(aq) \rightarrow Ag(s) \qquad \textit{Reduction}$$

Step 3. Balance each half-reaction, first for mass and then for charge (by adding electrons).

$$Cu(s) \rightarrow Cu^{2+}(aq) + 2\ e^- \qquad \textit{Add 2 electrons to products}$$

$$Ag^+(aq) + e^- \rightarrow Ag(s) \qquad \textit{Add 1 electron to reactants}$$

Step 4. Balance the two half-reactions so that the number of electrons gained equals the number of electrons lost.

$$Cu(s) \rightarrow Cu^{2+}(aq) + 2\ e^-$$

$$2\ Ag^+(aq) + 2\ e^- \rightarrow 2\ Ag(s) \qquad \textit{Multiply reaction by 2}$$

Step 5. Add the two half-reactions, eliminating any species that are identical in the reactants and products.

$$Cu(s) + 2\ Ag^+(aq) \rightarrow Cu^{2+}(aq) + 2\ Ag(s) \qquad \textit{Eliminate electrons from reactants and products}$$

The equation is now balanced both for mass and for charge.

Example Problem 20.1.2 Balance redox reactions using the half-reaction method.

Balance the following redox reaction using the half-reaction method.

$$Al(s) + Fe^{3+}(aq) \rightarrow Al^{3+}(aq) + Fe^{2+}(aq)$$

Solution:

You are asked to balance a redox reaction using the half-reaction method.

You are given an unbalanced redox reaction.

Follow the steps shown previously to balance the redox reaction.

Step 1. Assign oxidation numbers to all species in the reaction and identify the oxidized and reduced species.

$$\overset{0}{Al(s)} + \overset{+3}{Fe^{3+}(aq)} \rightarrow \overset{+3}{Al^{3+}(aq)} + \overset{+2}{Fe^{2+}(aq)}$$

oxidized   reduced

Step 2. Divide the reaction into oxidation and reduction half-reactions.

$$Al(s) \rightarrow Al^{3+}(aq) \qquad \textit{Oxidation}$$
$$Fe^{3+}(aq) \rightarrow Fe^{2+}(aq) \qquad \textit{Reduction}$$

Step 3. Balance each half-reaction, first for mass and then for charge (by adding electrons).

$$Al(s) \rightarrow Al^{3+}(aq) + 3\ e^- \qquad \textit{Add 3 electrons to products}$$
$$Fe^{3+}(aq) + e^- \rightarrow Fe^{2+}(aq) \qquad \textit{Add 1 electron to reactants}$$

Step 4. Balance the two half-reactions so that the number of electrons gained equals the number of electrons lost.

$$Al(s) \rightarrow Al^{3+}(aq) + 3\ e^-$$
$$3\ Fe^{3+}(aq) + 3\ e^- \rightarrow 3\ Fe^{2+}(aq) \qquad \textit{Multiply reaction by 3}$$

Step 5. Add the two half-reactions, eliminating any identical species in the reactants and products.

$$Al(s) + 3\ Fe^{3+}(aq) \rightarrow Al^{3+}(aq) + 3\ Fe^{2+}(aq) \qquad \textit{Eliminate electrons from reactants and products}$$

Video Solution

Tutored Practice
Problem 20.1.2

20.1c Balancing Redox Reactions in Acidic and Basic Solutions

Many redox reactions take place in acidic or basic solution. We can balance these types of reactions by expanding Step 3 of the half-reaction method:

Step 3. Balance each half-reaction, first for mass and then for charge (by adding electrons).

- Balance elements other than O and H.

- Balance O using H_2O.

- Balance H using H^+.

- If the solution is basic, eliminate H^+ by adding an equal number of OH^- ions to both sides of the equation. Combine equal numbers of H^+ and OH^- ions to form H_2O in reactants or products.

- Finally, balance charge by adding electrons.

Example Problem 20.1.3 Balance redox reactions in acidic solution.

Balance the reaction between IO_3^- and Cr^{2+} to form Cr^{3+} and I_2 in acidic solution.

Solution:

You are asked to balance a redox reaction in acidic solution using the half-reaction method.

You are given an unbalanced redox reaction.

Use the steps shown previously for balancing a redox reaction by the half-reaction method, including the modification to Step 3.

Step 1. Assign oxidation numbers to all species in the reaction and identify the oxidized and reduced species.

$$\overset{+5-2}{IO_3^-}(aq) + \overset{+2}{Cr^{2+}}(aq) \rightarrow \overset{+3}{Cr^{3+}}(aq) + \overset{0}{I_2}(aq)$$

I in IO_3^- is reduced and Cr^{2+} is oxidized.

Step 2. Divide the reaction into oxidation and reduction half-reactions.

$$IO_3^-(aq) \rightarrow I_2(aq) \qquad \textit{Reduction}$$
$$Cr^{2+}(aq) \rightarrow Cr^{3+}(aq) \qquad \textit{Oxidation}$$

Example Problem 20.1.3 (*continued*)

Step 3. Balance each half-reaction, first for mass and then for charge (by adding electrons).

First, balance elements other than O and H.

$$2\ IO_3^-(aq) \rightarrow I_2(aq)$$
$$Cr^{2+}(aq) \rightarrow Cr^{3+}(aq)$$

Balance O using H_2O.

$$2\ IO_3^-(aq) \rightarrow I_2(aq) + 6\ H_2O(\ell)$$
$$Cr^{2+}(aq) \rightarrow Cr^{3+}(aq)$$

Balance H using H^+.

$$2\ IO_3^-(aq) + 12\ H^+(aq) \rightarrow I_2(aq) + 6\ H_2O(\ell)$$
$$Cr^{2+}(aq) \rightarrow Cr^{3+}(aq)$$

Finally, balance charge by adding electrons.

$$2\ IO_3^-(aq) + 12\ H^+(aq) + 10\ e^- \rightarrow I_2(aq) + 6\ H_2O(\ell)$$
$$Cr^{2+}(aq) \rightarrow Cr^{3+}(aq) + e^-$$

Step 4. Balance the two half-reactions so that the number of electrons required in the reduction equals the number of electrons supplied by the oxidation.

$$2\ IO_3^-(aq) + 12\ H^+(aq) + 10\ e^- \rightarrow I_2(aq) + 6\ H_2O(\ell)$$
$$10\ Cr^{2+}(aq) \rightarrow 10\ Cr^{3+}(aq) + 10\ e^-$$

Step 5. Add the two half-reactions, eliminating any identical species in the reactants and products.

$$2\ IO_3^-(aq) + 12\ H^+(aq) + 10\ Cr^{2+}(aq) \rightarrow 10\ Cr^{3+}(aq) + I_2(aq) + 6\ H_2O(\ell)$$

Video Solution

Tutored Practice
Problem 20.1.3

Example Problem 20.1.4 Balance redox reactions in basic solution.

Balance the following reaction in basic solution:

$$CrO_4^{2-}(aq) + Pb(s) \rightarrow Cr(OH)_3(s) + HPbO_2^-(aq)$$

Solution:

You are asked to balance a redox reaction in basic solution using the half-reaction method.

You are given an unbalanced redox reaction.

Use the steps shown previously for balancing a redox reaction by the half-reaction method, including the modification to Step 3.

Example Problem 20.1.4 *(continued)*

Step 1. Assign oxidation numbers to all species in the reaction and identify the oxidized and reduced species.

$$\overset{+6\ -2}{CrO_4{}^{2-}}(aq) + \overset{0}{Pb}(s) \rightarrow \overset{+3\ -2\ +1}{Cr(OH)_3}(s) + \overset{+1\ +2\ -2}{HPbO_2{}^{-}}(aq)$$

Cr in $CrO_4{}^{2-}$ is reduced and Pb is oxidized.

Step 2. Divide the reaction into oxidation and reduction half-reactions.

$$CrO_4{}^{2-}(aq) \rightarrow Cr(OH)_3(s) \qquad \textit{Reduction}$$
$$Pb(s) \rightarrow HPbO_2{}^{-}(aq) \qquad \textit{Oxidation}$$

Step 3. Balance each half-reaction, first for mass and then for charge (by adding electrons).

First, balance elements other than O and H.

$$CrO_4{}^{2-}(aq) \rightarrow Cr(OH)_3(s)$$
$$Pb(s) \rightarrow HPbO_2{}^{-}(aq)$$

Balance O using H_2O.

$$CrO_4{}^{2-}(aq) \rightarrow Cr(OH)_3(s) + H_2O(\ell)$$
$$Pb(s) + 2\,H_2O(\ell) \rightarrow HPbO_2{}^{-}(aq)$$

Balance H using H^+.

$$CrO_4{}^{2-}(aq) + 5\,H^+(aq) \rightarrow Cr(OH)_3(s) + H_2O(\ell)$$
$$Pb(s) + 2\,H_2O(\ell) \rightarrow HPbO_2{}^{-}(aq) + 3\,H^+(aq)$$

Eliminate H^+ by adding an equal number of OH^- ions to both sides of the equation. Combine equal numbers of H^+ and OH^- ions to form H_2O in reactants or products.

$$CrO_4{}^{2-}(aq) + \underbrace{5\,H^+(aq) + 5\,OH^-(aq)}_{5\,H_2O(\ell)} \rightarrow Cr(OH)_3(s) + H_2O(\ell) + 5\,OH^-(aq)$$

$$Pb(s) + 2\,H_2O(\ell) + 3\,OH^-(aq) \rightarrow HPbO_2{}^{-}(aq) + \underbrace{3\,H^+(aq) + 3\,OH^-(aq)}_{3\,H_2O(\ell)}$$

Finally, balance charge by adding electrons.

$$CrO_4{}^{2-}(aq) + 5\,H_2O(\ell) + 3\,e^- \rightarrow Cr(OH)_3(s) + H_2O(\ell) + 5\,OH^-(aq)$$
$$Pb(s) + 2\,H_2O(\ell) + 3\,OH^-(aq) \rightarrow HPbO_2{}^{-}(aq) + 3\,H_2O(\ell) + 2\,e^-$$

◀- - - - -

Example Problem 20.1.4 *(continued)*

Step 4. Balance the two half-reactions so that the number of electrons gained equals the number of electrons lost.

$$2 \, CrO_4^{2-}(aq) + 10 \, H_2O(\ell) + 6 \, e^- \rightarrow 2 \, Cr(OH)_3(s) + 2 \, H_2O(\ell) + 10 \, OH^-(aq)$$

$$3 \, Pb(s) + 6 \, H_2O(\ell) + 9 \, OH^-(aq) \rightarrow 3 \, HPbO_2^-(aq) + 9 \, H_2O(\ell) + 6 \, e^-$$

Step 5. Add the two half-reactions, eliminating any identical species in the reactants and products.

$$2 \, CrO_4^{2-}(aq) + 3 \, Pb(s) + 5 \, H_2O(\ell) \rightarrow 2 \, Cr(OH)_3(s) + 3 \, HPbO_2^-(aq) + OH^-(aq)$$

Video Solution

Tutored Practice
Problem 20.1.4

20.1d Construction and Components of Electrochemical Cells

When an oxidizing agent such as Cu^{2+} is placed in direct contact with a reducing agent such as Zn, as shown in Interactive Figure 20.1.1, electrons can be transferred directly from the reducing agent to the oxidizing agent. We call this *direct electron transfer* because it occurs where the two species are in contact.

Electrons can also be transferred indirectly through an external conducting medium. For example, if we place a $CuSO_4$ solution and some Cu metal in one beaker and a $Zn(NO_3)_2$ solution and some Zn metal in another beaker and connect the metals by a piece of wire, electrons can move from Zn to the Cu^{2+} ions. The newly formed Zn^{2+} ions will move into the $Zn(NO_3)_2$ solution and new Cu will build up on the piece of copper metal. This is an example of *indirect electron transfer* (Figure 20.1.2).

When the electrons flow through the wire, reducing Cu^{2+} to Cu and oxidizing Zn to Zn^{2+}, negative charge builds up in the beaker where the reduction half-reaction occurs and positive charge builds up in the beaker where the oxidation half-reaction is taking place. This buildup of charge quickly stops the reaction from proceeding any further.

Charge buildup can be eliminated by adding a mechanism that allows counterions to flow between the two **half-cells**, the separated containers where oxidation and reduction take place. Two common options are the use of a porous disk or a **salt bridge**; a salt bridge is a stoppered tube containing an electrolyte in aqueous solution or suspended in a gel (Figure 20.1.3).

An **electrochemical cell** consists of two half-cells connected by a mechanism to allow the cell to maintain charge balance. There are two types of electrochemical cells. In a **voltaic** (or **galvanic**) **cell**, electrons flow spontaneously through the wire, and this electrical current can be used to do work. Batteries are examples of voltaic cells.

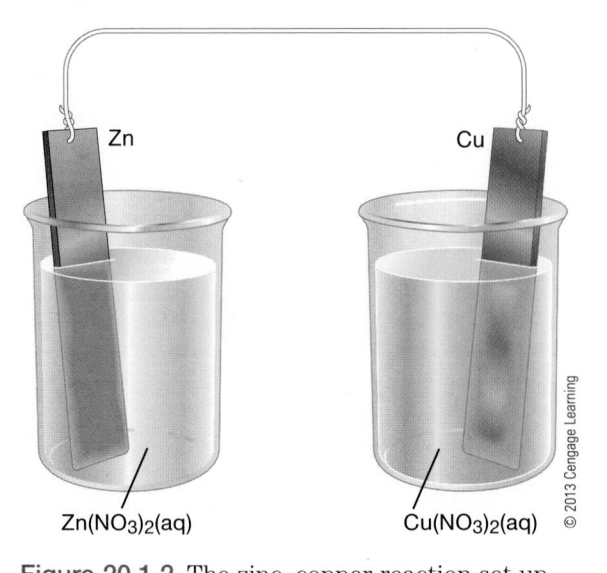

Figure 20.1.2 The zinc–copper reaction set up for indirect electron transfer

© 2013 Cengage Learning

Salt bridge

Porous disk

Figure 20.1.3 Examples of salt bridges

In an electrolytic cell, an external power source (such as a battery) is used to force the electrons to flow between the oxidation and reduction half-reactions in the nonspontaneous direction. We will discuss electrolytic cells later in this unit.

Electron transfer occurs at an **electrode**, which is typically a sample of metal involved in the redox reaction. In a typical voltaic cell (Interactive Figure 20.1.4),

- the electrode where oxidation takes place is called the **anode**;

- the electrode where reduction takes place is called the **cathode**;

- electrons flow through a wire between two half-cells, from the anode to the cathode;

- anions in the aqueous solution or salt bridge flow from the cathode to the anode and cations flow from the anode to the cathode; and

- the anode is assigned a negative charge, and the cathode is assigned a positive charge.

In the electrochemical cell shown in Interactive Figure 20.1.4,

- the Zn electrode is the anode, and the Cu electrode is the cathode;

- electrons flow from Zn to Cu;

- the concentration of Zn^{2+} increases and the concentration of Cu^{2+} decreases as the reaction proceeds;

Explore a voltaic electrochemical cell.

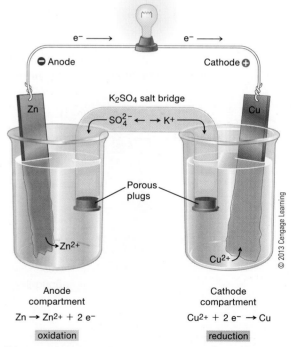

Diagram of a voltaic electrochemical cell

- the mass of the Zn electrode decreases and the mass of the Cu electrode increases as the reaction proceeds; and

- the anions in the salt bridge flow from the $CuSO_4$ solution to the $Zn(NO_3)_2$ solution, and the cations in the salt bridge flow from the $Zn(NO_3)_2$ solution to the $CuSO_4$ solution.

Not all electrochemical cells contain electrodes that are part of the oxidation or reduction half-reactions. For example, consider the oxidation of Fe^{2+} to Fe^{3+} or the reduction of MnO_4^- to Mn^{2+}. In these cases an inert, nonreactive metal is used as an electrode to allow electron transfer at the anode or cathode. A common inert electrode is platinum, which is used in the **standard hydrogen electrode (SHE)** (Figure 20.1.5). Here, a stream of hydrogen gas is bubbled through a chamber that contains a platinum electrode and an acidic solution. Depending on the setup of the electrochemical cell, the electrode can act as an anode or a cathode.

$$\text{Oxidation, anode:} \quad H_2(g) \rightarrow 2\,H^+(aq) + 2\,e^-$$

$$\text{Reduction, cathode:} \quad 2\,H^+(aq) + 2\,e^- \rightarrow H_2(g)$$

The platinum metal allows electrons to move to the H^+ ions or away from the H_2 molecules, but it does not participate in the chemical reaction.

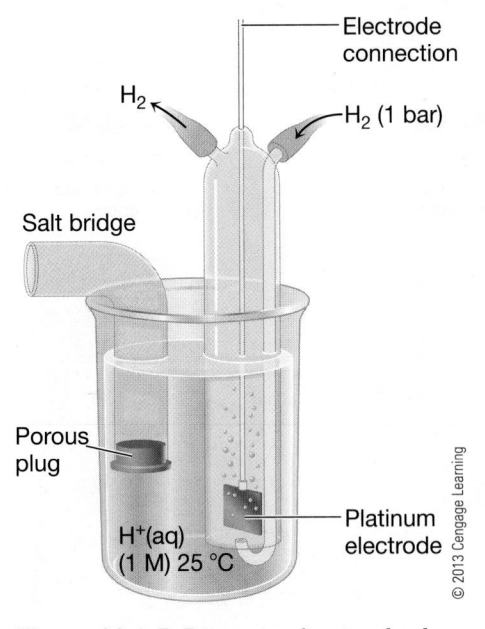

Figure 20.1.5 Diagram of a standard hydrogen electrode

Example Problem 20.1.5 Analyze the components of a voltaic electrochemical cell.

A voltaic electrochemical cell is constructed using the following reaction. The half-cell components are separated by a salt bridge.

$$Al(s) + 3\,Fe^{3+}(aq) \rightarrow Al^{3+}(aq) + 3\,Fe^{2+}(aq)$$

Identify the reactions that take place at the anode and at the cathode, the direction in which the electrons migrate in the external circuit, and the direction the anions in the salt bridge migrate.

Solution:

You are asked to identify the components of a voltaic electrochemical cell.

You are given the net reaction for the cell.

The oxidation number of aluminum increases from 0 in $Al(s)$ to +3 in Al^{3+}. Aluminum is oxidized. Oxidation occurs at the anode in a voltaic electrochemical cell.

$$Al(s) \rightarrow Al^{3+}(aq) + 3\,e^- \qquad \textit{Oxidation, anode}$$

Example Problem 20.1.5 *(continued)*

The oxidation number of iron decreases from +3 in Fe^{3+} to +2 in Fe^{2+}. Iron(III) is reduced. Reduction occurs at the cathode in a voltaic electrochemical cell.

$$Fe^{3+}(aq) + e^- \rightarrow Fe^{2+}(aq) \qquad \textit{Reduction, cathode}$$

Electrons are transferred from the reducing agent, Al(s), to the oxidizing agent, $Fe^{3+}(aq)$. The electrons flow from the anode half-cell to the cathode half-cell. To complete the circuit, anions migrate through the salt bridge from the cathode compartment to the anode compartment.

Video Solution

Tutored Practice Problem 20.1.5

20.1e Electrochemical Cell Notation

Chemists use a shorthand method for indicating the components of an electrochemical cell. Consider the electrochemical cell consisting of a zinc anode and a copper cathode. If the aqueous solutions of Zn^{2+} and Cu^{2+} each have a concentration of 1 mol/L, the shorthand notation for this electrochemical cell is

$$Zn(s)|Zn^{2+}(aq, 1\ M)||Cu^{2+}(aq, 1\ M)|Cu(s)$$

In this notation,

- the oxidation half-reaction (anode) is written on the left and the reduction half-reaction (cathode) is written on the right (the cell components are written in the order they appear in the half-reaction);

- physical states and solution concentrations are indicated in parentheses;

- a single vertical line separates species in different phases; and

- a double vertical line, which indicates a salt bridge, separates the oxidation and reduction half-cells.

For cells using inert electrodes, the electrode material is noted first for the anode and last for the cathode. For example, consider an electrochemical cell consisting of the following half-cells, where the iron is reduced at a platinum electrode.

$$Zn(s) \rightarrow Zn^{2+}(aq) + 2\ e^-$$

$$Fe^{3+}(aq) + e^- \rightarrow Fe^{2+}(aq)$$

Assuming all solutions have a concentration of 1 mol/L, the cell notation is

$$Zn(s)|Zn^{2+}(aq, 1\ M)||Fe^{3+}(aq, 1\ M), Fe^{2+}(aq, 1\ M)|Pt(s)$$

Example Problem 20.1.6 Write electrochemical cell notation.

Write the cell notation for an electrochemical cell consisting of an anode where H_2 is oxidized to H^+ at a platinum electrode and a cathode where Co^{3+} is reduced to Co. Assume all aqueous solutions have a concentration of 1 mol/L and gases have a pressure of 1 bar.

Solution:

You are asked to write cell notation for an electrochemical cell.

You are given the components of the electrochemical cell.

Follow the previously provided rules for writing the electrochemical cell notation. The two half-reactions taking place in this cell are

$$H_2(g) \rightarrow 2\, H^+(aq) + 2\, e^-$$
$$Co^{3+}(aq) + 3\, e^- \rightarrow Co(s)$$

Vertical lines separate the species with different physical states in the anode and cathode. Note that the inert electrode material comes first in the anode portion of the notation.

$$Pt(s)|H_2(g, 1\ bar)|H^+(aq, 1\ M)||Co^{3+}(aq, 1\ M)|Co(s)$$

Video Solution

Tutored Practice
Problem 20.1.6

Section 20.1 Mastery

20.2 Cell Potentials, Free Energy, and Equilibria

20.2a Cell Potentials and Standard Reduction Potentials

We know that reactions are, in part, controlled by thermodynamics. In particular, we've explored the concept of equilibrium in detail, where the extent to which a reaction proceeds is a function of the free energy change. Electrochemical reactions are no different—the extent to which they proceed and the energy they can provide depend on the nature of the reactants and products, as well as their concentrations at any point during the reaction.

As we saw earlier, a voltaic electrochemical cell is one where a spontaneous redox reaction "pushes" electrons through a wire. The rate of electron flow through a conductor is expressed in units of **amperes** (A), and it is related to the fundamental unit of electric charge (the **coulomb**) by the relationship 1 A = 1 C/s. The charge on a single electron is 1.602×10^{-19} C.

The driving force for the electron flow from the anode to the cathode is a drop in the potential energy of the electrons that are moving through the wire. This difference in potential energy, which is commonly called the **electromotive force (emf)**, is measured in units of **volts** (V), where 1 V = 1 J/C. In a voltaic cell, the difference in potential energy between the cathode and anode is called the **cell potential**, E_{cell}. Under standard conditions, where all solutions have concentrations of 1 mol/L and all gases have a pressure of 1 bar, the potential energy difference is the **standard cell potential, $E°_{cell}$**.

Consider the copper–zinc electrochemical cell described earlier. When a voltmeter is placed in series along the wire connecting the anode and cathode in this voltaic cell, a voltage of 1.10 V is measured. This means that the potential energy of the cathode is 1.10 V lower than that of the anode and that the electrons flow through the wire driven by a potential drop of 1.10 V (Figure 20.2.1).

Combinations of different oxidation and reduction half-cells result in different potential energy differences and therefore different cell potentials (Table 20.2.1). A positive cell potential indicates that the electrons are moving from a higher potential energy to a lower potential energy. The greater the cell potential, the more likely the reaction will occur spontaneously. A negative cell potential indicates that the reaction will not proceed spontaneously in the forward direction but will proceed spontaneously in the reverse direction.

Instead of tabulating cell potentials for every possible combination of oxidation and reduction half-reactions, it is useful to know the **standard reduction potential, $E°_{red}$**, for many reduction half-reactions, which can then be used to calculate $E°_{cell}$ values. The standard reduction potential for a half-reaction cannot be measured directly, though, because cell potentials are a measurement of a potential energy change and not an absolute amount of potential energy. Therefore, to report $E°_{red}$ for a half-reaction, we must arbitrarily define a zero-point standard half-reaction, against which all other half-reaction potentials can be measured. Chemists have defined the standard reduction potential of the reduction of H^+ to H_2, which is measured using the hydrogen electrode (SHE) described earlier, as 0 V. This half-reaction, which can be written as an oxidation or a reduction, can be used to measure the potential for other half-reactions.

$$2 \text{ H}^+(\text{aq}) + 2 \text{ e}^- \rightarrow \text{H}_2(\text{g}) \qquad E°_{red} = 0 \text{ V}$$

Consider an electrochemical cell consisting of a zinc half-reaction and the SHE (Figure 20.2.2). When the zinc electrode is the anode (it is connected to the negative terminal of the voltmeter) and the SHE is the cathode (connected to the positive terminal), this cell has a standard potential of 0.76 V. The net reaction for this cell can be written as the sum of two half-reactions, and the standard cell potential, the potential energy

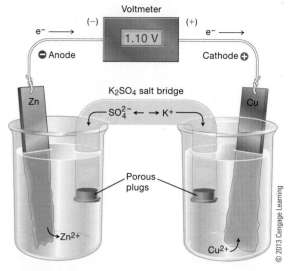

Figure 20.2.1 Cell potential for the copper–zinc voltaic cell

Table 20.2.1 Cell Potentials for Some Voltaic Cells

Anode	Cathode	$E°_{cell}$ (V)
Ni|Ni^{2+}(1 M)	Cd^{2+}(1 M)|Cd	0.15
Cd|Cd^{2+}(1 M)	Sn^{4+}(1 M), Sn^{2+}(1 M)|Pt	0.55
Ni|Ni^{2+}(1 M)	Ag$^+$(1 M)|Ag	1.05
Mg|Mg^{2+}(1 M)	Sn^{2+}(1 M)|Sn	2.23

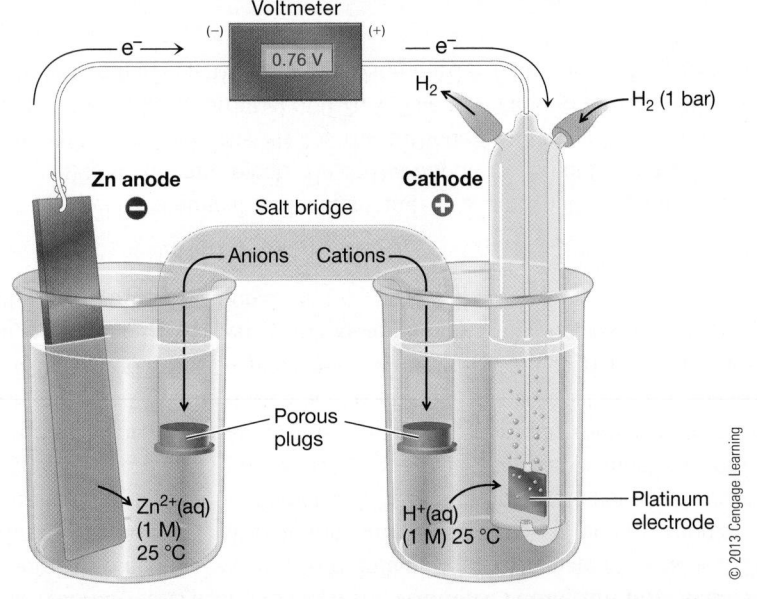

Figure 20.2.2 The cell potential for the zinc–SHE voltaic cell

difference between the two half-cells, is calculated from the standard reduction potentials (Equation 20.1).

$$E^{\circ}_{cell} = E^{\circ}_{cathode} - E^{\circ}_{anode} = E^{\circ}_{red} + E^{\circ}_{ox} \qquad \textbf{(20.1)}$$

In Equation 20.1, $E^{\circ}_{cathode}$ is the standard reduction potential for the reaction taking place at the cathode (in the reduction half-cell) and E°_{anode} is the standard reduction potential for the reaction taking place at the anode (in the oxidation half-cell). Reversing a half-reaction changes the sign of the standard potential, so we can also express the standard potential for the oxidation half-cell as a standard oxidation potential, E°_{ox}, where $E^{\circ}_{ox} = -E^{\circ}_{anode}$. Note that Equation 20.1 works only for balanced oxidation–reduction reactions where the number of electrons lost in the oxidation is equal to the number gained in the reduction.

We can use Equation 20.1 to calculate E°_{red} for the zinc half-reaction.

Oxidation, anode:	$Zn(s) \rightarrow Zn^{2+}(aq) + 2\ e^-$	$E^{\circ}_{anode} = ?$
Reduction, cathode:	$2\ H^+(aq) + 2\ e^- \rightarrow H_2(g)$	$E^{\circ}_{cathode} = 0\ V$
Net reaction:	$Zn(s) + 2\ H^+(aq) \rightarrow Zn^{2+}(aq) + H_2(g)$	$E^{\circ}_{cell} = 0.76\ V$

$$E^{\circ}_{cell} = E^{\circ}_{cathode} - E^{\circ}_{anode}$$

$$0.76\ V = 0\ V - E^{\circ}_{anode}$$

$$E^{\circ}_{anode} = -0.76\ V$$

The standard reduction potential for the reaction occurring in the oxidation half-cell is therefore -0.76 V. Similar experiments are used to calculate standard reduction potentials for other reduction half-reactions, and some selected E°_{red} values are shown in Interactive Table 20.2.2.

Notice in Interactive Table 20.2.2 that reduction potentials are listed in order of decreasing reduction potential. That is, reduction half-reactions with positive standard reduction potentials have reactants that are stronger oxidizing agents (more easily reduced) than H^+, and the products are poorer oxidizing agents (more difficult to oxidize) than H_2. Conversely, a negative standard reduction potential indicates that the reactant in the half-reaction is a poorer oxidizing agent (more difficult to reduce) than H^+ and that the product is a stronger reducing agent (is more easily oxidized) than H_2.

The position of elements in Interactive Table 20.2.2 reveals some trends in relative reducing and oxidizing ability. For example, F_2 is the most easily reduced species in the table, which makes sense because of fluorine's large electron affinity. However, because these reactions take place in aqueous solution, hydration enthalpies also play a role, which can result in some inconsistent trends. For example, Li is more easily oxidized than Na, but orbital energies and ionization energies suggest that Na should be more easily oxidized than Li. The difference in ionic radii and therefore hydration enthalpies result in Li being more easily oxidized than Na.

Interactive Table 20.2.2

Selected Standard Reduction Potentials in Aqueous Solution at 25 °C

Reduction Half-Reaction	E° (V)
$F_2(g) + 2\,e^- \rightarrow 2\,F^-(aq)$	2.87
$MnO_4^-(aq) + 8\,H^+(aq) + 5\,e^- \rightarrow Mn^{2+}(aq) + 4\,H_2O(\ell)$	1.51
$Cl_2(g) + 2\,e^- \rightarrow 2\,Cl^-(aq)$	1.36
$Cr_2O_7^{2-}(aq) + 14\,H^+(aq) + 6\,e^- \rightarrow 2\,Cr^{3+}(aq) + 7\,H_2O(\ell)$	1.33
$O_2(g) + 4\,H^+(aq) + 4\,e^- \rightarrow 2\,H_2O(\ell)$	1.229
$Br_2(\ell) + 2\,e^- \rightarrow 2\,Br^-(aq)$	1.08
$NO_3^-(aq) + 4\,H^+(aq) + 3\,e^- \rightarrow NO(g) + 2\,H_2O(\ell)$	0.96
$Hg^{2+}(aq) + 2\,e^- \rightarrow Hg(\ell)$	0.855
$O_2(g) + 4\,H^+(aq, 10^{-7}\,M) + 4\,e^- \rightarrow 2\,H_2O(\ell)$	0.82
$Ag^+(aq) + e^- \rightarrow Ag(s)$	0.799
$Hg_2^{2+}(aq) + 2\,e^- \rightarrow 2\,Hg(\ell)$	0.789
$Fe^{3+}(aq) + e^- \rightarrow Fe^{2+}(aq)$	0.771
$I_2(s) + 2\,e^- \rightarrow 2\,I^-(aq)$	0.535
$Cu^{2+}(aq) + 2\,e^- \rightarrow Cu(s)$	0.337
$2\,H^+(aq) + 2\,e^- \rightarrow H_2(g)$	0.0000
$Sn^{2+}(aq) + 2\,e^- \rightarrow Sn(s)$	−0.14
$Ni^{2+}(aq) + 2\,e^- \rightarrow Ni(s)$	−0.25
$Cr^{3+}(aq) + e^- \rightarrow Cr^{2+}(aq)$	−0.41
$2\,H_2O(\ell) + 2\,e^- \rightarrow H_2(g) + 2\,OH^-(aq, 10^{-7}\,M)$	−0.41
$Fe^{2+}(aq) + 2\,e^- \rightarrow Fe(s)$	−0.44
$Cr^{3+}(aq) + 3\,e^- \rightarrow Cr(s)$	−0.74
$Zn^{2+}(aq) + 2\,e^- \rightarrow Zn(s)$	−0.763
$2\,H_2O(\ell) + 2\,e^- \rightarrow H_2(g) + 2\,OH^-(aq)$	−0.83
$Al^{3+}(aq) + 3\,e^- \rightarrow Al(s)$	−1.66
$Mg^{2+}(aq) + 2\,e^- \rightarrow Mg(s)$	−2.37
$Na^+(aq) + e^- \rightarrow Na(s)$	−2.714
$K^+(aq) + e^- \rightarrow K(s)$	−2.925
$Li^+(aq) + e^- \rightarrow Li(s)$	−3.045

Example Problem 20.2.1 Explore relative reducing/oxidizing ability.

Use the half-reactions shown below to answer the following questions.

Half-Reaction	$E°$ (V)
$F_2(g) + 2\,e^- \rightarrow 2\,F^-(aq)$	2.870
$Pb^{2+}(aq) + 2\,e^- \rightarrow Pb(s)$	−0.126
$Al^{3+}(aq) + 3\,e^- \rightarrow Al(s)$	−1.660

a. Identify the strongest oxidizing agent and the weakest oxidizing agent.
b. Identify the strongest reducing agent and the weakest reducing agent.
c. Will $Al^{3+}(aq)$ oxidize $F^-(aq)$ to $F_2(g)$?
d. Identify all species that can be oxidized by $Pb^{2+}(aq)$.

Solution:

You are asked to characterize species using reduction potentials.

You are given half-reactions and standard reduction potentials.

a. The strongest oxidizing agent, F_2, has the largest (positive) standard reduction potential. The weakest oxidizing agent, Al^{3+}, has the most negative reduction potential.
b. The strongest reducing agent, Al, is the product in the half-reaction with the weakest oxidizing agent. The weakest reducing agent, F^-, is the product in the half-reaction with the strongest oxidizing agent.
c. The reaction $Al^{3+}(aq) + F^-(aq) \rightarrow F_2(g) + Al(s)$ is not product favored. Because F_2 is a stronger oxidizing agent than Al^{3+}, Al^{3+} will not oxidize F^- to F_2. The reaction is product favored when written in the reverse direction.
d. Of the species in the table, only Al^{3+} is a weaker oxidizing agent than Pb^{2+}. Therefore, Pb^{2+} will oxidize Al to Al^{3+}.

Video Solution

Tutored Practice
Problem 20.2.1

Standard reduction potentials can be used to predict $E°_{cell}$ for a redox reaction. Consider a voltaic cell consisting of Cu/Cu^{2+} and Ag/Ag^+ half-cells.

$$Cu^{2+} + 2\,e^- \rightarrow Cu \qquad E°_{red} = +0.34\text{ V}$$

$$Ag^+ + e^- \rightarrow Ag \qquad E°_{red} = +0.80\text{ V}$$

According to the standard reduction potentials, Ag^+ is *more favored* to be reduced than is Cu^{2+}. Therefore, when a voltaic cell is composed of these two half-reactions, Ag^+ will be reduced and Cu will be oxidized. That is, the half-cell more favored to undergo reduction will do so and will be the cell's cathode. The other half-cell will be the anode. The standard cell potential, the potential energy difference between the two half-cells, is calculated using Equation 20.1.

Oxidation, anode:	$Cu \rightarrow Cu^{2+} + 2\,e^-$	$E^{\circ}_{red} = +0.34$ V
Reduction, cathode:	$Ag^+ + e^- \rightarrow Ag$	$E^{\circ}_{red} = +0.80$ V
Net reaction:	$Cu(s) + 2\,Ag^+(aq) \rightarrow 2\,Ag(s) + Cu^{2+}(aq)$	$E^{\circ}_{cell} = 0.80$ V $- 0.34$ V $= 0.46$ V

In this cell, electrons will flow from the $Cu|Cu^{2+}$ half-cell to the $Ag^+|Ag$ half-cell, driven by a potential drop of 0.46 V.

Example Problem 20.2.2 Predict reaction and calculate E°_{cell} from standard reduction potentials.

A voltaic cell is constructed from a standard $Cr^{3+}|Cr^{2+}$ half-cell ($E^{\circ}_{red} = -0.410$ V) and a standard $Cu^{2+}|Cu$ half-cell ($E^{\circ}_{red} = 0.337$ V). What is the spontaneous reaction that takes place, and what is the standard cell potential?

Solution:

You are asked to identify the spontaneous reaction taking place in a voltaic cell and to calculate the standard cell potential.

You are given the components of the two half-cells.

The spontaneous cell reaction is the one in which the stronger oxidizing agent is reduced at the cathode and the stronger reducing agent is oxidized at the anode. The stronger oxidizing agent has the larger standard reduction potential.

$$Cu^{2+}(aq) + 2\,e^- \rightarrow Cu(s) \qquad E^{\circ}_{red} = 0.337 \text{ V}$$
$$Cr^{3+}(aq) + e^- \rightarrow Cr^{2+}(aq) \qquad E^{\circ}_{red} = -0.410 \text{ V}$$

In this cell, Cu^{2+} is more easily reduced than Cr^{3+}. The reaction taking place at the cathode is

$$Cu^{2+}(aq) + 2\,e^- \rightarrow Cu(s)$$

and the reaction occurring at the anode is

$$Cr^{2+}(aq) \rightarrow Cr^{3+}(aq) + e^-$$

The balanced equation for the spontaneous reaction is

Reduction, cathode:	$Cu^{2+}(aq) + 2\,e^- \rightarrow Cu(s)$
Oxidation, anode:	$2\,Cr^{2+}(aq) \rightarrow 2\,Cr^{3+}(aq) + 2\,e^-$
Net reaction:	$Cu^{2+}(aq) + 2\,Cr^{2+}(aq) \rightarrow Cu(s) + 2\,Cr^{3+}(aq)$

Use Equation 20.1 to calculate the standard cell potential for this reaction.

$$E^{\circ}_{cell} = E^{\circ}_{cathode} - E^{\circ}_{anode} = 0.337 \text{ V} - (-0.410 \text{ V}) = 0.747 \text{ V}$$

It is important to note that multiplying a half-reaction by a constant, as was done previously to the oxidation half-reaction, does not change the reduction potential for the half-cell.

Is your answer reasonable? A spontaneous reaction has a positive E°_{cell} value.

Video Solution

Tutored Practice
Problem 20.2.2

20.2b Cell Potential and Free Energy

Both cell potential and free energy (ΔG) tell us whether or not a reaction is spontaneous. Both are also related to the maximum amount of work (w_{max}) that a system can do on the surroundings. For a redox reaction, the maximum amount of work is equal to the maximum cell potential multiplied by the total charge transferred during the reaction (Equation 20.2), where the negative sign indicates the direction of energy flow (work is done on the surroundings).

$$w_{max} = -q \times E_{cell} \qquad \textbf{(20.2)}$$

The total charge transferred during a redox reaction, q, is equal to the moles of electrons transferred between oxidizing and reducing agents (n) multiplied by **Faraday's constant** (F), the charge on one mole of electrons.

$$1\ F = (1.60218 \times 10^{-19}\ C/e^-)(6.02214 \times 10^{23}\ e^-/1\ mol\ e^-)$$
$$= 96{,}485\ C/mol\ e^- = 96{,}485\ J/V \cdot mol\ e^-$$

Substituting this relationship into Equation 20.2,

$$w_{max} = -nFE_{cell} \qquad \textbf{(20.3)}$$

Because free energy is equal to the maximum amount of work that can be done on the surroundings, we can now relate cell potential to free energy (Equation 20.4).

$$w_{max} = -nFE_{cell} \qquad \text{and} \qquad w_{max} = \Delta G$$
$$\Delta G = -nFE_{cell} \qquad \textbf{(20.4)}$$

Under standard conditions,

$$\Delta G^\circ = -nFE^\circ_{cell} \qquad \textbf{(20.5)}$$

We have already seen that a positive cell potential indicates a spontaneous reaction, and we learned in Thermodynamics: Entropy and Free Energy (Unit 19) that a negative ΔG value also indicates a spontaneous reaction. Equations 20.4 and 20.5 confirm this relationship and give us a way to relate the two quantitatively.

Example Problem 20.2.3 Calculate free energy from cell potential.

Use standard reduction potentials to calculate the standard free energy change for the following reaction:

$$Hg^{2+}(aq) + 2\ Cr^{2+}(aq) \rightarrow Hg(\ell) + 2\ Cr^{3+}(aq)$$

Solution:

You are asked to use standard reduction potentials to calculate the standard free energy change for a redox reaction.

You are given the balanced redox reaction.

To use Equation 20.5, we must first calculate $E°_{cell}$ for the reaction. Separate the reaction into half-reactions and use a table of reduction potentials. The reduction potential of Hg^{2+} is greater than that of Cr^{3+}, so Hg^{2+} will be reduced and Cr^{2+} will be oxidized.

> Reduction, cathode: $Hg^{2+}(aq) + 2\ e^- \rightarrow Hg(\ell)$ $E°_{red} = 0.855\ V$
>
> Oxidation, anode: $Cr^{2+}(aq) \rightarrow Cr^{3+}(aq) + e^-$ $E°_{red} = -0.410\ V$
>
> $E°_{cell} = E°_{cathode} - E°_{anode} = 0.855\ V - (-0.410\ V) = 1.265\ V$

We also must know the total number of electrons transferred during the redox reaction. Balancing the equation, we see that $n = 2$.

> Reduction: $Hg^{2+}(aq) + 2\ e^- \rightarrow Hg(\ell)$
>
> Oxidation: $2\ Cr^{2+}(aq) \rightarrow 2\ Cr^{3+}(aq) + 2\ e^-$
>
> Net reaction: $Hg^{2+}(aq) + 2\ Cr^{2+}(aq) \rightarrow Hg(\ell) + 2\ Cr^{3+}(aq)$

Note that n is the number of electrons lost during oxidation or gained during reduction. It is equal to the number of electrons transferred during the reaction, not the total number of electrons that appear in the two half-reactions.

Finally, use Equation 20.5 to calculate the free energy change for this reaction.

$$\Delta G° = -nFE°_{cell} = -(2\ mol\ e^-)(96{,}485\ J/V \cdot mol\ e^-)(1.265\ V)(1\ kJ/10^3\ J) = -244.1\ kJ$$

Is your answer reasonable? A spontaneous reaction has a negative free energy change.

Video Solution

Tutored Practice
Problem 20.2.3

20.2c Cell Potential and the Equilibrium Constant

Now that we have related cell potential to free energy, we can relate it to the equilibrium constant. As you learned in Thermodynamics: Entropy and Free Energy (Unit 19),

$$\Delta G° = -RT\ln K$$

This relationship can be combined with Equation 20.5 to show the relationship between cell potential and the equilibrium constant.

$$-nFE°_{cell} = -RT\ln K$$

$$E°_{cell} = \frac{RT}{nF}\ln K \qquad (20.6)$$

At 298 K,

$$E°_{cell} = \frac{0.0257\text{ V}}{n}\ln K \qquad (20.7)$$

Although Equation 20.7 can be used to calculate a cell potential from an equilibrium constant, it is much more common to use it to calculate an equilibrium constant from an experimentally determined cell potential.

Example Problem 20.2.4 Use cell potential to calculate an equilibrium constant.

Calculate the cell potential and the equilibrium constant for the following reaction at 298 K:

$$2\text{ H}^+(aq) + Hg(\ell) \rightarrow H_2(g) + Hg^{2+}(aq)$$

Solution:

You are asked to calculate the cell potential and the equilibrium constant for a reaction at a given temperature.

You are given the balanced redox reaction.

First calculate $E°_{cell}$ for the reaction.

Reduction, cathode: $2\text{ H}^+(aq) + 2\text{ e}^- \rightarrow H_2(g)$
$E°_{red} = 0.000\text{ V}$

Oxidation, anode: $Hg(\ell) \rightarrow Hg^{2+}(aq) + 2\text{ e}^-$
$E°_{red} = 0.855\text{ V}$

Net reaction: $2\text{ H}^+(aq) + Hg(\ell) \rightarrow H_2(g) + Hg^{2+}(aq)$
$E°_{cell} = E°_{cathode} - E°_{anode} = -0.855\text{ V}$

The total number of electrons transferred during the reaction is 2 ($n = 2$). Use Equation 20.7 to calculate the equilibrium constant. (It is a good idea to carry extra significant figures through the calculations to avoid round-off errors when you do the anti-log.)

$$E°_{cell} = \frac{0.0257\text{ V}}{n}\ln K$$

$$\ln K = \frac{nE°_{cell}}{0.0257\text{ V}} = \frac{(2)(-0.855\text{ V})}{0.0257\text{ V}} = -66.537$$

$$K = e^{-66.537} = 1.27 \times 10^{-29}$$

Is your answer reasonable? The equilibrium constant is very small, which makes sense for this nonspontaneous reaction ($E°_{cell} < 0$).

Video Solution

Tutored Practice
Problem 20.2.4

Cell potential, free energy, and equilibrium constants are related as shown in Interactive Figure 20.2.3 and Table 20.2.3.

20.2d Cell Potentials Under Nonstandard Conditions

Standard reduction potentials such as those in Table 20.2.1 are valid under standard conditions where all solutions have concentrations of 1 mol/L. Many electrochemical cells do not operate under these conditions, however, so we must make adjustments for nonstandard conditions. The relationship between the standard cell potential (E°_{cell}) and the cell potential under nonstandard conditions (E_{cell}) can be derived from the relationship between free energy and the reaction quotient described in Thermodynamics: Entropy and Free Energy (Unit 19).

Recall that the free energy change for a reaction is equal to the sum of the standard free energy change for the reaction and the quantity $RT\ln Q$:

$$\Delta G = \Delta G^\circ + RT\ln Q$$

We have shown that $\Delta G^\circ = -nFE^\circ_{cell}$ and $\Delta G = -nFE_{cell}$. Substituting these relationships into the free energy–reaction quotient relationship gives the **Nernst equation** (Equation 20.8).

$$-nFE_{cell} = -nFE^\circ_{cell} + RT\ln Q$$

$$E_{cell} = E^\circ_{cell} - \frac{RT}{nF}\ln Q \qquad \textbf{(20.8)}$$

At 298 K, the Nernst equation can be simplified to

$$E_{cell} = E^\circ_{cell} - \frac{0.0257\ \text{V}}{n}\ln Q \qquad \textbf{(20.9)}$$

Notice that the magnitude of Q controls whether E_{cell} is greater or less than E°_{cell}. When the reaction quotient is greater than 1 (product concentrations are greater than reactant concentrations), $\ln Q > 0$ and $E_{cell} < E^\circ_{cell}$. Conversely, when the reaction quotient is less than 1 (reactant concentrations are greater than product concentrations), $\ln Q < 0$ and $E_{cell} > E^\circ_{cell}$.

The Nernst equation can be used in a variety of ways, some examples of which include predicting cell potential under nonstandard conditions, using a measured cell potential to determine concentration of a species in the reaction, using a measured cell potential to determine the pH at a hydrogen electrode, and relating cell potential to the pressure of a gaseous reactant.

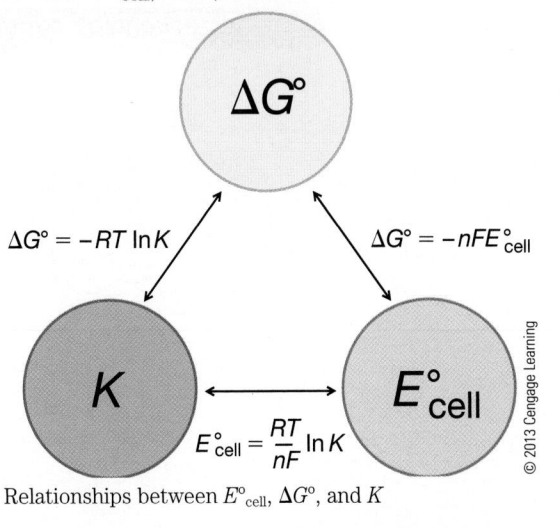

Interactive Figure 20.2.3

Relate E°_{cell}, ΔG°, and K.

$\Delta G^\circ = -RT\ln K$

$\Delta G^\circ = -nFE^\circ_{cell}$

$E^\circ_{cell} = \frac{RT}{nF}\ln K$

Relationships between E°_{cell}, ΔG°, and K

© 2013 Cengage Learning

Table 20.2.3 Nature of Chemical Reactions under Standard Conditions

ΔG°	K	E°_{cell}	Nature of Reaction
< 0	> 1	> 0	Spontaneous
0	1	0	At equilibrium
> 0	< 1	< 0	Nonspontaneous

Example Problem 20.2.5 Predict cell potential under nonstandard conditions.

What is the calculated value of the cell potential at 298 K for an electrochemical cell with the reaction shown below when $[Pb^{2+}] = 1.46$ M and $[Cr^{3+}] = 1.62 \times 10^{-4}$ M?

$$3\ Pb^{2+}(aq) + 2\ Cr(s) \rightarrow 3\ Pb(s) + 2\ Cr^{3+}(aq)$$

Solution:

You are asked to calculate the cell potential for an electrochemical cell under nonstandard conditions.

You are given the reaction occurring in the cell and the temperature and concentration of all species in the cell.

The cell potential under nonstandard conditions is calculated using the Nernst equation. To use this equation, we need to know E°_{cell} and n, the number of electrons transferred in the redox reaction. First, balance the equation and calculate E°_{cell}.

Reduction, cathode: $3\ Pb^{2+}(aq) + 6\ e^{-} \rightarrow 3\ Pb(s)$ $E^{\circ}_{red} = -0.126$V

Oxidation, anode: $2\ Cr(s) \rightarrow 2\ Cr^{3+}(aq) + 6\ e^{-}$ $E^{\circ}_{red} = -0.740$V

Net reaction: $3\ Pb^{2+}(aq) + 2\ Cr(s) \rightarrow 3\ Pb(s) + 2\ Cr^{3+}(aq)$

$$E^{\circ}_{cell} = E^{\circ}_{cathode} - E^{\circ}_{anode} = -0.126\text{ V} - (-0.740\text{ V}) = 0.614\text{ V}$$

Next, use Equation 20.9 to calculate the nonstandard cell potential:

$$n = 6 \qquad\qquad Q = \frac{[Cr^{3+}]^2}{[Pb^{2+}]^3}$$

$$[Pb^{2+}] = 1.46\text{ M} \qquad [Cr^{3+}] = 1.62 \times 10^{-4}\text{ M}$$

$$E_{cell} = E^{\circ}_{cell} - \frac{0.0257\text{ V}}{n} \ln Q$$

$$= 0.614\ V - \frac{0.0257\text{ V}}{6} \ln \left[\frac{(1.62 \times 10^{-4})^2}{(1.46)^3} \right]$$

$$= 0.614\ V + 0.080\ V$$

$$= 0.694\ V$$

Is your answer reasonable? The reactant concentration is greater than the product concentration, so $E_{cell} > E^{\circ}_{cell}$.

Video Solution

Tutored Practice
Problem 20.2.6

Example Problem 20.2.6 Use measured cell potential to calculate concentration.

When $[Cu^{2+}] = 1.07$ M, the observed cell potential at 298 K for a voltaic cell with the reaction shown below is 2.067 V. What is the Al^{3+} concentration in this cell?

$$3\ Cu^{2+}(aq) + 2\ Al(s) \rightarrow 3\ Cu(s) + 2\ Al^{3+}(aq)$$

Solution:

You are asked to use a measured cell potential to calculate the concentration of a species in the cell.

You are given the reaction occurring in the cell, the cell potential, and the temperature and concentration of all but one species in the cell.

The concentration of species in a voltaic cell under nonstandard conditions is calculated using the Nernst equation. To use this equation, we need to know E°_{cell} and n, the number of electrons transferred in the redox reaction. First, balance the equation and calculate E°_{cell}.

Reduction, cathode: $\quad 3\ Cu^{2+}(aq) + 6\ e^- \rightarrow 3\ Cu(s) \qquad E^\circ_{red} = 0.337\ V$

Oxidation, anode: $\quad 2\ Al(s) \rightarrow 2\ Al^{3+}(aq) + 6\ e^- \qquad E^\circ_{red} = -1.660V$

Net reaction: $\quad 3\ Cu^{2+}(aq) + 2\ Al(s) \rightarrow 3\ Cu(s) + 2\ Al^{3+}(aq)$

$$E^\circ_{cell} = E^\circ_{cathode} - E^\circ_{anode} = 0.337\ V - (-1.660\ V) = 1.997\ V$$

Next, use Equation 20.9 to calculate the Al^{3+} concentration:

$$n = 6 \qquad\qquad Q = \frac{[Al^{3+}]^2}{[Cu^{2+}]^3}$$

$$E_{cell} = 2.067\ V \qquad\qquad [Cu^{2+}] = 1.07 M$$

$$E_{cell} = E^\circ_{cell} - \frac{0.0257\ V}{n} \ln Q$$

$$2.067\ V = 1.997\ V - \frac{0.0257\ V}{6} \ln\left\{\frac{[Al^{3+}]^2}{(1.07)^3}\right\}$$

$$\ln\left\{\frac{[Al^{3+}]^2}{(1.07)^3}\right\} = -16.342$$

$$\frac{[Al^{3+}]^2}{(1.07)^3} = e^{-16.342} = 7.99 \times 10^{-8}$$

$$[Al^{3+}] = 3.13 \times 10^{-4} M$$

Is your answer reasonable? The cell potential is greater than the standard cell potential, so the reaction quotient is less than 1 (reactant concentrations are greater than product concentrations).

Video Solution

Tutored Practice
Problem 20.2.6

20.2e Concentration Cells

A **concentration cell** is an electrochemical cell where the two half-cells differ only in the concentration of species in solution (Figure 20.2.4). Because the half-reactions are the same in both cells, the standard cell potential is equal to zero ($E^{\circ}_{cell} = 0$ V for a concentration cell). The reaction proceeds because the cell potential is related to the reaction quotient, Q, the ratio of concentrations of species in the cell.

As the half-reactions that make up a concentration cell proceed, the higher concentration decreases and the lower concentration increases until the concentrations in both half-cells are the same; at that point, both E°_{cell} and E_{cell} are equal to zero.

As you will see in the following examples, it is important to identify which solution concentration is in the anode cell and which is in the cathode cell when working with concentration cells. Again, we use the Nernst equation to calculate cell potential or concentrations for these cells.

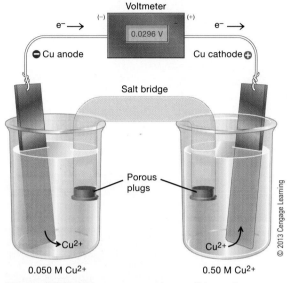

Figure 20.2.4 A concentration cell based on Cu/Cu²⁺ half-reactions

---

Example Problem 20.2.7 Calculate cell potential for a concentration cell.

Calculate the cell potential of a cell composed of a cathode consisting of a copper wire immersed in a 2.20 M CuSO₄ solution and an anode cell consisting of a copper wire immersed in a 1.50×10^{-4} M CuSO₄ solution. Assume the temperature of the solutions is 298 K.

Solution:

You are asked to calculate the cell potential for a concentration cell.

You are given the identity and concentration of the species in the concentration cell and the temperature.

The standard cell potential (E°_{cell}) is 0 V because the two standard half-cells are identical. The net reaction is

$$Cu(s) + Cu^{2+}(aq, cathode) \rightarrow Cu^{2+}(aq, anode) + Cu(s)$$

Notice that the higher [Cu²⁺] is at the cathode (its concentration will decrease as it is reduced to copper metal) and the lower [Cu²⁺] is at the anode (its concentration will increase as copper metal is oxidized to Cu²⁺).

Use Equation 20.9 to calculate the cell potential.

$$E_{cell} = 0 - \frac{0.0257 \text{ V}}{2} \ln\left(\frac{[Cu^{2+}]_{anode}}{[Cu^{2+}]_{cathode}}\right) = \frac{-0.0257 \text{ V}}{2} \ln\left(\frac{1.50 \times 10^{-4}}{2.20}\right) = 0.123 \text{ V}$$

Video Solution

Tutored Practice Problem 20.2.7

The pH of a solution can be measured using a concentration cell where both half-cells are hydrogen electrodes. As shown in the following example, one method uses a standard hydrogen electrode for one of the half-cells.

Example Problem 20.2.8 Calculate the concentration of an unknown solution in a concentration cell.

A concentration cell consisting of two hydrogen electrodes, where the cathode is a standard hydrogen electrode and the anode solution has an unknown pH, has a cell voltage of 0.371 V. What is the pH in the unknown solution? Assume the temperature of the solutions is 298 K.

Solution:

You are asked to calculate the pH in an unknown solution using a concentration cell.

You are given the identity and temperature of the half-cells, the concentration of the species in one of the half-cells, and the cell voltage.

The standard cell potential ($E°_{cell}$) is 0 V because the two standard half-cells are identical. The net reaction is

$$H_2(g) + 2\,H^+(aq, \text{cathode}) \rightarrow 2\,H^+(aq, \text{anode}) + H_2(g)$$

The concentration cell has a positive E_{cell}, which tells us that $[H^+]$ in the unknown solution is lower than $[H^+]$ in the standard hydrogen electrode (where $[H^+] = 1.0$ M).

Use Equation 20.9 to calculate $[H^+]$ in the unknown solution.

$$E_{cell} = E°_{cell} - \frac{0.0257\ \text{V}}{n}\ln\left(\frac{[H^+]^2_{\text{anode}}}{[H^+]^2_{\text{cathode}}}\right)$$

$$0.371 = 0 - \frac{0.0257\ \text{V}}{2}\ln\left(\frac{[H^+]^2_{\text{anode}}}{(1.0)^2}\right)$$

$$[H^+]_{\text{anode}} = 5.38 \times 10^{-7}\ \text{M}$$

$$pH = 6.27$$

Video Solution

Tutored Practice
Problem 20.2.8

Section 20.2 Mastery

20.3 Electrolysis

20.3a Electrolytic Cells and Coulometry

Electrolysis is the use of an external energy source, often a battery, to force a non-spontaneous reaction redox reaction to form products. An **electrolytic cell** is an electrochemical cell in which electrolysis takes place. Consider the zinc–copper voltaic cell we studied earlier.

$$Zn(s) + Cu^{2+}(aq) \rightarrow Zn^{2+}(aq) + Cu(s) \qquad E°_{cell} = +1.10 \text{ V}, \Delta G° = -212 \text{ kJ}$$

A voltaic cell consisting of the $Zn(s)|Zn^{2+}(aq)$ and $Cu(s)|Cu^{2+}(aq)$ half-cells will react in the forward direction, and the electrons will move from the zinc anode to the copper cathode. The cell potential under standard conditions is 1.10 V. By connecting a power source to the wire between the half-cells that produces an electrical voltage greater than 1.10 V, we can initiate the reverse reaction, which is thermodynamically unfavorable and nonspontaneous.

$$Zn^{2+}(aq) + Cu(s) \rightarrow Zn(s) + Cu^{2+}(aq) \qquad E°_{cell} = -1.10 \text{ V}, \Delta G° = +212 \text{ kJ}$$

As shown in Figure 20.3.1, this power source will "push" the electrons in the reverse direction, from the copper electrode to the zinc electrode.

The similarities and differences between voltaic and electrolytic cells are summarized in Table 20.3.1. As you can see, the primary difference, other than the need for an external power source in the electrolytic cells, is the +/− sign of the anode and cathode. In a voltaic cell, electrons are produced at the anode, so it is negative. In an electrolytic cell, the

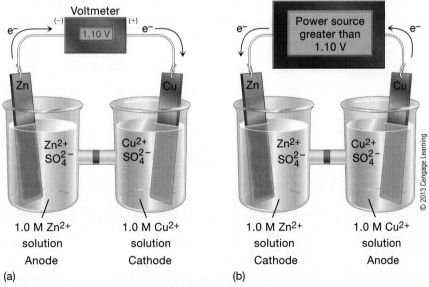

Figure 20.3.1 (a) A voltaic cell; (b) an electrolytic cell

Table 20.3.1 Voltaic and Electrolytic Cells

Type of Cell	Anode	Cathode	Direction of Anion Flow in Salt Bridge	Direction of Electron Flow
Voltaic	Oxidation, negative (−) electrode	Reduction, positive (+) electrode	Cathode to anode	Anode to cathode
Electrolytic	Oxidation, positive (+) electrode	Reduction, negative (−) electrode	Cathode to anode	Anode to cathode

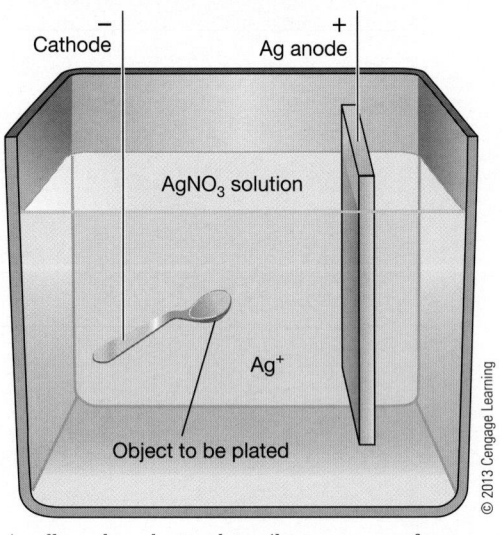

Interactive Figure 20.3.2

Explore electroplating.

A cell used to electroplate silver onto a surface

electrons are produced in the external power source and pushed toward the cathode, where reduction takes place. Therefore, the cathode in an electrolytic cell is negative.

Coulometry is the study of the amount of matter transformed during an electrolysis reaction. In coulometry, we relate charge, the moles of electrons transferred in a redox reaction, and the mass of material produced in the reaction. For example, consider the process of electroplating, where a thin coating of a valuable metal such as silver is plated onto a surface (Interactive Figure 20.3.2). The net reaction for this process is the reduction of silver ions.

$$Ag^+(aq) + e^- \rightarrow Ag(s)$$

The amount of metal plated onto the surface is controlled by the current, the time span during which plating occurs, and the number of electrons required to convert the metal ion to the neutral metal. Recall that 1 A = 1 C/s and 1 F = 96,485 C/mol e$^-$.

Example Problem 20.3.1 Calculate the mass of metal deposited during electrolysis.

What mass of silver metal will be deposited from a solution containing Ag^+ ions if a current of 0.756 A is applied for 69.5 minutes?

Solution:

You are asked to calculate the mass of metal deposited during an electrolysis experiment.

You are given the identity of the species undergoing electrolysis, the amount of current applied, and the experiment time.

First, calculate the total charge transferred in the reaction.

$$(69.5 \text{ min})\left(\frac{60 \text{ s}}{1 \text{ min}}\right)(0.756 \text{ A})\left(\frac{1 \text{ C}}{1 \text{ A} \cdot \text{s}}\right) = 3150 \text{ C}$$

Next, convert charge to moles of electrons using Faraday's constant.

$$(3150 \text{ C})\left(\frac{1 \text{ mol e}^-}{96,485 \text{ C}}\right) = 0.0327 \text{ mol e}^-$$

Finally, use the moles of electrons and the half-reaction for the reduction of silver ions to calculate the mass of silver produced in the reaction.

$$Ag^+(aq) + e^- \rightarrow Ag(s)$$

$$(0.0327 \text{ mol e}^-)\left(\frac{1 \text{ mol Ag}}{1 \text{ mol e}^-}\right)\left(\frac{107.87 \text{ g}}{1 \text{ mol Ag}}\right) = 3.52 \text{ g Ag}$$

Alternatively, the calculation can be done in a single step.

$$(69.5 \text{ min})\left(\frac{60 \text{ s}}{1 \text{ min}}\right)(0.756 \text{ A})\left(\frac{1 \text{ C}}{1 \text{ A} \cdot \text{s}}\right)\left(\frac{1 \text{ mol e}^-}{96,485 \text{ C}}\right)\left(\frac{1 \text{ mol Ag}}{1 \text{ mol e}^-}\right)\left(\frac{107.87 \text{ g}}{1 \text{ mol Ag}}\right) = 3.52 \text{ g Ag}$$

Video Solution

Tutored Practice
Problem 20.3.1

20.3b Electrolysis of Molten Salts

The simplest type of electrolytic cell involves the electrolysis of a molten salt, where the metal cation is reduced and the anion is oxidized. Consider the electrolysis of molten sodium chloride.

Reduction, cathode:	$2 \text{ Na}^+ + 2 \text{ e}^- \rightarrow 2 \text{ Na}(s)$
Oxidation, anode:	$2 \text{ Cl}^- \rightarrow \text{Cl}_2(g) + 2 \text{ e}^-$
Net reaction:	$2 \text{ NaCl}(\ell) \rightarrow 2 \text{ Na}(s) + \text{Cl}_2(g)$

The net reaction is nonspontaneous and is driven by the use of an external power source. Industrially, this reaction occurs in a Downs cell (Interactive Figure 20.3.3).

Explore electrolysis of a molten salt.

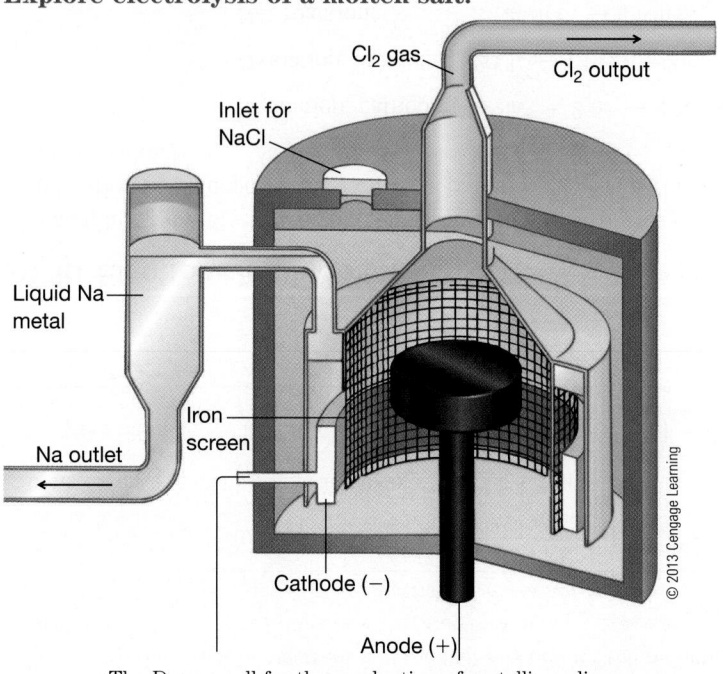

The Downs cell for the production of metallic sodium

In the cell, $CaCl_2$ is mixed with NaCl to lower the melting point from 801 °C to 580 °C. Both sodium and calcium metal are produced at a steel cathode, and gaseous chlorine is produced at a graphite anode. An iron screen separates the cathode from the anode, preventing the reaction products from undergoing an explosive reaction in the cell. The liquid sodium metal is separated from the calcium metal by taking advantage of the different densities and melting points of the metals. The Downs cell is expensive to construct and operate, but it is the most important industrial method for the production of metallic sodium.

What if an electrolytic cell contains a mixture of molten salts? Consider, for example, a cell containing molten NaCl and KCl. When an external power source is applied, which metal will be produced at the cathode, sodium or potassium? The answer lies in the reduction potentials for the species in the cell. For an electrolytic cell containing NaCl and KCl, consider the following reduction potentials:

$$Na^+ + e^- \rightarrow Na \qquad E^\circ_{red} = -2.71 \text{ V}$$

$$K^+ + e^- \rightarrow K \qquad E^\circ_{red} = -2.92 \text{ V}$$

In this case, the sodium ions are more easily reduced than the potassium ions. Therefore, Na^+ is more likely to be reduced at the cathode. For a mixture of anions, the ion with the least negative oxidation potential will be most likely to be oxidized at the anode.

Example Problem 20.3.2 Calculate quantity of materials produced from the electrolysis of a molten salt.

A current of 7.06 A is passed through an electrolysis cell containing molten $CaCl_2$ for 12.9 minutes.

a. Predict the products of the electrolysis and the reactions occurring at the cathode and anode.
b. Calculate the quantity or volume of products collected at the anode and cathode (assume gases are collected at 298 K and 1.00 atm).

Solution:

You are asked to predict the products in an electrolysis cell and calculate the quantity of products produced in the reaction.

You are given the identity of the material in the electrolysis cell, the current passed through the cell, the amount of time the current is passed, and the temperature and pressure at which any gaseous product is collected.

a. In the electrolysis of a molten salt, the metal cation is reduced and the anion is oxidized.

$$\text{Reduction, cathode:} \quad Ca^{2+} + 2\,e^- \rightarrow Ca(\ell)$$
$$\text{Oxidation, anode:} \quad 2\,Cl^- \rightarrow Cl_2(g) + 2\,e^-$$
$$\text{Net reaction:} \quad CaCl_2(\ell) \rightarrow Ca(\ell) + Cl_2(g)$$

b. Use the current, time, and moles of electrons transferred in the reaction to calculate the mass of calcium produced at the cathode. Recall that 1 A = 1 C/s.

$$(7.06 \text{ C/s})(12.9 \text{ min})\left(\frac{60 \text{ s}}{1 \text{ min}}\right)\left(\frac{1 \text{ mol e}^-}{96{,}485 \text{ C}}\right)\left(\frac{1 \text{ mol Ca}}{2 \text{ mol e}^-}\right)\left(\frac{40.08 \text{ g}}{1 \text{ mol Ca}}\right) = 1.13 \text{ g Ca}$$

Similarly, calculate moles of chlorine produced at the anode and use pressure and temperature to calculate the volume of the gas.

$$(7.06 \text{ C/s})(12.9 \text{ min})\left(\frac{60 \text{ s}}{1 \text{ min}}\right)\left(\frac{1 \text{ mol e}^-}{96{,}485 \text{ C}}\right)\left(\frac{1 \text{ mol Cl}_2}{2 \text{ mol e}^-}\right) = 0.0283 \text{ mol Cl}_2$$

$$V = \frac{nRT}{P} = \frac{(0.0283)(0.08206 \text{ L} \cdot \text{atm/K} \cdot \text{mol})(298 \text{ K})}{1.00 \text{ atm}} = 0.692 \text{ L}$$

Video Solution

Tutored Practice
Problem 20.3.2

20.3c Electrolysis of Aqueous Solutions

When an electrolytic cell involves aqueous solutions, it is necessary to consider the possibility of the oxidation or reduction of water. Consider the electrolysis of an aqueous sodium chloride solution. There are three species in solution: $Na^+(aq)$, $Cl^-(aq)$, and $H_2O(\ell)$. Possible reduction reactions:

$$Na^+(aq) + e^- \rightarrow Na(s) \qquad\qquad E^o_{red} = -2.71 \text{ V}$$

$$2\,H_2O(\ell) + 2\,e^- \rightarrow H_2(g) + 2\,OH^-(aq) \qquad\qquad E^o_{red} = -0.83 \text{ V}$$

Possible oxidation reactions:

$$2\,Cl^-(aq) \rightarrow Cl_2(g) + 2\,e^- \qquad\qquad E^o_{ox} = -E^o_{red} = -1.36 \text{ V}$$

$$2\,H_2O(\ell) \rightarrow O_2(g) + 4\,H^+(aq) + 4\,e^- \qquad\qquad E^o_{ox} = -E^o_{red} = -1.23 \text{ V}$$

Based on reduction and oxidation potentials, we would predict that water is reduced at the cathode and oxidized at the anode of this cell. In fact, while water is reduced at the cathode, chlorine gas is produced at the anode.

Reduction: $\quad 2\,H_2O(\ell) + 2\,e^- \rightarrow H_2(g) + 2\,OH^-(aq) \qquad\qquad E^o_{red} = -0.83 \text{ V}$

Oxidation: $\quad 2\,Cl^-(aq) \rightarrow Cl_2(g) + 2\,e^- \qquad\qquad E^o_{red} = 1.36 \text{ V}$

Net reaction: $\quad 2\,NaCl(aq) + 2\,H_2O(\ell) \rightarrow H_2(g) + 2\,NaOH(aq) + Cl_2(g) \qquad E^o_{cell} = -2.19 \text{ V}$

Why is chlorine produced at the anode when water is more easily reduced? The oxidation of water requires an *overvoltage,* a voltage greater than the amount predicted to cause the reaction to occur. The reason behind the overvoltage of the oxidation of water is complex, but it is related to the kinetics of the oxidation reaction. Overvoltages are difficult to predict, so the best way to know whether or not water is oxidized at the anode in an aqueous electrolytic cell is to perform the experiment.

Example Problem 20.3.3 Predict reactions in the electrolysis of an aqueous solution.

An aqueous CoI_2 solution is electrolyzed under 1 bar pressure using platinum electrodes. Write the half-reactions expected to occur at the anode and cathode, based on the cell potentials given.

Standard Reduction Potentials (V) at 25 °C

$I_2 + 2\ e^- \rightarrow 2\ I^-$	0.535
$O_2 + 4\ H_3O^+ + 4\ e^- \rightarrow 6\ H_2O$	1.229
$Co^{2+} + 2\ e^- \rightarrow Co$	−0.280
$2\ H_2O + 2\ e^- \rightarrow H_2 + 2\ OH^-$	−0.828

Solution:

You are asked to predict the products in an aqueous electrolysis cell.

You are given the identity of the material in the aqueous electrolysis cell.

The two species in the solution that can be oxidized are I^- and H_2O. Using standard reduction potentials, we predict that I^- will be oxidized first at the anode because $E^°_{ox}(I^-)$ is greater than $E^°_{ox}(H_2O)$.

$$2\ I^- \rightarrow I_2 + 2\ e^- \qquad E^°_{ox} = -0.535$$
$$6\ H_2O \rightarrow O_2 + 4\ H_3O^+ + 4\ e^- \qquad E^°_{ox} = -1.229$$

Therefore, the expected reaction at the anode is

$$2\ I^- \rightarrow I_2 + 2\ e^-$$

The two species in solution that can be reduced are Co^{2+} and H_2O. Using standard reduction potentials, we predict that Co^+ will be reduced first at the cathode because $E^°_{red}(Co^{2+})$ is greater than $E^°_{red}(H_2O)$. Therefore, the expected reaction at the cathode is

$$Co^{2+} + 2\ e^- \rightarrow Co$$

Video Solution

Tutored Practice
Problem 20.3.3

Section 20.3 Mastery

20.4 Applications of Electrochemistry: Batteries and Corrosion

20.4a Primary Batteries

A **battery** is a combination of one or more electrochemical cells that is used to convert chemical energy into electrical energy. Batteries are classified into two main types: **primary batteries**, which cannot be recharged, and **secondary batteries**, which can be recharged. All batteries have the following components:

- A cathode material that gets reduced

- An anode material that gets oxidized

- A separator that serves as a salt bridge

The main difference between batteries and the voltaic cells we have described so far is that batteries are portable and self-contained. Therefore, instead of aqueous solutions, most batteries use a moist paste to hold the electrolyte and a thin permeable plastic film separator to serve as a salt bridge. The only battery to differ from this method is the common lead-acid car battery, which uses an aqueous sulfuric acid solution.

A typical alkaline battery (Interactive Figure 20.4.1) uses zinc as the anode material. Zn is a good choice for three reasons:

- It is a reasonably strong metal and can help serve as part of the battery case.

- It is a reasonably good reducing agent.

- It does not quickly corrode when exposed to typical temperature and humidity conditions.

The cathode in an alkaline battery is manganese(IV) oxide, which is a good oxidizing agent. The reactions and cell potential for an alkaline battery are shown below.

Cathode: $2 \, MnO_2(s) + H_2O(\ell) + 2 \, e^- \rightarrow Mn_2O_3(s) + 2 \, OH^-(aq)$

Anode: $Zn(s) + 2 \, OH^-(aq) \rightarrow ZnO(s) + H_2O(\ell) + 2 \, e^-$

Net reaction: $2 \, MnO_2(s) + Zn(s) \rightarrow Mn_2O_3(s) + ZnO(s)$ $\quad E_{cell} = 1.5 \, V$

Another example of a primary battery is the common, small disk-shaped watch battery, also called a mercury battery (Figure 20.4.2). The mercury battery uses the same anode as the alkaline battery, but instead of manganese, the cathode in this battery consists of mercury(II) oxide. The cell reactions and cell potential are shown below.

Explore an alkaline battery.

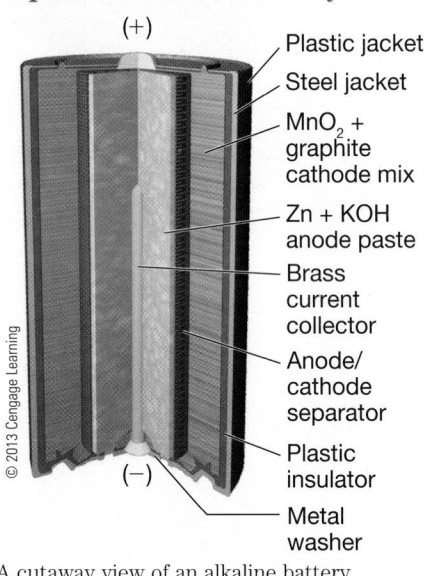

(+)

Plastic jacket
Steel jacket
MnO$_2$ + graphite cathode mix
Zn + KOH anode paste
Brass current collector
Anode/ cathode separator
Plastic insulator

(−)

Metal washer

© 2013 Cengage Learning

A cutaway view of an alkaline battery

Cathode:	$HgO(s) + H_2O(\ell) + 2e^- \rightarrow Hg(\ell) + 2\ OH^-(aq)$
Anode:	$Zn(s) + 2\ OH^-(aq) \rightarrow ZnO(s) + H_2O(\ell) + 2\ e^-$
Net reaction:	$HgO(s) + Zn(s) \rightarrow ZnO(s) + Hg(\ell) \qquad E_{cell} = 1.3\ V$

Of course, the presence of the heavy metal mercury means that these batteries should be recycled. Another variation is the silver battery, which uses Ag_2O as the cathode material and generates a cell potential of 1.6 V.

20.4b Secondary Batteries

Unlike primary batteries, secondary batteries can be recharged. The main design difference between primary and secondary batteries is that in a secondary battery the products of the redox reaction stay in contact with the electrodes, allowing the reaction to be reversed during the recharging process.

One of the most common secondary batteries is the lead-acid storage battery used in cars (Interactive Figure 20.4.3). This battery is composed of six lead-acid cells, each of which produces about 2 V, hooked in series to give a total cell potential of 12 V. In the charged state, each cell consists of a solid lead anode and a solid PbO_2 cathode held in place by a lead grid. The half-reactions and net reaction for this battery are shown below.

Cathode:	$PbO_2(s) + 2\ H_2SO_4(aq) + 2\ e^- \rightarrow PbSO_4(s) + 2\ H_2O(\ell)$
Anode:	$Pb(s) + SO_4^{2-}(aq) \rightarrow PbSO_4(s) + 2\ e^-$
Net reaction:	$PbO_2(s) + Pb(s) + H_2SO_4(aq) \rightarrow 2\ PbSO_4(s) + 2\ H_2O(\ell)$

When a lead-acid battery is discharged, the cathode and anode have been converted to solid $PbSO_4$. When the battery is recharged, an electric current, driven by a voltage applied against the voltage of the battery, is applied to the electrodes, forcing the reaction to move in the opposite direction (usually this external charge comes from the car's alternator). The electrodes are restored to their original states.

Recharging: $2\ PbSO_4(s) + 2\ H_2O(\ell) \rightarrow PbO_2(s) + Pb(s) + H_2SO_4(aq)$

The recharging reaction is an example of a *disproportionation reaction*, a redox reaction where one chemical species is both oxidized and reduced.

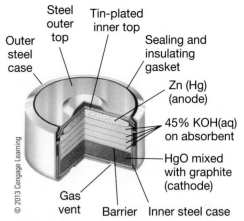

© 2013 Cengage Learning

Figure 20.4.2 A cutaway view of a mercury battery

Interactive Figure 20.4.3

Explore a lead-acid storage battery.

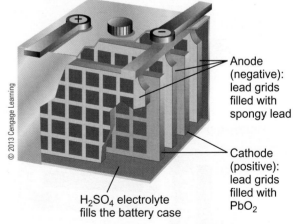

© 2013 Cengage Learning

A cutaway view of a lead-acid storage battery

A more portable secondary battery is the NiCd (or NiCad) battery (Figure 20.4.4). This rechargeable battery uses a nickel cathode and a cadmium anode.

Cathode: $NiO(OH)(s) + H_2O(\ell) + e^- \rightarrow Ni(OH)_2(s) + OH^-(aq)$

Anode: $Cd(s) + 2\ OH^-(aq) \rightarrow Cd(OH)_2(s) + 2\ e^-$

Net reaction: $2\ NiO(OH)(s) + Cd(s) + 2\ H_2O(\ell) \rightarrow 2\ Ni(OH)_2(s) + Cd(OH)_2(s)$

The product species, $Ni(OH)_2$ and $Cd(OH)_2$, remain in contact in the battery, thus allowing the battery to be recharged.

The nickel metal hydride (NiMH) battery is similar to the NiCad batteries in the use of a nickel cathode. The anode in these batteries is a metal hydride, where the metal is an intermetallic compound consisting of many different metals. The reactions in this type of rechargeable battery are shown below.

Cathode: $NiO(OH)(s) + H_2O(\ell) + e^- \rightarrow Ni(OH)_2(s) + OH^-(aq)$

Anode: $MH(s) + OH^-(aq) \rightarrow M(s) + H_2O(\ell) + e^-$

Net reaction: $MH(s) + NiO(OH)(s) \rightarrow M(s) + Ni(OH)_2(s)$

Most cell phones and laptop computers now use lithium-ion batteries. These batteries are especially useful because lithium is a very strong reducing agent and it has very low density, making the batteries lightweight. The lithium metal anode is encased in a matrix of graphite. The cathode is composed of the oxidizing agent CoO_2 and Li^+ ions, which become part of the product salt $LiCoO_2$.

Cathode: $CoO_2(s) + Li^+ + 1\ e^- \rightarrow LiCoO_2(s)$

Anode: $Li\text{-graphite} \rightarrow C(graphite) + Li^+ + e^-$

Net reaction: $Li(s) + CoO_2(s) \rightarrow LiCoO_2(s)$

Lithium-ion batteries should not be discharged completely because they will undergo an irreversible reaction forming Li_2O, which can ruin the battery. Most appliances that use lithium-ion batteries have electronic controllers that do not allow the battery to be either overcharged or overdischarged.

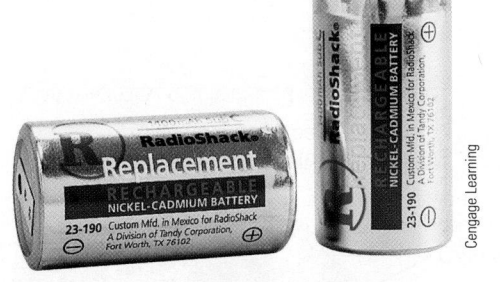

Figure 20.4.4 Rechargeable NiCad batteries

Example Problem 20.4.1 Calculate quantity of charge and maximum work from a battery.

The discharge reaction for a lithium–iron disulfide battery can be represented as

$$FeS_2(s) + 4\,Li(s) \rightarrow Fe(s) + 2\,Li_2S(s)$$

a. What quantity of charge (in coulombs) can be provided by a fully charged 1.50-V lithium–iron disulfide digital camera battery, if the mass of Li in the battery is 0.453 g?

b. What is the maximum amount of work (in joules) that can be obtained from this battery?

Solution:

You are asked to calculate the quantity of charge and the maximum amount of work that can be obtained from a battery.

You are given the balanced equation for the discharge reaction occurring in the battery, the battery voltage, and the mass of a reactant in the battery.

a. The quantity of charge is equal to the number of moles of electrons transferred when 0.453 g of Li reacts, multiplied by Faraday's constant, the charge on one mole of electrons. From the lithium oxidation states, it can be seen that one mole of electrons is transferred per mole of Li.

$$0.453 \text{ g Li} \times \frac{1 \text{ mol Li}}{6.941 \text{ g}} \times \frac{1 \text{ mol e}^-}{1 \text{ mol Li}} \times \frac{96{,}485 \text{ C}}{1 \text{ mol e}^-} = 6.30 \times 10^3 \text{ C}$$

b. Because voltage is the work per Coulomb of charge, multiply charge and cell voltage to determine the maximum amount of work obtainable when 6.30×10^3 C of charge passes through a potential difference of 1.50 V.

$$\text{work} = C \times V = (6.3 \times 10^3 \text{ C})(1.50 \text{ J/C}) = 9.45 \times 10^3 \text{ J}$$

The battery can perform at most 9.45×10^3 J of work.

Video Solution

Tutored Practice
Problem 20.4.1

20.4c Fuel Cells

A **fuel cell** is a voltaic cell where the reactants, a fuel and an oxidizer, are continuously supplied to the cell and the products are continuously removed from the cell. Unlike a battery, a fuel cell never "runs down." It will produce electricity as long as the fuel and the oxidizer are pumped through it.

Fuel cells were used to generate electricity in space by the U.S. space program as early as 1962. The space shuttles (program retired in July 2011) contained three fuel cell power plants, each of which contained 32 fuel cells that transformed hydrogen and oxygen into electrical power, water, and heat. The fuel cell power plants used in the space shuttle were 14 inches high, 15 inches wide, and 40 inches long and weighed 255 pounds. Each fuel cell on the space shuttle was capable of supplying 7 kW continuously and 12 kW at peak power output.

In an individual fuel cell (Interactive Figure 20.4.5), oxygen gas is fed into the cathode compartment, where it reacts with water to produce hydroxide ions.

$$\text{Cathode:} \quad O_2(g) + 2\,H_2O(\ell) + 4\,e^- \rightarrow 4\,OH^-(aq)$$

The hydroxide ions migrate through the electrolyte solution to the anode, where they combine with hydrogen to produce water vapor. (The reaction generates heat, which converts liquid water into water vapor.)

$$\text{Anode:} \quad H_2(g) + 2\,OH^-(aq) \rightarrow 2\,H_2O(g) + 2\,e^-$$

The anode and cathode are primarily composed of porous graphite, but both also contain small amounts of metal catalyst. Excess water vapor is removed by an internal circulating system that continuously pumps hydrogen gas into the anode compartment and the water vapor and excess hydrogen gas out of the cell. The gases leaving the fuel cell pass through a condenser, where water vapor is condensed and stored for later use.

The net reaction in the fuel cell is the combination of hydrogen and oxygen to make water vapor.

$$2\,H_2(g) + O_2(g) \rightarrow 2\,H_2O(g)$$

NASA research centers are researching new fuel cells and refining the fuel cell technology currently in use in the space program. Fuel cells are also being developed for use in generating power at stationary locations such as hospitals and for use as standby or backup power.

20.4d Corrosion

Not all electrochemical cells and redox reactions are useful. Some redox reactions are destructive and can have a significantly negative economic impact. One example is **corrosion**, the naturally occurring redox process by which metals are converted to oxides and sulfides. The corrosion of iron to form rust accounts for millions of dollars of damage every year to bridges, buildings, automobiles, and other iron-containing structures.

Iron corrodes in the presence of oxygen gas and water to form rust, a hydrated form of iron(III) oxide. There are many factors, such as pH, stress on the metal, and the presence of salts, that can affect the speed at which the metal corrodes. The process occurs at the surface of the metal, so iron serves as both the anode and the cathode. At the anode, iron is oxidized.

$$\text{Anode:} \quad Fe(s) \rightarrow Fe^{2+}(aq) + 2\,e^-$$

Explore a fuel cell.

A hydrogen fuel cell

© 2013 Cengage Learning

Explore the corrosion process.

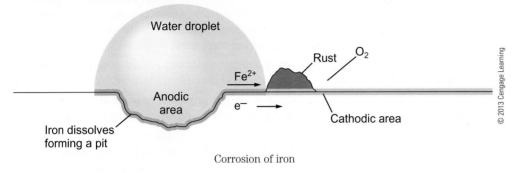

Corrosion of iron

The electrons travel through the iron to the cathodic area, a location where both oxygen and water are present.

$$\text{Cathode:} \quad O_2(g) + 4\,H^+(aq) + 4\,e^- \rightarrow 2\,H_2O(\ell)$$
$$\text{or} \quad O_2(g) + 2\,H_2O(\ell) + 4\,e^- \rightarrow 4\,OH^-(aq)$$

The iron(II) ions migrate toward the cathodic area where they are further oxidized by oxygen to Fe^{3+} and ultimately form rust, an iron(III) oxide compound with varying waters of hydration (Interactive Figure 20.4.6).

$$\text{Net reaction:} \quad 2\,Fe(s) + \tfrac{3}{2}\,O_2(g) + n\,H_2O(\ell) \rightarrow Fe_2O_3 \cdot nH_2O(s)$$

Iron surfaces can be protected by paint or by coating the surface with another metal. However, if the paint or metal surface cracks or flakes off, the iron can corrode. A better way to protect iron from rusting is to put it in contact with a metal that is easier to oxidize, a process called galvanizing.

In galvanized iron, the metal is coated with a layer of zinc. Zinc is easier to oxidize than iron; it is a stronger reducing agent.

$$Zn^{2+}(aq) + 2\,e^- \rightarrow Zn(s) \qquad E^\circ = -0.76\text{V}$$
$$Fe^{2+}(aq) + 2\,e^- \rightarrow Fe(s) \qquad E^\circ = -0.44\text{V}$$

When water and oxygen come in contact with galvanized iron, the iron no longer acts as the anode. Zinc is oxidized and acts as a **sacrificial anode** (Figure 20.4.7). The iron is protected and does not rust.

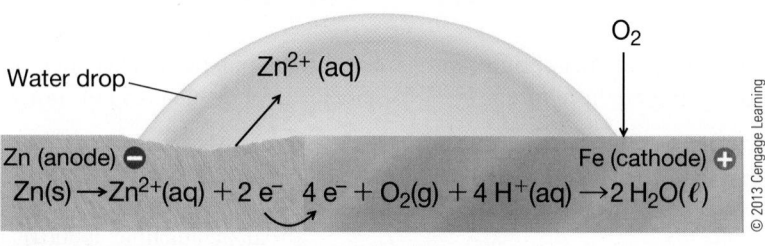

Water drop

Zn²⁺ (aq)

O₂

Zn (anode) ⊖   Fe (cathode) ⊕

$Zn(s) \rightarrow Zn^{2+}(aq) + 2\ e^-$ $4\ e^- + O_2(g) + 4\ H^+(aq) \rightarrow 2\ H_2O(\ell)$

© 2013 Cengage Learning

Figure 20.4.7 Zinc acts as a sacrificial anode.

Unit Recap

Key Concepts

20.1 Oxidation–Reduction Reactions and Electrochemical Cells

- In a redox reaction, the species that loses electrons has been oxidized and is the reducing agent and the species that gains electrons has been reduced and is the oxidizing agent (20.1a).

- Oxidation numbers are used to identify the species oxidized and reduced in a redox reaction (20.1a).

- The half-reaction method can be used to balance redox reactions in neutral, acidic, or basic solution (20.1b, 20.1c).

- An electrochemical cell consists of two half-cells connected by a mechanism to allow the cell to maintain charge balance (20.1d).

- In a voltaic cell, electrons flow spontaneously through the wire and the electrical current can be used to do work (20.1d).

- In an electrolytic cell, an external power source forces the electrons to flow between the oxidation and reduction half-reactions in the nonspontaneous direction (20.1d).

- In a voltaic cell, oxidation takes place at the anode, reduction takes place at the cathode, the anode is assigned a negative charge and the cathode is assigned a positive charge, electrons flow through a wire from the anode to the cathode, anions in the salt bridge flow from the cathode to the anode, and cations flow from the anode to the cathode (20.1d).

- Electrochemical cell notation shows the components of an electrochemical cell (20.1e).

20.2 Cell Potentials, Free Energy, and Equilibria

- Cell potential is the difference in potential energy between the cathode and the anode. Under standard conditions, this is the standard cell potential, E°_{cell} (20.2a).

- Standard reduction potentials are used to calculate standard cell potentials (20.2a).

- A positive E°_{red} indicates that the half-reaction reactants are more easily reduced than H^+ and that the products are more difficult to oxidize than H_2. A negative E°_{red} indicates that the half-reaction reactants are more difficult to reduce than H^+ and the products are more easily oxidized than H_2 (20.2a).

- A reaction with a negative standard free energy change has a positive E°_{cell} value and is spontaneous in the direction it is written (20.2b).

- A reaction with a positive E°_{cell} value has an equilibrium constant greater than one and is spontaneous in the direction it is written (20.2c).

- The Nernst equation relates cell potential, standard cell potential, temperature, the number of electrons transferred in the redox reaction, and the reaction quotient for the reaction (20.2d).

- In a concentration cell, the reactions in both half-cells are identical but the concentrations of species in solution differ. The reaction is driven by the magnitude of the reaction quotient for the reaction (20.2e).

20.3 Electrolysis

- In an electrolytic cell, oxidation takes place at the anode, reduction takes place at the cathode, the anode is assigned a positive charge and the cathode is assigned a negative charge, electrons flow through a wire from the anode to the cathode, anions in the salt bridge flow from the cathode to the anode, and cations flow from the anode to the cathode (20.3a).

- The amount of matter transformed during an electrolysis reaction is related to the number of electrons transferred in the redox reaction, the current, and the time span of the reaction (20.3a).

- In an electrolytic cell consisting of a molten salt, the cationic species is reduced and the anionic species is oxidized (20.3b).

- In an electrolytic cell consisting of an aqueous salt, the species reduced and oxidized depend on the relative E°_{red} and E°_{ox} values for the species in the solution (20.3c).

20.4 Applications of Electrochemistry: Batteries and Corrosion

- Batteries can be classified as primary or secondary (20.4a).

- Alkaline and mercury batteries are examples of primary batteries (20.4a).

- Secondary batteries can be recharged in part because the products of the redox reaction stay in contact with the electrodes during the redox reaction (20.4b).

- In a fuel cell, the reactants are continuously supplied to the cell and the products are continuously removed from the cell (20.4c).

- Corrosion is a costly, destructive process where metals are converted to oxides and sulfides (20.4d).

- Zinc is used to galvanize iron because it is more easily oxidized than iron and can act as a sacrificial anode (20.4d).

Key Equations

$$E^\circ_{\text{cell}} = E^\circ_{\text{cathode}} - E^\circ_{\text{anode}} = E^\circ_{\text{red}} + E^\circ_{\text{ox}} \quad \textbf{(20.1)}$$

$$w_{\text{max}} = -q \times E_{\text{cell}} \quad \textbf{(20.2)}$$

$$w_{\text{max}} = -nFE_{\text{cell}} \quad \textbf{(20.3)}$$

$$\Delta G = -nFE_{\text{cell}} \quad \textbf{(20.4)}$$

$$\Delta G^\circ = -nFE^\circ_{\text{cell}} \quad \textbf{(20.5)}$$

$$E^\circ_{\text{cell}} = \frac{RT}{nF} \ln K \quad \textbf{(20.6)}$$

$$E^\circ_{\text{cell}} = \frac{0.0257\ \text{V}}{n} \ln K \text{ at } 298\ \text{K} \quad \textbf{(20.7)}$$

$$E_{\text{cell}} = E^\circ_{\text{cell}} - \frac{RT}{nF} \ln Q \quad \textbf{(20.8)}$$

$$E_{\text{cell}} = E^\circ_{\text{cell}} - \frac{0.0257\ \text{V}}{n} \ln Q \text{ at } 298\ \text{K} \quad \textbf{(20.9)}$$

Key Terms

20.1 Oxidation–Reduction Reactions and Electrochemical Cells
electrochemistry
half-reaction method
half-cell
salt bridge
electrochemical cell
voltaic (galvanic) cell
electrode
anode
cathode
standard hydrogen electrode (SHE)

20.2 Cell Potentials, Free Energy, and Equilibria
ampere (A)
coulomb (C)
electromotive force (emf)
volt (V)
cell potential, E_{cell}
standard cell potential, E°_{cell}
standard reduction potential, E°_{red}
Faraday's constant
Nernst equation
concentration cell

20.3 Electrolysis
electrolysis
electrolytic cell
coulometry

20.4 Applications of Electrochemistry: Batteries and Corrosion
battery
primary battery
secondary battery
fuel cell
corrosion
sacrificial anode

Unit 20   Review and Challenge Problems

21 Organic Chemistry

Unit Outline

In This Unit...

Organic chemistry is the study of the compounds of carbon, specifically hydrocarbons and their derivatives. These compounds make up the bulk of living matter, plastics, and many of the most important materials we use in everyday life. Organic chemistry is one of the subdisciplines in chemistry, and it is a vast subject area that warrants entire courses of its own. In this entry-level unit on the subject we introduce you to the field of organic chemistry by exploring the variety in organic structures (we examine hydrocarbons and functional groups), revealing the order in organic reactivity (we examine the reactivity of alkenes, alcohols, and carbonyl compounds), and surveying the world of polymer chemistry (including plastics and biopolymers such as deoxyribonucleic acid [DNA]).

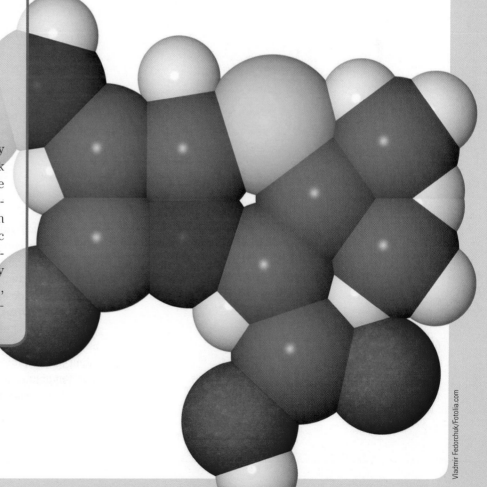

21.1 Hydrocarbons

21.1a Classes of Hydrocarbons

Organic chemistry is the study of the compounds of carbon. Hydrocarbons, which were introduced in Elements and Compounds (Unit 2), are binary compounds containing only carbon and hydrogen. While the simple chemical formulas of these compounds might imply a limited variety to their properties, there is great structural variation possible in bonding carbon atoms together and thus a virtually unlimited number of compounds are possible.

Hydrocarbons can be classified as either saturated or unsaturated. **Saturated hydrocarbons** include alkanes and cycloalkanes and are compounds containing only carbon–carbon single bonds. In saturated hydrocarbons, each carbon atom has the maximum number of hydrogen atoms bonded to it. **Unsaturated hydrocarbons** include alkenes, alkynes, and aromatic compounds and consist of compounds containing at least one carbon–carbon double or triple bond. As shown in Table 21.1.1, the carbon atoms in these compounds always have four bonds.

Table 21.1.1 Types of Hydrocarbons

	Saturated		Unsaturated		
	Alkanes	**Cycloalkanes**	**Alkenes**	**Alkynes**	**Aromatics**
Example	Ethane	Cyclopropane	Ethene	Acetylene	Benzene
Number of σ bonds to each C	4	4	3	2	3
Number of π bonds to each unsaturated C	—	—	1	2	1

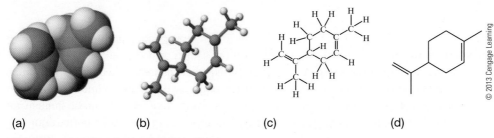

(a)　　　　　(b)　　　　　(c)　　　　　(d)

© 2013 Cengage Learning

Figure 21.1.1 Four representations of an organic molecule

Carbon is unique in its ability to form long chains bonded to other carbon atoms, which results in a wide variety of chemical structures and unique organic compounds. With the exception of CH_4, all hydrocarbons consist of a network of bonded C atoms. The structure of a hydrocarbon is often considered a *carbon skeleton* with H atoms added to fulfill the octet on carbon. Because of this, hydrocarbons are often drawn using simple line structures that show only the carbon network.

Consider limonene, a hydrocarbon found in citrus peels that is, in part, responsible for the smell of an orange. The structure of limonene is shown in Figure 21.1.1 in four ways: as (a) a space-filling molecular model, (b) a ball-and-stick molecular model, (c) a complete Lewis structure, and (d) an organic line structure.

All of the representations show that the compound has the formula $C_{11}H_{15}$, but only the representations in (a), (b), and (c) show all of the atoms in the molecule. In an organic line structure, each line represents a carbon–carbon bond, a bend in a line represents a carbon atom, and the end of a line represents a carbon atom.

Example Problem 21.1.1 Interpret organic line structures.

Write the molecular formula for the organic compound shown here:

Solution:

You are asked to write the molecular formula for an organic compound.

You are given the line structure for the compound.

Example Problem 21.1.1 *(continued)*

In an organic line structure, each line represents a carbon–carbon bond and the end of each line represents a carbon atom. This compound contains five carbon atoms. The number of hydrogen atoms attached to each carbon is determined using the fact that each carbon atom forms four bonds.

The molecular formula for the compound is C_5H_{10}.

Video Solution

Tutored Practice
Problem 21.1.1

21.1b Alkanes and Cycloalkanes

The simplest type of hydrocarbon is an **alkane**, a hydrocarbon containing only carbon–carbon single bonds. All alkanes have the general formula C_nH_{2n+2}; for example, propane, the three-carbon hydrocarbon, has the formula C_3H_8. In **cycloalkanes**, carbon atoms are joined to form a ring. Cycloalkanes have the general formula C_nH_{2n}.

The physical and chemical properties of alkanes and cycloalkanes are related to the fact that they are nonpolar, saturated hydrocarbons. For example, alkane boiling points are generally lower than the boiling points for other organic compounds with similar molecular weights. Alkanes with high molecular weights are white, waxy solids such as those found in paraffin wax and petroleum jelly and those used in candles. Because they are saturated, alkanes and cycloalkanes are generally unreactive toward most reagents, although they react vigorously with oxygen in combustion reactions to form carbon dioxide and water.

Naming Alkanes and Cycloalkanes

The International Union of Pure and Applied Chemistry (IUPAC) established specific rules for naming organic compounds. These rules allow chemists to name both simple and complex organic compounds. Here we will describe how to name only relatively simple organic compounds. The name of an alkane is based on the number of carbon atoms in the longest chain in the molecule. Specific names are used to indicate the number of carbon atoms in the chain, as shown in Table 21.1.2.

Table 21.1.2 Alkane Names for Chain Lengths up to 10 Carbons

Name	Number of C Atoms	Molecular Formula C_nH_{2n+2}	Condensed Structural Formula
Methane	1	CH_4	CH_4
Ethane	2	C_2H_6	CH_3CH_3
Propane	3	C_3H_8	$CH_3CH_2CH_3$
Butane	4	C_4H_{10}	$CH_3CH_2CH_2CH_3$
Pentane	5	C_5H_{12}	$CH_3CH_2CH_2CH_2CH_3$
Hexane	6	C_6H_{14}	$CH_3CH_2CH_2CH_2CH_2CH_3$
Heptane	7	C_7H_{16}	$CH_3CH_2CH_2CH_2CH_2CH_2CH_3$
Octane	8	C_8H_{18}	$CH_3CH_2CH_2CH_2CH_2CH_2CH_2CH_3$
Nonane	9	C_9H_{20}	$CH_3CH_2CH_2CH_2CH_2CH_2CH_2CH_2CH_3$
Decane	10	$C_{10}H_{22}$	$CH_3CH_2CH_2CH_2CH_2CH_2CH_2CH_2CH_2CH_3$

A fragment of a carbon chain that is missing a single hydrogen is named by changing the end of the carbon chain name to -*yl*. For example, the three-carbon fragment $CH_3CH_2CH_2$— is the propyl group.

The alkanes in Table 21.1.2 are examples of **unbranched hydrocarbons**, organic compounds where each carbon is bonded to no more than two other carbon atoms. The name of an unbranched hydrocarbon is based on the number of carbon atoms in the molecule, as shown in the table. In **branched hydrocarbons**, at least one carbon is connected to more than two carbon atoms, and the naming is more complex. The name of a branched hydrocarbon consists of a parent name, which indicates the longest carbon chain in the molecule, and substituent names, which indicate the fragments that are attached to the longest chain.

Rules for naming branched alkanes and cycloalkanes can be very complex. Interactive Table 21.1.3 shows a simplified set of rules for naming these compounds.

For example, consider the compounds shown in Figure 21.1.2. The first compound is named pentane because it is a five-carbon unbranched hydrocarbon. In the second structure, the longest continuous carbon chain contains four carbons, so it has the parent name butane. There is a methyl group (—CH_3) on the second carbon in the chain, so the full

Interactive Table 21.1.3

Rules for Naming Branched Alkanes and Cycloalkanes

1. Determine the parent name for the compound by identifying the number of carbon atoms in the longest continuous carbon chain in the molecule. For a cycloalkane, the number of atoms in the ring gives the parent name.

2. Name any alkane substituents attached to the chain by dropping the -*ane* from the alkane name and adding -*yl*.

3. Number the longest continuous carbon chain to place substituents on carbon atoms with the lowest possible numbers. For cycloalkanes, number the ring to locate the substituents at the lowest possible numbers, with substituents lowest in alphabetical order at lowest numbered carbons.

4. Name the substituents first, in alphabetical order. Use numbers to indicate the location of the branching and prefixes (*di-*, *tri-*, *tetra-*, etc.) to indicate multiple identical substituents.

5. Follow the substituent names with the parent name of the longest carbon chain.

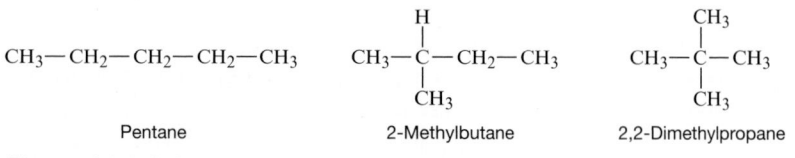

Pentane   2-Methylbutane   2,2-Dimethylpropane

Figure 21.1.2 Structures and names of three alkanes

Cyclopropane   Cyclobutane   1-Ethyl-2-methylcyclopentane

Figure 21.1.3 Structures and names of three cycloalkanes

name of the compound is 2-methylbutane. The third compound has three carbons (propane) in its longest chain and two methyl substituents on the second carbon in the three-carbon chain. Therefore, the compound name is 2,2-dimethylpropane.

Cycloalkanes are also named using the largest number of continuously connected carbon atoms. As shown in Figure 21.1.3, an unbranched cycloalkane is named using the prefix *cyclo-*, and a branched cycloalkane is named using the rules in Table 21.1.3.

Example Problem 21.1.2 Name alkanes and cycloalkanes.

Name the compounds shown:

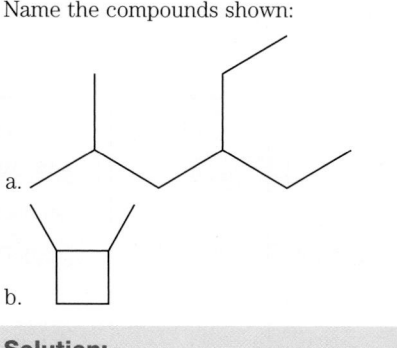

a.

b.

Solution:

You are asked to name an alkane or cycloalkane.

You are given the structure of the compound.

a. The longest carbon chain has six carbon atoms, so the parent name is hexane. There are two alkane substituents, one with one carbon (methyl) and one with two carbons (ethyl). The six-carbon chain is numbered to place the substituents on the lowest-numbered carbons.

Example Problem 21.1.2 *(continued)*

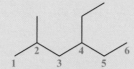

The compound name is 4-ethyl-2-methylhexane.

b. The parent name is the cycloalkane name, cyclobutane. The carbon atoms in the ring are numbered to give the substituents, the two methyl groups, the lowest possible numbers. The prefix *di-* is used to indicate the fact that there are two identical substituents.

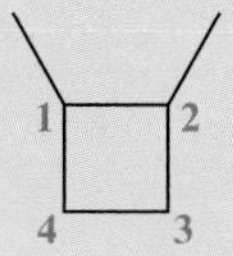

The compound name is 1,2-dimethylcyclobutane.

Video Solution

Tutored Practice Problem 21.1.2

21.1c Unsaturated Hydrocarbons

Unsaturated hydrocarbons contain at least one carbon–carbon double or triple bond and include alkenes, alkynes, and arenes, compounds derived from benzene. Alkenes and alkynes are generally more reactive than alkanes due to their unsaturation, but benzene, an unsaturated cyclic hydrocarbon, is actually quite unreactive.

Alkenes

Alkenes are hydrocarbons that contain at least one carbon–carbon double bond. Alkenes with one carbon–carbon double bond and no rings have the general formula C_nH_{2n}. The carbon atoms involved in the double bond are linked by one sigma bond and one pi bond, and each is sp^2-hybridized with trigonal planar geometry. Recall that molecules with pi bonds can exist as more than one isomer because of restricted rotation around the pi bond. Note that the two trigonal planes are coplanar due to the overlap of p orbitals (Figure 21.1.4).

Alkene names contain a parent alkane name with the *-ane* ending changed to *-ene*. In large alkanes a numbering system is used to indicate the position of the double bond. Branched alkenes are named using the same rules used for branched alkanes. For example, consider the alkenes in Figure 21.1.5.

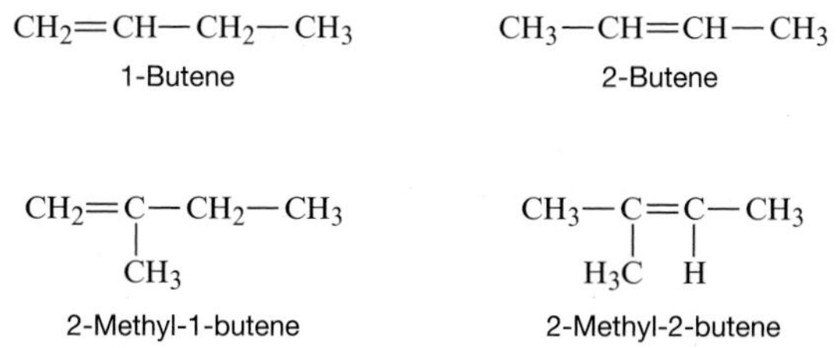

Figure 21.1.4 Sigma and pi bonding in ethene, an alkene

$CH_2{=}CH{-}CH_2{-}CH_3$ $CH_3{-}CH{=}CH{-}CH_3$
  1-Butene   2-Butene

$CH_2{=}C{-}CH_2{-}CH_3$ $CH_3{-}C{=}C{-}CH_3$
        | | |
       CH_3 H_3C H

  2-Methyl-1-butene   2-Methyl-2-butene

Figure 21.1.5 Structures and names of four alkenes

In all four compounds, the parent name is butene because the longest carbon chain contains four carbon atoms. The unbranched alkenes 1-butene and 2-butene differ in the position of the double bond, and the number indicates the position of the first carbon in the double bond. In general, we number the carbon atoms in an alkene to give the first carbon atom in the double bond the lowest number possible. In the branched alkenes, the position of the branched group is based on numbering that gives the carbon atoms in the double bond the lowest numbers possible. Cycloalkenes are named in a manner similar to the alkenes, but the parent name is based on the number of carbon atoms in the ring that contains the double bond and includes the *cyclo-* prefix.

Example Problem 21.1.3 Name alkenes and cycloalkenes.

Name the compounds shown:

a.

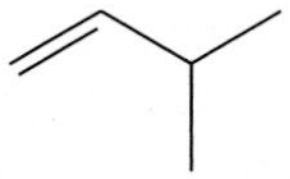

b.

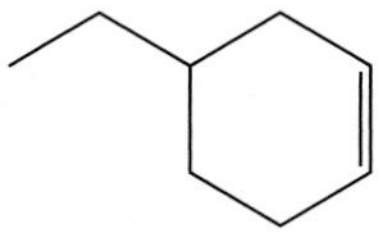

Solution:

You are asked to name an alkene or cycloalkene.

You are given the structure of the compound.

a. The longest carbon chain in the alkene has four carbon atoms, so the parent name is butene. The carbon chain is numbered to give the first carbon atom in the double bond the lowest possible number; the methyl group is therefore attached to carbon 3.

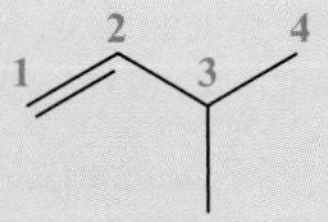

The compound name is 3-methyl-1-butene.

b. The parent name is the cycloalkene name, cyclohexene. The carbon atoms in the ring are numbered to place the first carbon atom in the double bond at carbon 1 and to give the substituent the lowest possible number; the ethyl group is therefore attached to carbon 4. Because a cycloalkene is always numbered to position the double bond at carbon 1, it is not necessary to indicate the position of the double bond in the compound name.

Example Problem 21.1.3 *(continued)*

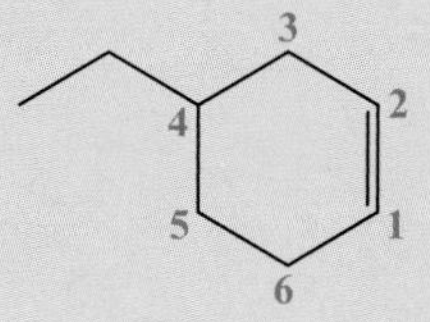

The compound name is 4-ethylcyclohexene.

Video Solution

Tutored Practice
Problem 21.1.3

Alkynes

Alkynes are hydrocarbons that contain at least one carbon–carbon triple bond. Alkynes with one carbon–carbon triple bond and no rings have the general formula C_nH_{2n-2}. The carbon atoms involved in the triple bond are linked by one sigma bond and two pi bonds, and each is *sp*-hybridized with linear geometry. Alkynes are very reactive and are generally not found in nature.

Like alkenes, alkynes are named using the parent alkane name with the *-ane* ending changed to *-yne*. The simplest alkyne, C_2H_2, is commonly named acetylene, but its IUPAC name is ethyne.

Arenes

Arenes, also called **aromatic hydrocarbons**, are compounds that contain a benzene ring, a six-member carbon ring of alternating single and double carbon–carbon bonds. This class of compounds was named *aromatic* because many of them have strong and distinctive odors. Now this term refers to a class of compounds that are unsaturated but are unexpectedly stable. Benzene, C_6H_6, the simplest aromatic compound, has two resonance structures (Figure 21.1.6). The unusual stability of this unsaturated cyclic hydrocarbon is due to the delocalized pi system in the ring, as discussed in Theories of Chemical Bonding (Unit 9). This delocalized bonding is often represented in line structures as a circle inside the six-membered carbon ring.

Substituted benzene rings are named using a numbering system or a prefix to indicate the substitution position on the ring. However, common names are still used for some benzene derivatives (Figure 21.1.7). When a benzene ring contains two substituents, the compound is named using a numbering system (where the numbers are assigned so that the substituent with the lower alphabetical ranking is carbon 1) or the term *ortho-* (*o-*),

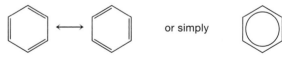

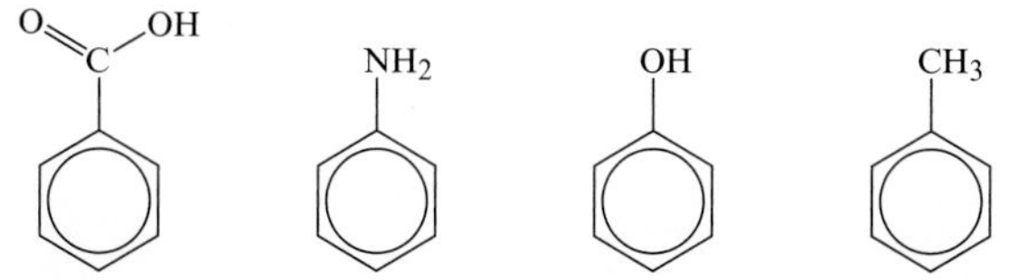

Figure 21.1.6 Representations of benzene

Figure 21.1.7 Common names for some mono-substituted benzene derivatives

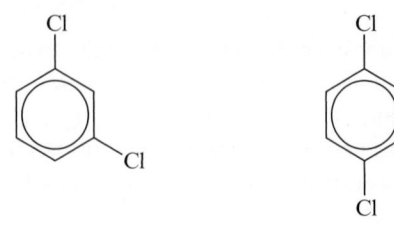

o-Dichlorobenzene   m-Dichlorobenzene   p-Dichlorobenzene

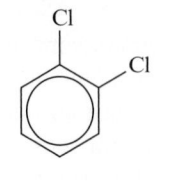

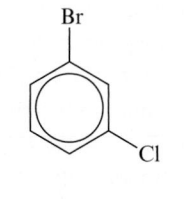

m-Chlorobromobenzene   2,4,6-Trinitrotoluene (TNT)

Figure 21.1.8 Structures and names of some di-substituted benzene derivatives

meta- (*m-*), or *para-* (*p-*) to indicate the position of the second substituent relative to the first. When the benzene ring contains three or more substituents, numbers indicate the positions of the substituents (Figure 21.1.8).

Example Problem 21.1.4 Name substituted arenes.

Name the following substituted arene:

Solution:

You are asked to name a substituted arene.

You are given the structure of the compound.

This is a derivative of toluene, one of the benzene derivatives given a special name. The methyl group is assigned to carbon 1, so the chloro substituent is on carbon 3.

Example Problem 21.1.4. *(continued)*

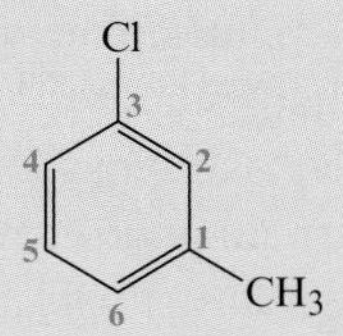

The compound is named 3-chlorotoluene, or *o*-chlorotoluene.

Video Solution

Tutored Practice
Problem 21.1.4

21.1d Hydrocarbon Reactivity

In general, alkenes and alkynes are more reactive than alkanes and aromatic compounds. Here we describe just a few of the more common reactions for these compounds.

Alkanes

Alkanes are saturated hydrocarbons, so they are fairly unreactive toward most reagents, with the exception of oxygen. These compounds are used mainly as fuels because they react vigorously with oxygen. For example, propane burns in oxygen in an exothermic reaction to produce carbon dioxide and water.

$$C_3H_8(g) + 5\,O_2(g) \rightarrow 3\,CO_2(g) + 4\,H_2O(g) \qquad \Delta H° = 2200 \text{ kJ/mol}$$

Alkanes also undergo **substitution reactions**, reactions where one group is substituted for another, with halogens when exposed to light of sufficient energy to break the X—X halogen bond (Figure 21.1.9).

Alkenes

Alkenes commonly undergo **addition reactions**, reactions where the pi bond is replaced by a new sigma bonded atom on each carbon. In the simplified example in Figure 21.1.10, the molecule AB is "added across" the alkene double bond.

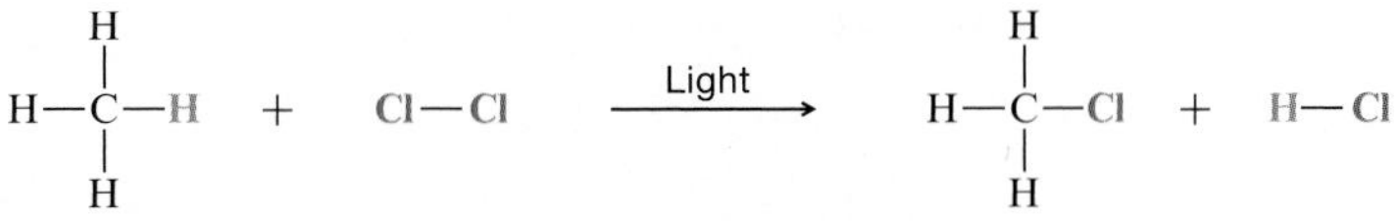

Figure 21.1.9 An alkane substitution reaction

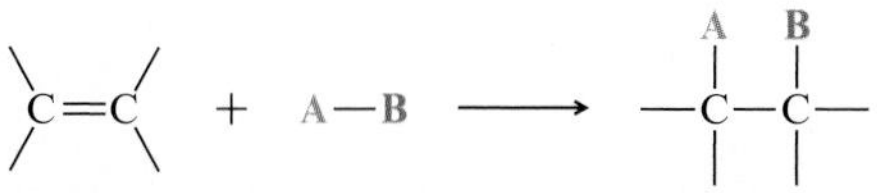

Figure 21.1.10 An alkene addition reaction

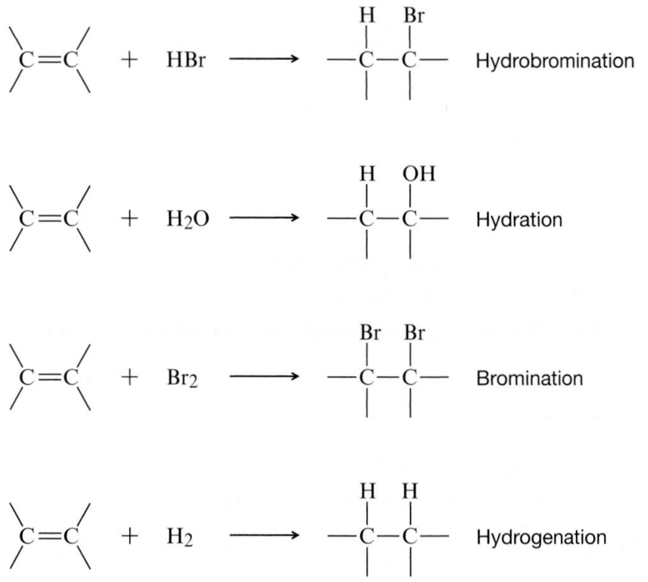

Figure 21.1.11 Some important alkene addition reactions

The driving force behind addition reactions is the fact that one sigma bond (between A and B) and one pi bond (between the two C atoms) are broken, but two sigma bonds are formed. Sigma bonds usually being stronger than pi bonds, the driving force of this reaction is the formation of stronger bonds. It is enthalpy favored, but entropy disfavored.

Many different small molecules can be added across an alkene double bond, as shown in Figure 21.1.11. **Hydrogenation** is the addition of an H_2 molecule across an alkene double bond, forming an alkane. This is a special case of addition reaction that occurs on a solid metal catalytic surface (Interactive Figure 21.1.12). In this reaction, H_2 molecules adsorb onto the surface, breaking the H—H bond and forming relatively weak metal—H bonds. The alkene does likewise, breaking the C—C pi bond and forming relatively weak metal—C bonds. The H atoms then migrate along the metal surface and bond to the C atoms, forming the alkane product, which is no longer bound to the metal

Interactive Figure 21.1.12

Explore a metal-catalyzed hydrogenation reaction.

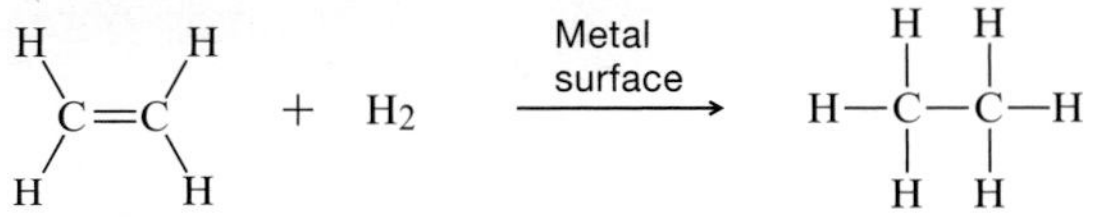

Hydrogenation of ethene

Investigate Markovnikov's rule.

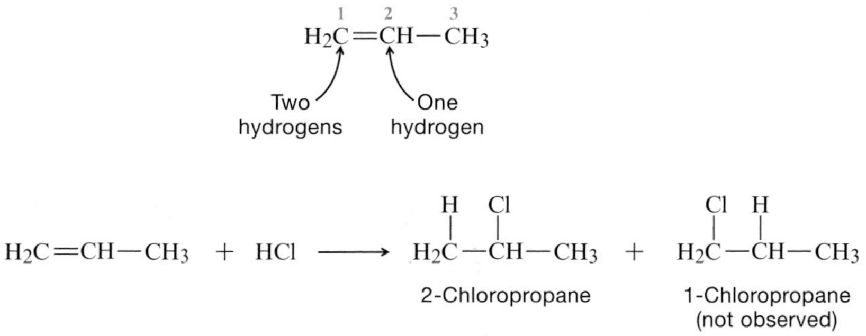

$$H_2\overset{1}{C}=\overset{2}{C}H-\overset{3}{C}H_3$$

Two hydrogens — One hydrogen

$$H_2C=CH-CH_3 \;+\; HCl \;\longrightarrow\; \underset{\text{2-Chloropropane}}{H_2\overset{\overset{\displaystyle H}{|}}{C}-\overset{\overset{\displaystyle Cl}{|}}{C}H-CH_3} \;+\; \underset{\substack{\text{1-Chloropropane}\\\text{(not observed)}}}{H_2\overset{\overset{\displaystyle Cl}{|}}{C}-\overset{\overset{\displaystyle H}{|}}{C}H-CH_3}$$

Reaction of HCl with propene

surface. The catalytic activity of the metal surface lies in its ability to facilitate bond-breaking steps by allowing weak, compensating bonds to form to the metal surface. The less energetic bond-breaking steps result in a lower activation energy and a faster reaction.

When the reagent added to the double bond has the general formula HX, the addition reaction is regioselective. **Regioselectivity** is reaction selectivity where a particular atom adds to one region of a molecule but not a different one. The regioselectivity in addition reactions of alkenes is summarized by **Markovnikov's rule**, which states that when an HX molecule adds to an alkene, the H atom adds to the carbon that has the greater number of H atoms.

For example, consider the reaction of HCl and propene (Interactive Figure 21.1.13). The reaction has two possible products: the Cl could add to the terminal carbon (carbon 1) or the central carbon (carbon 2). However, the only product of the reaction is 2-chloropropane. Markovnikov's rule predicts the product of this reaction because the hydrogen in HCl adds to the terminal carbon, the doubly bonded carbon in the reactant that had more hydrogen atoms.

Regioselectivity occurs for different reasons in different organic systems, and in the case of alkene addition reactions, it has to do with the relative stability of the possible intermediates in the reaction.

21.2 Isomerism

21.2a Constitutional Isomerism

Constitutional isomers are compounds with the same chemical formulas but different structural formulas. Alkanes with a small number of carbon atoms have a small number of possible constitutional isomers. Methane (CH_4), ethane (C_2H_6), and propane (C_3H_8) have no constitutional isomers; there are two constitutional isomers with four carbons (butane and 2-methylpropane); and there are three constitutional isomers with five carbons (pentane, 2-methylbutane, and 2,2-dimethylpropane; Figure 21.2.1).

As the number of carbon atoms increases, the number of possible constitutional isomers increases exponentially, as shown in Table 21.2.1.

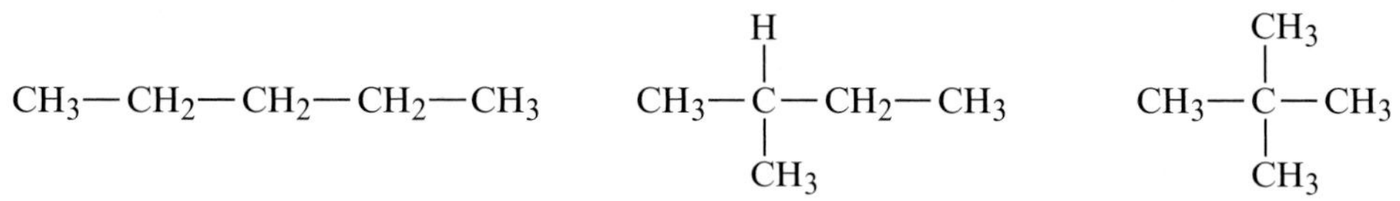

Table 21.2.1 The Number of Constitutional Isomers Possible for Some Alkanes

Formula	Number of Isomers
C_6H_{14}	5
C_7H_{16}	9
C_8H_{18}	18
C_9H_{20}	35
$C_{10}H_{22}$	75
$C_{15}H_{32}$	4,347
$C_{20}H_{42}$	366,319
$C_{30}H_{62}$	4,111,846,763

$$CH_3-CH_2-CH_2-CH_2-CH_3$$

Pentane

2-Methylbutane

2,2-Dimethylpropane

Figure 21.2.1 The three constitutional isomers of C_5H_{12}

Example Problem 21.2.1 Identify constitutional isomers.

Which of the following represent constitutional isomers of the molecule shown?

a.

b.

c.

Solution:

You are asked to identify the constitutional isomers of a compound.

You are given the structure of the compound and some possible constitutional isomers.

The compound in question is 2,3,4-trimethylhexane (C_9H_{20}). A constitutional isomer of this compound has the same chemical formula but a different structural formula.

a. This compound has the same chemical formula (C_9H_{20}) but a different structural formula (3-ethyl-2,4-dimethylpentane). This is a constitutional isomer of 2,3,4-trimethylhexane.
b. This compound is identical to 2,3,4-trimethylhexane, so it is not a constitutional isomer.
c. This compound has the chemical formula C_8H_{18}, which is not the same as the chemical formula for 2,3,4-trimethylhexane. It is not a constitutional isomer, it is a different compound.

21.2b Stereoisomerism

There are two types of stereoisomers: geometric and optical. As we saw in Theories of Chemical Bonding (Unit 9), compounds containing a carbon–carbon double bond can have geometric (*cis* and *trans*) isomers. For example, 1,2-dichloroethene has two **geometric isomers**, the *cis* isomer where the Cl atoms are on the same side of the molecule and the *trans* isomer where the Cl atoms are arranged on opposite sides of the molecule (Interactive Figure 21.2.2).

Optical isomers are two compounds that are nonsuperimposable (not identical) mirror images of each other. A compound that exhibits optical isomerism is **chiral**. One characteristic of a chiral compound is how it interacts with plane-polarized light, that is,

Video Solution

Tutored Practice
Problem 21.2.1

Interactive Figure 21.2.2

Name alkene stereoisomers.

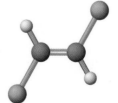

cis-1,2-Dichloroethylene   *trans*-1,2-Dichloroethylene

© 2013 Cengage Learning

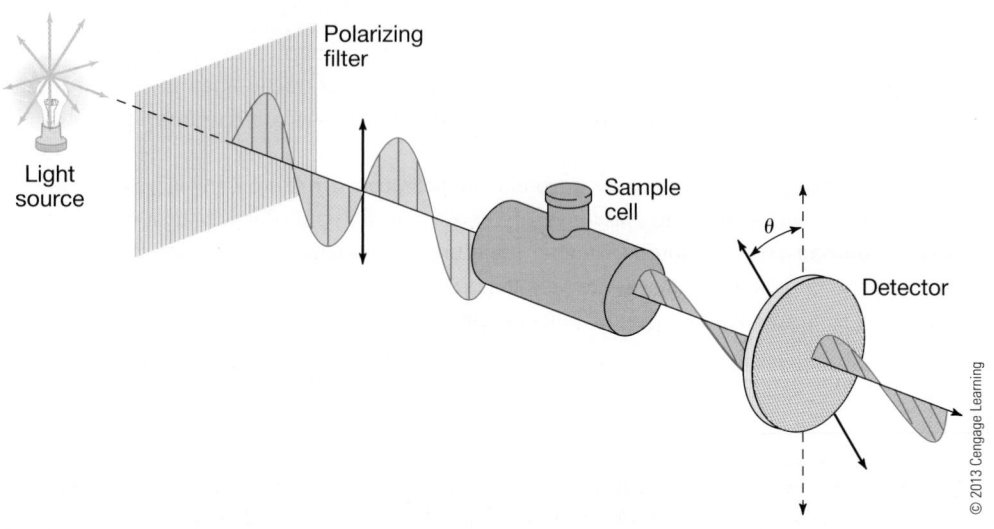

Figure 21.2.3 Rotation of plane-polarized light by a chiral molecule

© 2013 Cengage Learning

monochromatic light that has been passed through a polarizing filter (Figure 21.2.3). When plane-polarized light is passed through a solution containing a chiral compound, the plane rotates. The magnitude and angle of rotation are unique physical properties of the chiral compound. The mirror image of the chiral compound will also rotate plane-polarized light by the same angle, but in the opposite direction.

An organic molecule that has at least one carbon atom with four different groups bonded to it is chiral. A carbon atom attached to four different groups is called a **chiral center**. For example, lactic acid is a chiral compound because its mirror image is a unique (nonsuperimposable) compound (Interactive Figure 21.2.4).

The two unique stereoisomers are designated using the R,S system, which indicates the orientation of groups around the chiral center. The R,S system uses a set of priority rules to assign the configuration of a chiral center, but we will not cover that process here. In addition to its configuration, the direction that a chiral molecule rotates plane-polarized light is indicated by the symbol $(+)$ or $(-)$. Note that there is not a direct relationship between R and S configuration and $(+)$ and $(-)$ rotation; that is, for some molecules, the R isomer is $(+)$, and for others, the S isomer is $(-)$.

Section 21.2 Mastery

Identify chiral molecules.

H₃C—C—CO₂H (with H above and OH below)   X—C—Z (with H above and Y below)

Lactic acid, a molecule of general formula CHXYZ

(+)-Lactic acid   (–)-Lactic acid

Optical isomers of lactic acid

© 2013 Cengage Learning

21.3 Functional Groups

21.3a Identifying Functional Groups

The richness of organic chemistry lies in the great variety of physical and chemical properties offered by compounds with different functional groups. **Functional groups** are atoms or groups of atoms within a molecule that show characteristic physical and chemical properties.

The simplest functional group is the *alkyl halides*; in this group, a hydrogen atom in an alkane is replaced with a halogen. We can represent functional groups using a generic symbol, R, to represent the hydrocarbon parent attached to the functional group. The symbols R′ and R″ represent unique hydrocarbon groups attached to a functional group; in some cases, the R group can be a hydrogen atom. Thus, an alkyl halide is represented as R—X, where R is a hydrocarbon group and X is a halogen. An **amine**, a derivative of ammonia, is represented as RNR′R″, where R′ and R″ can be hydrocarbons or hydrogen atoms. The most common functional groups are shown in Interactive Table 21.3.1, which is organized by the type of atom attached to the hydrocarbon group.

Organic Functional Groups

Atom Attached to R	Functional Group	Structure
Halogen	Alkyl halide	R—X (X = F, Cl, Br, I)
Oxygen (single bond)	Alcohol	R—O—H
	Phenol	Ar—O—H (Ar = aromatic group)
	Ether	R—O—R′
Carbonyl (C=O) group	Carboxylic acid	$\begin{array}{c} O \\ \parallel \\ R—C—O—H \end{array}$
	Aldehyde	$\begin{array}{c} O \\ \parallel \\ R—C—H \end{array}$
	Ketone	$\begin{array}{c} O \\ \parallel \\ R—C—R′ \end{array}$
	Ester	$\begin{array}{c} O \\ \parallel \\ R—C—O—R′ \end{array}$
	Amide	$\begin{array}{cc} O & R′ \\ \parallel & \mid \\ R—C—N—R″ \end{array}$ (R, R′, R″ = H or hydrocarbon group)
	Acid halide	$\begin{array}{c} O \\ \parallel \\ R—C—X \end{array}$
Nitrogen	Amine	$\begin{array}{c} R′ \\ \mid \\ R—N—R″ \end{array}$ (R′, R″ = H or hydrocarbon group)
Sulfur	Thiol	R—S—H
	Thioether	R—S—R′

21.3b Alcohols

Alcohols are organic compounds containing the —OH functional group. Because they are polar, alcohols experience hydrogen bonding and generally have higher boiling points and solubility in water than the hydrocarbons from which they are derived. Alcohols are important organic molecules because they can be converted into many different types of compounds, such as alkyl halides, carboxylic acids, aldehydes, and esters.

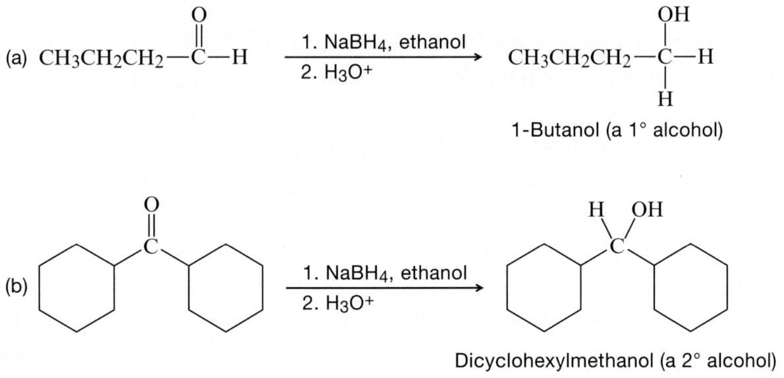

(a) $CH_3CH_2CH_2-\overset{\overset{\displaystyle O}{\|}}{C}-H$ $\xrightarrow[\text{2. }H_3O^+]{\text{1. NaBH}_4,\text{ ethanol}}$ $CH_3CH_2CH_2-\overset{\overset{\displaystyle OH}{|}}{\underset{\underset{\displaystyle H}{|}}{C}}-H$

1-Butanol (a 1° alcohol)

(b) $\xrightarrow[\text{2. }H_3O^+]{\text{1. NaBH}_4,\text{ ethanol}}$

Dicyclohexylmethanol (a 2° alcohol)

Figure 21.3.1 Synthesis of alcohols by (a) aldehyde reduction and (b) ketone reduction

Synthesis and Naming

Alcohols are synthesized from alkenes (hydration reactions) or by the reduction of aldehydes or ketones. For example, the reduction of the aldehyde butanal results in the formation of 1-butanol, and the reduction of the ketone dicyclohexyl ketone produces dicyclohexylmethanol (Figure 21.3.1).

Alcohols are categorized by the number of hydrogen atoms attached to the carbon where the —OH group is attached. In a primary (1°) alcohol, the —OH group is attached to a carbon with two H atoms; in a secondary (2°) alcohol, the —OH group is attached to a carbon atom with only one H atom; and in a tertiary (3°) alcohol, the —OH group is attached to a carbon atom with no H atoms (Interactive Figure 21.3.2).

Alcohols are named using the parent name of the longest carbon chain, where the carbon atoms are numbered to give the carbon attached to —OH the lowest possible number. Other substituents are named following the rules for naming branched hydrocarbons.

Interactive Figure 21.3.2

Explore alcohol structure and naming.

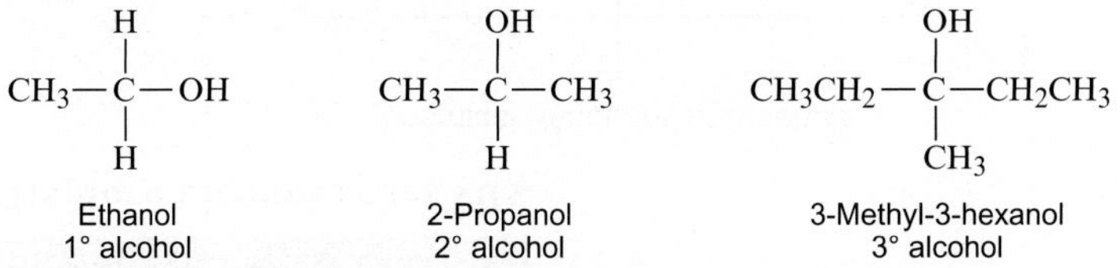

| Ethanol | 2-Propanol | 3-Methyl-3-hexanol |
| 1° alcohol | 2° alcohol | 3° alcohol |

Primary, secondary, and tertiary alcohols

Explore alcohol reactivity.

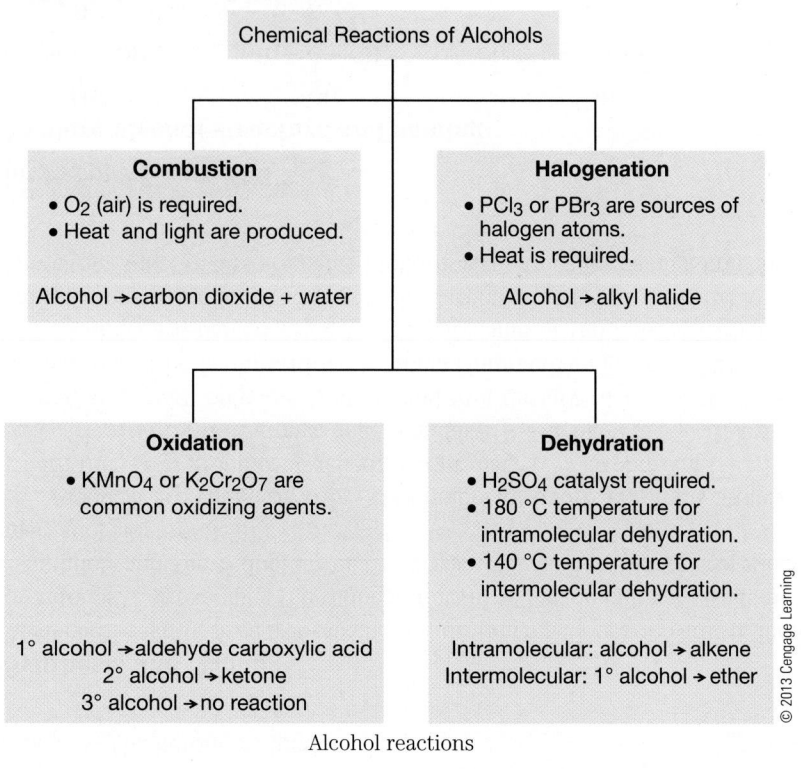

Alcohol reactions

Reactions

Alcohols undergo numerous reactions. The reactivity of alcohols follow trends based on whether the alcohol is primary, secondary, or tertiary (Interactive Figure 21.3.3).

The two most common alcohol reactions are dehydration and oxidation. The dehydration of an alcohol to form an alkene is an elimination reaction, which is defined as a reaction where a small molecule is separated from a larger molecule. As shown in Figure 21.3.4, an elimination reaction is the reverse of an alkene addition reaction.

In an alcohol dehydration reaction, H_2O is eliminated. The dehydration of a primary alcohol has only one possible product, an alkene with the double bond on a terminal carbon. However, a secondary alcohol dehydration has two possible products (Figure 21.3.5).

Figure 21.3.4 An elimination reaction

When 2-butanol undergoes dehydration, the carbon–carbon double bond can form between carbons 1 and 2 or between carbons 2 and 3. The major product formed is predicted using **Zaitsev's rule**, which states that the more stable product of an alcohol dehydration reaction is the alkene with the greatest number of substituents on the C=C carbons (Figure 21.3.6).

Alcohols also undergo intermolecular dehydration reactions, where H and OH from two alcohol molecules are lost. These reactions form water and an **ether** (R—O—R′) (Figure 21.3.7).

Oxidation of an alcohol can result in an aldehyde, a carboxylic acid, or a ketone. As shown in Figure 21.3.8, primary alcohols are oxidized to form aldehydes, which can undergo further oxidation to form carboxylic acids. Secondary alcohols can be oxidized to form ketones, and tertiary alcohols cannot be oxidized.

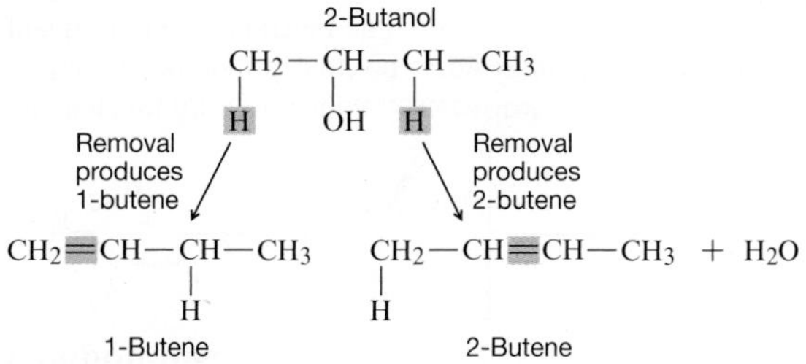
2-Butanol
Removal produces 1-butene
Removal produces 2-butene
1-Butene
2-Butene

Figure 21.3.5 Dehydration of the secondary alcohol 2-butanol

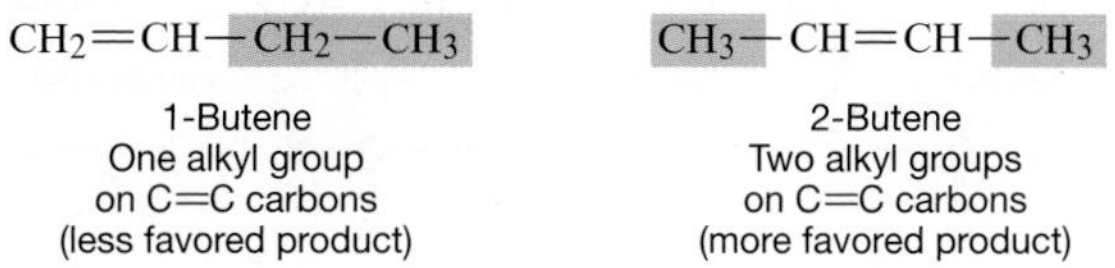
1-Butene
One alkyl group
on C=C carbons
(less favored product)
2-Butene
Two alkyl groups
on C=C carbons
(more favored product)

Figure 21.3.6 An application of Zaitsev's rule

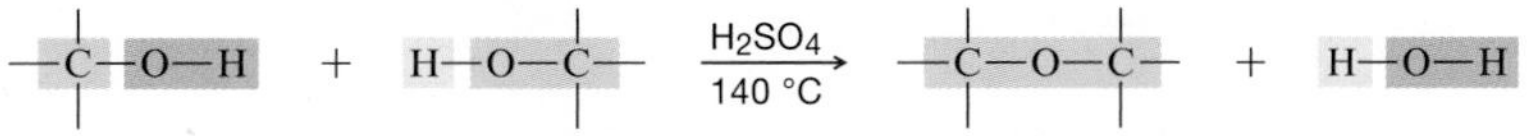
H2SO4
140 °C

Figure 21.3.7 Intermolecular dehydration of alcohols to form an ether

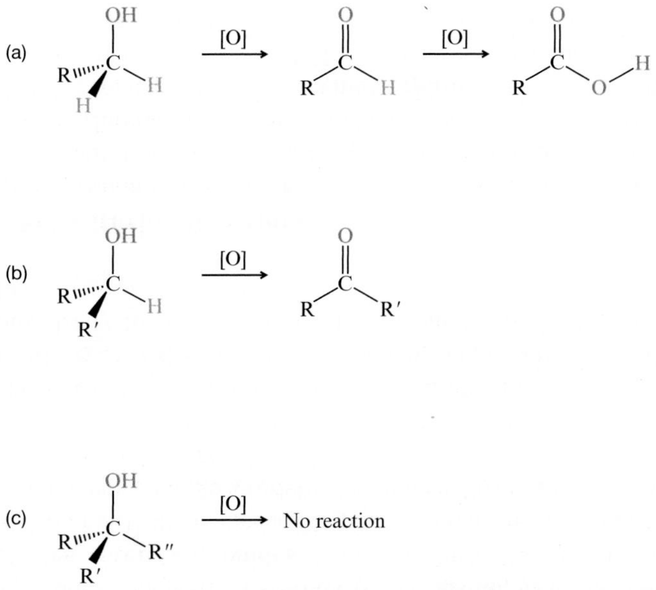
(a)
[O]
[O]
(b)
[O]
(c)
[O]
No reaction

Figure 21.3.8 Oxidation reactions of (a) primary, (b) secondary, and (c) tertiary alcohols ([O] = oxidizing agent)

21.3c Compounds Containing a Carbonyl Group

Many functional groups contain the **carbonyl** (C=O) **group**, including **carboxylic acids**, **ketones**, **aldehydes**, **esters**, and **amides** (Interactive Table 21.3.1). Here we describe some of the properties and reactivity of organic compounds containing these functional groups.

Carbonyl compounds undergo a variety of reactions. Figure 21.3.9 shows a diagram for only some of the reactions a carboxylic acid can undergo.

The nature of the groups attached to the carbonyl group influence the reactivity of the compound and can lead to the synthesis of a wide variety of compounds. For example, if we consider a carboxylic acid as a parent compound, simply substituting the —OH group for other groups leads to a wide range of compound types with different reactivity (Figure 21.3.10).

Nucleophilic Substitution Reactions

The carbonyl group contains a partial positively charged C atom and a partial negatively charged O atom. Substitution reactions of this group occur at the C atom and involve breaking the C—O pi bond (Interactive Figure 21.3.11). In this reaction, the carbonyl carbon acts as a Lewis acid and accepts electrons from a **nucleophile**, a species that can donate a pair of electrons to form a new covalent bond. Nucleophiles are by definition Lewis bases, but not all Lewis bases act as nucleophiles in nucleophilic substitution reactions.

Once the nucleophile bonds to the carbon atom of the carbonyl, the reverse process can occur: loss of one of the substituents bonded to the carbon, followed by reformation of the C—O pi bond. The net result is substitution of one group bonded to the carbonyl for another.

Section 21.3 Mastery

21.4 Synthetic Polymers

21.4a Addition Polymerization

Polymers are large molecules formed by the linking of many small molecules, called **monomers**. The two most common processes used to make synthetic polymers are addition polymerization and condensation polymerization. Synthetic polymers are ubiquitous in our modern world; they are used to make our electronic devices, the clothes we wear, and the containers that hold our food and drink.

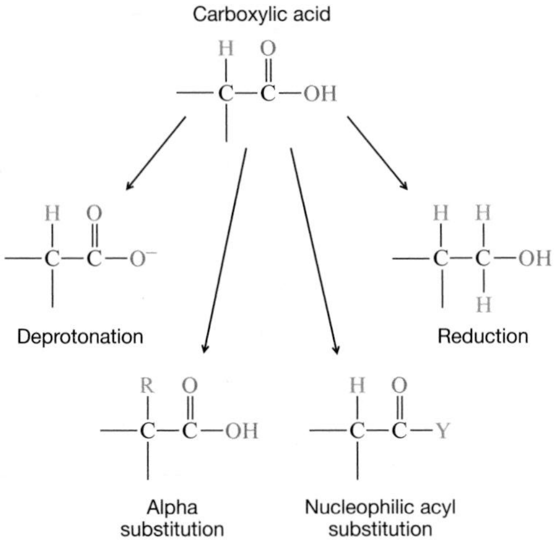

Figure 21.3.9 Selected reactions of a carboxylic acid

Figure 21.3.10 Derivatives of a carboxylic acid

Interactive Figure 21.3.11

Explore reactions of carbonyl-containing compounds.

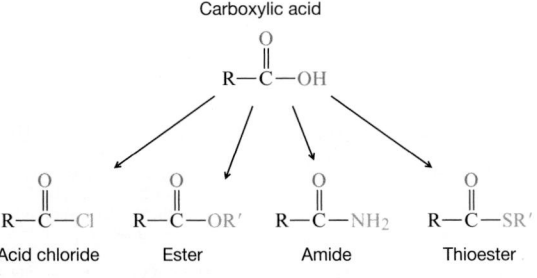

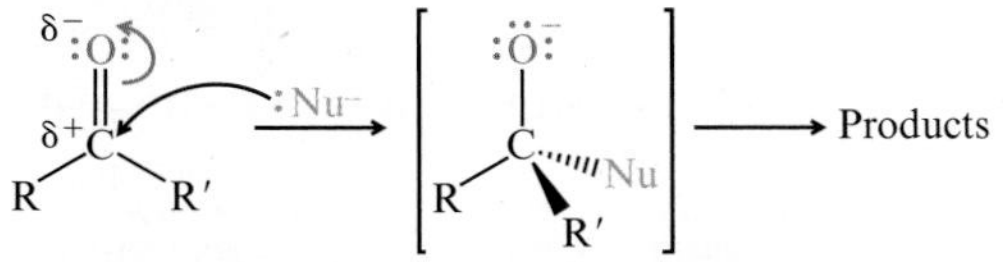

Mechanism for a nucleophilic substitution reaction of a carbonyl compound (curved arrow indicates change in position of an electron pair)

An **addition polymer** is a polymer that results from the combination of many monomers. In most addition polymerization reactions, monomers contain a $C=C$ double bond. These reactions are started using a free radical "initiator" (recall that a free radical is a compound that has an unpaired electron). The initiator reacts with the $C=C$ group, breaking the pi bond and forming a new bond between the initiator and one of the alkene carbons. The product of this reaction has an unpaired electron on a carbon atom; this step effectively adds a two-carbon unit to the free radical without losing the free radical functionality. The new free radical can now react with other alkene monomers, lengthening the carbon chain and forming a polymer (Figure 21.4.1). The result of an addition polymerization reaction is a long chain of sp^3-hybridized carbon atoms. This long-chained molecule is a polymer and can be thousands of carbons long.

Most addition polymers, with the exception of polyethylene and polytetrafluoroethylene (PFTE, Teflon®), have the same general structure where $—CH_2—$ groups alternate with substituted $—C(X)(Y)—$ groups (Table 21.4.1). Polyethylene consists of chains of $—CH_2—$ groups and in Teflon®, all of the groups attached to the carbon chain are fluorine atoms. The groups attached to carbon result in different chemical and physical properties of common plastics.

Polymer Processing

Most of the plastic objects in our lives are formed by melting plastic pellets and forming consumer products with useful shapes. The most important processes are injection molding, blow molding, film extrusion, and fiber spinning. Interactive Figure 21.4.2 shows the process of blow molding to make a plastic bottle.

21.4b Condensation Polymerization

Condensation polymers are formed when reactions between two monomers form a linked, larger structure, as well as a small molecule such as HCl or water. For the reaction to produce a long chain, each monomer must have two reactive groups. For example, **polyamides**, condensation polymers linked by amide functional groups, can be synthesized by reacting a *dicarboxylic acid*, a compound with two carboxylic acid functional groups, with a *diamine*, a compound with two amine functional groups.

Interactive Figure 21.4.3 shows the formation of a **polyester** from the reaction of a dicarboxylic acid with a dialcohol. In Step 1, the carboxylic acid and alcohol functional groups on the two monomers react to form an ester and water. The new molecule is a *dimer* because it is made up of two monomers. The dimer can react with additional monomer (dicarboxylic acid or dialcohol) in Step 2 to form a *trimer*. This reaction continues to occur, resulting in the formation of the polymer. Some examples of condensation polymers are shown in Table 21.4.2.

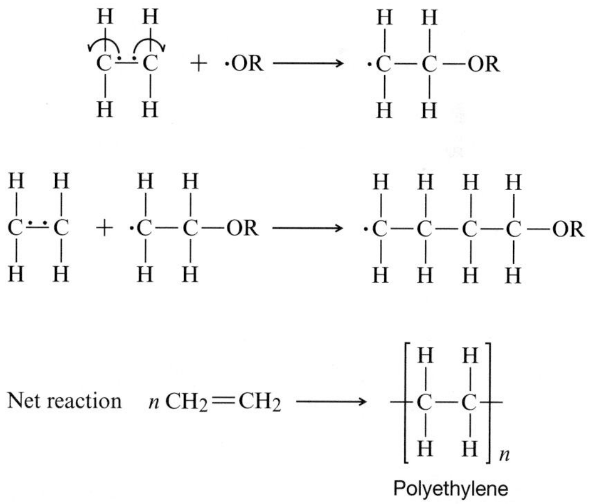

Figure 21.4.1 Addition polymerization (curved half-headed arrow indicates change in position of a single electron)

Table 21.4.1 Structures and Uses of Some Common Addition Polymers

$$\left[\begin{array}{cccccccc} H & X & H & X & H & X & H & X \\ | & | & | & | & | & | & | & | \\ -C & -C & -C & -C & -C & -C & -C & -C- \\ | & | & | & | & | & | & | & | \\ H & Y & H & Y & H & Y & H & Y \end{array}\right]$$

Polymer Name	X	Y	Common Uses
Polyethylene	H	H	Plastic bags, plastic soda bottles, wire and cable insulation
Polypropylene	H	CH_3	Clothing, carpets, ropes, reusable plastic containers
Polystyrene	H	C_6H_5	Styrofoam, plastic utensils, CD and DVD cases
Polyvinylchloride (PVC)	H	Cl	Pipes, hoses, tubing, wire insulation
Saran®	Cl	Cl	Plastic sheeting, shower curtains, tape, garden furniture
Polymethylmethacrylate (PMMA)	CH_3	CO_2CH_3	Plexiglass, acrylic, bone replacement, dentures
Polytetrafluoroethylene (Teflon®)	F	F	Nonstick cookware, nail polish additive, coating on clothing and rugs

Interactive Figure 21.4.2

Explore methods used to process plastics.

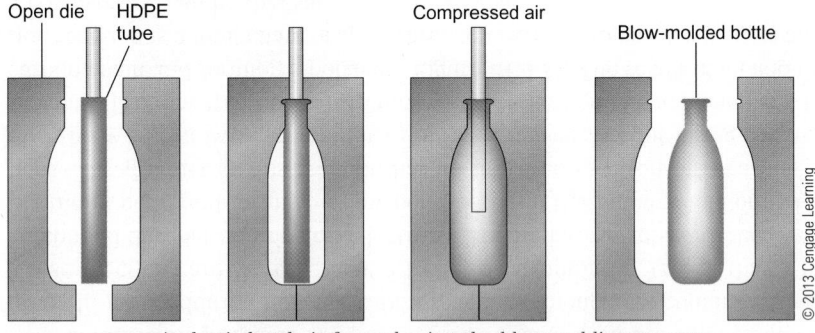

Open die   HDPE tube   Compressed air   Blow-molded bottle

© 2013 Cengage Learning

A plastic bottle is formed using the blow molding process.

Explore the synthesis of a condensation polymer.

Step 1

$$HO-\overset{\overset{\displaystyle O}{\|}}{C}-R-\overset{\overset{\displaystyle O}{\|}}{C}-OH \quad + \quad HO-R'-OH \longrightarrow$$

$$HO-\overset{\overset{\displaystyle O}{\|}}{C}-R-\overset{\overset{\displaystyle O}{\|}}{C}-O-R'-OH \; + \; H_2O$$

Step 2

$$HO-\overset{\overset{\displaystyle O}{\|}}{C}-R-\overset{\overset{\displaystyle O}{\|}}{C}-O-R'-OH \quad + \quad HO-\overset{\overset{\displaystyle O}{\|}}{C}-R-\overset{\overset{\displaystyle O}{\|}}{C}-OH \longrightarrow$$

$$HO-\overset{\overset{\displaystyle O}{\|}}{C}-R-\overset{\overset{\displaystyle O}{\|}}{C}-O-R'-O-\overset{\overset{\displaystyle O}{\|}}{C}-R-\overset{\overset{\displaystyle O}{\|}}{C}-OH \; + \; H_2O$$

A condensation polymer is formed from a dicarboxylic acid and a dialcohol.

Table 21.4.2 Structures and Uses of Some Common Condensation Polymers

Type	Example	Uses
Polyester	Polyethylene terephthalate (PET) $$\left[O-CH_2CH_2-O-\overset{\overset{\displaystyle O}{\|}}{C}-\!\!\left\langle\right\rangle\!\!-\overset{\overset{\displaystyle O}{\|}}{C} \right]_n$$	Beverage bottles, clothing, film, food packaging
Polyamide	Nylon-6,6 $$\left[\overset{\overset{\displaystyle H}{\|}}{N}-(CH_2)_6-\overset{\overset{\displaystyle H}{\|}}{N}-\overset{\overset{\displaystyle O}{\|}}{C}-(CH_2)_4-\overset{\overset{\displaystyle O}{\|}}{C} \right]_n$$	Textiles, bulletproof and fireproof garments
Polycarbonate	$$\left[\!\!\left\langle\right\rangle\!\!\overset{\overset{\displaystyle CH_3}{\|}}{\underset{\underset{\displaystyle CH_3}{\|}}{C}}\!\!\left\langle\right\rangle\!\!-O-\overset{\overset{\displaystyle O}{\|}}{C}-O \right]_n$$	Water bottles, electronics, safety helmets, household appliances
Polyurethane	$$\left[O-CH_2CH_2-O-\overset{\overset{\displaystyle O}{\|}}{C}-\underset{\underset{\displaystyle H}{\|}}{N}-\!\!\left\langle\right\rangle\!\!-\underset{\underset{\displaystyle H}{\|}}{N}-\overset{\overset{\displaystyle O}{\|}}{C} \right]_n$$	Flexible foams (furniture, clothing), rigid foams (refrigerators, freezers), moldings, footwear

21.4c Control of Polymer Properties

One of the hallmarks of polymer chemistry is the ability to make useful materials. All plastics we use are formed from polymers. The ability to control the properties of materials comes from the ability to control their structures. We examine three means that influence properties unique to polymers: chain length, branching, and cross-linking.

Chain Length

The length of polymer chains influences properties such as a polymer's flexibility and its ability to dissolve in a solvent. In addition polymerization, chain length is controlled by adjusting the ratio of initiator to monomer in the reaction mixture. When an addition polymerization reaction occurs, initiators start the reaction and monomers keep adding onto growing polymer chains until they are all reacted. If few initiator molecules are used, fewer chains will be formed, and each will add more monomers before the reaction stops. If many initiator molecules are used, more chains will form and the chains will have fewer monomer units each (Figure 21.4.4).

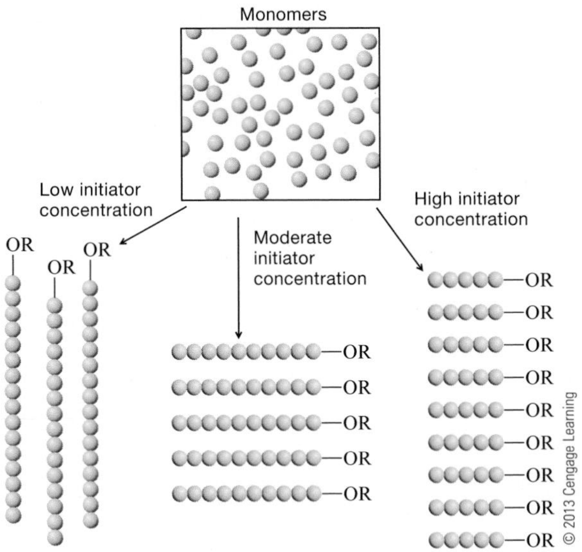

Branching

It is possible to use branching units in polymerization reactions that cause side chains to grow off the main polymer chain. This leads to bulky structures that do not pack closely together. Examples of this are high- and low-density polyethylene (HDPE and LDPE). HDPE is made up of straight-chain (not branched) polymer chains that pack closely together, resulting in a high-density solid. LDPE contains branched polyethylene, which cannot pack as tightly and results in a solid polymer with lower density and a more flexible structure (Figure 21.4.5).

Cross-Linking

It is possible to link one chain to another using branches off one chain that bond to branches off an adjacent chain. These linkages are called *cross-links* (Interactive Figure 21.4.6). Materials containing cross-linked polymers are generally much tougher and harder than those consisting of non–cross-linked polymers. The toughness and hardness can be controlled by the amount of cross-linking.

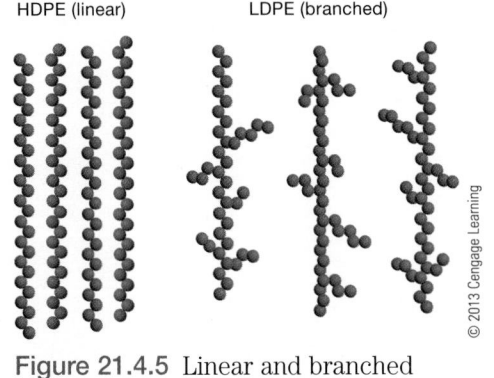

Figure 21.4.4 Chain length control

Figure 21.4.5 Linear and branched polyethylene

Section 21.4 Mastery

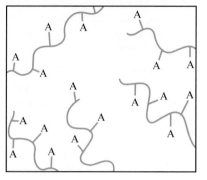

Interactive Figure 21.4.6

Explore cross-linking in polymers.

Not cross-linked   Cross-linked

Polymer cross-linking

© 2013 Cengage Learning

21.5 Biopolymers

21.5a Carbohydrates

Biopolymers, naturally occurring polymers produced by living organisms, are vital to the survival of all living things. Like synthetic polymers, biopolymers consist of monomer units that are linked together into a large molecule. In this unit we discuss the three main types of biopolymers: polysaccharides, polypeptides, and polynucleic acids.

Carbohydrates are the most abundant organic compounds on our planet. They store chemical energy, are important structural components in plants and connective tissues, and are an important building block in nucleic acids such as DNA. Most carbohydrates are composed of carbon, hydrogen, and oxygen (some also contain nitrogen), with empirical formulas equal or close to $C(H_2O)$.

Simple Carbohydrates

The simplest carbohydrates, called simple sugars, typically contain between three and seven carbon atoms. They are polyhydroxy compounds that contain either an aldehyde or ketone functional group. The linear forms of two common carbohydrates, glucose and fructose, are shown in Figure 21.5.1. Glucose is an *aldohexose* because it is a six-carbon carbohydrate that contains an aldehyde functional group. Fructose is a *ketohexose* because it is a six-carbon

Figure 21.5.1 The structures of glucose and fructose

D-Glucose   D-Fructose

carbohydrate that contains a ketone functional group. Both are simple sugars with the molecular formula $C_6H_{12}O_6$, or $C_6(H_2O)_6$.

The line structures in Figure 21.5.1 are called **Fisher projections** and are specifically used to show the conformation of carbohydrates. In a Fisher projection, all horizontal lines represent bonds directed toward the viewer and all vertical lines represent bonds directed away from the viewer. Note that carbohydrates contain at least one chiral center; glucose has four chiral carbons, and fructose has three. The configuration of chiral biomolecules such as carbohydrates and proteins is designated using the D,L system, the specifics of which we will not discuss here. However, it is interesting to note that the majority of carbohydrates found in nature are of the D form.

Example Problem 21.5.1 Identify aldoses and ketoses.

Identify each of the following carbohydrates as an aldose or ketose and as a triose, tetrose, pentose, or hexose.

a.
```
        CHO
   HO ──┼── H
   HO ──┼── H
    H ──┼── OH
    H ──┼── OH
        CH2OH
```

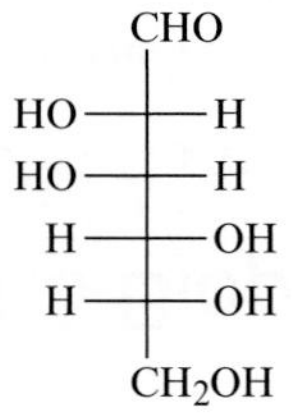

b.
```
        CH2OH
         |
         C═O
   HO ──┼── H
        CH2OH
```

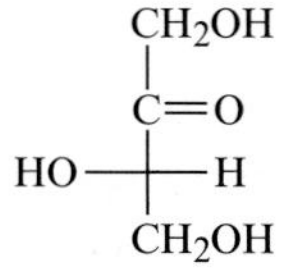

Solution:

You are asked to identify a carbohydrate as an aldose or ketose and as a triose, tetrose, pentose, or hexose.

You are given the structural formula of the carbohydrate.

a. This carbohydrate has an aldehyde functional group and six carbons. It is an aldose and a hexose, or an aldohexose.

b. This carbohydrate has a ketone functional group and four carbons. It is a ketose and a tetrose, or a ketotetrose.

Video Solution

Tutored Practice
Problem 21.5.1

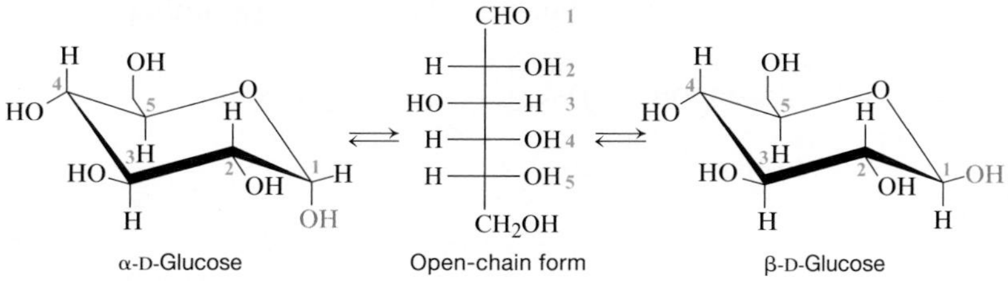

Figure 21.5.2 Cyclic forms of glucose

Cyclic Forms of Simple Carbohydrates

Most carbohydrates can undergo an intramolecular reaction between their alcohol and carbonyl functional groups to form a cyclic structure. Figure 21.5.2 shows the linear form of D-glucose and the two cyclic structures that result when this intramolecular reaction takes place.

The two products of the reaction differ only in the stereochemistry of the —OH group shown in red; in the alpha (α) form, the —OH group and the —CH₂OH group attached to carbon 5 lie on opposite sides of the ring, whereas in the beta (β) form, the two groups are on the same side of the ring. Both forms of D-glucose can be formed from the straight-chain form.

Complex Small Carbohydrates

An important property of carbohydrates is their ability to react with other carbohydrates to form larger structures. Figure 21.5.3 shows how two monosaccharides form a dimer (a disaccharide) by a condensation reaction that results in the formation of a new bond called a glycoside linkage.

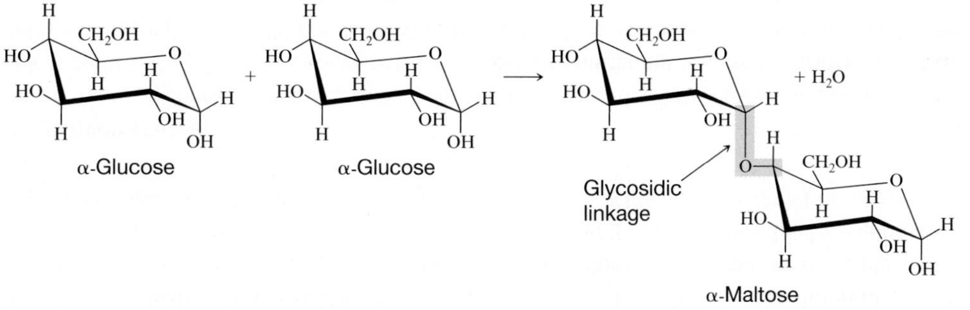

Figure 21.5.3 Formation of a disaccharide

Figure 21.5.4 shows two important disaccharides: sucrose, formed from glucose and fructose, and lactose, formed from galactose and glucose units. Notice that different conformations can result from linking two monosaccharides. Sucrose contains an α-1,2-glycosidic bond because carbon 1 of α-D-glucose is linked to carbon 2 of D-fructose. Lactose contains a β-1,4-glycosidic bond because carbon 1 on β-D-galactose is linked to carbon 4 of D-glucose.

Polysaccharides

Three monosaccharides link together to form a trisaccharide. When many monosaccharides link together, they form a polysaccharide, a polymer containing monosaccharides linked together by α or β glycosidic bonds. Two important polysaccharides are starch and cellulose, each consisting of thousands of glucose units linked together by α (amylose and amylopectin, both forms of starch) or β (cellulose) bonds (Figure 21.5.5).

Amylopectin and amylose have similar structures, but amylose does not have branching (the glucose units linked by α-1,6-glycosidic bonds in Figure 21.5.5). Both amylose and

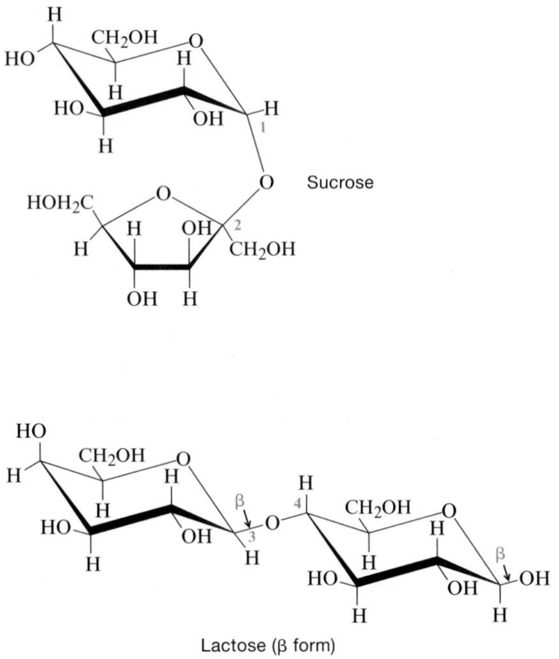

Figure 21.5.4 Structures of sucrose and lactose

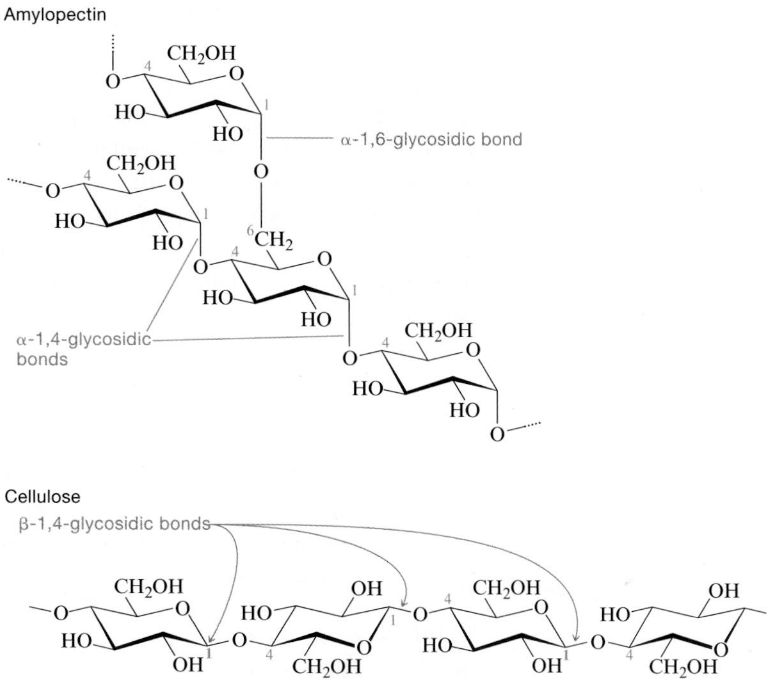

Figure 21.5.5 Amylopectin and cellulose

cellulose are straight-chain polysaccharides that differ only in the orientation of the glycosidic linkages.

21.5b Amino Acids

Proteins are biopolymers formed when many amino acids link together to form long chains. As you saw in Advanced Acid–Base Equilibria (Unit 17), amino acids contain an alpha carbon bonded to a hydrogen atom, an amine group, a carboxylic acid group, and a side-chain molecular fragment (R) (Figure 21.5.6).

In Figure 21.5.6, the amino group is shown in its basic form and the carboxylic acid group is shown in its acidic form. The nature of the side-chain molecular fragment, R, controls the properties of the amino acid. There are 20 common amino acids, each of which is chiral, with the exception of glycine. The D,L system used to designate the conformation of carbohydrates is also commonly used to indicate the conformation of amino acids. The majority of the amino acids found in the biological world are of the L form.

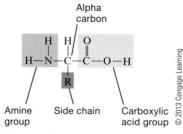

Figure 21.5.6 The structure of an amino acid

Acid–Base Properties of Amino Acids

Amino acids have acidic and basic functional groups and can therefore exist in different acid–base states with changing pH. For the simple amino acid alanine, there are three possible states, as shown in Interactive Figure 21.5.7.

At high pH, both the acidic and basic site are found in the basic (deprotonated) form. At low pH, both sites are in the acidic (protonated) form. The state most commonly found under neutral pH conditions has a protonated (acidic) ammonium group and deprotonated (basic) carboxylate group. This is the zwitterionic form of alanine. Amino acids exist in the zwitterionic form when dissolved in water, in bodily fluids, and in the solid state. The pH at which an aqueous amino acid solution contains only zwitterionic molecules is called its isoelectric point, or pI.

Interactive Figure 21.5.7

Investigate the relationship between pH and amino acid structure.

Acidic, neutral, and basic forms of the amino acid alanine

21.5c Proteins

A carboxylic acid can react with an amine in a condensation reaction, where the products are water and an amide (Figure 21.5.8). Because amino acids contain both carboxylic acid and amine functional groups, they can undergo condensation reactions. When two amino acids react to form an amide, the new bond between them is called a **peptide bond** and the amide is called a **dipeptide** (Figure 21.5.9). The dipeptide still has carboxylic acid and amine groups, so it can undergo further reaction to form additional peptide bonds, leading to larger and larger molecules. When many amino acids react to form peptide linkages, the result is a **polypeptide** (Figure 21.5.10).

Proteins are large polypeptides that have a biological function; they contain many linked amino acids. The smallest naturally occurring protein contains only 20 amino acids, whereas the largest known protein contains more than 30,000 amino acids. Figure 21.5.11 shows the amino acid sequence of the protein lysozyme, where each circle represents a single amino acid.

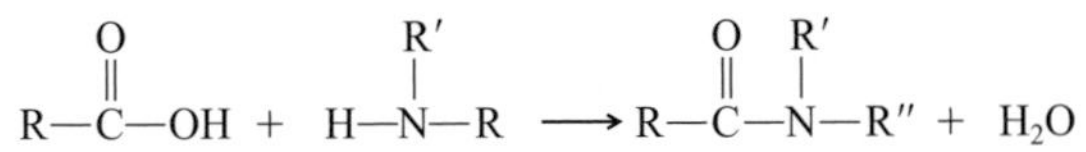

$$R-\overset{O}{\overset{\|}{C}}-OH + H-\overset{R'}{\underset{|}{N}}-R \longrightarrow R-\overset{O}{\overset{\|}{C}}-\overset{R'}{\underset{|}{N}}-R'' + H_2O$$

Carboxylic acid   Amine   Amide

Figure 21.5.8 A condensation reaction to form an amide

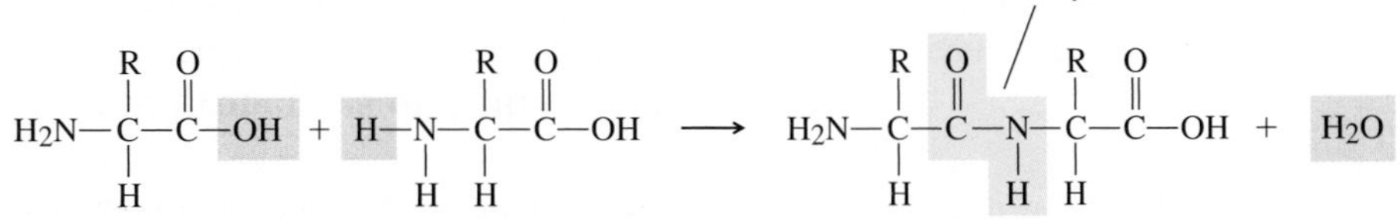

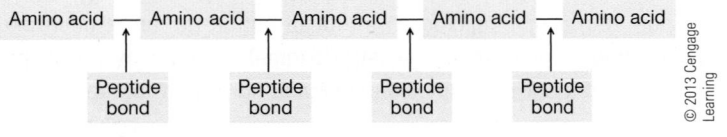

Figure 21.5.9 Two amino acids react to form a dipeptide.

Figure 21.5.10 A polypeptide contains amino acids linked by peptide bonds.

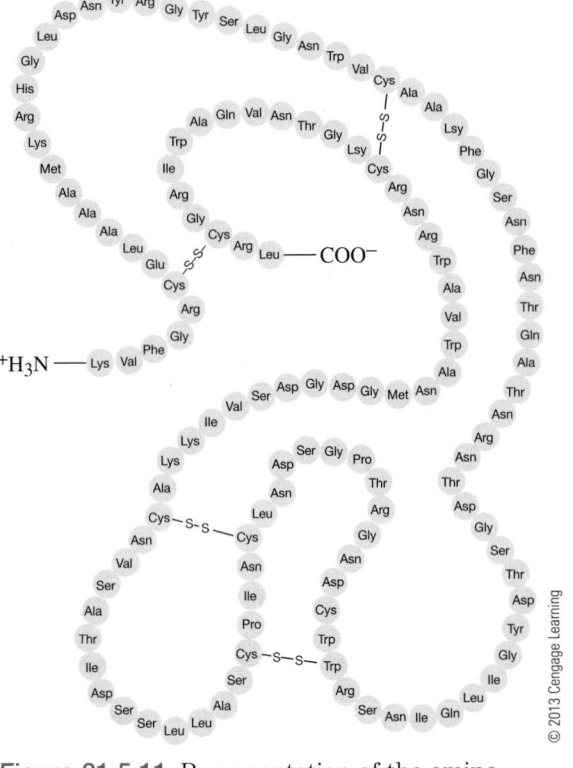

Figure 21.5.11 Representation of the amino acid linkage of lysozyme

© 2013 Cengage Learning

Interactive Figure 21.5.12

Explore an alpha helix.

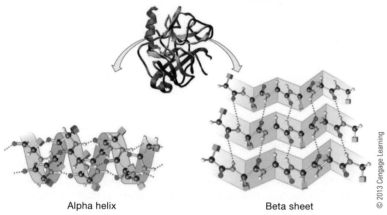

Alpha helix   Beta sheet

Secondary structure in a polypeptide chain

Protein Structure

The composition and sequence of amino acids making up a protein is called its primary structure, but the reactivity of a protein depends heavily on the way the different parts of the protein chain interact with one another. This longer-order structure is represented by the secondary, tertiary, and quaternary structures.

The secondary structure of a protein involves the way one part of the chain can bind (either strongly or weakly) with another part. Two structural motifs involving secondary structure are the alpha helix and beta sheets (Interactive Figure 21.5.12). In an alpha helix, carbonyl groups and amino groups located at short distances within the protein chain form intramolecular hydrogen bonds, which leads to the peptide chain forming a coil. Beta sheets are formed when sections of the polypeptide chain are aligned near one another and are held together by intramolecular hydrogen bonds between carbonyl and amino groups on adjacent chains.

Proteins also have higher-order structure controlling how different parts of the chain interact to form large units (tertiary structure) and the way those large units interact (quaternary structure). The structure of myoglobin involves a number of alpha helixes that "wrap around" a heme unit, a large group containing an iron atom surrounded by four nitrogen atoms. Figure 21.5.13 shows the structures of myoglobin and hemoglobin, each of which contains heme groups.

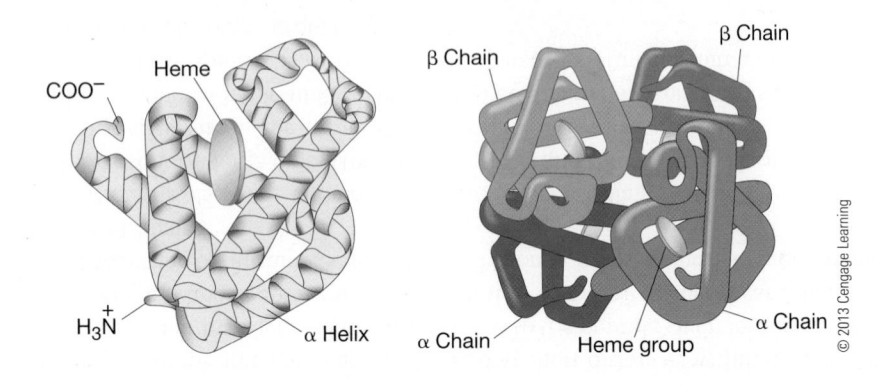

Figure 21.5.13 (a) Tertiary structure in myoglobin; (b) tertiary and quaternary structure in hemoglobin

The way the alpha helices in myoglobin wrap around the heme group represents tertiary structure. The way the four myoglobin-like units of hemoglobin bind together represents quaternary structure.

21.5d Nucleic Acids

Heredity and genetics arise through information carrying biopolymers called **nucleic acids**, including DNA and ribonucleic acid (RNA). Figure 21.5.14 shows the three structural components of a nucleotide, the repeating unit of a nucleic acid: a phosphate group, a five-carbon pentose sugar, and a nitrogen-containing base.

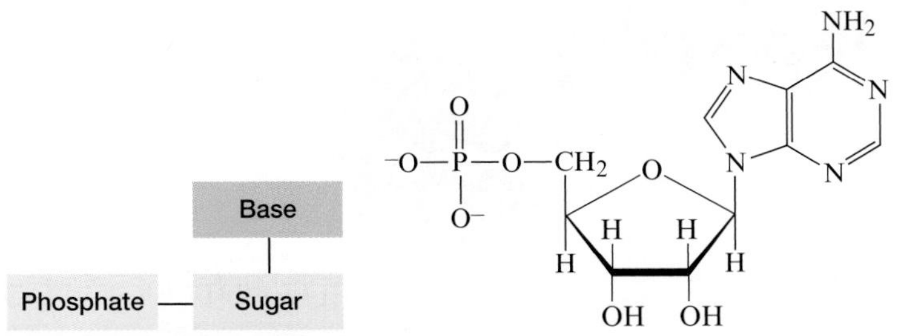

Figure 21.5.14 The structure of a nucleotide

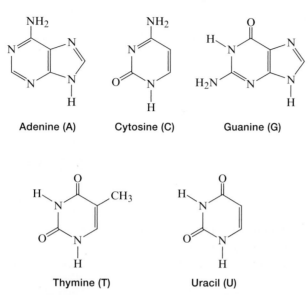

Figure 21.5.15 The bases found in nucleic acids

Adenine (A)   Cytosine (C)   Guanine (G)

Thymine (T)   Uracil (U)

Five different bases are found in nucleic acids; their structures are shown in Figure 21.5.15, along with the one-letter code used to identify them. The nucleic acids adenine, guanine, thymine, and cytosine are found in DNA. In RNA, the base uracil replaces thymine.

DNA and RNA are polynucleic acids in which the phosphate and sugar linkages form the polymeric backbone, and the base units are "side groups." In DNA, the sugar is deoxyribose and the bases are A, G, T, and C. RNA contains the sugar ribose and the bases A, G, C, and U. Figure 21.5.16 shows an expanded four-unit section of a DNA molecule. An individual DNA or RNA molecule can contain thousands of nucleotides.

The Double Helix

DNA consists of a double-stranded structure, whereas most RNA molecules are single stranded. The DNA structure, shown in Figure 21.5.17, is called a double helix. The double helix is formed by two intertwining polynucleic acid strands.

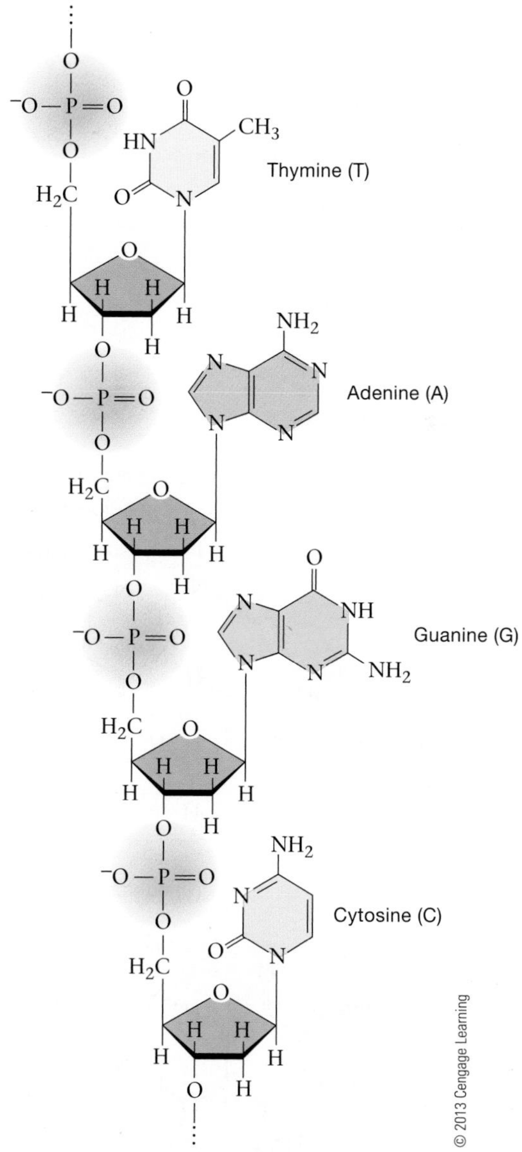

Thymine (T)

Adenine (A)

Guanine (G)

Cytosine (C)

© 2013 Cengage Learning

Figure 21.5.16 A portion of a DNA molecule

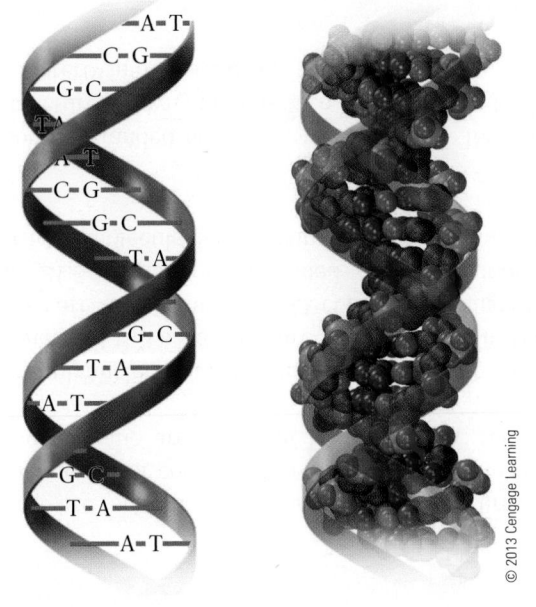

Figure 21.5.17 Representations of the DNA double helix

© 2013 Cengage Learning

Explore base pairing in DNA.

Thymine (T)

Adenine (A)

Cytosine (C)

sugar

Guanine (G)

Base pairing in DNA

The strands are held together by intermolecular forces (hydrogen bonds) between the base side groups. The principal guiding force in heredity is the specificity of these linkages. Each base has a specific "complementary base" to which it preferentially binds. Interactive Figure 21.5.18 shows these linkages.

Section 21.5 Mastery

Unit Recap

Key Concepts

21.1 Hydrocarbons

- Carbon and hydrogen form many different types of compounds, both saturated (alkanes and cycloalkanes) and unsaturated (alkenes, alkynes, and aromatics) (21.1a).

- Hydrocarbons are represented using space-filling models, ball-and-stick models, Lewis structures, and line structures (21.1a).

- Hydrocarbons are named using a systematic set of naming rules (21.1b, 21.1c).

- Alkanes and aromatics are generally less reactive than alkenes and alkynes (21.1d).

- Some common hydrocarbon reactions include substitution reactions and addition reactions (21.1d).

- Addition reactions can exhibit regioselectivity (21.1d).

21.2 Isomerism

- As the number of carbon atoms increases, the number of possible constitutional isomers for a given hydrocarbon increases quickly (21.2a).

- Organic compounds can exhibit both geometric and optical stereoisomerism (21.2b).

- Organic molecules that rotate plane-polarized light are chiral, have a chiral center, and have a nonsuperimposable mirror image (21.2b).

21.3 Functional Groups

- Organic compounds are characterized by the compound's functional groups (21.3a).

- Alcohols are named using a systematic set of naming rules and are characterized as primary, secondary, or tertiary (21.3b).

- Alcohols are generally synthesized by the hydration of an alkene or the reduction of an aldehyde or ketone (21.3b).

- Alcohols commonly undergo dehydration and oxidation reactions (21.3b).

- The dehydration of an alcohol can be regioselective (21.3b).

- Carbonyl compounds such as carboxylic acids also undergo numerous reactions, including nucleophilic substitution reactions (21.3c).

21.4 Synthetic Polymers

- Polymers are large molecules formed when many small molecules (monomers) link together (21.4).

- Addition polymers primarily form from alkene monomers in reactions that involve a free radical "initiator" (21.4a).

- Addition polymers such as polyethylene are further processed to form the items used in everyday life (21.4a).

- Condensation polymers form from monomers with at least two reactive groups such as alcohols, amines, and carboxylic acids (21.4b).

- Polymer properties are influenced by controlling the polymer chain length and the amount of branching and by cross-linking the polymer chains (21.4c).

21.5 Biopolymers

- Carbohydrates, also known as sugars, are chiral compounds composed primarily of carbon, hydrogen, and oxygen. They contain between three and seven carbon atoms and are polyhydroxy compounds that contain either an aldehyde or ketone functional group (21.5a).

- Carbohydrates exist in both linear and cyclic form (21.5a).

- Carbohydrates react in condensation reactions to form water and larger sugars, including disaccharides, trisaccharides, and polysaccharides, that are linked by glycosidic bonds (21.5a).

- The two most important polysaccharides are cellulose and starch (amylopectin and amylose) (21.5a).

- Amino acids are characterized by the side chain (the R group) attached to the molecule (21.5b).

- Amino acids act both as acids and bases and are typically found in their zwitterionic form (21.5b).

- Amino acids react to form dipeptides, tripeptides, and larger polypeptides that are linked by peptide bonds between the amino acids (21.5c).

- A protein is a polypeptide with a biological function and is characterized by its primary, secondary, tertiary, and quaternary structure (21.5c).

- DNA and RNA are polynucleic acids made up of nucleotide monomers (21.5d).

- DNA has a double-stranded helical structure, held together by hydrogen bonding between the complementary bases on the adjacent strands (21.5d).

Key Terms

21.1 Hydrocarbons
saturated hydrocarbon
unsaturated hydrocarbon
alkane
cycloalkane
unbranched hydrocarbon
branched hydrocarbon
alkene
alkyne
arene
aromatic hydrocarbon
substitution reaction
addition reaction
hydrogenation
regioselectivity
Markovnikov's rule

21.2 Isomerism
constitutional isomers
geometric isomers
optical isomers
chiral compound
chiral center

21.3 Functional Groups
functional group
amine
alcohol
Zaitsev's rule
ether
carbonyl group
carboxylic acid
ketone
aldehyde
ester

amide
nucleophile

21.4 Synthetic Polymers
polymer
monomer
addition polymer
condensation polymer
polyamide
polyester

21.5 Biopolymers
biopolymer
Fisher projections
peptide bond
dipeptide
polypeptide
protein
nucleic acid

Unit 21   Review and Challenge Problems

22 Applying Chemical Principles to the Main-Group Elements

Unit Outline

In This Unit...

The main-group elements, those in the *s* block and *p* block of the periodic table, vary greatly in their physical and chemical properties. They are metals, nonmetals, and metalloids and solids, liquids, and gases. This unit uses the chemical principles you have learned to describe some of the chemical and physical properties of these elements. We will use, for example, your knowledge of periodic trends, chemical structure, equilibria, and acid–base chemistry to understand the chemistry of some of the main-group elements.

22.1 Structures of the Elements

22.1a The Periodic Table

The periodic table is chemistry's organizational framework. Periodic trends in atomic size, orbital energy, and effective nuclear charge are important in understanding periodicity and the properties of the elements, especially the nonmetals. Recall from Electron Configurations and the Properties of Atoms (Unit 7) that atomic radii of elements generally decrease as you move from left to right across the periodic table and increase down a group (Figure 22.1.1).

The decrease in size across a group is related to the increase in effective nuclear charge and decrease in orbital energy (Figure 22.1.2). These differences result in significant differences in the properties of the elements and allow us to easily differentiate between metals and

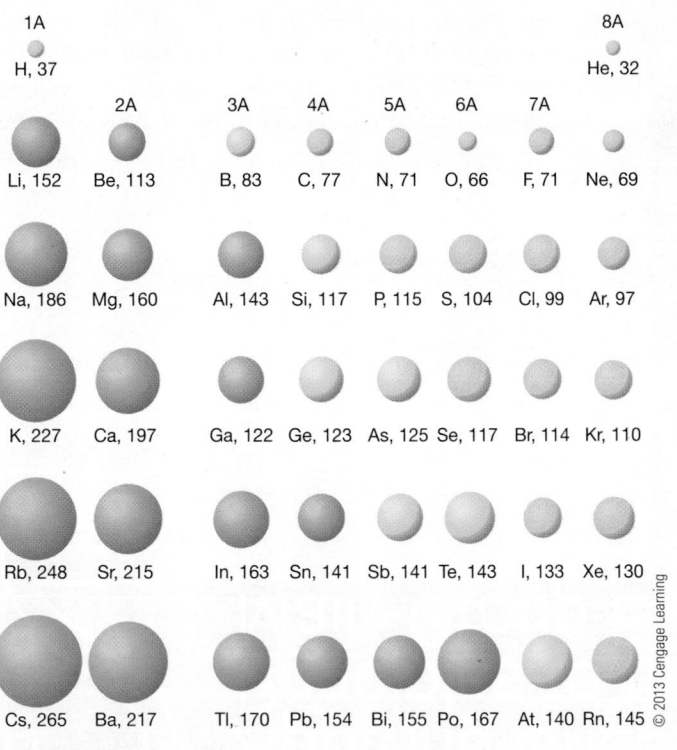

Figure 22.1.1 Atomic radii (pm) of the main-group elements

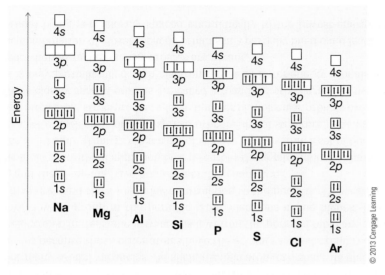

Figure 22.1.2 Orbital energies for elements in the third period

nonmetals. Main-group metals, located on the left and toward the bottom of the periodic table, have larger radii, low effective nuclear charge, and low first ionization energies and generally form cations when they react. Nonmetals, located on the right and toward the top of the periodic table, have smaller radii, large effective nuclear charge, and high first ionization energy and generally form anions when they react. Those along the diagonal line from B to Te display properties of both metals and nonmetals and are called metalloids (Interactive Figure 22.1.3).

22.1b Metals

Why are metals found to the left and toward the bottom of the periodic table? The answer lies in the properties of metals and how those properties come about based on orbital energies and the degree to which those orbitals are filled. Metals have important properties in common: their structure involves atoms packed closely together without discrete covalent bonds, they are good conductors of electricity, and they tend to lose electrons to form cations.

Bonding and Conduction in Metals

Elements on the left side of the periodic table have multiple vacancies in their valence orbitals, allowing metal atoms to form multiple bonding interactions with neighboring atoms. As

Investigate metals and nonmetals in the periodic table.

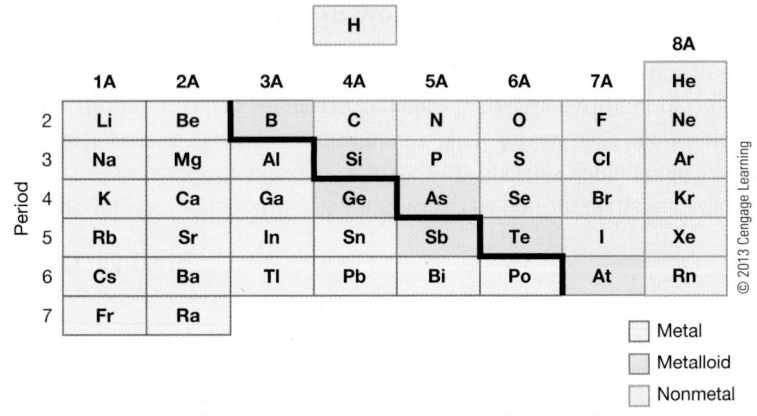

The main-group metals, nonmetals, and metalloids

a result, metals form closely packed structures, most often surrounded by 12 other like atoms. Although these interactions have some covalent character, the atoms in a metallic solid are not linked by traditional covalent bonds.

On the left side of the periodic table, atoms of a given element have few valence electrons but many orbital vacancies. As such, a single atom can share its electrons with many other atoms because those other atoms have the orbital vacancies required to accept those shared electrons. In the same way, atoms lower on the periodic table are more metallic. As you move down a group in the periodic table, electrons in the highest-filled orbitals are higher in energy and show much lower attraction to the nucleus. They can be shared more readily, and the elements have greater metallic character.

The molecular orbital approach best describes these interactions (Figure 22.1.4). Recall from The Solid State (Unit 12) that a set of valence orbitals from a set of bonded metal atoms leads to a set of molecular orbitals spread over the metal structure. When a sample of metal has many atoms, the structure has a large number of molecular orbitals over a range of energies. These closely spaced orbitals are called a "band" of orbitals, and in metals, the band of orbitals is only partially filled with electrons. That is, lower-energy orbitals in the band are occupied, whereas higher-energy orbitals are vacant.

The combination of closely spaced orbitals in a partially filled band leads to the highest-filled orbital being very closely spaced energetically to the lowest-energy vacant orbital.

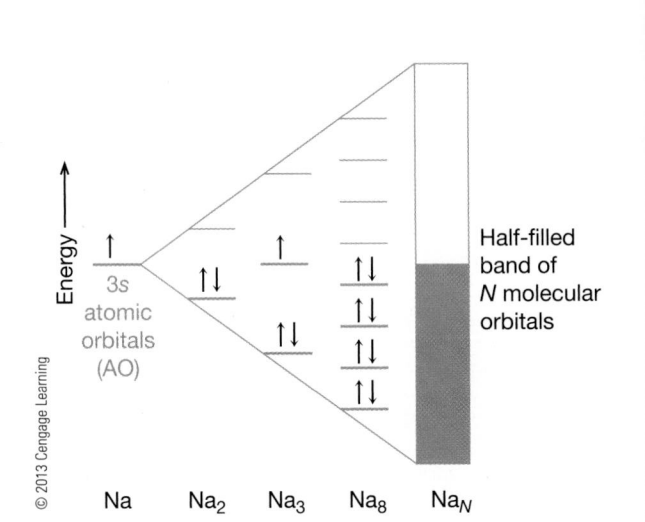

Figure 22.1.4 Formation of bands of molecular orbitals in a solid

This in turn means that electrons need gain only a very small amount of energy to move from one orbital to another. This small energy barrier to electron movement is the reason metals are good conductors of electricity.

Formation of Cations

Although metal atoms have multiple valence orbital vacancies, they tend to lose electrons in their interactions with other atoms, forming cations. This is due to orbital energy trends across each period in the periodic table. As you move across a period from left to right, the nuclear charge of the elements increases. As electrons are added to the valence shell, they repel one another, increasing shielding. The effective nuclear charge, the sum effects of nuclear positive charge attracting electrons and shielding electrons repelling valence electrons, increases as you move left to right because each added electron does not effectively shield the full +1 charge of the added proton. As the effective nuclear charge increases, the attraction of the outermost electrons to the nucleus increases and the energy of the valence orbital decreases. Therefore, the highest-energy valence orbitals are found toward the left side of each period. Electrons found in these orbitals are most easily lost, which is why metals tend to form cations (Interactive Figure 22.1.5).

22.1c Nonmetals

Unlike metals, elemental forms of the nonmetals consist of either monoatomic species (the noble gases) or atoms covalently bonded to one another. Some of the main-group elements form structures containing multiply bonded atoms, whereas others have only single bonds between atoms. Also, some elements form small, discrete molecules, and others form extended structures.

Two factors that control the molecular structure of the elements are the number of orbital vacancies and the ability to form pi bonds with neighboring atoms. Interactive Table 22.1.1 and Figure 22.1.6 show the number of orbital vacancies, the number of bonds needed to form to fill each atom's octet, for the main-group nonmetals.

The number of possible bonds an atom of a given element can form plays an important role in determining the element's structure. Starting with the Group 8A noble gases, we see that these elements do not form molecules in their elemental state; they exist as monatomic species because they have no valence orbital vacancies. The Group 7A elements, the halogens, have a single orbital vacancy and thus need only one bond to complete the octet. The elemental form of all halogens is diatomic molecules with a single X—X bond.

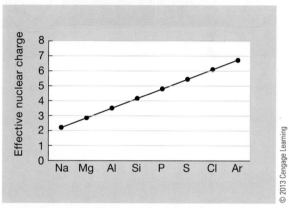

Interactive Figure 22.1.5

Explore the relationship between effective nuclear charge and the formation of ions.

Sodium has a low effective nuclear charge and therefore high orbital energies. Argon has a high effective nuclear charge and therefore low orbital energies.

Interactive Table 22.1.1

Orbital Vacancies for the Main-Group Nonmetals

Group Number	4A	5A	6A	7A	8A
Number of Orbital Vacancies	4	3	2	1	0
Number of Bonds Needed to Complete Octet	4	3	2	1	0

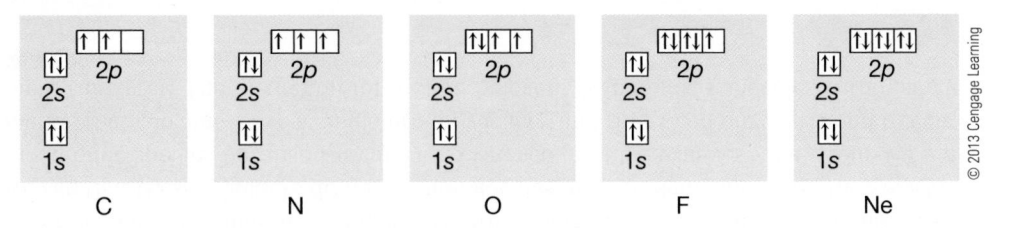

Figure 22.1.6 Electron configurations for the second-row main-group nonmetals

Element structure becomes more complex as we move to Groups 4–6A. For example, oxygen has two allotropes, O_2 and O_3, each of which consists of small molecules containing O=O double bonds. Sulfur, on the other hand, has multiple allotropes based on rings of singly bonded S atoms and a polymeric form containing long chains of singly bonded S atoms. Why does O form small molecules when S forms larger extended structures? The answer lies in the size of the atoms and their ability to form pi bonds.

Atomic Radius and Bonding in Nonmetals

Bonds form when orbitals on adjacent atoms overlap. Sigma bonds can form between atoms of any size because the bonds are the result of orbital overlap between the two atoms, along the internuclear axis. Pi bonds, on the other hand, are the result of side-to-side overlap of p or d orbitals. If the two atoms forming the pi bond are large, orbital overlap is not significant and pi bonds do not form.

Consider the Group 6A elements oxygen and sulfur. Recall that atomic radii increase moving down a group. Sulfur atoms are larger than oxygen atoms, and as a result, pi bonds form between oxygen atoms in elemental oxygen (O_2 and O_3) but not between sulfur atoms in elemental sulfur (Figure 22.1.7).

In terms of thermodynamics, we can consider the effect of both enthalpy (bond energies) and entropy (disorder) on the structure of an element. Entropy will favor the formation of a large number of molecules; thus the formation of many smaller molecules of an element is entropy favored. Enthalpy favors molecules with stronger bonds. In cases where relatively strong pi bonds can form, which results in smaller molecules, entropy "wins" and small, multiply bonded molecules are formed. When pi bonds are weak, enthalpy "wins" and extended structures containing only sigma bonds are formed. For oxygen and sulfur, the fact that oxygen forms O—O pi bonds results in small molecules. Sulfur cannot easily form strong pi bonds with other sulfur atoms due to its larger atomic

Figure 22.1.7 Elemental forms of sulfur: S_8 and polymeric sulfur

radius, so in its elemental state, it exists as larger extended structures with only S—S single bonds.

A similar situation exists in the elemental forms of the Group 5A elements. Nitrogen is small, can form N—N pi bonds, and thus exists as small triply bonded N_2 molecules. Phosphorus has a larger atomic radius, cannot easily form strong P—P pi bonds, and thus forms only P—P single bonds in its elemental structures. Phosphorus exists as multiple allotropes, the three most common of which are white (P_4 molecules), red (polymeric chains of P_4 monomers), and black (complex cross-linked polymeric structure) (Figure 22.1.8).

Finally, the Group 4A elements continue the relationship between elemental form and atomic radius. Carbon has a smaller atomic radius than silicon and can form C—C pi bonds. Elemental forms of carbon include graphite (which contains both C—C sigma and pi bonds), diamond (C—C sigma bonds only), and fullerenes such as buckyball (both C—C sigma and pi bonds) (Figure 22.1.9). Elemental silicon is a network solid similar to diamond that consists of a three-dimensional array of silicon atoms connected by Si—Si sigma bonds.

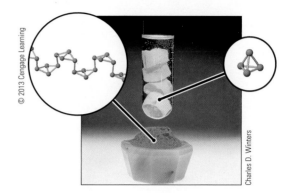

Figure 22.1.8 Two allotropes of phosphorus (white and red)

Section 22.1 Mastery

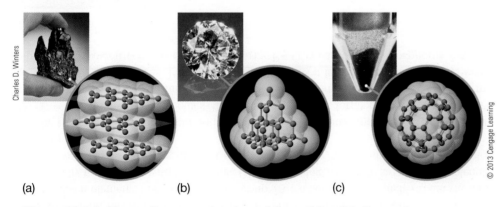

(a)   (b)   (c)

Figure 22.1.9 Three allotropes of carbon: (a) graphite, (b) diamond, and (c) buckyball

22.2 Oxides and Halides of the Nonmetals

22.2a Nonmetal Oxides

The oxides of the nonmetals (Interactive Table 22.2.1) follow structural trends similar to those found in the elements themselves. As we saw in the structures of the elements, atomic radius and orbital vacancy both have a significant effect on the types of bonds formed by atoms of a given element. For example, the Group 7A elements have a single orbital vacancy and generally form oxides with the formula OX_2 (X = halogen). These are all small molecule oxides, favored both in terms of bond energies (enthalpy) and multiple small molecules (entropy).

The Group 6A elements sulfur and selenium both form small molecule oxides, EO_2 and EO_3 (E = S, Se). In these molecules, the Group 6A element forms pi bonds to the smaller oxygen atoms. In Group 5A, nitrogen forms both sigma and pi bonds to O and thus forms many different small oxides. Phosphorus, however, has a larger radius and thus its compounds with oxygen tend to contain primarily P—O single bonds. Phosphorus forms a

Interactive Table 22.2.1

Selected Oxides of the Nonmetals

	Group 4A	Group 5A	Group 6A	Group 7A	Group 8A
Period 2	O=C=O	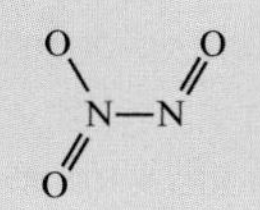		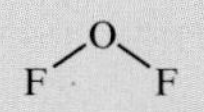	
Period 3	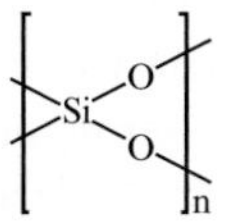	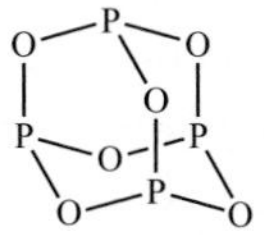	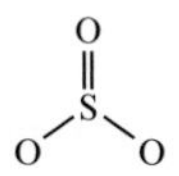	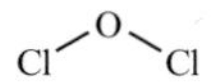	
Period 4			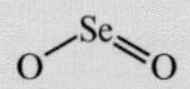	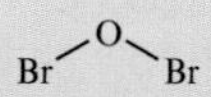	
Period 5					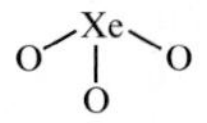

© 2013 Cengage Learning

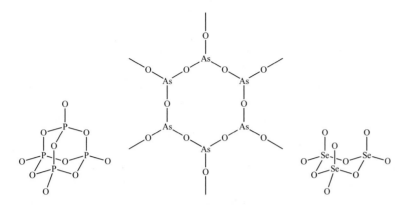

Figure 22.2.1 Structures of some nonmetal oxides containing larger nonmetals

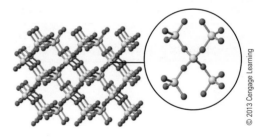

Figure 22.2.2 The structure of SiO_2

series of oxides that contain bridging oxygen atoms in P—O—P groups and terminal atoms in P—O bonds that have some double-bond character. Similar oxide structures are found for the larger nonmetals arsenic and selenium (Figure 22.2.1).

Finally, the Group 4A nonmetals carbon and silicon also follow this trend in oxide structure based on atomic radius. Carbon forms two molecular oxides, CO and CO_2, both of which contain C—O pi bonds and are entropy favored due to their small size. Silicon, however, forms an oxide containing only Si—O single bonds. Silicon dioxide, SiO_2, has a formula analogous to carbon dioxide, but it does not consist of individual SiO_2 molecules. Silicon dioxide (silica) is a covalently bonded network solid consisting of interlocking SiO_4 tetrahedra. The larger silicon atoms cannot effectively form pi bonds with oxygen (Figure 22.2.2).

Acid–Base Properties of Oxides

Metal oxides are generally **basic oxides** (they react with water to form bases or act as bases in chemical reactions), and nonmetal oxides are generally **acidic oxides** (they react with water to form acids or act as acids in chemical reactions). Again, we can explain these experimental observations using orbital energies.

For example, consider what happens when magnesium oxide reacts with water.

$$MgO(s) + H_2O(\ell) \rightarrow Mg(OH)_2(s) \rightleftharpoons Mg^{2+}(aq) + 2\,OH^-(aq)$$

Main-group metals have high-energy valence orbitals, so reaction with water leads to an ionic hydroxide salt (a base).

Nonmetal oxides react with water to form acids. For example, consider the reaction of carbon dioxide with water to form carbonic acid.

$$CO_2(g) + H_2O(\ell) \rightleftarrows H_2CO_3(aq)$$

$$H_2CO_3(aq) + H_2O(\ell) \rightleftarrows HCO_3^-(aq) + H_3O^+(aq)$$

There are two reasons this reaction results in the formation of carbonic acid. First, carbon is a nonmetal with orbital vacancies that are relatively low in energy. Oxygen is very electronegative and electron withdrawing, which pulls electron density away from the carbon, making it open to attack by the electronegative oxygen in the water molecule. Finally, in carbonic acid, the electronegative oxygen atoms pull electron density away from hydrogen, making it easy to remove (as the hydronium ion, H_3O^+).

22.2b Nonmetal Halides

All main-group elements form compounds with halogens, including the halogens themselves. (Compounds containing only halogens are called *interhalogen* compounds.) The stoichiometry of nonmetal halide compounds is related to orbital vacancies and the maximum oxidation state of the nonmetal combined with the halogen.

As shown in Table 22.2.2, the second-row nonmetals in Groups 4–6A form halide compounds with stoichiometries based on orbital vacancies. These compounds contain only sigma bonds to halogen atoms, and the structures of the compounds follow VSEPR theory. Larger elements, those in periods 3 and higher, can have expanded valence structures. For example, Table 22.2.3 shows some of the halides of the third-row nonmetals. The compounds formed are those expected based on orbital valencies, except that phosphorus, sulfur, and chlorine can also have an expanded valence. For example, phosphorus forms both PX_3 and PX_5 compounds with fluorine, chlorine, and bromine, and the PX_3 compound with iodine. In addition, compounds with the formula P_2X_4 are also known for X = F, Cl, Br, and I. Sulfur combines with fluorine to make SF_2, SF_4, SF_6, and S_2F_{10}. However, with the exception of SCl_2, it does not form analogous compounds with Cl, Br, or I (probably due to the weaker S—X bonds).

Recall that the tendency to form expanded valence structures begins in the third period of the periodic table because these elements can make use of vacant d orbitals to form bonds. The tendency increases as you move left to right in the periodic table because the energy of orbitals decreases, making those vacant orbitals more attractive to bonding electrons. Thus, phosphorus forms many compounds with an expanded valence, but Si does not (although it does form the ionic species SiF_5^- and SiF_6^{2-}).

Table 22.2.2 Halides of the Second-Row Group 4–6A Nonmetals

4A	5A	6A
CX_4 and other extended halogenated organics (e.g., C_3F_8)	NF_3, NCl_3, NBr_3, NI_3	OF_2, OCl_2, OBr_2

Table 22.2.3 Selected Halides of the Third-Row Group 4–7A Nonmetals

4A	5A	6A	7A
SiF_4	PF_3, PF_5	SF_2	ClF, BrCl, ICl
$SiCl_4$	PCl_3, PCl_5	SF_4, SF_6	ClF_3, ClF_5
$SiBr_4$	PBr_3, PBr_5	S_2F_{10}	
SiI_4	PI_3		

Melting Points of Selected Main-Group Halides

	1A	2A	3A	4A	5A	6A	7A
Compound	NaF	MgF_2	AlF_3	SiF_4	PF_3	SF_4	ClF
Melting point, °C	993	1263	1291	−90	−151	−121	−155

If we expand our view out further to the left in the periodic table, we see an interesting effect. Interactive Table 22.2.4 shows the melting points of the most common fluoride compound for the third-row main-group elements. Notice the large break between Group 3A and Group 4A when we move from Al to Si. This sharp change in melting point is due to the change from ionic compounds, which have strong ion–ion attractions and high melting points, to molecular covalent compounds, which have only relatively weak intermolecular forces. The compounds NaF, MgF_2, and AlF_3 are ionic, whereas those to the right in the table are covalent (molecular).

Section 22.2 Mastery

22.3 Compounds of Boron and Carbon

22.3a Boron Compounds

Boron has the electron configuration $[He]2s^22p^1$; it has five valence orbital vacancies. However, this does not translate into compounds with five bonds to boron. Instead, boron tends to form compounds with the general formula BY_3, where the boron atom is bonded to three other atoms or groups. For example, in BH_3, the central B atom has six valence electrons and therefore two vacancies. Because the boron is *electron deficient* in this compound, BH_3 and compounds like it can act as Lewis acids, accepting a pair of electrons from a Lewis base such as NH_3 (Figure 22.3.1).

Other compounds of boron show similar reactivity. For example, consider boric acid, $B(OH)_3$. The chemical formula of boric acid is similar to the ionic hydroxide salts of the Group 1A and 2A metals such as NaOH and $Mg(OH)_2$. These ionic compounds are basic; they dissociate in water to produce metal ions and hydroxide ions.

$$NaOH(s) \rightarrow Na^+(aq) + OH^-(aq)$$

$$Mg(OH)_2(s) \rightleftarrows Mg^{2+}(aq) + 2\ OH^-(aq)$$

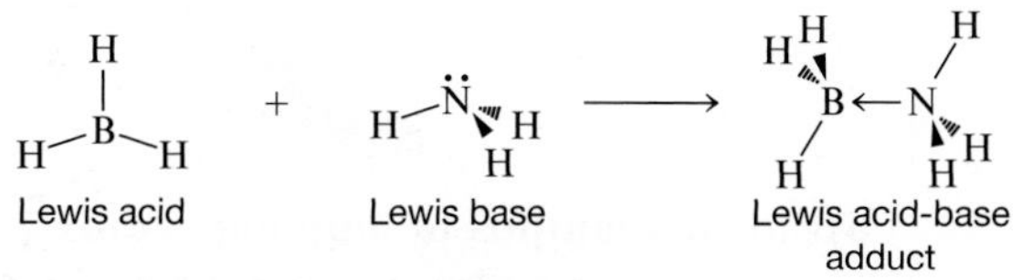

Figure 22.3.1 Reaction between BH_3 and NH_3

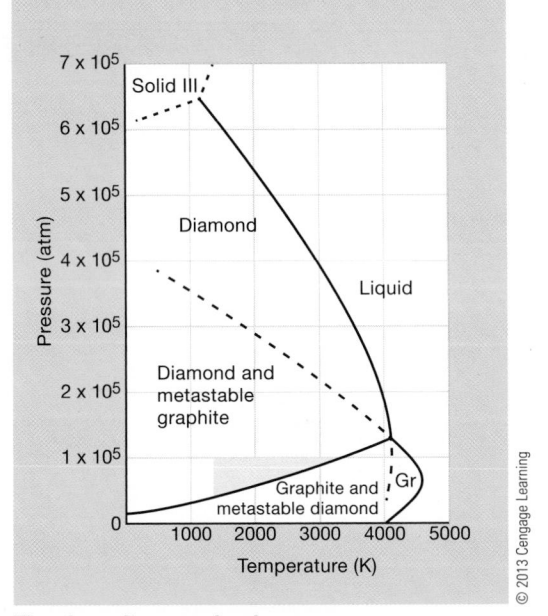

$$K_a = 7.3 \times 10^{-10}$$

Figure 22.3.2 Boric acid is a Lewis acid in water.

Boric acid, as its name implies, is an acid, not a base. It acts as a Lewis acid in water, forming borate and hydronium ions (Figure 22.3.2). Why does $B(OH)_3$ act differently than metal hydroxides? The answer is once again orbital energies. Recall that as you move from left to right across the periodic table, orbital energies decrease due to increasing effective nuclear charge. Group 1A elements have valence s orbitals of very high energy that can easily lose an electron to form a cation. Elements further to the right have lower orbital energies and decreased ability to form a cation. Boron has relatively low valence orbital energies and, instead of losing electrons, acts as an electron-pair acceptor.

Sodium Perborate

One of the most commercially important boron compounds is sodium perborate, $NaBO_3 \cdot 4H_2O$, which is the active ingredient in color-safe bleach. The structure of the perborate ion (Interactive Figure 22.3.3) shows that it is actually a dimer with peroxide groups linking the two boron atoms.

The active ingredient in bleach, sodium hypochlorite (NaClO), is added to laundry to kill bacteria and destroy dyes, chemical compounds that are used to color fabrics. Color-safe bleach contains the perborate ion because at high temperatures the ion decomposes to produce, among other products, hydrogen peroxide. Hydrogen peroxide can remove stains but, at the concentrations found in a typical wash containing color-safe bleach, it will not destroy dye molecules. Color-safe bleaches also contain brightening agents that adhere to fabrics and make them look brighter.

22.3b Elemental Carbon

Interactive Figure 22.3.4 shows a phase diagram for carbon under a wide range of temperatures and pressures. At room temperature and 1 bar, graphite is more stable than diamond.

$$C(s, \text{diamond}) \rightarrow C(s, \text{graphite}) \quad \Delta G^\circ = -2.9 \text{ kJ/mol}$$

Interactive Figure 22.3.3

Explore bonding in sodium perborate.

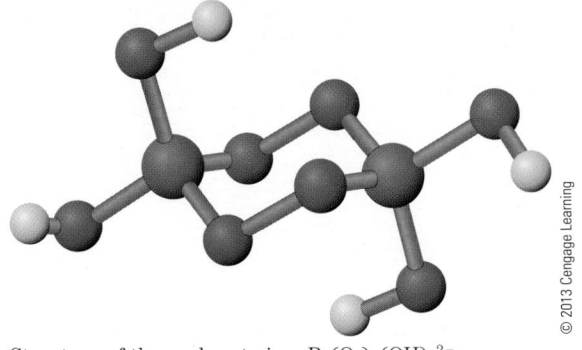

Structure of the perborate ion, $B_2(O_2)_2(OH)_4^{2-}$ (boron = purple, oxygen = red)

Interactive Figure 22.3.4

Explore stability and metastability.

The phase diagram of carbon

Under typical conditions, however, we know diamond does not convert into graphite. If it did, expensive jewels would be converted into pencil lead. The process does not occur because diamond is kinetically stable. That is, even though carbon is more stable as graphite, the activation energy for the conversion is so large that the reaction just does not occur. Species such as diamond are called metastable, meaning they are thermodynamically unstable but highly kinetically stable.

Synthetic diamonds can be made from graphite, and a large industry exists that is dedicated to this process. Two common methods are used to convert the more stable form of carbon, graphite, into diamond. One method puts carbon under conditions where diamond is more stable than graphite (high pressure, high temperature). The problem with this idea is that even under fairly extreme conditions where diamond is more stable than graphite, the graphite is metastable and reacts slowly, if at all. The key is to use a catalyst, commonly a metal such as chromium, iron, or nickel. The graphite dissolves into the molten metal, and experimental conditions are applied that favor the formation of diamond.

$$C(s, graphite) \rightleftarrows C(s, dissolved\ in\ molten\ metal) \rightleftarrows C(s, diamond)$$

A second method for synthesizing diamonds from graphite (or any carbon-containing material) is chemical vapor deposition, where a carbon-containing material is heated to about 1000 °C at relatively low pressures (~50 mbar). Excess hydrogen atoms are used to dilute the carbon and prevent the formation of graphite, and the gas-phase material condenses into a thin film of diamond on a surface.

22.3c Cave Chemistry

The chemistry of carbon compounds (specifically carbon–hydrogen compounds) was discussed in Organic Chemistry (Unit 21). Carbon also has interesting inorganic chemistry, particularly the chemistry of carbon–oxygen compounds. Carbonate chemistry, and particularly the chemistry involving carbonic acid in aqueous solution, is very important in biology, environmental chemistry, and the chemical industry. Here, we explore the carbonate equilibria that exist in caves such as the one shown in Interactive Figure 22.3.5.

Caves are formed when large bodies of limestone rock ($CaCO_3$) are dissolved. Calcium carbonate is an insoluble inorganic compound with a small K_{sp}.

$$CaCO_3(s) \rightleftarrows Ca^{2+}(aq) + CO_3^{2-}(aq) \qquad K_{sp} = 3.8 \times 10^{-9}$$

$$solubility = \sqrt{3.8 \times 10^{-9}} = 6.2 \times 10^{-5}\ mol/L = 0.0062\ g/L$$

However, the solubility of calcium carbonate in natural conditions is about 80 times greater than its solubility in pure water. We can explain this by exploring the effect of carbon dioxide on carbonate compound solubility.

Carbon dioxide reacts with water to form carbonic acid.

$$CO_2(g) + H_2O(\ell) \rightleftarrows H_2CO_3(aq)$$

When calcium carbonate is exposed to water with a high carbonic acid concentration, it dissolves.

$$H_2CO_3(aq) + CaCO_3(s) \rightleftarrows Ca^{2+}(aq) + 2\ HCO_3^-(aq)$$

The equilibrium constant for this reaction can be determined by adding reactions for which we know the equilibrium constants:

$CaCO_3(s) \rightleftarrows Ca^{2+}(aq) + CO_3^{2-}(aq)$	$K_{sp} = 3.8 \times 10^{-9}$
$CO_3^{2-}(aq) + H_2O(\ell) \rightleftarrows HCO_3^-(aq) + OH^-(aq)$	$K_b(CO_3^{2-}) = 2.1 \times 10^{-4}$
$H_2CO_3(aq) + H_2O(\ell) \rightleftarrows HCO_3^-(aq) + H_3O^+(aq)$	$K_a(H_2CO_3) = 4.2 \times 10^{-7}$
$H_3O^+(aq) + OH^-(aq) \rightleftarrows 2\ H_2O(\ell)$	$1/K_w = 1.0 \times 10^{14}$

$$H_2CO_3(aq) + CaCO_3(s) \rightleftarrows Ca^{2+}(aq) + 2\ HCO_3^-(aq) \qquad K = K_{sp} \times K_a \times K_b \times 1/K_w = 3.4 \times 10^{-5}$$

Although the equilibrium constant for the net reaction is not large (not greater than 1), it is about 10^4 times greater than K_{sp}. The equilibrium is pushed further to the right by large concentrations of carbonic acid (and other organic acids) in soil.

Cave formations such as stalactites and stalagmites result from the reverse of the dissolution of calcium carbonate in the presence of carbonic acid. Consider a drop of water, saturated with $CaCO_3$, hanging inside a cave. The precipitation of $CaCO_3$ is not due to water evaporation; the air inside most caves is saturated with water vapor, and evaporation happens little, if at all. Instead, deposition of $CaCO_3$ is again dependent on carbonate equilibria involving CO_2.

When a cave is forming, the partial pressure of CO_2 in the cave air is very high, much higher than in the outside air, where CO_2 pressure is nearly zero. The $CaCO_3$ is much more soluble in that acidic environment. When the cave is open to the outside air, the CO_2 pressure in the cave air drops to near zero. When this happens, the amount of H_2CO_3 dissolved in the cave water also decreases, as does the solubility of $CaCO_3$. Cave formations occur due to degassing of CO_2 out of droplets saturated with $CaCO_3$.

22.3d Carbon Dioxide and Global Warming

The greenhouse effect is caused by gases in Earth's atmosphere that trap energy that would otherwise escape Earth. All objects give off blackbody radiation, the wavelength of which depends on the temperature of the body. The sun gives off UV and visible radiation; Earth, being cooler, gives off infrared (IR) radiation (Figure 22.3.6).

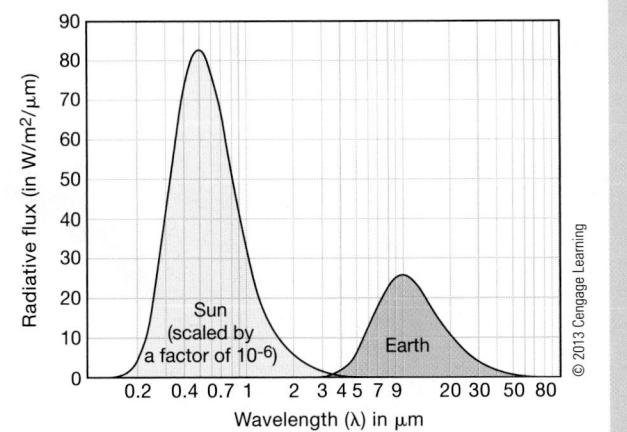

Figure 22.3.6 Blackbody radiation curves

Our planet is warmed by the radiation from the sun. Our atmosphere contains greenhouse gases, gases that can absorb IR radiation given off by Earth and send it back to Earth. This greenhouse effect warms the surface of our planet. Without greenhouse gases, the average temperature of Earth would be much lower than the actual average temperature.

We can quantify the impact of the greenhouse effect on the temperature of a planet by calculating the temperature of a planet such as Earth in the absence of greenhouse gases (Equation 22.1).

$$T_{\rm E} = \sqrt[4]{\frac{E_{\rm s}(1 - \alpha)}{4\sigma}} \qquad (22.1)$$

In Equation 22.1, $T_{\rm E}$ is the equilibrium average surface temperature, $E_{\rm s}$ is the solar constant (the amount of solar light that reaches the planet), α is the fraction of sunlight reaching the planet that is reflected before reaching the planet surface, and σ is a constant (5.67×10^{-8} W/m^{2}K^4).

For Earth, $E_{\rm s} = 1380$ W/m^2 and $\alpha = 0.28$. Solving for $T_{\rm E}$,

$$T_{\rm E} = \sqrt[4]{\frac{(1380 \text{ W/m}^2)(1 - 0.28)}{4(5.67 \times 10^{-8} \text{ W/m}^2\text{K}^4)}} = 257 \text{ K} \, (-16° \text{ C})$$

Earth's average temperature is about 15 °C, approximately 31 degrees warmer than what the average temperature would be without the greenhouse effect.

We can also calculate the fraction of radiation that leaves a planet but is trapped by greenhouse gases and reflected back to the planet surface. This quantity, called the IR transmission factor (f), is a measure of the fraction of IR light that escapes the planet's atmosphere (Equation 22.2).

$$f = \frac{\dfrac{E_{\rm s}}{4}(1 - \alpha)}{\sigma(T_{\rm E})^4} \qquad (22.2)$$

To calculate this quantity for our planet, we use the actual measured temperature for $T_{\rm E}$ (288 K).

$$f = \frac{\dfrac{1380 \text{ W/m}^2}{4}(1 - 0.28)}{(5.67 \times 10^{-8} \text{ W/m}^2\text{K}^4)(288 \text{ K})^4} = 0.64$$

This means that about 36% of the IR radiation leaving Earth's surface is trapped by greenhouse gases and radiated back to Earth.

Atmospheric compounds that are "greenhouse gases" absorb IR radiation when they can excite a vibration in the molecule that leads to a change in the molecule's dipole moment, changing its polarity. The three most abundant gases in the atmosphere, N_2, O_2, and Ar, are nonpolar and do not absorb IR light; they are not greenhouse gases. The two most important atmospheric gases that act as greenhouse gases are water, H_2O, and carbon dioxide, CO_2. Although CO_2 is linear and nonpolar, it has vibrational modes that cause it to become polar. The three vibrational modes of CO_2 are shown in Interactive Figure 22.3.7. Because two of these modes (bending and asymmetric stretching) result in a change in dipole moment, CO_2 absorbs IR radiation and is a greenhouse gas.

Section 22.3 Mastery

22.4 Silicon

22.4a Silicon Semiconductors

Silicon is the basis of the huge semiconductor industry. Interestingly, in a typical room, the Si in a silicon computer chip is the purest substance present, but it functions as a semiconductor precisely because it is impure. The key to the function of a silicon semiconductor is that the impurities are very specific and have very low concentrations.

Semiconductor chips are formed by first purifying Si to a very high purity of 99.9999%. The first step is to obtain $SiCl_4$ from relatively pure silicon.

$$Si(s) + 2 Cl_2(g) \rightarrow SiCl_4(\ell)$$

The reason for forming $SiCl_4$ is that it has a relatively low boiling point and can be purified by distillation to a purity of greater than 99%.

Next, the $SiCl_4$ is reduced with zinc.

$$SiCl_4(\ell) + 2 Zn(s) \rightarrow Si(s) + 2 ZnCl_2(s)$$

Finally, the Si is cast into a cylindrical shape and highly purified by zone refining (Interactive Figure 22.4.1). The zone refining process is based on the same principle that causes the solvent in a solution to solidify in its pure form as temperature is lowered. When a solution freezes, the solvent forms a solid that does not contain any particles of any solute. Because of this, the solute (in this case the impurities in the Si) stay in solution. Zone refining works by using radio waves to heat a thin layer of Si in the cylinder until it melts. The region of melted Si is slowly moved down the cylinder. As this happens, each area of the Si cylinder melts and then, as the liquid layer passes, solidifies. The key is that as the liquid layer passes, it takes the impurities away with it. The impurities dissolve in the solution when the solid melts but stay in solution when it refreezes.

Interactive Figure 22.3.7

Explore molecular vibrations.

$\longleftarrow O{=}C{=}O \longrightarrow$ Symmetric stretching

Bending

$\longleftarrow O{=}C{=}O \longleftarrow$ Asymmetric stretching (one bond stretches while the other compresses)

© 2013 Cengage Learning

Vibrational modes for CO_2

Interactive Figure 22.4.1

Explore the zone refining process.

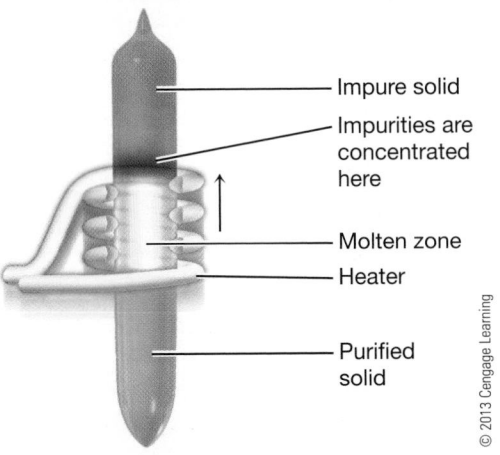

— Impure solid
— Impurities are concentrated here
— Molten zone
— Heater
— Purified solid

© 2013 Cengage Learning

Diagram showing the zone refining process for silicon

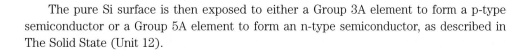

The pure Si surface is then exposed to either a Group 3A element to form a p-type semiconductor or a Group 5A element to form an n-type semiconductor, as described in The Solid State (Unit 12).

22.4b Silicates

Most of the minerals in Earth's crust are silicates, silicon compounds that contain extensive networks of silicon–oxygen bonds. These compounds form the basis of all rocks and rock chemistry. Unlike carbon, whose chemistry is dominated by compounds containing chains and rings of carbon atoms, the chemistry of silicon is dominated by Si—O bonds. The relatively weak Si—Si bond energy (Si—Si, 222 kJ/mol; Si—O, 452 kJ/mol) means that the same types of structures found in carbon compounds are just not found in stable silicon compounds.

The structures of silicates follow the following rules:

1. Every Si is bonded to four O atoms.

2. Oxygen atoms can either be bonded to a single Si atom (terminal O) or bridge between two Si atoms (bridging O).

3. The ratio of O atoms to Si atoms in a structure controls how branched and interlinked the structure is.

4. The charge on a silicon–oxygen unit is determined using oxidation numbers of +4 for Si and −2 for O.

For example, consider Mg_2SiO_4. The silicon oxygen unit is the silicate ion, SiO_4^{4-}. The charge on the ion is the sum of the silicon and oxygen oxidation states [Si + 4 × (O) = (+4) + 4 × (−2) = −4]; the structure of the ion is tetrahedral, and all oxygen atoms are terminal (Figure 22.4.2).

Figure 22.4.2 The SiO_4^{4-} ion

Now consider the $Si_2O_7^{6-}$ ion, which consists of two [SiO_4] tetrahedra sharing a single (bridging) oxygen. The charge on the ion is the sum of the silicon and oxygen oxidation states [2 × (Si) + 7 × (O) = 2 × (+4) + 7 × (−2) = −6]; the geometry is tetrahedral around each silicon atom, and all but one oxygen atom are terminal (Figure 22.4.3).

As the O:Si ratio decreases, more O atoms bridge between Si atoms in order to maintain four oxygen atoms around each silicon atom. Interactive Table 22.4.1 shows the types of formulas and structures of the various silicates as a function of O:Si ratio.

As the ratio decreases from 4 O:1 Si to 2 O:1 Si, the silicate structures change from monomeric [SiO_4] units to dimers, to chains and rings, to double chains, to two-dimensional sheets, and finally to three-dimensional networks.

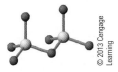

Figure 22.4.3 The $Si_2O_7^{6-}$ ion

Silicate Structures

Structure	Corners Shared at Each Si Atom	Repeating Unit	Si:O Ratio	Example
Tetrahedra	0	SiO_4^{4-}	1:4	Olivines
Pairs of tetrahedra	1	$Si_2O_7^{6-}$	1:3.5	Thortveitite
Closed rings	2	SiO_3^{2-}	1:3	Beryl
Infinite single chains	2	SiO_3^{2-}	1:3	Pyroxenes
Infinite double chains	2.5	$Si_4O_{11}^{6-}$	1:2.75	Amphiboles
Infinite sheets	3	$Si_2O_5^{2-}$	1:2.5	Talc
Infinite network	4	SiO_2	1:2	Quartz

Because of the wide variation in structure and arrangement of the [SiO_4] units, silicates have a wide range of physical and chemical properties. For example, the single- and double-chain structures result in fibrous minerals such as asbestos (Figure 22.4.4).

Because all silicates (with the exception of SiO_2) are anions, they form ionic compounds with various metal ions. Some of these, those with relatively high O:Si ratios, are traditional ionic compounds with relatively small silicate anions whose negative charge is balanced by the positive charges of the cations. Those with lower O:Si ratios have extended, negatively charged Si—O chains or networks, along with accompanying cations. These compounds tend to have layered structures such as those found in clays such as kaolinite and montmorillonite (Figure 22.4.5).

Silica, the neutral silicon–oxygen compound with the empirical formula SiO_2, is a network solid consisting of a three-dimensional network of interlocking [SiO_4] tetrahedra (Figure 22.4.6).

22.4c Silicones

Most Americans first encounter silicones when they play with Silly Putty as a child. Silly Putty, a compound made by mistake during research into synthetic rubber during World War II, has as its main ingredient a silicone polymer. Modern uses of silicones include

Figure 22.4.4 A sample of the mineral asbestos

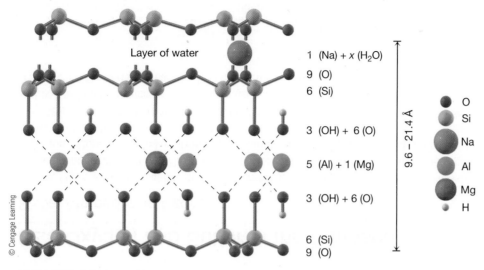

Layer of water

1	(Na) + x (H$_2$O)
9	(O)
6	(Si)
3	(OH) + 6 (O)
5	(Al) + 1 (Mg)
3	(OH) + 6 (O)
6	(Si)
9	(O)

9.6 – 21.4 Å

- O
- Si
- Na
- Al
- Mg
- H

Figure 22.4.5 A small portion of the montmorillonite structure

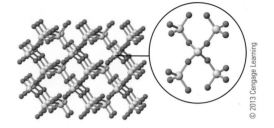

Figure 22.4.6 A small portion of the silica (SiO$_2$) structure

lubricants, adhesives, insulation, and medicine (implants); their wide use is related to the desirable properties of silicones such as chemical inertness, lack of reactivity toward UV light, high gas permeability, and low toxicity.

Silicones combine the high stability of silicates with the hydrophobic nature of hydrocarbons. Silicones are small molecules or polymers with a silicon–oxygen backbone and carbon-containing side groups. The simplest silicone contains only two silicon atoms.

$$
\begin{array}{ccc}
& CH_3 & CH_3 \\
& | & | \\
H_3C-&Si-O-Si&-CH_3 \\
& | & | \\
& CH_3 & CH_3
\end{array}
$$

Notice that each silicon atom is bonded to both carbon and oxygen. Larger silicones exist, as do silicone polymers (polysilicones).

$$
\left[
\begin{array}{c}
R \\
| \\
Si-O \\
| \\
R
\end{array}
\right]_n
$$

Silicones are synthesized from compounds containing relatively reactive Si—Cl bonds.

$$2\ (CH_3)_3SiCl + H_2O \rightarrow (CH_3)_3Si—O—Si(CH_3)_3 + 2\ HCl$$

As shown in Interactive Table 22.4.2, three types of reagents are used when forming silicones.

Unbranched silicones are formed by the reaction between a capping reagent such as $(CH_3)_3SiCl$ and a chain-extending reagent such as $(CH_3)_2SiCl_2$. As the reaction progresses, each Cl is replaced by a bridging oxygen atom. Therefore, each extending unit lengthens the growing chain by a $(CH_3)_2Si—O—$ unit. Each time a capping unit reacts, that end of the chain loses any active Si sites and growth in that direction is halted. The length of the chain can be controlled by the ratio of capping to extending reagents used. When [extending] >> [capping], the structures will be relatively long because many extending reactions will occur before the less concentrated capping unit halts growth. When [extending] is not much greater than [capping], the chain lengths will be shorter.

Likewise, the use of branching reagents leads to branches and cross-links that form in the growing structure. The greater the amount of branching reagent used, the more cross-linked the resulting structure. In this way, a wide variety of structures with predictable properties can be formed. Liquid silicone oils can be produced by using no branching reagents, and their boiling points can be controlled by controlling the chain lengths by altering the [extending]:[capping] ratio. Solid silicones with varying toughness and hardness can be formed by controlling the amount of branching reagent used. This way, very stable solids ranging from very soft to very hard and tough can be formed.

Interactive Table 22.4.2

Reagents Used to Synthesize Silicones

Reagent	Example	Result
Chain capping	$(CH_3)_3SiCl$	$(CH_3)_3Si—O—$
Chain extending	$(CH_3)_2SiCl_2$	$Si—O—\underset{\underset{CH_3}{\vert}}{\overset{\overset{CH_3}{\vert}}{Si}}—O—$
Chain branching	$(CH_3)SiCl_3$	$Si—O—\underset{\underset{O}{\vert}}{\overset{\overset{CH_3}{\vert}}{Si}}—O—$

© 2013 Cengage Learning

Section 22.4 Mastery

22.5 Oxygen and Sulfur in the Atmosphere

22.5a Atmospheric Ozone

Ozone, O_3, is a highly reactive form of oxygen. In the upper atmosphere (stratosphere), ozone is beneficial to life on Earth because it blocks short-wavelength, highly energetic UV radiation. Ozone in the lower atmosphere is toxic and is the principal pollutant that defines harmful air quality alerts in cities.

Ozone is less stable than dioxygen, O_2. Ozone is formed in the upper atmosphere when highly energetic photons from the sun cause an O_2 molecule to split. The free oxygen atoms (O radicals) formed can react with O_2 molecules to form ozone.

$$O_2(g) + photon \rightarrow 2\ O(g)$$

$$2\ O(g) + 2\ O_2(g) \rightarrow 2\ O_3(g)$$

Net reaction: $3\ O_2(g) + photon \rightarrow 2\ O_3(g)$

Atmospheric ozone is measured in units called the Dobson, with the symbol DU. One DU is equivalent to a 10-mm-thick layer of pure ozone at 1 atm pressure and a temperature of 0 °C. In the 1980s, scientists discovered that the amount of ozone in the atmosphere had decreased significantly, especially over Antarctica around October each year (Interactive Figure 22.5.1). This area with significantly decreased ozone is called the ozone hole (even though it is not really a hole, as the ozone concentration is not zero within the "ozone hole"). The decrease in stratospheric ozone is primarily due to the effect of free radicals, compounds that react with stratospheric ozone and convert it to oxygen, specifically the chlorine free radical.

Chlorine free radicals enter the atmosphere in the form of chlorofluorocarbons, CFCs. CFCs, compounds containing carbon, fluorine, and chlorine, are a class of compounds developed in the 1890s and put into use as refrigerants, industrial cleaners, and firefighting materials in the 20th century. They are extraordinarily nontoxic and have just the right boiling points and enthalpy of vaporization values to make them useful in refrigeration and as propellants in spray cans.

Long after CFCs were put into commercial use, it was found that they are so stable that they move into the upper atmosphere without being destroyed. In the upper atmosphere, a CFC such as Freon-11 (CF_3Cl) absorbs high-energy photons from the sun, breaking the carbon–chlorine bond.

$$CCl_3F(g) + photon \rightarrow CCl_2F(g) + Cl(g)$$

The Cl radicals are highly reactive, and they react with ozone in a three-step mechanism where Cl is a catalyst.

$$Cl + O_3 \rightarrow ClO + O_2$$

$$O_3 \rightarrow O_2 + O \text{ (in presence of light)}$$

$$\underline{ClO + O \rightarrow Cl + O_2}$$

Net reaction $\quad 2\ O_3 \rightarrow 2\ O_2$

A single chlorine free radical can destroy up to 10^5 ozone molecules before it is converted to a more inert form.

In Antarctic winter, there is no light from the sun hitting the pole, and because of weather patterns, there is a vortex around the South Pole that prevents the atmosphere over the pole from mixing with the air at lower latitudes. During winter, the chlorine radicals are captured in "reservoir compounds" such as HCl and chlorine nitrate $ClONO_2$. These compounds hold the chlorine until the sun returns, which it does each year in the southern hemisphere spring, in October. At that point a large amount of chlorine radicals are produced and a marked decrease in ozone occurs.

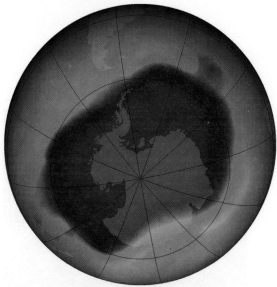

Interactive Figure 22.5.1

Investigate the mechanism of ozone depletion.

Courtesy of NASA

Total Ozone (Dobson Units)
110 220 330 440 550

Ozone concentrations over Antarctica on September 24, 2006. The size of the hole was the largest ever observed.

Ozone depletion has decreased in recent years, as evidenced by a decrease in the size of the "ozone hole" over Antarctica. This decrease is due primarily to the international agreement on phasing out substances that deplete the ozone layer (the Montreal Protocol, first signed by participating nations in 1987) and research into the creation of CFC substitutes.

22.5b Sulfur and Acid Rain

Acid rain is an environmental problem that was identified as early as the 17th century and was addressed by government regulations in the 1970s. As we saw earlier in this unit oxides of nonmetals are acidic. In particular, all coal contains sulfur, sometimes in significant proportions, in the form of metal sulfide minerals. When coal is burned, largely in coal-fired power plants, the sulfur is also burned.

$$S(s) + O_2(g) \rightarrow SO_2(g)$$

If the SO_2 is ejected into the atmosphere, it slowly, over a span of days, is further oxidized.

$$SO_2(g) + \tfrac{1}{2} O_2(g) \rightarrow SO_3(g)$$

Sulfur trioxide then reacts with water in clouds to form sulfuric acid.

$$SO_3(g) + H_2O(g) \rightarrow H_2SO_4(aq)$$

The sulfuric acid dissolves in cloud droplets and when those droplets fall to Earth as rain, that rain is acidic. As shown in Interactive Figure 22.5.2, acid rain can damage statues, buildings, and plants, and it is also very damaging to lakes, streams, and the species that live in these bodies of water.

The solutions to this problem are conceptually simple: use a fuel other than coal (natural gas contains very little sulfur), remove the sulfur from the coal prior to burning, or capture the SO_2 formed after burning but before it escapes the smokestack. This latter process is called scrubbing.

Scrubbing can be categorized as wet or dry. Both methods often involve the use of limestone ($CaCO_3$), slaked lime [$Ca(OH)_2$], soda lime [a mixture of $Ca(OH)_2$, KOH, and NaOH], or soda ash (Na_2CO_3). In all cases, the scrubber reacts with the SO_2 gas to form a sulfate or sulfite salt that is removed as a solid. For example,

$$CaCO_3(s) + SO_2(g) \rightarrow CaSO_3(s) + CO_2(g)$$

$$Ca(OH)_2(s) + SO_2(g) \rightarrow CaSO_3(s) + H_2O(\ell)$$

The calcium sulfite is further oxidized to calcium sulfate, which can be sold as gypsum.

$$CaSO_3(s) + H_2O(\ell) + \tfrac{1}{2} O_2(g) \rightarrow CaSO_4 \cdot H_2O(s)$$

Interactive Figure 22.5.2

Explore the acid-base properties of oxides.

© Vanessa Miles/Alamy

Acid rain can damage limestone statues.

Section 22.5 Mastery

Unit Recap

Key Concepts

22.1 Structures of the Elements

- The physical and chemical properties of metals are related to the fact that they have multiple orbital vacancies in valence orbitals, low effective nuclear charge, and high-energy valence orbitals (22.1b).

- The physical and chemical properties of nonmetals are related to the fact that they form covalent bonds with other nonmetals, have high effective nuclear charge, and have low-energy valence orbitals (22.1c).

- The structure of elemental nonmetals is related to atomic radius and the ability of atoms of an element to form pi bonds (22.1c).

22.2 Oxides and Halides of the Nonmetals

- The type of oxide formed by a nonmetal is related to orbital vacancies and atomic radius (ability to form pi bonds) (22.2a).

- Metal oxides are basic because the metals have high-energy valence orbitals; nonmetal oxides are acidic because of the low-energy vacant orbitals on the nonmetal (22.2a).

- The type of halide formed by a nonmetal is related to orbital vacancies, atomic radius (ability to form pi bonds), and ability to form compounds with expanded octets (22.2b).

22.3 Compounds of Boron and Carbon

- Boron forms a wide variety of compounds with other nonmetals, many of which are electron deficient (22.3a).

- The most stable elemental form of carbon is graphite (at 1 bar and 25 °C), but synthetic diamond can be synthesized by a number of techniques (22.3b).

- The creation of caves and cave formations is controlled by the effect carbon dioxide has on the solubility of calcium carbonate in water (22.3c).

- Carbon dioxide is an important greenhouse gas, a gas that reflects infrared radiation back to Earth's surface (22.3d).

22.4 Silicon

- Silicon is purified using the zone refining process to a very high purity for use in the semiconductor industry (22.4a).

- Silicates form the basis of all rocks and rock chemistry and are found in most of the minerals in Earth's crust (22.4b).

- Silicates contain structures based on SiO_4 tetrahedra connected in chains, rings, sheets, and three-dimensional networks (22.4b).

- Silicones are silicon–oxygen polymers with hydrophobic hydrocarbon side groups (22.4c).

- The structure of a silicone can be controlled by using capping, extending, and branching reagents in the synthesis process (22.4c).

22.5 Oxygen and Sulfur in the Atmosphere

- Ozone in the stratosphere is beneficial to life on Earth because it shields us from harmful UV radiation (22.5a).

- Stratospheric ozone concentrations have decreased due to free radicals such as those formed by CFCs when they enter the upper atmosphere (22.5a).

- Burning coal ejects large amounts of sulfur into the atmosphere, which reacts with water vapor to form sulfuric acid and falls to Earth in the form of acid rain (22.5b).

- Scrubbing is used to remove sulfur oxides from plant emissions before they can enter the atmosphere (22.5b).

Key Equations

$$T_E = \sqrt[4]{\frac{E_s(1 - \alpha)}{4\sigma}} \qquad (22.1)$$

$$f = \frac{\frac{E_s}{4}(1 - \alpha)}{\sigma(T_E)^4} \qquad (22.2)$$

Key Terms

22.2 Oxides and Halides of the Nonmetals
basic oxide
acidic oxide

Unit 22 Review and Challenge Problems

23

The Transition Metals

Unit Outline

In This Unit...

We have used our understanding of basic principles of chemistry to undertake brief explorations of the properties and reactivity of organic compounds and main-group elements, particularly those of the *p*-block elements. We now turn to the more complex chemistry of the transition metals, which consist of the metals found in Groups 3B–2B (Groups 3–12 in the IUPAC numbering system) of the periodic table. These elements play crucial roles in the manufacture of modern materials and in many important biological processes. In this unit we will examine how the metals are formed from their ores, used in alloys, and used to form coordination complexes, which are special types of Lewis acid–base compounds.

Vasilyev/Shutterstock.com

23.1 Properties of the Transition Metals

23.1a General Characteristics of Transition Metals

Transition metals are traditionally defined as the elements in Groups 3B–2B (3–12) of the periodic table. However, IUPAC defines transition elements as those elements whose atoms have partially filled d orbitals in their neutral or cationic state. Using this definition, the elements in Group 2B—zinc, cadmium, mercury, and copernicium—are not considered transition metals. However, in this unit we will occasionally include these elements in our discussion of the chemistry of the transition metals. Interactive Figure 23.1.1 shows the elements traditionally considered to be transition metals.

The chemical and physical properties of the transition metals are closely linked to the presence of ns electrons outside the $(n-1)d$ orbitals; although transition metals have different numbers of d electrons, they all have either 1 or 2 ns electrons in the valence shell. As a result, transition metals show great similarity in physical and chemical properties. Unlike the main-group elements, the properties do not differ greatly when moving between groups. In general, most transition metals

- are found in multiple oxidation states;

- have a silvery, metallic appearance (with the exception of copper and gold);

- have high melting points (with the exception of mercury, a liquid at room temperature);

- are highly conductive;

- form highly colored compounds; and

- are ductile (able to be drawn into a wire) and malleable (able to be hammered into thin sheets).

In this unit we will primarily consider the chemistry of the first-row transition metals (Sc–Zn). The electron configurations for these elements are shown in Figure 23.1.2.

23.1b Atomic Size and Electronegativity

Effective nuclear charge and the size of the highest-energy occupied atomic orbitals both influence the size of atoms. Consider the atomic radii for the transition metals, shown in Interactive Figure 23.1.3. In general, the atomic size of the transition metals decreases the

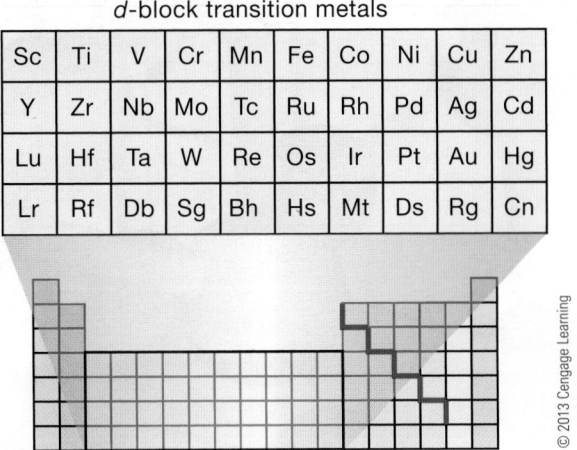

Interactive Figure 23.1.1

Investigate the properties of the coinage metals.

d-block transition metals

Sc	Ti	V	Cr	Mn	Fe	Co	Ni	Cu	Zn
Y	Zr	Nb	Mo	Tc	Ru	Rh	Pd	Ag	Cd
Lu	Hf	Ta	W	Re	Os	Ir	Pt	Au	Hg
Lr	Rf	Db	Sg	Bh	Hs	Mt	Ds	Rg	Cn

© 2013 Cengage Learning

The transition metals

farther right the element appears in a period, which matches the trend we saw for the main-group elements. However, there are some interesting details in the data shown in Interactive Figure 23.1.3.

- Atomic size decreases as you move across a period, but the change is not as significant as was observed for the main-group elements.

- Atomic size decreases from Sc to Mn but then levels out.

- Atomic size increases slightly at the right side of the d-block.

- Within groups, the period 5 transition metals are larger than the period 4 transition metals but are the same size or only slightly smaller than the period 6 transition metals.

The size of a transition metal is determined by the size of the ns orbital, the highest-energy filled atomic orbital for a transition metal. As you move across a period, both the number of protons in the nucleus and the number of electrons are increasing, but electrons are being added to $(n-1)d$ orbitals. These electrons screen the outer ns electrons from the nuclear charge, effectively offsetting the effect of an increase in nuclear charge. As a result, the atomic size of transition metals generally decreases across a period, but not by a significant amount. At the end of the transition metal group, atomic radius increases slightly due to a number of factors that include electron–electron repulsion.

The general trend in size across a period is similar for all three groups of transition metals, but the trend in size moving down a group does not follow the trend of the main-group elements. Although La is larger than Y, Hf and Zr are very similar in size, a trend that continues across the period. This phenomenon is called the **lanthanide contraction**, which is due to the inefficient shielding of nuclear charge by the $4f$ electrons. Because the $4f$ electrons do not effectively shield outer (valence) electrons from the full nuclear charge, the valence electrons are more strongly attracted to the nucleus than predicted. As a result, the atomic radius for elements with a filled $4f$ subshell is smaller than we would predict based on periodic trends. The lanthanide contraction effectively offsets the increase in size expected when moving down a group in the periodic table.

Electronegativity

Electronegativity continues the trend of minimal changes moving across the transition metal series (Figure 23.1.4). The similarity in the electronegativity values for the transition

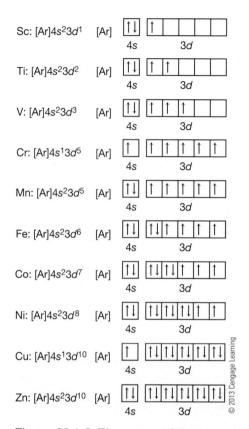

Figure 23.1.2 Electron configurations for the first-row transition metals

metals is, as described earlier, related to the similarity in the electron configurations of these elements, all of which have one or two $(n-1)s$ electrons in their valence shell. Consider the increase in electronegativity within a period for main-group elements (Li–F) when compared with the increase within the first row of transition metals (Sc–Zn). In the case of the main-group elements, electronegativity increases from 1.0 to 4.0 within the second period. In the first-row transition metals, the range of values is much smaller (1.3–1.8). Notice that the electronegativity of transition metals is higher than that of most main-group metals, suggesting that they are more likely to form covalent compounds than the main-group metals, particularly those in Groups 1A and 2A.

23.1c Ionization Energy and Oxidation States

One of the most interesting properties of transition metals is the occurrence of multiple oxidation states for each element, a property that results in extensive redox chemistry. The reactivity of these elements is related to the generally low ionization energy values.

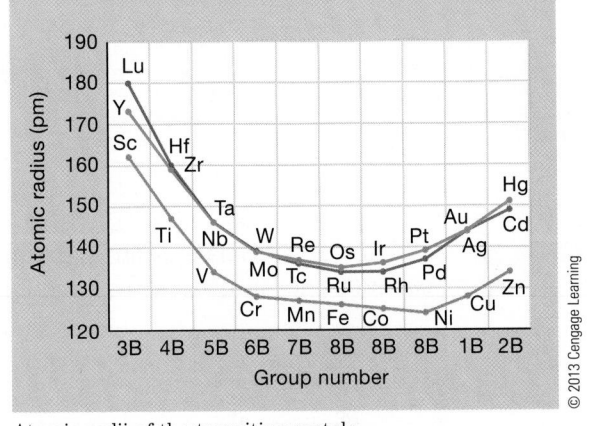

Atomic radii of the transition metals

1A	2A						8B			1B	2B	3A	4A	5A	6A	7A
																H 2.1
Li 1.0	Be 1.5											B 2.0	C 2.5	N 3.0	O 3.5	F 4.0
Na 1.0	Mg 1.2	3B	4B	5B	6B	7B				1B	2B	Al 1.5	Si 1.8	P 2.1	S 2.5	Cl 3.0
K 0.9	Ca 1.0	Sc 1.3	Ti 1.4	V 1.5	Cr 1.6	Mn 1.6	Fe 1.7	Co 1.7	Ni 1.8	Cu 1.8	Zn 1.6	Ga 1.7	Ge 1.9	As 2.1	Se 2.4	Br 2.8
Rb 0.9	Sr 1.0	Y 1.2	Zr 1.3	Nb 1.6	Mo 1.6	Tc 1.7	Ru 1.8	Rh 1.8	Pd 1.8	Ag 1.6	Cd 1.6	In 1.6	Sn 1.8	Sb 1.9	Te 2.1	I 2.5
Cs 0.8	Ba 1.0	La 1.1	Hf 1.3	Ta 1.4	W 1.5	Re 1.7	Os 1.9	Ir 1.9	Pt 1.8	Au 1.9	Hg 1.7	Tl 1.6	Pb 1.7	Bi 1.8	Po 1.9	At 2.1

■ <1.0   □ 1.5–1.9   ▨ 2.5–2.9   ▨ 4.0
□ 1.0–1.4   ▨ 2.0–2.4   □ 3.0–3.9

Figure 23.1.4 Electronegativities of the elements

Ionization Energy

Ionization energy, the energy required to remove an electron from a species in the gas phase, generally increases as you move from left to right across the periodic table, because of increasing effective nuclear charge, and decreases moving down a group, because of higher principal quantum number. Interactive Figure 23.1.5 shows the first $[M(g) \rightarrow M^+(g) + e^-]$, second $[M^+(g) \rightarrow M^{2+}(g) + e^-]$, and third $[M^{2+}(g) \rightarrow M^{3+}(g) + e^-]$ ionization energies for the first-row transition metals.

The first ionization energies increase slightly across the period, following the trend expected for a slight increase in effective nuclear charge across the period. Recall that each first-row transition metal has at least one $4s$ electron and one or more $3d$ electrons. When these elements form cations, $4s$ electrons are removed before $3d$ electrons, so the first ionization energy is the energy required to remove a $4s$ electron. As we described in our discussion of the trend in atomic radii, effective nuclear charge does not increase significantly across the period in the transition metals, so the first ionization energies of these elements also does not increase very much.

The second ionization energy is greater than the first ionization energy, as expected, but the trend across the transition metals still shows only a slight increase. Notice the higher values for Cr and Cu; both of these elements have a single $4s$ electron, so the second electron is removed from a lower-energy $3d$ orbital. The third ionization energy of the transition metals is a measure of the energy required to remove a $3d$ electron (both $4s$ electrons have been removed). In this case the anticipated result of increasing effective nuclear charge as you move left to right across a period is observed.

Oxidation States

Most transition metals have at least two stable oxidation states (Figure 23.1.6).

For the Group 3B–7B elements, the maximum oxidation state is equal to the group number. In general, the maximum oxidation state for a transition metal is observed only when the metal is combined with the highly electronegative elements oxygen and fluorine. For example, manganese has an oxidation state of +7 in the permanganate ion (MnO_4^-) and chromium has a +6 oxidation state in the chromate ion (CrO_4^{2-}). The high oxidation state on the metal means these ions are strong oxidizing agents.

Section 23.1 Mastery

Explore trends in ionization energies for the transition metals.

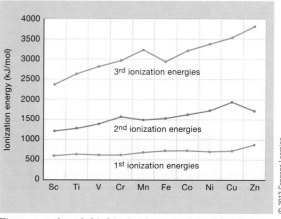

First, second, and third ionization energies of the first-row transition metals

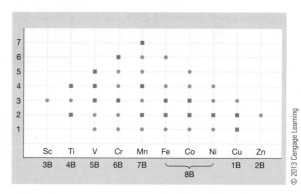

Figure 23.1.6 Common (blue circles) and less common (red squares) oxidation states for the first-row transition metals

23.2 Isolation from Metal Ores

23.2a Common Ores

Almost all metals are found in nature in compounds, not as the pure element. A rock body rich in a particular element is an **ore** of that element. However, the particular ore form of an element varies with its chemical properties and thus its position in the periodic table. The metallic character of an element, a measure of how metallic an element behaves when forming compounds, is a measure of the tendency of a metallic element to form ionic bonds as opposed to covalent bonds when bonding with a nonmetal. Elements with greater metallic character tend to form stable oxide compounds; those with less metallic character form sulfide ores, or even exist in the elemental (native) state in nature.

Metals are classified in terms of what kinds of ores they form (Interactive Figure 23.2.1). The elements on the left of the transition metals tend to form oxide ores; these metals are called **lithophiles**. Those on the right side are called **chalcophiles** and form sulfide ores. **Siderophiles** have an affinity for iron, exhibit high solubility in molten iron, and are most often found in their native, metallic state. These classifications are generally the result of the tendency for an element to form ionic or covalent bonds, and as such, they are linked to the electronegativity of the transition metals. Some elements can belong to more than one category; for example, iron is a lithophile but is often found in pyrite-based ores such as pyrite (FeS_2) and chalcopyrite ($CuFeS_2$). Some common ores of the transition metals are shown in Table 23.2.1.

23.2b Extraction of Metals from Ores

Most metals exist as positively charged ions in ionic compounds. **Extraction**, the process of chemically reacting an ore to give the neutral metal, requires reduction of the metal cation to form the neutral metal. The type of reduction process used depends on how difficult it is to reduce the metal (Interactive Figure 23.2.2). Metals that are very difficult to reduce (mostly main-group metals such as sodium, potassium, and aluminum) are produced by electrolytic reduction or by chemical reduction using a metal that was formed electrolytically. Transition

Interactive Figure 23.2.1

Explore the relationship between electronegativity and common ore types.

Phosphates   Sulfides

Oxides   Native metals

3B	4B	5B	6B	7B	8B	8B	8B	1B	2B
Sc	Ti	V	Cr	Mn	Fe	Co	Ni	Cu	Zn
Y	Zr	Nb	Mo	Tc	Ru	Rh	Pd	Ag	Cd
Lu	Hf	Ta	W	Re	Os	Ir	Pt	Au	Hg

Types of ores formed by transition metals

© 2013 Cengage Learning

metals tend to be more easily reduced and are extracted from the ore by reaction with coke, which consists mostly of elemental carbon. Toward the right side of the transition metal block, many elements are particularly easy to reduce and can be formed from the ore by simple heating. One example is mercury, which can be formed by roasting (heating) its sulfide ore, HgS (cinnabar).

Notice the trend shown in Interactive Figure 23.2.2 as you move left to right across the periodic table, where elements to the left are produced via reduction by electrolysis, those in the middle and toward the top are formed via reduction with carbon, and those in the lower right are found free in nature or can be formed by simple heating. This trend follows the periodic trend in electronegativity of the metals. Metals on the left are the least electronegative and the most difficult to reduce to the metallic state; these elements require electrolytic reduction. Metals on the lower right are the most electronegative and the easiest to reduce; these elements are often found in nature in the metallic state. Metals in the middle of the periodic table can be reduced to elemental form by the more moderate carbon reduction process.

Pyrometallurgy

Metals such as iron can be extracted from ores using **pyrometallurgy**, a technique that involves the use of high-temperature processes. Iron is found in a large variety of ores, both as an oxide and as a sulfide. (Iron sulfides are converted to iron oxides by *roasting*, heating the ore in the presence of air, before the iron is extracted from the ore.) Iron oxides most commonly found in iron ores are hematite (Fe_2O_3, iron in the +3 oxidation state) and magnetite (Fe_3O_4, iron in +2 and +3 oxidation states). The iron oxides are reduced to iron metal in a blast furnace (Interactive Figure 23.2.3).

The conversion of iron ore into molten iron takes place in a series of steps. First, iron ore, coke (primarily carbon), and limestone ($CaCO_3$) are added to the top of the blast furnace. The carbon is oxidized to carbon monoxide in an exothermic reaction that adds heat to the furnace, speeding up the reactions in the furnace.

$$C(s) + \tfrac{1}{2} O_2(g) \rightarrow CO(g) \qquad \text{Exothermic}$$

The carbon monoxide then reduces the iron in the iron ore.

Reaction	Change in Fe Oxidation State
$3\ Fe_2O_3(s) + CO(g) \rightarrow 2\ Fe_3O_4(s) + CO_2(g)$	$+3 \rightarrow +2, +3$
$Fe_3O_4(s) + CO(g) \rightarrow 3\ FeO(s) + CO_2(g)$	$+2, +3 \rightarrow +2$
$FeO(s) + CO(g) \rightarrow Fe(\ell) + CO_2(g)$	$+2 \rightarrow 0$

Table 23.2.1 Ores of Selected Transition Metals

Metal Classification	Element	Ore
Lithophile	Ti	Ilmenite, $FeTiO_3$ Rutile, TiO_2
	Mn	Pyrolusite, MnO_2
	Fe	Magnetite, Fe_3O_4 Hematite, Fe_2O_3
	W	Wolframite, $FeWO_4$ Scheelite, $CaWO_4$
	Ta	Tantalite, $FeTa_2O_6$
Chalcophile	Cu	Chalcocite, Cu_2S Chalcopyrite, $CuFeS_2$ Covellite, CuS
	Mo	Molybdenite, MoS_2
	Hg	Cinnabar, HgS
	Ag	Acanthite, Ag_2S
	Fe	Pyrite, FeS_2
	Co	Linnaeite, Co_3S_4

The iron produced in a blast furnace is called *pig iron* and is further refined in other processes to produce steel and other iron-containing products. The gases produced in the reactions move up and exit at the top of the furnace, although some of the carbon dioxide reacts with coke to produce additional carbon monoxide.

$$C(s) + CO_2(g) \rightarrow 2\ CO(g) \qquad \text{Endothermic}$$

Note that although solid carbon (coke) can also reduce the solid iron ore to iron metal, the reaction between a solid and a gas (in this case, carbon monoxide) is much more kinetically favored because the gas can envelop all of the available surface area of the solid.

Limestone, $CaCO_3$, is added to remove silica-containing impurities in the ore. At the high temperatures found in the furnace, the solid decomposes to lime (CaO) and carbon dioxide.

$$CaCO_3(s) \rightarrow CaO(s) + CO_2(g) \qquad \text{Endothermic}$$

The lime then reacts with the silica impurities (typically SiO_2 and silicate minerals) to produce a molten metal silicate mixture called *slag*, which is primarily composed of calcium silicate.

$$CaO(s) + SiO_2(s) \rightarrow CaSiO_3(\ell) \qquad \text{Endothermic}$$

The slag is less dense than the molten iron, so it forms a layer on top of the iron and is drawn off as it builds up in the furnace. The slag protects the molten iron from reaction with oxygen in the furnace.

Electrometallurgy

Titanium forms ionic compounds that are too difficult to reduce using carbon. Instead the neutral metal is produced by chemical reduction using electrolytically produced Mg metal at high temperatures. One principal ore of titanium is rutile, TiO_2. The first step in forming metallic titanium is to convert TiO_2 into $TiCl_4$.

$$TiO_2(s) + 2\ Cl_2(g) \rightarrow TiCl_4(\ell) + O_2(g)$$

Although this reaction does not reduce the titanium, which remains in the +4 oxidation state, it does form a more reactive chloride compound that can be used in the reduction step.

$$2\ Mg(\ell) + TiCl_4(g) \rightarrow 2\ MgCl_2(\ell) + Ti(s)$$

Hydrometallurgy

Some ores, most importantly those of gold, contain the native metallic element. In these cases the most difficult task is to separate the metal from the rest of the rock. For gold,

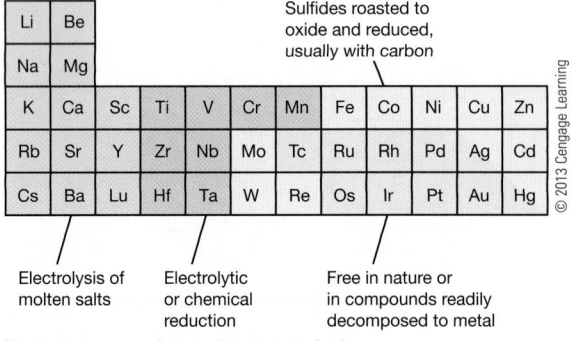

Interactive Figure 23.2.2

Predict the extraction method for transition metals.

Techniques used to extract metals from ores

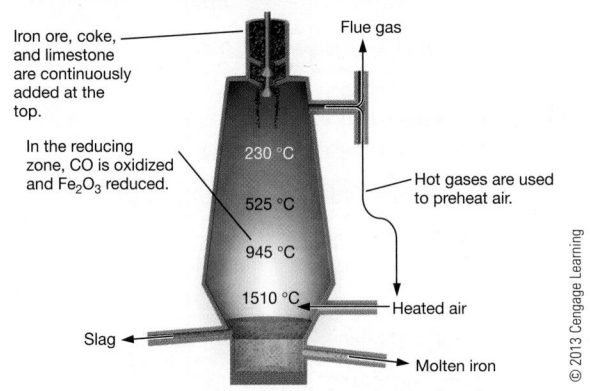

Interactive Figure 23.2.3

Explore a blast furnace.

A blast furnace is used to extract iron from its ore.

the difficulty is that the metal has a very low concentration in the ore—often as little as 25 parts per million—so only a tiny amount of gold can be separated from a large amount of rock. One method for extracting gold involves **hydrometallurgy**, a technique that makes use of reactions in aqueous solution. The most common method used today for gold extraction is to chemically react the gold with cyanide ion. This process dissolves the gold into an aqueous solution, leaching it from the rest of the rock. To maximize the contact between the chemical reagents and the gold, the gold ore is crushed into a powder before it is treated.

The reaction for the dissolution of the gold metal is a redox reaction, where gold is oxidized and oxygen gas is the oxidizing agent.

$$4 \text{ Au(s)} + 8 \text{ CN}^-\text{(aq)} + \text{O}_2\text{(g)} + 2 \text{ H}_2\text{O}(\ell) \rightarrow 4 \text{ Au(CN)}_2^-\text{(aq)} + 4 \text{ OH}^-\text{(aq)}$$

Once free of the rock, metallic gold is produced by another redox reaction. Here, the gold is reduced using an inexpensive metal that is easily oxidized such as zinc.

$$\text{Zn(s)} + 2 \text{ Au(CN)}_2^-\text{(aq)} \rightarrow \text{Zn(CN)}_4^{2-}\text{(aq)} + 2 \text{ Au(s)}$$

The gold-depleted rock is disposed of on-site in piles called *mine tailings*. Many mine tailings are contaminated with cyanide salts that are not recycled during the extraction process. The cyanide can then enter water reservoirs such as groundwater, lakes, and streams, compromising the health of the surrounding environment.

Section 23.2 Mastery

23.3 Coordination Compounds: Structure and Isomerism

23.3a Composition of Coordination Compounds

Transition metals form Lewis acid–base adducts where a single Lewis acid (a metal atom or ion) bonds with multiple Lewis bases. These compounds are called coordination compounds. A large branch of chemistry, called *coordination chemistry,* involves synthesizing compounds with just the right set of molecules or ions around just the right metal atom or ion, in just the right arrangement to lead to a compound with particularly desired properties. We also use this field of chemistry to help understand the structure and behavior of transition metals in biomolecules such as enzymes and proteins.

A typical **coordination compound** consists of two or three components: (1) a metal (typically a transition metal, but coordination compounds can contain main-group metals); (2) one or more Lewis bases (called *ligands*) covalently bonded to the metal; and (3) if needed, counterions that balance the charge on the other two components in the coordination compound. Recall that the bond between a ligand and a metal is often called a *coordinate–covalent bond* because both bonding electrons come from a single species, the Lewis base. If the combination of the metal and the Lewis bases results in a charged species, it is called a **complex ion**. If it does not, no counterions are needed. For example, the coordination compound $[Co(NH_3)_6]Cl_3$ contains a complex ion, $Co(NH_3)_6^{3+}$, and chloride (Cl^-) counterions.

Square brackets are traditionally used to enclose the formula of a complex ion or a neutral coordination compound. It is important to note that counterions in a coordination compound are ionic species that do not form covalent bonds with the transition metal; they simply balance the charge on the complex ion. Other species that can be written outside the square brackets include *waters of hydration*, water molecules that are trapped in the solid crystal lattice but are not directly bonded to the transition metal.

Ligands

The Lewis bases that form coordinate covalent bonds with a transition metal in a coordination compound are called **ligands**. (The term *ligand* is also used in biology but is pronounced differently. In chemistry a short "i" is used, as in *linen*; in biology a long "i" is used, as in *lion*.) Ligands can be neutral or anionic species; cationic ligands are unusual.

Ligands are classified by the number of attachments they form to a central metal atom. **Monodentate ligands** such as Cl^- and NH_3 have one attachment point to a metal. Ligands that have more than one potential attachment point are called **chelating ligands** or **multidentate ligands**. Chelating ligands are identified by the number of potential attachment points. For example, a **bidentate ligand** such as ethylenediamine ($H_2NCH_2CH_2NH_2$, abbreviated "en") has two attachment points, and a **hexadentate ligand** has six attachment points. Due to their complex formulas, many multidentate ligands are given abbreviations that are used in coordination compound formulas. Some examples include "en" (for ethylenediamine, $H_2NCH_2CH_2NH_2$), and "ox" (for the oxalate ion, $^-O_2CCO_2^-$).

An **ambidentate ligand** is a special type of ligand that has more than one possible attachment site but does not generally act as a multidentate ligand. For example, the ambidentate ligand SCN^- typically attaches to a metal at S or N, but not both at the same time. Some common ligands are shown in Figure 23.3.1.

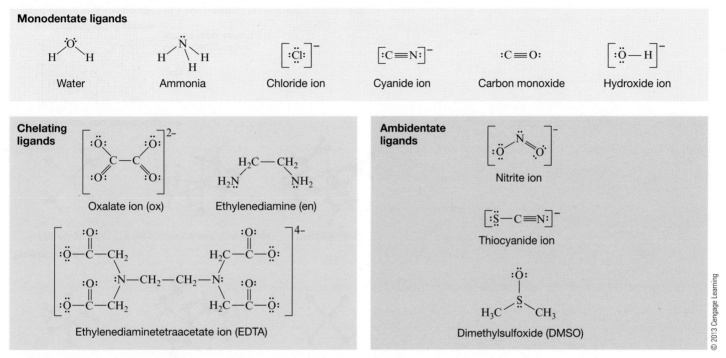

Figure 23.3.1 Some monodentate, chelating, and ambidentate ligands

General Structure of Coordination Compounds

The composition and structure of coordination compounds vary widely, but there are some terms that are commonly used to describe the components and general structure of most coordination compounds.

- **Coordination number** (CN): The number of atoms covalently bonded to the metal in the coordination compound.

- **Coordination sphere** (or **inner coordination sphere**): The portion of the coordination compound where species are directly bonded to the metal.

- **Outer coordination sphere:** The portion of the coordination compound that does not include the metal and any ligands covalently bonded to the metal. Includes counterions and waters of hydration.

Interactive Figure 23.3.2 shows how these terms, and some of the terms used to describe ligands, apply to coordination compounds and complex ions.

23.3b Naming Coordination Compounds

Coordination compounds are named according to an established system of rules.

1. If the coordination compound contains a complex ion, name the cation first, followed by the anion name.
2. In the name of the complex ion or neutral coordination compound, name the ligands first and then name the transition metal.

Interactive Figure 23.3.2

Explore the components of a coordination compound.

Coordination number

CN = 6   CN = 6   CN = 4

 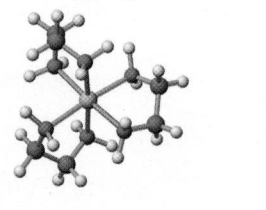

Type of ligand

4 monodentate   2 monodentate and 2 bidentate   1 hexadentate

 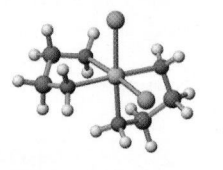 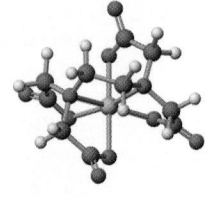

Inner and outer coordination sphere

$$\left[\begin{array}{c} NH_3 \\ H_3N \diagdown \; | \diagup Cl \\ Co \\ H_3N \diagup \; | \diagdown NH_3 \\ NH_3 \end{array} \right]^{+}$$

NO_2^-

Outer coordination
sphere (counterion)

Inner coordination sphere

Structures and components of some common coordination compounds

3. Ligands names:
   a. Name the ligands in alphabetical order. When using prefixes (as described later), do not alphabetize the prefix—alphabetize only the names of the ligands.
   b. If there is more than one of a particular ligand, use a prefix to indicate the number of ligands. Monodentate ligands use the prefixes *di-*, *tri-*, *tetra-*, *penta-*, and *hexa-* to indicate 2, 3, 4, 5, and 6 ligands, respectively. For complex ligand names or names that contain a prefix such as *di-* or *tri-*, place the ligand name in parentheses and use the prefixes *bis-*, *tris-*, *tetrakis-*, *pentakis-*, and *hexakis-* to indicate 2, 3, 4, 5, and 6 ligands, respectively.
   c. If the ligand is an anion, change the name to end in *-o*. (See Table 23.3.1.)
   d. If the ligand is neutral, name it using its common name. There are some exceptions to this rule, as shown in Table 23.3.1.
4. If the complex ion is an anion, the suffix *-ate* is added to the metal name.
   a. If the metal is Fe, Cu, Ag, or Au, use the Latin name:
      Fe   ferrate   Cu   cuprate
      Ag   argentate   Au   aurate
5. The charge on the metal is added after its name, in parentheses and in Roman numerals.

Table 23.3.1 Names of Some Common Ligands

Neutral Ligands		Anionic Ligands	
NH_3	Ammine	Cl^-	Chloro
H_2O	Aqua	Br^-	Bromo
CO	Carbonyl	CN^-	Cyano
NO	Nitrosyl	OH^-	Hydroxo
$H_2NCH_2CH_2NH_2$	Ethylenediamine	$C_2O_4{}^{2-}$	Oxalato
$P(C_6H_5)_3$	Triphenylphosphine	SCN^- (S-bonded)	Thiocyanato
		NCS^- (N-bonded)	Isothiocyanato
		$NO_2{}^-$ (N-bonded)	Nitro
		ONO^- (O-bonded)	Nitrito

Metal Oxidation State

The charge on the metal in a coordination compound is the oxidation state of the metal. As you learned in Chemical Reactions and Solution Stoichiometry (Unit 4), the sum of the oxidation states for all species in a compound is equal to the overall charge on the compound. When determining the oxidation state of a metal in a coordination compound, however, it is simpler to use the charges on the ligands and counterions instead of the oxidation states of the individual atoms in the compound. For example, consider the coordination compound $[Co(NH_3)_6]Cl_3$. Each of the NH_3 ligands is neutral, and the overall charge on the compound is zero. Therefore, the charge on the cobalt must be +3 to balance the charge on the three Cl^- ions.

$$\text{(charge on Co)} + 6 \times \text{(charge on } NH_3\text{)} + 3 \times \text{(charge on Cl)} = 0$$

$$\text{(charge on Co)} + 6(0) + 3(-1) = 0$$

$$\text{charge on Co} = +3$$

Consider the following examples that demonstrate how to name some simple coordination compounds.

$$[Co(NH_3)_6]Cl_3$$

This compound has six ammonia ligands (prefix = *hexa-*, ligand name = *ammine*) and a cobalt ion with a +3 oxidation state. The counterion is the chloride ion. The compound name is hexaamminecobalt(III) chloride. Notice the space in the name between the cation (the complex ion) and the anion.

$$K_4[Fe(CN)_6]$$

The cation in this compound is the potassium ion, and the anion is a complex ion. The complex ion has six cyanide ligands (prefix = *hexa-*, ligand name = *cyano*), an iron ion with a +2 oxidation state, and an overall negative charge (metal name = *ferrate*). The compound name is potassium hexacyanoferrate(II).

Example Problem 23.3.1 Name a coordination compound.

Name the coordination compound $[Pt(en)Cl_2]$.

Solution:

You are asked to name a coordination compound.

You are given the formula of the coordination compound.

In this neutral coordination compound there are two chloride ligands (prefix = *di-*, ligand name = *chloro*), an ethylenediamine ligand (abbreviated "en" in the compound formula), and a platinum ion with a +2 oxidation state. The compound name is dichloroethylenediamineplatinum(II).

Video Solution

Tutored Practice
Problem 23.3.1

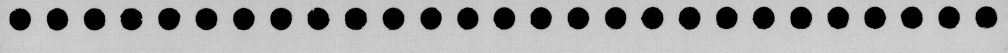

The formula of a coordination compound can be determined from its name. For example, consider the compound tetraaquadichlorochromium(III) chloride. The compound has four water ligands (tetraaqua), two chloride ion ligands (dichloro), a Cr^{3+} ion, and chloride ion counterions. The combination of the metal and ligands gives a complex ion with an overall +1 charge $[4(0) + 2(-1) + (+3) = +1]$, so only one Cl^- counterion is needed to balance the charge on the complex ion. The compound formula is $[Cr(H_2O)_4Cl_2]Cl$.

Example Problem 23.3.2 Write the formula of a coordination compound.

Write the formula of the coordination compound sodium tetrachlorocuprate(II).

Solution:

You are asked to write the name of a coordination compound.

You are given the name of the coordination compound.

The compound has four chloride ligands (tetrachloro), a Cu^{2+} ion, and sodium ion counterions. The combination of the metal and ligands gives a complex ion with an overall -2 charge $[4(-1) + (+2) = -2]$, so two Na^+ counterions are needed to balance the charge on the complex ion. The compound formula is $Na_2[CuCl_4]$.

Video Solution

Tutored Practice
Problem 23.3.2

23.3c Stability and the Chelate Effect

The **chelate effect** states that coordination compounds containing chelating ligands are more thermodynamically stable than similar compounds containing only monodentate ligands. For example, consider the formation constants for the similar nickel compounds $Ni(NH_3)_6^{2+}$ and $Ni(en)_3^{2+}$.

$$Ni^{2+}(aq) + 6\ NH_3(aq) \rightarrow Ni(NH_3)_6^{2+}(aq) \qquad K_f = 5.5 \times 10^8$$

$$Ni^{2+}(aq) + 3\ H_2NCH_2CH_2NH_2(aq) \rightarrow Ni(en)_3^{2+}(aq) \qquad K_f = 2.1 \times 10^{18}$$

The two compounds have the same coordination number (6) and each has six Ni—N bonds. However, the compound containing the chelating ligand is much more stable than the one containing only monodentate ligands.

The relative stability of the two compounds can be explained by examining the free energy change for the reactions, ΔG. Recall that we can calculate free energy from the enthalpy change and entropy change for a reaction at a given temperature ($\Delta G = \Delta H - T\Delta S$). First, consider the enthalpy change (ΔH) for the reactions. If we rewrite the reactions to include the water molecules associated with the nickel(II) ion, it is clear that the enthalpy change for the two reactions is probably very similar in magnitude due to the similar energy of the bonds being broken (six Ni—OH_2 bonds) and formed (six Ni—N bonds).

$$Ni(H_2O)_6^{2+}(aq) + 6\ NH_3(aq) \rightarrow Ni(NH_3)_6^{2+}(aq) + 6\ H_2O(\ell)$$

$$Ni(H_2O)_6^{2+}(aq) + 3\ H_2NCH_2CH_2NH_2(aq) \rightarrow Ni(en)_3^{2+}(aq) + 6\ H_2O(\ell)$$

The magnitude of ΔS for the two reactions is not the same, however. In the reaction involving the monodentate NH_3 ligand, there is no significant change in entropy; seven moles of reactants form seven moles of products. In the reaction involving the multidentate ligand, the entropy change has a larger positive value; 4 moles of reactants form 7 moles of products. Because the reaction with the multidentate ligand has a larger positive ΔS value, ΔG for this reaction is a larger negative value, and thus the formation of this compound is more favored thermodynamically.

Another factor that can explain the favored formation of compounds involving multidentate ligands is known as the "high local concentration" effect (Interactive Figure 23.3.3). Again consider the formation of $Ni(en)_3^{2+}$. When one nitrogen atom of an ethylenediamine ligand attaches to a nickel(II) ion, the second coordinating nitrogen atom is placed in close proximity to the nickel(II) ion, which could help drive the formation of the product. Coordination of a monodentate ligand such as NH_3 has no effect on the likelihood that additional ammonia ligands will form a bond with the metal ion, so monodentate ligands do not experience this effect.

The chelate effect can be used to synthesize a desired compound. For example, the chelate effects makes the conversion of $Ni(NH_3)_6^{2+}$ to $Ni(en)_3^{2+}$ favorable ($K \gg 1$).

$$Ni(NH_3)_6^{2+}(aq) + 3\ H_2NCH_2CH_2NH_2(aq) \rightarrow Ni(en)_3^{2+}(aq) + 6\ NH_3(aq) \qquad K = 4.9 \times 10^9$$

23.3d Isomerism

Coordination compound geometry, the arrangement in space of the ligands attached to a central metal, can vary, but the most common geometries are octahedral (CN = 6), tetrahedral (CN = 4), and square planar (CN = 4). Compounds with octahedral and square planar geometries exhibit a variety of isomers, which we can categorize as structural isomers or stereoisomers.

Structural Isomers

Structural isomers are compounds with the same composition that differ in the number and types of chemical bonds. Two of the most important examples of structural isomers for coordination compounds are *linkage isomers* and *ionization isomers*.

© 2013 Cengage Learning

Interactive Figure 23.3.3

Explore the chelate effect.

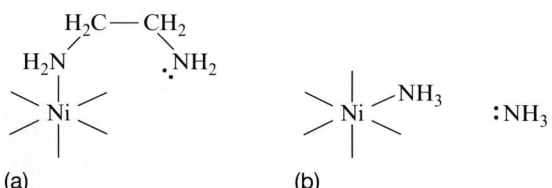

(a) (b)

(a) A chelating ligand creates a high local ligand concentration. (b) Coordination of a monodentate ligand has no effect on the concentration of other ligands near the coordination sphere.

Linkage isomers are two coordination compounds that have the same metal center and ligands but differ in how one or more of the ligands are attached to the metal. For example, consider the complex ion $[Co(NH_3)_5NO_2]^{2+}$. The nitrite ion, NO_2^- is an ambidentate ligand that can attach to a metal through two different atoms, but not both at the same time. In Figure 23.3.4, the nitrite ion is bonded to the cobalt ion at the nitrogen atom in the structure on the left and through the oxygen atom in the structure on the right. The thiocyanate ion (SCN^-) is another example of an ambidentate ligand.

Ionization isomers are two coordination compounds in which two different anions switch positions between the inner and outer coordination sphere. Figure 23.3.5 shows two ionization isomers of $[Co(NH_3)_5Cl]NO_2$. The coordination compound on the left has a Cl^- ligand bonded to Co^{2+} inside the coordination sphere while NO_2^- is a counterion. In the compound on the right, the ions have changed positions; the NO_2^- ion is bonded to Co^{2+} inside the coordination sphere (it is acting as a ligand) and Cl^- is a counterion.

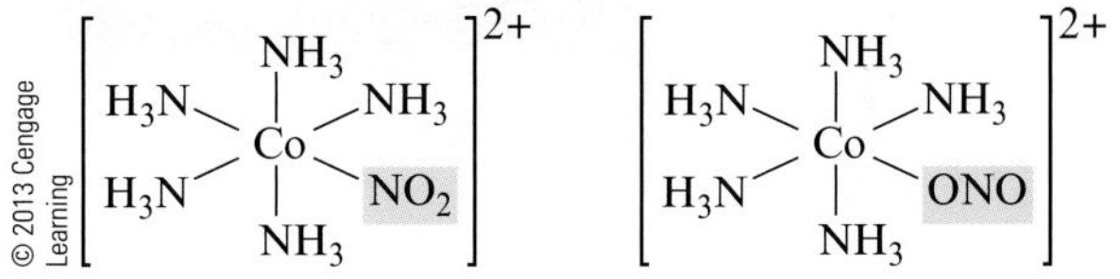

Stereoisomers

Stereoisomers are coordination compounds that have the same chemical composition and number and types of chemical bonds but differ in the spatial arrangement of the atoms (or ligands). There are two types of stereoisomers: *geometric isomers* and *optical isomers.*

Cis–trans and **fac–mer** isomers are examples of geometric isomers. In a *cis* isomer, two identical ligands are arranged adjacent to each other; in a *trans* isomer, two identical ligands are arranged across from each other (Figure 23.3.6). *Cis* and *trans* isomers are common in both octahedral and square planar geometries.

In a *fac* (facial) isomer, three identical ligands are arranged at 90° to each other, on a triangular face of an octahedron. In a *mer* (meridional) isomer, the three identical ligands are arranged in a line, or are coplanar, in an octahedral compound (Figure 23.3.7).

Optical isomers are two compounds that have the same composition and arrangement of ligands but differ in that the two isomers are mirror images of each other that are not identical (are not superimposable). A coordination compound or complex ion that exhibits optical isomerism is *chiral.* One characteristic of a chiral compound is how it interacts with plane-polarized light, monochromatic light that has been passed through a polarizing filter (Figure 23.3.8). When plane-polarized light is passed through a solution containing a chiral compound, the plane rotates. The magnitude and angle of rotation are unique physical properties of the chiral compound. The mirror image of the chiral compound will also rotate plane-polarized light by the same angle, but in the opposite direction. For example, the compound *cis*-$[Cr(en)_2Cl_2]$ is chiral because its mirror image is a unique (nonsuperimposable) compound (Figure 23.3.9).

Figure 23.3.4 Linkage isomers of $[Co(NH_3)_5NO_2]^{2+}$

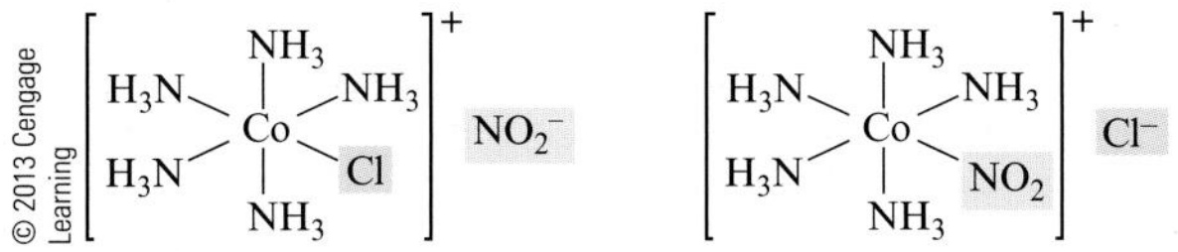

Figure 23.3.5 Ionization isomers of $[Co(NH_3)_5Cl]NO_2$

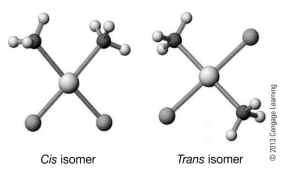

Cis isomer   *Trans* isomer

Figure 23.3.6 *Cis*- and *trans*-$[Pt(NH_3)_2Cl_2]$

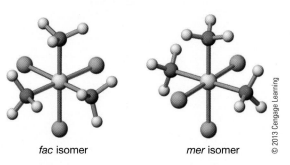

fac isomer   *mer* isomer

Figure 23.3.7 *Fac*- and *mer*-$[Co(NH_3)_3Cl_3]$

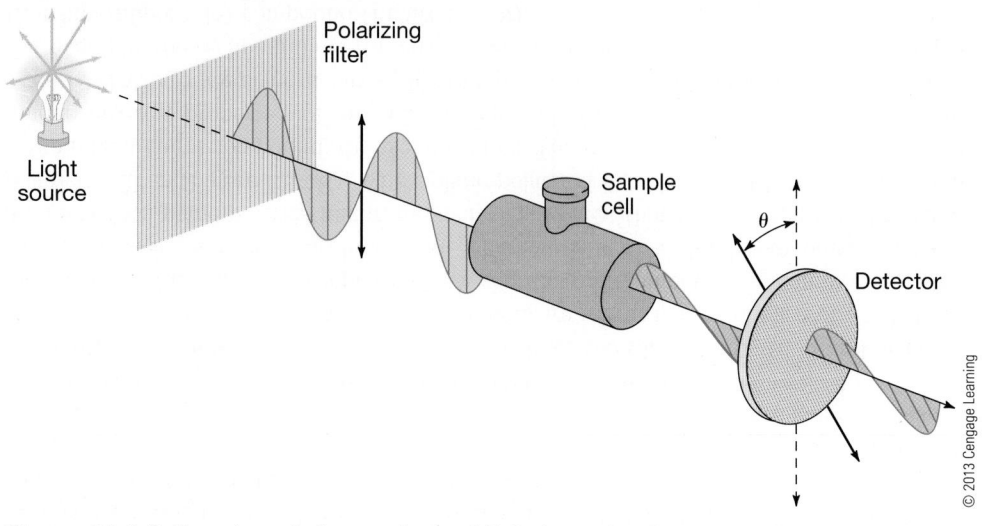

Figure 23.3.8 Rotation of plane-polarized light by a chiral compound

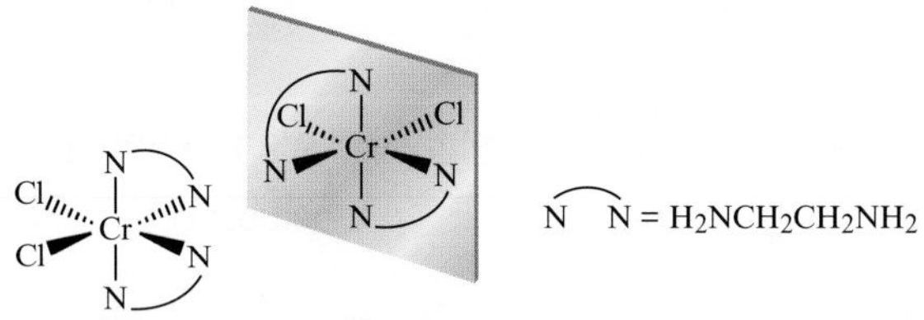

Figure 23.3.9 Optical isomers of *cis*-[Cr(en)$_2$Cl$_2$]

Interactive Table 23.3.2 summarizes the general characteristics of the different structural and stereoisomers described here.

Interactive Table 23.3.2

Structural and Stereoisomers of Coordination Compounds

Classification	Type of Isomer	Characteristics
Structural isomers	Linkage	Contain at least one ambidentate ligand. Two isomers differ in the atom in the ambidentate ligand attached to the metal.
Structural isomers	Ionization	Contains at least one counterion that can also act as a ligand. Two isomers differ in the ligands inside and outside the coordination sphere of the metal in the coordination compound.
Stereoisomers	Geometric *Cis* and *trans*	Two isomers differ in the positions of two identical ligands with respect to each other.
Stereoisomers	Geometric *Fac* and *mer*	Two isomers differ in the positions of three identical ligands with respect to each other.
Stereoisomers	Optical	Two isomers are nonsuperimposable mirror images of each other.

Section 23.3 Mastery

23.4 Coordination Compounds: Bonding and Spectroscopy

23.4a Crystal Field Theory

To this point we have described only the metal–ligand interaction in coordination compounds as a Lewis acid–base interaction, where the metal is the electron-pair acceptor (Lewis acid) and the ligands act as Lewis bases (lone-pair donors). This simple, localized model does not allow us to explain the wide range of colors and the magnetic properties of coordination compounds. To explain these properties we must use a different bonding model, **crystal field theory**, which focuses on the energy of the metal d orbitals. In crystal field theory, we assume that ligands are *negative point charges*, small, localized regions of negative charge.

To understand how ligands (point charges) affect the energy of the metal d orbitals, we will begin by orienting the d orbitals in an octahedral field of ligands. As shown in Figure 23.4.1, if the six ligands in an octahedral compound are placed along the x-, y-, and z-axes, the d orbitals fall into two different groups.

Two of the d orbitals have electron density oriented along the x-, y-, and z-axes (pointed directly at the point charges), and three of the d orbitals have electron density oriented between the x-, y-, and z-axes (pointed between the point charges). As a result, when the point charges get close to the metal d orbitals, repulsive forces increase the energy of the $d_{x^2-y^2}$ and d_{z^2} orbitals (known as the e_g orbitals) more than the energy of the d_{xy}, d_{xz}, and d_{yz} orbitals (known as the t_{2g} orbitals) (Figure 23.4.2).

The difference in energy between the t_{2g} and e_g sets in an octahedral field of point charges is the **crystal field splitting energy**, symbolized by Δ_o. The magnitude of Δ_o is a

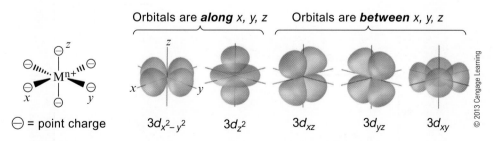

Orbitals are **along** x, y, z Orbitals are **between** x, y, z

⊖ = point charge $3d_{x^2-y^2}$ $3d_{z^2}$ $3d_{xz}$ $3d_{yz}$ $3d_{xy}$

© 2013 Cengage Learning

Figure 23.4.1 d orbitals oriented in an octahedral field of point charges

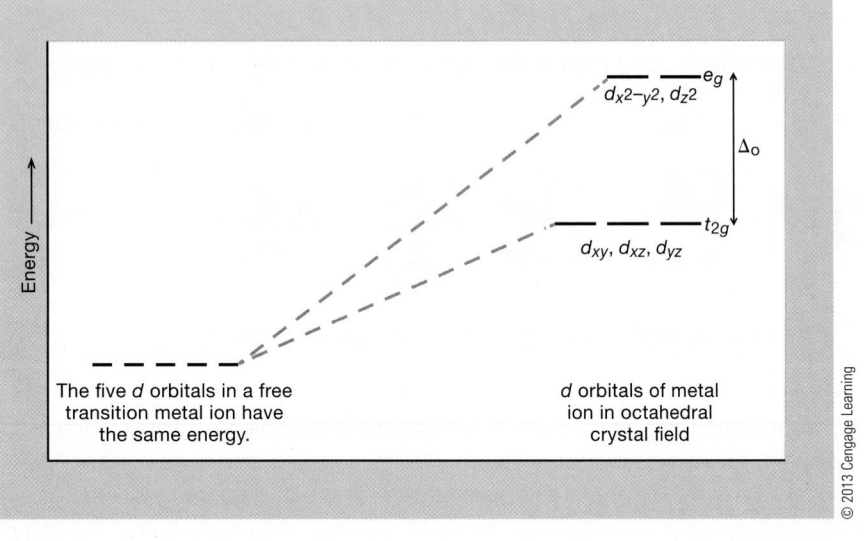

Figure 23.4.2 Splitting of the d orbitals into t_{2g} and e_g sets in an octahedral field

function of the strength of the interaction between ligands that form sigma bonds to a metal; the stronger the interaction, the higher the energy of the e_g orbitals and the larger the crystal field splitting energy.

The crystal field splitting energy for a compound can be used to account for its color and magnetic properties. We can also apply crystal field theory to other geometries such as tetrahedral and square planar. In these geometries, the metal d orbitals also split but in different ways due to the different orientation of the point charges (Figure 23.4.3).

In a tetrahedral ligand field, none of the d orbitals point directly at the ligand point charges, so the energy difference between the sets of d orbitals (Δ_t) is always very small, much smaller than Δ_o. In a square planar complex, the energy gap between the two highest-energy d orbitals (Δ_{sp}) can be large or small.

Consequences of d-Orbital Splitting

The energy difference between the t_{2g} and e_g orbitals in an octahedral compound (Δ_o) can vary with the strength of the metal–ligand interaction. For example, consider the two complex ions $[Fe(H_2O)_6]^{2+}$ and $[Fe(CN)_6]^{4-}$. In both complexes, iron has a +2 oxidation state

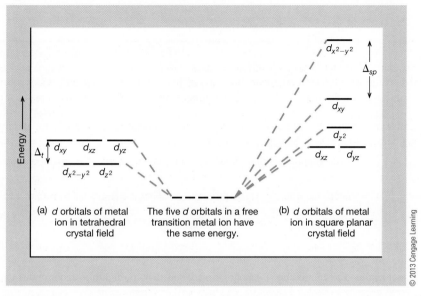

Figure 23.4.3 Splitting of the d orbitals in (a) tetrahedral and (b) square planar ligand fields

$$d_{x^2-y^2}, \overline{d_{z^2}} \quad \Delta_o(H_2O)$$

$$\underline{\uparrow\downarrow}\ \underline{\uparrow}\ \underline{\uparrow}$$
$$d_{xy},\ d_{xz},\ d_{yz}$$

high spin
$[Fe(H_2O)_6]^{2+}$

$$d_{x^2-y^2}, \overline{d_{z^2}} \quad \Delta_o(CN^-)$$

$$\underline{\uparrow\downarrow}\ \underline{\uparrow\downarrow}\ \underline{\uparrow\downarrow}$$
$$d_{xy},\ d_{xz},\ d_{yz}$$

low spin
$[Fe(CN)_6]^{4-}$

Figure 23.4.4 High-spin and low-spin arrangements for an octahedral d^6 metal complex

Interactive Figure 23.4.5

Explore crystal field theory

	d^4	d^5	d^6	d^7

High spin

High-spin and low-spin arrangements for octahedral d^4–d^7 configurations

and the electron configuration $[Ar]3d^6$. However, $[Fe(H_2O)_6]^{2+}$ is paramagnetic (has unpaired electrons) and $[Fe(CN)_6]^{4-}$ is diamagnetic (all electrons are paired). We can account for this difference in magnetism by arranging the electrons in the t_{2g} and e_g orbitals in two different ways (Figure 23.4.4).

If the metal–ligand interaction is weak (a **weak-field** case), Δ_o is small and it is smaller than the energy required to pair electrons in an orbital (the **pairing energy**). In this situation a **high-spin** configuration results where electrons occupy all five d orbitals before any are paired. If the metal–ligand interaction is strong (a **strong-field** case), Δ_o is larger than the pairing energy and a **low-spin** configuration results where the electrons fill the lower-energy d orbitals before filling the higher-energy d orbitals.

High- and low-spin configurations for octahedral compounds are possible only for metals with 4, 5, 6, or 7 d electrons (Interactive Figure 23.4.5). Tetrahedral complexes are almost always high spin due to the small magnitude of Δ_t.

Spectrochemical Series

The **spectrochemical series** shows the effect different ligands have on the magnitude of d-orbital splitting. The series is the result of experiments on many octahedral coordination compounds. A small portion of the series is shown here:

$$CO > CN^- > P(C_6H_5)_3 > NO_2^- > en > NH_3 > H_2O > OH^-$$
$$> F^- > NO_3^- > SCN^- > Cl^- > Br^- > I^-$$

Ligands at the "high" end of the series, such as CO and CN^-, are called **strong-field ligands** because they have a large effect on Δ_o. Coordination compounds containing these ligands have a large Δ_o and can be low spin. Ligands at the "low" end of the series, such as OH^- and Cl^-, are called **weak-field ligands** because they have a small effect on Δ_o. Coordination compounds containing these ligands have a small Δ_o and can be high spin.

In the middle of the series are ligands such as en and NH_3, which are not considered to be strong-field or weak-field ligands. For compounds containing these ligands, the magnitude of Δ_o depends on the metal charge (a high metal oxidation state results in a large Δ_o) and the size of the metal (Δ_o for first-row transition metals is generally smaller than Δ_o for second- and third-row transition metals).

Example Problem 23.4.1 Predict the number of unpaired electrons for a transition metal complex.

Predict the number of unpaired electrons for the following compounds or complex ions.

a. $CoCl_6^{4-}$
b. $Ni(H_2O)_6^{2+}$

Solution:

You are asked to predict the number of unpaired electrons for a coordination compound or complex ion.

You are given the formula of the coordination compound or complex ion.

a. The cobalt ion in $CoCl_6^{4-}$ has an oxidation state of $+2$, so it has seven d electrons. The chloride ion is a weak-field ligand, so Δ_o is small and the compound is high spin with three unpaired electrons.

b. The nickel ion in $Ni(H_2O)_6^{2+}$ has an oxidation state of $+2$, so it has eight d electrons. There is only one way to arrange eight electrons in an octahedral coordination compound: six electrons in the t_{2g} orbitals and two electrons in the e_g orbitals. The compound has two unpaired electrons.

Video Solution

Tutored Practice
Problem 23.4.1

23.4b Molecular Orbital Theory

Crystal field theory, which is based on electrostatic interactions between charged species, is very useful for explaining the relative energies of metal d orbitals in a coordination compound. However, it does not explain the stability of coordination compounds or the effect ligands have on the magnitude of the energy gap (Δ) between the d orbitals. For example, as shown in Table 23.4.1, coordination compounds tend to have very large formation constants (K_f), which suggest the formation of these compounds is very thermodynamically favorable.

Also, when comparing Δ_o for octahedral coordination compounds containing only NH_3, H_2O, or OH^- ligands, crystal field theory predicts the negatively charged OH^- will have the strongest metal–ligand interaction, and NH_3, which is less polar than H_2O, will have the weakest metal–ligand interaction. As we saw in the spectrochemical series, the actual effect of these ligands on Δ_o is $NH_3 > H_2O > OH^-$. By supplementing crystal field theory with molecular orbital theory, we can better understand how the interactions between ligands and metals affects the relative energy of the metal d orbitals in coordination compounds.

Metal–Ligand Sigma Bonding

In molecular orbital theory, atomic orbitals overlap to form molecular orbitals, which can be delocalized over many atoms. When building a molecular orbital diagram to describe sigma bonding in a coordination compound, we consider the overlap of the valence atomic orbitals on the transition metal ($3d$, $4s$, and $4p$ for a first-row transition metal), with six filled orbitals representing the electrons that ligands donate to the metal when forming sigma bonds (Figure 23.4.6). Notice the following features of the molecular orbital diagram.

- The orbitals that have electron density along the x-, y-, and z-axes (the $4s$, $4p$, $3d_{x^2-y^2}$, and $3d_{z^2}$ valence orbitals on the metal and the six ligand sigma donor orbitals) overlap to form six bonding (σ) and six antibonding (σ^*) molecular orbitals.

- The remaining orbitals (the transition metal $3d_{xy}$, $3d_{zy}$, and $3d_{yz}$ orbitals) become nonbonding molecular orbitals (the t_{2g} orbitals in Figure 23.4.6).

- The sigma donor ligand electrons fill the low-energy bonding molecular orbitals, lowering the overall energy of the compound.

- The metal valence electrons fill the five next higher energy orbitals, the t_{2g} (nonbonding) and e_g (antibonding) molecular orbitals.

- The energy gap between the t_{2g} and e_g orbitals is Δ_o. *Molecular orbital theory predicts the splitting of the* d *orbitals into two sets just as crystal field theory does.*

Table 23.4.1 Formation Constants for Some Complex Ions

Formation Equilibrium	K_f
$Ag^+ + 2\,NH_3 \rightleftarrows [Ag(NH_3)_2]^+$	1.1×10^7
$Al^{3+} + 4\,OH^- \rightleftarrows [Al(OH)_4]^-$	1.1×10^{33}
$Co^{2+} + 6\,NH_3 \rightleftarrows [Co(NH_3)_6]^{2+}$	1.3×10^5
$Fe^{2+} + 6\,CN^- \rightleftarrows [Fe(CN)_6]^{4-}$	1.0×10^{35}

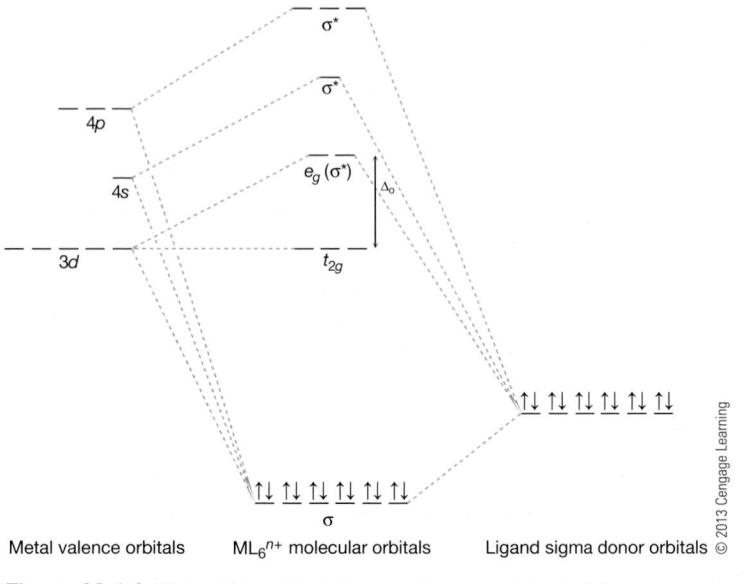

Figure 23.4.6 Molecular orbital diagram for an octahedral first-row transition metal complex ion

Labels on figure: $4p$, $4s$, $3d$, σ^*, $\overline{\sigma^*}$, $e_g\,(\sigma^*)$, Δ_o, t_{2g}, σ, Metal valence orbitals, $ML_6{}^{n+}$ molecular orbitals, Ligand sigma donor orbitals

© 2013 Cengage Learning

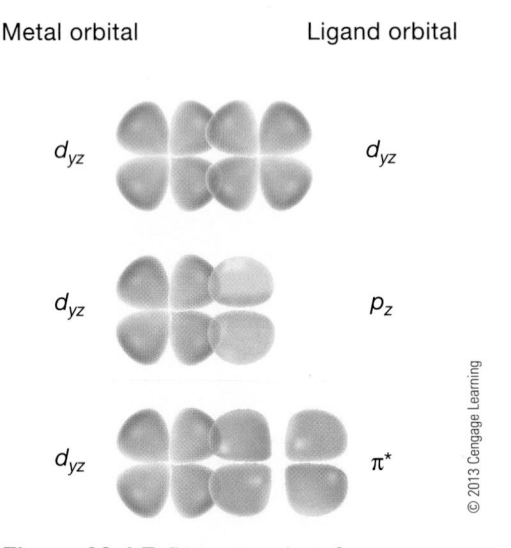

Metal orbital   Ligand orbital

d_{yz} d_{yz}

d_{yz} p_z

d_{yz} π^*

© 2013 Cengage Learning

Figure 23.4.7 Pi interactions between metal d orbitals and ligand orbitals

Metal–Ligand Pi Bonding

All ligands act as sigma electron-pair donors to metals in coordination compounds. In addition, some ligands also form pi bonds with metals, and it is these pi bonds that help explain the impact ligands have on the magnitude of Δ_o. Recall from Theories of Chemical Bonding (Unit 9) that pi bonds form when two orbitals overlap to form a bond where the bonding region is above and below the internuclear axis. (◄ Flashback to Section 9.3a Formation of Pi Bonds) In octahedral coordination compounds, the metal orbitals that have electron density between the x-, y-, and z-axes (the t_{2g} orbitals, d_{xy}, d_{xz}, and d_{yz}) can form pi bonds with ligand p, d, or π^* (pi antibonding) orbitals (Figure 23.4.7).

There are two possible types of pi interactions between a metal and a ligand: ligand-to-metal pi donation and metal-to-ligand pi donation.

Ligand-to-Metal Pi Donation

Ligands such as OH^-, Cl^-, and H_2O have filled p orbitals (lone pairs or filled orbitals that have significant p character) that are lower in energy than the metal t_{2g} orbitals in an octahedral complex. These ligand orbitals can interact with the metal t_{2g} orbitals, forming new pi bonding and antibonding molecular orbitals (Figure 23.4.8). (◄ Flashback to Section 9.4b Pi Bonding and Antibonding Molecular Orbitals)

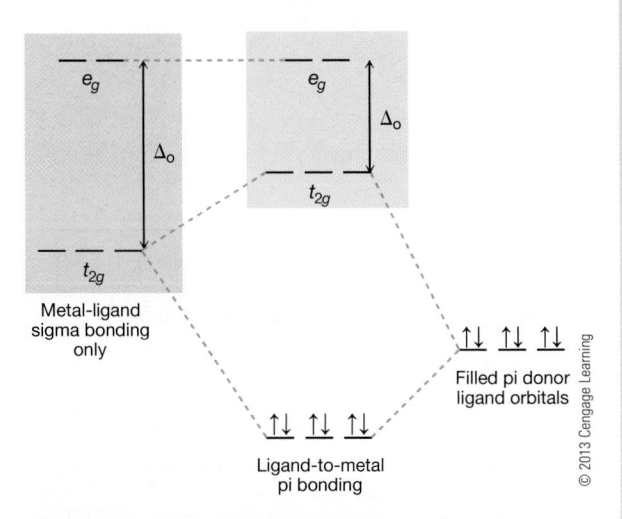

Labels on figure: e_g, e_g, Δ_o, Δ_o, t_{2g}, t_{2g}, Metal-ligand sigma bonding only, Ligand-to-metal pi bonding, Filled pi donor ligand orbitals

© 2013 Cengage Learning

Figure 23.4.8 Interaction between filled pi donor ligand orbitals and metal d orbitals

As shown in Figure 23.4.8, the result of the formation of these new molecular orbitals is *a decrease in the magnitude of* Δ_o. Therefore, ligands that can donate pi electron density to a metal (pi donor ligands such as Cl^- and H_2O) are generally weak-field ligands; compounds containing these ligands have smaller Δ_o values and are often high-spin complexes.

Metal-to-Ligand Pi Donation

Ligands such as CO, CN^-, and $P(C_6H_5)_3$ have empty π^* or d orbitals that are higher in energy than the metal e_g orbitals in an octahedral complex. These ligand orbitals can interact with the metal t_{2g} orbitals, forming new pi bonding and antibonding molecular orbitals (Figure 23.4.9).

As shown in Figure 23.4.9, the result of the formation of these new molecular orbitals is *an increase in the magnitude of* Δ_o. Therefore, ligands that can accept pi electron density from a metal (pi acceptor ligands) are generally strong-field ligands; compounds containing these ligands have larger Δ_o values and are often low-spin complexes. This type of metal-to-ligand pi donation is commonly called *pi backbonding*.

It is important to remember that these pi interactions occur in addition to the sigma bonding between ligands and metals. The two different types of metal–ligand interactions we have discussed and their effect on the magnitude of Δ_o are summarized in Interactive Table 23.4.2.

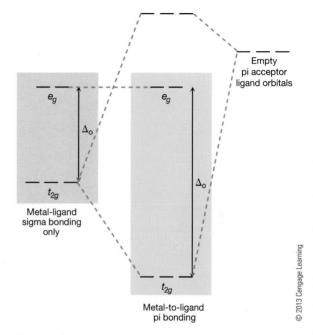

Figure 23.4.9 Interaction between empty pi acceptor ligand orbitals and metal d orbitals

---

Interactive Table 23.4.2

Metal–Ligand Bonding Interactions and Δ_o

Type of Ligand	Examples	Relative Magnitude of Δ_o	Example
Sigma donor only	NH_3, en	Intermediate	$Co(NH_3)_6^{3+}$ $\Delta_o = 274$ kJ/mol
Sigma donor and pi donor	Cl^-, H_2O	Smaller	$Co(H_2O)_6^{3+}$ $\Delta_o = 218$ kJ/mol
Sigma donor and pi acceptor	CO, CN^-	Larger	$Co(CN)_6^{3-}$ $\Delta_o = 401$ kJ/mol

23.4c Spectroscopy

Spectroscopy is a technique that analyzes how much light a solution absorbs at a given wavelength. Most chemical compounds absorb light in the visible and/or ultraviolet region by promoting an electron from a lower-energy orbital to one of higher energy. Measurements of the energy of the light absorbed tell us about the spacing between the different energy levels of the orbitals in a compound.

For coordination compounds, these transitions have energies that typically correspond to wavelengths in the visible region of the electromagnetic spectrum. When light of one wavelength is absorbed, the observed color of the solution corresponds to the color of the complementary wavelength. For example, when an aqueous solution containing a coordination compound absorbs red–violet light, we see its complementary color, green (Figure 23.4.10).

In a typical teaching laboratory, Spec-20 spectrometers are used to measure the light absorbed by solutions containing coordination compounds. In a Spec-20, the source light in the instrument passes through a diffraction grating, which separates the light into its constituent wavelengths. After setting the desired wavelength on the instrument, a sample containing a coordination compound (typically in aqueous solution) is placed in the instrument and the amount (intensity) of light that passes through the sample is measured by a detector.

In a more sophisticated version of the instrument, called a Photodiode Array UV-Vis Spectrophotometer, the source light first passes through the sample and then encounters the diffraction grating. The light then falls upon a series (also called an array) of photodiodes. The position of each diode determines what wavelength it measures. The amount of current produced in the diode indicates how much light of that wavelength is striking the diode. The more light that strikes the diode, the less light was absorbed by the sample.

In both instruments, the amount of light hitting the detector with the sample in place (I) is compared with that in the absence of the sample (I_0). The sample's **absorbance** at a given wavelength is calculated from these two values (Equation 23.1).

$$\text{absorbance } (A) = -\log \frac{I}{I_0} \qquad \textbf{(23.1)}$$

For example, if the sample absorbs 90% of the light, 10% is transmitted and $A = -\log(10/100) = 1.0$; if it absorbs 99% of the light, $A = -\log(1/100) = 2.0$.

d–d Transitions

Many transition metal compounds absorb visible light in a process that promotes a d electron from lower-energy d orbitals to higher-energy d orbitals. These electronic transitions are called d–d transitions, and the colors of many transition metal compounds can be

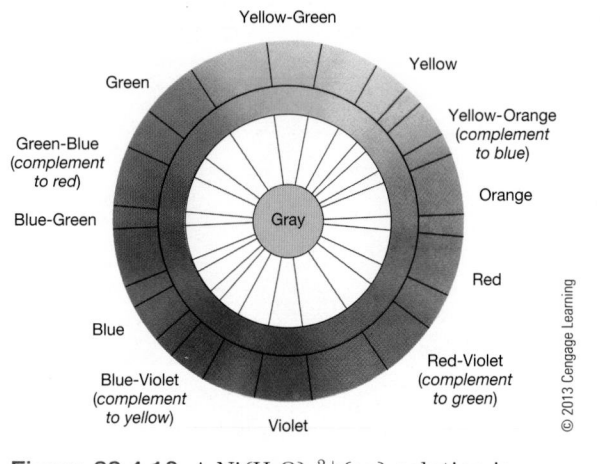

Figure 23.4.10 A $Ni(H_2O)_6^{2+}$(aq) solution is green because it absorbs red–violet light.

explained by this transition, the electron configuration of the metal, and the magnitude of Δ. It is important to note, however, that other electronic transitions are possible that absorb energy outside the visible range and that not all possible d–d transitions are "allowed." A d–d transition that results in a change in the number of unpaired electrons is "spin forbidden" (for reasons that go beyond the scope of the material covered in this text), and therefore the observed color for these transitions is not very intense.

The general relationship between the wavelength of light absorbed, the types of ligands, and Δ_o for octahedral coordination compounds is summarized in Table 23.4.3 and demonstrated in the examples that follow. Note that there will be exceptions to these general relationships because the geometry of the compound and the charge on the metal and its size also impact the magnitude of Δ.

Example 1 Why is an aqueous solution of $Ni(NH_3)_6^{2+}$ deep blue while an aqueous solution of $Ni(H_2O)_6^{2+}$ is green (Interactive Figure 23.4.11)?

Both complex ions contain the Ni^{2+} ion and have an octahedral geometry, but they have different ligands. Ammonia is a sigma-only donor ligand, whereas water can both sigma donate and pi donate (see Table 23.4.2). Therefore, Δ_o for $Ni(H_2O)_6^{2+}$ is less than Δ_o for $Ni(NH_3)_6^{2+}$. The compound with the smaller Δ_o will absorb light with lower energy (toward the red end of the visible spectrum). Aqueous solutions of $Ni(H_2O)_6^{2+}$ are green because they absorb low-energy (red) light. Aqueous solutions of $Ni(NH_3)_6^{2+}$ are blue because Δ_o is larger and the compound absorbs higher-energy (orange–yellow) light.

Example 2 Why are aqueous solutions of Mn^{2+} nearly colorless?

The manganese(II) ion has five d electrons. In an octahedral field of ligands, a d^5 configuration can be either high spin or low spin. The fact that the solution is nearly colorless tells us the ion is high spin because d–d transitions are spin forbidden for this electron configuration.

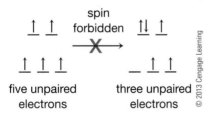

Table 23.4.3 Ligand Type, Δ_0, and Wavelength of Light Absorbed

Type of Ligand	Relative Magnitude of Δ_o	Type of Wavelength Absorbed
Weak sigma donor only	Smaller	Longer (lower energy)
Strong sigma donor only	Larger	Shorter (higher energy)
Sigma donor and pi donor	Smaller	Longer (lower energy)
Sigma donor and pi acceptor	Larger	Shorter (higher energy)

Interactive Figure 23.4.11

Explore the colors of transition metal compounds.

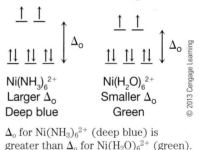

Δ_o for $Ni(NH_3)_6^{2+}$ (deep blue) is greater than Δ_o for $Ni(H_2O)_6^{2+}$ (green).

Section 23.4 Mastery

Unit Recap

Key Concepts

23.1 Properties of the Transition Metals

- Transition metals generally have similar chemical and physical properties (23.1a).
- Atomic size decreases only slightly as you move left to right across the transition metals in the periodic table (23.1b).
- The lanthanide contraction explains why the size of the transition metals does not increase significantly moving down a group (23.1b).
- First and second ionization energy values for the transition metals are generally much smaller than third ionization energies due to the similar valence electron configurations of these elements (23.1c).
- Most transition metals have at least two oxidation states (23.1c).

23.2 Isolation from Metal Ores

- Metals can be classified by the kinds of ores they form. Most transition metals are lithophiles, chalcophiles, or siderophiles (23.2a).
- Metals are extracted from ores by a number of different methods that include electrolysis, reaction with a reducing agent such as carbon (coke), and roasting (23.2b).

23.3 Coordination Compounds: Structure and Isomerism

- Coordination compounds typically consist of a metal, some ligands, and counterions (if needed) (23.3a).
- Ligands are classified by the number of attachments they form to a metal (23.3a).
- Coordination compounds and complex ions vary in their coordination number, the number and type of species inside and outside the coordination sphere, and the types of ligands attached to the metal (23.3a).
- Coordination compounds are named by a standard set of rules (23.3b).
- The chelate effect explains why coordination compounds containing chelating ligands are more thermodynamically stable than similar compounds containing only monodentate ligands (23.3c).
- Octahedral and square planar coordination compounds exhibit a variety of structural isomers (linkage isomers and ionization isomers) and stereoisomers (geometric isomers and optical isomers) (23.3d).

23.4 Coordination Compounds: Bonding and Spectroscopy

- Crystal field theory explains why the d orbitals in a coordination compound are not equal in energy (23.4a).

- The octahedral crystal field splitting energy, Δ_o, can be larger than electron-pairing energy (a strong-field case) or smaller than electron-pairing energy (a weak-field case) (23.4a).

- For coordination compounds with four to seven d electrons, high-spin (small Δ_o) and low-spin (large Δ_o) electron configurations are possible (23.4a).

- The spectrochemical series shows the effect ligands have on the magnitude of Δ_o; compounds containing strong-field ligands generally have a large Δ_o, whereas compounds containing weak-field ligands generally have a small Δ_o (23.4a).

- Molecular orbital theory is applied to coordination compounds to help explain why these compounds have large K_f values and the effect ligands have on the magnitude of Δ_o (23.4b).

- Metal–ligand sigma bonding results in the formation of low-energy bonding molecular orbitals (23.4b).

- Ligand-to-metal pi bonding results in a decrease in the magnitude of Δ_o (23.4b).

- Metal-to-ligand pi bonding (pi backbonding) results in an increase in the magnitude of Δ_o (23.4b).

- Coordination compounds are colored because of d–d transitions, where electrons move from lower-energy d orbitals to higher-energy d orbitals (23.4d).

- The color of a coordination compound is the complementary color of the light absorbed by the compound (23.4d).

- Not all d–d transitions are allowed or absorb light in the visible region of the electromagnetic spectrum (23.4d).

Key Equations

$$\text{absorbance } (A) = -\log \frac{I}{I_0} \tag{23.1}$$

Key Terms

23.1 Properties of the Transition Metals
transition metal
lanthanide contraction

23.2 Isolation from Metal Ores
ore
lithophile
chalcophile
siderophile
extraction
pyrometallurgy
hydrometallurgy

23.3 Coordination Compounds: Structure and Isomerism
coordination compound
complex ion
ligand

monodentate ligand
chelating ligand
multidentate ligand
bidentate ligand
hexadentate ligand
ambidentate ligand
coordination number
coordination sphere
    (or inner coordination sphere)
outer coordination sphere
chelate effect
coordination compound geometry
structural isomer
linkage isomer
ionization isomer
stereoisomer
cis–trans isomers

fac–mer isomers
optical isomers

23.4 Coordination Compounds: Bonding and Spectroscopy
crystal field theory
crystal field splitting energy (Δ)
strong field
low spin
weak field
pairing energy
high spin
spectrochemical series
strong-field ligand
weak-field ligand
spectroscopy
absorbance

Unit 23   Review and Challenge Problems

24 Nuclear Chemistry

Unit Outline

In This Unit...

Your study of chemistry to this point has been overwhelmingly based on the study of electrons and how they interact. Strikingly absent from the conversation to this point has been the nucleus of the atom. We discussed the nucleus in Elements and Compounds (Unit 2) when describing the components of an atom, how an atom gets its identity (the number of protons in the nucleus), and isotopes. We also talked about how the nucleus contains more than 99.9% of the mass of an atom but less than 0.00000000000001% of its total volume. Aside from that, the nucleus has been considered unchangeable throughout our study of chemistry. This is for a good reason; most of the time, nuclei are perfectly stable and do not change. This unit explores situations when nuclei do react and undergo change.

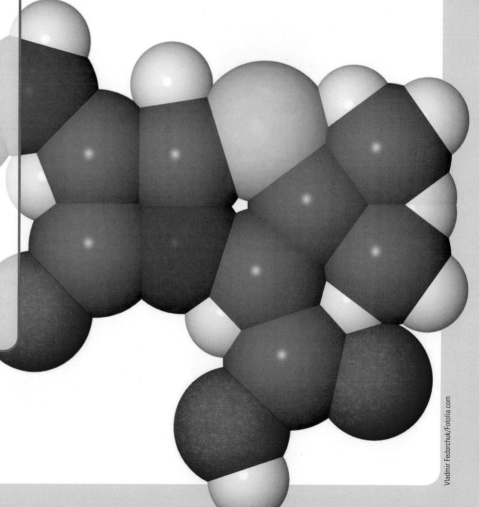

24.1 Nuclear Reactions

24.1a Nuclear vs. Chemical Reactions

In a chemical reaction, electrons move from one environment to another, and almost all reactions are accompanied by energy being absorbed or released when those electrons move. In a **nuclear reaction**, the particles in the nucleus, not the electrons, undergo change, and these changes are also accompanied by an exchange of energy with the surroundings.

Unlike chemical reactions, which involve changes to the number, arrangement, or environment of electrons, nuclear reactions involve changes in the number and types of particles in the nucleus. The exchange of energy in a nuclear reaction is similar to what happens during chemical reactions, but the scale of the energy exchanged is not at all similar. The energy change that happens during most chemical reactions is on the order of hundreds to thousands of kJ/mol. For example, propane, commonly used in barbecue grills and for home heating, releases 2200 kJ/mol when burned, or about 50 kJ/g. The reaction of uranium-235 (U-235), the fuel for the first nuclear explosion, released 1.95×10^{10} kJ/mol, or 8.31×10^7 kJ/g. This is over 10^6 times more energy per gram of reactant than what we consider a vigorous chemical explosion. Interactive Figure 24.1.1 illustrates this difference, showing the relative amounts of water that can be boiled using a single gram of propane and a single gram of the most commonly used nuclear energy fuel, U-235.

Another difference between chemical and nuclear reactions is the types of electromagnetic radiation they emit. Just as the transition of electrons from one energy level to another in an atom is quantized and occurs with the transfer of a quantum of energy, nucleons exhibit a similar transition in the nucleus. When an excited state nucleus "relaxes" to the ground state, it too gives off a quantum of energy. When electrons relax, we see some of these transitions as visible light because they are in the visible spectrum (about 1 to 3 eV/photon). When the nucleus relaxes, the transitions are on the order of MeV/photon or millions of electron volts per photon. These high-energy photons are the gamma rays in the high-energy region of the electromagnetic spectrum. (◄Flashback to Section 6.1b The Electromagnetic Spectrum)

Interactive Figure 24.1.1

Compare energy released by nuclear and chemical reactions.

Burning 1 g of propane generates enough heat to boil the water in the beaker. The heat generated by the fission of 1 g of uranium-235 will boil the water held in the large tank.

24.1b Natural Radioactive Decay

Radioactive decay is the process where an unstable nucleus emits energy and a small particle and is thereby transformed into a more stable nucleus. All elements with an atomic number greater than 83 ($Z > 83$) are naturally radioactive, and as you will see later in this unit, isotopes of many lighter elements also undergo radioactive decay. Interactive Table 24.1.1 lists the six particles that can be emitted from nuclei during radioactive decay.

An **alpha (α) particle** is a helium nucleus ($^4\text{He}^{2+}$), which consists of two protons and two neutrons. Alpha particles have a relatively large mass for a nuclear particle (relative mass 4.0026 u), carry a positive charge ($+2$), move relatively slowly ($\leq 10\%$ of the speed of light), and have low penetrating ability (an alpha particle can be stopped by a sheet of paper). The particle is represented using the Greek letter α ($^4_2\alpha$) or by the full atomic symbol for helium-4 (^4_2He). Note that although the particle formally has a $+2$ charge, by convention we do not include charge when describing nuclear reactions.

A **beta (β) particle** is an electron that is generated in the nucleus. Beta particles have very little mass (relative mass 0.0055 u), carry a negative charge (-1), move relatively quickly ($\leq 90\%$ of the speed of light), and have a higher penetrating ability than an alpha

Common Types of Radioactive Emissions

Type and Symbol	Identity	Mass (u)	Charge	Velocity	Penetration
Alpha (α, $^4_2\alpha$, ^4_2He)	Helium nucleus	4.0026	$+2$	$\leq 10\%$ speed of light	Low
Beta (β, $^0_{-1}\beta$, $^0_{-1}e$)	Electron	0.00055	-1	$\leq 90\%$ speed of light	Low to moderate, depending on energy
Positron ($^0_{+1}\beta$, $^0_{+1}e$)	Positively charged electron	0.00055	$+1$	$\leq 90\%$ speed of light	Low to moderate, depending on energy
Gamma ray (γ, $^0_0\gamma$)	High-energy electromagnetic radiation	0	0	Speed of light	High
Neutron (^1_0n)	Neutron	1.008665	0	$\leq 10\%$ speed of light	Very high
Proton (^1_1p, ^1_1H)	Proton, hydrogen nucleus	1.007276	$+1$	$\leq 10\%$ speed of light	Low to moderate, depending on energy

particle (a beta particle can pass through paper but can be stopped by a 0.5-cm-thick sheet of lead). The particle is represented using the Greek letter β ($_{-1}^{0}\beta$) or the letter e ($_{-1}^{0}e$), and is assigned a mass number of 0 and an atomic number of -1 (not a true atomic number, the -1 subscript helps in balancing nuclear reactions). A positron is the antimatter counterpart to an electron and is effectively a positively charged electron. Positrons have the same properties as electrons, with the exception of the charge on the particle. The particle is represented using the Greek letter β ($_{+1}^{0}\beta$) or the letter e ($_{+1}^{0}e$) and is assigned a mass number of 0 and an atomic number of $+1$ (not a true atomic number, the $+1$ subscript helps in balancing nuclear reactions).

Gamma rays are high-energy photons emitted from a nucleus when it decays from an excited, high-energy state. Gamma rays have no mass or charge, move very fast (at the speed of light), and have a very high penetrating ability (a gamma ray can be stopped by 10-cm-thick piece of lead). Gamma rays are represented using the Greek symbol γ ($_{0}^{0}\gamma$) and are assigned mass and atomic numbers of zero.

24.1c Radioactive Decay and Balancing Nuclear Reactions

A radioactive **nuclide**, a nucleus with a specific makeup of neutrons and protons, can undergo radioactive decay by ejecting a nuclear particle, capturing an electron, or giving off energy in the form of gamma radiation. In a radioactive decay reaction, the nucleus undergoing a transformation is called the **parent nucleus** and the nucleus formed during the transformation is called the **daughter nucleus**. As you will see in this unit, some nuclear transformations involve multiple daughter nuclei.

When writing an equation to represent a decay process or any other nuclear process, the reaction must be balanced. Just as a balanced chemical equation must have the same number of atoms of each element on each side of the reaction arrow, a balanced nuclear reaction must have the same number of nuclear particles on each side of the reaction arrow. We balance nuclear equations by balancing both mass and charge; the sum of the mass numbers of the reactants (the sum of the protons and neutrons in the nuclei) is equal to the sum of the mass numbers of the products, and charge balance is obtained by balancing the atomic numbers of the reactants and products.

For example, consider the alpha decay of uranium-238.

	$^{238}_{92}\text{U}$	$\rightarrow$	$^{234}_{90}\text{Th}$ +	$^{4}_{2}\alpha$
Number of protons	92		90	2
Number of neutrons	146		144	2
Total mass	238		234	4

The mass number of U-238 is equal to the sum of the mass numbers of thorium-234 (Th-234) and an alpha particle. In addition, charge is balanced because the number of protons is balanced (92 = 90 + 2). Both mass and charge are balanced in this reaction.

When balancing reactions involving charged species such as electrons, beta particles, positrons, and protons, the atomic number assigned to the particle changes the proton count for the reaction.

Consider this electron capture process:

	$^{83}_{37}\text{Rb}$ +	$^{0}_{-1}\text{e}$	$\rightarrow$	$^{83}_{36}\text{Kr}$
Number of protons	37	−1		36
Number of neutrons	46	0		47
Total mass	83	0		83

An electron really does not have a "−1" number of protons; however, assigning electrons and beta particles an atomic number of −1 maintains charge balance in the reaction. In the preceding reaction, there is a total charge of 37 + (−1) = 36 in the reactants, which balances the charge in the product nuclide.

Decay Processes

When a nucleus emits an alpha particle, it loses two protons and two neutrons in a process called **alpha decay**. For example, uranium-238 undergoes alpha decay to form thorium-234.

$$^{238}_{92}\text{U} \rightarrow {}^{234}_{90}\text{Th} + {}^{4}_{2}\alpha$$

Example Problem 24.1.1 Write balanced nuclear equations involving alpha decay.

Write a balanced nuclear equation for the alpha decay of radium-226.

Solution:

You are asked to write a balanced nuclear equation for a reaction involving alpha decay.

You are given the identity of the particle undergoing alpha decay.

$$^{226}_{88}\text{Ra} \rightarrow {}^{4}_{2}\alpha + {}^{222}_{86}\text{Rn}$$

Is your answer reasonable? When a nuclide undergoes alpha decay, the mass number decreases by four and the atomic number decreases by two.

Video Solution

Tutored Practice
Problem 24.1.1

Beta decay (or beta emission) occurs when a neutron in an unstable nucleus is converted to a proton and an electron and the nuclear electron is ejected from the nucleus.

$$^{1}_{0}\text{n} \rightarrow {}^{1}_{1}\text{p} + {}^{0}_{-1}\beta$$

When an unstable nucleus emits a beta particle, the number of neutrons in the nucleus decreases by one and the number of protons in the nucleus increases by one, but there is no change in mass number. For example, cesium-137 is a beta emitter.

$$^{137}_{55}\text{Cs} \rightarrow {}^{137}_{56}\text{Ba} + {}^{0}_{-1}\beta$$

Cesium-137 is converted to barium-137 when it undergoes beta decay. The mass number does not change, but the number of protons increases by one, thus changing the identity of the element.

Example Problem 24.1.2 Write balanced nuclear equations involving beta decay.

Write a balanced nuclear equation for the beta decay of carbon-14.

Solution:

You are asked to write a balanced nuclear equation for a reaction involving beta decay.

You are given the identity of the particle undergoing beta decay.

$$^{14}_{6}\text{C} \rightarrow {}^{0}_{-1}\beta + {}^{14}_{7}\text{N}$$

Is your answer reasonable? When a nuclide undergoes beta decay, the mass number does not change and the atomic number increases by one.

Video Solution

Tutored Practice
Problem 24.1.2

Similar to beta decay, **positron emission** occurs when a proton in an unstable nucleus is converted to a neutron and a positron and the positron is ejected from the nucleus.

$$^{1}_{1}\text{p} \rightarrow {}^{1}_{0}\text{n} + {}^{0}_{+1}\beta$$

When an unstable nucleus emits a positron, the number of protons in the nucleus decreases by one and the number of neutrons in the nucleus increases by one, but there is no change in mass number. For example, sodium-22 is a positron emitter.

$$^{22}_{11}\text{Na} \rightarrow {}^{22}_{10}\text{Ne} + {}^{0}_{+1}\beta$$

In this nuclear reaction, the mass number does not change but the atomic number decreases because a proton is converted to a neutron and the emitted positron. Positrons do not have a long half-life. They typically collide with an electron within microseconds of being generated, annihilating both particles and generating energy in the form of gamma rays.

Example Problem 24.1.3 Write balanced nuclear equations involving positron emission.

Write a balanced nuclear equation for the reaction when aluminum-25 undergoes positron emission.

Solution:

You are asked to write a balanced nuclear equation for a reaction involving positron emission.

You are given the identity of the particle undergoing positron emission.

$$^{25}_{13}\text{Al} \rightarrow {}^{0}_{+1}\beta + {}^{25}_{12}\text{Mg}$$

Is your answer reasonable? When a nuclide undergoes positron emission decay, the mass number does not change and the atomic number decreases by one.

Video Solution

Tutored Practice
Problem 24.1.3

One of the few instances where orbital electrons are involved in a nuclear reaction is in **electron capture**, where the nucleus captures an electron from its electron cloud. When an electron is captured by the nucleus, it combines with a proton to form a neutron. Thus, as we saw in positron emission, electron capture results in the decrease (by one) in the number of protons in the nucleus and an increase (by one) in the number of neutrons in the nucleus, but no change in the mass number. For example, note the decrease in atomic number when rubidium-83 undergoes electron capture to form krypton-83.

$$^{83}_{37}\text{Rb} + {}^{0}_{-1}\text{e} \rightarrow {}^{83}_{36}\text{Kr}$$

In the nuclear equation electron capture process, notice that the electron appears as a reactant, not a product of the reaction.

Gamma rays are often emitted after other nuclear events such as one of the decay processes discussed previously. For example, when cesium-137 undergoes beta decay, it forms a metastable isotope, barium-137m. A *metastable isotope* is in a high-energy state but is stable enough to be detected before it undergoes further decay, often by gamma emission. Metastable isotopes are indicated by adding the symbol "m" to the mass number.

$$^{137}_{55}\text{Cs} \rightarrow {}^{137\text{m}}_{56}\text{Ba} + {}^{0}_{-1}\beta$$

The metastable, high-energy state isotope decays to the ground state isotope by undergoing gamma emission.

$$^{137\text{m}}_{56}\text{Ba} \rightarrow {}^{137}_{56}\text{Ba} + {}^{0}_{0}\gamma$$

Many nuclear reactions produce excited state daughter nuclei that decay to the ground state by gamma emission. Note that in gamma emission there is no change in atomic number or mass number because this process is only releasing energy.

Section 24.1 Mastery

24.2 Nuclear Stability

24.2a Band of Stability

There are more than 250 naturally occurring nuclides. A plot of these nuclides by increasing numbers of protons and neutrons is shown in Interactive Figure 24.2.1, where the red dots represent unstable nuclei and the black dots represent stable (nonradioactive) nuclei.

From Interactive Figure 24.2.1, we can make the following observations:

- For elements with $Z \leq 20$, stable nuclei have approximately equal numbers of protons and neutrons. That is, the neutron:proton ratio (n/p) is very close to 1.

- As Z increases, stable nuclei have an n/p > 1.

- There are no stable nuclei with $Z > 83$.

- Elements with an even atomic number have a greater number of stable isotopes than those with an odd atomic number.

- Most stable isotopes have an even number of neutrons.

Interactive Figure 24.2.1

Explore the band of stability.

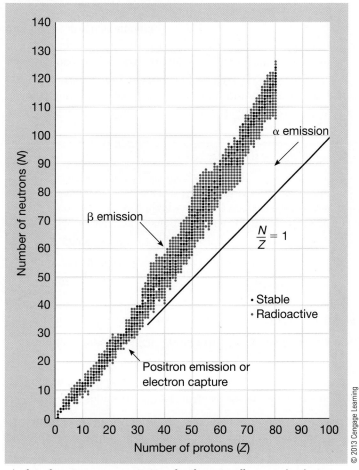

A plot of neutrons versus protons for the naturally occurring isotopes

The stable nuclei in Interactive Figure 24.2.1 lie along what is called the **band of stability**. Nuclides to the left of the band of stability have an n/p ratio that is too high, whereas those to the right of the band have an n/p ratio that is too low. We can use this information to predict the type of radioactive decay an unstable isotope undergoes in order to become more stable.

- Nuclides with a high n/p ratio often undergo beta decay because this will decrease the number of neutrons in the nucleus and increase the number of protons.

- Nuclides with a low n/p ratio often undergo positron emission or electron capture because either process will increase the number of neutrons in the nucleus and decrease the number of protons.

- Nuclides with $Z > 83$ often undergo alpha decay because this will decrease the mass number by four and the atomic number by two.

Figure 24.2.2 shows the effect of these decay processes on the number of neutrons and protons in a nuclide.

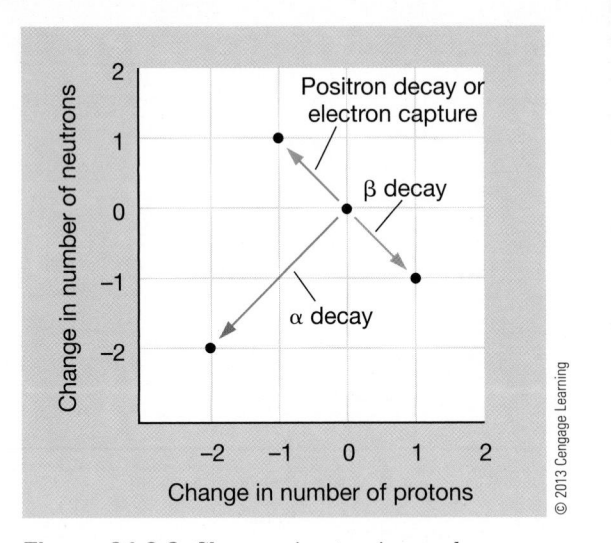

Figure 24.2.2 Changes in atomic number and mass number for different decay processes

---

Example Problem 24.2.1 Predict the decay product for a nuclide.

Based on nuclear stability, what is the most likely product nuclide when phosophorus-28 undergoes decay?

Solution:

You are asked to predict the decay product for a nuclide.

You are given the identity of the nuclide.

The atomic weight of phosphorus is 31.0 u and the atomic weight of phosphorus-28 is about 28 u. Therefore, ^{28}P has a lower mass than the stable isotopes of phosphorus. This means that the isotope has too few neutrons (its n/p ratio is too low), and it will decay by positron emission or electron capture to increase this ratio.

$$^{28}_{15}\text{P} \rightarrow {}^{0}_{+1}\beta + {}^{28}_{14}\text{Si}$$

Video Solution

Tutored Practice
Problem 24.2.1

For many unstable nuclides, one radioactive decay does not result in a stable nuclide; instead, further decay occurs. We can represent the decay path of a radioactive isotope in a **radioactive decay series** (Figure 24.2.3). In Figure 24.2.3, ^{238}U undergoes a series of decay processes before forming a stable nuclide, ^{206}Pb. Notice that most of the nuclides undergo alpha decay, the decay process that decreases both atomic mass and atomic number by the greatest amount.

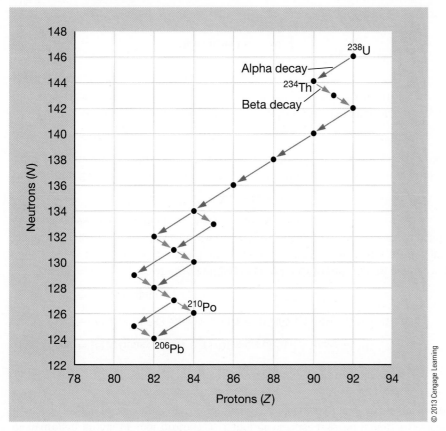

Figure 24.2.3 The ^{238}U decay series

24.2b Binding Energy

The nucleus of an atom is very small, about 1/10,000 the diameter of the atom. As more and more protons are packed into a tiny volume, repulsive forces increase, and as shown in Interactive Figure 24.2.1, increasing numbers of neutrons are required for a stable nucleus. **Binding energy**, $E_\mathbf{b}$, is a measure of the force holding the nuclear particles together in a nucleus. We can think of binding energy either as the amount of energy required to separate a nuclide's nuclear particles or the negative energy change that would occur if a nuclide's nuclear particles were combined to form a nucleus.

The nucleus is so much more stable than its separate nuclear particles that the energy difference can be observed as a mass difference. For example, consider phosphorus-31, which has 15 protons and 16 neutrons in its nucleus. We can calculate the binding energy for this isotope by first recognizing that the mass of a nucleus is always less than the mass of the individual particles that make up that nucleus. The **mass defect**, Δm, is the difference between the mass of the nuclear particles that make up a nucleus and the mass of an atom of the isotope of interest (Equation 24.1).

$$\Delta m = \Sigma(\text{mass of nuclear particles}) - (\text{mass of isotope}) \qquad \textbf{(24.1)}$$

The mass of one mole of neutrons is 1.00867 g/mol, and we use the mass of one mole of hydrogen atoms (1 mol protons + 1 mol electrons, 1.00783 g/mol) to calculate mass defect in order to account for the mass of the electrons included in the isotope mass. For ^{31}P,

$$^{31}_{15}P \rightarrow 15\,^{1}_{1}H + 16\,^{1}_{0}n$$

$$\Delta m = [(15)(\text{mass of }^{1}H) + (16)(\text{mass of }^{1}n)] - \text{mass of }^{31}P$$

$$= [(15)(1.00783 \text{ g/mol}) + (16)(1.00867 \text{ g/mol})] - 30.97376 \text{ g/mol}$$

$$= 0.28241 \text{ g/mol}$$

The mass defect is related to binding energy by Einstein's theory of special relativity, which states that the energy of a body is equivalent to its mass times the square of the speed of light, $E = mc^2$. To calculate the energy change for the process of combining nuclear particles to form a nucleus, this equation becomes

$$\Delta E = (\Delta m)c^2 \qquad \textbf{(24.2)}$$

In Equation 24.2, energy has units of joules and mass is expressed in units of kg (because $1 \text{ J} = 1 \text{ kg} \cdot \text{m}^2/\text{s}^2$). We can now calculate binding energy for one mole of phosphorus-31 nuclei.

$$\Delta E = (2.8241 \times 10^{-4} \text{ kg})(2.998 \times 10^8 \text{ m/s})^2 = 2.538 \times 10^{13} \text{ J}$$

Binding energy is usually reported in units of kJ per mol of nucleons, where a **nucleon** is a nuclear particle. The ^{31}P nucleus has 15 protons and 16 neutrons, or 31 nuclear particles.

$$E_{\text{b}} \text{ per mol nucleons} = \frac{2.538 \times 10^{13} \text{ J}}{31 \text{ mol nucleons}} \times \frac{1 \text{ kJ}}{10^3 \text{ J}} = 8.188 \times 10^8 \text{ kJ/mol nucleons}$$

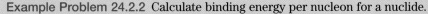

Example Problem 24.2.2 Calculate binding energy per nucleon for a nuclide.

Calculate the binding energy (in kJ/mol nucleons) for silver-107, which has a mass of 106.90509 g/mol.

Solution:

You are asked to calculate the binding energy for a nuclide.

You are given the identity of the nuclide.

Silver-107 has 47 protons and 60 neutrons. First, calculate the mass defect for this isotope.

$$^{107}_{47}Ag \rightarrow 47\,^1_1H + 60\,^1_0n$$

$$\Delta m = [(47)(\text{mass of }^1H) + (60)(\text{mass of }^1n)] - \text{mass of }^{107}Ag$$

$$= [(47)(1.00783\ \text{g/mol}) + (60)(1.00867\ \text{g/mol})] - 106.90509\ \text{g/mol}$$

$$= 0.98312\ \text{g/mol}$$

Use the mass defect to calculate binding energy for one mole of silver-107 nuclei.

$$\Delta E = (9.8312 \times 10^{-4}\ \text{kg})(2.998 \times 10^8\ \text{m/s})^2 = 8.836 \times 10^{13}\ \text{J}$$

$$E_b \text{ per mol nucleons} = \frac{8.836 \times 10^{13}\ \text{J}}{107\ \text{mol nucleons}} \times \frac{1\ \text{kJ}}{10^3\ \text{J}} = 8.258 \times 10^8\ \text{kJ/mol nucleons}$$

Video Solution

**Tutored Practice
Problem 24.2.2**

24.2c Relative Binding Energy

Binding energy per nucleon is an indication of the stability of the nucleus of an atom, so we can use it to compare the relative stability of nuclei. Interactive Figure 24.2.4 shows a plot of binding energy per mole of nucleons as a function of mass number. The isotopes with the greatest relative binding energy have a mass number between 50 and 75 u, with a maximum at ^{56}Fe.

For elements lighter than ^{56}Fe, adding more protons and neutrons results in an increase in nuclear attractions that outweighs the added electrostatic repulsions caused by having more protons near one another. Moving past ^{56}Fe toward heavier nuclei, adding more protons and neutrons leads to an increase in electrostatic repulsions that outweighs the additional nuclear attractions. This means that all nuclei are thermodynamically unstable compared with ^{56}Fe. As a result, iron is the most abundant of the heavier elements in the universe.

To increase stability, lighter nuclei can undergo nuclear **fusion**, the process where two or more nuclei combine to form a heavier nucleus. Nuclei heavier than Fe can become more stable by undergoing **fission**, where one heavy nucleus splits into smaller, more stable nuclei. The processes of fusion and fission are important in many applications of radioactivity such as power generation and the synthesis of new elements.

Section 24.2 Mastery

Investigate relative binding energy.

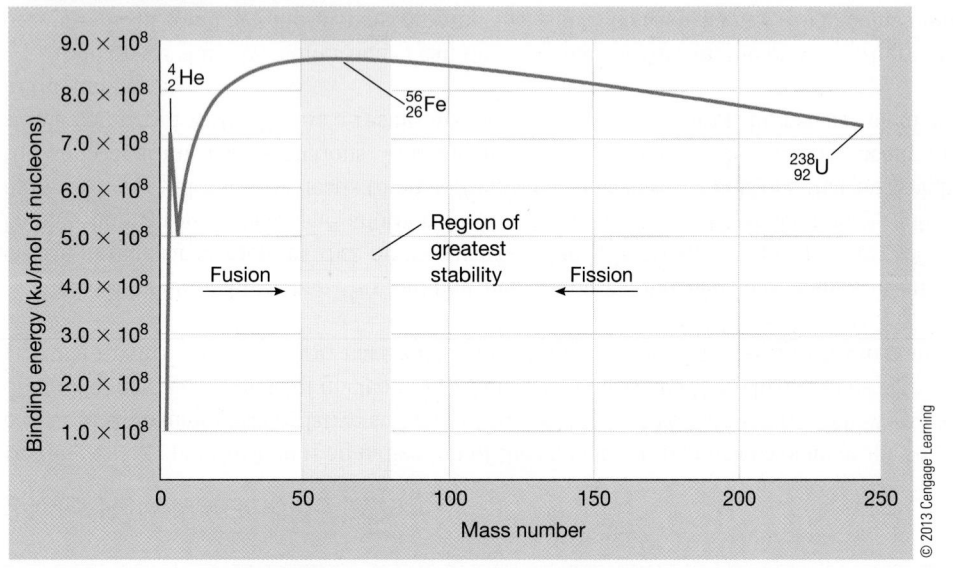

A plot of binding energy per nucleon versus atomic number

24.3 Kinetics of Radioactive Decay

24.3a Rate of Decay

Several applications of radioactive isotopes are linked to the speed at which the isotopes undergo decay. A medical procedure involving radioisotopes is likely to use ones that decay quickly so as to minimize the patient's exposure to radiation. Determining the age of ancient artifacts, however, relies on radioisotopes with long decay times.

As you learned in Chemical Kinetics (Unit 14), all nuclear decay processes follow first-order kinetics. The integrated first-order rate law can be rewritten for radioactive decay by substituting N, the number of nuclei, for [A], the concentration of a reactant (Equation 24.3).

$$\ln\left(\frac{N}{N_0}\right) = -kt \qquad \textbf{(24.3)}$$

The rate constant, k, can be calculated from the half-life of the radioactive decay for the isotope (Equation 24.4).

$$t_{1/2} = \frac{\ln 2}{k}$$ (24.4)

Half-lives for radioactive isotopes vary widely, from fractions of a second to billions of years (Table 24.3.1). We can use the first-order rate law to calculate the amount of material remaining at any time after $t = 0$, as shown in the following example problem.

Table 24.3.1 Decay Characteristics of Some Radioactive Nuclei

Nuclide	$t_{1/2}$	Decay Mode
^{3}H (tritium)	12.26 years	Beta
^{8}Be	$\sim 1 \times 10^{-16}$ s	Alpha
^{14}C	5730 years	Beta
^{32}P	14.28 days	Beta
^{109}Cd	453 days	Electron capture
^{131}I	8.04 days	Beta
^{226}Ra	1600 years	Alpha
^{238}U	4.47×10^9 years	Alpha
^{239}Pu	2.411×10^4 years	Alpha

Example Problem 24.3.1 Calculate the amount of radioactive material remaining after a given period of time.

The half-life of radon-222 is 3.82 days. If you begin with 37.4 mg of this isotope, what mass remains after 6.04 days have passed?

Solution:

You are asked to calculate the amount of radioactive material remaining after a given period of time.

You are given the half-life and initial mass of the radioactive material.

First, calculate the rate constant from the half-life.

$$k = \frac{\ln 2}{t_{1/2}} = \frac{\ln 2}{3.82 \text{ days}} = 0.181 \text{ d}^{-1}$$

Use half-life and the integrated first-order rate law to calculate the mass of radon-222 remaining after 6.04 days. In Equation 24.2, N is the number of nuclei, which is directly proportional to the mass of the sample.

$$\ln\left(\frac{N}{37.4 \text{ mg}}\right) = -(0.181 \text{ d}^{-1})(6.04 \text{ days})$$

$$\frac{N}{37.4 \text{ mg}} = 0.335$$

$$N = 12.5 \text{ mg}$$

Is your answer reasonable? The amount of time passed, 6.04 days, is a little less than two half-lives, so the amount remaining is between one half and one fourth of the original mass.

Video Solution

Tutored Practice
Problem 24.3.1

The half-life of a radioactive element can be determined by measuring its rate of decay using a **Geiger counter**, a device that detects ionizing radiation (Figure 24.3.1).

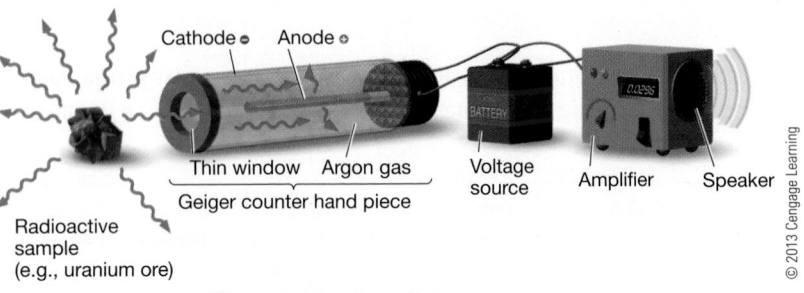

Figure 24.3.1 A Geiger counter

The Geiger counter allows us to measure the **activity**, A, of the sample, which is directly related to the rate constant (k, also called the *activity constant*) and the number of radioactive atoms present in the sample (N). Because radioactive decay follows first-order kinetics,

$$A = kN$$

and the ratio N/N_0 is equal to the ratio of activity at time t to that at the beginning of the experiment, A/A_0 (Equation 24.5).

$$\ln\left(\frac{A}{A_0}\right) = -kt \qquad \textbf{(24.5)}$$

Radioactive activity is typically reported in units of **curies** (Ci), millicuries (mCi), disintegrations per second (s^{-1}), or **becquerels** (Bq).

$$1\ \text{Ci} = 3.7 \times 10^{10}\ \text{s}^{-1} = 3.7 \times 10^{10}\ \text{Bq}$$

24.3b Radioactive Dating

Carbon dating uses the radioactive decay of a naturally occurring radioactive isotope of carbon, ^{14}C, to determine the age of artifacts that contain carbon. Carbon has two nonradioactive isotopes, ^{12}C (98.9%) and ^{13}C (1.1%). Radioactive ^{14}C is found in very small amounts (about 1 part per trillion) on Earth. It is formed in the upper atmosphere when ^{14}N reacts with high-energy solar radiation. Because ^{14}C is constantly decaying and being reformed in the upper atmosphere, there is a relatively constant concentration of ^{14}C present as $^{14}CO_2$ in the atmosphere. Like all CO_2, $^{14}CO_2$ can be taken up by plants and converted to organic material during photosynthesis. Any plant (or anything that eats plants) contains a constant concentration of ^{14}C (equal to the concentration in the atmosphere) as long as the plant (or animal or human) is alive.

Once the plant or animal or human dies, however, its ^{14}C content starts to decrease by radioactive decay. If we compare the amount of ^{14}C present in the nonliving object today with the amount that would be present if the object were living, we can estimate the amount of time that has passed since the object stopped taking up ^{14}C.

To calculate the age of an object that used to be alive, the ^{14}C radioactivity is measured (typically in counts per minute per gram of C) and compared with the ^{14}C radioactivity in living objects (the half-life of ^{14}C is 5730 years). This technique has been found to be effective for dating items up to 60,000 years old, but the process of dating objects using ^{14}C dating is complicated by the fact that the amount of ^{14}C in the atmosphere varies over time. Careful use of calibration curves and atmospheric measurements of atmospheric ^{14}C concentration allows scientists to use this method to obtain rather accurate dates for objects that contain previously living material.

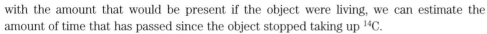

Example Problem 24.3.2 Use carbon dating to determine the age of an object.

A wooden bowl found in a cave in France is found to have a ^{14}C radioactivity of 10.2 counts per minute per gram of carbon. Living wood has an activity of 13.6 counts per minute per gram of carbon. How long ago did the tree containing the wood used to make the bowl die?

Solution:

You are asked to determine how long ago a tree used to make a wooden bowl died.

You are given the ^{14}C radioactivity of the wooden bowl and of living wood.

First, we will assume that the fraction of ^{14}C in the atmosphere has been constant over time. Therefore, the activity at $t = 0$ is assumed to have been 13.6 counts/min · g. Because the radioactivity of a sample is directly proportional to the amount of ^{14}C present, the ratio of ^{14}C present now and when the tree used to make the wood bowl stopped taking up $^{14}CO_2$ is equal to the ratio of activities. Thus, we can use the ^{14}C activity to determine the age of the artifact.

First, calculate the rate constant for the radioactive decay of ^{14}C.

$$t_{1/2} = \frac{0.693}{k}$$

$$k = \frac{0.693}{t_{1/2}} = \frac{0.693}{5730 \text{ y}} = 1.21 \times 10^{-4} \text{ y}^{-1}$$

Next, use the integrated first-order rate law to determine the age of the artifact.

$$\ln \frac{N_t}{N_0} = -kt$$

$$\ln\left(\frac{10.2 \text{ counts/min} \cdot \text{g}}{13.6 \text{ counts/min} \cdot \text{g}}\right) = -(1.21 \times 10^{-4} \text{ y}^{-1})t$$

$$t = 2380 \text{ y}$$

The artifact is about 2380 years old.

Video Solution

Tutored Practice
Problem 24.3.2

Table 24.3.2 Some Radioisotopes Used in Dating Nonorganic Objects

Parent Isotope	Stable Daughter Product	$t_{1/2}$
^{238}U	^{206}Pb	4.47×10^9 y
^{235}U	^{207}Pb	7.04×10^8 y
^{232}Th	^{208}Pb	1.4×10^{10} y
^{87}Rb	^{87}Sr	4.75×10^{10} y
^{40}K	^{40}Ar	1.277×10^9 y
^{147}Sm	^{143}Nd	1.06×10^{11} y

Other radioactive isotopes can be used to date other types of nonorganic materials, such as rocks. The technique is similar to that used for carbon dating but uses isotopes that are much longer lived. Some important radioactive isotopes used in dating nonorganic objects and the stable daughter products are shown in Table 24.3.2.

Section 24.3 Mastery

24.4 Fission and Fusion

24.4a Types of Fission Reactions

Fission was first discovered in 1938, when radioactive barium was detected among the products formed when uranium was bombarded with neutrons.

$$^{235}_{92}U + ^{1}_{0}n \rightarrow ^{141}_{56}Ba + ^{92}_{36}Kr + 3\,^{1}_{0}n$$

This reaction releases a large amount of energy, about 8.3×10^7 kJ. Shortly after this discovery, Albert Einstein, at the request of several scientists, wrote a letter to President Franklin Roosevelt urging him to begin research to develop a bomb using the energy from fission reactions so that Germany would not develop one first. This letter ultimately resulted in the beginning of the Manhattan Project, a top-secret research effort to develop an atomic bomb.

Neutron-Induced Fission

The preceding reaction is an example of neutron-induced fission, where the fission event is initiated by the collision of a neutron with a heavy fissionable (*fissile*) nucleus. As shown in the following reactions, neutron bombardment of ^{235}U can produce other nuclides.

$$^{235}_{92}U + ^{1}_{0}n \rightarrow ^{135}_{50}Sn + ^{99}_{42}Mo + 2\,^{1}_{0}n$$

$$^{235}_{92}U + ^{1}_{0}n \rightarrow ^{139}_{52}Te + ^{94}_{40}Zr + 3\,^{1}_{0}n$$

More than 370 different nuclides have been detected from the neutron-induced fission of ^{235}U. For reasons that are not well understood, fission is unsymmetrical (ratio of mass numbers for the product nuclides is approximately 1.4) (Figure 24.4.1).

On average, neutron-induced fission of ^{235}U produces 2.5 neutrons per uranium nucleus. These neutrons can then collide with other fissile nuclides, creating a **chain reaction** (Interactive Figure 24.4.2).

Interactive Figure 24.4.2

Explore a nuclear chain reaction.

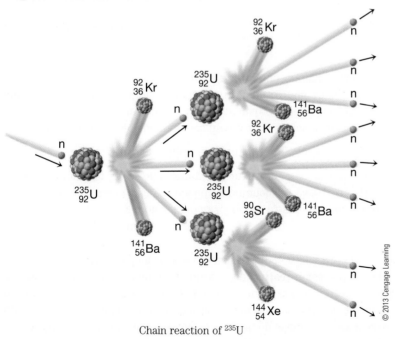

Chain reaction of ^{235}U

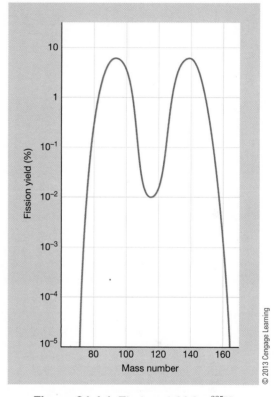

Figure 24.4.1 Fission yield for ^{235}U

A chain reaction is described as *critical* if, on average, every neutron produced in the fission reaction collides with another fissile nuclide and induces fission. A *subcritical* chain reaction produces less than one fission per neutron, and a *supercritical* chain reaction produces more than one fission per neutron. The **critical mass** is the minimum mass of fissile material needed to sustain a critical chain reaction. For ^{235}U, this critical mass is 880 g.

Spontaneous Fission

A number of very heavy radioactive nuclei decay by spontaneous fission. As the name suggests, this occurs when a parent nuclide spontaneously cleaves into two smaller more stable nuclei. Californium-252 is one example of an isotope that undergoes spontaneous fission:

$$^{252}_{98}\text{Cf} \rightarrow {}^{141}_{56}\text{Ba} + {}^{107}_{42}\text{Mo} + 4\,{}^{1}_{0}\text{n}$$

Nuclei that undergo spontaneous fission are used as neutron sources for a number of applications including the detection of trace materials in airline luggage by screening equipment and remote detection of moisture in grain storage units. The daughter nuclides of spontaneous fission have a very high n/p ratio and undergo beta decay to increase their stability.

24.4b Nuclear Fuel

Elemental uranium consists primarily of the nonfissile isotope ^{238}U (99.3%); the fissile isotope ^{235}U is only 0.7% abundant. For the uranium in nuclear power plants to sustain a nuclear chain reaction, the concentration of fissile ^{235}U must be increased to about 3% to 5%. Nuclear weapons require a much higher enrichment, in excess of 85% due to the need for supercritical explosive kinetics (Interactive Table 24.4.1).

Interactive Table 24.4.1

Types of Enriched Uranium

Classification	^{235}U Concentration	Application
Slightly enriched uranium (SEU)	0.9–2.0%	Fuel for nuclear power plants
Low-enriched uranium (LEU)	<20%	Fuel for nuclear power plants
Highly enriched uranium (HEU)	>20%	Nuclear weapons Fuel for nuclear power plants

Enriched Uranium

Uranium ore contains small amounts of UO_2 and UO_3. Before the enrichment process begins, the uranium is extracted from the ore and converted to UF_6, a compound that sublimes at a relatively low temperature (56.5 °C). The process involves four steps:

Step 1. Oxidize UO_2 to UO_3.

Step 2. Solubilize the uranium by reaction with H_2SO_4. The solution containing dissolved uranium is separated from the rest of the crushed rock.

$$UO_3(s) + 2\,H^+(aq) \rightarrow UO_2{}^{2+}(\text{solid salt}) + H_2O(\ell)$$

$$UO_2{}^{2+}(\text{solid salt}) + 3\,SO_4{}^{2-}(aq) \rightarrow UO_2(SO_4)_3{}^{4-}(aq)$$

Step 3. Convert leached uranium into pure UO_3 using a soluble base (R_3N).

$$2\,R_3N(aq) + H_2SO_4(aq) \rightarrow (R_3NH)_2SO_4(aq)$$

$$2\,(R_3NH)_2SO_4(aq) + UO_2(SO_4)_3{}^{4-}(aq) \rightarrow (R_3NH)_4UO_2(SO_4)_3(aq) + 2\,SO_4{}^{2-}(aq)$$

$$(R_3NH)_4UO_2(SO_4)_3(aq) + 2\,(NH_4)_2SO_4(aq) \rightarrow 4\,R_3N(aq) + (NH_4)_4UO_2(SO_4)_3(aq) + 2\,H_2SO_4(aq)$$

$$2\,NH_3(aq) + 2\,UO_2(SO_4)_3{}^{4-}(aq) \rightarrow (NH_4)_2U_2O_7(s) + 4\,SO_4{}^{2-}(aq)$$

$$(NH_4)_2U_2O_7(s) \rightarrow U_3O_8(s) + H_2O(\ell) + NH_3(g)$$

U_3O_8 is then used to make pure UO_3.

Step 4. Convert UO_3 into UF_6.

$$UO_3(s) + H_2(g) \rightarrow UO_2(s) + H_2O(\ell)$$

$$UO_2(s) + 4\,HF(g) \rightarrow UF_4(s) + 2\,H_2O(\ell)$$

$$UF_4(s) + F_2(g) \rightarrow UF_6(\ell \text{ or } g)$$

The final product, UF_6, is a mixture of $^{238}UF_6$ (99.3%) and $^{235}UF_6$ (0.7%). The two compounds differ only in molar mass, so gaseous diffusion or a gas centrifuge is used to increase the concentration of $^{235}UF_6$ in the sample. It is the components for gas centrifuges that nuclear inspectors look for when searching installations suspected of secretly creating nuclear weapons. The uranium is easy to hide; the giant machines needed to enrich the uranium are not. A new and much more efficient process using laser excitement of selective isotopes is presently in testing.

Breeder Reactors

Another fissile nuclide used for nuclear fuel is ^{239}Pu. Plutonium-239 is not a naturally occurring isotope, but it is formed in large amounts inside nuclear reactors using uranium as a fuel. A **breeder reactor** is a nuclear reactor that produces fissile material at a greater rate than fissile material is consumed. The production of ^{239}Pu uses the non-fissile isotope of uranium, ^{238}U.

$$^{238}_{92}\text{U} + {}^{1}_{0}\text{n} \rightarrow {}^{239}_{92}\text{U} + {}^{0}_{-1}\beta$$

$$^{239}_{92}\text{U} \rightarrow {}^{239}_{93}\text{Np} + {}^{0}_{-1}\beta$$

$$^{239}_{93}\text{Np} \rightarrow {}^{239}_{94}\text{Pu} + {}^{0}_{-1}\beta$$

24.4c Nuclear Power

The energy produced in a neutron-induced fission reaction can be harnessed in a nuclear power plant and used to create electricity. In 2010, there were 432 operable nuclear power reactors worldwide; in the United States, 104 reactors were operating and producing 20% of the nation's electricity. France generates the highest percentage of its electricity from nuclear reactors (76% of its electricity was generated from 58 reactors in 2010).

A nuclear power plant consists of three main components: the reactor, the heat exchanger, and the steam turbine (Interactive Figure 24.4.3). Key to the design of a nuclear power plant is the **nuclear reactor**, the container in which the controlled nuclear reaction occurs, and the control of the chain reaction; it is important that the chain reaction run so that it is barely critical. A variety of methods are used to control the speed of the neutrons being produced in the fission reaction to control both the speed of the chain reaction and its efficiency. **Moderators**, substances such as nonenriched water, heavy water (enriched in deuterium), and graphite, slow the neutrons and increase reaction efficiency. **Control rods**, which consist of elements that absorb neutrons, such as boron, silver, indium, and cadmium, can be inserted between the fuel rods to slow or stop the chain reaction.

In a water-cooled nuclear reactor, the heat generated by the fission reaction heats the water surrounding the fuel assembly. The water does not boil, however, because it is kept at very high pressure. The heated water passes through a heat exchanger, generating steam, and then is pumped back to the reactor. The steam turns a turbine, generating electricity; is condensed into water by another heat exchanger, typically using lake or river water; and is then pumped back to the original heat exchange unit.

Explore a nuclear power plant.

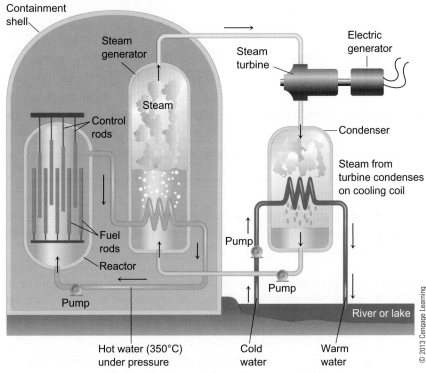

Schematic diagram of a nuclear power plant

Nuclear Fusion

Nuclear fusion, the combination of two or more light nuclides into a larger nuclide, generates a large amount of energy. Fusion is the source of energy in the stars, where large amounts of hydrogen nuclei are fused to form helium nuclei.

$$4\,{}^{1}_{1}\text{H} \rightarrow {}^{4}_{2}\text{He} + 2\,{}^{0}_{+1}\beta \qquad \Delta E = -2.5 \times 10^{9}\,\text{kJ}$$

Other small nuclei undergo fusion, such as the fusion of deuterium nuclei to form ^{3}He. The large amounts of energy generated in fusion reactions led scientists to study fusion as a potential energy source. However, generating the temperature required to initiate a fusion reaction (approximately 1×10^{8} K for the fusion of deuterium and tritium) is extremely difficult. At these high temperatures, the reactants exist as a **plasma**, an ionized gas consisting of unbound nuclei and electrons. It has proved difficult in fusion reactors to contain the plasma at the correct temperature long enough to generate more energy than has been used to create the plasma.

Even with these challenges, fusion as a potential energy source continues to be studied worldwide. A number of countries are currently cooperating on the design and construction of the International Thermonuclear Experimental Reactor (ITER), an experimental fusion reactor located in the south of France that will allow scientists to explore the feasibility of using fusion as an energy source on Earth.

Section 24.4 Mastery

24.5 Applications and Uses of Nuclear Chemistry

24.5a Stellar Synthesis of Elements

When the universe was first created, after the big bang, it was initially composed almost solely of hydrogen atoms. Most of the other elements in the universe have been created in stars from nuclear reactions that built up from hydrogen. Nuclear synthesis happens in stars because that is where the temperature is high enough to allow nuclear fusion to take place.

A logarithmic plot of relative element abundance in the universe versus atomic number shows some interesting features, many of which we can explain by exploring the reactions that take place in stars. (See Interactive Figure 24.5.1, where the relative abundance of Si is set at 10^6.)

- Hydrogen and helium are by far the most abundant elements in the universe.

- Iron has an unusually high abundance compared with the other elements near it in the periodic table.

- Elements with even atomic numbers are more abundant than elements with odd atomic numbers.

- The heavier elements are not very abundant, and some elements (such as Li, Be, and B) have unusually low abundance compared with the elements they are near in the periodic table.

Interactive Figure 24.5.1

Explore cosmic elemental abundance.

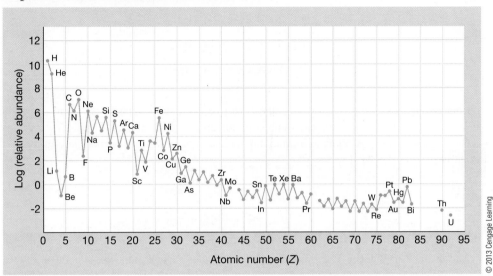

Relative cosmic abundance of the elements

Hydrogen Burning

Stars are large collections of hydrogen and helium held together by gravity. When the temperature in a star reaches about 1×10^7 K, the process called hydrogen burning begins. The overall reaction is the conversion of four protons into a helium nucleus plus two positrons and two **neutrinos** (ν_e, neutral particles with very small mass that travel at speeds close to the speed of light).

$$4\,{}^{1}_{1}\text{H} \rightarrow {}^{4}_{2}\text{He} + 2\,{}^{0}_{+1}\beta + 2\,\nu_e$$

The reaction actually occurs in multiple steps, the most important of which are

$${}^{1}_{1}\text{H} + {}^{1}_{1}\text{H} \rightarrow {}^{2}_{1}\text{H} + {}^{0}_{+1}\beta + \nu_e$$

$${}^{2}_{1}\text{H} + {}^{1}_{1}\text{H} \rightarrow {}^{3}_{2}\text{He} + \gamma$$

$${}^{3}_{2}\text{He} + {}^{3}_{2}\text{He} \rightarrow {}^{4}_{2}\text{He} + 2\,{}^{1}_{1}\text{H}$$

In stars that contain the elements N, C, and O, catalytic cycles, an example of which is shown here, can also result in hydrogen burning. Note that the net reaction for the CNO cycle is also the conversion of four protons to a helium nucleus.

$$^{12}_{6}C + ^{1}_{1}H \rightarrow ^{13}_{7}N + \gamma$$

$$^{13}_{7}N \rightarrow ^{13}_{6}C + ^{0}_{+1}\beta$$

$$^{13}_{6}C + ^{1}_{1}H \rightarrow ^{14}_{7}N + \gamma$$

$$^{14}_{7}N + ^{1}_{1}H \rightarrow ^{15}_{8}O + \gamma$$

$$^{15}_{8}O \rightarrow ^{15}_{7}N + ^{0}_{+1}\beta$$

$$^{15}_{7}N + ^{1}_{1}H \rightarrow ^{12}_{6}C + ^{4}_{2}He$$

Net reaction: $4^{1}_{1}H \rightarrow ^{4}_{2}He + 2\,^{0}_{+1}\beta + 3\,\gamma$

Helium and Carbon Burning

When a large star gets older and the temperature increases, helium burning begins. The net reaction for helium burning is the fusion of three ^{4}He nuclei to form ^{12}C. However, the reaction actually occurs in two steps, where the first is a rapid equilibrium forming ^{8}Be. Beryllium-8 is very unstable at the temperatures found in stars, so it has a very short half-life and is converted to ^{12}C in an excited state (^{12}C*) that gives off a gamma ray, forming ground state ^{12}C. Beryllium-8 is essentially "skipped" in nuclear synthesis in stars.

$$^{4}_{2}He + ^{4}_{2}He \rightleftarrows ^{8}_{4}Be$$

$$^{8}_{4}Be + ^{4}_{2}He \rightarrow ^{12}_{6}C* \rightarrow ^{12}_{6}C + \gamma$$

Net reaction: $3\,^{4}_{2}He \rightarrow ^{12}_{6}C + \gamma$

Additional helium-burning reactions combine helium nuclei with heavier nuclei to synthesize ^{16}O, ^{20}Ne, and ^{24}Mg. Carbon burning also takes place as the star's temperature increases, and ^{12}C nuclei combine to form heavier nuclei and also to regenerate ^{1}H and ^{4}He.

$$^{12}_{6}C + ^{12}_{6}C \rightarrow ^{24}_{12}Mg + \gamma$$

$$^{12}_{6}C + ^{12}_{6}C \rightarrow ^{23}_{11}Na + ^{1}_{1}H$$

$$^{12}_{6}C + ^{12}_{6}C \rightarrow ^{20}_{10}Ne + ^{4}_{2}He$$

The fact that most of the elements up through about calcium are built up from ^{4}He nuclei explains why elements in this range with even atomic numbers and mass numbers divisible by 4 are more abundant than those with odd atomic numbers.

Other Stellar Synthesis Reactions

The reactions that take place in stars after helium and carbon burning depend on the size and temperature of the star. The most important types of reactions that occur are

- the alpha process, which is responsible for the synthesis of nuclei with even atomic numbers up to ^{40}Ca;

- the equilibrium process, which takes places just before a short-lived star explodes and accounts for the relative abundance of elements from Ti to Cu and the unusually high abundance of Fe; and

- neutron capture-beta decay (both slow and rapid), processes that are responsible for most of the elements beyond atomic number 20.

The x-process is the single stellar synthesis process that does not occur in a star. Instead, this process occurs in interstellar spaces, where cosmic rays consisting of heavier elements come in contact with the hydrogen and helium in interstellar gases. When nuclei collide in interstellar space, *spallation* reactions occur, resulting in fragmentation of the heavier nuclei. The products of the spallation reactions include the elements Li, Be, and B, elements that were skipped during stellar synthesis reactions.

24.5b Induced Synthesis of Elements

Elements can also be synthesized in a laboratory setting. In fact, most of the elements above uranium, the **trans-uranium elements**, can only be synthesized artificially.

Early researchers synthesized low-mass isotopes by bombarding elements with alpha particles.

$$^{14}_{7}N + ^{4}_{2}\alpha \rightarrow ^{17}_{8}O + ^{1}_{1}p$$

These reactions are very difficult to perform because of the high energy needed to overcome the repulsive forces between positively charged nuclei. For elements with higher mass, even more energy is required to induce a nuclear reaction of this type. Several techniques, including the cyclotron and the linear particle accelerator, have been developed that allow scientists to perform these bombardment reactions.

In a cyclotron, alternating voltage is applied to two hollow, D-shaped electrodes. A particle source for the charged particle sits at the center of the device. The charged particle is accelerated by the electric field when it passes through the gap between the two electrodes (Interactive Figure 24.5.2). Neptunium, the first of the trans-uranium elements to be synthesized, was synthesized using neutrons produced using a cyclotron.

Interactive Figure 24.5.2

Explore a cyclotron.

Magnetic field

Spiral path
of ions

Hollow D-shaped
electrodes

High-frequency
voltage

+
−

−
+

Emerging
ion beam

Negative electrode
to deflect beam
to target

Positive
ion source

Target
material

© 2013 Cengage Learning

A linear accelerator works in a similar manner. An ion source is placed at one end of the accelerator, and the charged particle passes through tubes with reversing voltage (Figure 24.5.3). The particle accelerates as it passes between tubes; by repeating the process multiple times, the resulting particle beam has very high energy.

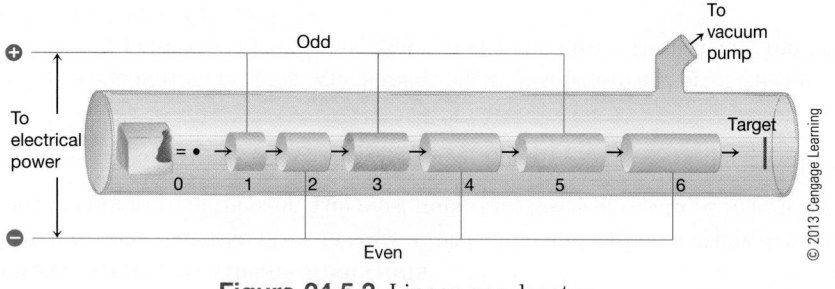

Odd

To
vacuum
pump

⊕

To
electrical
power

= •

Target

0   1   2   3   4   5   6

⊖

Even

© 2013 Cengage Learning

Figure 24.5.3 Linear accelerator

A third method used to synthesize heavy isotopes is neutron bombardment. Because neutrons carry no charge, repulsive forces are not an issue and the process occurs at lower energy. Many radioisotopes are synthesized by neutron capture, including the ^{14}C used in carbon dating.

$$^{14}_{7}N + ^{1}_{0}n \rightarrow ^{14}_{6}C + ^{1}_{1}p$$

24.5c Nuclear Medicine

Radioisotopes are used for a variety of applications in medicine. Some are used for medical imaging, whereas others are used for diagnosis or treatment of a medical problem.

Medical Imaging

Positron emitters are used in the medical imaging technique called positron emission tomography (PET). In this application, a positron-emitting isotope such as ^{11}C, ^{13}N, or ^{15}O is chemically bound to a tracer molecule and injected into a patient. The positrons are annihilated when they encounter an electron, emitting two gamma rays that move in opposite directions.

$$^{0}_{+1}\beta + ^{0}_{-1}e \rightarrow 2\,\gamma$$

The gamma rays are detected by detectors in the PET scanner that are located 180° apart. This technique is widely used for brain scans (Figure 24.5.4).

Other types of imaging take advantage of the fact that a radioisotope will accumulate in a specific location in the body. For example, technetium-99m can be bound to the pyrophosphate ion ($P_4O_7^{2-}$) and then injected into the bloodstream. The complex accumulates in the bones and kidneys. As ^{99m}Tc decays, it ejects gamma rays that are detected and used to create an image that can then be used to detect bone tumors. Table 24.5.1 shows some of the radioisotopes used for location-specific medical imaging.

Medical Treatment

Radioisotopes can also be used to treat medical problems. For example, iodine-131 is a beta emitter that is used to treat hyperthyroidism or thyroid cancer. Because iodine naturally concentrates in the thyroid, a patient can be treated with an oral dose of sodium iodide containing a relatively high concentration of iodine-131. The radioisotope concentrates in the thyroid and the high-energy beta particles emitted during decay cause the thyroid or tumor to shrink.

Other examples of medical treatment of cancer using radioisotopes include the external use of a high-energy beam of gamma radiation (from a gamma emission source) or the internal use of gamma-emitting radioisotopes delivered in gold- or platinum-coated delivery vessels.

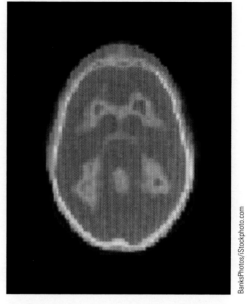

BanksPhotos/iStockphoto.com

Figure 24.5.4 PET scan of an axial section through a healthy human brain

Table 24.5.1 Diagnostic Radioisotopes		
Radioisotope	$t_{1/2}$ **(hours)**	**Site for Diagnosis**
^{99m}Tc	6.0	As $^{99m}TcO_4^-$ to the thyroid
^{201}Tl	72.9	To the heart
^{131}I	193	To the thyroid
^{67}Ga	78.2	To various tumors and abscesses

Example Problem 24.5.1 Investigate applications of nuclear medicine.

a. What type of radioactive emission is used in a PET scan?
b. Are radioactive isotopes used in medicine for imaging, treating disease, or both?

Solution:

You are asked to answer questions about the use of radioactive isotopes in nuclear medicine.

You are given a specific application of radioactive isotopes.

a. PET stands for positron emission tomography. It uses a positron-emitting radioactive isotope that is embedded as part of a "tracer" molecule that is injected in a patient and will concentrate in certain parts of the body.
b. Radioactive isotopes are used in imaging, such as the PET technique, and they are also used to treat disease such as cancer, where emitted radiation is used to kill cancer cells.

Video Solution

Tutored Practice
Problem 24.5.1

24.5d Radioactivity in the Home

We are exposed to ionizing radiation every day of our lives. Many of the atoms that make up our body are naturally radioactive. In addition, everything we eat and all the matter that surrounds us emits low levels of ionizing radiation. In a typical home, radioactivity can be both helpful and harmful.

Smoke Detectors

Smoke detectors save lives by alerting people to the presence of smoke, giving them a chance to exit the building safely. In a smoke detector, a small sample of an alpha emitter is used to ionize air, completing an electrical circuit (Interactive Figure 24.5.5). When smoke is in the air, it interferes with the ionizing radiation and the circuit is broken, triggering the alarm.

Food Irradiation

Radioactivity can be used to prevent food spoilage and food-borne illnesses such as *Salmonella* or *Escherichia coli.* This method is used for many foods because it reduces the need for chemical preservatives and improves shelf life. Food irradiation works by damaging the organism's DNA beyond its ability to repair itself. Most food is irradiated by gamma rays from ^{60}Co or ^{137}Cs sources, although x-ray radiation is also used.

Public perception of this technique has been an issue, because some mistakenly worry that the food becomes radioactive when exposed to ionizing radiation. Others worry that the food quality is degraded by the exposure. However, food spoilage is dramatically reduced with use of this method (Figure 24.5.6).

Interactive Figure 24.5.5

Explore a smoke detector.

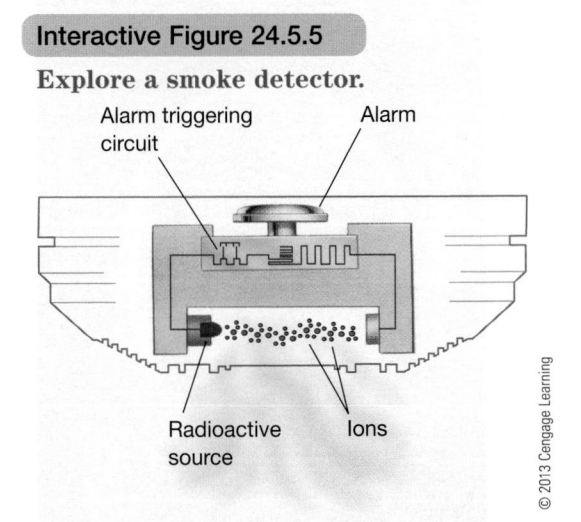

Schematic of a smoke detector

© 2013 Cengage Learning

Figure 24.5.6 A sample of strawberries labeled with the Radura logo, indicating that the food has been irradiated

Courtesy of Food Technologies Service, Inc.

Radon

When buying a home, many people insist on doing a radon test, which might seem unusual because radon is a colorless, odorless, chemically inert gas. However, radon is also radioactive and can potentially cause lung cancer.

Radon is a naturally occurring element in our environment; approximately 55% of the natural radiation you are exposed to is due to radon. Radon-222 is produced by the decay of naturally occurring uranium-238.

$$^{238}_{92}\text{U} \xrightarrow{^{4}_{2}\alpha} \xrightarrow{^{0}_{-1}\beta} \xrightarrow{^{0}_{-1}\beta} \xrightarrow{^{4}_{2}\alpha} \xrightarrow{^{4}_{2}\alpha} \xrightarrow{^{4}_{2}\alpha} \, ^{222}_{86}\text{Rn}$$

As radon gas is produced in the ground, it escapes through cracks and fissures into the atmosphere. However, it can also become trapped in underground mines and in homes (by passing through porous concrete blocks or foundation cracks), as shown in Interactive Figure 24.5.7.

Gaseous radon-222 is an alpha emitter that undergoes alpha decay to form polonium-218, a solid. If radon-222 decays when it is in the lungs, solid polonium-218 adheres to lung tissues, where it undergoes further alpha decay.

$$^{222}_{86}\text{Rn} \rightarrow \, ^{4}_{2}\alpha + \, ^{218}_{84}\text{Po}$$

$$^{218}_{84}\text{Po} \rightarrow \, ^{4}_{2}\alpha + \, ^{214}_{82}\text{Pb}$$

The alpha particles produced in the lungs do not travel very far, but they can destroy epithelial cells in the lungs, which can lead to the formation of lung cancer.

Radon can be detected using an in-home commercial radon test. The U.S. Environmental Protection Agency (EPA) suggests addressing the radon in a home if its level is greater than 4 picocuries per liter (pCi/L). Often, the radon problem in a home can be solved by simply sealing cracks in basement floors and walls, closing gaps around pipes, sealing cavities in walls, or installing a ventilation system to remove the radon.

Section 24.5 Mastery

Interactive Figure 24.5.7

Investigate radon in the home.

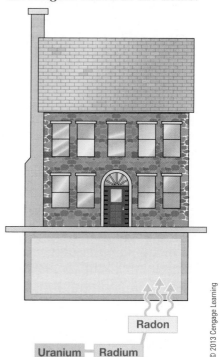

Radon enters a house through openings in basement floors and walls.

Unit Recap

Key Concepts

24.1 Nuclear Reactions

- Nuclear reactions differ from chemical reactions in that nuclear reactions involve changes to the particles in the nucleus (24.1a).

- Natural radioactive decay is the process where an unstable nucleus emits energy and a small particle and is thereby transformed into a more stable nucleus (24.1b).

- The most common decay products are alpha particles, beta particles, positrons, and gamma rays (24.1b).

- Nuclear reactions are always balanced for charge and mass (24.1c).

- The most common decay processes are alpha decay, beta emission, positron emission, and electron capture (24.1c).

24.2 Nuclear Stability

- A plot of stable and unstable isotopes shows that for elements with $Z > 20$, a stable nucleus has more neutrons than protons (24.2a).

- No nuclei with $Z > 83$ are stable (24.2a).

- Nuclides with $Z > 83$ are most likely to decay by emitting an alpha particle (24.2a).

- Nuclides with too many neutrons are most likely to decay by emitting a beta particle (24.2a).

- Nuclides with too few neutrons are most likely to decay by emitting a positron or by electron capture (24.2a).

- The mass of an isotope is always less than the sum of the mass of its component particles (24.2b).

- The mass defect for an isotope is directly related to its binding energy (24.2b).

- A plot of relative binding energies shows that ^{56}Fe has the most stable relative binding energy (24.2c).

24.3 Kinetics of Radioactive Decay

- Radioactive decay processes follow first-order kinetics (24.3a).

- The integrated first-order rate law relates the initial amount or activity of the radioactive material, the amount or activity at a specific time, the rate constant, and time (24.3a).

- Carbon dating is based on the natural abundance of ^{14}C and can be used to date objects containing organic material (24.3b).

24.4 Fission and Fusion

- Fission can occur naturally or be induced by neutron bombardment (24.4a).
- Chain reactions, fission reactions that produce neutrons that can cause additional fission reactions, can be subcritical, critical, or supercritical (24.4a).
- The nuclear fuel used in most nuclear reactors and weapons, a mixture of ^{235}U and ^{238}U, must be enriched in ^{235}U before it is used in any application (24.4b).
- A nuclear power plant uses the energy produced by a fission reaction to generate electricity (24.4c).
- Nuclear fusion is a potential energy source, but it has proved difficult to initiate and sustain controlled fusion in a laboratory setting (24.4c).

24.5 Applications and Uses of Nuclear Chemistry

- Nuclear reactions that take place in stars and in interstellar gases help us understand the relative abundance of elements in the universe (24.5a).
- Elements and isotopes that do not occur naturally can be synthesized in bombardment reactions using particles generated in a cyclotron or a linear particle accelerator (24.5b).
- Radioisotopes are used in the medical imaging and treatment of diseases (24.5c).
- Household applications of nuclear chemistry include the use of radioisotopes in smoke detectors, the gamma irradiation of food, and the naturally occurring radon gas that can enter a home through a basement or foundation (24.5d).

Key Equations

$\Delta m = \Sigma(\text{mass of nuclear particles}) - (\text{mass of isotope})$ **(24.1)**

$\Delta E = (\Delta m)c^2$ **(24.2)**

$\ln\left(\dfrac{N}{N_0}\right) = -kt$ **(24.3)**

$t_{1/2} = \dfrac{\ln 2}{k}$ **(24.4)**

$\ln\left(\dfrac{A}{A_0}\right) = -kt$ **(24.5)**

Key Terms

24.1 Nuclear Reactions
nuclear reaction
alpha (α) particle
beta (β) particle
positron
gamma ray
nuclide
parent nucleus
daughter nucleus
alpha decay
beta decay
positron emission
electron capture

24.2 Nuclear Stability
band of stability
radioactive decay series
binding energy (E_b)
mass defect
nucleon
fusion
fission

24.3 Kinetics of Radioactive Decay
Geiger counter
activity
curie
becquerel

24.4 Fission and Fusion
chain reaction
critical mass
breeder reactor
nuclear reactor
moderator
control rod
plasma

24.5 Applications and Uses of Nuclear Chemistry
neutrino
trans-uranium element

Unit 24   Review and Challenge Problems

Reference Tables

Fundamental Constants

Name	Symbol	Value
Avogadro's Number	N	$6.022 \times 10^{23} \text{ mol}^{-1}$
Boltzmann's Constant	k_b	$1.381 \times 10^{-23} \text{ J K}^{-1}$
Electron Charge	e	$1.602 \times 10^{-19} \text{ C}$
Faraday's Constant	$F = Ne$	$9.649 \times 10^4 \text{ C mol electrons}^{-1}$
Ideal Gas Constant	$R = Nk_b$	$0.08206 \text{ L atm mol}^{-1} \text{ K}^{-1}$ $8.314 \text{ L kPa mol}^{-1} \text{ K}^{-1}$
Planck's Constant	h	$6.626 \times 10^{-34} \text{ J s}$
Rydberg Constant	R	$1.097 \times 10^7 \text{ m}^{-1}$
Speed of Light (in vacuum)	c	$2.998 \times 10^8 \text{ m s}^{-1}$
Standard Acceleration (due to gravity)	g	9.807 m s^{-2}
Universal Gravitational Constant	G	$6.674 \times 10^{-11} \text{ m}^3 \text{ kg}^{-1} \text{ s}^{-2}$
Vacuum Permittivity	ε_o	$8.854 \times 10^{-12} \text{ C}^2 \text{ J}^{-1} \text{ m}^{-1}$
Mass of Electron	m_e	$9.109 \times 10^{-31} \text{ kg}$
Mass of Neutron	m_n	$1.675 \times 10^{-27} \text{ kg}$
Mass of Proton	m_p	$1.673 \times 10^{-27} \text{ kg}$

Vapor Pressure of Water at Various Temperatures

Temperature (°C)	Vapor Pressure (torr)	Temperature (°C)	Vapor Pressure (torr)
−10	2.1	23	21.1
−5	3.2	24	22.4
0	4.6	25	23.8
1	4.9	26	25.2
2	5.3	27	26.7
3	5.7	28	28.3
4	6.1	29	30.0
5	6.5	30	31.8
6	7.0	35	42.2
7	7.5	40	55.3
8	8.0	45	71.9
9	8.6	50	92.5
10	9.2	55	118.0
11	9.8	60	149.4
12	10.5	65	187.5
13	11.2	70	233.7
14	12.0	75	289.1
15	12.8	80	355.1
16	13.6	85	433.6
17	14.5	90	525.8
18	15.5	95	633.9
19	16.5	100	760.0
20	17.5	105	906.1
21	18.7	110	1074.6
22	19.8		

Average Bond Enthalpies (kJ/mol) at 25 °C

Single Bonds

	H	C	N	O	F	Si	P	S	Cl	Br	I
H	436	413	391	463	565	318	322	347	432	366	299
C		346	305	358	485	---	---	272	339	285	213
N			163	201	283	---	---	---	192	243	---
O				146	184	452	335	---	218	201	201
F					155	565	490	284	253	249	278
Si						222	---	293	381	310	234
P							201	---	326	---	184
S								226	255	213	---
Cl									242	216	208
Br										193	175
I											151

Average Bond Enthalpies (kJ/mol) at 25 °C

Multiple Bonds

Double Bonds (kJ/mol)		Triple Bonds (kJ/mol)	
C=C	602	C≡C	835
O=O	498		
C=O	732	C≡O	1072
N=O	607		
N=N	418	N≡N	945
C=N	615	C≡N	887

Acid Ionization Constants at 25 °C

Acid	Formula	K_{a1}	K_{a2}	K_{a3}
Acetic acid	CH_3COOH	1.8×10^{-5}		
Acetylsalicylic acid (aspirin)	$HC_9H_7O_4$	3.0×10^{-4}		
Aluminum ion	$Al(H_2O)_4^{3+}$	1.2×10^{-5}		
Arsenic acid	H_3AsO_4	2.5×10^{-4}	5.6×10^{-8}	3.0×10^{-13}
Ascorbic acid	$H_2C_6H_6O_6$	7.9×10^{-5}	1.6×10^{-12}	
Benzoic acid	C_6H_5COOH	6.3×10^{-5}		
Carbonic acid	H_2CO_3	4.2×10^{-7}	4.8×10^{-11}	
Ferric ion	$Fe(H_2O)_6^{3+}$	4.0×10^{-3}		
Formic acid	$HCOOH$	1.8×10^{-4}		
Hydrocyanic acid	HCN	4.0×10^{-10}		
Hydrofluoric acid	HF	7.2×10^{-4}		
Hydrogen peroxide	H_2O_2	2.4×10^{-12}		
Hydrosulfuric acid	H_2S	1.0×10^{-7}	1.0×10^{-19}	
Hypochlorous acid	$HClO$	3.5×10^{-8}		
Nitrous acid	HNO_2	4.5×10^{-4}		
Oxalic acid	$H_2C_2O_4$	5.9×10^{-2}	6.4×10^{-5}	
Phenol	C_6H_5OH	1.0×10^{-10}		
Phosphoric acid	H_3PO_4	7.5×10^{-3}	6.2×10^{-8}	3.6×10^{-13}
Sulfuric acid	H_2SO_4	very large	1.2×10^{-2}	
Sulfurous acid	H_2SO_3	1.7×10^{-2}	6.4×10^{-8}	
Zinc ion	$Zn(H_2O)_4^{2+}$	2.5×10^{-10}		

Base Ionization Constants at 25 °C

Base	Formula	K_b
Ammonia	NH_3	1.8×10^{-5}
Aniline	$C_6H_5NH_2$	7.4×10^{-10}
Caffeine	$C_8H_{10}N_4O_2$	4.1×10^{-4}
Codeine	$C_{18}H_{21}O_3N$	8.9×10^{-7}
Diethylamine	$(C_2H_5)_2NH$	6.9×10^{-4}
Dimethylamine	$(CH_3)_2NH$	5.9×10^{-4}
Ethylamine	$C_2H_5NH_2$	4.3×10^{-4}
Hydroxylamine	NH_2OH	9.1×10^{-9}
Isoquinoline	C_9H_7N	2.5×10^{-9}
Methylamine	CH_3NH_2	4.2×10^{-4}
Morphine	$C_{17}H_{19}O_3N$	7.4×10^{-7}
Piperidine	$C_5H_{11}N$	1.3×10^{-3}
Pyridine	C_5H_5N	1.5×10^{-9}
Quinoline	C_9H_7N	6.3×10^{-10}
Triethanolamine	$C_6H_{15}O_3N$	5.8×10^{-7}
Triethylamine	$(C_2H_5)_3N$	5.2×10^{-4}
Trimethylamine	$(CH_3)_3N$	6.3×10^{-5}
Urea	N_2H_4CO	1.5×10^{-14}

Selected Thermodynamic Data (at 298.15 K)

Aluminum	ΔH_f^o(kJ/mol)	ΔG_f^o(kJ/mol)	S^o(J/mol K)
Al(s)	0	0	28.3
AlCl$_3$(s)	–704.2	–628.8	110.7
Al$_2$O$_3$(s)	–1675.7	–1582.3	50.9
Al(OH)$_3$(s)	–1277.0		
Barium			
Ba(s)	0	0	67.0
BaCl$_2$(s)	–858.6	–810.4	123.7
BaCO$_3$(s)	–1219.0	–1139.0	112.0
BaO(s)	–553.5	–525.1	70.4
Ba(OH)$_2$(s)	–946.0		
BaSO$_4$(s)	–1473.2	–1362.2	132.2
Beryllium			
Be(s)	0	0	9.5
BeO(s)	–599.0	–569.0	14.0
Be(OH)$_2$(s)	–902.5	–815.0	51.9
Bromine			
Br(g)	111.9	82.4	175.0
Br$_2$(ℓ)	0	0	152.2
Br$_2$(g)	30.9	3.1	245.5
Br$_2$(aq)	–3.0	4.0	130.0
Br$^-$(aq)	–121.0	–175.0	82.0
BrF$_3$(g)	–255.6	–229.4	292.5
HBr(g)	–36.3	–53.5	198.7
Cadmium			
Cd(s)	0	0	52.0
CdO(s)	–258.0	–228.0	55.0

Selected Thermodynamic Data (at 298.15 K)

	ΔH_f^o(kJ/mol)	ΔG_f^o(kJ/mol)	S^o(J/mol K)
Cd(OH)$_2$(s)	–561.0	–474.0	96.0
CdS(s)	–162.0	–156.0	65.0
CdSO$_4$(s)	–935.0	–823.0	123.0
Calcium			
Ca(s)	0	0	41.4
Ca(g)	178.2	144.3	158.9
Ca^{2+}(g)	1925.9		
CaC$_2$(s)	–59.8	–64.9	70.0
CaCO$_3$(s, calcite)	–1206.9	–1128.8	92.9
CaCl$_2$(s)	–795.8	–748.1	104.6
CaF$_2$(s)	–1219.6	–1167.3	68.9
CaH$_2$(s)	–186.2	–147.2	42.0
CaO(s)	–635.1	–604.0	39.8
CaS(s)	–482.4	–477.4	56.5
Ca(OH)$_2$(s)	–986.1	–898.5	83.4
Ca(OH)$_2$(aq)	–1002.8	–868.1	–74.5
Ca$_3$(PO$_4$)$_2$(s)	–4126.0	–3890.0	241.0
CaSO$_4$(s)	–1434.1	–1321.8	106.7
CaSiO$_3$(s)	–1630.0	–1550.0	84.0
Carbon			
C(s, graphite)	0	0	5.7
C(s, diamond)	1.9	2.9	2.4
C(g)	716.7	671.3	158.1
CCl$_4$(ℓ)	–135.4	–65.2	216.4
CCl$_4$(g)	–102.9	–60.6	309.9
CHCl$_3$(ℓ)	–134.5	–73.7	201.7

continued

Selected Thermodynamic Data (at 298.15 K)

	ΔH_f° (kJ/mol)	ΔG_f° (kJ/mol)	S° (J/mol K)
$CHCl_3(g)$	–103.1	–70.3	295.7
$CH_4(g)$	–74.8	–50.7	186.3
$CH_3OH(g)$	–200.7	–162.0	239.8
$CH_3OH(\ell)$	–238.7	–166.3	126.8
$H_2CO(g)$	–116.0	–110.0	219.0
$HCOOH(g)$	–363.0	–351.0	249.0
$HCN(g)$	135.1	125.0	202.0
$C_2H_2(g)$	226.7	209.2	200.9
$C_2H_4(g)$	52.3	68.2	219.6
$CH_3CHO(g)$	–166.0	–129.0	250.0
C_2H_4O(g, ethylene oxide)	–53.0	–13.0	242.0
$CH_3CH_2OH(\ell)$	–277.7	–174.8	160.7
$CH_3CH_2OH(g)$	–235.1	–168.5	282.7
$CH_3COOH(\ell)$	–484.0	–389.0	160.0
$C_2H_6(g)$	–84.7	–32.8	229.6
$C_3H_6(g)$	20.9	62.7	266.9
$C_3H_8(g)$	–103.8	–23.5	269.9
$CH_2{=}CHCN(\ell)$	152.0	190.0	274.0
$C_6H_6(\ell)$	49.0	124.5	172.8
$C_6H_{12}O_6(s)$	–1275.0	–911.0	212.0
$CO(g)$	–110.5	–137.2	197.7
$CO_2(g)$	–393.5	–394.4	213.7
$CS_2(g)$	117.4	67.1	237.8
$COCl_2(g)$	–218.8	–204.6	283.5
Chlorine			
$Cl(g)$	121.7	105.7	165.2

Selected Thermodynamic Data (at 298.15 K)

	ΔH_f° (kJ/mol)	ΔG_f° (kJ/mol)	S° (J/mol K)
$Cl_2(g)$	0	0	223.1
$Cl_2(aq)$	–23.0	7.0	121.0
$Cl^-(aq)$	–167.0	–131.0	57.0
$Cl^-(g)$	–233.1		
$HCl(g)$	–92.3	–95.3	186.9
$HCl(aq)$	–167.2	–131.2	56.5
Chromium			
$Cr(s)$	0	0	23.8
$Cr_2O_3(s)$	–1139.7	–1058.1	81.2
$CrO_3(s)$	–579.0	–502.0	72.0
$CrCl_3(s)$	–556.5	–486.1	123.0
Copper			
$Cu(s)$	0	0	33.2
$CuCl_2(s)$	–220.1	–175.7	108.1
$CuCO_3(s)$	–595.0	–518.0	88.0
$Cu_2O(s)$	–170.0	–148.0	93.0
$CuO(s)$	–157.3	–129.7	42.6
$Cu(OH)_2(s)$	–450.0	–372.0	108.0
$CuS(s)$	–49.0	–49.0	67.0
Fluorine			
$F_2(g)$	0	0	202.8
$F(g)$	79.0	61.9	158.8
$F^-(g)$	–255.4		
$F^-(aq)$	–332.6	–278.8	–13.8
$HF(g)$	–271.1	–273.2	173.8

continued

	ΔH_f° (kJ/mol)	ΔG_f° (kJ/mol)	S° (J/mol K)
HF(aq)	–332.6	–278.8	88.7
Hydrogen			
H_2(g)	0	0	130.7
H(g)	218.0	203.2	114.7
H^+(g)	1536.2		
H^+(aq)	0		
OH^-(aq)	–230.0	–157.0	–11.0
H_2O(ℓ)	–285.8	–237.1	69.9
H_2O(g)	–241.8	–228.6	188.8
H_2O_2(ℓ)	–187.8	–120.4	109.6
Iodine			
I_2(s)	0	0	116.1
I_2(g)	62.4	19.3	260.7
I_2(aq)	23.0	16.0	137
I(g)	106.8	70.3	180.8
I^-(g)	–197.0		
I^-(aq)	–55.0	–52.0	106.0
ICl(g)	17.8	–5.5	247.6
Iron			
Fe(s)	0	0	27.8
Fe_3C(s)	21.0	15.0	108.0
$FeCl_2$(s)	–341.8	–302.3	118.0
$FeCl_3$(s)	–399.5	–333.9	142.3
$Fe_{0.95}O$(s)(wustite)	–264.0	–240.0	59.0
FeO(s)	–272.0		

	ΔH_f° (kJ/mol)	ΔG_f° (kJ/mol)	S° (J/mol K)
Fe_3O_4(s, magnetite)	–1118.4	–1015.4	146.4
Fe_2O_3(s, hematite)	–824.2	–742.2	87.4
FeS(s)	–95.0	–97.0	67.0
FeS_2(s, pyrite)	–178.2	–166.9	52.9
$FeSO_4$(s)	–929.0	–825.0	121.0
$Fe(CO)_5$(ℓ)	–774.0	–705.3	338.1
Lead			
Pb(s)	0	0	64.8
$PbCl_2$(s)	–359.4	–314.1	136.0
PbO(s, yellow)	–217.3	–187.9	68.7
PbO_2(s)	–277.0	–217.0	69.0
PbS(s)	–100.4	–98.7	91.2
$PbSO_4$(s)	–920.0	–813.0	149.0
Magnesium			
Mg(s)	0	0	32.7
$MgCl_2$(s)	–641.3	–591.8	89.6
$MgCO_3$(s)	–1095.8	–1012.1	65.7
MgO(s)	–601.7	–569.4	26.9
$Mg(OH)_2$(s)	–924.5	–833.5	63.2
MgS(s)	–346.0	–341.8	50.3
Manganese			
Mn(s)	0	0	32.0
MnO(s)	–385.0	–363.0	60.0
Mn_3O_4(s)	–1387.0	–1280.0	149.0
Mn_2O_3(s)	–971.0	–893.0	110.0

continued

Selected Thermodynamic Data (at 298.15 K)	ΔH_f^o(kJ/mol)	ΔG_f^o(kJ/mol)	S^o(J/mol K)
$MnO_2(s)$	–521.0	–466.0	53.0
$MnO_4^-(aq)$	–543.0	–449.0	190.0
Mercury			
$Hg(\ell)$	0	0	75.9
$HgCl_2(s)$	–224.3	–178.6	146.0
$Hg_2Cl_2(s)$	–265.4	–210.7	191.7
$HgO(s, red)$	–90.8	–58.5	70.3
$HgS(s, red)$	–58.2	–50.6	82.4
Nickel			
$Ni(s)$	0	0	29.9
$NiCl_2(s)$	–305.3	–259.0	97.7
$NiO(s)$	–239.7	–211.7	38.0
$Ni(OH)_2(s)$	–538.0	–453.0	79.0
$NiS(s)$	–93.0	–90.0	53.0
Nitrogen			
$N_2(g)$	0	0	191.6
$N(g)$	472.7	455.6	153.3
$NH_3(g)$	–46.1	–16.5	192.5
$NH_3(aq)$	–80.0	–27.0	111.0
$NH_4^+(aq)$	–132.0	–79.0	113.0
$NO(g)$	90.3	86.6	210.8
$NOCl(g)$	51.7	66.1	261.8
$NO_2(g)$	33.2	51.3	240.1
$N_2O(g)$	82.1	104.2	219.9
$N_2O_4(g)$	9.2	97.9	304.3
$N_2O_4(\ell)$	–20.0	97.0	209.0

Selected Thermodynamic Data (at 298.15 K)	ΔH_f^o(kJ/mol)	ΔG_f^o(kJ/mol)	S^o(J/mol K)
$N_2O_5(s)$	–42.0	134.0	178.0
$N_2H_4(\ell)$	50.6	149.3	121.2
$N_2H_3CH_3(\ell)$	54.0	180.0	166.0
$HNO_3(aq)$	–207.4	–111.3	146.4
$HNO_3(\ell)$	–174.1	–80.7	155.6
$HNO_3(g)$	–135.1	–74.7	266.4
$NH_4ClO_4(s)$	–295.0	–89.0	186.0
$NH_4Cl(s)$	–314.4	–202.9	94.6
$NH_4Cl(aq)$	–299.7	–210.5	169.9
$NH_4NO_3(s)$	–365.6	–183.9	151.1
$NH_4NO_3(aq)$	–339.9	–190.6	259.8
Oxygen			
$O_2(g)$	0	0	205.1
$O(g)$	249.2	231.7	161.1
$O_3(g)$	142.7	163.2	238.9
Phosphorus			
$P(s, white)$	0	0	164.4
$P(s, red)$	–70.4	–48.4	91.2
$P(s, black)$	–39.0	–33.0	23.0
$P(g)$	314.6	278.3	163.2
$P_4(s, white)$	0	0	41.1
$P_4(s, red)$	–17.6	–12.1	22.8
$P_4(g)$	59.0	24.0	280.0
$PF_5(g)$	–1578.0	–1509.0	296.0
$PH_3(g)$	5.4	13.4	210.2
$PCl_3(g)$	–287.0	–267.8	311.8

continued

Selected Thermodynamic Data (at 298.15 K)

	ΔH_f^o(kJ/mol)	ΔG_f^o(kJ/mol)	S^o(J/mol K)
$H_3PO_4(\ell)$	–1279.0	–1119.1	110.5
$H_3PO_4(aq)$	–1288.0	–1143.0	158.0
$P_4O_{10}(s)$	–2984.0	–2697.7	228.9
Potassium			
$K(s)$	0	0	64.2
$KCl(s)$	–436.7	–409.1	82.6
$KClO_3(s)$	–397.7	–296.3	143.1
$KClO_4(s)$	–433.0	–304.0	151.0
$KI(s)$	–327.9	–324.9	106.3
$K_2O(s)$	–361.0	–322.0	98.0
$K_2O_2(s)$	–496.0	–430.0	113.0
$KO_2(s)$	–283.0	–238.0	117.0
$KOH(s)$	–424.8	–379.1	78.9
$KOH(aq)$	–482.4	–440.5	91.6
Silicon			
$Si(s)$	0	0	18.3
$SiBr_4(\ell)$	–457.3	–443.9	277.8
$SiC(s)$	–65.3	–62.8	16.6
$SiCl_4(g)$	–657.0	–617.0	330.7
$SiH_4(g)$	34.3	56.9	204.6
$SiF_4(g)$	–1614.9	–1572.7	282.5
$SiO_2(s, quartz)$	–910.9	–856.6	41.8
Silver			
$Ag(s)$	0	0	42.6
$Ag^+(aq)$	105.0	77.0	73.0
$AgBr(s)$	–100.0	–97.0	107.0

Selected Thermodynamic Data (at 298.15 K)

	ΔH_f^o(kJ/mol)	ΔG_f^o(kJ/mol)	S^o(J/mol K)
$AgCN(s)$	146.0	164.0	84.0
$AgCl(s)$	–127.1	–109.8	96.2
$Ag_2CrO_4(s)$	–712.0	–622.0	217.0
$AgI(s)$	–62.0	–66.0	115.0
$Ag_2O(s)$	–31.1	–11.2	121.3
$AgNO_3(s)$	–124.4	–33.4	140.9
$Ag_2S(s)$	–32.0	–40.0	146.0
Sodium			
$Na(s)$	0	0	51.2
$Na(g)$	107.3	76.8	153.7
$Na^+(g)$	609.4		
$Na^+(aq)$	–240.0	–262.0	59.0
$NaBr(s)$	–361.0	–349.0	86.8
$Na_2CO_3(s)$	–1130.7	–1044.4	135.0
$NaHCO_3(s)$	–948.0	–852.0	102.0
$NaCl(s)$	–411.2	–384.1	72.1
$NaCl(g)$	–176.7	–196.7	229.8
$NaCl(aq)$	–407.3	–393.1	115.5
$NaH(s)$	–56.0	–33.0	40.0
$NaI(s)$	–288.0	–282.0	91.0
$NaNO_2(s)$	–359.0		
$NaNO_3(s)$	–467.0	–366.0	116.0
$Na_2O(s)$	–416.0	–377.0	73.0
$Na_2O_2(s)$	–515.0	–451.0	95.0
$NaOH(s)$	–425.6	–379.5	64.5
$NaOH(aq)$	–470.1	–419.2	48.1

continued

Selected Thermodynamic Data (at 298.15 K)			
	ΔH_f^o(kJ/mol)	ΔG_f^o(kJ/mol)	S^o(J/mol K)
Sulfur			
S(s, rhombic)	0	0	31.8
S(s, monoclinic)	0.3	0.1	33.0
S(g)	278.8	238.3	167.8
S_2^-(aq)	33.0	86.0	−15.0
S_8(g)	102.0	50.0	431.0
S_2Cl_2(g)	−18.4	−31.8	331.5
SF_6(g)	−1209.0	−1105.3	291.8
H_2S(g)	−20.6	−33.6	205.8
SO_2(g)	−296.8	−300.2	248.2
SO_3(g)	−395.7	−371.1	256.8
$SOCl_2$(g)	−212.5	−198.3	309.8
SO_4^{2-}(aq)	−909.0	−745.0	20.0
$H_2SO_4(\ell)$	−814.0	−690.0	156.9
H_2SO_4(aq)	−909.3	−744.5	20.1
Tin			
Sn(s, white)	0	0	51.6
Sn(s, gray)	−2.1	0.1	44.1
$SnCl_4(\ell)$	−511.3	−440.1	258.6
$SnCl_4$(g)	−471.5	−432.2	365.8
SnO(s)	−285.0	−257.0	56.0
SnO_2(s)	−580.7	−519.6	52.3
$Sn(OH)_2$(s)	−561.0	−492.0	155.0
Titanium			
Ti(s)	0	0	30.6

Selected Thermodynamic Data (at 298.15 K)			
	ΔH_f^o(kJ/mol)	ΔG_f^o(kJ/mol)	S^o(J/mol K)
$TiCl_4(\ell)$	−804.2	−737.2	252.3
$TiCl_4$(g)	−763.2	−726.7	354.9
TiO_2(s)	−939.7	−884.5	49.9
Uranium			
U(s)	0	0	50.0
UF_6(s)	−2137.0	−2008.0	228.0
UF_6(g)	−2113.0	−2029.0	380.0
UO_2(s)	−1084.0	−1029.0	78.0
U_3O_8(s)	−3575.0	−3393.0	282.0
UO_3(s)	−1230.0	−1150.0	99.0
Xenon			
Xe(g)	0	0	170.0
XeF_2(g)	−108.0	−48.0	254.0
XeF_4(s)	−251.0	−121.0	146.0
XeF_6(g)	−294.0		
XeO_3(s)	402.0		
Zinc			
Zn(s)	0	0	41.6
$ZnCl_2$(s)	−415.1	−369.4	111.5
ZnO(s)	−348.3	−318.3	43.6
$Zn(OH)_2$(s)	−642.0		
ZnS(s, wurtzite)	−193.0		
ZnS(s, zinc blende)	−206.0	−201.3	57.7
$ZnSO_4$(s)	−983.0	−874.0	120.0

Standard Reduction (Electrode) Potentials at 25 °C

Half–Cell Reaction	E^o (volts)	Half–Cell Reaction	E^o (volts)
$F_2(g) + 2\ e^- \rightarrow 2\ F^-(aq)$	2.87	$2\ H^+(aq) + 2\ e^- \rightarrow H_2(g)$	0.0000
$Ce^{4+}(aq) + e^- \rightarrow Ce^{3+}(aq)$	1.61	$Pb^{2+}(aq) + 2\ e^- \rightarrow Pb(s)$	−0.126
$MnO_4^-(aq) + 8\ H^+(aq) + 5\ e^- \rightarrow Mn^{2+}(aq) + 4\ H_2O(\ell)$	1.51	$Sn^{2+}(aq) + 2\ e^- \rightarrow Sn(s)$	−0.14
$Cl_2(g) + 2\ e^- \rightarrow 2\ Cl^-(aq)$	1.36	$Ni^{2+}(aq) + 2\ e^- \rightarrow Ni(s)$	−0.25
$Cr_2O_7^{2-}(aq) + 14\ H^+(aq) + 6\ e^- \rightarrow 2\ Cr^{3+}(aq) + 7\ H_2O(\ell)$	1.33	$Co^{2+}(aq) + 2\ e^- \rightarrow Co(s)$	−0.28
$O_2(g) + 4\ H^+(aq) + 4e^- \rightarrow 2\ H_2O(\ell)$	1.229	$Cd^{2+}(aq) + 2\ e^- \rightarrow Cd(s)$	−0.403
$Br_2(\ell) + 2\ e^- \rightarrow 2\ Br^-(aq)$	1.08	$Cr^{3+}(aq) + e^- \rightarrow Cr^{2+}(aq)$	−0.41
$NO_3^-(aq) + 4\ H^+(aq) + 3\ e^- \rightarrow NO(g) + 2\ H_2O(\ell)$	0.96	$2\ H_2O(\ell) + 2\ e^- \rightarrow H_2(g) + 2\ OH^-\ (aq,\ 10^{-7}\ M)$	−0.41
$2\ Hg^{2+}(aq) + 2\ e^- \rightarrow Hg_2^{2+}(aq)$	0.920	$Fe^{2+}(aq) + 2\ e^- \rightarrow Fe(s)$	−0.44
$Hg^{2+}(aq) + 2\ e^- \rightarrow Hg(\ell)$	0.855	$Cr^{3+}(aq) + 3\ e^- \rightarrow Cr(s)$	−0.74
$O_2(g) + 4\ H^+\ (aq,\ 10^{-7}\ M) + 4\ e^- \rightarrow 2\ H_2O(\ell)$	0.82	$Zn^{2+}(aq) + 2\ e^- \rightarrow Zn(s)$	−0.763
$Ag^+(aq) + e^- \rightarrow Ag(s)$	0.799	$2\ H_2O(\ell) + 2\ e^- \rightarrow H_2(g) + 2\ OH^-(aq)$	−0.83
$Hg_2^{2+}(aq) + 2\ e^- \rightarrow 2\ Hg(\ell)$	0.789	$Mn^{2+}(aq) + 2\ e^- \rightarrow Mn(s)$	−1.18
$Fe^{3+}(aq) + e^- \rightarrow Fe^{2+}(aq)$	0.771	$Al^{3+}(aq) + 3\ e^- \rightarrow Al(s)$	−1.66
$I_2(s) + 2\ e^- \rightarrow 2\ I^-(aq)$	0.535	$Mg^{2+}(aq) + 2\ e^- \rightarrow Mg(s)$	−2.37
$Fe(CN)_6^{3-}(aq) + e^- \rightarrow Fe(CN)_6^{4-}(aq)$	0.48	$Na^+(aq) + e^- \rightarrow Na(s)$	−2.714
$Cu^{2+}(aq) + 2\ e^- \rightarrow Cu(s)$	0.337	$K^+(aq) + e^- \rightarrow K(s)$	−2.925
$Cu^{2+}(aq) + e^- \rightarrow Cu^+(aq)$	0.153	$Li^+(aq) + e^- \rightarrow Li(s)$	−3.045
$S(s) + 2\ H^+(aq) + 2\ e^- \rightarrow H_2S(aq)$	0.14		

Solubility Product Constants (K_{sp}) at 25 °C

	Formula	K_{sp}		Formula	K_{sp}
Bromides	$PbBr_2$	6.3×10^{-6}	**Fluorides**	CaF_2	3.9×10^{-11}
	$AgBr$	3.3×10^{-13}		PbF_2	3.7×10^{-8}
Carbonates	$BaCO_3$	8.1×10^{-9}		MgF_2	6.4×10^{-9}
	$CaCO_3$	3.8×10^{-9}	**Hydroxides**	$AgOH$	2.0×10^{-8}
	$CoCO_3$	8.0×10^{-13}		$Al(OH)_3$	1.9×10^{-33}
	$CuCO_3$	2.5×10^{-10}		$Ca(OH)_2$	7.9×10^{-6}
	$FeCO_3$	3.5×10^{-11}		$Cr(OH)_3$	6.7×10^{-31}
	$PbCO_3$	1.5×10^{-13}		$Co(OH)_2$	2.5×10^{-16}
	$MgCO_3$	4.0×10^{-5}		$Cu(OH)_2$	1.6×10^{-19}
	$MnCO_3$	1.8×10^{-11}		$Fe(OH)_2$	7.9×10^{-15}
	$NiCO_3$	6.6×10^{-9}		$Fe(OH)_3$	6.3×10^{-38}
	Ag_2CO_3	8.1×10^{-12}		$Pb(OH)_2$	2.8×10^{-16}
	$ZnCO_3$	1.5×10^{-11}		$Mg(OH)_2$	1.5×10^{-11}
Chlorides	$PbCl_2$	1.7×10^{-5}		$Mn(OH)_2$	4.6×10^{-14}
	$AgCl$	1.8×10^{-10}		$Ni(OH)_2$	2.8×10^{-16}
Chromates	$BaCrO_4$	2.0×10^{-10}		$Zn(OH)_2$	4.5×10^{-17}
	$CaCrO_4$	7.1×10^{-4}	**Iodides**	PbI_2	8.7×10^{-9}
	$PbCrO_4$	1.8×10^{-14}		AgI	1.5×10^{-16}
	Ag_2CrO_4	9.0×10^{-12}	**Oxalates**	BaC_2O_4	1.1×10^{-7}
Cyanides	$Ni(CN)_2$	3.0×10^{-23}		CaC_2O_4	2.3×10^{-9}
	$AgCN$	1.2×10^{-16}		MgC_2O_4	8.6×10^{-5}
	$Zn(CN)_2$	8.0×10^{-12}	**Phosphates**	$AlPO_4$	1.3×10^{-20}
Fluorides	BaF_2	1.7×10^{-6}		$Ba_3(PO_4)_2$	1.3×10^{-29}

Solubility Product Constants (K_{sp}) at 25 °C

	Formula	K_{sp}		Formula	K_{sp}
Phosphates	$Ca_3(PO_4)_2$	1.0×10^{-25}	**Sulfides**	CuS	7.9×10^{-37}
	$CrPO_4$	2.4×10^{-23}		FeS	4.9×10^{-18}
	$Pb_3(PO_4)_2$	3.0×10^{-44}		Fe_2S_3	1.4×10^{-88}
	Ag_3PO_4	1.3×10^{-20}		PbS	3.2×10^{-28}
	$Zn_3(PO_4)_2$	9.1×10^{-33}		MnS	5.1×10^{-15}
Sulfates	$BaSO_4$	1.1×10^{-10}		NiS	3.0×10^{-21}
	$CaSO_4$	2.4×10^{-5}		Ag_2S	1.0×10^{-49}
	$PbSO_4$	1.8×10^{-8}		ZnS	2.0×10^{-25}
	Ag_2SO_4	1.7×10^{-5}	**Sulfites**	$BaSO_3$	8.0×10^{-7}
Sulfides	CaS	8×10^{-6}		$CaSO_3$	1.3×10^{-8}
	CoS	5.9×10^{-21}		Ag_2SO_3	1.5×10^{-14}

Glossary

absolute zero The lowest possible temperature, equivalent to -273.15 °C, used as the zero point of the Kelvin scale (1.3b)

absorbance The negative logarithm of the transmittance (23.4c)

accuracy The agreement between the measured quantity and the accepted value (1.3c)

acid A substance that, when dissolved in pure water, increases the concentration of hydrogen ions (4.3b); *see also* **Arrhenius acid, Brønsted–Lowry acid, Lewis acid**

acid dissociation constant (K_a) The equilibrium constant for the ionization of an acid in aqueous solution (16.3a)

acid–base indicator A dye that shows by a change in color when the acid–base reaction is complete (4.5d, 17.3d)

acid–base reaction An exchange reaction between an acid and a base producing a salt and water (4.1b)

acid–base titration An analytical method used to accurately determine the concentration of an acid or base solution or the molar mass of an unknown acid or base (4.5d)

acidic oxide An oxide of a nonmetal that acts as an acid (22.2a)

acidic solution A solution in which the concentration of hydronium ions is greater than the concentration of hydroxide ions (16.2a)

actinides The series of elements between actinium and rutherfordium in the periodic table (2.2a)

activated complex A high-energy transition state (14.5a)

activation energy (E_a) The minimum amount of energy that must be absorbed by a system to cause it to react (14.1b, 14.5a)

activity (A) A measure of the rate of nuclear decay, the number of disintegrations observed in a sample per unit time (24.3a)

actual yield *see* **experimental yield**

addition polymer A synthetic organic polymer formed by directly joining monomer units (21.4a)

addition reaction A reaction where the pi bond in an alkene is replaced by a new sigma-bonded atom on each carbon (21.1d)

adhesive force A force of attraction between molecules of two different substances (11.3c)

aerosol A type of emulsion consisting of a mixture of small solid or liquid particles in a gas (13.5b)

alcohol Any of a class of organic compounds characterized by the presence of a hydroxyl group bonded to a saturated carbon atom (21.3a)

aldehyde Any of a class of organic compounds characterized by the presence of a carbonyl group, in which the carbon atom is bonded to at least one hydrogen atom (21.3a)

alkali metal Any of the metals in Group 1A of the periodic table (2.2a)

alkaline earth metal Any of the elements in Group 2A of the periodic table (2.2a)

alkane Any of a class of hydrocarbons in which each carbon atom is bonded to four other atoms (21.1b)

alkene Any of a class of hydrocarbons in which there is at least one carbon–carbon double bond (21.1c)

alkyne Any of a class of hydrocarbons in which there is at least one carbon–carbon triple bond (21.1c)

allotropes Different forms of the same element that exist in the same physical state under the same conditions of temperature and pressure (2.2a)

alloy A mixture of a metal with one or more other elements that retains metallic characteristics (13.5a)

alpha (α) particle A helium nucleus ejected from certain radioactive substances (24.1b)

alpha (α) plot A graphical representation of the relationship between pH and solution composition (17.2c)

alpha decay The process where a nucleus emits an alpha particle (24.1c)

ambidentate ligand A ligand that has more than one possible attachment site but does not generally act as a multidentate ligand (23.3a)

amide Any of a class of organic compounds characterized by the presence of an amido (—NRCO—) group (21.3a)

amine A derivative of ammonia in which one or more of the hydrogen atoms are replaced by organic groups (21.3a)

amino acid A compound containing an amine group and a carboxyl group, both attached to the same carbon atom (17.4b)

amorphous solid A solid that lacks long-range regular structure and displays a melting range instead of a specific melting point (12.1a)

ampere (A) The unit of electric current (20.2a)

amphiprotic substance A substance that can behave as either a Brønsted–Lowry acid or a Brønsted–Lowry base (16.1c)

amplitude The maximum height of a wave, as measured from the axis of propagation (6.1a)

angular momentum quantum number (ℓ) The quantum number that indicates the shape of the orbital (6.5a)

anion An ion with a negative electric charge (2.1a)

anode The electrode of an electrochemical cell at which oxidation occurs (20.1d)

antibonding molecular orbital A molecular orbital in which the energy of the electrons is higher than that of the parent orbital electrons (9.4a)

aqueous solution A solution in which the solvent is water (4.2a)

arene A general term used for aromatic hydrocarbons (21.1c)

aromatic hydrocarbon Any of a class of hydrocarbons characterized by the presence of a benzene ring or related structure (21.1c)

Arrhenius acid A substance containing hydrogen that, when dissolved in water, increases the concentration of H^+ ions (16.1a)

Arrhenius base A substance containing the hydroxide group that, when dissolved in water, increases the concentration of OH^- ions (16.1a)

Arrhenius equation A mathematical expression that relates reaction rate to activation energy, collision frequency, and molecular orientation (14.5b)

atom The smallest particle of an element that retains the characteristic chemical properties of that element (1.2a)

atomic mass *see* **atomic weight**

atomic mass unit (u) The unit of a scale of relative atomic masses of the elements; $1\ u = 1/12$ of the mass of a carbon atom with six protons and six neutrons (2.1a)

atomic nucleus The core of an atom; made up of protons and neutrons (2.1a)

atomic number (Z) The number of protons found in the nucleus of an atom of an element (2.1b)

atomic scale A scale of measurement used to describe individual atoms or molecules (1.1a)

atomic symbol The one- or two-letter symbol that represents the element along with the atomic number, written as a subscript number, and the mass, written as a superscript number (2.1b)

atomic weight The average mass of an atom in a natural sample of the element (2.1c)

autoionization of water Proton transfer between two water molecules to produce a hydronium ion and a hydroxide ion (16.2a)

average reaction rate The change in concentration of a reactant or product over a defined time interval (14.2a)

Avogadro's law Equal volumes of gases under the same conditions of temperature and pressure have equal numbers of particles (10.2c)

Avogadro's number (N_A) The number of particles in one mole of any substance (6.0221415×10^{23}) (3.1a)

balanced equation A chemical equation showing the relative amounts of reactants and products (3.3a)

ball-and-stick model A molecular model that shows atoms as colored spheres connected by sticks that represent covalent bonds (2.3c)

band of stability A graphical representation of nuclei that are stable with respect to radioactive decay (24.2a)

band theory A formal approach to the bonding in crystalline metallic solids (12.4a)

bar A unit of pressure; 1 bar = 100 kPa (10.1a)

barometer An apparatus used to measure atmospheric pressure (10.1b)

base A substance that, when dissolved in pure water, increases the concentration of hydroxide ions (4.3b); *see also* **Arrhenius base, Brønsted–Lowry base, Lewis base**

base dissociation constant (K_b) The equilibrium constant for the ionization of a base in aqueous solution (16.3a)

basic oxide An oxide of a metal that acts as a base (22.2a)

basic solution A solution in which the concentration of hydronium ions is less than the concentration of hydroxide ions (16.2a)

battery A device consisting of two or more electrochemical cells (20.4a)

becquerel (Bq) The SI unit of radioactivity; 1 decomposition per second (24.3a)

beta (β) particle An electron ejected at high speed from certain radioactive substances (24.1b)

beta decay A nuclear decay process that occurs when a neutron in an unstable nucleus is converted to a proton and an electron and the nuclear electron is ejected from the nucleus (24.1c)

bidentate ligand A chelating ligand with two attachment points (23.3a)

bimolecular An elementary step where two reacting species collide (14.6a)

binary nonmetal A compound formed from two elements, both nonmetals (2.3d)

binding energy (E_b) The energy required to separate a nucleus into individual protons and neutrons (24.2b)

biopolymer A naturally occurring polymer produced by a living organism (21.5)

body-centered cubic unit cell A unit cell with one lattice point on each corner of a cube and a single lattice point in the center of the unit cell (12.2b)

boiling point The temperature at which the vapor pressure of a liquid is equal to the external pressure on the liquid (1.2b, 11.2c)

Boltzmann distribution The distribution of speeds in a collection of moving molecules (10.5b)

bond An interaction between two or more atoms that holds them together by reducing the potential energy of their electrons (8.1a)

bond angle The angle between two atoms bonded to a central atom (8.5a)

bond energy The enthalpy change for breaking a bond in a molecule, with the reactants and products in the gas phase at standard conditions (8.3a)

bond length The distance between the nuclei of two bonded atoms (8.3a)

bond order The number of bonding electron pairs shared by two atoms in a molecule (8.3a, 9.4c)

bonding molecular orbital A molecular orbital in which the energy of the electrons is lower than that of the parent orbital electrons (9.4a)

bonding pair Two electrons, shared by two atoms, that contribute to the bonding attraction between the atoms (8.2a)

Boyle's law The pressure and volume of a gas sample are inversely related when the amount of gas and temperature are held constant; $PV = k_B$ (10.2a)

Bragg's law The condition for observing a reflection by diffraction from a crystal; $n\lambda = 2d \sin \theta$ (12.2d)

branched hydrocarbon A hydrocarbon in which at least one carbon is connected to more than two carbon atoms (21.1b)

Bravais lattices The 14 observed repeat patterns for lattice points in three-dimensional crystalline solids (12.1b)

breeder reactor A nuclear reactor that produces fissile material at a greater rate than fissile material is consumed (24.4b)

Brønsted–Lowry acid A proton donor (16.1a)

Brønsted–Lowry base A proton acceptor (16.1a)

buffer capacity The amount of strong acid or base that can be added to a buffer without a drastic change in pH (17.2b)

buffer solution A solution that resists a change in pH when hydroxide or hydronium ions are added (17.2b)

calorie (cal) The quantity of energy required to raise the temperature of 1.00 g of pure liquid water from 14.5 °C to 15.5 °C (1.3b)

calorimeter A measurement device used in experiments that measure heat exchange (5.4c)

calorimetry The experimental determination of the energy changes of reactions (5.4c)

capillary action The result of the cohesive forces within a bulk sample of liquid being overcome by adhesive forces (11.3c)

carbonyl group The functional group that characterizes aldehydes and ketones, consisting of a carbon atom doubly bonded to an oxygen atom (21.3a)

carboxylic acids Any of a class of organic compounds characterized by the presence of a carboxyl group (21.3a)

catalysis The technique of using a catalyst to influence the rate of a reaction (14.6e)

catalyst A substance that increases the rate of a reaction while not being consumed in the reaction (14.3a)

cathode The electrode of an electrochemical cell at which reduction occurs (20.1d)

cation An ion with a positive electrical charge (2.1a)

cell potential (E_{cell}) The difference in potential energy between the cathode and anode in a voltaic cell (20.2a)

Celsius temperature scale A scale defined by the freezing and boiling points of pure water, defined as 0 °C and 100 °C (1.3b)

ceramics Solid inorganic compounds that combine metal and nonmetal atoms (12.3d)

cesium chloride structure A 1:1 ionic crystal structure that can be described as a simple cubic arrangement of the larger ions surrounding cubic holes occupied by the smaller ions (12.3b)

chain reaction A self-sustaining fission reaction (24.4a)

chalcophile A transition metal that forms sulfide ores (23.2a)

change in enthalpy (ΔH) The energy as heat transferred at constant pressure (5.2a)

change in enthalpy for a reaction (ΔH_{rxn}) *see* **enthalpy change for a reaction**

change in internal energy (ΔE_{system}) The sum of the energy in the form of heat exchanged between system and surroundings and the work done by or on the system; $\Delta E_{system} = q + w$ (5.1c)

Charles's law The temperature and volume of a gas sample are directly related when the pressure and the amount of gas are held constant; $V = k_C T$ (10.2b)

chelate effect Coordination compounds containing chelating ligands are more thermodynamically stable than similar compounds containing only monodentate ligands (23.3c)

chelating ligand A ligand that forms more than one coordinate covalent bond with the central metal ion in a complex (23.3a)

chemical analysis The determination of the amounts or identities of the components of a mixture (3.5)

chemical change A change that involves the transformation of one or more substances into one or more different substances (1.2b); *see also* **reaction**

chemical compound Matter that is composed of two or more kinds of atoms, chemically combined in definite proportions (1.2a)

chemical energy A form of potential energy that can be released when new chemical bonds are formed (5.1a)

chemical equation A written representation of a chemical reaction, showing the reactants and products, their physical states, and the direction in which the reaction proceeds (3.3a)

chemical equilibrium The state of a chemical system when the rate of the forward reaction is equal to the rate of the reverse reaction (15.1b)

chemical kinetics The study of the rates of chemical reactions under various conditions and of reaction mechanisms (14.1a)

chemical properties The properties of a substance that involve a chemical change in the material and often involve a substance interacting with other chemicals (1.2b)

chemistry The study of matter, its transformations, and how it behaves (1.1a)

chiral center A carbon atom attached to four different groups (21.2b)

chiral compound A molecule that is not superimposable on its mirror image (21.2b); *see also* **enantiomers**

cis–trans isomers An example of geometric isomers; in a *cis* isomer, two identical ligands are arranged adjacent to each other; in a *trans* isomer, two identical ligands are arranged across from each other (9.3e, 23.3d)

closed system A system in which energy but not matter can be passed to or from the surroundings (5.1a)

closest-packed structure One of two ways of packing identical spheres with minimum unoccupied volume; each sphere is in contact with six others in a single layer and with three in each of the layers above and below (12.2c)

cohesive force A force of attraction between molecules of a single substance (11.3)

colligative properties The properties of a solution that depend only on the number of solute particles per solvent molecule and not on the nature of the solute or solvent (13.4)

collision theory of reaction rates A theory of reaction rates that assumes that molecules must collide in order to react (14.1b)

colloid A state of matter intermediate between a solution and a suspension, in which solute particles are large enough to scatter light but too small to settle out (13.5b)

combination reaction A chemical reaction in which two or more reactants, usually elements or compounds, combine to form one product, usually a compound (4.1a)

combined gas law An equation that allows calculation of pressure, temperature, and volume when a given amount of gas undergoes a change in conditions (10.3a)

combustion analysis A common technique used to determine the chemical formula of a compound that contains carbon and hydrogen, where a weighed sample of the compound is burned in the presence of excess oxygen (3.5a)

combustion reaction The reaction of a compound with molecular oxygen to form products in which all elements are combined with oxygen (3.5a)

common ion effect The limiting of acid (or base) ionization caused by addition of its conjugate base (or conjugate acid) (17.2b)

complex ion A charged species consisting of a transition metal and Lewis bases (18.3a, 23.3a)

concentration The amount of solute dissolved in a given amount of solution (4.5a)

concentration cell An electrochemical cell where the two half-cells differ only in the concentration of species in solution (20.2e)

concerted process A reaction that involves more than one chemical process happening simultaneously (14.6a)

condensation The movement of molecules from the gas to the liquid phase (11.1b)

condensation polymer A synthetic organic polymer formed by combining monomer units in such a way that a small molecule, usually water, is split out (21.4b)

condensed phase A phase in which the particles are packed in close proximity to one another (11.1a)

condensed structural formula A variation of a molecular formula that shows groups of atoms (2.3b)

conduction band The next-lowest-energy band of orbitals above the valence band (12.4a)

conductor A substance in which either the valence band and the conduction band are contiguous or the valence band and the conduction band overlap, so there is no energy gap between the two (12.4a)

conformations The different three-dimensional arrangements of atoms in a molecule that can be interconverted by rotation around single bonds (9.3d)

conjugate acid–base pair A pair of compounds or ions that differs by the presence of one hydrogen ion (16.1b)

constitutional isomers Compounds with the same chemical formula but different structural formulas (21.2a)

continuous spectrum The spectrum of white light emitted by a heated object that consists of light of all wavelengths (6.3a)

control rods Rods that are inserted between the fuel rods in a nuclear reactor to slow or stop the chain reaction by absorbing neutrons (24.4c)

conversion factor A multiplier that relates the desired unit to the starting unit (1.4a)

coordinate–covalent bond Interatomic attraction resulting from the sharing of a lone pair of electrons from one atom with another atom (18.3a)

coordination compound A compound in which a metal ion or atom is bonded to one or more molecules or anions to define a structural unit (23.3a)

coordination compound geometry The arrangement in space of the central metal ion and the ligands attached to it (23.3d)

coordination number The number of nearest neighbors an atom has in an extended crystal lattice or the number of ligands attached to the central metal ion in a coordination compound (12.2a, 23.3a)

coordination sphere The portion of a coordination compound where species are directly bonded to the metal (23.3a)

core electrons The electrons in an atom's completed set of shells (7.3b)

corrosion The deterioration of metals by oxidation–reduction reactions (20.4d)

coulomb (C) The quantity of charge that passes a point in an electric circuit when a current of 1 ampere flows for 1 second (20.2a)

Coulomb's law The force of attraction between the oppositely charged ions of an ionic compound is directly proportional to their charges and inversely proportional to the square of the distance between them (8.1a)

coulometry The study of the amount of matter transformed during an electrolysis reaction (20.3a)

covalent bond An interatomic attraction resulting from the sharing of electrons between the atoms (8.1a)

covalent compound A compound formed by atoms that are covalently bonded to each other (2.3a)

covalent radius The distance between the nuclei of two atoms of an element when they are held together by a single bond (7.4b)

covalent solid A crystalline solid consisting of a three-dimensional extended network of atoms held together by covalent bonds (12.1a)

critical mass The minimum mass of fissile material needed to sustain a critical chain reaction (24.4a)

critical point The upper end of the curve of vapor pressure versus temperature (12.5b)

critical pressure The pressure at the critical point (12.5b)

critical temperature The temperature at the critical point; above this temperature the vapor cannot be liquefied at any pressure (12.5b)

crystal field splitting energy (Δ) The difference in potential energy between sets of d orbitals in a metal atom or ion surrounded by ligands (23.4a)

crystal field theory A bonding model for coordination compounds that focuses on the energy of the metal d orbitals and assumes that ligands are negative point charges (23.4a)

crystal lattice A solid, regular array of positive and negative ions (12.1b)

crystalline solid A solid in which the particles are arranged in a regular way (12.1a)

cubic closest-packed structure A closest-packed structure where the layers are arranged in an ABCABC…repeating pattern (12.2c)

cubic hole The hole at the center of a unit cell defined by a simple cubic lattice with an atom at each corner (12.3a)

curie (Ci) A unit of radioactivity (24.3a)

cycloalkane An alkane in which the carbon atoms are joined to form a ring (21.1b)

Dalton's law of partial pressures The total pressure of a mixture of gases is the sum of the pressures of the components of the mixture (10.4a)

daughter nucleus The nucleus (or nuclei) produced in a radioactive decay reaction (24.1c)

decomposition reaction A chemical reaction in which one compound breaks down into simpler elements and/or compounds (4.1a)

density The ratio of the mass of an object to its volume (1.3b)

deposition The physical process in which a gas is converted to a solid (12.5a)

diamagnetism The physical property of being repelled by a magnetic field (7.1b)

diffusion The gradual mixing of the molecules of two or more substances by random molecular motion (10.5c)

dilution A method used to prepare a very dilute solution from a more concentrated solution (4.5b)

dimensional analysis A general problem-solving approach that uses the dimensions or units of each value to guide the calculations done (1.4a)

dipeptide An amide consisting of two amino acids (21.5c)

dipole A separation of partial positive and partial negative charge within a bond or a molecule

dipole moment The product of the magnitude of the partial charges in a molecule and the distance by which they are separated (8.4b)

dipole–dipole intermolecular force The electrostatic force between two neutral molecules that have permanent dipole moments (11.4a)

dipole–induced dipole intermolecular force The electrostatic force between two neutral molecules, one having a permanent dipole and the other having an induced dipole (11.4b)

diprotic acid An acid that can dissociate to form two moles of H^+ ions per mole of acid (4.3b, 16.1c)

direct addition A method used to prepare solutions with relatively high solute concentrations (4.5b)

dispersion forces *see* **London dispersion forces**

displacement reaction A chemical reaction that proceeds by the interchange of reactant cation–anion partners (4.1b)

dissociation constant (K_d) The equilibrium constant for the dissociation of a complex ion (18.3b)

dopant An impurity added to a semiconductor to change its properties (12.4a)

double bond A bond formed by sharing two pairs of electrons, one pair in a sigma bond and the other in a pi bond (8.2a)

double displacement reaction A type of displacement reaction during which two atoms, ions, or molecular fragments exchange (4.1b)

dynamic equilibrium A condition in which the forward and reverse reaction rates in a physical or chemical system are equal (11.2a)

effective nuclear charge (Z^*) The nuclear charge experienced by an electron in a multielectron atom, as modified by the other electrons (7.4a)

effusion The movement of gas molecules through a membrane or other porous barrier by random molecular motion (10.5c)

electrochemical cell A device that produces an electric current as a result of an electron transfer reaction (20.1d)

electrochemistry The area of chemistry that studies electron-transfer (oxidation–reduction) reactions (20.1a)

electrode A device, such as a metal plate or wire, for conducting electrons into and out of solutions in electrochemical cells (20.1d)

electrolysis The use of electrical energy to produce chemical change (20.3a)

electrolyte A substance that ionizes in water or on melting to form an electrically conducting solution (4.2a)

electrolytic cell An electrochemical cell in which electrolysis takes place (20.3a)

electromagnetic radiation Radiation that consists of wavelike electric and magnetic fields, including light, microwaves, radio signals, and x-rays (6.1)

electromagnetic spectrum The array of different types of electromagnetic radiation arranged by wavelength (6.1b)

electromotive force (emf) The drop in potential energy of the electrons moving through a wire from anode to cathode (20.2a)

electron A negatively charged subatomic particle found in the space about the nucleus (2.1a)

electron affinity The energy change occurring when an atom of the element in the gas phase gains an electron (7.4d)

electron capture A nuclear process in which an inner-shell electron is captured (24.1c)

electron configuration A representation of how electrons are distributed in orbitals for an atom or ion (7.3a)

electron deficient A compound in which an element has an incomplete octet (8.2c)

electron density The probability of finding an atomic electron within a given region of space, related to the square of the electron's wave function (6.4b)

electron spin The negatively charged electron spinning on an axis (7.1a)

electron-pair geometry The geometry determined by all the bond pairs and lone pairs in the valence shell of the central atom (8.5a)

electromagnetic radiation Energy that travels through space as waves (6.1a)

electronegativity (χ) A measure of the ability of an atom in a molecule to attract electrons to itself (8.4b)

electrostatic forces Forces of attraction or repulsion caused by electric charges (11.4)

element Matter that is composed of only one kind of atom (1.2a)

elementary step A simple event in which some chemical transformation occurs; one of a sequence of events that form the reaction mechanism (14.6a)

empirical formula A molecular formula showing the simplest possible ratio of atoms in a molecule (2.3b, 3.2c)

emulsion A colloidal dispersion of one liquid in another (16.5b)

endothermic process A thermodynamic process in which energy as heat flows into a system from its surroundings (5.2a)

energy The capacity to do work and transfer heat (1.3b, 5.1a); *see also* **enthalpy, heat**

enthalpy (H) The sum of the internal energy of the system and the product of its pressure and volume (5.2a)

enthalpy change (ΔH) *see* **change in enthalpy**

enthalpy change for a reaction (ΔH_{rxn}) The energy transferred during a chemical reaction, measured at constant pressure (5.4a)

enthalpy of dissolution ($\Delta H_{dissolution}$) The amount of energy as heat involved in the process of solution formation (5.4c, 13.3b)

enthalpy of fusion (ΔH_{fus}) The energy as heat required to convert one mole of a substance from a solid to a liquid at constant temperature (5.3c)

enthalpy of hydration (ΔH_{hyd}) The enthalpy change when one mole of a gaseous ion dissolves in water, forming a hydrated ion (13.2d)

enthalpy of sublimation ($\Delta H_{sublimation}$) The energy as heat required to convert one mole of a substance from a solid to a gas (12.5a)

enthalpy of vaporization (ΔH_{vap}) The quantity of energy as heat required to convert one mole of a liquid to a gas at constant temperature (5.3c)

entropy (S) A measure of the dispersal of energy in a system (13.2a, 19.1b)

equilibrium constant (K) The constant in the equilibrium constant expression (15.2a)

equilibrium constant expression A mathematical expression that relates the concentrations of the reactants and products at equilibrium at a particular temperature to a numerical constant (15.2a)

equilibrium vapor pressure The pressure of the vapor of a substance at equilibrium in contact with its liquid or solid phase in a sealed container (11.2a)

equivalence point The point in a titration at which one reactant has been exactly consumed by addition of the other reactant (4.5d)

ester Any of a class of organic compounds structurally related to carboxylic acids but in which the hydrogen atom of the carboxyl group is replaced by a hydrocarbon group (21.3a)

ether Any of a class of organic compounds characterized by the presence of an oxygen atom singly bonded to two carbon atoms (21.3a)

excess reactant A reactant in a chemical reaction present in an amount greater than is required for complete reaction with other reactants (3.4a)

exchange reaction *see* **displacement reaction**

excited state The state of an atom in which at least one electron is not in the lowest possible energy level (6.3b)

exothermic process A thermodynamic process in which energy as heat flows from a system to its surroundings (5.2a)

expanded valence Elements with more than eight electrons in a Lewis structure (8.2c)

experimental yield The measured amount of product obtained from a chemical reaction (3.4b)

extensive variable A variable or property that depends on the amount of matter present (5.4b)

extraction The process of chemically reacting an ore to give the neutral metal (23.2b)

fac–mer isomers Geometric isomers in which three identical ligands are arranged on a triangular face of an octahedron (*fac*) or in a line (*mer*) (23.3d)

face-centered cubic unit cell A cubic unit cell in which there are lattice points on each corner of the cube and a lattice point centered on each cube face (12.2c)

Fahrenheit temperature scale A scale defined by the freezing and boiling points of pure water, defined as 32 °F and 212 °F (1.3b)

Faraday's constant (F) The proportionality constant that relates standard free energy of reaction to standard potential; the charge carried by one mole of electrons (20.2b)

ferromagnetism A form of paramagnetism, seen in some metals and their alloys, in which the magnetic effect is greatly enhanced (7.1b)

first law of thermodynamics The total energy of the universe is constant (5.1c, 19.1a)

Fisher projection A line structure specifically used to show the conformation of carbohydrates (21.5a)

fission The highly exothermic process by which very heavy nuclei split to form lighter nuclei (24.2c)

formal charge The charge on an atom in a molecule or ion calculated by assuming equal sharing of the bonding electrons (8.4a)

formation constant (K_f) An equilibrium constant for the formation of a complex ion (18.3b)

formula weight The sum of the atomic weights of the elements that make up a substance multiplied by the number of atoms of each element in the formula for the substance (3.1b)

free radical A neutral atom or molecule containing an unpaired electron (8.2c)

freezing point The temperature at which the solid and liquid phases are in equilibrium at 1 atm (1.2b)

frequency (ν) The number of complete waves passing a point in a given amount of time (6.1a)

fuel cell A voltaic cell in which reactants are continuously added (20.4c)

functional group A structural fragment found in all members of a class of compounds (21.3)

fusion The state change from solid to liquid (11.1b)

fusion, nuclear The highly exothermic process by which comparatively light nuclei combine to form heavier nuclei (24.2c)

galvanic cell *see* **voltaic cell**

gamma ray High-energy electromagnetic radiation (24.1b)

gas The phase of matter in which a substance has no definite shape and a volume defined only by the size of its container (1.2b)

gas-forming reaction A double displacement reaction that results in the formation of a gas (4.1b)

Geiger counter A device that detects ionizing radiation (24.3a)

gel A colloidal dispersion with a structure that prevents it from flowing (13.5b)

geometric isomers Isomers in which the atoms of the molecule are arranged in different geometric relationships (21.2b)

Gibbs free energy (G) A thermodynamic state function relating enthalpy, temperature, and entropy (19.3a)

Graham's law of effusion The rate of effusion of a gas is inversely related to the square root of its molar mass (10.5c)

ground state The state of an atom in which all electrons are in the lowest possible energy levels (6.3b, 7.3a)

groups The vertical columns in the periodic table (2.2a)

half-cell A compartment of an electrochemical cell in which a half-reaction occurs (20.1d)

half-equivalence point The midpoint of a titration, where half of the species being titrated has been consumed (17.3b)

half-life ($t_{1/2}$) The time required for the concentration of one of the reactants to reach half of its initial value (14.4c)

half-reaction method A systematic procedure for balancing oxidation–reduction reactions (20.1b)

half-reactions The two chemical equations into which the equation for an oxidation–reduction reaction can be divided, one representing the oxidation process and the other the reduction process (4.4a)

halogen Any of the elements in Group 7A of the periodic table (2.2a)

heat of fusion *see* **enthalpy of fusion**

heat of sublimation *see* **enthalpy of sublimation**

heat of vaporization *see* **enthalpy of vaporization**

Henderson–Hasselbalch equation An equation used to calculate the pH of a buffer solution (17.2b)

Henry's law The concentration of a gas dissolved in a liquid at a given temperature is directly proportional to the partial pressure of the gas above the liquid (13.3a)

hertz The unit of frequency, or cycles per second; 1 Hz = 1 s^{-1} (6.1a)

Hess's law If a reaction is the sum of two or more other reactions, the enthalpy change for the overall process is the sum of the enthalpy changes for the constituent reactions (5.5a)

heterogeneous catalyst A catalyst that is not in the same phase as the compounds undergoing reaction (14.6e)

heterogeneous mixture A mixture in which the properties in one region or sample are different from those in another region or sample (1.2c)

heteronuclear diatomic molecule A molecule composed of two atoms of different elements (9.4e)

hexadentate ligand A chelating ligand with six attachment points (23.3a)

hexagonal closest-packed structure A closest-packed structure where the layers are arranged in an ABABAB...repeating pattern (12.2c)

high-spin configuration The electron configuration for a coordination complex with the maximum number of unpaired electrons (23.4a)

homogeneous catalyst A catalyst that is in the same phase as the reaction mixture (14.6e)

homogeneous mixture A mixture in which the properties are the same throughout, regardless of the optical resolution used to examine it (1.2c)

homonuclear diatomic molecule A molecule composed of two identical atoms (9.4d)

Hund's rule of maximum multiplicity The most stable arrangement of electrons is that with the maximum number of unpaired electrons, all with the same spin direction (7.3b)

hybrid orbital An orbital formed by mixing two or more atomic orbitals (9.2a)

hybrid orbitals A set of equal-energy orbitals that are the combination of an atom's atomic orbitals (9.2a)

hydrated compound A compound in which molecules of water are associated with ions (3.2e)

hydrated ion An ion surrounded by water molecules (4.2a)

hydrocarbon A compound that contains only carbon and hydrogen (2.3c)

hydrogen bonding Attraction between a hydrogen atom and a very electronegative atom to produce an unusually strong dipole–dipole attraction (11.4a)

hydrogenation An addition reaction in which the reagent is molecular hydrogen (21.1d)

hydrolysis reaction A reaction with water in which a bond to oxygen is broken (16.3a)

hydrometallurgy Recovery of metals from their ores by reactions in aqueous solution (23.2b)

hypertonic A solution with a higher solute concentration than that of normal body fluids (13.4a)

hypotonic A solution with a lower solute concentration than that of normal body fluids (13.4a)

ideal gas A simplification of real gases in which it is assumed that there are no forces between the molecules and that the molecules occupy no volume (10.3b)

ideal gas constant (R) The proportionality constant in the ideal gas law; 0.082057 L · atm/mol · K or 8.314510 J/mol · K (10.3b)

ideal gas law A law that relates pressure, volume, number of moles, and temperature for an ideal gas (10.3b)

ideal solution A solution that obeys Raoult's law (13.4b)

immiscible A term used to describe two liquids that do not intermix (13.1a)

indicator A substance used to signal the equivalence point of a titration by a change in some physical property such as color (4.5d)

induced dipole Separation of charge in a normally non-polar molecule, caused by the approach of a polar molecule (11.4b)

induced dipole–induced dipole intermolecular force *see* **London dispersion forces**

initial rate The instantaneous reaction rate at the start of a reaction (14.2b)

inner coordination sphere *see* **coordination sphere**

inorganic acid An acid that produces the hydrogen ion (H^+) when dissolved in water and that contains hydrogen and one or more nonmetals (2.3d)

insoluble A term describing a compound that does not dissolve to an appreciable extent in a specific solvent (4.2b, 13.1a)

instantaneous rate The rate of a reaction at any given point in time, equal to the slope of a line tangent to the concentration–time curve at a given point in time (14.2b)

insulator A material that does not conduct electricity because the valence band is full and the energy gap is so large that motion of electrons from it to the empty conduction band is prohibited (12.4c)

integrated rate law A mathematical equation derived by integration of a rate law equation (14.4a)

intermediate A species that is produced in one step of a reaction mechanism and completely consumed in a later step (14.6b)

intermolecular forces (IMFs) Interactions between molecules, between ions, or between molecules and ions (8.1a, 11.1a)

internal energy The sum of the potential and kinetic energies of the particles in the system (5.1c, 19.1a)

intrinsic semiconductor A pure, undoped material that conducts electricity under certain conditions due to a small energy gap between a full valence band and an empty conduction band (12.4a)

ion An atom or group of atoms that has lost or gained one or more electrons so that it is no longer electrically neutral (2.1a); *see also* **anion, cation**

ion–dipole intermolecular force The electrostatic force between an ion and a neutral molecule that has a permanent dipole moment (13.2d)

ionic bonding The attraction between a positive ion and a negative ion, resulting from the complete (or nearly complete) transfer of one or more electrons from one atom to another (8.1a)

ionic compound A compound formed by the combination of positive and negative ions (2.4)

ionic solid A solid formed by the condensation of anions and cations (12.1a)

ionization constant for water (K_w) The equilibrium constant for the autoionization of water (16.2a)

ionization energy The energy change required to remove an electron from an atom or ion in the gas phase (7.4c)

ionization isomers Two or more complexes in which a coordinated ligand and a noncoordinated counter-ion are exchanged (23.3d)

isoelectric point (pI) The pH at which an amino acid has equal numbers of positive and negative charges (17.3d)

isoelectronic ions Ions that have the same number of electrons but different numbers of protons (7.5c)

isolated system A system in which neither matter nor energy can be passed to or from the surroundings (5.1a)

isomers Two or more compounds with the same molecular formula but different arrangements of atoms (9.3d)

isotonic A solution with the same solute concentration as that of normal body fluids (13.4a)

isotopes Atoms with the same atomic number but different mass numbers because of a difference in the number of neutrons (2.1c)

joule (J) The SI unit of energy (1.3b)

Kelvin temperature scale A scale in which the unit is the same size as the Celsius degree but the zero point is the lowest possible temperature (1.3b); *see also* **absolute zero**

ketone Any of a class of organic compounds characterized by the presence of a carbonyl group, in which the carbon atom is bonded to two other carbon atoms (21.3a)

kilogram (kg) The SI base unit of mass (1.3b)

kinetic energy The energy of a moving object, dependent on its mass and velocity (5.1a)

kinetic-molecular theory A theory of the behavior of matter at the molecular level (10.5)

lanthanide contraction The decrease in ionic radius that results from the filling of the 4*f* orbitals (23.1b)

lanthanides The series of elements between lanthanum and hafnium in the periodic table (2.2a)

lattice energy (U) The energy of formation of one mole of a solid crystalline ionic compound from ions in the gas phase (12.4b)

lattice points The corners of the unit cell in a crystal lattice (12.1b)

law of conservation of matter When a chemical reaction takes place, matter is neither created nor destroyed (3.3a)

Le Chatelier's principle A change in any of the factors determining an equilibrium will cause the system to adjust to reduce the effect of the change (15.4)

length The longest dimension of an object (1.3b)

Lewis acid A substance that can accept a pair of electrons to form a new bond (16.1a, 18.3a)

Lewis acid–base adduct The product of a Lewis acid–base reaction (18.3a)

Lewis base A substance that can donate a pair of electrons to form a new bond (16.1a, 18.3a)

Lewis symbol/structure A notation for the electron configuration of an atom or molecule (8.2a)

ligands The molecules or anions bonded to the central metal atom in a coordination compound (23.3a)

limiting reactant The reactant present in limited supply that determines the amount of product formed (3.4a)

line spectrum The spectrum of light emitted by excited atoms in the gas phase, consisting of discrete wavelengths (6.3a)

linkage isomers Two or more complexes in which a ligand is attached to the metal atom through different atoms (23.3d)

liquid The phase of matter in which a substance has no definite shape but a definite volume (1.2b)

liter (L) A unit of volume convenient for laboratory use; $1 \text{ L} = 1000 \text{ cm}^3$ (1.3b)

lithophile A transition metal that tends to form oxide ores (23.2a)

London dispersion forces Intermolecular attractions between two neutral molecules, both having induced dipoles (11.4c)

lone (electron) pairs Pairs of valence electrons that do not contribute to bonding in a covalent molecule (8.2a)

low-spin configuration The electron configuration for a coordination complex with the minimum number of unpaired electrons (23.4a)

macroscopic scale Processes and properties on a scale large enough to be observed directly (1.1a)

magnetic quantum number (m_l) The quantum number related to an orbital's orientation in space (6.5a)

main-group element An element in the A groups in the periodic table (2.2a)

Markovnikov's rule When an HX molecule adds to an alkene, the H atom adds to the carbon that has the greater number of H atoms (21.1d)

mass A measure of the quantity of matter in an object (1.3b)

mass defect The difference between the mass of the nuclear particles that make up a nucleus and the mass of an atom of the isotope of interest (24.2b)

mass number (A) The sum of the number of protons and neutrons in the nucleus of an atom of an element (2.1b)

matter Anything that has mass and occupies space (1.1a)

mechanical energy The sum of the kinetic and potential energy of an object as a whole (5.1a)

melting point The temperature at which the crystal lattice of a solid collapses and solid is converted to liquid (1.2b, 12.5a)

metal An element characterized by a tendency to give up electrons and by good thermal and electrical conductivity (2.2a)

metallic bonding The attractive forces that exist between the electrons and nuclei in metal atoms (8.1a)

metallic radius The distance between the nuclei of two atoms in a metallic crystal (7.4b)

metallic solid A solid consisting of positively charged metal atom cores held together by attractions to their valence electrons, which are delocalized over the entire crystal (12.1a)

metalloid An element with properties of both metals and nonmetals (2.2a)

meter (m) The SI base unit of length (1.3b)

method of initial rates An experimental method used to determine the order of a reaction with respect to a reacting species (14.3b)

microstate One possible distribution of the energy in a material at the molecular level (19.1c)

milliliter (mL) A unit of volume equivalent to one thousandth of a liter; 1 mL = 1 cm^3 (1.3b)

millimeter of mercury (mm Hg) A common unit of pressure, defined as the pressure that can support a 1-millimeter column of mercury; 760 mm Hg = 1 atm (10.1a)

miscible A term used to describe two liquids that intermix completely (13.1a)

mixture A combination of two or more substances in which each substance retains its identity (1.2c)

moderator A substance used in a nuclear reactor to slow the neutrons and increase reaction efficiency (24.4c)

molality (m) The number of moles of solute per kilogram of solvent (13.1b)

molar mass (M) The mass in grams of one mole of particles of any substance

molarity (M) The number of moles of solute per liter of solution

mole (mol) The SI base unit for amount of substance

mole fraction (χ) The ratio of the number of moles of one substance to the total number of moles in a mixture of substances (10.4b)

molecular covalent compound A compound formed by the combination of atoms without significant ionic character (2.3a); *see also* **covalent compound**

molecular formula A written formula that expresses the number of atoms of each type within one molecule of a compound (2.3b, 3.2d)

molecular geometry *see* **shape**

molecular orbital diagram An energy diagram that shows both the energy of the atomic orbitals from the atoms that are combining and the energy of the molecular orbitals (9.4c)

molecular orbital theory A model of bonding in which pure atomic orbitals combine to produce molecular orbitals that are delocalized over two or more atoms (9.1a)

molecular solid A solid formed by the condensation of covalently bonded molecules (12.1a)

molecular weight The formula weight of a substance that exists as individual molecules (3.1b)

molecularity The number of particles colliding in an elementary step (14.6a)

molecule The smallest unit of a compound that retains the composition and properties of that compound (1.2a)

monoatomic ion A single atom that has gained or lost one or more electrons (2.4a)

monodentate ligand A ligand with one attachment point to a metal (23.3a)

monomer The small units from which a polymer is constructed (21.4)

monoprotic acid A Brønsted–Lowry acid that can donate one proton (4.3b, 16.1c)

multidentate ligand A ligand with more than one potential attachment point to a metal (23.3a)

n-type semiconductor A semiconductor that has been doped with a substance that introduces extra electrons (12.4a)

Nernst equation A mathematical expression that relates the potential of an electrochemical cell to the concentrations of the cell reactants and products (20.2d)

net ionic equation A chemical equation involving only those substances undergoing chemical changes in the course of the reaction (4.3a)

network covalent compound A compound made up of a network of covalently bonded atoms (2.3a, 12.1a); *see also* **covalent solid**

neutral solution A solution in which the concentrations of hydronium ion and hydroxide ion are equal (16.2a)

neutralization reaction An acid–base reaction that produces a neutral solution of a salt and water (4.1b)

neutrino A massless, chargeless particle emitted by some nuclear reactions (24.5a)

neutron An electrically neutral subatomic particle found in the nucleus (2.1a)

newton (N) The SI unit of force, 1 N = 1 kg · m/s^2 (5.1a)

noble gas notation An abbreviated form of *spdf* notation that replaces the completed electron shells with the symbol of the corresponding noble gas in brackets (7.3b)

noble gas Any of the elements in Group 8A of the periodic table (2.2a)

node A point of zero amplitude of a wave (6.5b)

nonelectrolyte A substance that dissolves in water to form an electrically nonconducting solution (4.2a)

nonmetal An element characterized by a lack of metallic properties (2.2a)

nonpolar bond A bond in which electrons experience the same attractive force to both nuclei (8.4b)

normal boiling point The boiling point when the external pressure is 1 atm (11.2c)

normal melting point The melting temperature at a pressure of 1 atm (12.5a)

nuclear reaction A reaction involving one or more atomic nuclei, resulting in a change in the identities of the isotopes (24.1)

nuclear reactor A container in which a controlled nuclear reaction occurs (24.4c)

nucleic acids A class of polymers, including RNA and DNA, that are the genetic material of cells (21.5c)

nucleon A nuclear particle, either a neutron or a proton (24.2b)

nucleophile A species that can donate a pair of electrons to form a new covalent bond (21.3c)

nuclide A nucleus with a specific makeup of neutrons and protons (24.1c)

octahedral hole An empty space in a closest-packed structure that is surrounded by six atoms or ions whose centers define the vertices of an octahedron (12.3a)

octet rule When forming bonds, atoms of main-group elements gain, lose, or share electrons to achieve a stable configuration having eight valence electrons (8.2b)

optical isomers Two compounds that are nonsuperimposable (not identical) mirror images of each other (21.2b, 23.3c)

orbital The matter wave for an allowed energy state of an electron in an atom or molecule (6.4b)

orbital box notation A representation of the electron configuration in an atom or ion that uses boxes or horizontal lines to represent orbitals and arrows to represent electrons (7.3b)

orbital overlap Partial occupation of the same region of space by orbitals from two atoms (9.1a)

ore A sample of matter containing a desired mineral or element, usually with large quantities of impurities (23.2)

organic acids Compounds made up mostly of carbon, hydrogen, and oxygen that also contain the —C(O)OH structural group (4.3b)

ore A sample of matter containing a desired mineral or element, usually with large quantities of impurities (23.2)

osmosis The movement of solvent molecules through a semipermeable membrane from a region of lower solute concentration to a region of higher solute concentration (13.4a)

osmotic pressure (Π) The pressure exerted by osmosis in a solution system at equilibrium (13.4a)

outer coordination sphere The portion of the coordination compound that does not include the metal and any ligands covalently bonded to the metal (23.3a)

oxidation The loss of electrons by an atom, ion, or molecule (4.4a)

oxidation number A number assigned to each element in a compound in order to keep track of the electrons during a reaction (4.4b)

oxidation–reduction reaction A reaction involving the transfer of one or more electrons from one species to another (4.4a)

oxidation–reduction titration A titration in which an oxidation–reduction reaction is used to analyze a solution of unknown concentration (4.5d)

oxidizing agent The substance that accepts electrons and is reduced in an oxidation–reduction reaction (4.4a)

oxoacids Groups of acids that differ only in the number of oxygen atoms (2.3d)

p-type semiconductor A semiconductor that has been doped with a substance that has atoms with fewer electrons than those of the pure semiconductor (12.4a)

pairing energy The additional potential energy due to the electrostatic repulsion between two electrons in the same orbital (23.4a)

paramagnetism The physical property of being attracted by a magnetic field (7.1b)

parent nucleus The nucleus undergoing a transformation in a radioactive decay reaction (24.1c)

partial charge The charge on an atom in a molecule or ion calculated by assuming sharing of the bonding electrons proportional to the electronegativity of the atom (8.4c)

partial pressure The pressure exerted by one gas in a mixture of gases (10.4a)

parts per billion (ppb) The mass of solute (g) in 10^9 g of solution (13.1b)

parts per million (ppm) The mass of solute (g) in 10^6 g of solution (13.1b)

pascal (Pa) The SI unit of pressure; 1 Pa = 1 N/m^2 (10.1a)

Pauli exclusion principle No two electrons in an atom can have the same set of four quantum numbers (7.3a)

peptide bond The chemical bond between amino acids in a polypeptide (21.5c)

percent abundance The percentage of the atoms of a natural sample of the pure element represented by a particular isotope (2.1c)

percent composition The percentage of the mass of a compound represented by each of its constituent elements (3.2b)

percent yield The actual yield of a chemical reaction as a percentage of its theoretical yield (3.4b)

periodic table of the elements A table of elements organized by increasing atomic number (2.2a)

periods The horizontal rows in the periodic table (2.2a)

pH The negative of the base-10 logarithm of the hydrogen ion concentration; a measure of acidity (16.2b)

phase diagram A graph showing which phases of a substance exist at various temperatures and pressures (12.5b)

photoelectric effect The ejection of electrons from a metal bombarded with light of at least a minimum frequency (6.2a)

photon A "particle" of electromagnetic radiation having zero mass and an energy given by Planck's law (6.2a)

physical change A change that involves only physical properties (1.2b)

physical properties Properties of a substance that can be observed and measured without changing the composition of the substance (1.2b)

pi (π) bond The second (and third, if present) bond in a multiple bond; results from sideways overlap of p atomic orbitals (9.3a)

pK_a The negative of the base-10 logarithm of the acid ionization constant (16.3b)

Planck's constant (h) The proportionality constant that relates the frequency of radiation to its energy (6.2a)

plasma A gaslike phase of matter that consists of charged particles (24.4d)

pOH The negative of the base-10 logarithm of the hydroxide ion concentration; a measure of basicity (16.2b)

polar bond A chemical bond in which there is an uneven distribution of bond electron density (8.4b)

polar covalent bond *see* **polar bond**

polarizability The extent to which the electron cloud of an atom or molecule can be distorted by an external electric charge (11.4b)

polyamide A condensation polymer formed by linking monomers by amide bonds (21.4b)

polyatomic ion An ion consisting of more than one atom (2.4b)

polyester A condensation polymer formed by linking monomers by ester groups (21.4b)

polymer A large molecule composed of many smaller repeating units, usually arranged in a chain (21.4)

polypeptide A polymer consisting of many amino acids connected by peptide bonds (21.5c)

polyprotic acid A Brønsted–Lowry acid that can donate more than one proton (16.1c)

positron A particle having the same mass as an electron but a positive charge (24.1b)

positron emission A type of radioactive decay that occurs when a proton in an unstable nucleus is converted to a neutron and a positron and the positron is ejected from the nucleus (24.1c)

potential energy The energy that results from an object's position (5.1a)

precipitate A water-insoluble solid product of a reaction, usually of water-soluble reactants (4.2b)

precipitation reaction An exchange reaction that produces an insoluble salt, or precipitate, from soluble reactants (4.1b)

precision The agreement of repeated measurements of a quantity with one another (1.3c)

pressure The force exerted on an object divided by the area over which the force is exerted (10.1b)

primary battery A battery that cannot be recharged (20.4a)

primary standard A pure, solid acid or base that can be accurately weighed for preparation of a titrating reagent (4.5d)

primitive cubic unit cell *see* **simple cubic unit cell**

principal quantum number (n) The quantum number that describes the size and energy of the shell in which the orbital resides (6.3b, 6.5a)

principle of microscopic reversibility The elementary steps in a reaction mechanism are reversible (15.1a)

product A substance formed in a chemical reaction (3.3a)

product-favored reaction A reaction in which reactants are completely or largely converted to products at equilibrium (15.2a)

properties A collection of characteristics that describe matter (1.2a)

protein A polymer formed by condensation of amino acids, sometimes conjugated with other groups (21.5c)

proton A positively charged subatomic particle found in the nucleus (2.1a)

pure substance A substance that contains only one type of element or compound and has fixed chemical composition (1.2b)

pyrometallurgy Recovery of metals from their ores by high-temperature processes (23.2b)

qualitative analysis A process used to separate mixtures of ions and to identify the components of a mixture (18.4c)

quantum mechanics A general theoretical approach to atomic behavior that describes the electron in an atom as a matter wave (6.4a)

quantum number A set of numbers with integer values that define the properties of an atomic orbital (6.5a)

radioactive decay series A series of nuclear reactions by which a radioactive isotope decays to form a stable isotope (24.2a)

Raoult's law The vapor pressure of the solvent is proportional to the mole fraction of the solvent in a solution (13.4b)

rate constant (k) The proportionality constant in the rate equation

rate law The mathematical relationship between reactant concentration and reaction rate (14.3a)

rate-determining step The slowest elementary step of a reaction mechanism (14.6c)

reactant A starting substance in a chemical reaction (3.3a)

reactant-favored reaction A reaction in which only a small amount of reactants is converted to products at equilibrium (15.2a)

reaction A process in which substances are changed into other substances by rearrangement, combination, or separation of atoms (4.1a)

reaction coordinate diagram A plot that shows energy as a function of the progress of the reaction from reactants to products (14.5a)

reaction mechanism The sequence of events at the molecular level that controls the speed and outcome of a reaction (14.6)

reaction order The exponent of a concentration term in the reaction's rate equation (14.3a)

reaction quotient (Q) The product of concentrations of products divided by the product of concentrations of reactants, each raised to the power of its stoichiometric coefficient in the chemical equation (15.3b); *see also* **equilibrium constant**

reaction quotient expression A mathematical expression that relates the concentrations of the reactants and products at a particular temperature to a numerical constant (15.3b); *see also* **equilibrium constant expression**

reaction rate The change in concentration of a reagent per unit time (14.2a)

reducing agent The substance that donates electrons and is oxidized in an oxidation–reduction reaction (4.4a)

reduction The gain of electrons by an atom, ion, or molecule (4.4a)

regioselectivity Reaction selectivity where a particular atom adds to one region of a molecule but not to a different one (21.1d)

resonance hybrid The actual electron arrangement for a molecule or ion that is intermediate between resonance structures but not represented by any individual resonance structure (8.2d)

resonance structure The possible structures of a molecule for which more than one Lewis structure can be written, differing by the number of bond pairs between a given pair of atoms (8.2d)

reverse osmosis The application of pressure greater than the osmotic pressure of impure solvent to force solvent through a semipermeable membrane to the region of lower solute concentration (13.4a)

reversible process A process for which it is possible to return to the starting conditions along the same path without altering the surroundings (15.1a)

root mean square (rms) speed The square root of the average of the squares of the speeds of the molecules in a sample (10.5b)

sacrificial anode A metal, such as zinc, that is easily oxidized and protects another material, such as iron, from oxidation (20.4d)

salt An ionic compound whose cation comes from a base and whose anion comes from an acid (4.3c)

salt bridge A device for maintaining the balance of ion charges in the compartments of an electrochemical cell (20.1d)

saturated hydrocarbon A hydrocarbon containing only carbon–carbon single bonds (21.1a); *see also* **alkane**

saturated solution A solution in which the solute concentration is at the solubility limit at a given temperature (13.1a)

scientific notation Notation used to represent very large or very small numbers (1.3a)

second law of thermodynamics The total entropy of the universe is continually increasing (19.1b)

secondary battery A battery that can be recharged (20.4a)

selectivity A measure of the tendency of a reaction to form one set of products over another (14.6e)

semiconductors Substances that can conduct small quantities of electric current (12.4a)

semimetal An element with properties of both metals and nonmetals (2.2a); *see also* **metalloid**

semipermeable membrane A thin sheet of material through which only certain types of molecules can pass (13.4a)

shape The arrangement of atoms around the central atom in a molecule or ion (8.5a)

SI units Abbreviation for *Système International d'Unités,* a uniform system of measurement units in which a single base unit is used for each measured physical quantity (1.3a)

siderophile A transition metal that has an affinity for iron, exhibits high solubility in molten iron, and is most often found in a native, metallic state (23.2a)

sigma (σ) bond A bond formed by the overlap of orbitals head to head and with bonding electron density concentrated along the axis of the bond (9.1a)

significant figures The digits in a measured quantity that are known exactly, plus one digit that is inexact to the extent of ± 1 (1.3c)

simple cubic unit cell A unit cell with one lattice point on each corner of a cube (12.2a)

single bond A bond formed by sharing one pair of electrons; a sigma bond (8.2a)

single displacement reaction A displacement reaction in which one molecular fragment is exchanged for another (4.1b)

sodium chloride structure A 1:1 ionic crystal structure that can be viewed as a face-centered cubic arrangement of the larger ions with the smaller ions occupying all of the octahedral holes (12.3b)

sol A colloidal dispersion of a solid substance in a fluid medium (13.5b)

solid The phase of matter in which a substance has both definite shape and definite volume (1.2b)

solubility The concentration of solute in equilibrium with undissolved solute in a saturated solution (13.1a)

solubility product constant (K_{sp}) An equilibrium constant relating the concentrations of the ionization products of a dissolved substance (18.1b)

soluble A compound that dissolves in a solvent to an appreciable extent (4.2b, 13.1a)

solute The substance dissolved in a solvent to form a solution (4.2a)

solution A homogeneous mixture in a single phase (4.2a)

solvated A species dissolved in a solvent that is surrounded by solvent molecules (4.2a)

solvent The medium in which a solute is dissolved to form a solution (4.2a)

sp hybrid orbital A hybrid orbital made up of one part s and one part p atomic orbital (9.2c)

sp^2 hybrid orbital A hybrid orbital made up of one part s and two parts p atomic orbital (9.2b)

sp^3 hybrid orbital A hybrid orbital made up of one part s and three parts p atomic orbital (9.2a)

sp^3d hybrid orbital A hybrid orbital made up of one part s, three parts p, and one part d atomic orbital (9.2d)

sp^3d^2 hybrid orbital A hybrid orbital made up of one part s, three parts p, and two parts d atomic orbital (9.2e)

space-filling model A representation of a molecule where interpenetrating spheres represent the relative amount of space occupied by each atom in the molecule (2.3c)

$spdf$ notation A notation for the electron configuration of an atom in which the number of electrons assigned to a subshell is shown as a superscript after the subshell's symbol (7.3b)

specific heat capacity (C) The quantity of heat required to raise the temperature of 1.00 g of a substance by 1.00 °C (5.3a)

spectator ion An ion that is present in a solution in which a reaction takes place but that is not involved in the net process (4.3a)

spectrochemical series An ordering of ligands by the magnitudes of the splitting energies they cause (23.4a)

spectroscopy An analytical method based on the absorption and transmission of specific wavelengths of light (23.4c)

speed of light The speed at which all electromagnetic radiation travels in a vacuum, 2.998×10^8 m/s (6.1a)

spin quantum number (m_s) The quantum number that describes the spin state of an electron (7.1a)

spin state One of the two possible orientations of the magnetic field generated by a spinning electron (7.1a)

spontaneous process A process that occurs without a continual external source of energy (19.1a)

standard atmosphere (atm) A unit of pressure; 1 atm = 760 mm Hg (10.1b)

standard cell potential ($E°_{cell}$) The potential of an electrochemical cell measured under standard conditions (20.2a)

standard enthalpy of reaction ($\Delta H°_{rxn}$) The enthalpy change of a reaction that occurs with all reactants and products in their standard states (5.6a)

standard entropy of reaction ($\Delta S°_{rxn}$) The entropy change of a reaction that occurs with all reactants and products in their standard states (19.2b)

standard free energy of formation ($\Delta G_f°$) The free energy change for the formation of one mole of a compound from its elements, all in their standard states (19.3b)

standard free energy of reaction ($\Delta G°_{rxn}$) The free energy change of a reaction that occurs with all reactants and products in their standard states (19.3b)

standard heat of formation ($\Delta H_f°$) The enthalpy change of a reaction for the formation of one mole of a compound directly from its elements, all in their standard states (5.6a)

standard hydrogen electrode (SHE) An electrode in which hydrogen gas is bubbled through a chamber that contains a platinum electrode and an acidic solution and that can act as an anode or cathode (20.1d)

standard molar entropy ($S°$) The entropy of a substance in its most stable form at a pressure of 1 bar (19.1f)

standard molar volume The volume occupied by one mole of gas at standard temperature and pressure; 22.414 L (10.3c)

standard reduction potential ($E°_{red}$) The cell potential for a reduction half-reaction measured under standard conditions (20.2a)

standard state The most stable form of an element or compound in the physical state in which it exists at 1 bar and the specified temperature (5.6a)

standard temperature and pressure (STP) A temperature of 0 °C and a pressure of exactly 1 atm (10.3c)

standardization The accurate determination of the concentration of an acid, base, or other reagent for use in a titration (4.5d)

state function A quantity whose value is determined only by the state of the system (5.5a)

states of matter The three physical states: solid, liquid, and gas (1.2b)

stereoisomers Two or more compounds with the same molecular formula and the same atom-to-atom bonding but with different arrangements of the atoms in space (23.3d)

stoichiometric coefficients The multiplying numbers assigned to the species in a chemical equation in order to balance the equation (3.3a)

stoichiometric factor A conversion factor relating moles of one species in a reaction to moles of another species in the same reaction (3.3c)

stoichiometry The study of the quantitative relations between amounts of reactants and products (3.2a)

strong acid An acid that ionizes completely in aqueous solution (4.3b)

strong base A base that ionizes completely in aqueous solution (4.3b)

strong electrolyte A substance that dissolves in water to form a good conductor of electricity (4.2a)

strong field Strong metal–ligand bonding in a coordination compound, which results in a low-spin electron configuration (23.4a)

strong-field ligands Any of the ligands at the "high" end of the spectrochemical series that have a large effect on the crystal field splitting energy (Δ_o) (23.4a)

structural electron pair Nonbonding or bonding electrons that repel one another and are arranged around the central atom in a molecule or ion so as to avoid one another as best as possible (8.5a)

structural formula A variation of a molecular formula that expresses how the atoms in a compound are connected (2.3b)

structural isomers Two or more compounds with the same molecular formula but with different atoms bonded to each other (23.3d)

subatomic particles A collective term for protons, neutrons, and electrons (2.1a)

sublimation The direct conversion of a solid to a gas (12.5a)

subshell One or more orbitals in the electron shell of an atom, defined by a value of ℓ and designated by a different letter (s, p, d, and f are the letters assigned to the first four subshells) (6.5b)

substitution reaction A reaction where one group is substituted for another (21.1d)

superconductor A material that has no resistance to the flow of electric current (12.3d)

supercritical fluid A substance at or above the critical temperature and pressure (12.5b)

supersaturated solution A solution in which the solute concentration is greater than the solubility limit at a given temperature (13.1a)

surface tension The energy required to disrupt the surface of a liquid (11.3a)

surfactant A substance that changes the properties of a surface, typically in a colloidal suspension (13.5b)

surroundings Everything outside the system in a thermodynamic process (5.1c)

system The substance being evaluated for energy content in a thermodynamic process (5.1c)

temperature A physical property that determines the direction of heat flow in an object on contact with another object (1.3b)

termolecular An elementary step in a reaction mechanism where three species collide simultaneously (14.6a)

tetrahedral hole An empty space in a closest-packed structure that is surrounded by four atoms or ions whose centers define the vertices of a tetrahedron (12.3a)

theoretical yield The maximum amount of product that can be obtained from the given amounts of reactants in a chemical reaction (3.4b)

thermal energy The sum of the kinetic and potential energies of all the atoms, molecules, or ions within a system (5.1a)

thermal equilibrium A condition in which the system and its surroundings are at the same temperature (5.3b)

thermochemistry The study of how energy in the form of heat is involved in chemical change (5.1a)

thermodynamics The study of the relationships between heat, energy, and work and the conversion of one into the other (5.1c)

third law of thermodynamics The entropy of a pure, perfectly formed crystal at 0 K is zero (19.1f)

titrant The substance being added during a titration (17.3)

titration A procedure for quantitative analysis of a substance by an essentially complete reaction in solution with a measured quantity of a reagent of known concentration (4.5d); *see also* **acid–base titration**

torr A unit of pressure equivalent to 1 mm of mercury (10.1b)

trans-uranium elements The elements in the periodic table with atomic numbers greater than 92 (24.5b)

transition metals The elements in Groups 3B–2B (3–12) of the periodic table, or those elements whose atoms have partially filled d orbitals in their neutral or cationic state (2.2a, 23.1a)

transition state The arrangement of reacting molecules and atoms at the point of maximum potential energy (14.5a)

translational symmetry A property of a pattern in which translation-related objects appear at regular intervals in certain directions (12.1b)

trigonal hole An empty space in a closest-packed structure that is surrounded by three equidistant atoms or ions (12.3a)

triple bond A bond formed by sharing three pairs of electrons, one pair in a sigma bond and the other two in pi bonds (8.2a)

triple point The temperature and pressure at which the solid, liquid, and vapor phases of a substance are in equilibrium (12.5b)

unbranched hydrocarbon An organic compound where each carbon is bonded to no more than two other carbon atoms (21.1b)

uncertainty principle It is not possible to know with great certainty both an electron's position and its momentum at the same time (6.4a)

unimolecular An elementary step in a reaction mechanism involving a single reacting species (14.6a)

unit cell The smallest unit in a crystal lattice that reflects the full symmetry of the solid and from which the entire crystal can be built by repeated translation (12.1b)

unsaturated hydrocarbon A hydrocarbon that contains at least one carbon–carbon double or triple bond (21.1a)

unsaturated solution A solution in which the solute concentration is less than the solubility limit at a given temperature (13.1a)

valence band The highest-energy occupied orbitals in a solid metal, semiconductor, or insulator (12.4a)

valence bond theory A model of bonding in which a bond arises from the overlap of atomic orbitals on two atoms to give a bonding orbital with electrons localized between the atoms (9.1a)

valence electrons The outermost and most reactive electrons of an atom (7.3d)

valence shell electron-pair repulsion (VSEPR) model A model for predicting the shapes of molecules in which structural electron pairs are arranged around each atom to minimize electron–electron repulsions (8.5a)

van der Waals equation A mathematical expression that describes the behavior of nonideal gases (10.5d)

van der Waals solid *see* **molecular solid**

van't Hoff equation The equation used to estimate the new value for an equilibrium constant when there is a change in temperature (15.4c)

van't Hoff factor (i) The ratio of the experimentally measured freezing point depression of a solution to the value calculated from the apparent molality (13.4a)

vapor pressure The pressure of the vapor of a substance in contact with its liquid or solid phase in a sealed container (10.4a, 11.2a)

vaporization The state change from liquid to gas (11.1b)

viscosity The resistance of a liquid to flow (11.3b)

volatility The tendency of the molecules of a substance to escape from the liquid phase into the gas phase (13.4b)

volt (V) The electric potential through which 1 coulomb of charge must pass in order to do 1 joule of work (20.2a)

voltaic cell An electrochemical cell where electrons flow spontaneously through a wire (20.1d)

volume The amount of space a substance occupies (1.3b)

volumetric glassware Laboratory glassware that has been carefully calibrated to contain very accurate volumes or allow for the measurement of very accurate volumes (4.5b)

water of hydration The water associated with a hydrated compound (3.2e)

wave function (ψ) An equation that characterizes the electron as a matter wave (6.4b)

wave–particle duality The idea that the electron has properties of both a wave and a particle (6.4a)

wavelength (λ) The distance between successive crests (or troughs) in a wave (6.1a)

weak acid An acid that is only partially ionized in aqueous solution (4.3b)

weak base A base that is only partially ionized in aqueous solution (4.3b)

weak electrolyte A substance that dissolves in water to form a poor conductor of electricity (4.2a)

weak field Weak metal–ligand bonding in a coordination compound, which results in a high-spin electron configuration (23.4a)

weak-field ligand Any of the ligands at the "low" end of the spectrochemical series that have a small effect on the crystal field splitting energy (Δ_o) (23.4a)

wedge-and-dash model A two-dimensional representation of a three-dimensional structure that can easily be drawn on paper, where bonds are represented by lines, wedges, or dashes (2.3c)

work Energy transfer that occurs as a mass is moved through a distance against an opposing force (5.1a)

Zaitsev's rule The more stable product of an alcohol dehydration reaction is the alkene with the greatest number of substituents on the C=C carbons (21.3b)

zeolite A solid inorganic material with an aluminosilicate framework that results in a porous structure (12.3d)

zinc blende structure A 1:1 ionic crystal structure that can be described as a face-centered cubic arrangement of the larger ions with the smaller ions occupying half of the tetrahedral holes (12.3c)

zwitterion An amino acid in which both the amine group and the carboxyl group are ionized (17.4b)

Index

Page numbers followed by the letter *f* indicate figures
Page numbers followed by the letter *t* indicate tables

atomic radius and bonding in, 740–741, 740–741*f*

overview, 738*f*, 739–741, 739*t*, 740*f*, 741*f*

periodic table and, 35

nonpolar bonds, 247

nonstandard conditions, 674–676

normal boiling point, 335

normal melting point, 385, 385*t*

n-type semiconductors, 382

nuclear chemistry, 789–822

  fission, 806–811

  fusion, 811–812

  kinetics of radioactive decay, 802–806. *See also* radioactive decay

  radioactivity applications, 812–819. *See also* radioactivity

  reactions, 790–796. *See also* nuclear reactions

  stability, 796–802. *See also* nuclear stability

nuclear fuel, 808–810

nuclear medicine, 817–818

nuclear power plants, 810–811, 811*f*

nuclear reactions, 790–796

  chemical reactions vs., 790, 790*f*

  decay processes, 793–796

  defined, 790

  natural radioactive decay, 791–792, 791*t*

  radioactive decay and balancing nuclear reactions, 792–796

nuclear reactor, 809–810

nuclear stability, 796–802

  band of stability, 796–798, 797–799*f*

  binding energy, 799–801

  relative binding energy, 801, 802*f*

nuclear symbol, 29–30

nuclear synthesis, 812–817

  carbon burning, 814

  elemental abundance, 813*f*

  helium burning, 814

  hydrogen burning, 813–814

  induced synthesis of elements, 815–817, 816*f*

  stellar synthesis reactions, 812–815

nuclear weapons, 808–809, 808*t*

nucleic acids, 728–729*f*, 728–730

nucleon, 800

nucleophiles, 716

nucleophilic substitution reactions, 716, 716*f*

nucleotide, 728, 728*f*

nucleus. *See* atomic nucleus

nuclide, 792

O

octahedral hole, 371–372*f*

octet rule, 231, 234–236

optical isomers, 709–710, 710–711*f*, 775, 776*f*

orbital box notation, 198

orbital energies

  in single- and multielectron species, 196, 196–197*f*

  trends in, 207–209, 208*f*, 213, 737, 737*f*, 739, 739*f*, 746

orbital energy diagrams, 187–188, 188*f*

orbital overlap, 266

orbitals

  defined, 182, 182*f*

  nodes, 184, 185*f*, 187, 187*f*

  shapes of, 184–187, 185*f*, 186*t*

ores, 764–767

  common, 764, 764*f*, 765*t*

  defined, 764

  electrometallurgy, 766

  extraction of metals from, 764–767, 766*f*

  hydrometallurgy, 766–767

  pyrometallurgy, 765–766, 766*f*

organic acids, 105, 105*f*

organic chemistry, 695–733

  biopolymers, 721–730

  functional groups, 711–716

  hydrocarbons, 696–708

  isomerism, 708–711

  synthetic polymers, 716–720, 721*f*

osmosis, 416–421, 417*f*, 419–420*f*

osmotic pressure, 416–421, 417*f*, 419*f*, 420*f*

outer coordination sphere, 769, 770*f*

overall reaction order, 444

overvoltage, 684

oxidation, 110–111, 715, 715*f*

oxidation numbers

  overview, 111–113, 112*t*

  rules for assigning oxidation numbers, 112*t*, 653*t*

  transition metals, 762–763, 763*f*

oxidation–reduction reactions

  balancing reactions, 654–656

  balancing reactions in acidic and basic solutions, 657–660

  electrochemical cells and, 660–663

overview, 110–114, 110*f*, 652–654, 653*t*

  recognition of, 113–114, 114*t*

  rules for assigning oxidation numbers, 652, 653*t*

oxidation–reduction titrations, 129–130

oxidation states. *See* oxidation numbers

oxidizing agents, 110, 652

oxoacids, 42, 42*t*

ozone hole, 755–756, 755*f*

P

pairing energy, 779

paramagnetic, 195, 195*f*, 287, 287*f*

parent nucleus, 792

partial charge, 252

partial pressure, 307–312

  Dalton's law of, 307–309, 308*f*, 309*t*

  and mole fractions, 309–310

parts per billion (ppb), 402–403

parts per million (ppm), 402–403

pascal (Pa), 295

Pauli exclusion principle, 196–197, 197*f*

peptide bond, 726

percent abundance, 31

percent composition

  determination of, 62–63

  empirical formulas from, 63–65

percent yield, 82–83

periodic table of the elements

  atomic radii, 209–210, 210*f*, 736, 736*f*

  atomic size and, 209–210, 760–762

  electron affinity and, 212–213, 212*f*

  electron configurations and, 205–206*f*, 205–207

  introduction to, 33–37, 33*t*, 34*f*, 36*f*

  ionization energy and, 211–212, 763

  main-group elements, 735–758

  transition metals, 759–788

  trends in orbital energies and, 207–209, 208*f*, 213

periods, 33

permanent gas, 388

PET (positron emission tomography), 817, 817*f*

pH

  and simultaneous equilibria, 609–610, 610*f*

  of buffers, 560–565

  pH and pOH calculations, 522–523*f*, 522–524

of salt solutions, 544–545

of strong acid–strong base solutions, 532–533

of weak acid solutions, 533–538

of weak base solutions, 538–542

phase changes

  energy and, 148–151, 149*f*

  kinetic molecular theory, 328–329, 328*t*

  of solids, 385–386, 385*t*, 386*f*

  standard entropy change for, 628–629, 629*t*

  standard molar entropy, 627–628*t*, 627–629, 627*f*

phase diagrams, 386–391, 387*f*, 388*t*, 389*f*

phenolphthalein, 125, 583, 583*f*

pH indicators, 582–583, 583*f*

pH meter, 522, 522*f*

photoelectric effect, 174–175, 175*f*

photoelectron spectroscopy, 290, 290*f*

photon, 175, 175*f*, 188, 188*f*

pH titration plots. *See also* acid–base titrations

  as indicator of acid or base strength, 580–582

  polyprotic acid titration plots, 586–587, 586*f*

physical change, 7

physical properties

  hydrogen bonding effect on, 348–350, 349–350*f*

  overview, 5–6

  polarizability effect on, 347–348, 347*t*

physical states

  entropy and, 622, 627

pi backbonding, 783

pi (π) bonds, 276–282

  antibonding molecular orbitals and, 284, 284*f*

  atomic radius and, 740, 741*f*

  conformations, 281–282, 281*f*

  formation of, 276–277, 276*f*

  hybridization and number of possible *p–p* pi bonds relationship, 280, 280*t*

  isomers, 281–282, 281*f*

  metal–ligand bonding interactions, 782, 782*f*, 783*t*

pig iron, 766

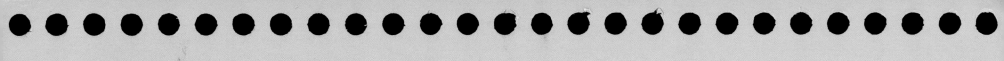